W9-CYN-146

ADVANCED CALCULUS

WILFRED KAPLAN
Department of Mathematics
University of Michigan

ADVANCED CALCULUS

SECOND EDITION

ADDISON-WESLEY PUBLISHING COMPANY
Reading, Massachusetts
Menlo Park, California · London · Amsterdam · Don Mills, Ontario · Sydney

 Printed in the United States of America. Published simultaneously in Canada. Library of Congress Catalog Card No. 77-184161.

ISBN 0-201-03611-8
IJKLMNOPQR-MA-89876543210

Is mathematical analysis . . . only a vain play of the mind? It can give to the physicist only a convenient language; is this not a mediocre service, which, strictly speaking, could be done without; and even is it not to be feared that this artificial language may be a veil interposed between reality and the eye of the physicist? Far from it; without this language most of the intimate analogies of things would have remained forever unknown to us; and we should forever have been ignorant of the internal harmony of the world, which is . . . the only true objective reality.

HENRI POINCARÉ

Preface to the Second Edition

Since the appearance of the first edition of this text, the freshman and sophomore courses in calculus have expanded in scope and now normally include a considerable amount of material on vectors. Furthermore, along with the calculus, these courses often include matrices and linear algebra. These developments have motivated the principal modifications in this new edition.

In particular, the previous Chapter 1 on 3-dimensional vector geometry has been replaced by a summary of this topic, included in the introductory chapter. A new Chapter 1 covers matrices and linear algebra. This chapter can be used for review and reference or to study the topic afresh. A very few sections are important for the chapters to follow; as in the previous edition, the sections which are not essential are marked by an asterisk. It is thus possible to treat this chapter lightly and give little weight to applications of linear algebra, or to study the chapter in depth and take advantage of the algebra to clarify such topics as maxima and minima, eigenvalues, normal modes of vibration.

The ninth chapter of the first edition was a lengthy treatment of analytic functions of a complex variable. This chapter became the basis of a separate book by the author (*Introduction to Analytic Functions*, Reading, Mass., Addison-Wesley, 1966). In view of the appearance of this book, it was decided to replace the old Chapter 9 by a much shorter one on the same topic, derived in large part from a chapter in the author's book *Operational Methods for Linear Systems* (Reading, Mass., Addison-Wesley, 1962). In particular, the elaborate discussion of conformal mapping has been omitted; for this topic, one is referred to the separate book cited above.

As with the first edition, the purpose of this book is to provide sufficient material for a course in advanced calculus up to a year in length. It is hoped that the great variety of topics covered will also make this work useful as a reference book.

The background assumed is that usually obtained in freshman and sophomore courses in algebra, analytic geometry, and calculus. The introductory chapter provides a concise review of these subjects; it also serves as a handy reference list of basic definitions and formulas.

The subject matter of the book includes all the topics usually to be found in texts on advanced calculus. However, there is more than the usual emphasis on applications and on physical motivation. Vectors are introduced at the outset and serve at many points to indicate the intrinsic geometrical and physical significance of mathematical relations. Numerical methods of integration and of solving differential equations are emphasized, both because of their practical value and because of the insight they give into the limit process.

A high level of rigor is maintained throughout. Definitions are clearly labeled as such and all important results are formulated as theorems. A few of the finer points concerning the real number system (the Heine-Borel theorem, Weierstrass-Bolzano theorem, and related notions) are omitted. The theorems whose proofs depend on these tools are stated without proof, with references to more advanced treatises. A competent teacher can easily fill in these gaps, if so desired, and thereby present a complete course in real analysis.

A large number of problems, with answers, are distributed throughout the text. They include simple exercises of the "drill" type and more elaborate ones planned to stimulate critical reading. Some of the finer points of the theory are relegated to the problems, with hints given where appropriate.

Generous references to the literature are given and each chapter concludes with a list of books for supplementary reading.

Topical summary. Chapter 1 takes up the simplest properties of matrices, develops vectors in n-dimensional space V^n and linear mappings from V^n to V^m and, in starred sections, goes more deeply into these topics. Partial derivatives are taken up in the second chapter, at first without reference to vectors and matrices, and then with the aid of the linear algebra, especially for geometric applications. The third chapter introduces the divergence and curl and the basic identities; orthogonal coordinates are treated concisely; a new final section provides an introduction to tensors in n-dimensional space.

The fourth chapter, on integration, has as its main goal a clarification of the concept of definite and indefinite integrals. To this end, numerical methods receive special attention. Improper single and multiple integrals are studied and treated in the same way as infinite series, with which they are coordinated at the end of Chapter 6. Chapter 5 is devoted to line and surface integrals. While the notions are first presented without vectors, it very soon becomes clear how natural the vector approach is for this subject. At the end of the chapter an unusually complete treatment of transformation of variables in a multiple integral is given.

Chapter 6 studies infinite series, without assumption of previous knowledge. The notions of upper and lower limits are introduced and used sparingly as a simplifying device; with their aid the theory is given in almost complete form. The usual tests are given, in particular, the root test. With its aid, the treatment of power series is greatly simplified. Uniform convergence is presented with great care and applied to power series. Final sections point out the parallel with improper integrals; in particular, power series are shown to correspond to the Laplace transform.

The seventh chapter is a complete treatment of Fourier series at an elementary level. The first sections give a simple introduction, with many examples; the approach is gradually deepened and a convergence theorem is proved with a minimum of formal work. Orthogonal functions are then studied, with the aid of inner product, norm, and vector procedures. A general theorem on complete systems enables one to deduce completeness of the trigonometric system and Legendre polynomials as a corollary. In closing sections, the treatments of Bessel functions and the Fourier integral are expanded beyond those given in the first edition, and there is new material on Laplace transforms and on generalized functions.

Chapter 8 is a fairly concise treatment of ordinary differential equations with emphasis on the linear equation and its applications. Problems of forced motion are treated from the "input-output" point of view. Matrices are applied in the discussion of systems of linear equations.

Chapter 9 develops the theory of analytic functions with emphasis on power series, Laurent series and residues, and their applications.

The final chapter, on partial differential equations, lays great stress on the relationship between the problem of forced vibrations of a spring (or a system of springs) and the partial differential equation $\rho u_{tt} + hu_t - k^2\nabla^2 u = F(x, y, z, t)$. By pursuing this idea vigorously the physical meaning of the partial differential equation is made transparent and the mathematical tools used become natural. Numerical methods are also motivated on a physical basis.

Throughout a number of references are made to the text *Calculus and Linear Algebra* by Wilfred Kaplan and Donald J. Lewis (2 vols., New York, John Wiley & Sons, 1970–1971), cited simply as *CLA*.

SUGGESTIONS ON THE USE OF THIS BOOK AS THE TEXT FOR A COURSE. It is recommended that the introductory chapter either be omitted from a course outline or else be taken up very briefly. Its main purpose is for reference and as a "refresher" for the student.

The chapters are independent of each other in the sense that each one can be started with a knowledge of only the simplest notions of the previous ones. The later portions of the chapter may depend on some of the later portions of earlier ones. It is thus possible to construct a course using just the earlier portions of several chapters. The following is an illustration of a plan for a 1-semester course, meeting four hours a week: 0–6, 0–7, 1–1 to 1–4, 1–8, 1–10, 2–1 to 2–12, 2–14 to 2–18, 3–1 to 3–6, 4–1 to 4–4, 4–6 to 4–9, 4–11, 5–1 to 5–13, 5–15, 6–1 to 6–7, 6–11 to 6–19.

If it is desired that one topic be stressed, then the corresponding chapters can be taken up in full detail. For example, Chapters 1, 3, and 5 together provide a very substantial training in vector analysis; Chapters 7 and 10 together contain sufficient material for a one-semester course in partial differential equations.

The author expresses his appreciation to the many colleagues who have given advice and encouragement in the preparation of this book. Professors R. C. F. Bartels, F. E. Hohn, and J. Lehner deserve special thanks and recognition for their thorough criticisms of the manuscript; a number of improvements were made on the basis of their suggestions. Others whose counsel has been of value are Professors R. V. Churchill, C. L. Dolph, G. E. Hay, M. Morkovin, G. Piranian, G. Y. Rainich, L. L. Rauch, M. O. Reade, E. Rothe, H. Samelson, R. Büchi, A. J. Lohwater, W. Johnson, Dr. G. Béguin. To his wife the author expresses his deeply felt appreciation for her aid and counsel in every phase of the arduous task.

To the Addison-Wesley Publishing Company the author expresses his appreciation for their unfailing cooperation and for the high standards of publishing which they have set and maintained.

WILFRED KAPLAN

June 1972

Contents

Chapter 5 Vector Integral Calculus

Part I. Two-Dimensional Theory

Part II. Three-Dimensional Theory and Applications

Chapter 6 Infinite Series

ADVANCED CALCULUS

CHAPTER 0

Review of Algebra, Analytic Geometry and Calculus

In this chapter a review of the fundamentals of algebra, analytic geometry and calculus is presented. The ideas discussed here will serve as the basis of all the theory to follow and reference will often be made to them. Thus this chapter serves both as preparation for the later ones and as a convenient list of formulas and theorems for reference.

0–1 The real number system. The system of real numbers can be thought of as composed of the following:

(a) the *rational numbers:* the positive and negative integers $1, 2, 3, \ldots,$ $-1, -2, -3, \ldots$ and the number 0; the fractions p/q, where p and q are integers;

(b) the *irrational numbers:* numbers expressible as infinite decimals (e.g., $-3.14159\ldots$) but not as ratios of integers.

Together these form a collection of numbers any two of which can be added, subtracted, multiplied, or divided (except for division by zero), subject to the basic rules of algebra:

$$\begin{gathered} a+b=b+a, \quad a\cdot b=b\cdot a, \\ a+(b+c)=(a+b)+c, \quad a\cdot(bc)=(ab)\cdot c, \\ a(b+c)=a\cdot b+a\cdot c, \quad a+0=a, \quad a\cdot 1=a. \end{gathered} \tag{0–1}$$

The real numbers can be identified with the points of an infinite straight line as in Fig. 0–1. On this line a point O has been selected as origin, a unit of length has been chosen and a positive direction assigned. To each number x is then assigned a point P on the line; P is at O if x is 0 and is in the positive or negative direction from O according as x is positive or negative, with the distance OP equal to $|x|$, the absolute value of x (equal to x when x is positive and to $-x$ when x is negative). In this way every number x is represented by one point P and, conversely, each P represents just one number x.

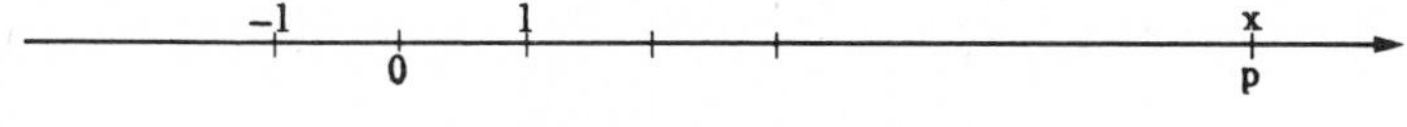

FIG. 0–1. Real numbers.

This geometric picture suggests that the numbers can be ordered: $a > b$ or $b < a$ means simply that $a - b$ is positive or that a lies to the right of b on the above number axis. The $=$ sign, the $>$ or $<$ signs, and the $|\ |$ signs obey the following laws:

Rules of equality:

$$\begin{aligned}&\text{if } a = b \text{ and } b = c, \text{ then } a = c; \quad \text{if } a = b, \text{ then } b = a;\\ &a = a \text{ always}; \quad \text{if } a = a' \text{ and } b = b', \text{ then}\\ &a + b = a' + b' \text{ and } a \cdot b = a' \cdot b'.\end{aligned} \tag{0–2}$$

Rules of inequality:

$$\begin{aligned}&\text{if } a < b \text{ and } b < c, \text{ then } a < c; \quad \text{if } a < b, \text{ then } b > a;\\ &a < a \text{ is impossible}; \quad \text{if } a < a' \text{ and } b < b', \text{ then}\\ &a + b < a' + b' \text{ and, if } a \text{ and } b \text{ are positive, } a \cdot b < a' \cdot b'.\end{aligned} \tag{0–3}$$

Rules for absolute values:

$$\begin{aligned}&|a| \geqq 0; \quad |a| = 0 \text{ if and only if } a = 0;\\ &|a \cdot b| = |a| \cdot |b|; \quad |a + b| \leqq |a| + |b|.\end{aligned} \tag{0–4}$$

0–2 The complex number system. Unless otherwise indicated, the numbers in this text are real. However, the study of the solution of algebraic equations such as

$$x^2 + 1 = 0, \quad x^2 + 2x + 2 = 0$$

leads one to introduce complex numbers of the form $a + bi$, where a and b are real and i, the imaginary unit, has the property: $i^2 = -1$. To each pair of real numbers a, b there corresponds one complex number $a + bi$ and conversely.

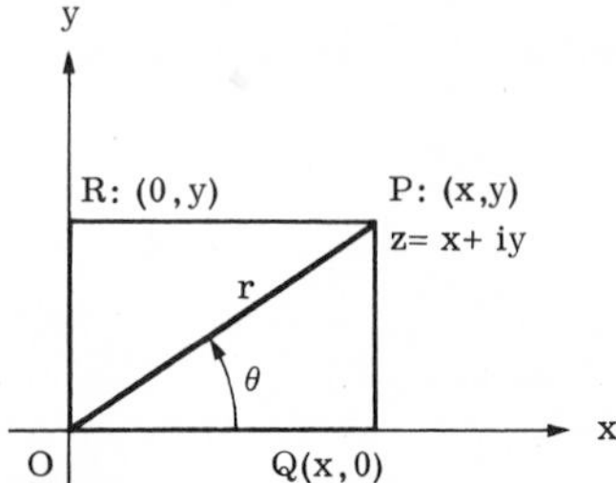

FIG. 0–2. Complex numbers.

The complex numbers can be identified with the points of a plane, the xy plane, in which two perpendicular axes (directed lines) have been chosen and a unit of length has been selected, as in Fig. 0–2. To the complex number $z = x + iy$ corresponds the point P with rectangular coordinates (x, y). The number $0 + 0i = 0$ is represented by the origin O, the point of intersection of the x and y axes. The numbers $x + 0i = x$ (*real* numbers) are represented by the points $(x, 0)$ on the x axis just as on the above number axis; the numbers $0 + iy = iy$ (*pure imaginary* numbers) are represented similarly by the points $(0, y)$ on the y axis. The general complex number $z = x + iy$ is represented by the point P whose projection Q on the x axis is $(x, 0)$ and whose projection R on the y axis is $(0, y)$; x is termed the *real part* of z and y the *imaginary part* of z. If $z = x + iy$, then we write: $\bar{z} = x - iy$ and call $\bar{z}$ the *complex conjugate* of z.

The distance $r = OP$ is termed the *absolute value* or *modulus* of the complex number z and is denoted by $|z|$. By the Pythagorean theorem,

$$|z| = r = \sqrt{x^2 + y^2}. \tag{0–5}$$

The signed angle $\theta = \measuredangle XOP$, measured from the positive x axis to OP, is the *argument* or *amplitude* of $x + iy$; angles will be measured positively in the counterclockwise direction and in terms of radians, unless otherwise indicated, so that a complete cycle is 2π. One has then

$$\arg z = \theta = \text{arc sin}\,\frac{y}{r} = \text{arc cos}\,\frac{x}{r} = \text{arc tan}\,\frac{y}{x}, \tag{0–6}$$

as in trigonometry, θ being determined for given z only up to multiples of 2π. The numbers r, θ are the *polar coordinates* of P.

The complex numbers can be added, subtracted, multiplied, and divided (except for division by zero) and obey the same algebraic rules (0–1) as the real numbers. One has further

$$\begin{aligned}(x_1 + iy_1) + (x_2 + iy_2) &= (x_1 + x_2) + i(y_1 + y_2),\\ (x_1 + iy_1)\cdot(x_2 + iy_2) &= x_1x_2 - y_1y_2 + i(x_1y_2 + x_2y_1).\end{aligned} \tag{0–7}$$

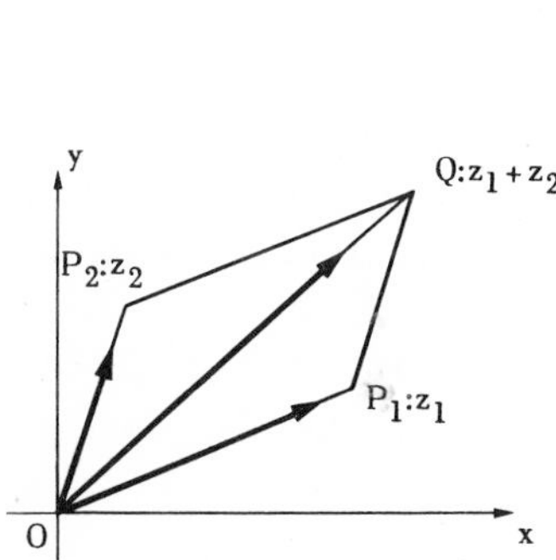

FIG. 0–3. Addition of complex numbers.

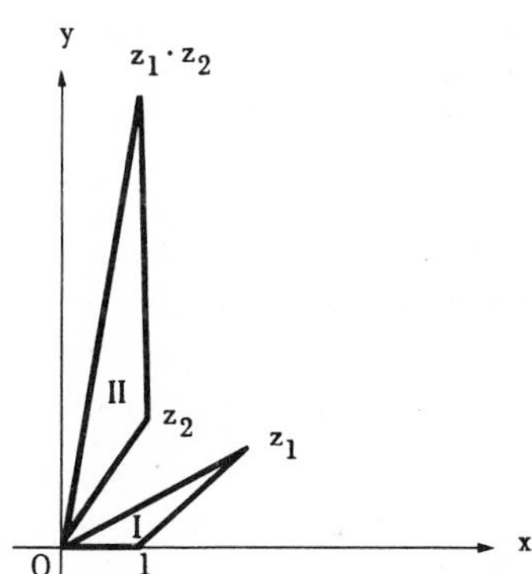

FIG. 0–4. Multiplication of complex numbers.

The first part of (0–7) shows that addition of complex numbers is in accordance with the "parallelogram law" by which forces are added in mechanics, as illustrated in Fig. 0–3. The second part of (0–7), when written in polar coordinates, is the basis of a graphical construction of the product of two complex numbers:

$$\begin{aligned}z_1 \cdot z_2 &= (r_1 \cos\theta_1 + ir_1 \sin\theta_1)\cdot(r_2\cos\theta_2 + ir_2\sin\theta_2)\\ &= r_1r_2[\cos\theta_1\cos\theta_2 - \sin\theta_1\sin\theta_2 + i(\sin\theta_1\cos\theta_2 + \cos\theta_1\sin\theta_2)]\\ &= r_1r_2[\cos(\theta_1+\theta_2) + i\sin(\theta_1+\theta_2)].\end{aligned} \tag{0–8}$$

Accordingly,

$$|z_1\cdot z_2| = |z_1|\cdot|z_2|, \quad \arg(z_1\cdot z_2) = \arg z_1 + \arg z_2; \tag{0–9}$$

from this it follows that the triangles I and II of Fig. 0–4 are similar. Thus $z_1 \cdot z_2$ can be constructed graphically from z_1 and z_2.

The rules of equality for complex numbers are the same as for real numbers, but inequalities between complex numbers have no meaning. The rules for absolute values are the same as for real numbers. The rule

$$|z_1 + z_2| \leqq |z_1| + |z_2| \tag{0–10}$$

expresses the geometric condition: $OQ \leqq OP_1 + P_1Q$ of Fig. 0–3.

0–3 Algebra of real and complex numbers. If x is a real number and n is a positive integer, then x^n, the nth power of x, is defined as $x \cdot x \cdots x$ (n factors). One then verifies the rules

$$x^m \cdot x^n = x^{m+n}, \quad (x^m)^n = x^{mn}. \tag{0–11}$$

A polynomial in x of degree n is an expression of form

$$a_0x^n + a_1x^{n-1} + \cdots + a_{n-1}x + a_n,$$

where $a_0, a_1, \ldots, a_n$ are real numbers and $a_0 \neq 0$. Thus $x^2 + 2x - 3$ is a polynomial in x of degree 2.

An algebraic equation in x of degree n is a polynomial in x of degree n set equal to 0, e.g.,

$$x^2 - 4x - 5 = 0.$$

It is shown in algebra that such an equation has at most n real roots. In particular, if $a > 0$, the equation $x^n = a$ has precisely one positive real root, denoted by $\sqrt[n]{a}$ or by $a^{\frac{1}{n}}$.

The preceding definitions extend at once to powers, polynomials, and algebraic equations involving complex numbers. Thus

$$z^2 + 1 = 0$$

is an algebraic equation of degree 2; so also is

$$(1 - i)z^2 + iz - 1 = 0,$$

in which the coefficients are complex. It is shown in advanced mathematics that an equation of degree n has n complex roots, some of which may coincide; if the coefficients are real, then the imaginary roots come in conjugate pairs.

A linear equation is one of first degree, e.g.,

$$3x - 2 = 0, \quad 5z + 4i = 0.$$

This has always just one solution: $ax + b = 0$ has the root $x = -(b/a)$.

A quadratic equation is one of second degree, e.g.,

$$x^2 - 5x + 6 = 0.$$

The general equation

$$az^2 + bz + c = 0$$

has the roots

$$z = \frac{-b \pm \sqrt{b^2 - 4ac}}{2a}. \tag{0–12}$$

If a, b, c are real, then the character of the roots is determined by the *discriminant:* $b^2 - 4ac$. If $b^2 - 4ac > 0$, then the roots are real and distinct; if $b^2 - 4ac = 0$, then the roots are real and equal; if $b^2 - 4ac < 0$, the roots are the conjugate complex numbers $x \pm iy$.

Explicit formulas are available for solving the general third and fourth degree equations. [See L. E. Dickson, *First Course in the Theory of Equations* (New York: Wiley, 1922) Chapter IV.] For equations of higher degree there are no explicit formulas. Numerical methods are available for finding real and complex roots of equations of high degree; see Chapter 10 of *Introduction to Numerical Analysis* by F. B. Hildebrand (New York: McGraw-Hill, 1956).

Equations of form

$$z^n = a$$

can be solved for complex z, with a real or complex. The roots are the "nth roots of a." The solution is based on the Demoivre formula

$$(\cos\theta + i\sin\theta)^n = \cos n\theta + i\sin n\theta, \tag{0–13}$$

which follows from the rule (0–8) for multiplication. From this one concludes that, if a has polar coordinates r, θ, so that $a = r(\cos\theta + i\sin\theta)$, then

$$\sqrt[n]{a} = \sqrt[n]{r}\left[\cos\left(\frac{\theta}{n} + k\frac{2\pi}{n}\right) + i\sin\left(\frac{\theta}{n} + k\frac{2\pi}{n}\right)\right], \tag{0–14}$$

where k ranges over the values $0, 1, 2, \ldots, n-1$ and $\sqrt[n]{r}$ is the positive real nth root of r. This is illustrated in Fig. 0–5, with $n = 5$.

A system of *simultaneous linear equations* is a system such as

$$a_1x + b_1y + c_1z = k_1, \quad a_2x + b_2y + c_2z = k_2, \quad a_3x + b_3y + c_3z = k_3. \tag{0–15}$$

Here there are three equations in the three unknowns x, y, z; in general one may have n equations in m unknowns. These can in general be solved by combining equations to successively eliminate variables.

For n equations in n unknowns the solution can be expressed in terms of determinants, the theory of which is discussed below. One denotes by D the "determinant of the coefficients"; for (0–15) this is the determinant

$$D = \begin{vmatrix} a_1 & b_1 & c_1 \\ a_2 & b_2 & c_2 \\ a_3 & b_3 & c_3 \end{vmatrix}. \tag{0–16}$$

In general D is an nth order determinant. One denotes by $D_1, D_2, \ldots$

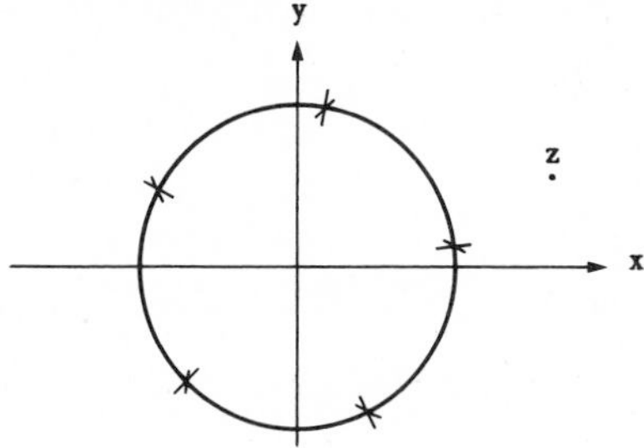

FIG. 0–5. Fifth roots of $z = 2 + i$.

the determinants obtained from D by replacing the first, second, . . . column of D by the numbers $k_1, k_2, \ldots, k_n$. Thus for (0–15)

$$D_2 = \begin{vmatrix} a_1 & k_1 & c_1 \\ a_2 & k_2 & c_2 \\ a_3 & k_3 & c_3 \end{vmatrix}. \tag{0–17}$$

The solution of the system of equations is then given by

$$x = \frac{D_1}{D}, \quad y = \frac{D_2}{D}, \quad \ldots; \tag{0–18}$$

this is Cramer's Rule. If $D \neq 0$, (0–18) gives the one and only solution. If $k_1 = k_2 = k_3 = 0$ (*homogeneous case*) and $D \neq 0$, (0–18) gives the trivial solution $x = y = z = 0$; if $D = 0$, there are infinitely many solutions.

For n equations in m unknowns, with $n < m$, one can try to solve as above for n unknowns in terms of the remaining $m - n$ unknowns. This will be possible, provided the corresponding determinant of coefficients D for these n unknowns is not 0. For $n > m$, there are more equations than unknowns; the equations are contradictory unless certain determinants are 0. For a full discussion the reader is referred to Chapter VIII of the book by Dickson cited above. See also Section 1–13 below.

For linear equations such as (0–15) the coefficients and unknowns may be real or complex numbers; the form of the solutions is the same in either case.

The properties of determinants needed in this book are listed here:

$$\begin{vmatrix} a & b \\ c & d \end{vmatrix} = ad - bc; \tag{0–19}$$

$$\begin{vmatrix} a_1 & b_1 & c_1 \\ a_2 & b_2 & c_2 \\ a_3 & b_3 & c_3 \end{vmatrix} = a_1 \begin{vmatrix} b_2 & c_2 \\ b_3 & c_3 \end{vmatrix} - b_1 \begin{vmatrix} a_2 & c_2 \\ a_3 & c_3 \end{vmatrix} + c_1 \begin{vmatrix} a_2 & b_2 \\ a_3 & b_3 \end{vmatrix}; \tag{0–20}$$

$$\begin{vmatrix} a_1 & b_1 & c_1 & d_1 \\ a_2 & b_2 & c_2 & d_2 \\ a_3 & b_3 & c_3 & d_3 \\ a_4 & b_4 & c_4 & d_4 \end{vmatrix} = a_1 \begin{vmatrix} b_2 & c_2 & d_2 \\ b_3 & c_3 & d_3 \\ b_4 & c_4 & d_4 \end{vmatrix} - b_1 \begin{vmatrix} a_2 & c_2 & d_2 \\ a_3 & c_3 & d_3 \\ a_4 & c_4 & d_4 \end{vmatrix} + \cdots. \tag{0–21}$$

Thus, in general, a determinant of nth order is defined in terms of determinants of $(n - 1)$st order; the coefficients of $a_1, b_1, \ldots$ on the right side of (0–20) and (0–21) are the "cofactors" of these elements.

$$\begin{vmatrix} a_1 & b_1 & c_1 \\ a_2 & b_2 & c_2 \\ a_3 & b_3 & c_3 \end{vmatrix} = \begin{vmatrix} a_1 & a_2 & a_3 \\ b_1 & b_2 & b_3 \\ c_1 & c_2 & c_3 \end{vmatrix}; \tag{0–22}$$

in general, rows and columns can be interchanged.

$$\begin{vmatrix} a_1 & b_1 & c_1 \\ a_2 & b_2 & c_2 \\ a_3 & b_3 & c_3 \end{vmatrix} = - \begin{vmatrix} a_2 & b_2 & c_2 \\ a_1 & b_1 & c_1 \\ a_3 & b_3 & c_3 \end{vmatrix}; \tag{0–23}$$

in general, interchanging two rows (or columns) multiplies the determinant by -1.

$$\begin{vmatrix} ka_1 & kb_1 & kc_1 \\ a_2 & b_2 & c_2 \\ a_3 & b_3 & c_3 \end{vmatrix} = k \begin{vmatrix} a_1 & b_1 & c_1 \\ a_2 & b_2 & c_2 \\ a_3 & b_3 & c_3 \end{vmatrix}; \tag{0–24}$$

a factor of any row (or column) can be placed before the determinant.

$$\begin{vmatrix} ka_1 & kb_1 & kc_1 \\ a_1 & b_1 & c_1 \\ a_2 & b_2 & c_2 \end{vmatrix} = 0; \tag{0–25}$$

if two rows (or columns) are proportional, the determinant equals 0.

$$\begin{vmatrix} a_1 & b_1 & c_1 \\ a_2 & b_2 & c_2 \\ a_3 & b_3 & c_3 \end{vmatrix} + \begin{vmatrix} A_1 & b_1 & c_1 \\ A_2 & b_2 & c_2 \\ A_3 & b_3 & c_3 \end{vmatrix} = \begin{vmatrix} a_1 + A_1 & b_1 & c_1 \\ a_2 + A_2 & b_2 & c_2 \\ a_3 + A_3 & b_3 & c_3 \end{vmatrix}; \tag{0–26}$$

this indicates how two determinants differing only in one row or column can be added.

$$\begin{vmatrix} a_1 & b_1 & c_1 \\ a_2 & b_2 & c_2 \\ a_3 & b_3 & c_3 \end{vmatrix} = \begin{vmatrix} a_1 + ka_2 & b_1 + kb_2 & c_1 + kc_2 \\ a_2 & b_2 & c_2 \\ a_3 & b_3 & c_3 \end{vmatrix}; \tag{0–27}$$

the value of the determinant is unchanged if the elements of one row are multiplied by the same quantity k and added to the corresponding elements of another row. By suitable choice of k, one can use this rule to introduce zeros; by repetition of the process, one can reduce all elements but one in a chosen row to zero. This procedure is basic for numerical evaluation of determinants.

Proofs of these rules and further properties are given in Chapter VIII of the book by Dickson mentioned above.

Three other formulas of algebra will be useful:

$$a + (a + d) + (a + 2d) + \cdots + [a + (n - 1)d] = n\frac{a + [a + (n - 1)d]}{2}; \tag{0–28}$$

$$a + ar + ar^2 + \cdots + ar^{n-1} = a\frac{1 - r^n}{1 - r} \quad (r \neq 1); \tag{0–29}$$

$$(a + b)^n = a^n + na^{n-1}b + \frac{n(n - 1)}{2!}a^{n-2}b^2 + \cdots + \binom{n}{r}a^{n-r}b^r + \cdots + b^n, \tag{0–30}$$

where

$$\binom{n}{r} = \frac{n(n - 1)\cdots(n - r + 1)}{r!} = \frac{n!}{r!(n - r)!}. \tag{0–31}$$

These give respectively the sum of an *arithmetic progression*, the sum of a *geometric progression*, and the *binomial theorem*. Throughout n is a positive integer and $n!$ (read "n factorial") is defined as follows:

$$0! = 1, \qquad n! = 1\cdot 2 \cdots n \quad \text{for} \quad n = 1, 2, 3, \ldots. \tag{0–32}$$

These formulas can be proved by the *principle of induction:* If a theorem concerning integers n is known to be true for $n = 1$ and if the assumed truth of this theorem for $n = k$ implies its truth for $n = k + 1$, then the theorem is true for all positive integers n.

0–4 Plane analytic geometry. The rectangular coordinate system is introduced in the plane as in Section 0–2 above. Each point has then coordinates x (abscissa) and y (ordinate). The distance d between the points $P_1\colon(x_1, y_1)$ and $P_2\colon(x_2, y_2)$ is given by the formula

$$d = \sqrt{(x_2 - x_1)^2 + (y_2 - y_1)^2}. \tag{0–33}$$

This follows from the Pythagorean theorem.

The set of points which satisfy a linear equation

$$Ax + By + C = 0 \tag{0–34}$$

(A, B, C any real numbers, A and B not both 0) is a straight line and every straight line can be so represented. The *slope* of the line is defined as

$$m = \tan\omega, \tag{0–35}$$

where ω is the angle from the positive x axis to the line, as shown in Fig. 0–6; the slope is undefined (equal to ∞) for lines parallel to the y axis. If (x_1, y_1) and (x_2, y_2) are two distinct points on a line, then the slope is

$$m = \frac{y_2 - y_1}{x_2 - x_1}. \tag{0–36}$$

Two lines are parallel precisely when their slopes are equal; they are perpendicular when the slope of one is the negative reciprocal of the slope of the other.

The line through (x_1, y_1) with slope m has the equation

$$y - y_1 = m(x - x_1). \tag{0–37}$$

After division by $\sqrt{A^2 + B^2}$, the general equation (0–34) can be written in the *normal form:*

$$x\cos\alpha + y\sin\alpha = p, \tag{0–38}$$

where $\alpha = \omega \pm \pi/2$ is the angle from the positive x axis to a line through O perpendicular to the line and p is the perpendicular distance from O to the line, as in Fig. 0–6.

The *equation* of *second degree* in x and y:

$$Ax^2 + Bxy + Cy^2 + Dx + Ey + F = 0 \tag{0–39}$$

has as locus a circle (if $A = C, B = 0$), an ellipse (if $B^2 - 4AC < 0$), a hyperbola (if $B^2 - 4AC > 0$), or a parabola (if $B^2 - 4AC = 0$). These loci can be degenerate: e.g., point ellipse, pair of lines.

The points of the xy plane can also be described by polar coordinates r, θ, as in Section 0–2 above. In terms of polar coordinates, the equation (0–38) of the straight line becomes

$$r \cos (\theta - \alpha) = p \tag{0–40}$$

and the conic sections (0–39) with focus at O become

$$r = \frac{l}{1 + e \cos (\theta - \beta)}, \tag{0–41}$$

where e (the eccentricity), β, and l are constants.

0–5 Solid analytic geometry. If three mutually perpendicular directed lines Ox, Oy, Oz, meeting at O, are chosen in space, then a coordinate system is determined in space. Points $(x, 0, 0)$, $(0, y, 0)$, and $(0, 0, z)$ are located on the corresponding axes as in the plane and a general triple (x, y, z) corresponds to the point P whose projections on the three axes are $(x, 0, 0)$, $(0, y, 0)$, and $(0, 0, z)$ as illustrated in Fig. 0–7.

A repeated application of the Pythagorean theorem shows that the distance d between two points (x_1, y_1, z_1) and (x_2, y_2, z_2) of space is given by

$$d = \sqrt{(x_2 - x_1)^2 + (y_2 - y_1)^2 + (z_2 - z_1)^2}. \tag{0–42}$$

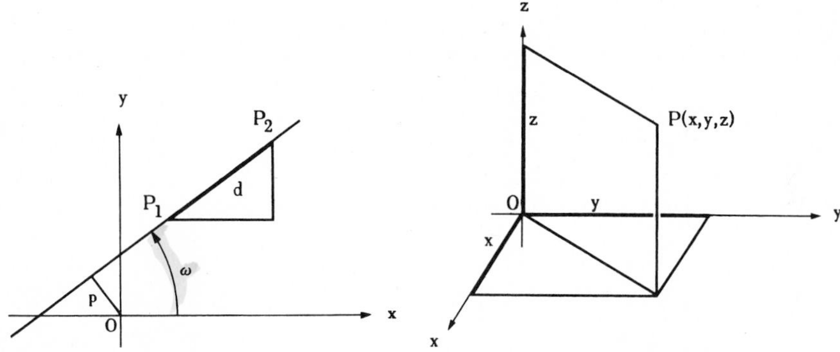

FIG. 0–6. FIG. 0–7. Coordinates in space.

A straight line L in space is assigned sets of *direction numbers* or *direction components* by the following procedure. Let $P_1: (x_1, y_1, z_1)$ and $P_2: (x_2, y_2, z_2)$ be two distinct points on L. Then the three numbers $a = x_2 - x_1$, $b = y_2 - y_1$, $c = z_2 - z_1$ (in that order) form a set of direction numbers. Other sets are obtained by varying the two points on L. If a', b', c' is a second set, then these numbers are in the same ratio as a, b, c; i.e., one has

$$a:b:c = a':b':c'. \tag{0–43}$$

Further, if a', b', c' is any triple of numbers, other than 0, 0, 0, such that (0–43) holds, then a', b', c' is a set of direction numbers of L.

It follows that L_1 and L_2 are parallel or coincident lines if and only if

$$a_1:b_1:c_1 = a_2:b_2:c_2 \tag{0–44}$$

for corresponding sets of direction numbers.

If L passes through the origin: $(0, 0, 0)$, then P_1 can be chosen as the origin; one concludes that, in this case, the coordinates (x, y, z) of any other point on L form a set of direction numbers of L.

If L is a line through (x_1, y_1, z_1) with direction numbers a, b, c, then every point (x, y, z) on L must satisfy the condition:

$$x - x_1:y - y_1:z - z_1 = a:b:c \tag{0–45}$$

or (if none of the direction numbers is 0)

$$\frac{x - x_1}{a} = \frac{y - y_1}{b} = \frac{z - z_1}{c}, \tag{0–46}$$

which are symmetric equations for L. If the common value of the three ratios is denoted by t, one obtains

$$x = x_1 + at, \quad y = y_1 + bt, \quad z = z_1 + ct, \tag{0–47}$$

which are parametric equations for L; as the "parameter" t ranges over all real numbers, (x, y, z) moves along L and to each point of L corresponds precisely one t.

The (unsigned) angle θ between two intersecting directed lines L_1, L_2 is defined as in plane geometry. If L_1 and L_2 are nonintersecting directed lines, parallel and similarly directed lines L_1' and L_2' can be drawn through a point P' of space; the angle θ between L_1 and L_2 is defined to be the same as that between L_1' and L_2'. This does not depend on the choice of P'.

The angles α, β, γ between a directed line L and the (directed) x, y, and z axes are called the direction angles of L. The quantities

$$l = \cos \alpha, \quad m = \cos \beta, \quad n = \cos \gamma \tag{0–48}$$

are termed direction cosines of L. If the direction on L is reversed, l, m, and n all change sign. It can be shown that l, m, n form a set of direction numbers for L and that

$$l^2 + m^2 + n^2 = 1; \tag{0–49}$$

conversely, if (a, b, c) is any set of direction numbers for L and

$$a^2 + b^2 + c^2 = 1, \tag{0–50}$$

then $a = l$, $b = m$, $c = n$ for appropriate direction on L. From this it follows that, if a, b, c is a set of direction numbers, then

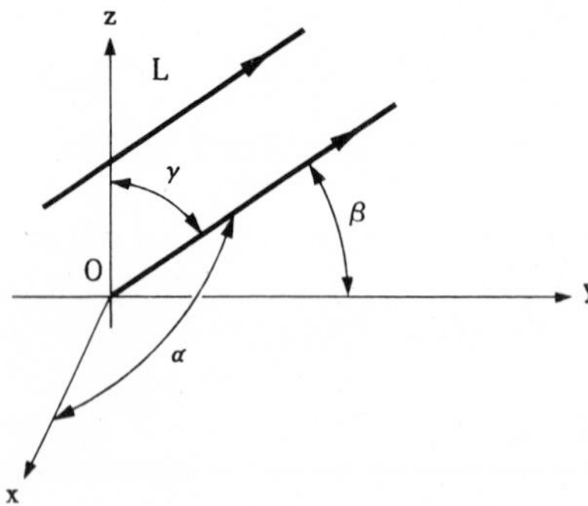

Fig. 0–8. Direction angles.

$$l = \frac{a}{\sqrt{a^2 + b^2 + c^2}}, \quad m = \frac{b}{\sqrt{a^2 + b^2 + c^2}}, \quad n = \frac{c}{\sqrt{a^2 + b^2 + c^2}} \tag{0–51}$$

is a set of direction cosines for L.

If L_1 and L_2 are two directed lines in space and θ is the angle between them, then

$$\cos\theta = l_1 l_2 + m_1 m_2 + n_1 n_2, \tag{0–52}$$

where l_1, m_1, n_1 and l_2, m_2, n_2 are direction cosines for L_1 and L_2 respectively. This can be proved by applying the law of cosines to the triangle with vertices at $(0, 0, 0)$, (l_1, m_1, n_1), (l_2, m_2, n_2), two of whose sides have unit length and are parallel to L_1 and L_2 respectively. From (0–52) it follows that L_1 and L_2 are perpendicular if and only if

$$l_1 l_2 + m_1 m_2 + n_1 n_2 = 0 \tag{0–53}$$

and that L_1 and L_2 are perpendicular if and only if

$$a_1 a_2 + b_1 b_2 + c_1 c_2 = 0, \tag{0–54}$$

where a_1, b_1, c_1 and a_2, b_2, c_2 are sets of direction numbers for L_1 and L_2.

Every equation of first degree:

$$Ax + By + Cz + D = 0 \tag{0–55}$$

represents a plane in space and every plane can be so represented. If (x_1, y_1, z_1) is a point of the plane, then the equation can be written thus:

$$A(x - x_1) + B(y - y_1) + C(z - z_1) = 0; \tag{0–56}$$

this shows that A, B, C is a set of direction numbers of a line perpendicular to every line in the plane, i.e., of a *normal* to the plane. It follows that two planes

$$A_1x + B_1y + C_1z + D_1 = 0,$$
$$A_2x + B_2y + C_2z + D_2 = 0$$

are parallel or coincident if and only if

$$A_1 : B_1 : C_1 = A_2 : B_2 : C_2 \tag{0–57}$$

and are perpendicular if and only if

$$A_1A_2 + B_1B_2 + C_1C_2 = 0. \tag{0–58}$$

After division by $\pm\sqrt{A^2 + B^2 + C^2}$, the general equation (0–55) can be put in the normal form:

$$lx + my + nz = p, \tag{0–59}$$

where l, m, n are the direction cosines of a normal to the plane through the origin and p is the perpendicular distance from the origin to the plane.

The equations of second degree:

$$Ax^2 + By^2 + Cz^2 + Dxy + Exz + Fyz + Gx + Hy + Jz + K = 0 \tag{0–60}$$

represent quadric surfaces in space. These include the ellipsoid (with the sphere a special case), the hyperboloid of one sheet, the hyperboloid of two sheets, the elliptic paraboloid, the hyperbolic paraboloid, the elliptic cone, and quadric cylinders.

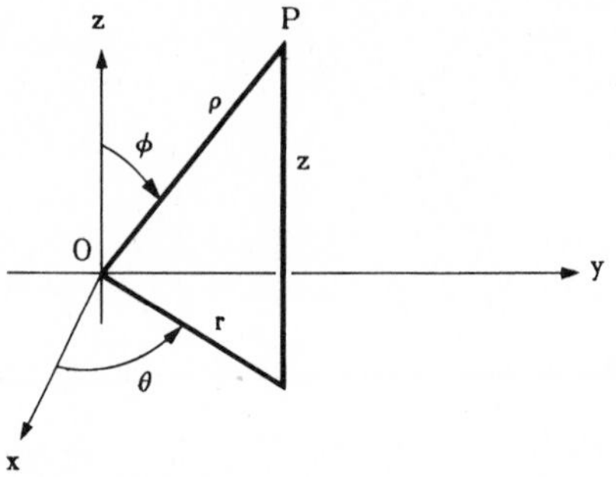

Two other coordinate systems are commonly used in space: cylindrical coordinates and spherical coordinates. Cylindrical coordinates are obtained by replacing two of the rectangular coordinates by the corresponding polar coordinates, e.g., as in Fig. 0–9, (x, y) by (r, θ); thus (r, θ, z) form the cylindrical coordinates of P. Spherical coordinates use a polar coordinate angle in one coordinate plane, the distance ρ from the origin O to P and the angle between OP and the third axis; e.g., as in Fig. 0–9, one can use the angles θ and ϕ and the distance ρ. Relations between the rectangular, cylindrical, and spherical coordinate systems can be obtained from the following formulas:

FIG. 0–9. Cylindrical and spherical coordinates.

$$x = r\cos\theta, \quad y = r\sin\theta, \qquad z = \rho\cos\phi, \quad r = \rho\sin\phi; \tag{0–61}$$

thus one has immediately the equations:

$$x = \rho\sin\phi\cos\theta, \quad y = \rho\sin\phi\sin\theta, \quad z = \rho\cos\phi, \tag{0–62}$$

connecting spherical and rectangular coordinates.

0–6 Vectors. We consider vectors in 3-dimensional space. A vector is a combination of a *magnitude* (positive real number) and a *direction*. Each vector can be represented by a directed line segment PQ, but can equally well be represented by a directed line segment $P'Q'$, parallel to PQ and having the same direction as PQ. It will be convenient to denote by $\overrightarrow{PQ}$ both the directed line segment and the vector which it represents. Accordingly, we write $\overrightarrow{PQ} = \overrightarrow{P'Q'} = \mathbf{a}$, where $\mathbf{a}$ is the vector. In general, vectors are denoted by boldface letters: $\mathbf{a}$, $\mathbf{b}$, $\mathbf{u}$, $\mathbf{v}$, $\mathbf{F}$, $\mathbf{M}$, ... The magnitude (also called *length* or *norm*) of $\mathbf{a}$ is denoted by $|\mathbf{a}|$ or by a.

It is convenient to define a zero vector $\mathbf{0}$ having magnitude 0 and ambiguous direction, so that the vector $\mathbf{0}$ is both parallel to and perpendicular to every vector $\mathbf{a}$.

Addition of vectors is defined by the parallelogram rule of Fig. 0–3. Thus in general, $\overrightarrow{PQ} + \overrightarrow{QR} = \overrightarrow{PR}$. Given $\mathbf{a}$ and $\mathbf{b}$, there is a unique vector $\mathbf{c}$ such that $\mathbf{b} + \mathbf{c} = \mathbf{a}$; we write $\mathbf{c} = \mathbf{a} - \mathbf{b}$, thereby defining *subtraction*.

In the theory of vectors one refers to real numbers as *scalars*. If k is a scalar and $\mathbf{a}$ a vector, then $k\mathbf{a}$ is defined to be $\mathbf{0}$ if k is 0 or $\mathbf{a}$ is $\mathbf{0}$, and otherwise to be a vector whose magnitude is $k|\mathbf{a}|$ and whose direction is the same as that of $\mathbf{a}$ or opposite to that of $\mathbf{a}$ according as $k > 0$ or $k < 0$.

From these definitions one has the rules:

$$\mathbf{a} + \mathbf{b} = \mathbf{b} + \mathbf{a}, \qquad (\mathbf{a} + \mathbf{b}) + \mathbf{c} = \mathbf{a} + (\mathbf{b} + \mathbf{c}); \tag{0–63}$$
$$\mathbf{a} + \mathbf{0} = \mathbf{a}, \qquad \mathbf{a} - \mathbf{a} = \mathbf{0}; \tag{0–64}$$
$$1\mathbf{a} = \mathbf{a}, \; 0\mathbf{a} = \mathbf{0}, \qquad (h_1h_2)\mathbf{a} = h_1(h_2\mathbf{a}); \tag{0–65}$$
$$h_1(\mathbf{a} + \mathbf{b}) = h_1\mathbf{a} + h_1\mathbf{b}, \qquad \mathbf{c} - \mathbf{a} = \mathbf{c} + (-\mathbf{a}), \tag{0–66}$$
$$|\mathbf{a}| \geqq 0, \qquad |\mathbf{a}| = 0 \text{ if and only if } \mathbf{a} = \mathbf{0}; \tag{0–67}$$
$$|\mathbf{a} + \mathbf{b}| \leqq |\mathbf{a}| + |\mathbf{b}| \qquad \text{(Triangle inequality)}. \tag{0–68}$$

Vectors $\mathbf{a}$, $\mathbf{b}$ are said to be *collinear* or *linearly dependent* if there are scalars h_1, h_2, not both zero, such that

$$h_1\mathbf{a} + h_2\mathbf{b} = \mathbf{0}.$$

This is equivalent to asserting that $\mathbf{a}$ and $\mathbf{b}$ can be represented by line segments on the same line.

Vectors $\mathbf{a}$, $\mathbf{b}$, $\mathbf{c}$ are said to be *coplanar* or *linearly dependent* if there are scalars h_1, h_2, h_3, not all zero, such that

$$h_1\mathbf{a} + h_2\mathbf{b} + h_3\mathbf{c} = \mathbf{0}.$$

In this case $\mathbf{a}$, $\mathbf{b}$, $\mathbf{c}$ can be represented by line segments in the same plane.

If $\mathbf{a}$, $\mathbf{b}$, or $\mathbf{a}$, $\mathbf{b}$, $\mathbf{c}$ are not linearly dependent, then they are said to be *linearly independent.*

If $\mathbf{a}$, $\mathbf{b}$ are linearly independent, then every vector $\mathbf{c}$ coplanar with $\mathbf{a}$, $\mathbf{b}$ can be represented uniquely as a linear combination of $\mathbf{a}$ and $\mathbf{b}$:

$$\mathbf{c} = h_1\mathbf{a} + h_2\mathbf{b}.$$

The *angle* θ between two nonzero vectors $\mathbf{a}$ and $\mathbf{b}$ is defined to be $\measuredangle AOB$, where O is any point of space and A, B are chosen so that

$$\overrightarrow{OA} = \mathbf{a}, \qquad \overrightarrow{OB} = \mathbf{b}.$$

It will be assumed that θ is measured in radians and that $0 \leqq \theta \leqq \pi$. We write: $\theta = \measuredangle(\mathbf{a}, \mathbf{b})$.

If $\mathbf{a}$, $\mathbf{b}$, $\mathbf{c}$ are linearly independent, then every vector $\mathbf{d}$ in space can be represented uniquely as a linear combination of $\mathbf{a}$, $\mathbf{b}$, and $\mathbf{c}$:

$$\mathbf{d} = h_1\mathbf{a} + h_2\mathbf{b} + h_3\mathbf{c}.$$

The *scalar product* (*dot product* or *inner product*) of $\mathbf{a}$ and $\mathbf{b}$, denoted by $\mathbf{a} \cdot \mathbf{b}$, is defined to be 0 if $\mathbf{a}$ or $\mathbf{b}$ is $\mathbf{0}$, and otherwise to be the scalar

$$\mathbf{a} \cdot \mathbf{b} = ab \cos\theta \qquad (a = |\mathbf{a}|,\; b = |\mathbf{b}|,\; \theta = \measuredangle(\mathbf{a}, \mathbf{b})). \tag{0–69}$$

Here one calls $b \cos\theta$ the component of $\mathbf{b}$ in the direction of $\mathbf{a}$ and denotes this number by $\text{comp}_a\,\mathbf{b}$. Thus one can write

$$\mathbf{a} \cdot \mathbf{b} = a\,\text{comp}_a\,\mathbf{b} = b\,\text{comp}_b\,\mathbf{a}. \tag{0–70}$$

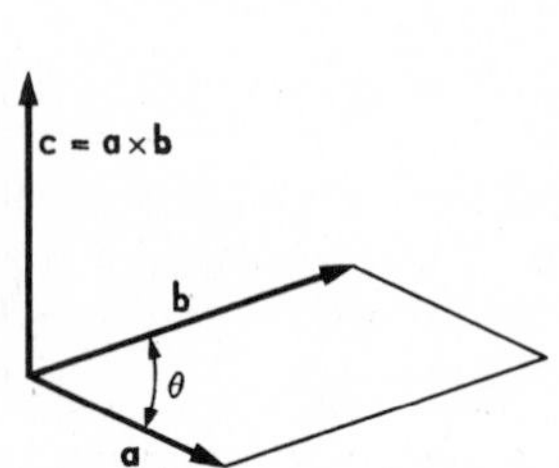

FIG. 0–10. The cross product (vector product).

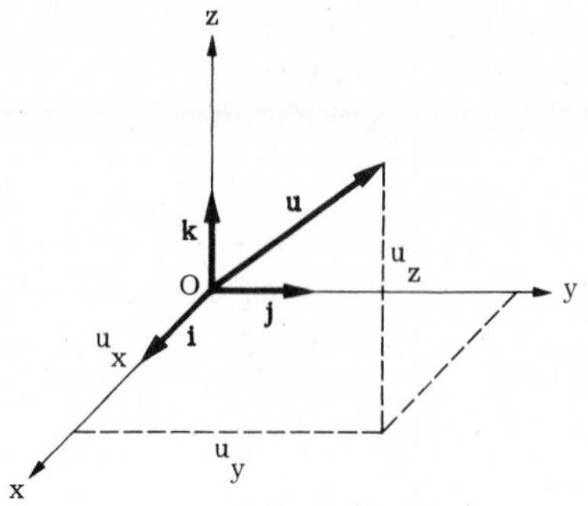

FIG. 0–11. Basis.

In mechanics the scalar product $\mathbf{F}\cdot\overrightarrow{AB}$ of force $\mathbf{F}$ and displacement vector $\overrightarrow{AB}$ is the *work* done by the constant force $\mathbf{F}$ acting on a particle moving from A to B. By (0–69), $\mathbf{F}\cdot\overrightarrow{AB}$ equals the length of $\overrightarrow{AB}$ times the component of $\mathbf{F}$ in the direction of motion, and this agrees with the usual definition of work.

The *cross product* or *vector product* of two vectors $\mathbf{a}$, $\mathbf{b}$, denoted by $\mathbf{a}\times\mathbf{b}$, is defined to be $\mathbf{0}$ if $\mathbf{a}$ or $\mathbf{b}$ is $\mathbf{0}$, and otherwise to be the vector $\mathbf{c}$ perpendicular to both $\mathbf{a}$ and $\mathbf{b}$, such that $c = ab\sin\theta$ and such that $\mathbf{a}$, $\mathbf{b}$, $\mathbf{c}$ form a *positively oriented triple of vectors.* For the usual righthanded coordinate system, such a triple is suggested by the thumb, index finger, and middle finger of the right hand; the corresponding fingers of the left hand, in the order named, form a *negatively* oriented triple. The cross product is illustrated in Fig. 0–10.

From the definition it follows that $c = |\mathbf{a}\times\mathbf{b}|$ equals the *area* of the parallelogram whose sides are $\mathbf{a}$ and $\mathbf{b}$, as in Fig. 0–10.

The scalar product and vector product have the following properties:

$$\mathbf{a}\cdot\mathbf{b} = \mathbf{b}\cdot\mathbf{a}, \qquad \mathbf{a}\cdot(\mathbf{b}+\mathbf{c}) = (\mathbf{a}\cdot\mathbf{b}) + (\mathbf{a}\cdot\mathbf{c}); \tag{0–71}$$
$$\mathbf{a}\cdot(h\mathbf{b}) = (h\mathbf{a})\cdot\mathbf{b} = h(\mathbf{a}\cdot\mathbf{b}); \tag{0–72}$$
$$\mathbf{a}\cdot\mathbf{a} = a^2, \qquad \mathbf{a}\cdot\mathbf{b} = 0 \ \text{ if and only if } \ \mathbf{a}\perp\mathbf{b}; \tag{0–73}$$
$$\mathbf{a}\times\mathbf{b} = -\mathbf{b}\times\mathbf{a}, \qquad \mathbf{a}\times(\mathbf{b}+\mathbf{c}) = \mathbf{a}\times\mathbf{b} + \mathbf{a}\times\mathbf{c}; \tag{0–74}$$
$$\mathbf{a}\times(h\mathbf{b}) = (h\mathbf{a})\times\mathbf{b} = h(\mathbf{a}\times\mathbf{b}); \tag{0–75}$$
$$\mathbf{a}\times\mathbf{a} = \mathbf{0}, \qquad \mathbf{a}\times\mathbf{b} = \mathbf{0} \ \text{ if and only if } \ \mathbf{a}\parallel\mathbf{b}. \tag{0–76}$$

Basis vectors, Cartesian components. Let a righthanded Cartesian xyz-coordinate system be introduced in space. Let $\mathbf{i}$ be a unit vector (vector of length 1) having the direction of the positive x-axis; let $\mathbf{j}$, $\mathbf{k}$ be chosen similarly for the y- and z-axes (Fig. 0–11).

We call $\mathbf{i}$, $\mathbf{j}$, $\mathbf{k}$ the *standard basis* for vectors in space. These vectors are linearly independent; hence for each vector $\mathbf{u}$ in space we have a representation:

$$\mathbf{u} = u_x\mathbf{i} + u_y\mathbf{j} + u_z\mathbf{k} \tag{0–77}$$

for unique scalars u_x, u_y, u_z, called the (Cartesian) *components* of $\mathbf{u}$. We also write, more concisely:

$$\mathbf{u} = (u_x, u_y, u_z), \tag{0–78}$$

and thus can identify the vectors in space with the ordered triples of numbers. In particular,

$$\mathbf{0} = (0, 0, 0), \qquad \mathbf{i} = (1, 0, 0), \qquad \mathbf{j} = (0, 1, 0), \qquad \mathbf{k} = (0, 0, 1).$$

We find that

$$\mathbf{i}\cdot\mathbf{i} = 1,\ \mathbf{j}\cdot\mathbf{j} = 1,\ \mathbf{k}\cdot\mathbf{k} = 1;\quad \mathbf{i}\cdot\mathbf{j} = 0,\ \mathbf{j}\cdot\mathbf{k} = 0,\ \mathbf{i}\cdot\mathbf{k} = 0; \tag{0–79}$$

$$\mathbf{i}\times\mathbf{i} = \mathbf{0},\ \mathbf{j}\times\mathbf{j} = \mathbf{0},\ \mathbf{k}\times\mathbf{k} = \mathbf{0};\quad \mathbf{i}\times\mathbf{j} = \mathbf{k},\ \mathbf{j}\times\mathbf{k} = \mathbf{i},\ \mathbf{k}\times\mathbf{i} = \mathbf{j}. \tag{0–80}$$

Vector operations can now be carried out in terms of components:

$$\mathbf{u} \pm \mathbf{v} = (u_x \pm v_x,\ u_y \pm v_y,\ u_z \pm v_z); \tag{0–81}$$

$$h\mathbf{u} = (hu_x,\ hu_y,\ hu_z); \tag{0–82}$$

$$\mathbf{u}\cdot\mathbf{v} = u_xv_x + u_yv_y + u_zv_z; \tag{0–83}$$

$$\mathbf{u}\times\mathbf{v} = (u_yv_z - u_zv_y)\mathbf{i} + (u_zv_x - u_xv_z)\mathbf{j} + (u_xv_y - u_yv_x)\mathbf{k} = \begin{vmatrix} \mathbf{i} & \mathbf{j} & \mathbf{k} \\ u_x & u_y & u_z \\ v_x & v_y & v_z \end{vmatrix}, \tag{0–84}$$

where the determinant is expanded formally by minors of the first row.

Unit vectors, direction cosines. If $\mathbf{u}$ is a unit vector, then for any nonzero vector $\mathbf{v}$, $\mathbf{v}\cdot\mathbf{u} = v\cos\theta$, the component of $\mathbf{v}$ in the direction of $\mathbf{u}$. In particular,

$$\mathbf{v}\cdot\mathbf{i} = (v_x\mathbf{i} + v_y\mathbf{j} + v_z\mathbf{k})\cdot\mathbf{i} = v_x = v\cos\alpha,$$

where $\alpha = \measuredangle(\mathbf{v}, \mathbf{i})$. Similarly,

$$\mathbf{v}\cdot\mathbf{j} = v_y = v\cos\beta, \qquad \mathbf{v}\cdot\mathbf{k} = v_z = v\cos\gamma.$$

We call α, β, γ the direction angles of $\mathbf{v}$ and $\cos\alpha$, $\cos\beta$, $\cos\gamma$ the direction cosines of $\mathbf{v}$. When $\mathbf{v}$ is itself a unit vector, then

$$\mathbf{v} = v_x\mathbf{i} + v_y\mathbf{j} + v_z\mathbf{k} = \cos\alpha\mathbf{i} + \cos\beta\mathbf{j} + \cos\gamma\mathbf{k}.$$

Otherwise,

$$\mathbf{v} = v(\cos\alpha\mathbf{i} + \cos\beta\mathbf{j} + \cos\gamma\mathbf{k}),$$

and

$$\mathbf{v}/v = \cos\alpha\mathbf{i} + \cos\beta\mathbf{j} + \cos\gamma\mathbf{k}$$

is a unit vector having the same direction as $\mathbf{v}$.

The *scalar triple product* of $\mathbf{u}$, $\mathbf{v}$, $\mathbf{w}$ is the scalar $\mathbf{u}\times\mathbf{v}\cdot\mathbf{w}$. One has $\mathbf{u}\times\mathbf{v}\cdot\mathbf{w} = \mathbf{w}\times\mathbf{u}\cdot\mathbf{v} = \mathbf{v}\times\mathbf{w}\cdot\mathbf{u}$, so that the value is unchanged when the vectors are permuted cyclically. Also, dot and cross can be interchanged: $\mathbf{u}\times\mathbf{v}\cdot\mathbf{w} = \mathbf{u}\cdot\mathbf{v}\times\mathbf{w}$. These properties all follow from the fact that

$$\mathbf{u}\times\mathbf{v}\cdot\mathbf{w} = \begin{vmatrix} u_x & u_y & u_z \\ v_x & v_y & v_z \\ w_x & w_y & w_z \end{vmatrix}. \tag{0–85}$$

The scalar triple product also equals $\pm V$, where V is the volume of the parallelepiped whose edges are $\mathbf{u}$, $\mathbf{v}$, $\mathbf{w}$. The $+$ sign is used when the triple $\mathbf{u}$, $\mathbf{v}$, $\mathbf{w}$ is positively oriented; the $-$ sign is used when it is negatively oriented. Thus, the sign of the determinant in (0–85) can be used to determine the orientation. When the vectors are coplanar, $V = 0$ and the determinant is 0; otherwise, $V > 0$.

The *vector triple products are the expressions*

$$\mathbf{u} \times (\mathbf{v} \times \mathbf{w})$$

and

$$(\mathbf{u} \times \mathbf{v}) \times \mathbf{w}.$$

These are generally unequal. One has the identities

$$\mathbf{u} \times (\mathbf{v} \times \mathbf{w}) = (\mathbf{u} \cdot \mathbf{w})\mathbf{v} - (\mathbf{u} \cdot \mathbf{v})\mathbf{w}, \tag{0–86}$$

$$(\mathbf{u} \times \mathbf{v}) \times \mathbf{w} = (\mathbf{u} \cdot \mathbf{w})\mathbf{v} - (\mathbf{v} \cdot \mathbf{w})\mathbf{u}. \tag{0–87}$$

The right sides can be remembered in words as "outer dot remote adjacent minus outer dot adjacent remote."

Lines and planes in space can be analyzed by vectors. A point P is on the line L through points P_1, P_2 precisely when $\overrightarrow{P_1P} \times \overrightarrow{P_1P_2} = \mathbf{0}$, and this equation is a concise form of the equation of L. One can also give a vector counterpart of the parametric equations (0–47):

$$\overrightarrow{OP} = \mathbf{a} + t\mathbf{b}, \tag{0–88}$$

where O is the origin and P is an arbitrary point on the line.

The equation $A(x - x_1) + B(y - y_1) + C(z - z_1) = 0$ for a plane through P_1: (x_1, y_1, z_1) can be replaced by the vector equation

$$\overrightarrow{P_1P} \cdot \mathbf{n} = 0, \tag{0–89}$$

where $\mathbf{n} = A\mathbf{i} + B\mathbf{j} + C\mathbf{k}$ is a nonzero normal vector of the plane.

0–7 Sets, functions, limits, continuity. In mathematics a *set* or a *class* means a well-defined collection of objects: for example, the set of all positive integers, the set of all straight lines in the plane. Sets are commonly denoted by capital letters: $A, B, C, \ldots$ An object a in a set A is said to be an *element* of A: one writes $a \in A$. If A and B are sets and every element of A is also an element of B, then one says that A is *contained in* B or that A is a *subset* of B, and writes $A \subset B$. If A and B are subsets of a set C, then one writes $A \cup B$ for the *union* of A and B: the set consisting of all objects in A or B, or *both*. One writes $A \cap B$ for the *intersection* of A and B: the set of all objects which are in *both* A and B. The intersection may contain *no* objects. This leads one to define the *empty set* (often denoted by $\emptyset$) as the set having *no elements*. The empty set is regarded as a *subset of every set*. Also, $A \cup \emptyset = A$ and $A \cap \emptyset = \emptyset$.

Let a and b be given real numbers, with $a < b$. The set of all real numbers x such that $a < x < b$ is called an *open interval;* similarly, a *closed interval* is formed of all x such that $a \leqq x \leqq b$. A *half-open interval* is formed of all x such that $a \leqq x < b$ or such that $a < x \leqq b$. One also considers *infinite intervals* defined by corresponding inequalities: the intervals $a < x$ (or $a < x < \infty$), $x < b$ (or $-\infty < x < b$) and $-\infty < x < \infty$ (all real x) are called *infinite open intervals*. The intervals $a \leqq x$ (or $a \leqq x < \infty$) and $x \leqq b$ (or $-\infty < x \leqq b$) are called *half-open infinite intervals*.

Let X and Y be sets. If to each element x of X there is associated an element y of Y, then one says that there is a *function* whose *domain* is X and whose values lie in Y. Typically, one denotes a function by a letter such as $f, g, \varphi, \psi, F, G$. If f is a function with domain X and having values in Y, then for each x of X one denotes by $f(x)$ the value of f—that is, the element y of Y assigned to x. Thus $y = f(x)$. The *range* of f is the set of all values of the function; that is, it is the set of all y such that $y = f(x)$ for some x in X. The function f is also called a *mapping* of X into Y. If the range of f is *all* of Y, f is said to map X *onto* Y, and one calls f an *onto* mapping. If different x's always correspond to different y's—that is, $f(x_1) \neq f(x_2)$ for $x_1 \neq x_2$—then f is said to be a *one-to-one mapping* or a *one-to-one correspondence*. For a one-to-one correspondence f, for each y in the range of f there is exactly one x in X such that $f(x) = y$. One writes $x = f^{-1}(y)$ and thereby defines the *inverse function* or *inverse mapping* f^{-1}.

One often writes, somewhat loosely, "the function $y = f(x)$" or "the function $f(x)$" or "the function $y(x)$." Usually the context makes the meaning clear. However, for greatest precision the function should be denoted by a single letter f, and $f(x)$ should denote the particular value assigned to a chosen x.

When X and Y are sets of real numbers, the function f is said to be a real-valued function of a real variable or, more briefly, a function of a real variable. Such a function is commonly defined by an equation: $y = x^2 - 1$ defines a function f whose domain is the set X of all real numbers, and $f(x) = x^2 - 1$ for each x. For example, $f(0) = -1, f(1) = 0, f(2) = 3$. The range of this function is the infinite interval $-1 \leqq y < \infty$. The function is not one-to-one, since $f(1) = f(-1) = 0$.

A function of a real variable is often defined *implicitly*. For example, the conditions

$$x^2 + y^2 = 1 \qquad (y > 0)$$

assign a unique y to each x such that $-1 < x < 1$.

Let $y = f(x)$ for $a \leqq x < x_0$, where $a < x_0$. Then the equation

$$\lim_{x \to x_0 -} f(x) = c \tag{0–90}$$

means the following: given any positive number ϵ, there exists a positive number δ such that for all x satisfying $x_0 - \delta < x < x_0$ one has

$$|f(x) - c| < \epsilon. \tag{0–91}$$

In words, for x sufficiently close to x_0 (and less than x_0), $f(x)$ is as close to c as desired. The equation

$$\lim_{x \to x_0+} f(x) = c \tag{0-92}$$

is defined similarly for $f(x)$ given in an interval $x_0 < x \leqq b$, so that only values of x greater than x_0 are considered. Finally, if $f(x)$ is given both to the left and right of x_0, i.e., for $a \leqq x < x_0$ and for $x_0 < x \leqq b$, then

$$\lim_{x \to x_0} f(x) = c \tag{0-93}$$

means that for each positive ϵ there is a positive δ such that (0–91) holds for all x such that $0 < |x - x_0| < \delta$. Thus (0–93) is equivalent to (0–90) and (0–92) combined. It is to be remarked that if $f(x_0)$ is also defined this fact is ignored in all three definitions.

Now let $f(x)$ be defined for $a \leqq x \leqq b$. Then $f(x)$ is said to be continuous at the value x_0, $a < x_0 < b$, if $\lim_{x \to x_0} f(x)$ exists and

$$\lim_{x \to x_0} f(x) = f(x_0). \tag{0-94}$$

The function $f(x)$ is said to be continuous to the left at $x = b$ if

$$\lim_{x \to b-} f(x) = f(b) \tag{0-95}$$

and to be continuous to the right at $x = a$ if

$$\lim_{x \to a+} f(x) = f(a). \tag{0-96}$$

Finally, $f(x)$ is said to be continuous over the interval $a \leqq x \leqq b$ if it is continuous for each x_0 inside the interval ($a < x_0 < b$) and continuous to the left at b, continuous to the right at a.

It may happen that a limit (0–93) fails to exist in such a manner that, as x approaches x_0, $f(x)$ increases without bound, so that for each number M there is a positive δ such that $f(x) > M$ for $0 < |x - x_0| < \delta$. In this case one writes

$$\lim_{x \to x_0} f(x) = +\infty. \tag{0-97}$$

Similar definitions are given for

$$\lim_{x \to x_0} f(x) = -\infty, \quad \lim_{x \to x_0+} f(x) = +\infty, \text{ etc.}$$

If $f(x)$ is defined for all x greater than a, then the equation

$$\lim_{x \to +\infty} f(x) = c \tag{0-98}$$

is defined to mean: given any $\epsilon > 0$, there is a number K such that $|f(x) - c| < \epsilon$ for $x > K$. Similar definitions are given for

$$\lim_{x \to -\infty} f(x) = c, \quad \lim_{x \to +\infty} f(x) = +\infty, \text{ etc.}$$

THEOREM. *If* $\lim f(x) = c$, $\lim g(x) = d$, *then* $\lim [f(x) + g(x)] = c + d$, $\lim [f(x) \cdot g(x)] = c \cdot d$, $\lim \dfrac{f(x)}{g(x)} = \dfrac{c}{d}$, *provided* $d \neq 0$; *the sum, product, and quotient of continuous functions are continuous, provided there is no division by* 0.

The limits here may be of any one of the types (0–90), (0–92), (0–93), (0–98). The continuity also applies to the left, to the right, or over an interval.

If $y = f(u)$ and $u = g(x)$ are given functions, a new function, the "function of the function" or *composite function*, is defined by setting

$$h(x) = f[g(x)].$$

One also writes $h = f \circ g$. The function h will be defined in an interval $a \leqq x \leqq b$, provided $g(x)$ is defined in such an interval and has values in an interval $c \leqq u \leqq d$ within which $f(u)$ is defined. The continuity of f and g then implies that of h:

THEOREM. *Let* $y = f(u)$ *and* $u = g(x)$ *be defined and continuous for* $c \leqq u \leqq d$ *and* $a \leqq x \leqq b$ *respectively; let* $c \leqq g(x) \leqq d$ *for* $a \leqq x \leqq b$. *Then* $h(x) = f[g(x)]$ *is defined and continuous for* $a \leqq x \leqq b$.

If c is a constant, then $f(x) = c$ is continuous for all x; if $f(x) = x$, then $f(x)$ is continuous for all x. In both cases one can verify directly that the definition of continuity is satisfied. From these two functions one can, by multiplication alone, construct the functions:

$$x^2, x^3, \ldots, x^n; cx, cx^2, \ldots, cx^n \quad (n \text{ a positive integer}).$$

The above theorem on products of continuous functions ensures that all these functions are continuous for all x. Applying the theorem for the sum and quotient of two functions, one concludes that a polynomial

$$f(x) = c_0x^n + c_1x^{n-1} + \cdots + c_{n-1}x + c_n$$

is continuous for all x, as is any rational function

$$f(x) = \frac{c_0x^n + c_1x^{n-1} + \cdots + c_{n-1}x + c_n}{a_0x^m + a_1x^{m-1} + \cdots + a_{m-1}x + a_m}$$

in every interval containing no root of the denominator.

Vector functions. If to each value of the real variable t in an interval $t_1 \leqq t \leqq t_2$ there is assigned a vector $\mathbf{u}$ in space, then $\mathbf{u}$ is said to be given as a vector function of t over that interval. For example, one may have

$$\mathbf{u} = t\mathbf{a} + (1 - t)\mathbf{b}, \quad 0 \leqq t \leqq 1, \tag{0–99}$$

where $\mathbf{a}$ and $\mathbf{b}$ are given vectors, or one may have

$$\mathbf{u} = t^2\mathbf{i} + t^3\mathbf{j} + \sin t\,\mathbf{k}, \quad 0 \leqq t \leqq 2\pi, \tag{0–100}$$

where **i**, **j**, **k** form a triple of mutually perpendicular unit vectors as above. If such a function is given, a notation such as

$$\mathbf{u} = \mathbf{F}(t), \quad t_1 \leqq t \leqq t_2, \tag{0–101}$$

or more simply,

$$\mathbf{u} = \mathbf{u}(t), \quad t_1 \leqq t \leqq t_2, \tag{0–102}$$

is used. If a coordinate system is chosen in space, then the vector **u** can always be expressed in the form

$$\mathbf{u} = u_x\mathbf{i} + u_y\mathbf{j} + u_z\mathbf{k}, \tag{0–103}$$

where u_x, u_y, u_z are the corresponding components. These components will themselves depend on t; it will be assumed that the axes are fixed, independent of t. One can then write:

$$u_x = f(t), \quad u_y = g(t), \quad u_z = h(t), \quad t_1 \leqq t \leqq t_2. \tag{0–104}$$

Thus a vector function of t determines three scalar functions of t. Conversely, if $f(t)$, $g(t)$ and $h(t)$ are three scalar functions of t defined for $t_1 \leqq t \leqq t_2$, then the vector

$$\mathbf{u} = f(t)\mathbf{i} + g(t)\mathbf{j} + h(t)\mathbf{k} \tag{0–105}$$

is a vector function of t. In the triple notation of Section 0–6, the function (0–105) can be written:

$$\mathbf{u} = \mathbf{F}(t) = (f(t), g(t), h(t)). \tag{0–105′}$$

A vector function of t can be represented graphically as a curve in space. Thus let O be a fixed reference point and let P be chosen so that $\overrightarrow{OP} = \mathbf{u}$. As t varies, P will trace out a curve, as shown in Fig. 0–21. If axes are chosen with origin at O and **u** is written as in (0–105), then the equations

$$x = f(t), \quad y = g(t), \quad z = h(t) \tag{0–106}$$

are simply parametric equations for the curve traced by P; the parameter t can be interpreted as *time*.

While this representation of the vector function is very useful, a vector function can arise in other ways, for example as the velocity vector of the moving point P.

The vector function $\mathbf{u} = \mathbf{u}(t)$ is said to have a *limit* **v** as t approaches t_0:

$$\lim_{t \to t_0} \mathbf{u}(t) = \mathbf{v}, \tag{0–107}$$

if

$$\lim_{t \to t_0} |\mathbf{u}(t) - \mathbf{v}| = 0, \tag{0–108}$$

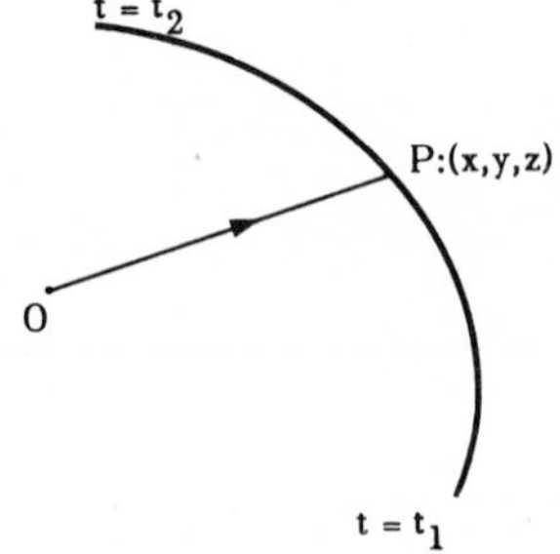

Fig. 0–12. Graph of a vector function.

i.e., if the difference between $\mathbf{u}(t)$ and $\mathbf{v}$ can be made arbitrarily small (as a vector) for t sufficiently close to t_0. The function $\mathbf{u} = \mathbf{u}(t)$ is said to be *continuous* at the value t_0 if one has

$$\lim_{t \to t_0} \mathbf{u}(t) = \mathbf{u}(t_0). \tag{0–109}$$

From the definition one proves that $\mathbf{u}(t) = f(t)\mathbf{i} + g(t)\mathbf{j} + h(t)\mathbf{k}$ has limit $a\mathbf{i} + b\mathbf{j} + c\mathbf{k}$ if and only if $f(t) \to a$, $g(t) \to b$, and $h(t) \to c$ as $t \to t_0$. From this we deduce that $\mathbf{u}(t)$ *is continuous at a value* t_0 *if and only if its components* u_x, u_y, u_z *are continuous at* t_0. It then follows, as above, that if $\mathbf{u}_1(t)$ and $\mathbf{u}_2(t)$ are two vector functions of t, both defined and continuous for $t_1 \leqq t \leqq t_2$, then the functions

$$\mathbf{u}_1(t) + \mathbf{u}_2(t), \quad \mathbf{u}_1(t) \cdot \mathbf{u}_2(t), \quad \mathbf{u}_1(t) \times \mathbf{u}_2(t) \tag{0–110}$$

are continuous functions of t over that interval.

0–8 The elementary transcendental functions. If a is a positive number, a^x is defined in algebra to have a meaning for every rational x; thus $a^{-\frac{3}{2}}$ means the reciprocal of the positive square root of a^3. It can be shown (see Section 5–7 of the book by Kaplan and Lewis listed at the end of the chapter) that a^x can be defined also for irrational values of x, so that the resulting function is defined and continuous for all values of x. Because of the continuity, a number such as a^π can be computed approximately by using a rational number close to π, such as 3.1416.

Further properties of this function, the "exponential function of base a," are the following:

$$a^x \cdot a^y = a^{x+y}; \quad \frac{a^x}{a^y} = a^{x-y}; \tag{0–111}$$

$$(a^x)^y = a^{xy}; \tag{0–112}$$

$$\begin{gathered}\text{if } a > 1, \text{ then } a^x \text{ increases as } x \text{ increases,}\\ \lim_{x \to -\infty} a^x = 0 \text{ and } \lim_{x \to +\infty} a^x = +\infty; \quad 1^x = 1 \text{ for all } x;\end{gathered} \tag{0–113}$$

$$\text{if } a \neq 1 \text{ and } x_2 > 0, \text{ then } a^{x_1} = x_2 \text{ can be solved uniquely for } x_1. \tag{0–114}$$

Property (0–114) leads to the definition of the logarithm: $x_1 = \log_a x_2$ if $a^{x_1} = x_2$. If a is positive and not equal to 1, then the function $\log_a x$ is defined for all positive values of x. It can be shown to be continuous for these values of x and to have the following properties, corresponding to (0–111) through (0–114):

$$\left.\begin{aligned}\log_a (x_1x_2) &= \log_a x_1 + \log_a x_2;\\ \log_a \frac{x_1}{x_2} &= \log_a x_1 - \log_a x_2;\end{aligned}\right\} \quad (0\text{–}115)$$

$$\log_a x_1^{x_2} = x_2 \log_a x_1; \quad (0\text{–}116)$$

$$\begin{aligned}&\text{if } a > 1, \text{ then } \log_a x \text{ increases as } x \text{ increases,}\\ &\lim_{x\to 0+} \log_a x = -\infty \quad \text{and} \quad \lim_{x\to +\infty} \log_a x = +\infty;\end{aligned} \quad (0\text{–}117)$$

$$\text{the equation } \log_a x_2 = x_1 \text{ can be solved uniquely for } x_2. \quad (0\text{–}118)$$

The logarithmic and exponential functions are related by the equations:

$$\log_a a^x = x; \quad a^{\log_a x} = x; \quad (0\text{–}119)$$

these express the fact that the two functions are inverses of each other.

The functions $\sin x$ and $\cos x$ are defined in trigonometry for all values of x. Throughout calculus, angles are measured in radians, so that both functions have period 2π. From the geometrical meaning of these functions, it can be shown that they are continuous for all x. Some of the basic identities satisfied by $\sin x$ and $\cos x$ are collected here for reference:

$$\left.\begin{aligned}
&\sin (x \pm y) = \sin x \cos y \pm \cos x \sin y,\\
&\cos (x \pm y) = \cos x \cos y \mp \sin x \sin y,\\
&\sin x + \sin y = 2 \sin \frac{x+y}{2} \cos \frac{x-y}{2},\\
&\cos x + \cos y = 2 \cos \frac{x+y}{2} \cos \frac{x-y}{2},\\
&\sin x \sin y = -\tfrac{1}{2}[\cos (x+y) - \cos (x-y)],\\
&\cos x \cos y = \tfrac{1}{2}[\cos (x+y) + \cos (x-y)],\\
&\sin x \cos y = \tfrac{1}{2}[\sin (x+y) + \sin (x-y)],\\
&\sin 2x = 2 \sin x \cos x, \quad \cos 2x = \cos^2 x - \sin^2 x, \quad \sin^2 x + \cos^2 x = 1,\\
&\sin (-x) = -\sin x, \quad \cos (-x) = \cos x, \quad \sin\left(\frac{\pi}{2} - x\right) = \cos x,\\
&\sin (\pi - x) = \sin x, \quad \sin^2 \frac{x}{2} = \frac{1-\cos x}{2}, \quad \cos^2 \frac{x}{2} = \frac{1+\cos x}{2},\\
&c^2 = a^2 + b^2 - 2ab \cos C, \text{ for triangle of sides } a, b, c.
\end{aligned}\right\} \quad (0\text{–}120)$$

The four other trigonometric functions are defined by the identities:

$$\tan x = \frac{\sin x}{\cos x}, \quad \cot x = \frac{\cos x}{\sin x}, \quad \csc x = \frac{1}{\sin x}, \quad \sec x = \frac{1}{\cos x}. \quad (0\text{–}121)$$

They satisfy further identities, related to those for $\sin x$ and $\cos x$.

For given x, the equation $\sin y = x$ has infinitely many solutions y, provided $-1 \leqq x \leqq 1$. The symbol arc $\sin x$ or $\sin^{-1} x$ denotes this "multiple-valued function" of x; thus arc $\sin \frac{1}{2}$ has the values $\pi/6 + 2n\pi$, $5\pi/6 + 2n\pi$ ($n = 0, \pm 1, \pm 2, \ldots$). To make this function single-valued, the restriction

$$-\frac{\pi}{2} \leqq \text{arc} \sin x \leqq \frac{\pi}{2}, \quad -1 \leqq x \leqq 1 \tag{0–122}$$

is imposed and the resulting function is called the principal value of arc $\sin x$. In this book, the symbol arc $\sin x$ will mean this principal value unless otherwise indicated. Similar definitions apply to the functions arc $\cos x$ and arc $\tan x$, with the restrictions:

$$0 \leqq \text{arc} \cos x \leqq \pi, \quad -1 \leqq x \leqq 1; \quad -\frac{\pi}{2} < \text{arc} \tan x < \frac{\pi}{2}, \quad -\infty < x < \infty. \tag{0–123}$$

All three functions as thus restricted are continuous where defined. They are the respective inverses of $\sin x$, $\cos x$, $\tan x$, as is indicated by the identities:

$$\begin{aligned}
&\sin(\text{arc} \sin x) = x, \quad \cos(\text{arc} \cos x) = x, \quad \tan(\text{arc} \tan x) = x,\\
&\text{arc} \sin(\sin x) = x, \quad \text{arc} \tan(\tan x) = x, \quad -\tfrac{1}{2}\pi \leqq x \leqq \tfrac{1}{2}\pi,\\
&\text{arc} \cos(\cos x) = x, \quad 0 \leqq x \leqq \pi.
\end{aligned} \tag{0–124}$$

The inverse functions satisfy other identities, e.g.,

$$\text{arc} \tan x + \text{arc} \tan y = \text{arc} \tan \frac{x + y}{1 - xy}$$

(not necessarily principal values), paralleling those for the direct functions.

The functions $\sin x$, $\cos x$ and e^x (where $e = 2.71828\ldots$, as defined in the next section) and their inverses are usually referred to as the elementary transcendental functions. Functions built up from these and polynomials by a finite number of applications of the operations of arithmetic, of raising to powers, and of substitutions (function of function) are termed elementary functions. Thus

$$y = \log_e (1 + \sqrt{1 - x^2 \cos x})$$

is an elementary function.

The nature of the functions $\sin x$, $\cos x$, $\tan x$, e^x, $\log_e x$, arc $\sin x$, arc $\tan x$ is portrayed graphically in Fig. 0–10. These graphs can also be used for rough numerical work.

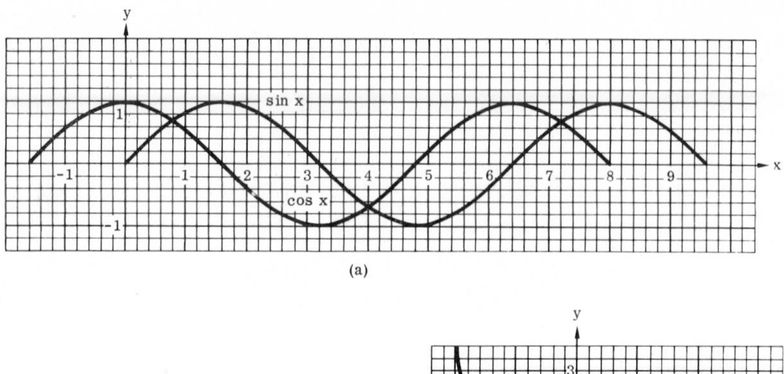

(a)

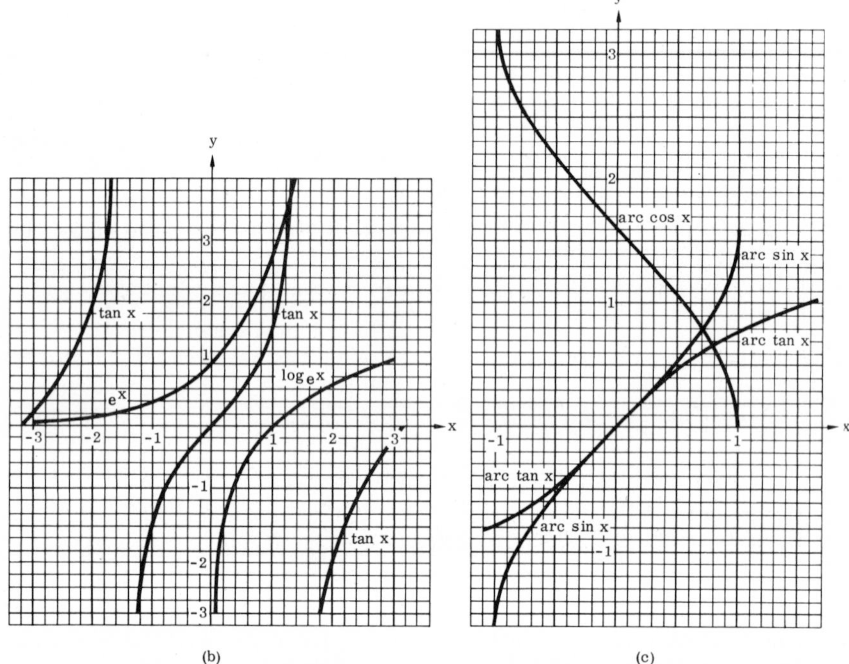

(b) (c)

FIG. 0–13. Graphs of elementary transcendental functions.

0–9 The differential calculus. Let the function $f(x)$ be defined for $a < x < b$. For each x of this interval, the derivative $y' = f'(x)$ is defined by the equation

$$f'(x) = \lim_{\Delta x \to 0} \frac{f(x + \Delta x) - f(x)}{\Delta x}, \tag{0–125}$$

provided this limit exists. It is to be remarked that the limit cannot exist if $f(x)$ is discontinuous at the value x considered; however, continuity alone does not imply existence of the derivative.

The differential dy of the function $y = f(x)$ is defined as follows:

$$dy = f'(x)\,\Delta x. \tag{0–126}$$

In this equation Δx can be replaced by dx, for a reason to be noted below, so that one has

$$dy = f'(x)\,dx \quad \text{or} \quad \frac{dy}{dx} = f'(x). \tag{0–127}$$

The derivative can be interpreted geometrically as the slope of the tangent to the curve $y = f(x)$ at the point (x, y) considered. The equation of the tangent at (x_1, y_1) is then, by (0–37),

$$y - y_1 = f'(x_1)(x - x_1). \tag{0–128}$$

Thus dx and dy can be interpreted as changes in x and y following the tangent to the curve, as in Fig. 0–14.

The second derivative $f''(x)$ is defined as the derivative of the first derivative and higher derivatives are correspondingly defined. The notations

$$\frac{d^2y}{dx^2}, \quad \frac{d^3y}{dx^3}, \quad \ldots, \quad \frac{d^ny}{dx^n}, \quad \ldots$$

or y'', y''', y^{iv}, y^{v}, $\ldots$, $y^{(n)}$,

are also used for higher derivatives.

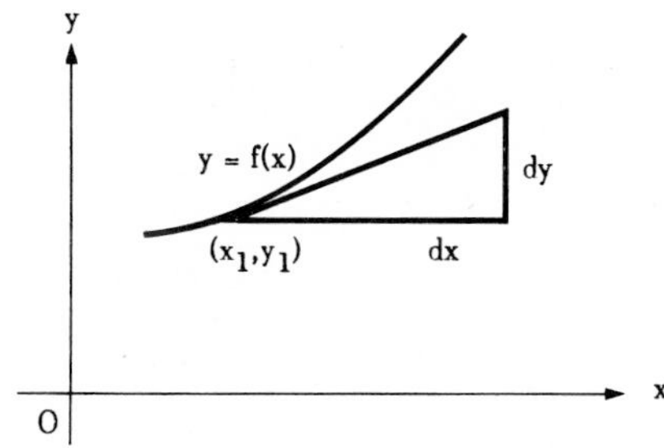

FIG. 0–14. Differentials.

The basic properties of the derivative are summarized in the following theorem.

THEOREM. *Let $u = f(x)$, $v = g(x)$ be defined for $a < x < b$. Then, for each x of this interval for which u' and v' exist, the functions $u + v$, $u \cdot v$, u/v also have derivatives, given by the formulas*

$$(u + v)' = u' + v', \quad (u \cdot v)' = uv' + vu', \quad \left(\frac{u}{v}\right)' = \frac{vu' - uv'}{v^2} \tag{0–129}$$

(except, in the last case, when $v = 0$).

If $w = h(u)$ and $u = f(x)$ are such that $w = h[f(x)]$ is defined for $a < x < b$, then, if u' exists at a particular x and $h'(u)$ exists for the corresponding value of u, $w = h[f(x)]$ has a derivative for this value of x, given by the chain rule:

$$\frac{dw}{dx} = h'(u)\,\frac{du}{dx}. \tag{0–130}$$

Repeated application of the rule for the derivative of the product $u \cdot v$ gives the rule:

$$\text{if } y = x^n, \quad \text{then} \quad y' = nx^{n-1} \tag{0–131}$$

for n any positive integer. This rule can be extended to all real values of n. The case $n = 0$ is included in the rule:

$$\text{if } y = c \ (c = \text{constant}), \quad \text{then} \quad y' = 0. \tag{0–132}$$

The successive derivatives of $y = u \cdot v$ are given by the formulas

$$(u \cdot v)' = u'v + uv', \quad (u \cdot v)'' = u''v + 2u'v' + uv'',$$
$$(u \cdot v)''' = u'''v + 3u''v' + 3u'v'' + uv''', \ldots ;$$

the general case is given by the Leibniz rule:

$$(u \cdot v)^{(n)} = u^{(n)} \cdot v + nu^{(n-1)}v' + \frac{n \cdot (n-1)}{1 \cdot 2} u^{(n-2)}v'' + \cdots + \binom{n}{r} u^{(n-r)}v^{(r)} + \cdots + uv^{(n)}, \quad (0\text{–}133)$$

which parallels the binomial theorem (0–30).

The derivative of a quotient of two products is given by the rule:

$$\left(\frac{u \cdot v}{w \cdot z}\right)' = \left(\frac{u \cdot v}{w \cdot z}\right)\left(\frac{u'}{u} + \frac{v'}{v} - \frac{w'}{w} - \frac{z'}{z}\right);$$

this includes the usual product and quotient rules as special cases. In general one has

$$\left(\frac{u_1 \cdot u_2 \cdots u_n}{v_1 \cdot v_2 \cdots v_m}\right)' = \left(\frac{u_1 \cdot u_2 \cdots u_n}{v_1 \cdot v_2 \cdots v_m}\right)\left(\frac{u_1'}{u_1} + \frac{u_2'}{u_2} + \cdots + \frac{u_n'}{u_n} - \frac{v_1'}{v_1} - \frac{v_2'}{v_2} - \cdots - \frac{v_m'}{v_m}\right). \quad (0\text{–}134)$$

This can be proved by induction or by "logarithmic differentiation," to be explained in Prob. 33 below.

Let y be defined as an implicit function of x by an equation such as

$$x^3 + x^2y - xy^2 - 2x - y - 1 = 0; \quad (0\text{–}135)$$

then replacing y by the corresponding function of x leads to an identity in x. One can now differentiate both sides of the equation with respect to x and thereby obtain a relation satisfied by y'. Thus in the above example

$$3x^2 + x^2y' + 2xy - 2xyy' - y^2 - 2 - y' = 0.$$

This relation can be differentiated again to find y'', and so on.

Let $w = h(u)$ and $u = f(x)$, so that one may apply the chain rule (0–130). Then $dw = h'(u)\,\Delta u$ is the differential of w in terms of u and $du = f'(x)\,\Delta x$ is the differential of u in terms of x. From (0–130) one concludes that the differential of w in terms of x is given by

$$dw = h'(u)f'(x)\,\Delta x = h'(u)\,du; \quad (0\text{–}136)$$

that is, if x is the independent variable, then dw and du are related by the same formula as are dw and Δu when u is independent. Thus replacing Δu by du leads to a correct result, no matter which variable is independent. Accordingly, as far as differentials are concerned, dependent and independent variables can be treated on an equal basis. One has, for example, the following: if $y = f(x)$, then $dy = f'(x)\,dx$, whence

$$\frac{dx}{dy} = \frac{1}{f'(x)} \quad [f'(x) \neq 0]; \quad (0\text{–}137)$$

this gives the derivative of the inverse function. In the above implicit equation (0–135) one can treat the two variables in the same way by taking differentials instead of derivatives, obtaining

$$3x^2\,dx + x^2\,dy + 2xy\,dx - 2xy\,dy - y^2\,dx - 2dx - dy = 0,$$

from which either derivative dy/dx or dx/dy can be obtained. Extension of this reasoning permits one to conclude that, for parametric equations:

$$x = f(t), \quad y = g(t), \tag{0–138}$$

one has

$$\frac{dy}{dx} = \frac{g'(t)}{f'(t)} \quad [f'(t) \neq 0]. \tag{0–139}$$

It should be remarked that the condition $f'(t) \neq 0$ [as also the analogous one for (0–137)] is precisely the condition needed to guarantee that the inverse function $t = t(x)$ [or, for (0–137), $x = x(y)$] is well-defined and differentiable. This can easily be seen from a graphical representation of the functions.

The rules above permit one to evaluate explicitly the derivatives of all rational functions of x and, more generally, of all functions built up from polynomials by repeated use of the operations of arithmetic and that of raising to a constant power. Implicit functions, defined by an equation formed by such operations on x and y, can also be differentiated (to give y' as a function of x and y), as can parametric functions (0–138) of this type (to give y' as a function of t).

For the basic transcendental functions one has the rules:

$$(\sin x)' = \cos x, \quad (\cos x)' = -\sin x, \quad (a^x)' = a^x \log_e a,$$
$$(\log_a x)' = \frac{1}{\log_e a}\frac{1}{x}. \tag{0–140}$$

The rules for $\sin x$ and $\cos x$ are a consequence of the relation

$$\lim_{\Delta x \to 0} \frac{\sin \Delta x}{\Delta x} = 1, \tag{0–141}$$

which holds when angles are measured in radians. The rules for a^x and $\log_a x$ are a consequence of the limit relation:

$$\lim_{\Delta x \to 0} (1 + \Delta x)^{\frac{1}{\Delta x}} = e = 2.71828\ 18285\ldots. \tag{0–142}$$

Because of (0–140) it is natural to use e as a base for exponential and logarithmic functions in calculus and that will be done in this text. The rules (0–140) then reduce to

$$(e^x)' = e^x, \quad (\log x)' = \frac{1}{x} \quad (\log x = \log_e x). \tag{0–143}$$

The change to base e is facilitated by the rules

$$a^x = e^{x \log a}, \quad \log_a x = \frac{\log x}{\log a}. \tag{0–144}$$

A basic theorem of differential calculus is the Law of the Mean:

$$f(b) - f(a) = f'(x_1)(b - a), \quad a < x_1 < b. \tag{0–145}$$

Here $f(x)$ is assumed to be continuous for $a \leqq x \leqq b$ and to be differentiable (i.e., to possess a derivative) for $a < x < b$. The theorem then asserts the existence of a number x_1 such that (0–145) holds. This has a simple geometric meaning: at some point (x_1, y_1) on the graph of $f(x)$, between a and b, the tangent is parallel to the chord joining the end points of the graph. This theorem implies that, if $f'(x) = 0$ throughout an interval, then $f(x)$ is constant throughout that interval.

A special case of the Law of the Mean is Rolle's Theorem: *Let $f(x)$ be continuous for $a \leqq x \leqq b$ and differentiable for $a < x < b$; if $f(a) = f(b)$, then $f'(x_1) = 0$ for at least one x_1 such that $a < x_1 < b$.*

If $f(x)$ is defined for $a \leqq x \leqq b$, one defines continuity of $f(x)$ over this interval with the aid of the right and left limits at a and b, as in the definition following Eq. (0–96) above. In the same way one can define differentiability of $f(x)$ over the interval $a \leqq x \leqq b$ in terms of appropriate right and left limits. More precisely, one defines the derivative of $f(x)$ to the right, at $x = a$, as the limit:

$$\lim_{\Delta x \to 0+} \frac{f(a + \Delta x) - f(a)}{\Delta x}$$

and the derivative of $f(x)$ to the left at $x = b$ as the limit:

$$\lim_{\Delta x \to 0-} \frac{f(b + \Delta x) - f(b)}{\Delta x},$$

provided the limits exist. The function $f(x)$ is then said to be differentiable over the interval $a \leqq x \leqq b$ provided $f(x)$ has a derivative at each x inside the interval $(a < x < b)$ and a derivative to the right at a, a derivative to the left at b. The derivatives at a and b can be denoted by $f'(a)$ and $f'(b)$, if the context makes clear that these are defined as right and left limits.

Derivative of a vector function. The velocity and acceleration vectors. The *derivative* of the vector function $\mathbf{u} = \mathbf{u}(t)$ is defined as a limit:

$$\frac{d\mathbf{u}}{dt} = \lim_{\Delta t \to 0} \frac{\mathbf{u}(t + \Delta t) - \mathbf{u}(t)}{\Delta t} = \lim_{\Delta t \to 0} \frac{\Delta \mathbf{u}}{\Delta t}, \tag{0–146}$$

provided the limit exists. This is illustrated in Fig. 0–15, where $\mathbf{u}$ is represented by the vector $\overrightarrow{OP}$, with O fixed. The numerator $\mathbf{u}(t + \Delta t) - \mathbf{u}(t) = \Delta \mathbf{u}$ represents the vector $\overrightarrow{PP'}$, which is the displacement of the moving point P in the interval t to $t + \Delta t$. The quantity $\Delta \mathbf{u}$ is a vector and $\Delta \mathbf{u}/\Delta t$ is the vector $\Delta \mathbf{u}$ times the scalar $1/\Delta t$. Thus $\Delta \mathbf{u}/\Delta t$ is a vector and its limit is a vector $d\mathbf{u}/dt$.

In terms of components (0–105) one has

$$\mathbf{u}(t + \Delta t) - \mathbf{u}(t) = [f(t + \Delta t) - f(t)]\mathbf{i} + [g(t + \Delta t) - g(t)]\mathbf{j} + [h(t + \Delta t) - h(t)]\mathbf{k}. \quad (0\text{–}147)$$

Hence, on dividing by Δt and letting Δt approach 0, one finds

$$\frac{d\mathbf{u}}{dt} = f'(t)\mathbf{i} + g'(t)\mathbf{j} + h'(t)\mathbf{k} = \frac{du_x}{dt}\mathbf{i} + \frac{du_y}{dt}\mathbf{j} + \frac{du_z}{dt}\mathbf{k}; \quad (0\text{–}148)$$

i.e., to differentiate a vector function, one differentiates each component separately.

If one defines the tangent to a curve as the limiting position of a secant (if the limit exists), then one concludes that, except when $d\mathbf{u}/dt = \mathbf{0}$, the vector $d\mathbf{u}/dt$ represents the tangent to the curve traced by P at the point P, as shown in Fig. 0–15.

It can be shown, as in Section 0–10 that, if s is the distance traversed by P from time $t = t_1$ up to time t, then

$$\frac{ds}{dt} = \sqrt{f'(t)^2 + g'(t)^2 + h'(t)^2} = \sqrt{\left(\frac{dx}{dt}\right)^2 + \left(\frac{dy}{dt}\right)^2 + \left(\frac{dz}{dt}\right)^2}, \quad (0\text{–}149)$$

i.e., that

$$ds = \sqrt{dx^2 + dy^2 + dz^2}. \quad (0\text{–}150)$$

Thus, if $\mathbf{u} = \overrightarrow{OP}$ is the position vector of the moving point P, then the vector $\mathbf{v} = (d/dt)\overrightarrow{OP}$ is tangent to the curve traced by P and has at each point a magnitude

$$|\mathbf{v}| = \left|\frac{d\mathbf{u}}{dt}\right| = \sqrt{f'(t)^2 + g'(t)^2 + h'(t)^2} = \frac{ds}{dt}. \quad (0\text{–}151)$$

One concludes that $\mathbf{v}$ is precisely the velocity vector of the moving point P; for $\mathbf{v}$ is tangent to the path, has the magnitude $v = ds/dt$ (the "speed"), and clearly points in the direction of motion. One has thus the rule:

$$\frac{d}{dt}\overrightarrow{OP} = \text{velocity of } P, \quad (0\text{–}152)$$

when O is a fixed reference point.

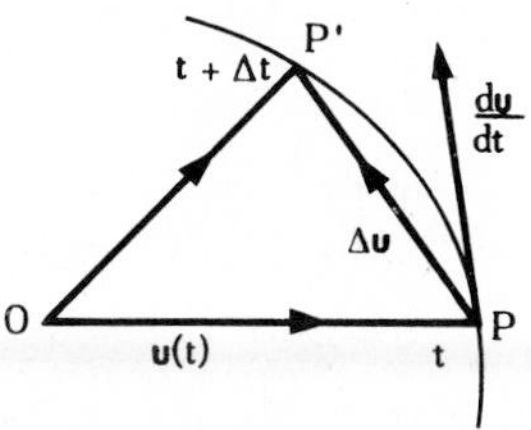

Fig. 0–15. Derivative of a vector function.

The rules for derivative of sum and product have analogues for vectors:

$$\frac{d}{dt}(\mathbf{u}+\mathbf{v}) = \frac{d\mathbf{u}}{dt} + \frac{d\mathbf{v}}{dt}; \tag{0–153}$$

$$\frac{d}{dt}(\mathbf{u}\cdot\mathbf{v}) = \mathbf{u}\cdot\frac{d\mathbf{v}}{dt} + \frac{d\mathbf{u}}{dt}\cdot\mathbf{v}; \tag{0–154}$$

$$\frac{d}{dt}(\mathbf{u}\times\mathbf{v}) = \mathbf{u}\times\frac{d\mathbf{v}}{dt} + \frac{d\mathbf{u}}{dt}\times\mathbf{v}; \tag{0–155}$$

$$\frac{d}{dt}(f\mathbf{u}) = f\frac{d\mathbf{u}}{dt} + \frac{df}{dt}\mathbf{u} \quad [f = f(t) = \text{scalar function of } t]; \tag{0–156}$$

$$\frac{d\mathbf{a}}{dt} = \mathbf{0} \quad (\mathbf{a} = \text{constant vector}). \tag{0–157}$$

These include as special cases the rules:

$$\frac{d}{dt}(\mathbf{a}\cdot\mathbf{v}) = \mathbf{a}\cdot\frac{d\mathbf{v}}{dt} \qquad (\mathbf{a} = \text{constant vector}); \tag{0–158}$$

$$\frac{d}{dt}(\mathbf{a}\times\mathbf{v}) = \mathbf{a}\times\frac{d\mathbf{v}}{dt} \qquad (\mathbf{a} = \text{constant vector}); \tag{0–159}$$

$$\frac{d}{dt}(c\mathbf{u}) = c\frac{d\mathbf{u}}{dt} \qquad (c = \text{constant scalar}); \tag{0–160}$$

$$\frac{d}{dt}(f\mathbf{a}) = \frac{df}{dt}\mathbf{a} \qquad (\mathbf{a} = \text{constant vector}). \tag{0–161}$$

These rules can all be proved by use of components and (0–148) or directly from the definition (0–146).

There is also a chain rule for the derivative of a vector:

$$\frac{d\mathbf{u}}{dt} = \frac{d\alpha}{dt}\frac{d\mathbf{u}}{d\alpha}, \tag{0–162}$$

if $\mathbf{u}$ is a function of α, which is in turn a function of t. The proof is similar to that for ordinary functions; in fact, by taking components, one reduces (0–162) to the rule (0–130).

The *second derivative* of a vector function $\mathbf{u}$ is defined as the derivative of the derivative; higher derivatives are defined in analogous fashion:

$$\frac{d^2\mathbf{u}}{dt^2} = \frac{d}{dt}\left(\frac{d\mathbf{u}}{dt}\right), \quad \frac{d^3\mathbf{u}}{dt^3} = \frac{d}{dt}\left(\frac{d^2\mathbf{u}}{dt^2}\right), \quad \ldots . \tag{0–163}$$

By repeated application of (0–148) one concludes that these can be computed by a corresponding differentiation of components:

$$\frac{d^2\mathbf{u}}{dt^2} = \frac{d^2u_x}{dt^2}\mathbf{i} + \frac{d^2u_y}{dt^2}\mathbf{j} + \frac{d^2u_z}{dt^2}\mathbf{k}, \quad \ldots . \tag{0–164}$$

If $\mathbf{u}$ is interpreted as position vector $\overrightarrow{OP}$ of a point P, then the second derivative of $\mathbf{u}$ with respect to time t is defined to be the *acceleration vector* of P:

$$\mathbf{a} = \frac{d\mathbf{v}}{dt} = \frac{d^2\overrightarrow{OP}}{dt^2} = \frac{d^2x}{dt^2}\mathbf{i} + \frac{d^2y}{dt^2}\mathbf{j} + \frac{d^2z}{dt^2}\mathbf{k}. \tag{0–165}$$

The magnitude of the acceleration is

$$|\mathbf{a}| = \sqrt{\left(\frac{d^2x}{dt^2}\right)^2 + \left(\frac{d^2y}{dt^2}\right)^2 + \left(\frac{d^2z}{dt^2}\right)^2}; \tag{0–166}$$

it should be noted that this is not in general equal to the rate of change of the speed v.

0–10 The integral calculus. The definite integral $\int_a^b f(x)\,dx$ of a function $f(x)$, defined for $a \leqq x \leqq b$, is defined by the limiting process:

$$\int_a^b f(x)\,dx = \lim_{\substack{n\to\infty \\ \max \Delta_i x \to 0}} \sum_{i=1}^{n} f(x_i^*)\,\Delta_i x. \tag{0–167}$$

Here one denotes by x_i $(i = 0, 1, \ldots, n)$ a sequence of values of x on the interval, such that $a = x_0 < x_1 < x_2 < \cdots < x_n = b$ and by $\Delta_i x$ the difference $x_i - x_{i-1}$. The value x_i^* is any value of x such that

$$x_{i-1} \leqq x_i^* \leqq x_i,$$

as illustrated in Fig. 0–16. The limiting process is understood as follows: there exists a number I such that, for n sufficiently large and the largest $\Delta_i x$ sufficiently small, the sum

$$\sum_{i=1}^{n} f(x_i^*)\,\Delta_i x = f(x_1^*)\,\Delta_1 x + \cdots + f(x_n^*)\,\Delta_n x$$

differs from I by as little as desired, no matter how the values x_i^* are chosen. The number I is then the value of the integral on the left of (0–167). The existence of this limit can be established if $f(x)$ is continuous for $a \leqq x \leqq b$.

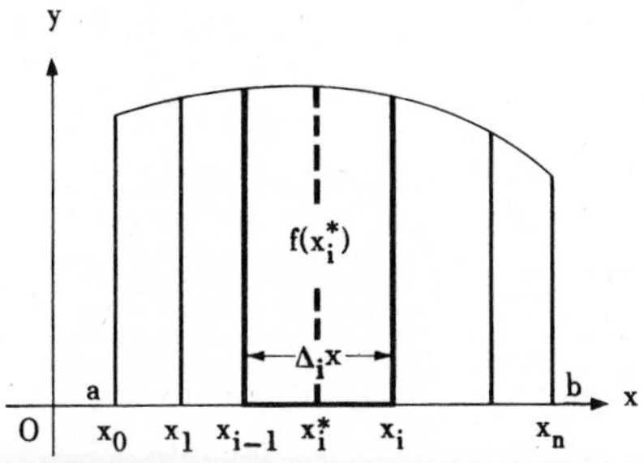

Fig. 0–16. The definite integral.

If $f(x)$ is continuous and nonnegative throughout the interval, the integral can be interpreted as the area of the part of the plane bounded by the x axis, the graph of $y = f(x)$ and the lines $x = a$, $x = b$. If A denotes this area, one has then

$$A = \int_a^b f(x)\,dx \qquad [f(x) \geqq 0]. \tag{0–168}$$

The definite integral satisfies certain basic laws:

$$\int_a^b [f(x) + g(x)]\,dx = \int_a^b f(x)\,dx + \int_a^b g(x)\,dx; \tag{0–169}$$

$$\int_a^b c\,f(x)\,dx = c\int_a^b f(x)\,dx \quad (c = \text{constant}); \tag{0–170}$$

$$\int_a^b f(x)\,dx + \int_b^c f(x)\,dx = \int_a^c f(x)\,dx; \tag{0–171}$$

$$\int_a^b f(x)\,dx = f(x_1)(b-a) \quad (a < x_1 < b); \tag{0–172}$$

if $M_1 \leqq f(x) \leqq M_2$ for $a \leqq x \leqq b$, then

$$M_1(b-a) \leqq \int_a^b f(x)\,dx \leqq M_2(b-a). \tag{0–173}$$

The functions $f(x)$ and $g(x)$ are here assumed to be continuous over the intervals concerned; equations (0–169), (0–170), (0–171), and the inequality (0–173) hold under somewhat more general conditions. Equation (0–172) is the Law of the Mean for integrals; it can be deduced from (0–145) above.

If $a > b$, $\int_a^b f(x)\,dx$ is defined as $-\int_b^a f(x)\,dx$, while $\int_a^a f(x)\,dx$ is defined to be 0. Consequently, in (0–171), the numbers a, b, c can be any three numbers of an interval in which $f(x)$ is defined.

The variable x in a definite integral $\int_a^b f(x)\,dx$ is a "dummy variable;" that is, x can be replaced by another symbol without affecting the value:

$$\int_a^b f(x)\,dx = \int_a^b f(t)\,dt, \quad \int_0^\pi \sin t\,dt = \int_0^\pi \sin u\,du.$$

The existence of an indefinite integral of $f(x)$, i.e., of a function $F(x)$ whose derivative is $f(x)$, is shown by the rule

$$\frac{d}{dx}\int_a^x f(u)\,du = f(x), \tag{0–174}$$

that is,

$$F(x) = \int_a^x f(u)\,du$$

is such a function. It is assumed that $f(x)$ is continuous for $a \leqq x \leqq b$

and (0–174) then holds for $a < x < b$; it holds also for $x = a$ and $x = b$ if one considers derivatives to the right and left.

The rule (0–174) is often referred to as the Fundamental Theorem of Calculus, for it provides the vital connecting link between the two tools of calculus, differentiation and integration.

The difference of two indefinite integrals of $f(x)$ has derivative 0 and must therefore be a constant. Thus all indefinite integrals of $f(x)$ are given by the formula

$$\int f(x)\,dx = F(x) + C, \tag{0–175}$$

where $F(x)$ is any one function whose derivative is $f(x)$, and C is an arbitrary constant.

The definite integral can be evaluated by means of a known indefinite integral by the formula

$$\int_a^b f(x)\,dx = G(b) - G(a), \quad \text{if } G'(x) = f(x). \tag{0–176}$$

For let

$$F(x) = \int_a^x f(u)\,du,$$

so that, by (0–174), $F'(x) = f(x)$. Hence $F(x)$ and $G(x)$ are two indefinite integrals of $f(x)$ and one has

$$G(x) = F(x) + C$$

for some choice of C. From this one concludes:

$$G(b) - G(a) = F(b) - F(a) = \int_a^b f(x)\,dx - \int_a^a f(x)\,dx = \int_a^b f(x)\,dx.$$

Thus (0–176) is proved.

While the existence of indefinite integrals (of continuous functions) is established, the technique of finding them is far from simple. Except in special cases, the integrals of the elementary functions considered above cannot be expressed in terms of such elementary functions. These special cases include integrals of rational functions and rational functions of $\sin x$ and $\cos x$. [See Prob. 37(r) below.] Certain other integrals can be reduced to these by the use of the rule for substitution:

$$\int f(x)\,dx = \int f[x(u)]\,\frac{dx}{du}\,du \tag{0–177}$$

and for integration by parts:

$$\int u(x)v'(x)\,dx = u(x)v(x) - \int v(x)u'(x)\,dx, \tag{0–178}$$

which are available under appropriate hypotheses. Problem 37 illustrates the use of these rules and itself provides a compact table of integrals. For a more complete list one is referred to tables such as the following: H. B. Dwight, *Mathematical Tables* (New York: McGraw-Hill Co., 1941); B. O. Peirce, *A Short Table of Integrals* (Boston: Ginn and Co., 1929).

The definite integral finds a wide application in physical problems. Besides the application to plane area mentioned above, it is used for volumes, lengths of curves, area of surfaces, mass, center of mass, moment of inertia, electrostatic potentials, gravitational potentials, and many other quantities. These will be referred to in the following chapters. The length of a curve $y = f(x)$, $a \leqq x \leqq b$, is given by the formula:

$$s = \int_a^b \sqrt{1 + f'(x)^2}\, dx, \tag{0–179}$$

provided $f(x)$ has a continuous derivative $f'(x)$ over this interval. From this, by (0–174), one deduces that, if s is the length of the curve from $x = a$ up to a variable x, then

$$\frac{ds}{dx} = \sqrt{1 + y'^2} \tag{0–180}$$

or

$$ds^2 = dx^2 + dy^2. \tag{0–181}$$

Thus ds is the hypotenuse of the differential right triangle shown in Fig. 0–17.

If a curve is given in parametric form:

$$x = g(t), \quad y = h(t), \quad \alpha \leqq t \leqq \beta, \tag{0–182}$$

then t can be interpreted as time and the length s of the part of the curve from α to β as the distance the point (x, y) has moved in this interval. From (0–181) one concludes that

$$\frac{ds}{dt} = \sqrt{\left(\frac{dx}{dt}\right)^2 + \left(\frac{dy}{dt}\right)^2} = \sqrt{g'(t)^2 + h'(t)^2}; \tag{0–183}$$

this is simply the *speed* of the moving point (x, y). The total distance moved is then given by the integral

$$s = \int_\alpha^\beta \sqrt{\left(\frac{dx}{dt}\right)^2 + \left(\frac{dy}{dt}\right)^2}\, dt. \tag{0–184}$$

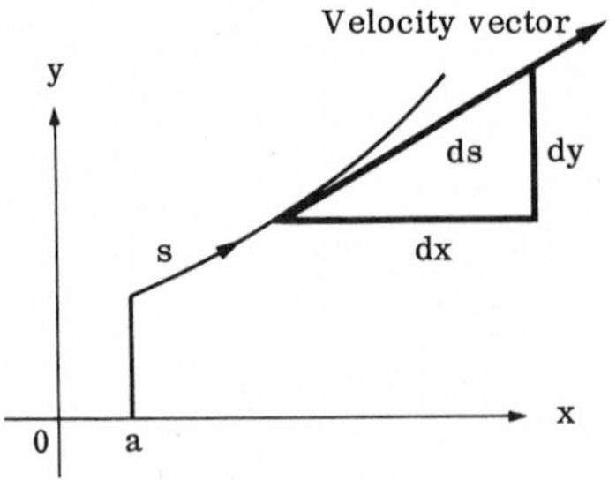

FIG. 0–17. Differential of arc length.

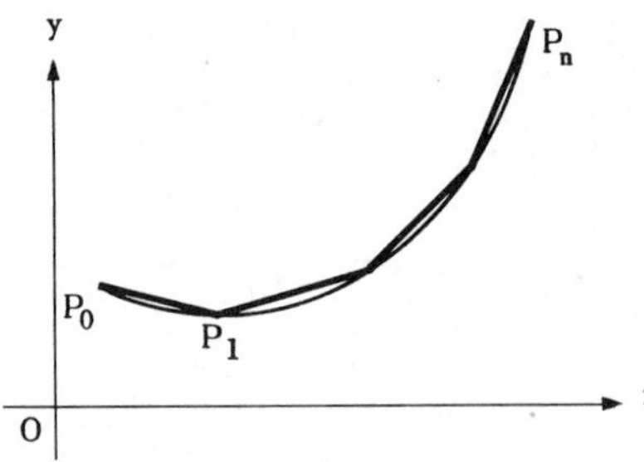

FIG. 0–18. Arc length.

This integral can be defined directly as the limit of a sum representing the length of a broken line $P_0P_1 \ldots P_n$ inscribed in the path, as shown in Fig. 0–18. The points P_i are the positions (x_i, y_i) corresponding to t_i, where $t_0 = \alpha < t_1 < t_2 < \cdots < t_n = \beta$. The limit is taken as n becomes infinite and the maximum difference $t_i - t_{i-1}$ approaches 0. The quantities dx/dt and dy/dt in (0–184) can be interpreted as the x and y components of the velocity vector of the moving point (x, y).

Integration of vector functions. Let $\mathbf{u} = \mathbf{F}(t) = f(t)\mathbf{i} + g(t)\mathbf{j} + h(t)\mathbf{k}$ be given for $a \leqq t \leqq b$. Then one defines the definite integral of $\mathbf{F}(t)$ from a to b as a limit of a sum:

$$\int_a^b \mathbf{F}(t)\,dt = \lim_{\substack{n\to\infty \\ \max\Delta_i t\to 0}} \sum_{i=1}^n \Delta_i t\,\mathbf{F}(t_i), \tag{0–185}$$

exactly as for the integral of a function $f(x)$. It follows that the integral exists when $\mathbf{F}$ is continuous in the interval $a \leqq t \leqq b$. From the properties of limits one finds that

$$\int_a^b \mathbf{F}(t)\,dt = \int_a^b f(t)\,dt\,\mathbf{i} + \int_a^b g(t)\,dt\,\mathbf{j} + \int_a^b h(t)\,dt\,\mathbf{k}. \tag{0–186}$$

For example,

$$\begin{aligned}\int_0^\pi (3t^2\mathbf{i} + e^t\mathbf{j} + \sin t\mathbf{k})\,dt &= (t^3\mathbf{i} + e^t\mathbf{j} - \cos t\mathbf{k})\Big|_0^\pi \\ &= \pi^3\mathbf{i} + (e^\pi - 1)\mathbf{j} + 2\mathbf{k}.\end{aligned}$$

In general, we conclude from (0–186) that

$$\int_a^b \mathbf{G}'(t)\,dt = \mathbf{G}(t)\Big|_a^b = \mathbf{G}(b) - \mathbf{G}(a). \tag{0–187}$$

We can interpret $\mathbf{v} = \mathbf{G}'(t)$ as the velocity vector of the point P, where $\overrightarrow{OP} = \mathbf{G}(t)$. Then (0–187) asserts that the integral of the velocity vector, over a certain time interval, is the total displacement of the point P in that interval.

The indefinite integral of $\mathbf{u} = \mathbf{F}(t)$ is defined as a vector function whose derivative is $\mathbf{F}(t)$. From the properties of the derivative, we conclude that if $\mathbf{F}(t)$ is continuous in an interval, then the indefinite integral exists in that interval and is given by

$$\int \mathbf{F}(t)\,dt = \int f(t)\,dt\mathbf{i} + \int g(t)\,dt\mathbf{j} + \int h(t)\,dt\mathbf{k} + \mathbf{C}, \tag{0–188}$$

where $\mathbf{C}$ is an arbitrary constant vector.

The definite and indefinite integrals of vector functions obey rules analogous to those for real functions (rules (0–169) through (0–172), (0–174), (0–177), and (0–178)).

Problems

Inequalities and absolute values (Section 0–1)

1. Indicate graphically the parts of the x axis for which each of the following inequalities is satisfied:

(a) $x > 3$ (e) $-1 < x < 2$ (i) $|x| > 3$
(b) $x < 2$ (f) $0 \leqq x \leqq 1$ (j) $|x - 2| < 1$
(c) $x > -2$ (g) $-1 \leqq x \leqq 1$ (k) $|x + 1| > 2$
(d) $x \geqq 1$ (h) $|x| \leqq 1$ (l) $x^2 - 1 > 0$

(m) $x^2 - x \geqq 0$ (o) $\dfrac{x}{1-x} > 0$

(n) $(x - 1)(x - 2)(x - 3) > 0$ (p) $\dfrac{x}{1-x} > \dfrac{1-x}{x}$

2. Graph the following functions:
(a) $y = |x|$, (b) $y = |x + 1|$, (c) $y = |x^2|$, (d) $y = x|x|$.

Complex numbers (Section 0–2)

3. Reduce to the form $x + iy$:

(a) $(3 + i) + (2 - 5i)$ (d) $\dfrac{1+i}{1-i}$ (g) $i^{15} + i^9 + i$

(b) $(2 - i)(2 + i)$ (e) $(1 + i)^3$ (h) $\sqrt{-9}$

(c) $(2 + i)(3 - 2i)$ (f) $\dfrac{1}{i}$ (i) $(x + iy)^3$

4. Plot the complex numbers: i, $-i$, $-1 + i$, $-1 - i$, $-2i$; also determine the absolute value and argument of each.

Solution of equations and determinants (Section 0–3)

5. Solve the equations for z:

(a) $z^2 - 4 = 0$ (d) $z^2 - z + 1 = 0$ (g) $z^2 + 2bz + b^2 + c^2 = 0$

(b) $z^2 = 9z$ (e) $2z - \dfrac{1}{z} = 1$ (h) $z^3 - 1 = 0$

(c) $z^2 = 4z - 4$ (f) $\dfrac{1}{1-z} + \dfrac{1}{1+z} = 1$ (i) $z^5 = -1 - i$

6. Evaluate the determinants:

(a) $\begin{vmatrix} 1 & -3 \\ 5 & +1 \end{vmatrix}$ (b) $\begin{vmatrix} 1 & 0 & 0 \\ 3 & 5 & 5 \\ 0 & 7 & 2 \end{vmatrix}$ (c) $\begin{vmatrix} a & b & c \\ a & 2b & 4c \\ a & 3b & 5c \end{vmatrix}$

(d) $\begin{vmatrix} a+1 & a & a & a \\ b & 2b & b & 2b \\ c & 2c & 3c & 4c \\ 1 & 1 & 1 & 1 \end{vmatrix}$

7. Solve the simultaneous equations:

(a) $\begin{cases} x - y + z = 1 \\ x + y - z = 2 \\ 3x + y + z = 0 \end{cases}$ (b) $\begin{cases} x - y + z = 0 \\ 3x - y + 2z = 0 \\ 6x - 4y + 5z = 0 \end{cases}$ (c) $\begin{cases} x - y + u = 0 \\ x + z - u = 1 \\ y - z + u = 0 \\ x + y + z = 1 \end{cases}$

Plane analytic geometry (Section 0–4)

8. Graph the curves:

(a) $3x^2 + 3y^2 - 6x + 12y - 1 = 0$
(b) $x^2 - 2y^2 + 2x - 4y + 5 = 0$
(c) $r = 3 \cos \theta$
(d) $r = 2 + \sin \theta$
(e) $xy = 1$
(f) $xy - x - y = 2$

9. Find the equation of a line perpendicular to the line $3x + y = 3$ and through the point $(-2, 1)$.

10. Find the distance from the origin to the line $x + y = 2$.

Solid analytic geometry (Section 0–5)

11. Find the equation of a plane through (0, 0, 0) and perpendicular to the line through (2, 1, 3) and (5, 3, 0).

12. Find the foot of the perpendicular from (1, 1, 1) to the line

$$\frac{x-1}{3} = \frac{y+2}{2} = \frac{z-1}{-1}.$$

13. Show that the points (1, 0, 4), (3, 2, 5) and (6, 0, 3) are vertices of a right triangle and find its area.

14. Draw a sketch of each of the following surfaces:

(a) $2x^2 + 2y^2 + z^2 = 4$, (b) $x^2 + y^2 - 4z = 0$, (c) $z = xy$.

15. Transform the equations of problem 14 to cylindrical and spherical coordinates.

16. Describe the loci: $r =$ constant, $\theta =$ constant for cylindrical coordinates and the loci: $\rho =$ constant, $\phi =$ constant for spherical coordinates.

Vectors (Section 0–6)

17. Let $\mathbf{a} = \mathbf{i} - 2\mathbf{j}$, $\mathbf{b} = 2\mathbf{i} + \mathbf{j} - \mathbf{k}$, $\mathbf{c} = \mathbf{i} - 7\mathbf{j} + \mathbf{k}$. Evaluate:
(a) $2\mathbf{a} + \mathbf{b}$, $3\mathbf{b} - \mathbf{c}$, $\mathbf{a} - 2(5\mathbf{b} + 3\mathbf{c})$
(b) $\mathbf{a} \cdot \mathbf{b}$, $\mathbf{b} \cdot (2\mathbf{c} - \mathbf{a})$, $|\mathbf{a} + \mathbf{b}|$
(c) $\measuredangle(\mathbf{a}, \mathbf{c})$ (d) $\text{comp}_{\mathbf{b}}\, \mathbf{c}$
(e) $\mathbf{a} \times \mathbf{b}$, $\mathbf{a} \times \mathbf{c}$, $\mathbf{a} \times (3\mathbf{b} + \mathbf{c})$ (f) $\mathbf{a} \times \mathbf{b} \cdot \mathbf{c}$, $\mathbf{c} \cdot \mathbf{b} \times \mathbf{a}$
(g) $\mathbf{a} \times (\mathbf{b} \times \mathbf{c})$, $(\mathbf{a} \times \mathbf{b}) \times \mathbf{c}$

18. Let $\mathbf{a}$, $\mathbf{b}$, $\mathbf{c}$ be as in Problem 17.
(a) Show that $\mathbf{a}$, $\mathbf{b}$ are linearly independent.
(b) Show that $\mathbf{a}$, $\mathbf{b}$, $\mathbf{c}$ are linearly dependent.

19. (a) Find the equation of a line in space through the point (3, 4, 7), and parallel to the line through (5, 2, 3) and (1, 4, 3).

(b) Show that the lines $\overrightarrow{OP} = (1, 0, 1) + t(1, 5, 6)$ and $\overrightarrow{OP} = (3, -4, -1) + t(-1, 2, 1)$ lie in a plane and find the equation of the plane.

20. Use vectors to prove the following theorems of geometry:

(a) The medians of a triangle intersect in a point which is a trisection point of each median.

(b) Let $ABCD$ be a quadrilateral (not necessarily planar) in space. Let M_1, M_2, M_3, M_4 be the successive mid-points of its sides. Then M_1M_3 and M_2M_4 meet and bisect each other (and are hence the diagonals of a parallelogram).

21. Let the points A: (1, 0, 1), B: (2, 1, −1), C: (0, 1, 0), D: (1, 3, −2) in space be given.
(a) Find the area of the triangle ABC.
(b) Find the volume of the parallelepiped three of whose edges are AB, AC and AD.
(c) Find the volume of the tetrahedron $ABCD$.

Functions, limits and continuity (Sections 0–7 and 0–9)

22. Let the functions $f(x) = 1/x^2$ and $g(x) = 2^x$ be given.
(a) Find $f(1)$, $f(-1)$, $g(0)$, $g(1)$.
(b) Find the domain and range of f.
(c) Find the domain and range of g.
(d) Is f one-to-one? If so, find its inverse.
(e) Is g one-to-one? If so, find its inverse.

23. Given that $\lim_{x\to 0} \dfrac{\sin x}{x} = 1$, evaluate $\lim_{x\to 0} \dfrac{x^2 e^x}{\sin x}$.

24. Evaluate the following limits:

(a) $\lim_{x\to\infty} \dfrac{\sin x}{x}$, (b) $\lim_{x\to\infty} \dfrac{2x-1}{3x+5}$, (c) $\lim_{x\to\infty} \log \log x$.

25. l'Hôpital's Rule states that, if $\lim_{x\to a} f(x) = 0$ and $\lim_{x\to a} g(x) = 0$, then

$$\lim_{x\to a} \frac{f(x)}{g(x)} = \lim_{x\to a} \frac{f'(x)}{g'(x)},$$

provided the limit on the right side exists. Apply this rule to evaluate the following limits:

(a) $\lim_{x\to 0} \dfrac{e^x - 1}{x}$, (b) $\lim_{x\to 0} \dfrac{1-\cos x}{x^2}$, (c) $\lim_{x\to 0} \dfrac{\sin^2 x}{1 - \sec x}$.

The rule also applies when f/g is of the "indeterminate form" ∞/∞ at $x = a$, also when the limit refers to x approaching infinity. Evaluate

(d) $\lim_{x\to 0+} x \log x$, (e) $\lim_{x\to\infty} \dfrac{e^x}{x^3}$, (f) $\lim_{x\to\infty} \dfrac{x}{\log^2 x}$.

26. Determine for what values of x the following functions are defined and for what values they are continuous:

(a) $y = \dfrac{1}{x}$ (c) $y = \dfrac{x-1}{x^3 - 2x + 1}$ (e) $y = \dfrac{|x|}{x}$

(b) $y = \log x$ (d) $y = |x|$ (f) $y = e^{\frac{1}{x}}$

Elementary functions (Sections 0–8 and 0–9)

27. Graph the following functions, with the aid of the first and second derivatives:

(a) $y = x^3 - x$ (c) $y = e^{-2x}$ (e) $y = e^x \sin x$

(b) $y = \dfrac{x}{1+x^2}$ (d) $y = \sin x + \sin 2x$ (f) $y = \log(1 + \cos x)$

28. (a) evaluate 2^π (b) simplify: $e^{2 \log x}$
(c) solve for x: $y = \sin \log (1 + \sqrt{1 - x^2})$

29. The functions cosh x (hyperbolic cosine) and sinh x (hyperbolic sine) are defined by the formulas:

$$\cosh x = \frac{e^x + e^{-x}}{2}, \quad \sinh x = \frac{e^x - e^{-x}}{2}.$$

Graph these functions and the function

$$\tanh x = \frac{\sinh x}{\cosh x}.$$

Prove the rules:

(a) $\cosh^2 x - \sinh^2 x = 1$ (b) $\dfrac{d}{dx} \cosh x = \sinh x$

(c) $\dfrac{d}{dx} \sinh x = \cosh x$ (d) $\sinh 2x = 2 \sinh x \cosh x$

(e) $\cosh 2x = \cosh^2 x + \sinh^2 x$

(f) $\sinh (x + y) = \sinh x \cosh y + \cosh x \sinh y$

Differential calculus (Section 0–9)

30. Differentiate:

(a) $y = \dfrac{x - 1}{\sqrt{x^2 + 1}}$, (b) $y = \log \sin (1 - 2x)$, (c) $y = x^2 e^{x^2} \sin 3x$.

31. Find y' and y'' if

(a) $x = e^t$, $y = t - t^3$, (b) $x^2 - y^2 + x - 2y + 1 = 0$.

32. Establish the formulas:

(a) $d \tan x = \sec^2 x \, dx$ (d) $d \operatorname{arc} \sin x = \dfrac{1}{\sqrt{1 - x^2}} dx$

(b) $d \sec x = \sec x \tan x \, dx$ (e) $d \operatorname{arc} \cos x = -\dfrac{1}{\sqrt{1 - x^2}} dx$

(c) $d \csc x = - \csc x \cot x \, dx$ (f) $d \operatorname{arc} \tan x = \dfrac{1}{1 + x^2} dx$

33. *Logarithmic differentiation.* Differentiate the following functions by first taking the logarithm of both sides of the equation:

(a) $y = (x^2 - 1)^3(x^2 + 2x)^4$ (c) $y = x^x$

(b) $y = \sqrt{(x - 1)(2x + 3)(x - 5)}$ (d) $y = (\sin x)^{\cos x}$

(e) $y = \dfrac{u_1(x) \cdot u_2(x) \cdots u_n(x)}{v_1(x) \cdot v_2(x) \cdots v_m(x)}$ [(0–95) above]

34. Use the differential to compute approximately and check the accuracy of the results by comparison with 5-place tables.

(a) $\sqrt{101}$, (b) $\sin (0.05)$, (c) $\log (1.02)$.

Integral calculus (Section 0–10)

35. Find the area between the curves:

(a) $y = 0$ and $y = 1 - x^2$,
(b) $y = x^3$ and $y = x^{\frac{1}{3}}$,
(c) $y = 6 \operatorname{arc} \sin x$ and $y = \pi \sin \pi x$.

36. A point moves in the xy plane according to the law

$$x = \sin t, \quad y = \cos^2 t.$$

Find:
(a) the equation of the curve traced,
(b) the x and y velocity components at time $t = 0$,
(c) the speed at time $t = 0$,
(d) the distance moved between times $t = 0$ and $t = \frac{\pi}{2}$.

37. Verify the following integration formulas. Formulas (a) through (g) are essentially restatements of basic differentiation rules. The others should be reduced to these seven by appropriate use of substitution, integration by parts, partial fractions, and trigonometric identities; they can then be checked by differentiation.

(a) $\int x^n\,dx = \frac{x^{n+1}}{n+1} + C \quad (n \neq -1)$

(b) $\int \frac{dx}{x} = \log |x| + C$

(c) $\int \sin x\,dx = -\cos x + C$

(d) $\int \cos x\,dx = \sin x + C$

(e) $\int e^{ax}\,dx = \frac{e^{ax}}{a} + C$

(f) $\int \frac{dx}{\sqrt{a^2 - x^2}} = \arcsin \frac{x}{a} + C$

(g) $\int \frac{dx}{a^2 + x^2} = \frac{1}{a} \arctan \frac{x}{a} + C$

(h) $\int \log x\,dx = x \log x - x + C$

(i) $\int e^{ax} \sin bx\,dx = \frac{e^{ax}}{a^2 + b^2} [a \sin bx - b \cos bx] + C$

(j) $\int e^{ax} \cos bx\,dx = \frac{e^{ax}}{a^2 + b^2} [a \cos bx + b \sin bx] + C$

(k) $\int x^n e^{ax}\,dx = \frac{e^{ax}}{a} \left[x^n - \frac{nx^{n-1}}{a} + \frac{n(n-1)x^{n-2}}{a^2} - \cdots \pm \frac{n!}{a^n} \right] + C$

(l) $\int \frac{dx}{a^2 - x^2} = \frac{1}{2a} \log \left| \frac{a+x}{a-x} \right| + C$

(m) $\int \frac{dx}{x^2(x^2+1)^2} = -\frac{1}{x} - \frac{x}{2(x^2+1)} - \frac{3}{2} \arctan x + C$

(n) $\int \sin^2 x\,dx = \frac{1}{2}x - \frac{1}{2} \sin x \cos x + C$

(o) $\int \cos^2 x\,dx = \frac{1}{2}x + \frac{1}{2} \sin x \cos x + C$

(p) $\int \tan x\,dx = \log |\sec x| + C$

(q) $\int \cot x\,dx = \log |\sin x| + C$

(r) $\int \csc x\,dx = \log \left| \tan \frac{x}{2} \right| + C \quad$ [Hint: set $t = \tan \frac{x}{2}$. Show that $\cos x = \frac{1-t^2}{1+t^2}$, $\sin x = \frac{2t}{1+t^2}$, $dx = \frac{2\,dt}{1+t^2}$.]

(s) $\int \sec x\,dx = \log \left| \tan \left(\frac{\pi}{4} + \frac{x}{2} \right) \right| + C$

(t) $\displaystyle\int \frac{dx}{a+\cos x} = \frac{2}{\sqrt{a^2-1}} \tan^{-1}\left(\frac{\sqrt{a^2-1}\tan\frac{x}{2}}{a+1}\right) + C, \quad a > 1$

(u) $\displaystyle\int \frac{dx}{a+\sin x} = \frac{2}{\sqrt{a^2-1}} \tan^{-1}\left(\frac{a\tan\frac{x}{2}+1}{\sqrt{a^2-1}}\right) + C, \quad a > 1$

(v) $\displaystyle\int \frac{dx}{\sqrt{x^2 \pm a^2}} = \log|x + \sqrt{x^2 \pm a^2}| + C$

(w) $\displaystyle\int \frac{dx}{x\sqrt{x^2-a^2}} = \frac{1}{a}\operatorname{arc\,cos}\frac{a}{x} + C$

(x) $\displaystyle\int \frac{dx}{x\sqrt{a^2 \pm x^2}} = -\frac{1}{a}\log\left|\frac{a+\sqrt{a^2 \pm x^2}}{x}\right| + C$

(y) $\displaystyle\int \sqrt{a^2-x^2}\,dx = \tfrac{1}{2}\left[x\sqrt{a^2-x^2} + a^2\sin^{-1}\frac{x}{a}\right] + C$

(z) $\displaystyle\int \sin x^2\,dx = C + \frac{x^3}{3\cdot 1!} - \frac{x^7}{7\cdot 3!} + \frac{x^{11}}{11\cdot 5!} + \cdots$

Calculus of vector functions (Sections 0–7, 0–9, 0–10)

38. A point P moves in space according to the law

$$\overrightarrow{OP} = \cos t\mathbf{i} + \sin t\mathbf{j} + t\mathbf{k}.$$

(a) Sketch the path traced.
(b) Find the velocity vector for any t and plot for $t = 0$, $t = \frac{\pi}{2}$.
(c) Find the acceleration vector for any t and plot for $t = 0$, $t = \frac{\pi}{2}$.

39. Let the point P move in the xy-plane and have polar coordinates r, θ, which then become functions of time t. Write $\dot{r}$, $\dot{\theta}$ or ω for dr/dt, $d\theta/dt$, $\ddot{r}$ and $\ddot{\theta}$ or α for d^2r/dt^2, $d^2\theta/dt^2$. Write $\mathbf{r} = \overrightarrow{OP} = r(\cos\theta\mathbf{i} + \sin\theta\mathbf{j}) = r\mathbf{e}$, where $\mathbf{e} = \cos\theta\mathbf{i} + \sin\theta\mathbf{j}$. Also write $\mathbf{h} = \mathbf{k}\times\mathbf{e} = -\sin\theta\mathbf{i} + \cos\theta\mathbf{j}$. Prove:

(a) $\mathbf{v} = d\mathbf{r}/dt = \dot{r}\mathbf{e} + r\dot{\theta}\mathbf{h}$, so that $\mathbf{v}$ has radial component $v_{\text{rad}} = dr/dt$ and transverse component $v_{\text{trans}} = r\omega$.

(b) $\mathbf{a} = d\mathbf{v}/dt = (\ddot{r} - r\dot{\theta}^2)\mathbf{e} + (r\ddot{\theta} + 2\dot{r}\dot{\theta})\mathbf{h}$, so that $a_{\text{rad}} = \ddot{r} - r\omega^2$, $a_{\text{trans}} = r\alpha + 2\dot{r}\omega$.

40. Let P move in the xy-plane with velocity $\mathbf{v} \neq \mathbf{0}$. Write $\mathbf{v} = v\mathbf{T} = v(\cos\alpha\mathbf{i} + \sin\alpha\mathbf{j})$, so that $\mathbf{T}$ is a unit tangent vector in the direction of motion, and let $\mathbf{n} = \mathbf{k}\times\mathbf{T} = -\sin\alpha\mathbf{i} + \cos\alpha\mathbf{j}$, so that $\mathbf{n}$ is a unit normal vector to the path.

(a) Prove: $\mathbf{a} = \dfrac{d\mathbf{v}}{dt} = \dfrac{dv}{dt}\mathbf{T} + v\dfrac{d\alpha}{dt}\mathbf{n}$.

(b) Let $\rho = |ds/d\alpha|$, so that ρ is the *radius of curvature* of the path and $\kappa = 1/\rho$ is the *curvature* of the path. Show from the result of part (a) that

$$\mathbf{a} = \frac{dv}{dt}\mathbf{T} + \frac{v^2}{\rho}\mathbf{N},$$

where $\mathbf{N}$ is an appropriate unit normal vector. Thus $\mathbf{a}$ has tangential component $a_{\text{tang}} = dv/dt$ and normal component $a_{\text{norm}} = v^2/\rho$.

(c) Show that $|\mathbf{v}\times\mathbf{a}| = v^3/\rho$.

41. Let a point P move in space with speed 1, so that the time parameter t can be identified with arc length s and $\mathbf{v} = d\mathbf{r}/ds = \mathbf{T}$, a unit tangent vector. The radius of curvature of the path is then defined as ρ, where

$$\frac{1}{\rho} = |\mathbf{a}| = \left|\frac{d\mathbf{v}}{ds}\right| = \left|\frac{d\mathbf{T}}{ds}\right|.$$

Thus $\mathbf{T}$ and $\rho(d\mathbf{T}/ds) = \mathbf{N}$ are a pair of unit vectors; $\mathbf{T}$ is tangent to the curve at P, $\mathbf{N}$ is normal to the curve and is termed the *principal normal*. The vector $\mathbf{B} = \mathbf{T} \times \mathbf{N}$ is known as the *binormal*. Establish the following relations:

(a) $\mathbf{B}$ is a unit vector and $(\mathbf{T}, \mathbf{N}, \mathbf{B})$ is a positive triple of unit vectors;

(b) $\dfrac{d\mathbf{B}}{ds} \cdot \mathbf{B} = 0$ and $\dfrac{d\mathbf{B}}{ds} \cdot \mathbf{T} = 0$;

(c) there is a scalar $-\tau$ such that $\dfrac{d\mathbf{B}}{ds} = -\tau\mathbf{N}$; τ is known as the *torsion;*

(d) $\dfrac{d\mathbf{N}}{ds} = -\dfrac{1}{\rho}\mathbf{T} + \tau\mathbf{B}$.

The equations

$$\frac{d\mathbf{T}}{ds} = \frac{1}{\rho}\mathbf{N}, \quad \frac{d\mathbf{N}}{ds} = -\frac{1}{\rho}\mathbf{T} + \tau\mathbf{B}, \quad \frac{d\mathbf{B}}{ds} = -\tau\mathbf{N}$$

are known as the *Frenet formulas*. For further properties of curves, see D. J. Struik, *Differential Geometry* (1961).

42. Evaluate the integrals:

(a) $\displaystyle\int_0^1 [(te^{t^2}\mathbf{i} + \cos^2 t\mathbf{j} + (t + 1)^5\mathbf{k}]\, dt$

(b) $\displaystyle\int_0^1 [(3\mathbf{i} + 2\mathbf{j} - 5\mathbf{k})] \cdot [(1 + t^2)^{1/3}\mathbf{i} + (1 + t^2)^{1/3}\mathbf{j} + (1 + t^2)^{1/3}\mathbf{k}]\, dt.$

43. With the aid of (0–186), prove the following integration rules. Assume $\mathbf{u}$ to be a constant vector and $\mathbf{F}(t)$ to be a vector function continuous for $a \leqq t \leqq b$.

(a) $\displaystyle\int_a^b \mathbf{u} \cdot \mathbf{F}(t)\, dt = \mathbf{u} \cdot \int_a^b \mathbf{F}(t)\, dt$ (b) $\displaystyle\int_a^b \mathbf{u} \times \mathbf{F}(t)\, dt = \mathbf{u} \times \int_a^b \mathbf{F}(t)\, dt.$

Proof by induction (Section 0–3)

44. Prove the Demoivre formula (0–13) by induction, carrying out the following steps: (a) verify the truth of the formula for $n = 1$; (b) show that, if the formula is true for one value of n, then it must be true for $n + 1$.

45. Prove by induction formula (0–29) for the sum of a geometric progression.

46. Prove by induction that $y = x^n$ is continuous for every positive integer n.

47. Find other examples above of theorems which appear to require proof by induction.

Answers

3. (a) $5 - 4i$, (b) 5, (c) $8 - i$, (d) i, (e) $-2 + 2i$, (f) $-i$, (g) i, (h) $\pm 3i$, (i) $x^3 - 3xy^2 + i(3x^2y - y^3)$.

4. Absolute values: 1, 1, $\sqrt{2}$, $\sqrt{2}$, 2; arguments: $\frac{\pi}{2}, \frac{3\pi}{2}, \frac{3\pi}{4}, \frac{5\pi}{4}, \frac{3\pi}{2}$.

5. (a) ± 2, (b) 0, 9, (c) 2, 2, (d) $\frac{1}{2}(1 \pm \sqrt{3}\,i)$, (e) $-\frac{1}{2}$, 1, (f) $\pm i$, (g) $-b \pm ci$, (h) 1, $\frac{1}{2}(-1 \pm \sqrt{3}\,i)$, (i) $2^{0.1}\left[\cos\left(\frac{\pi}{4} + \frac{2n\pi}{5}\right) + i \sin\left(\frac{\pi}{4} + \frac{2n\pi}{5}\right)\right]$, $n = 0, 1, 2, 3, 4$.

6. (a) 16, (b) -25, (c) $-2abc$, (d) $-2bc$.

7. (a) $x = \frac{3}{2}$, $y = -2$, $z = -\frac{5}{2}$, (b) $y = -x$, $z = -2x$ (infinitely many solutions), (c) $x = \frac{2}{3}$, $y = \frac{1}{3}$, $z = 0$, $u = -\frac{1}{3}$.

9. $x - 3y = -5$. 10. $\sqrt{2}$. 11. $3x + 2y - 3z = 0$. 12. $(\frac{16}{7}, -\frac{8}{7}, \frac{4}{7})$.

13. $\frac{3\sqrt{17}}{2}$. 15. Cylindrical: (a) $2r^2 + z^2 = 4$, (b) $r^2 - 4z = 0$, (c) $2z = r^2 \sin 2\theta$. Spherical: (a) $\rho^2(1 + \sin^2\phi) = 4$, (b) $\rho \sin^2\phi - 4\cos\phi = 0$, (c) $2\cos\phi = \rho \sin^2\phi \sin 2\theta$.

17. (a) $4\mathbf{i} - 3\mathbf{j} - \mathbf{k}$, $5\mathbf{i} + 10\mathbf{j} - 4\mathbf{k}$, $-25\mathbf{i} + 30\mathbf{j} + 4\mathbf{k}$; (b) 0, -12, $\sqrt{11}$; (c) $\cos^{-1} 15/\sqrt{255}$, about 0.35 radians; (d) $-5/\sqrt{6}$; (e) $2\mathbf{i} + \mathbf{j} + 5\mathbf{k}$, $-2\mathbf{i} - \mathbf{j} - 5\mathbf{k}$, $4\mathbf{i} + 2\mathbf{j} + 10\mathbf{k}$; (f) 0, 0; (g) $30\mathbf{i} + 30\mathbf{j} - 30\mathbf{k}$, $36\mathbf{i} + 3\mathbf{j} - 15\mathbf{k}$.

19. (a) $\overrightarrow{OP} = (3, 4, 7) + t(-4, 2, 0)$, (b) $x + y - z = 0$.

21. (a) $\sqrt{\frac{14}{2}}$, (b) 3, (c) $\frac{1}{2}$.

22. (a) 1, 1, 1, 2; (b) Domain: all real numbers except 0; range: all positive numbers; (c) Domain: all real numbers; range: all positive numbers; (d) no; (e) yes, $x = \log_2 y$.

23. 0. 24. (a) 0, (b) $\frac{2}{3}$, (c) ∞. 25. (a) 1, (b) $\frac{1}{2}$, (c) -2, (d) 0, (e) ∞, (f) ∞. 26. The functions are defined and continuous for the following ranges of x: (a) $x \neq 0$, (b) $x > 0$, (c) all x except 1 and $\frac{1}{2}(-1 \pm \sqrt{5})$, (d) all x, (e) $x \neq 0$, (f) $x \neq 0$. 28. (a) 8.8252, (b) x^2, (c) $x = \pm\sqrt{1 - (e^{\arcsin y} - 1)^2}$. 30. (a) $(x + 1)(x^2 + 1)^{-\frac{3}{2}}$, (b) $-2 \cot (1 - 2x)$, (c) $xe^{x^2}[3x \cos 3x + 2 \sin 3x(x^2 + 1)]$.

31. (a) $y' = e^{-t}(1 - 3t^2)$, $y'' = e^{-2t}(3t^2 - 6t - 1)$, (b) $y' = \frac{2x + 1}{2(y + 1)}$, $y'' = \frac{1}{y + 1} - \frac{(2x + 1)^2}{4(y + 1)^3}$.

33. (a) $y' = (x^2 - 1)^3(x^2 + 2x)^4\left[\frac{6x}{x^2 - 1} + \frac{8(x + 1)}{x^2 + 2x}\right]$, (b) $y' = \frac{1}{2}\sqrt{(x - 1)(2x + 3)(x - 5)}\left[\frac{1}{x - 1} + \frac{2}{2x + 3} + \frac{1}{x - 5}\right]$ (c) $y' = x^x(1 + \log x)$, (d) $y' = \sin x^{\cos x - 1}\,[\cos^2 x - \sin^2 x \log \sin x]$.

35. (a) $\frac{4}{3}$, (b) 1, (c) $14 - \pi - 6\sqrt{3}$. 36. (a) $y = 1 - x^2$, $-1 \leqq x \leqq 1$, (b) $v_x = 1$, $v_y = 0$, (c) $v = 1$, (d) $\frac{1}{2}\sqrt{5} + \frac{1}{4}\log(2 + \sqrt{5})$.

38. (b) $\mathbf{v} = -\sin t\mathbf{i} + \cos t\mathbf{j} + \mathbf{k}$, (c) $\mathbf{a} = -\cos t\mathbf{i} - \sin t\mathbf{j}$.

42. (a) $\frac{e - 1}{2}\mathbf{i} + \frac{2 + \sin 2}{4}\mathbf{j} + \frac{21}{2}\mathbf{k}$, (b) 0.

Suggested References

BARDELL, R. H., AND SPITZBART, A., *College Algebra*, 2nd ed. Reading, Mass.: Addison-Wesley Pub. Co., 1966.

BERS, LIPMAN, *Calculus*. New York: Holt, Rinehart and Winston, Inc., 1969.

COURANT, RICHARD AND ROBBINS, HERBERT E., *What Is Mathematics?* New York: Oxford University Press, 1941. This is an elementary introduction to the methods of higher mathematics, includes algebra, analytic geometry, calculus, and a variety of other topics.

DICKSON, LEONARD E., *First Course in the Theory of Equations*. New York: John Wiley and Sons, Inc., 1922. This includes algebra of real and complex numbers, solution of cubic and quartic equations, determinants, and linear equations.

KAPLAN, WILFRED, AND LEWIS, DONALD J., *Calculus and Linear Algebra*, 2 vols. New York: John Wiley and Sons, Inc., 1970–1971.*

THOMAS, GEORGE B., *Calculus and Analytic Geometry*, 4th ed. Reading, Mass.: Addison-Wesley, 1968.

VANCE, E. P., *Modern Algebra and Trigonometry*, 2nd ed. Reading, Mass.: Addison-Wesley, 1968.

* This work will be cited in the text as *CLA*.

CHAPTER 1

Matrices and n-Dimensional Geometry

Introduction. A matrix is a rectangular array of numbers such as

$$\begin{bmatrix} 3 & 5 & 7 \\ 1 & 2 & -3 \end{bmatrix}.$$

Such arrays appear naturally in the study of simultaneous linear equations (Section 0–3); for example, the coefficients of the unknowns form such an array. They appear in many other contexts: in differential equations, in geometry of n-dimensional space, in physical problems in mechanics and electrical networks, and in many other applications.

In the first part of this chapter we shall study matrices and the ways in which they can be combined. In the second part we shall develop n-dimensional geometry and the theory of vectors in n-dimensional space; in this theory matrices play an important role in describing linear mappings.

PART I. MATRICES

1–1 Terminology concerning matrices. By a *matrix* we mean a rectangular array of m rows and n columns:

$$\begin{bmatrix} a_{11} & \dots & a_{1n} \\ \vdots & & \vdots \\ a_{m1} & \dots & a_{mn} \end{bmatrix}. \tag{1–1}$$

For this chapter (with a very few exceptions), the objects $a_{11}, a_{12}, \dots, a_{mn}$ will be real numbers. In some applications they are complex numbers, and in some they are functions of one or more variables. We call each a_{ij} an *entry* of the matrix; more specifically, a_{ij} is the ij-entry.

We can denote a matrix by a single letter such as $A, B, C, X, Y, \dots$ If A denotes the matrix (1–1), then we write also, concisely, $A = (a_{ij})$.

Let A be the matrix (a_{ij}) of (1–1). We say that A is an $m \times n$ matrix. When $m = n$, we say that A is a *square matrix of order* n. The following are examples of matrices:

$$A = \begin{bmatrix} 2 & 3 & 5 \\ 1 & 2 & 3 \end{bmatrix}, \quad B = \begin{bmatrix} 1 & 2 \\ 4 & 3 \end{bmatrix}, \quad C = \begin{bmatrix} 1 & 4 \\ 2 & -3 \end{bmatrix}. \tag{1–2}$$

Here A is 2×3, B and C are 2×2; B and C are *square matrices of order* 2.

An important square matrix is the *identity matrix* of order n, denoted by I:

$$I = \begin{bmatrix} 1 & 0 & \dots & 0 \\ 0 & 1 & \dots & 0 \\ \vdots & & & \vdots \\ 0 & 0 & \dots & 1 \end{bmatrix} = (\delta_{ij}), \quad \delta_{ij} = \begin{cases} 1 & \text{for } i = j, \\ 0 & \text{for } i \neq j. \end{cases} \tag{1–3}$$

We call δ_{ij} the *Kronecker delta symbol.* One sometimes write I_n to indicate the order of I, but normally the context makes this unnecessary.

For each m and n we define the $m \times n$ *zero matrix*

$$O = \begin{bmatrix} 0 & \dots & 0 \\ \vdots & & \vdots \\ 0 & \dots & 0 \end{bmatrix}. \tag{1–4}$$

One sometimes denotes this matrix by O_{mn} to indicate the *size*—that is, the number of rows and columns.

In general, two matrices $A = (a_{ij})$ and $B = (b_{ij})$ are said to be equal, $A = B$, when A and B have the same size and $a_{ij} = b_{ij}$ for all i and j.

A $1 \times n$ matrix A is formed of one row: $A = (a_{11}, \dots, a_{1n})$. We call such a matrix a *row vector*. In a general $m \times n$ matrix (1–1), each of the successive rows forms a row vector. We often denote a row vector by a boldface symbol: **u**, **v**, ... (or in handwriting, by an arrow). Thus the matrix A in (1–2) has the row vectors $\mathbf{u}_1 = (2, 3, 5)$ and $\mathbf{u}_2 = (1, 2, 3)$.

Similarly, an $m \times 1$ matrix A is formed of one column:

$$A = \begin{bmatrix} a_{11} \\ \vdots \\ a_{m1} \end{bmatrix}. \tag{1–5}$$

We call such a matrix a *column vector*. For typographical reasons, we sometimes denote this matrix by col $(a_{11}, \dots, a_{m1})$ or even by $(a_{11}, \dots, a_{m1})$, if the context makes clear that a column vector is intended. We also denote column vectors by boldface letters: **u**, **v**, ... The matrix B in (1–2) has the column vectors $\mathbf{v}_1 = \text{col}\,(1, 4)$ and $\mathbf{v}_2 = \text{col}\,(2, 3)$.

We denote by **0** the row vector or column vector $(0, \dots, 0)$. The context will make clear whether **0** is a row vector or a column vector and the number of entries.

The vectors occurring here can be interpreted geometrically as vectors in k-dimensional space, for appropriate k. For example, the row vectors or column vectors with three entries are simply ordered triples of numbers and, as in Section 0–6, they can be represented as vectors $a\mathbf{i} + b\mathbf{j} + c\mathbf{k}$ in 3-dimensional space. This interpretation is discussed in Part II of the present chapter.

As mentioned earlier, matrices often arise in connection with simultaneous linear equations. Let such a set of equations be given:

$$\begin{aligned} a_{11}x_1 + \cdots\ a_{1n}x_n &= y_1, \\ &\vdots \\ a_{m1}x_1 + \cdots\ a_{mn}x_n &= y_m. \end{aligned} \tag{1–6}$$

Here we may think of $y_1, \ldots, y_m$ as given numbers and $x_1, \ldots, x_n$ as unknown numbers, to be found; however, we may also think of $x_1, \ldots, x_n$ as variable numbers and $y_1, \ldots, y_m$ as "dependent variables" whose values are determined by the values chosen for the "independent variables" $x_1, \ldots, x_n$. Both points of view will be important in this chapter. In either case, we call $A = (a_{ij})$ the *coefficient matrix* of the set of equations. The numbers $y_1, \ldots, y_m$ can be considered as the entries in a column vector $\mathbf{y} = \text{col}\,(y_1, \ldots, y_m)$. The numbers $x_1, \ldots, x_n$ can be thought of as the entries in a row vector or column vector $\mathbf{x}$; in this chapter, usually we write $\mathbf{x} = \text{col}\,(x_1, \ldots, x_n)$.

1–2 Addition of matrices. Scalar times matrix. Let $A = (a_{ij})$ and $B = (b_{ij})$ be matrices of the *same size*, both $m \times n$. Then one defines the sum $A + B$ to be the $m \times n$ matrix $C = (c_{ij})$ such that $c_{ij} = a_{ij} + b_{ij}$ for all i and j; that is, *one adds two matrices by adding corresponding entries.* For example,

$$\begin{bmatrix} 3 & 5 & 1 \\ 1 & 0 & -2 \end{bmatrix} + \begin{bmatrix} 0 & 1 & 0 \\ 2 & 3 & 5 \end{bmatrix} = \begin{bmatrix} 3 & 6 & 1 \\ 3 & 3 & 3 \end{bmatrix}.$$

Let c be a number (scalar); let $A = (a_{ij})$ be an $m \times n$ matrix. Then one defines cA to be the $m \times n$ matrix $B = (b_{ij})$ such that $b_{ij} = ca_{ij}$ for all i and j; that is, *cA is obtained from A by multiplying each entry of A by c.* For example,

$$5\begin{bmatrix} 2 & 0 \\ 1 & -2 \end{bmatrix} = \begin{bmatrix} 10 & 0 \\ 5 & -10 \end{bmatrix}.$$

We denote $(-1)A$ by $-A$ and $B + (-A)$ by $B - A$.

From these definitions we can deduce the following rules governing the two operations:

1. $A + B = B + A$.

2. $A + (B + C) = (A + B) + C$.

3. $c(A + B) = cA + cB$.

4. $(a + b)C = aC + bC$. (1–7)

5. $a(bC) = (ab)C$. 6. $1A = A$.

7. $0A = O$. 8. $A + O = A$.

9. $A + C = B$ if and only if $C = B - A$.

Throughout we assume that the sizes of the matrices are such that the operations have meaning. The proofs are obtained by simply applying the definitions. For example, for Rule 1 we write

$$A + B = (a_{ij}) + (b_{ij}) = (a_{ij} + b_{ij}) = (b_{ij} + a_{ij}) = (b_{ij}) + (a_{ij}) = B + A.$$

Problems

In these problems, the following matrices are given:

$$A = \begin{bmatrix} 1 \\ 3 \end{bmatrix}, \quad B = \begin{bmatrix} 2 \\ 0 \end{bmatrix}, \quad C = \begin{bmatrix} 2 & 3 \\ 4 & 1 \end{bmatrix}, \quad D = \begin{bmatrix} 1 & -1 \\ 2 & 0 \end{bmatrix}, \quad E = \begin{bmatrix} 1 & 2 \\ 2 & 4 \end{bmatrix},$$

$$F = \begin{bmatrix} 1 & 4 & 5 \\ 2 & 0 & 7 \end{bmatrix}, \quad G = \begin{bmatrix} 3 & 1 & 4 \\ -1 & 0 & -1 \end{bmatrix}, \quad H = (1, 0, 1), \quad J = (3, 5, 2),$$

$$K = (3, 5), \quad L = \begin{bmatrix} 3 & 1 & 0 \\ 2 & 5 & 6 \\ 1 & 4 & 3 \end{bmatrix}, \quad M = \begin{bmatrix} 2 & -1 & 0 \\ 1 & 2 & 1 \\ 3 & 2 & -1 \end{bmatrix}, \quad N = \begin{bmatrix} 1 & 4 \\ 0 & 3 \\ 7 & 1 \end{bmatrix},$$

$$P = \begin{bmatrix} 2 & 2 \\ -1 & -1 \\ 3 & 3 \end{bmatrix}.$$

1. (a) Give the number of rows and columns for each of the matrices A, F, H, L, and P.

(b) Writing $A = (a_{ij})$, $B = (b_{ij})$ and so on, give the values of the following entries: a_{11}, a_{21}, c_{21}, c_{22}, d_{12}, e_{21}, f_{11}, g_{23}, g_{21}, h_{12}, m_{23}.

(c) Give the row vectors of C, G, L, and P.

(d) Give the column vectors of D, F, L, and N.

2. Evaluate each expression which is meaningful:

(a) $A + B$. (b) $C + D$. (c) $E + F$. (d) $L + M$. (e) $N - P$. (f) $G - F$. (g) $5C$. (h) $2E$. (i) $3E + 4D$. (j) $2C + D - E$. (k) $3L - N$.

3. Solve for X: (a) $C + X = D$. (b) $F - 5X = G$.

4. Solve for X and Y: (a) $X + Y = N$, $X - Y = P$.
(b) $2X - 3Y = L$, $X - 2Y = M$.

5. Prove each of the following rules of (1–7): (a) Rule 2. (b) Rule 3. (c) Rule 4. (d) Rule 5. (e) Rule 6. (f) Rule 7. (g) Rule 8. (h) Rule 9.

Answers

1. (a) Number of rows first: A: 2 and 1, F: 2 and 3, H: 1 and 3, L: 3 and 3, P: 3 and 2. (b) $a_{11} = 1$, $a_{21} = 3$, $c_{21} = 4$, $c_{22} = 1$, $d_{12} = -1$, $e_{21} = 2$, $f_{11} = 1$, $g_{23} = -1$, $g_{21} = -1$, $h_{12} = 0$, $m_{23} = 1$. (c) C: (2, 3) and (4, 1), G: (3, 1, 4) and (−1, 0, −1), L: (3, 1, 0), (2, 5, 6) and (1, 4, 3), P: (2, 2), (−1, −1), (3, 3). (d) D: col (1, 2) and col (−1, 0), F: col (1, 2), col (4, 0) and col (5, 7), L: col (3, 2, 1), col (1, 5, 4) and col (0, 6, 3), N: col (1, 0, 7) and col (4, 3, 1).

2. (a) $\begin{bmatrix} 3 \\ 3 \end{bmatrix}$, (b) $\begin{bmatrix} 3 & 2 \\ 6 & 1 \end{bmatrix}$, (c) meaningless, (d) $\begin{bmatrix} 5 & 0 & 0 \\ 3 & 7 & 7 \\ 4 & 6 & 2 \end{bmatrix}$,

(e) $\begin{bmatrix} -1 & 2 \\ 1 & 4 \\ 4 & -2 \end{bmatrix}$, (f) $\begin{bmatrix} 2 & -3 & -1 \\ -3 & 0 & -8 \end{bmatrix}$, (g) $\begin{bmatrix} 10 & 15 \\ 20 & 5 \end{bmatrix}$, (h) $\begin{bmatrix} 2 & 4 \\ 4 & 8 \end{bmatrix}$,

(i) $\begin{bmatrix} 7 & 2 \\ 14 & 12 \end{bmatrix}$, (j) $\begin{bmatrix} 4 & 3 \\ 8 & -2 \end{bmatrix}$, (k) meaningless.

3. (a) $D - C = \begin{bmatrix} -1 & -4 \\ -2 & -1 \end{bmatrix}$, (b) $(\frac{1}{5})(F - G) = \begin{bmatrix} -\frac{2}{5} & \frac{3}{5} & \frac{1}{5} \\ \frac{3}{5} & 0 & \frac{8}{5} \end{bmatrix}$.

4. (a) $X = \frac{1}{2}(N + P) = \begin{bmatrix} \frac{3}{2} & 3 \\ -\frac{1}{2} & 1 \\ 5 & 2 \end{bmatrix}$, $Y = \frac{1}{2}(N - P) = \begin{bmatrix} -\frac{1}{2} & 1 \\ \frac{1}{2} & 2 \\ 2 & -1 \end{bmatrix}$.

(b) $X = 2L - 3M = \begin{bmatrix} 0 & 5 & 0 \\ 1 & 4 & 9 \\ -7 & 2 & 9 \end{bmatrix}$, $Y = L - 2M = \begin{bmatrix} -1 & 3 & 0 \\ 0 & 1 & 4 \\ -5 & 0 & 5 \end{bmatrix}$.

1–3 Multiplication of matrices. In order to motivate the definition of the product AB of two matrices A and B, we consider two systems of simultaneous equations:

$$\left.\begin{aligned} a_{11}u_1 + \cdots + a_{1p}u_p &= y_1, \\ &\vdots \\ a_{m1}u_1 + \cdots + a_{mp}u_p &= y_m. \end{aligned}\right\} \tag{1–8}$$

$$\left.\begin{aligned} b_{11}x_1 + \cdots + b_{1n}x_n &= u_1, \\ &\vdots \\ b_{p1}x_1 + \cdots + b_{pn}x_n &= u_p. \end{aligned}\right\} \tag{1–9}$$

Such pairs of systems arise in many practical problems. A typical situation is that in which $x_1, \ldots, x_n$ are known numbers and $y_1, \ldots, y_m$ are sought, all coefficients b_{ij} and a_{ij} being known. The second set of equations allows us to compute $u_1, \ldots, u_p$; if we substitute the values found in the first set of equations, we can then find $y_1, \ldots, y_m$. We carry this out for the general case:

$$\begin{aligned} y_1 &= a_{11}u_1 + \cdots + a_{1p}u_p \\ &= a_{11}(b_{11}x_1 + \cdots + b_{1n}x_n) + \cdots + a_{1p}(b_{p1}x_1 + \cdots + b_{pn}x_n) \\ &= (a_{11}b_{11} + \cdots + a_{1p}b_{p1})x_1 + \cdots + (a_{11}b_{1n} + \cdots + a_{1p}b_{pn})x_n; \end{aligned}$$

and, in general, for $i = 1, \ldots, m$,

$$y_i = (a_{i1}b_{11} + \cdots + a_{ip}b_{p1})x_1 + \cdots + (a_{i1}b_{1n} + \cdots + a_{ip}b_{pn})x_n$$

or

$$y_i = c_{i1}x_1 + \cdots + c_{in}x_n, \quad i = 1, \ldots, m$$

where

$$c_{ij} = a_{i1}b_{1j} + \cdots + a_{ip}b_{pj} \qquad \text{for } i = 1, \ldots, m, \quad j = 1, \ldots, n. \tag{1–10}$$

Thus, from the coefficient matrix $A = (a_{ij})$ and the coefficient matrix $B = (b_{ij})$, we obtain the coefficient matrix $C = (c_{ij})$ by the rule (1–10). We write $C = AB$ and have thereby defined the *product of the matrices A and B*.

We observe that $(a_{i1}, \ldots, a_{ip})$ is the ith *row* vector of A and that col $(b_{1j}, \ldots, b_{pj})$ is the jth *column* vector of B. Hence, to form the product $AB = C = (c_{ij})$, we obtain each c_{ij} by multiplying corresponding entries of the ith row of A and the jth column of B and adding. The process is suggested in Fig. 1–1.

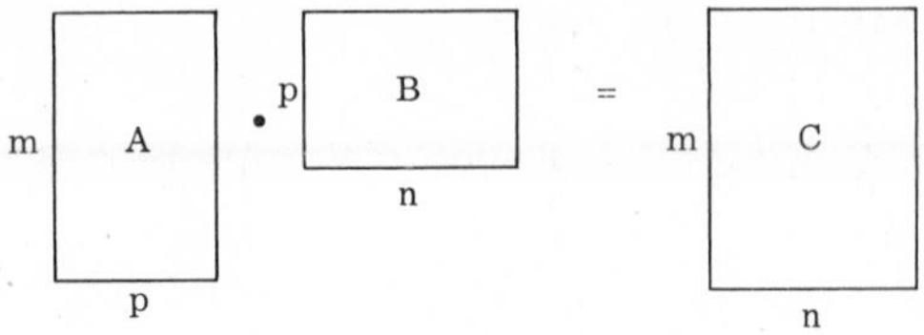

Fig. 1–1. Product of two matrices.

We remark that the product AB is defined only when the number of *columns* of A equals the number of *rows* of B; that is, when A is $m \times p$ and B is $p \times n$, AB is defined and is $m \times n$. Also, when AB is defined, BA need not be defined and even when it is, AB is generally *not equal* to BA; that is, there is *no commutative law* for multiplication.

Example 1. $\begin{bmatrix} a & b \\ c & d \end{bmatrix}\begin{bmatrix} e & f & g \\ h & i & j \end{bmatrix} = \begin{bmatrix} ae + bh & af + bi & ag + bj \\ ce + dh & cf + di & cg + dj \end{bmatrix}.$

Here $m = 2$, $p = 2$, $n = 3$.

Example 2. $\begin{bmatrix} 1 & 2 & 1 \\ 3 & 0 & 5 \\ 2 & 1 & -7 \end{bmatrix}\begin{bmatrix} 1 \\ 4 \\ 2 \end{bmatrix} = \begin{bmatrix} 1 + 8 + 2 \\ 3 + 0 + 10 \\ 2 + 4 - 14 \end{bmatrix} = \begin{bmatrix} 11 \\ 13 \\ -8 \end{bmatrix}$

Here $m = 3$, $p = 3$, $n = 1$.

The second example illustrates the important case of the product $A\mathbf{v}$, where A is an $m \times n$ matrix and $\mathbf{v}$ is an $n \times 1$ column vector. The product $A\mathbf{v}$ is again a column vector $\mathbf{u}$, $m \times 1$.

In the general product $AB = C$, as defined above, we note that the jth column vector of C is formed from A and the jth column vector of B, for the jth column vector of C is

$$\begin{bmatrix} a_{11}b_{1j} + & \cdots & + a_{1p}b_{pj} \\ a_{21}b_{1j} + & \cdots & + a_{2p}b_{pj} \\ & \vdots & \\ a_{m1}b_{1j} + & \cdots & + a_{mp}b_{pj} \end{bmatrix} = \begin{bmatrix} a_{11} & \cdots & a_{1p} \\ a_{21} & \cdots & a_{2p} \\ \vdots & & \vdots \\ a_{m1} & \cdots & a_{mp} \end{bmatrix} \cdot \begin{bmatrix} b_{1j} \\ \vdots \\ b_{pj} \end{bmatrix}.$$

Hence, if we denote the successive column vectors of B by $\mathbf{u}_1, \ldots, \mathbf{u}_n$, then the column vectors of $C = AB$ are $A\mathbf{u}_1, \ldots, A\mathbf{u}_n$. Symbolically,

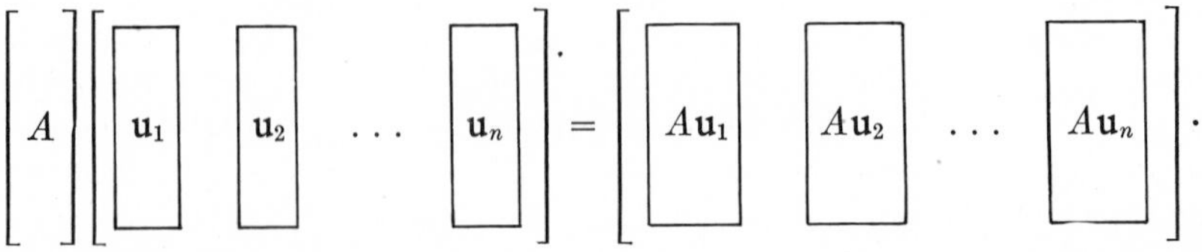

$$\begin{bmatrix} A \end{bmatrix} \begin{bmatrix} \boxed{\mathbf{u}_1} & \boxed{\mathbf{u}_2} & \cdots & \boxed{\mathbf{u}_n} \end{bmatrix} = \begin{bmatrix} \boxed{A\mathbf{u}_1} & \boxed{A\mathbf{u}_2} & \cdots & \boxed{A\mathbf{u}_n} \end{bmatrix}.$$

EXAMPLE 3. To calculate AB, where

$$A = \begin{bmatrix} 3 & 1 & 2 \\ 1 & 0 & 5 \end{bmatrix}, \quad B = \begin{bmatrix} 1 & 0 & 3 & 1 & 3 \\ 5 & 2 & 1 & 5 & 1 \\ -1 & 2 & 4 & -1 & 4 \end{bmatrix},$$
$$\qquad\qquad\qquad \mathbf{u}_1 \quad \mathbf{u}_2 \quad \mathbf{u}_3 \quad \mathbf{u}_4 \quad \mathbf{u}_5$$

we calculate

$$A\mathbf{u}_1 = \begin{bmatrix} 6 \\ -4 \end{bmatrix}, \quad A\mathbf{u}_2 = \begin{bmatrix} 6 \\ 10 \end{bmatrix}, \quad A\mathbf{u}_3 = \begin{bmatrix} 18 \\ 23 \end{bmatrix},$$

and then note that $\mathbf{u}_4 = \mathbf{u}_1$ and $\mathbf{u}_5 = \mathbf{u}_3$, so that $A\mathbf{u}_4 = A\mathbf{u}_1$ and $A\mathbf{u}_5 = A\mathbf{u}_3$. Therefore,

$$AB = \begin{bmatrix} 6 & 6 & 18 & 6 & 18 \\ -4 & 10 & 23 & -4 & 23 \end{bmatrix}.$$

EXAMPLE 4. The simultaneous equations

$$\begin{aligned} 3x + 2y + 5z &= u, \\ 4x - 5y - 8z &= v, \\ 7x + 2y + 9z &= w \end{aligned}$$

are equivalent to the matrix equation

$$\begin{bmatrix} 3 & 2 & 5 \\ 4 & -5 & -8 \\ 7 & 2 & 9 \end{bmatrix} \begin{bmatrix} x \\ y \\ z \end{bmatrix} = \begin{bmatrix} u \\ v \\ w \end{bmatrix},$$

for the product on the left equals the column vector

$$\begin{bmatrix} 3x + 2y + 5z \\ 4x - 5y - 8z \\ 7x + 2y + 9z \end{bmatrix}$$

and this equals col (u, v, w) precisely when the given simultaneous equations hold.

In the same way, the two sets of simultaneous equations (1–8) and (1–9) can be replaced by the equations

$$A\mathbf{u} = \mathbf{y} \quad \text{and} \quad B\mathbf{x} = \mathbf{u}.$$

The elimination process at the beginning of this section is equivalent to replacing $\mathbf{u}$ by $B\mathbf{x}$ in the first equation to obtain $\mathbf{y} = A(B\mathbf{x})$. *Our definition of the product AB is then such that* $\mathbf{y} = A(B\mathbf{x}) = (AB)\mathbf{x}$. Therefore, *for every column vector* $\mathbf{x}$,

$$A(B\mathbf{x}) = (AB)\mathbf{x}.$$

Powers of a square matrix. If A is a square matrix of order n, then the product AA has meaning and is again $n \times n$; we write A^2 for this product. Similarly, $A^3 = A^2A$, $A^4 = A^3A, \ldots, A^{s+1} = A^sA, \ldots$ We also define

A^0 to be the $n \times n$ identity matrix I. Negative powers can also be defined for certain square matrices A; see Section 1–4 below.

Rules for multiplication. Multiplication of matrices obeys a set of rules, which we adjoin to those of the preceding section:

$$
\begin{aligned}
&10.\ A(BC) = (AB)C. \quad 11.\ AI = A. \quad 12.\ IA = A.\\
&13.\ A(B + C) = AB + AC. \quad 14.\ c(AB) = A(cB).\\
&15.\ AO = O. \quad 16.\ OA = O. \quad 17.\ A^0 = I. \quad 18.\ A^kA^l = A^{k+l}.\\
&19.\ (A^k)^l = A^{kl}. \qquad 20.\ A\mathbf{x} = B\mathbf{x} \text{ for all } \mathbf{x} \text{ if and only if } A = B.
\end{aligned}
\tag{1–11}
$$

Here, the sizes of the matrices must again be such that the operations are defined. For example, in Rule 13, if A is $m \times p$, then B and C must be $p \times n$.

To prove Rule 10 (associative law), we let C have the column vectors $\mathbf{u}_1, \ldots, \mathbf{u}_k$. Then BC has the column vectors $B\mathbf{u}_1, \ldots, B\mathbf{u}_k$, and hence $A(BC)$ has the column vectors $A(B\mathbf{u}_1), \ldots, A(B\mathbf{u}_k)$. But, as remarked above, $A(B\mathbf{x}) = (AB)\mathbf{x}$ for every $\mathbf{x}$. Therefore $A(BC)$ has the column vectors $(AB)\mathbf{u}_1, \ldots, (AB)\mathbf{u}_k$. But these are the column vectors of $(AB)C$. Hence $A(BC) = (AB)C$.

For Rule 11, A is, say, $m \times p$ and I is $p \times p$, so that

$$
AI = \begin{bmatrix} a_{11} & \cdots & a_{1p} \\ \vdots & & \vdots \\ a_{m1} & \cdots & a_{mp} \end{bmatrix}
\begin{bmatrix} 1 & 0 & \cdots & 0 \\ 0 & 1 & \cdots & 0 \\ \vdots & \vdots & & \vdots \\ 0 & 0 & \cdots & 1 \end{bmatrix}
= \begin{bmatrix} a_{11} & \cdots & a_{1p} \\ \vdots & & \vdots \\ a_{m1} & \cdots & a_{mp} \end{bmatrix} = A.
$$

We can also write $AI = C = (c_{ij})$, where

$$c_{ij} = a_{i1}\delta_{1j} + \cdots + a_{ip}\delta_{pj} = a_{ij},$$

since $\delta_{jj} = 1$ but $\delta_{ij} = 0$ for $i \neq j$. Rule 12 is proved similarly. For Rule 13 we have $A(B + C) = D$, where

$$
\begin{aligned}
d_{ij} &= a_{i1}(b_{1j} + c_{1j}) + \cdots + a_{ip}(b_{pj} + c_{pj})\\
&= (a_{i1}b_{1j} + \cdots + a_{ip}b_{pj}) + (a_{i1}c_{1j} + \ldots + a_{ip}c_{pj}),
\end{aligned}
$$

and hence $D = AB + AC$. Rule 14 is proved similarly. Rules 15 and 16 follow from the fact that all entries of O are 0; here again the size of O must be such that the products have meaning.

In Rules 17, 18 and 19, A is a square matrix and k and l are nonnegative integers. Rule 17 is true by definition of A^0; and Rules 18 and 19 are true for $l = 0$ and $l = 1$ by definition. They can be proved for general l by induction (Problem 4 below).

For Rule 20, let A and B be $m \times n$ and let $\mathbf{e}_1, \ldots, \mathbf{e}_n$ be the column vectors of the identity matrix I of order n. Then AI is a matrix whose columns are $A\mathbf{e}_1, \ldots, A\mathbf{e}_n$, and BI is a matrix whose column vectors are $B\mathbf{e}_1, \ldots, B\mathbf{e}_n$. If $A\mathbf{x} = B\mathbf{x}$ for all $\mathbf{x}$, then we have

$$A\mathbf{e}_1 = B\mathbf{e}_1, \ldots, A\mathbf{e}_n = B\mathbf{e}_n.$$

Therefore $AI = BI$ or, by Rule 11, $A = B$. Conversely, if $A = B$, then $A\mathbf{x} = B\mathbf{x}$ for all $\mathbf{x}$, by the definition of equality of matrices.

Remark. Because of the associative law, Rule 10, we can generally drop parentheses in multiple products of matrices. For example, we replace $[A(BC)]D$ by $ABCD$. No matter how we group the factors, the same result is obtained.

Problems

Let the matrices $A, \ldots, P$ be given as at the beginning of the set of problems following Section 1–2.

1. Evaluate each expression which is meaningful:

(a) AB.
(b) CA.
(c) AC.
(d) CD and DC.
(e) CE and EC.
(f) AI.
(g) IL.
(h) $2I + 3GL$.
(i) $C^2 - 3C - 10C^0$.
(j) $E(E - 5I)$.
(k) LNI.
(l) MPG.
(m) $HL + J$.
(n) KA.
(o) $OC + N$.
(p) E^2.
(q) E^3.
(r) E^4.
(s) $HELP$.

2. Calculate RS for each of the following choices of R and S:

(a) $R = \begin{bmatrix} 3 & 1 \\ 5 & 2 \end{bmatrix}, \quad S = \begin{bmatrix} 1 & 0 & 2 & 0 & 2 \\ 0 & 1 & 3 & 1 & 3 \end{bmatrix}.$

(b) $R = \begin{bmatrix} 1 & 4 \\ 2 & 1 \\ 5 & 0 \end{bmatrix}, \quad S = \begin{bmatrix} 2 & 1 & 1 & 2 & 1 \\ 3 & 2 & 2 & 3 & 2 \end{bmatrix}.$

3. Consider each of the following pairs of simultaneous equations as cases of (1–8) and (1–9) and express $y_1, \ldots$ in terms of $x_1, \ldots$ (i) by eliminating $u_1, \ldots$ and (ii) by multiplying the coefficient matrices:

(a) $\begin{cases} 3u_1 + 2u_2 = y_1 \\ 5u_1 + 6u_2 = y_2 \end{cases}$ and $\begin{cases} 5x_1 - x_2 = u_1 \\ x_1 + 2x_2 = u_2 \end{cases}$

(b) $\begin{cases} 2u_1 - u_2 = y_1 \\ 5u_1 + u_2 = y_2 \end{cases}$ and $\begin{cases} x_1 + 2x_2 - x_3 = u_1 \\ 2x_1 + 3x_2 + x_3 = u_2 \end{cases}$

4. Prove each of the following rules of Section 1–3:
(a) Rule 12. (b) Rule 14.
(c) Rule 18, by induction with respect to l.
(d) Rule 19, by induction with respect to l and Rule 18.

5. Let A be a square matrix. Prove:
(a) $A^2 - I = (A + I)(A - I)$.
(b) $A^3 - I = (A - I)(A^2 + A + I)$.
(c) $A^2 - 2A - 3I = (A - 3I)(A + I)$.
(d) $6A^2 - A - 2I = (2A + I)(3A - 2I)$.

6. If A and B are $n \times n$ matrices, is $A^2 - B^2$ necessarily equal to $(A - B) \times (A + B)$? When must this be true?

7. Find nonzero 2×2 matrices A and B such that $A^2 + B^2 = O$.

Answers

1. (a) meaningless, (b) $\begin{bmatrix} 11 \\ 7 \end{bmatrix}$, (c) meaningless, (d) $\begin{bmatrix} 8 & -2 \\ 6 & -4 \end{bmatrix}$ and $\begin{bmatrix} -2 & 2 \\ 4 & 6 \end{bmatrix}$, (e) $\begin{bmatrix} 8 & 16 \\ 6 & 12 \end{bmatrix}$ and $\begin{bmatrix} 10 & 5 \\ 20 & 10 \end{bmatrix}$, (f) A, (g) L, (h) undefined, (i) O, (j) O, (k) $\begin{bmatrix} 3 & 15 \\ 44 & 29 \\ 22 & 19 \end{bmatrix}$, (l) $\begin{bmatrix} 10 & 5 & 15 \\ 6 & 3 & 9 \\ 2 & 1 & 3 \end{bmatrix}$, (m) (7, 10, 5), (n) (18), (o) N (O is 3×2), (p) $\begin{bmatrix} 5 & 10 \\ 10 & 20 \end{bmatrix} = 5E$, (q) $25E$, (r) $125E$, (s) meaningless.

2. (a) $\begin{bmatrix} 3 & 1 & 9 & 1 & 9 \\ 5 & 2 & 16 & 2 & 16 \end{bmatrix}$, (b) $\begin{bmatrix} 14 & 9 & 9 & 14 & 9 \\ 7 & 4 & 4 & 7 & 4 \\ 10 & 5 & 5 & 10 & 5 \end{bmatrix}$

3. (a) $y_1 = 17x_1 + x_2$, $y_2 = 31x_1 + 7x_2$,
(b) $y_1 = x_2 - 3x_3$, $y_2 = 7x_1 + 13x_2 - 4x_3$.

6. It is true precisely when $AB = BA$.

1–4 Inverse of a square matrix. Let A be an $n \times n$ matrix. If an $n \times n$ matrix B exists such that $AB = I$, then we call B an *inverse* of A. We shall see below that A can have at most one inverse. Hence, if $AB = I$, we call B *the* inverse of A and write $B = A^{-1}$.

For a general $n \times n$ matrix A we denote by det A the determinant formed from A; that is,

$$\det A = \begin{vmatrix} a_{11} & \cdots & a_{1n} \\ \vdots & & \vdots \\ a_{n1} & \cdots & a_{nn} \end{vmatrix}. \tag{1–12}$$

We stress that det A is a number, whereas A itself is a square array—that is, a matrix. If A and B are $n \times n$ matrices, then one has the rule

$$\det A \det B = \det(AB). \tag{1–13}$$

For a proof, see *CLA*, Sections 10–13 and 10–14. From this rule it follows that if A has an inverse, then $\det A \neq 0$. For $AB = I$ implies

$$\det A \det B = \det I = 1,$$

so that $\det A \neq 0$; also $\det B = \det A^{-1} \neq 0$ and, in fact,

$$\det B = \det A^{-1} = \frac{1}{\det A}. \tag{1–14}$$

Conversely, *if* $\det A \neq 0$, *then* A *has an inverse.* For if $\det A \neq 0$, then the simultaneous linear equations

$$\begin{aligned} a_{11}x_1 + \cdots + a_{1n}x_n &= y_1 \\ \vdots \qquad\qquad & \\ a_{n1}x_1 + \cdots + a_{nn}x_n &= y_n \end{aligned} \tag{1–15}$$

can be solved for $x_1, \ldots, x_n$ by Cramer's Rule (Section 0–3). For example,

$$x_1 = \begin{vmatrix} y_1 & a_{12} & \ldots & a_{1n} \\ \vdots & \vdots & & \vdots \\ y_n & a_{n2} & \ldots & a_{nn} \end{vmatrix} \div D,$$

where $D = \det A$. Upon expanding the first determinant on the right, we obtain an expression of the form

$$x_1 = b_{11}y_1 + \cdots + b_{1n}y_n,$$

with appropriate constants $b_{11}, \ldots, b_{1n}$. In general,

$$x_i = b_{i1}y_1 + \cdots + b_{in}y_n, \quad i = 1, \ldots, n. \tag{1–16}$$

Now our given equations (1–15) are equivalent to the matrix equation

$$A\mathbf{x} = \mathbf{y},$$

and the solution (1–16) is given by

$$\mathbf{x} = B\mathbf{y}.$$

The fact that this is a solution is expressed by the relation

$$AB\mathbf{y} = \mathbf{y} \quad \text{or} \quad AB\mathbf{y} = I\mathbf{y}.$$

This relation holds for all $\mathbf{y}$. Hence by Rule 20 of Section 1–3, we must have $AB = I$, so that B is an inverse of A.

The reasoning just given also provides a constructive way of finding A^{-1}. One simply forms the equations (1–15) and solves for $x_1, \ldots, x_n$. The solution can be written as $\mathbf{x} = B\mathbf{y}$, where $B = A^{-1}$.

Example 1. $A = \begin{bmatrix} 2 & 5 \\ 1 & 3 \end{bmatrix}$. The simultaneous equations are

$$2x_1 + 5x_2 = y_1, \quad x_1 + 3x_2 = y_2.$$

We solve by elimination, and find

$$x_1 = 3y_1 - 5y_2, \quad x_2 = -y_1 + 2y_2.$$

Therefore, $A^{-1} = \begin{bmatrix} 3 & -5 \\ -1 & 2 \end{bmatrix}$. We check by verifying that $AA^{-1} = I$.

Nonsingular matrices. A matrix A having an inverse is said to be *nonsingular*. Hence we have shown that A is nonsingular precisely when $\det A \neq 0$. A square matrix having *no* inverse is said to be *singular*.

Now let A have an inverse B, so that $AB = I$. Then as remarked above, also $\det B \neq 0$, so that B also has an inverse B^{-1}, and $BB^{-1} = I$. We can now write

$$BA = BAI = BABB^{-1} = B(AB)B^{-1} = BIB^{-1} = BB^{-1} = I.$$

Therefore, also, $BA = I$. Furthermore, if $AC = I$, then

$$C = IC = BAC = BI = B.$$

This shows that the inverse of A is unique. Furthermore, if $CA = I$, then

$$C = CI = CAB = IB = B.$$

These results can be summarized as follows: *the inverse* $B = A^{-1}$ *of* A *is unique and* B *satisfies the two equations*

$$AB = I \quad \textit{and} \quad BA = I;$$

furthermore, if a matrix B *satisfies either one of these two equations, then* B *must satisfy the other equation and* $B = A^{-1}$.

The inverse satisfies several additional rules:

21. $(AD)^{-1} = D^{-1}A^{-1}$. 22. $(cA)^{-1} = c^{-1}A^{-1}$ $(c \neq 0)$.

23. $(A^{-1})^{-1} = A$.

Here A and D are assumed to be nonsingular $n \times n$ matrices. To prove Rule 21, we write

$$(AD)(D^{-1}A^{-1}) = A(DD^{-1})A^{-1} = AIA^{-1} = AA^{-1} = I.$$

Therefore $D^{-1}A^{-1}$ must be the inverse of AD. The proof of Rule 22 is left as an exercise (Problem 5 below). For Rule 23, we reason that A^{-1} is nonsingular and hence A^{-1} has an inverse. But $A^{-1}A = I$, so that A is the inverse of A^{-1}; that is, $A = (A^{-1})^{-1}$.

Rule 21 extends to more than two factors: for example,

$$(ABCD)^{-1} = D^{-1}C^{-1}B^{-1}A^{-1}.$$

The proof is as above. In this way we see that the product of two or more nonsingular matrices is nonsingular.

Negative powers of a square matrix. Let A be nonsingular, so that A^{-1} exists. For each positive integer p, we now define A^{-p} to mean $(A^{-1})^p$. Since A is nonsingular, A^p is also nonsingular; in fact, A^p has the inverse

$$(AA \cdots A)^{-1} = A^{-1}A^{-1} \cdots A^{-1} = (A^{-1})^p.$$

Therefore,

$$A^{-p} = (A^p)^{-1} \quad \text{or} \quad A^pA^{-p} = I. \tag{1–17}$$

Rules 18 and 19,

$$A^pA^q = A^{p+q}, \tag{1–18}$$

$$(A^p)^q = A^{pq}, \tag{1–19}$$

are now satisfied, for A nonsingular, for arbitrary integers p and q, positive, negative, or zero. These rules are proved for p and q nonnegative in Section 1–3.

To prove (1–18) for general p and q, we first reason that, since

$$(A^{p+q})^{-1} = A^{-p-q},$$

(1–18) is equivalent to $A^pA^qA^{-p-q} = I$. This in turn is equivalent to the statement

$$(*) \qquad A^pA^qA^r = I \quad \text{whenever} \quad p + q + r = 0.$$

Now for s and t nonnegative, we know that $A^sA^t = A^{s+t}$. This implies

$$A^sA^tA^{-s-t} = I \quad \text{and} \quad A^{-s}A^{-t}A^{s+t} = I;$$

and hence

$$A^{-s-t}A^sA^t = I, \quad A^tA^{-s-t}A^s = I, \quad A^{s+t}A^{-s}A^{-t} = I, \quad A^{-t}A^{s+t}A^{-s} = I.$$

The last six equations state that (*) holds in each of the cases: $p \geqq 0$, $q \geqq 0$, $r \leqq 0$; $p \leqq 0$, $q \leqq 0$, $r \geqq 0$; $p \leqq 0$, $q \geqq 0$, $r \geqq 0$; $p \geqq 0$, $q \leqq 0$, $r \geqq 0$; $p \geqq 0$, $q \leqq 0$, $r \leqq 0$; $p \leqq 0$, $q \geqq 0$, $r \leqq 0$. These are *all possible cases*. Hence (*) is proved and (1–18) follows.

We note that, by (1–18), $A^pA^q = A^qA^p$. Hence A^p and A^q commute (that is, obey the commutative law) under multiplication.

The proof of (1–19) is left as an exercise (Problem 6 below).

The procedure of Example 1 for finding the inverse of a matrix can be much improved. The calculation of inverses of matrices is very important in numerical work, and various procedures have been devised for this purpose, especially adapted to digital computers (see *CLA*, Section 10–16).

Inverses can be used to solve matrix equations—for example, equations of the form $AX = B$ or $XA = B$, where A and B are known and X is sought; if A is $n \times n$ and nonsingular, then we find, respectively,

$$X = A^{-1}B \quad \text{and} \quad X = BA^{-1},$$

and verify in each case that the equation is satisfied. As a special case, we solve the equation $A\mathbf{x} = \mathbf{y}$ (A being $n \times n$) to obtain $\mathbf{x} = A^{-1}\mathbf{y}$.

Remark. If $k < n$, A is an $n \times k$ matrix, and B is a $k \times n$ matrix, then AB is an $n \times n$ matrix and AB is *singular*. To see this, we complete A and B to $n \times n$ matrices by adding extra columns and rows of zeros. Let A_1 and B_1 be the expanded matrices obtained. Then $AB = A_1B_1$, since the added zeros contribute nothing to the product. But A_1 and B_1 are singular, so that $A_1B_1 = AB$ is singular, since det A_1B_1 is 0 by (1–13).

EXAMPLE 2.

$$\begin{bmatrix} 2 & 1 \\ 4 & 1 \\ 5 & 3 \end{bmatrix}\begin{bmatrix} 1 & 2 & 4 \\ 2 & 1 & 5 \end{bmatrix} = \begin{bmatrix} 2 & 1 & 0 \\ 4 & 1 & 0 \\ 5 & 3 & 0 \end{bmatrix}\begin{bmatrix} 1 & 2 & 4 \\ 2 & 1 & 5 \\ 0 & 0 & 0 \end{bmatrix} = \begin{bmatrix} 4 & 5 & 13 \\ 6 & 9 & 21 \\ 11 & 13 & 35 \end{bmatrix}.$$

Problems

1. Verify that each of the following matrices is nonsingular and find the inverse of each:

(a) $A = \begin{bmatrix} 3 & 5 \\ 2 & 4 \end{bmatrix}$.

(b) $B = \begin{bmatrix} 4 & 7 \\ 1 & 6 \end{bmatrix}$.

(c) $C = \begin{bmatrix} 1 & 0 & 1 \\ 2 & 2 & 1 \\ 0 & 1 & -1 \end{bmatrix}$.

(d) $D = \begin{bmatrix} 2 & 0 & 1 \\ 3 & 1 & 2 \\ 4 & 0 & 3 \end{bmatrix}$.

2. Let $A, \ldots, D$ be as in Problem 1. With the aid of the answers to Problem 1, solve for X or $\mathbf{x}$:

(a) $AX = \begin{bmatrix} 0 & 1 \\ 1 & 0 \end{bmatrix}$.

(b) $AX = \begin{bmatrix} 1 & 5 \\ 6 & 2 \end{bmatrix}$.

(c) $BXA = \begin{bmatrix} 7 & 2 \\ 0 & 5 \end{bmatrix}$.

(d) $B^2XA^2 = \begin{bmatrix} 3 & 4 \\ 2 & 1 \end{bmatrix}$.

(e) $A\mathbf{x} = \text{col}\,(4, 3)$.

(f) $B\mathbf{x} = \text{col}\,(1, 7)$.

(g) $BX = \begin{bmatrix} 1 & 6 & 2 \\ 5 & 0 & -3 \end{bmatrix}$.

(h) $XC = \begin{bmatrix} 0 & 5 & 4 \\ 5 & 0 & 1 \end{bmatrix}$.

(i) $\mathbf{x}C = (1, 2, 3)$.

(j) $\mathbf{x}D = (2, 1, 0)$.

3. Simplify:
(a) $[(AB)^{-1}A^{-1}]^{-1}$.
(b) $(ABC)^{-1}(C^{-1}B^{-1}A^{-1})^{-1}$.
(c) $\{[(A^{-1})^2B]^{-1}A^{-2}B^{-1}\}^{-2}$.

4. Prove:
(a) If A and B are nonsingular and $AB = BA$, then $A^{-1}B^{-1} = B^{-1}A^{-1}$.
(b) If $ABC = I$, then $BCA = I$ and $CAB = I$.

5. Prove the Rule 22.

6. Prove the rule (1–19) for arbitrary integers p, q. [Hint: For $s \geqq 0, t \geqq 0$, show by (1–17) that $(A^s)^{-t} = (A^{-s})^t = A^{-st}$ and that $(A^{-s})^{-t} = A^{st}$.]

7. Let $A = (a_{ij})$ be a nonsingular square matrix of order n Prove:

(a) If $n = 2$, then $A^{-1} = \dfrac{1}{\det A}\begin{bmatrix} a_{22} & -a_{12} \\ -a_{21} & a_{11} \end{bmatrix}$.

(b) If $n = 3$, then A^{-1} is the matrix

$$\frac{1}{\det A}\begin{bmatrix} \begin{vmatrix} a_{22} & a_{23} \\ a_{32} & a_{33} \end{vmatrix} & \begin{vmatrix} a_{13} & a_{12} \\ a_{33} & a_{32} \end{vmatrix} & \begin{vmatrix} a_{12} & a_{13} \\ a_{22} & a_{23} \end{vmatrix} \\ \begin{vmatrix} a_{23} & a_{21} \\ a_{33} & a_{31} \end{vmatrix} & \begin{vmatrix} a_{11} & a_{13} \\ a_{31} & a_{33} \end{vmatrix} & \begin{vmatrix} a_{13} & a_{11} \\ a_{23} & a_{21} \end{vmatrix} \\ \begin{vmatrix} a_{21} & a_{22} \\ a_{31} & a_{32} \end{vmatrix} & \begin{vmatrix} a_{12} & a_{11} \\ a_{32} & a_{31} \end{vmatrix} & \begin{vmatrix} a_{11} & a_{12} \\ a_{21} & a_{22} \end{vmatrix} \end{bmatrix}.$$

(c) Let A_{ij} denote the minor determinant of A obtained by deleting the ith row and jth column from A. Let $b_{ij} = (-1)^{i+j}A_{ji}$. The matrix $B = (b_{ij})$

is called the *adjoint* of A and is denoted by adj A. Show that

$$A^{-1} = \frac{1}{\det A}\,\text{adj}\,A.$$

8. Let all matrices occurring in the following equations be *square* of order n. Solve for X and Y, stating which matrices are assumed to be nonsingular:
(a) $X + Y = A,\quad X - Y = B.$ (d) $AX + BY = C,\quad DX + EY = F.$
(b) $X + Y = A,\quad X + BY = C.$ (e) $XA + YB = C,\quad XD + YE = F.$
(c) $X + AY = B,\quad X + CY = D.$

Answers

1. (a) $\frac{1}{2}\begin{bmatrix} 4 & -5 \\ -2 & 3 \end{bmatrix}$, (b) $\frac{1}{17}\begin{bmatrix} 6 & -7 \\ -1 & 4 \end{bmatrix}$, (c) $\begin{bmatrix} 3 & -1 & 2 \\ -2 & 1 & -1 \\ -2 & 1 & -2 \end{bmatrix}$,
(d) $\frac{1}{2}\begin{bmatrix} 3 & 0 & -1 \\ -1 & 2 & -1 \\ -4 & 0 & 2 \end{bmatrix}$.

2. (a) $\frac{1}{2}\begin{bmatrix} -5 & 4 \\ 3 & -2 \end{bmatrix}$, (b) $\begin{bmatrix} -13 & 8 \\ 8 & -2 \end{bmatrix}$, (c) $\frac{1}{34}\begin{bmatrix} 214 & -279 \\ -64 & 89 \end{bmatrix}$,
(d) $\frac{1}{1156}\begin{bmatrix} -1714 & 1553 \\ 1458 & -1981 \end{bmatrix}$, (e) $\frac{1}{2}$ col $(1, 1)$, (f) $(\frac{1}{17})$ col $(-43, 27)$,
(g) $\frac{1}{17}\begin{bmatrix} -29 & 36 & 33 \\ 19 & -6 & -14 \end{bmatrix}$, (h) $\begin{bmatrix} -18 & 9 & -13 \\ 13 & -4 & 8 \end{bmatrix}$, (i) $(-7, 4, -6)$,
(j) $\frac{1}{2}(5, 2, -3)$.

3. (a) A^2B, (b) I, (c) B^4.

8. All matrices whose inverses appear in the answers are assumed to be nonsingular.

(a) $X = \frac{1}{2}(A + B),\quad Y = \frac{1}{2}(A - B).$
(b) $X = (B - I)^{-1}(BA - C),\quad Y = (B - I)^{-1}(C - A).$
(c) $X = B - A(A - C)^{-1}(B - D),\quad Y = (A - C)^{-1}(B - D).$
(d) $X = (B^{-1}A - E^{-1}D)^{-1}(B^{-1}C - E^{-1}F),$
$Y = (A^{-1}B - D^{-1}E)^{-1}(A^{-1}C - D^{-1}F).$
(e) $X = (CB^{-1} - FE^{-1})(AB^{-1} - DE^{-1})^{-1},$
$Y = (CA^{-1} - FD^{-1})(BA^{-1} - ED^{-1})^{-1}.$

***1–5 Eigenvalues of a square matrix.** Let A be an $n \times n$ matrix. For some *nonzero* column vector $\mathbf{v} = \text{col}\,(v_1, \ldots, v_n)$ it may happen that, for some scalar λ,

$$A\mathbf{v} = \lambda\mathbf{v}. \tag{1–20}$$

If this occurs, we say that λ is an *eigenvalue* of A and that $\mathbf{v}$ is an *eigenvector* of A, associated with the eigenvalue λ. The concept of eigenvalue has important applications in many branches of physics. An important example is the *spectrum*—of light, of an atom, of a nucleus. The frequencies occurring in the spectrum correspond to the eigenvalues of a matrix (or of a suitable generalization of a matrix).

We can write Eq. (1–20) in the form $A\mathbf{v} = \lambda I\mathbf{v}$ or in the form

$$(A - \lambda I)\mathbf{v} = \mathbf{0}.$$

Thus $\mathbf{v} = \text{col}\,(v_1, \ldots, v_n)$ is an eigenvector of A associated with the eigenvalue λ precisely when $v_1, \ldots, v_n$ form a nontrivial solution of the set of homogeneous linear equations

$$\begin{aligned}(a_{11} - \lambda)v_1 + a_{12}v_2 + \cdots + \quad & a_{1n}v_n = 0,\\ a_{21}v_1 + (a_{22} - \lambda)v_2 + \cdots + \quad & a_{2n}v_n = 0,\\ \vdots \qquad\qquad & \qquad\quad \vdots\\ a_{n1}v_1 + a_{n2}v_2 \quad + \cdots + (a_{nn} & - \lambda)v_n = 0.\end{aligned} \tag{1–21}$$

Now we know (Section 0–3) that the Eqs. (1–21) have a nontrivial solution precisely when the determinant of the coefficients is 0; that is, when $\det\,(A - \lambda I) = 0$ or

$$\begin{vmatrix} a_{11} - \lambda & a_{12} & \ldots & a_{1n} \\ a_{21} & a_{22} - \lambda & \ldots & a_{2n} \\ \vdots & \vdots & & \vdots \\ a_{n1} & a_{n2} & \ldots & a_{nn} - \lambda \end{vmatrix} = 0. \tag{1–22}$$

When expanded, (1–22) becomes an algebraic equation of degree n for λ, called the *characteristic equation* of the matrix A. The eigenvalues of A are simply the real roots of the characteristic equation. (The complex roots of the characteristic equation can also be interpreted as eigenvalues of A; see below.)

Example 1. Let $A = \begin{bmatrix} 1 & 2 \\ 3 & 2 \end{bmatrix}$. Then the characteristic equation is

$$\begin{vmatrix} 1 - \lambda & 2 \\ 3 & 2 - \lambda \end{vmatrix} = 0 \quad \text{or} \quad \lambda^2 - 3\lambda - 4 = 0 \quad \text{or} \quad (\lambda - 4)(\lambda + 1) = 0.$$

The roots are $\lambda_1 = 4$ and $\lambda_2 = -1$. To find $\mathbf{v}$ for $\lambda = \lambda_1 = 4$, we form the equations

$$(1 - 4)v_1 + 2v_2 = 0, \quad 3v_1 + (2 - 4)v_2 = 0,$$

and find that the solutions are all vectors $k(2, 3)$. Hence the eigenvectors associated with the eigenvalue $\lambda_1 = 4$ are all vectors $k(2, 3)$ for nonzero k. In the same way we find that all eigenvectors associated with the eigenvalue $\lambda_2 = -1$ are all vectors $k(1, -1)$ for nonzero k.

Example 2. Let

$$B = (\lambda_i \delta_{ij}) = \begin{bmatrix} \lambda_1 & 0 & \ldots & 0 \\ \vdots & \vdots & & \vdots \\ 0 & 0 & \ldots & \lambda_n \end{bmatrix}.$$

We call B a *diagonal* matrix and write $B = \text{diag}(\lambda_1, \ldots, \lambda_n)$. Then B has the characteristic equation

$$\begin{vmatrix} \lambda_1 - \lambda & \ldots & 0 \\ \vdots & & \\ 0 & \ldots & \lambda_n - \lambda \end{vmatrix} = 0 \quad \text{or} \quad (\lambda_1 - \lambda)(\lambda_2 - \lambda)\ldots(\lambda_n - \lambda) = 0.$$

Hence, B has the eigenvalues $\lambda_1, \ldots, \lambda_n$. We leave to Problem 5 below the discussion of the eigenvectors of B. We remark here that, when $\lambda_1, \ldots, \lambda_n$ are n different numbers, the eigenvectors associated with the eigenvalue λ_k are the vectors $c(0, \ldots, 0, 1, 0, \ldots, 0)$, with 1 as kth entry and c nonzero.

Similar matrices. Let A and B be $n \times n$ matrices. We say that B is *similar* to A if

$$B = C^{-1}AC$$

for some nonsingular $n \times n$ matrix C. If this holds, then

$$A = CBC^{-1} = (C^{-1})^{-1}BC^{-1},$$

so that A is also similar to B. Hence we speak of similar matrices A, B.

If A and B are similar, then A and B have the same characteristic equation. For

$$\begin{aligned} \det(B - \lambda I) &= \det(C^{-1}AC - \lambda I) = \det(C^{-1}AC - \lambda C^{-1}IC) \\ &= \det C^{-1}(A - \lambda I)C \\ &= \det C^{-1} \det(A - \lambda I) \det C = \det(A - \lambda I), \end{aligned}$$

since $\det C^{-1} \det C = 1$. It follows also that A and B have the same eigenvalues.

Matrices with n distinct real eigenvalues. Now let the $n \times n$ matrix A have n distinct (real) eigenvalues $\lambda_1, \ldots, \lambda_n$ and let $\mathbf{v}_1, \ldots, \mathbf{v}_n$ be eigenvectors associated with $\lambda_1, \ldots, \lambda_n$, respectively: $A\mathbf{v}_i = \lambda_i\mathbf{v}_i$, $i = 1, \ldots, n$. Now let C be the matrix whose column vectors are $\mathbf{v}_1, \ldots, \mathbf{v}_n$, respectively. Write $\mathbf{v}_j = \text{col}(v_{1j}, \ldots, v_{nj})$ for $j = 1, \ldots, n$. Then

$$\begin{aligned} AC &= A\begin{bmatrix} v_{11} & \ldots & v_{1n} \\ \vdots & & \vdots \\ v_{n1} & \ldots & v_{nn} \end{bmatrix} = \begin{bmatrix} \lambda_1 v_{11} & \ldots & \lambda_n v_{1n} \\ \vdots & & \vdots \\ \lambda_1 v_{n1} & \ldots & \lambda_n v_{nn} \end{bmatrix} \\ &= \begin{bmatrix} v_{11} & \ldots & v_{1n} \\ \vdots & & \vdots \\ v_{n1} & \ldots & v_{nn} \end{bmatrix}\begin{bmatrix} \lambda_1 & \ldots & 0 \\ \vdots & & \vdots \\ 0 & \ldots & \lambda_n \end{bmatrix} = CB, \end{aligned}$$

where B is a diagonal matrix, $B = \text{diag}(\lambda_1, \ldots, \lambda_n)$. It can be shown that C must be nonsingular (see Problem 12 following Section 1–10 below). Hence $B = C^{-1}AC$ and *A is similar to the diagonal matrix* $\text{diag}(\lambda_1, \ldots, \lambda_n)$.

EXAMPLE 3. $A = \begin{bmatrix} 1 & 2 & 2 \\ 2 & 3 & -2 \\ -5 & 3 & 8 \end{bmatrix}$

Here A has the characteristic equation

$$\begin{vmatrix} 1-\lambda & 2 & 2 \\ 2 & 3-\lambda & -2 \\ -5 & 3 & 8-\lambda \end{vmatrix} = 0 \quad \text{or} \quad (\lambda-3)(\lambda-4)(\lambda-5) = 0.$$

For $\lambda = 3$, the eigenvector $\mathbf{v} = (v_1, v_2, v_3)$ must satisfy the equations

$$-2v_1 + 2v_2 + 2v_3 = 0, \quad 2v_1 + 0v_2 - 2v_3 = 0, \quad -5v_1 + 3v_2 + 5v_3 = 0.$$

Hence $v_1 = (1, 0, 1)$ is an associated eigenvector. Similarly, for $\lambda_2 = 4$ an associated eigenvector is $(2, 2, 1)$ and for $\lambda_3 = 5$ an associated eigenvector is $(0, 1, -1)$. Hence $C^{-1}AC = B$, with

$$C = \begin{bmatrix} 1 & 2 & 0 \\ 0 & 2 & 1 \\ 1 & 1 & -1 \end{bmatrix}, \quad B = \begin{bmatrix} 3 & 0 & 0 \\ 0 & 4 & 0 \\ 0 & 0 & 5 \end{bmatrix} = \text{diag}\,(3, 4, 5).$$

We check by verifying that $AC = CB$. Thus A is similar to the diagonal matrix B.

In general, some of the eigenvalues may coincide (multiple roots of the characteristic equation), and there may be less than n real roots of the characteristic equation (1–22). To handle the general case, it is best to extend the theory of matrices to the case of *complex matrices*, whose elements are complex numbers. All concepts generalize easily to this case and, in particular, one can define complex eigenvalues and complex eigenvectors as above. For example, the matrix

$$\begin{bmatrix} 1 & -2 \\ 1 & -1 \end{bmatrix}$$

has the characteristic equation $\lambda^2 + 1 = 0$. The eigenvalues are $\pm i$, and associated eigenvectors are $(2, 1 \mp i)$. One proves, exactly as above, that when the matrix A has n distinct complex eigenvalues, A is similar to a diagonal matrix.

Problems

1. Find all eigenvalues and asociated eigenvectors:

(a) $\begin{bmatrix} 3 & 1 \\ 4 & 3 \end{bmatrix}$, (b) $\begin{bmatrix} 1 & 3 \\ 2 & 6 \end{bmatrix}$, (c) $\begin{bmatrix} 0 & 1 & -2 \\ 2 & 1 & 0 \\ 4 & -2 & 5 \end{bmatrix}$, (d) $\begin{bmatrix} 5 & -2 & 8 \\ -4 & 0 & -5 \\ -4 & 2 & -7 \end{bmatrix}$.

2. (a), . . . , (d). For each of the matrices of Problem 1, call the matrix A and find a nonsingular matrix C and a diagonal matrix B such that $A = C^{-1}BC$.

3. (*Complex case*) For each of the following choices of matrix A, find all eigenvalues and associated eigenvectors, and find a diagonal matrix B such that $A = C^{-1}BC$:

(a) $\begin{bmatrix} 1 & -1 \\ 4 & 1 \end{bmatrix}$, (b) $\begin{bmatrix} 3 & -2 \\ 1 & 5 \end{bmatrix}$, (c) $\begin{bmatrix} 0 & 0 & 0 \\ 0 & 1 & -2 \\ 0 & 1 & -1 \end{bmatrix}$.

4. (*Repeated roots*) Find all eigenvalues and associated eigenvectors:

(a) I_3, (b) O_{44}, (c) $\begin{bmatrix} 3 & -4 \\ 4 & -5 \end{bmatrix}$, (d) $\begin{bmatrix} 0 & 1 & -2 \\ -6 & 5 & -4 \\ 0 & 0 & 3 \end{bmatrix}$.

5. (*Diagonal matrices*) (a) Let $B = \operatorname{diag}(\lambda, \mu)$. Show that, if $\lambda \neq \mu$, then the eigenvectors associated with λ are all nonzero vectors $c(1, 0)$ and those associated with μ are all nonzero vectors $c(0, 1)$; show that, if $\lambda = \mu$, then the eigenvectors associated with λ are all nonzero vectors (v_1, v_2).

(b) Let $B = \operatorname{diag}(\lambda_1, \lambda_2, \lambda_3)$ and let $\mathbf{e}_1 = (1, 0, 0)$, $\mathbf{e}_2 = (0, 1, 0)$, $\mathbf{e}_3 = (0, 0, 1)$ (column vectors). Show that if λ_1, λ_2, λ_3 are distinct, then for each λ_k the associated eigenvectors are the vectors $c\mathbf{e}_k$ for $c \neq 0$; show that if $\lambda_1 = \lambda_2 \neq \lambda_3$, then the eigenvectors associated with λ_1 are all nonzero vectors $c_1\mathbf{e}_1 + c_2\mathbf{e}_2$ and those associated with λ_3 are all nonzero vectors $c\mathbf{e}_3$; show that if $\lambda_1 = \lambda_2 = \lambda_3$, then the eigenvectors associated with λ_1 are all nonzero vectors $\mathbf{v} = (v_1, v_2, v_3)$.

(c) Let $B = \operatorname{diag}(\lambda_1, \ldots, \lambda_n)$. Show that the eigenvectors associated with the eigenvalue λ_k are all nonzero vectors $\mathbf{v} = (v_1, \ldots, v_n)$ such that $v_i = 0$ for all i such that $\lambda_i \neq \lambda_k$.

6. Let A and B be similar $n \times n$ matrices with $A = C^{-1}BC$.

(a) Prove that $\det A = \det B$.

(b) The trace of a square matrix is defined as the sum of the diagonal terms: the trace of $A = (a_{ij})$ is $a_{11} + a_{22} + \cdots + a_{nn}$. Show that A and B have equal traces. [Hint: Consider the coefficient of λ^{n-1} in the characteristic equation.]

(c) Prove: If $\mathbf{v}$ is an eigenvector of A associated with the eigenvalue λ, then $C\mathbf{v}$ is an eigenvector of B associated with the eigenvalue λ.

7. Prove that the matrix $A = \begin{bmatrix} 1 & 1 \\ 0 & 1 \end{bmatrix}$ is not similar to a diagonal matrix. [Hint: Show by the results of Problem 6 that the diagonal matrix would have to be I.]

8. (a) Prove: every square matrix is similar to itself. (b) If A is similar to B and B is similar to C, then A is similar to C.

Answers

Throughout, k is an arbitrary nonzero real scalar; c is an arbitrary nonzero complex scalar.

1. (a) $\lambda = 1$, $k(1, -2)$ and $\lambda = 5$, $k(1, 2)$. (b) $\lambda = 0$, $k(3, -1)$ and $\lambda = 7$, $k(1, 2)$. (c) $\lambda = 1$, $k(0, 2, 1)$; $\lambda = 2$, $k(1, 2, 0)$; $\lambda = 3$, $k(1, 1, -1)$. (d) $\lambda = 1$, $k(3, -2, -2)$; $\lambda = -1$, $k(-1, 1, 1)$; $\lambda = -2$, $k(2, -1, -2)$.

2. (a) $C = \dfrac{1}{4}\begin{bmatrix} 2 & -1 \\ 2 & 1 \end{bmatrix}$, $B = \begin{bmatrix} 1 & 0 \\ 0 & 5 \end{bmatrix}$.

(b) $C = \dfrac{1}{7}\begin{bmatrix} 2 & -1 \\ 1 & 3 \end{bmatrix}$, $B = \begin{bmatrix} 1 & 0 \\ 0 & 7 \end{bmatrix}$.

(c) $C = \begin{bmatrix} -2 & 1 & -1 \\ 3 & -1 & 2 \\ -2 & 1 & -2 \end{bmatrix}$, $B = \begin{bmatrix} 1 & 0 & 0 \\ 0 & 2 & 0 \\ 0 & 0 & 3 \end{bmatrix}$.

(d) $C = \begin{bmatrix} 1 & 0 & 1 \\ 2 & 2 & 1 \\ 0 & 1 & -1 \end{bmatrix}$, $B = \begin{bmatrix} 1 & 0 & 0 \\ 0 & -1 & 0 \\ 0 & 0 & -2 \end{bmatrix}$.

3. (a) $\lambda = 1 + 2i$, $c(1, -2i)$ and $\lambda = 1 - 2i$, $c(1, 2i)$, $C = \frac{1}{4i}\begin{bmatrix} 2i & -1 \\ 2i & 1 \end{bmatrix}$, $B = \begin{bmatrix} 1+2i & 0 \\ 0 & 1-2i \end{bmatrix}$. (b) $\lambda = 4 + i$, $c(2, -1 - i)$ and $\lambda = 4 - i$, $c(2, -1 + i)$, $C = \frac{1}{4i}\begin{bmatrix} -1+i & -2 \\ 1+i & 2 \end{bmatrix}$, $B = \begin{bmatrix} 4+i & 0 \\ 0 & 4-i \end{bmatrix}$.
(c) $\lambda = 0$, $c(1, 0, 0)$; $\lambda = i$, $c(0, 2, 1 - i)$; $\lambda = -i$, $c(0, 2, 1 + i)$;
$C = \frac{1}{4i}\begin{bmatrix} 1 & 0 & 0 \\ 0 & 1+i & -2 \\ 0 & -1+i & 2 \end{bmatrix}$, $B = \begin{bmatrix} 0 & 0 & 0 \\ 0 & i & 0 \\ 0 & 0 & -i \end{bmatrix}$.

4. (a) $\lambda = 1$, all nonzero vectors. (b) $\lambda = 0$, all nonzero vectors. (c) $\lambda = -1, k(1, 1)$. (d) $\lambda = 2, k(1, 2, 0)$ and $\lambda = 3, a(1, 3, 0) + b(0, 2, 1)$, with a and b not both 0.

***1–6 The transpose.** Let $\mathbf{A} = (a_{ij})$ be an $m \times n$ matrix. We denote by A' the $n \times m$ matrix $B = (b_{ij})$ such that $b_{ij} = a_{ji}$ for $i = 1, \ldots, n$, $j = 1, \ldots, m$. Thus $B = A'$ is obtained from A by interchanging rows and columns. The following pair is an illustration:

$$A = \begin{bmatrix} 3 & 1 & 2 \\ 5 & 0 & 7 \end{bmatrix}, \quad A' = \begin{bmatrix} 3 & 5 \\ 1 & 0 \\ 2 & 7 \end{bmatrix}.$$

The first row of A becomes the first column of A'; the second row of A becomes the second column of A'. In general, we call A' the *transpose* of A. We observe that $I' = I$. The transpose of a matrix obeys several rules, which we adjoin to our list:

$$24.\ (A + B)' = A' + B'. \quad 25.\ (cA)' = cA'. \quad 26.\ (A')' = A. \quad 27.\ (AB)' = B'A'. \quad 28.\ \text{If } A \text{ is nonsingular, then } (A^{-1})' = (A')^{-1}. \tag{1–23}$$

To prove Rule 24, we write $D = (A + B) = (d_{ij})$, so that

$$d_{ij} = a_{ij} + b_{ij}$$

for all i and j. Then $D' = E = (e_{ij})$, where $e_{ij} = d_{ji}$ for all i and j, or

$$e_{ij} = a_{ji} + b_{ji}.$$

Thus $E = A' + B'$ or $D' = A' + B'$. The proofs of Rules 25 and 26 are left as exercises (Problem 4 below).

To prove Rule 27, we let A be $m \times p$, B be $p \times n$. We then write $C = AB = (c_{ij})$, $D = A' = (d_{ij})$, $E = B' = (e_{ij})$. Then

$$B'A' = ED = F = (f_{ij}),$$

where

$$\begin{aligned} f_{ij} &= e_{i1}d_{1j} + \cdots + e_{ip}d_{pj} = b_{1i}a_{j1} + \cdots + b_{pi}a_{jp} \\ &= a_{j1}b_{1i} + \cdots + a_{jp}b_{pj} = c_{ji} \quad (i = 1, \ldots, n, j = 1, \ldots, m). \end{aligned}$$

Hence, $F = C'$ or $B'A' = (AB)'$.

To prove Rule 28, we write: $AA^{-1} = I$. Then by Rule 27,

$$(A^{-1})' A' = I' = I.$$

From this equation it follows, as in Section 1–4, that $(A')^{-1} = (A^{-1})'$.

A matrix A such that $A = A'$ is called a *symmetric* matrix. Here A must be a square matrix. The matrix I is symmetric, as are the following matrices:

$$\begin{bmatrix} 1 & 2 \\ 2 & 3 \end{bmatrix}, \quad \begin{bmatrix} 3 & -1 & 0 \\ -1 & 7 & 2 \\ 0 & 2 & 4 \end{bmatrix}$$

Also, every diagonal matrix is symmetric.

Symmetric matrices are useful in discussing quadratic forms; that is, algebraic expressions of the form

$$\sum_{i=1}^{n} \sum_{j=1}^{n} a_{ij} x_i x_j. \tag{1–24}$$

For $n = 2$, the expression is

$$a_{11} x_1^2 + a_{12} x_1 x_2 + a_{21} x_2 x_1 + a_{22} x_2^2.$$

Here $x_1 x_2$ is the same as $x_2 x_1$, so that we could combine the second and third terms. However, it is preferable to split the combined term into two equal terms, each having as coefficient the average of a_{12} and a_{21}. For example, $3x_1^2 + 5x_1x_2 + 7x_2x_1 + 4x_2^2$ is replaced by

$$3x_1^2 + 6x_1x_2 + 6x_2x_1 + 4x_2^2. \tag{1–25}$$

By proceeding similarly for the general quadratic form (1–24), we can always assume that the *coefficient matrix* (a_{ij}) is symmetric, and it is standard practice to write quadratic forms in this way. For the example just considered, this matrix is $\begin{bmatrix} 3 & 6 \\ 6 & 4 \end{bmatrix}$.

Now let the $n \times n$ symmetric matrix $A = (a_{ij})$ be given, and consider the quadratic form (1–24). Here we can consider $x_1, \ldots, x_n$ as variables. For each assignment of numerical values to $x_1, \ldots, x_n$, the form (1–24) has a numerical value Q. Hence Q is a function of the n variables $x_1, \ldots, x_n$. However, it is simpler to think of Q as a function of the vector $\mathbf{x} = \text{col}\,(x_1, \ldots, x_n)$:

$$Q(\mathbf{x}) = \sum_{i=1}^{n} \sum_{j=1}^{n} a_{ij} x_i x_j. \tag{1–26}$$

Furthermore, for each $\mathbf{x}$, we can compute the number $Q(\mathbf{x})$ by matrix multiplications:

$$Q(\mathbf{x}) = \mathbf{x}' A \mathbf{x}. \tag{1–27}$$

Here $\mathbf{x}'$ is the transpose of the column vector $\mathbf{x}$, and is therefore the row vector $(x_1, \ldots, x_n)$. Accordingly, (1–27) is the same as

$$Q(\mathbf{x}) = (x_1, \ldots, x_n) \begin{bmatrix} a_{11} & \ldots & a_{1n} \\ \vdots & & \vdots \\ a_{n1} & \ldots & a_{nn} \end{bmatrix} \begin{bmatrix} x_1 \\ \vdots \\ x_n \end{bmatrix}.$$

The product of the last two factors is an $n \times 1$ column vector whose ith entry is $a_{i1}x_1 + \cdots + a_{in}x_n$. The product of the $1 \times n$ row vector $(x_1, \ldots, x_n)$ and this $n \times 1$ column vector is a 1×1 matrix—that is, a number; in fact, it is precisely the number on the right of (1–26). Therefore, (1–27) is indeed another way of writing (1–26).

As an illustration, we write the quadratic form $Q(\mathbf{x})$ of (1–25) as follows:

$$Q(\mathbf{x}) = \mathbf{x}' \begin{bmatrix} 3 & 6 \\ 6 & 4 \end{bmatrix} \mathbf{x} \quad (\mathbf{x} = \operatorname{col}(x_1, x_2)).$$

When expanded, this becomes

$$\begin{aligned} Q(\mathbf{x}) &= (x_1, x_2) \begin{bmatrix} 3 & 6 \\ 6 & 4 \end{bmatrix} \begin{bmatrix} x_1 \\ x_2 \end{bmatrix} = (x_1, x_2) \begin{bmatrix} 3x_1 + 6x_2 \\ 6x_1 + 4x_2 \end{bmatrix} \\ &= x_1(3x_1 + 6x_2) + x_2(6x_1 + 4x_2) = 3x_1^2 + 6x_1x_2 + 6x_2x_1 + 4x_2^2, \end{aligned}$$

as expected.

***1–7 Orthogonal matrices.** Let A be a real $n \times n$ matrix. Then A is said to be *orthogonal* if

$$AA' = I. \tag{1–28}$$

Hence, A is orthogonal if and only if $A^{-1} = A'$; that is, if and only if the inverse of A equals the transpose of A. Thus, every orthogonal matrix is nonsingular. The following are examples of orthogonal matrices:

$$A = \begin{bmatrix} \frac{3}{5} & \frac{4}{5} \\ -\frac{4}{5} & \frac{3}{5} \end{bmatrix}, \quad B = \begin{bmatrix} \frac{2}{3} & \frac{2}{3} & \frac{1}{3} \\ \frac{2}{3} & -\frac{1}{3} & -\frac{2}{3} \\ \frac{1}{3} & -\frac{2}{3} & \frac{2}{3} \end{bmatrix}.$$

Let us consider the row vectors $\mathbf{u}_1, \mathbf{u}_2$ of A as vectors in the xy-plane:

$$\mathbf{u}_1 = \tfrac{3}{5}\mathbf{i} + \tfrac{4}{5}\mathbf{j}, \quad \mathbf{u}_2 = -\tfrac{4}{5}\mathbf{i} + \tfrac{3}{5}\mathbf{j}.$$

Then we observe that $\mathbf{u}_1$ and $\mathbf{u}_2$ are both unit vectors and that $\mathbf{u}_1 \cdot \mathbf{u}_2 = 0$, so that $\mathbf{u}_1, \mathbf{u}_2$ are perpendicular. A similar statement applies to the column vectors of A:

$$\mathbf{v}_1 = \tfrac{3}{5}\mathbf{i} - \tfrac{4}{5}\mathbf{j}, \quad v_2 = \tfrac{4}{5}\mathbf{i} + \tfrac{3}{5}\mathbf{j}.$$

We can proceed similarly with the row vectors or column vectors of B, regarding them as vectors in space: $\mathbf{u}_1 = \frac{2}{3}\mathbf{i} + \frac{2}{3}\mathbf{j} + \frac{1}{3}\mathbf{k}, \ldots$ Again we verify that the row vectors, or the column vectors, are mutually perpendicular unit vectors.

The geometrical concepts used here can be generalized to n dimensions (Section 1–8). Here we phrase them algebraically. For an $n \times n$ matrix

$A = (a_{ij})$, the crucial conditions are as follows:

$$
\begin{aligned}
a_{i1}^2 + \cdots + a_{in}^2 &= 1,\ i = 1, \ldots, n, & (1\text{–}29)\\
a_{i1}a_{j1} + \cdots + a_{in}a_{jn} &= 0,\ i \neq j,\ i = 1, \ldots, n,\ j = 1, \ldots, n, & (1\text{–}30)\\
a_{1j}^2 + \cdots + a_{nj}^2 &= 1,\ j = 1, \ldots, n, & (1\text{–}31)\\
a_{1j}a_{1k} + \cdots + a_{nj}a_{nk} &= 0,\ j \neq k,\ j = 1, \ldots, n,\ k = 1, \ldots, n. & (1\text{–}32)
\end{aligned}
$$

(Here (1–29) states that the row vectors are unit vectors; (1–30) states that different row vectors are orthogonal; (1–31) and (1–32) express the analogous conditions on the column vectors.)

THEOREM. (a) *Let A be an $n \times n$ orthogonal matrix. Then conditions* (1–29), . . . (1–32) *all hold.*

(b) *If A is an $n \times n$ matrix such that* (1–29) *and* (1–30) *hold or such that* (1–31) *and* (1–32) *hold, then A is orthogonal.*

Proof. (a) Let A be orthogonal, so that (1–28) holds. By the definition of matrix multiplication, (1–28) asserts that

$$a_{i1}a_{j1} + \cdots + a_{in}a_{jn} = \delta_{ij} = \begin{cases} 1, & i = j, \\ 0, & i \neq j. \end{cases} \qquad (1\text{–}33)$$

Hence, (1–29) and (1–30) follow. From (1–28) we have also, by the properties of inverses (Section 1–4),

$$A'A = I, \qquad (1\text{–}34)$$

and hence

$$a_{1i}a_{1j} + \cdots + a_{ni}a_{nj} = \delta_{ij}; \qquad (1\text{–}35)$$

and this implies (1–31) and (1–32).

(b) If (1–29) and (1–30) hold, then (1–33) holds, so that $AA' = I$; if (1–31) and (1–32) hold, then (1–35) holds, so that $A'A = I$. In either case, A is orthogonal.

Orthogonal matrices are important in studying changes of coordinates (Section 1–14 below).

If two $n \times n$ matrices B, C are such that

$$B = A^{-1}CA \qquad (1\text{–}36)$$

for some orthogonal matrix A, then B is said to be *orthogonally congruent* to C. It then follows that C is also orthogonally congruent to B, so that we refer to B, C as *orthogonally congruent matrices* (Problem 9 below). We observe that orthogonally congruent matrices are similar.

If C is symmetric, then C is orthogonally congruent to a diagonal matrix B, $B =$ diag $(\lambda_1, \lambda_2, \ldots, \lambda_n)$, where $\lambda_1, \ldots, \lambda_n$ are the (necessarily real) eigenvalues of C (with multiple eigenvalues repeated in accordance with their multiplicities). For a proof of this theorem, see Section 5–3 of the book by Perlis listed at the end of the chapter.

This result is very important for the study of quadratic forms. For let C be a symmetric matrix and let A be an orthogonal matrix such that

$A^{-1}CA = B = \text{diag}(\lambda_1, \ldots, \lambda_n)$. Then let us consider the quadratic form

$$Q(\mathbf{x}) = \sum_{i=1}^{n}\sum_{j=1}^{n} c_{ij}x_ix_j = \mathbf{x}'C\mathbf{x}, \quad \mathbf{x} = \text{col}(x_1, \ldots, x_n).$$

We express $x_1 \ldots, x_n$ in terms of new variables $y_1, \ldots, y_n$ by writing

$$\mathbf{x} = A\mathbf{y}, \quad \mathbf{y} = \text{col}(y_1, \ldots, y_n).$$

Then $\mathbf{x}' = \mathbf{y}'A' = \mathbf{y}'A^{-1}$, and hence $Q(\mathbf{x})$ becomes a quadratic form $Q_1(\mathbf{y})$:

$$Q(\mathbf{x}) = Q_1(\mathbf{y}) = \mathbf{y}'A^{-1}CA\mathbf{y} = \mathbf{y}'B\mathbf{y} = \sum_{i=1}^{n}\sum_{j=1}^{n} b_{ij}y_iy_j$$
$$= \lambda_1y_1^2 + \cdots + \lambda_ny_n^2.$$

Hence, $Q(\mathbf{x})$ *can be written in terms of* $y_1, \ldots, y_n$ *as a quadratic form containing only the squares of the unknowns.* The coefficients $\lambda_1, \ldots, \lambda_n$ are the eigenvalues of B. But B and C have the same eigenvalues. Hence, once we know the eigenvalues of C (with their multiplicities), we can write Q in terms of $y_1, \ldots, y_n$. It is not necessary to find the matrix A.

We can interpret $x_1, \ldots, x_n$ as Cartesian coordinates in n-dimensional space (Section 1–8). Then $y_1, \ldots, y_n$ are simply new Cartesian coordinates in n-dimensional space with the same origin (Section 1–14). For $n = 2$, the equation $Q(\mathbf{x}) = 1$—that is, the equation

$$c_{11}x_1^2 + c_{12}x_1x_2 + c_{21}x_2x_1 + c_{22}x_2^2 = 1,$$

—represents a conic section in the x_1x_2-plane. In terms of the new coordinates y_1, y_2, this equation becomes $Q_1(\mathbf{y}) = 1$, or

$$\lambda_1y_1^2 + \lambda_2y_2^2 = 1,$$

and hence represents an ellipse or hyperbola. (See Section 0–4.) In fact, in the language of plane analytic geometry, the theorem on symmetric matrices for the case $n = 2$ is equivalent to the familiar statement that every second-degree equation $Ax^2 + Bxy + Cy^2 = 1$ can be reduced to the standard form $Ax^2 + By^2 = 1$ by an appropriate rotation of axes in the xy-plane.

Problems

1. Find the transpose of each of the matrices:

(a) $\begin{bmatrix} 1 & 2 & 3 \\ 3 & 0 & 5 \end{bmatrix}$ (b) $\begin{bmatrix} 3 & 1 \\ 0 & 2 \\ 1 & 0 \end{bmatrix}$ (c) $(1, 5, 0, 4)$ (d) $\begin{bmatrix} 1 \\ 0 \\ 7 \end{bmatrix}$

2. Choose a and b so that each of the following matrices becomes symmetric:

(a) $\begin{bmatrix} 1 & 3a-1 \\ 2a & 3 \end{bmatrix}$, (b) $\begin{bmatrix} 2 & a & 3 \\ b-a & 0 & 4+a \\ 3 & b & 5 \end{bmatrix}$

3. For each of the following quadratic forms obtain the coefficient matrix when the form is written in such a way that the coefficient matrix is symmetric:

(a) $5x_1^2 + 4x_1x_2 + 3x_2^2$, (b) $7x_1^2 + 2x_1x_2 - x_2^2$,

(c) $x_1^2 + 3x_2^2 - x_3^2 + 4x_1x_2 + 6x_1x_3 + 2x_2x_3$,

(d) $2x_1^2 + x_2^2 + x_3^2 + 2x_1x_3 - 4x_2x_3$.

4. Prove each of the following parts of (1–23): (a) Rule 25, (b) Rule 26.

5. Show that each of the following matrices is orthogonal:

(a) $\dfrac{1}{13}\begin{bmatrix} 5 & 12 \\ -12 & 5 \end{bmatrix}$, (c) $\dfrac{1}{7}\begin{bmatrix} 2 & 3 & 6 \\ 6 & 2 & -3 \\ 3 & -6 & 2 \end{bmatrix}$,

(b) $\begin{bmatrix} \cos\omega & \sin\omega \\ -\sin\omega & \cos\omega \end{bmatrix}$, (d) $\dfrac{1}{2}\begin{bmatrix} 1 & 1 & 1 & 1 \\ 1 & -1 & -1 & 1 \\ 1 & 1 & -1 & -1 \\ 1 & -1 & 1 & -1 \end{bmatrix}$.

6. (a), (b), (c). Represent the row vectors of each of the matrices in Problem 5 (a), (b), (c) as vectors in the plane or in space, graph them, and verify that they are mutually perpendicular unit vectors.

7. (a), (b), (c). Proceed as in Problem 6 with column vectors instead of row vectors.

8. Let A and B be $n \times n$ orthogonal matrices. Prove: (a) $\det A = \pm 1$. (b) AB is also orthogonal. (c) A' and A^{-1} are orthogonal.

9. Prove: (a) Every square matrix is orthogonally congruent to itself.
(b) If B is orthogonally congruent to C, then C is congruent to B.
(c) If B is orthogonally congruent to C and C is orthogonally congruent to D, then B is orthogonally congruent to D.

10. We consider *complex matrices:* that is, matrices with complex entries. If $A = (a_{ij})$ is such a matrix, we denote by $\bar{A}$ the *conjugate* of A: that is, the matrix $(\bar{a}_{ij})$, where $\bar{z}$ is the conjugate of the complex number z (Section 0–3). Prove the following:
(a) If z_1 and z_2 are complex numbers, then $\overline{(z_1 + z_2)} = \bar{z}_1 + \bar{z}_2$ and $\overline{z_1z_2} = \bar{z}_1\bar{z}_2$.
(b) If A and B are matrices and c is a complex scalar, then $\overline{(A + B)} = \bar{A} + \bar{B}$, $\overline{AB} = \bar{A}\,\bar{B}$, $\overline{(cA)} = \bar{c}\,\bar{A}$.
(c) Every complex matrix A can be written uniquely as $A_1 + iA_2$, where A_1 and A_2 are real matrices, and $A_1 = \frac{1}{2}(A + \bar{A})$, $A_2 = (2i)^{-1}(A - \bar{A})$. We call A_1 the real part of A, A_2 the imaginary part of A.
(d) If A is a square matrix, then $(\bar{A})' = \overline{(A')}$ and, if A is nonsingular, then $\overline{(A^{-1})} = (\bar{A})^{-1}$.
(e) A is a real matrix if and only if $A = \bar{A}$.

11. Let A be a real square matrix, let λ be real, and let $A\mathbf{v} = \lambda\mathbf{v}$ for a nonzero *complex* column vector $\mathbf{v}$. Show that $A\mathbf{u} = \lambda\mathbf{u}$ for a real nonzero vector $\mathbf{u}$, so that λ is an eigenvalue of A considered as a real matrix. [Hint: Let $\mathbf{v} = \mathbf{p} + i\mathbf{q}$, where $\mathbf{p}$ and $\mathbf{q}$ are real and not both zero. Show, with the aid of the results of Problem 10, that $A\mathbf{p} = \lambda\mathbf{p}$ and $A\mathbf{q} = \lambda\mathbf{q}$ and hence that $\mathbf{u}$ can be chosen as one of $\mathbf{p}$, $\mathbf{q}$.]

12. Let A be a real symmetric matrix. Show, with the aid of the results of Problem 10, that all eigenvalues of A are real. [Hint: Let $A\mathbf{v} = \lambda\mathbf{v}$ for some

complex λ and some nonzero complex vector $\mathbf{v} = \text{col}(v_1, \ldots, v_n)$. Consider the product $Q = \mathbf{v}' A \bar{\mathbf{v}}$. Show that Q is real and that

$$Q = \lambda \mathbf{v}' \bar{\mathbf{v}} = \lambda(|v_1|^2 + \cdots + |v_n|^2).$$

Conclude that λ is real.]

Remarks. By Problem 11, λ is actually an eigenvalue of A as a real matrix. The expression $\mathbf{v}' A \bar{\mathbf{v}}$ is a special case of a Hermitian quadratic form $Q(\mathbf{v})$. See Section 5–7 of the book by Perlis listed at the end of the chapter.

Answers

1. (a) $\begin{bmatrix} 1 & 3 \\ 2 & 0 \\ 3 & 5 \end{bmatrix}$ (b) $\begin{bmatrix} 3 & 0 & 1 \\ 1 & 2 & 0 \end{bmatrix}$, (c) col (1, 5, 0, 4), (d) (1, 0, 7).

2. (a) $a = 1$, (b) $a = 4$, $b = 8$.

3. (a) $\begin{bmatrix} 5 & 2 \\ 2 & 3 \end{bmatrix}$, (b) $\begin{bmatrix} 7 & 1 \\ 1 & -1 \end{bmatrix}$, (c) $\begin{bmatrix} 1 & 2 & 3 \\ 2 & 3 & 1 \\ 3 & 1 & -1 \end{bmatrix}$, (d) $\begin{bmatrix} 2 & 0 & 1 \\ 0 & 1 & -2 \\ 1 & -2 & 1 \end{bmatrix}$.

PART II. n-DIMENSIONAL GEOMETRY AND LINEAR MAPPINGS

1–8 Analytic geometry and vectors in n-dimensional space. The formal operations on vectors and coordinates in Section 0–6 suggest that the restriction to 3-dimensional space, and hence to three coordinates or components, is not necessary.

We define *Euclidean n-dimensional space* E^n to be a space having n coordinates $x_1, \ldots, x_n$. Throughout this section n will be a fixed positive integer. A *point* P of the space is by definition an ordered n-tuple $(x_1, \ldots, x_n)$; all points of the space are obtained by allowing $x_1, \ldots, x_n$ to take on all real values. The point $(0, \ldots, 0)$ is the *origin*, O. The *distance* between two points A: $(a_1, \ldots, a_n)$ and B: $(b_1, \ldots, b_n)$ is defined to be the number

$$d = \sqrt{(a_1 - b_1)^2 + \cdots + (a_n - b_n)^2}\,. \tag{1–37}$$

A vector $\mathbf{v}$ in n-dimensional space is defined to be an ordered n-tuple $(v_1, \ldots, v_n)$ of real numbers; $v_1, \ldots, v_n$ are the *components* of $\mathbf{v}$ (with respect to the given coordinates). In particular, we define a zero vector

$$\mathbf{0} = (0, \ldots, 0). \tag{1–38}$$

To the pair of points A, B in that order corresponds the vector

$$\overrightarrow{AB} = (b_1 - a_1, \ldots, b_n - a_n). \tag{1–39}$$

Remarks. Both points and vectors are represented by n-tuples, but this will be seen to cause no confusion. In fact, there are advantages in being able to go back and forth freely between the two points of view: that of points and that of vectors. To each point P: $(x_1, \ldots, x_n)$ we can associate the vector $\overrightarrow{OP} = (x_1, \ldots, x_n) = \mathbf{x}$. Conversely, to each vector

$\mathbf{x} = (x_1, \ldots, x_n)$ we can assign the point P whose coordinates are $(x_1, \ldots, x_n)$, and then $\mathbf{x} = \overrightarrow{OP}$. Below we shall also interpret our vectors as matrices, either as row vectors or as column vectors.

The *sum* of two vectors, *multiplication* of a vector by a scalar, and the *scalar product* (or *inner product* or *dot product*) of two vectors are defined by the equations:

$$\mathbf{u} + \mathbf{v} = (u_1 + v_1, \ldots, u_n + v_n), \tag{1–40}$$

$$h\mathbf{u} = (hu_1, \ldots, hu_n), \tag{1–41}$$

$$\mathbf{u} \cdot \mathbf{v} = u_1v_1 + \cdots + u_nv_n. \tag{1–42}$$

The *length* or *norm* of $\mathbf{v} = (v_1, \ldots, v_n)$ is defined as the scalar

$$|\mathbf{v}| = \sqrt{\mathbf{v} \cdot \mathbf{v}} = \sqrt{v_1^2 + \cdots + v_n^2}\,. \tag{1–43}$$

If $|\mathbf{v}| = 1$, $\mathbf{v}$ is called a *unit vector.*

The vector product of two vectors can be generalized to n dimensions only with the aid of *tensors* (see, however, *CLA*, Section 11–9). The set of all vectors $(v_1, \ldots, v_n)$ with the operations (1–40) through (1–43) will be denoted by V^n.

In V^n the following properties can then be verified (Problem 6 below):

$$\begin{array}{ll} \text{I. } \mathbf{u} + \mathbf{v} = \mathbf{v} + \mathbf{u}. & \text{II. } (\mathbf{u} + \mathbf{v}) + \mathbf{w} = \mathbf{u} + (\mathbf{v} + \mathbf{w}). \\ \text{III. } h(\mathbf{u} + \mathbf{v}) = h\mathbf{u} + h\mathbf{v}. & \text{IV. } (a + b)\mathbf{u} = a\mathbf{u} + b\mathbf{u}. \\ \text{V. } (ab)\mathbf{u} = a(b\mathbf{u}). & \text{VI. } 1\mathbf{u} = \mathbf{u}. \\ \text{VII. } 0\mathbf{u} = \mathbf{0}. & \text{VIII. } \mathbf{u} \cdot \mathbf{v} = \mathbf{v} \cdot \mathbf{u}. \\ \text{IX. } (\mathbf{u} + \mathbf{v}) \cdot \mathbf{w} = \mathbf{u} \cdot \mathbf{w} + \mathbf{v} \cdot \mathbf{w}. & \text{X. } (a\mathbf{u}) \cdot \mathbf{v} = a(\mathbf{u} \cdot \mathbf{v}). \\ \text{XI. } \mathbf{u} \cdot \mathbf{u} \geqq 0. & \text{XII. } \mathbf{u} \cdot \mathbf{u} = 0 \text{ if and only if } \mathbf{u} = \mathbf{0}. \end{array} \tag{1–44}$$

A set of k vectors $\mathbf{v}_1, \ldots, \mathbf{v}_k$ of V^n is said to be *linearly independent* if an equation

$$c_1\mathbf{v}_1 + \cdots + c_k\mathbf{v}_k = \mathbf{0} \tag{1–45}$$

can hold only if $c_1 = \cdots = c_k = 0$. If a relation (1–45) does hold with not all c's equal to 0, then the vectors are said to be *linearly dependent.*

Remarks. We observe that for every $n \times k$ matrix A,

$$A \text{ col } (c_1, \ldots, c_k) = c_1\mathbf{v}_1 + \cdots + c_k\mathbf{v}_k, \tag{1–46}$$

where $\mathbf{v}_1, \ldots, \mathbf{v}_k$ are the column vectors of A. For

$$\begin{bmatrix} a_{11} & \cdots & a_{1k} \\ \vdots & & \\ a_{n1} & \cdots & a_{nk} \end{bmatrix} \begin{bmatrix} c_1 \\ \vdots \\ c_k \end{bmatrix} = \begin{bmatrix} c_1a_{11} + \cdots + c_ka_{1k} \\ \vdots \\ c_1a_{n1} + \cdots + c_ka_{nk} \end{bmatrix} = c_1\mathbf{v}_1 + \cdots + c_k\mathbf{v}_k.$$

Thus, the linear combinations of $\mathbf{v}_1, \ldots, \mathbf{v}_k$ can be expressed as $A\mathbf{c}$. Accordingly, to state that $\mathbf{v}_1, \ldots, \mathbf{v}_k$ are linearly independent is the same as to state that $A\mathbf{c} = \mathbf{0}$ is satisfied only for $\mathbf{c} = \mathbf{0}$. In particular, for $k = n$, A is a square matrix and $A\mathbf{c} = \mathbf{0}$ is equivalent to n homogeneous linear equations in n unknowns; thus the column vectors $\mathbf{v}_1, \ldots, \mathbf{v}_n$ of A are linearly independent precisely when these equations have only the trivial solution $c_1 = 0, \ldots, c_n = 0$—that is, when $\det A \neq 0$, or A is nonsingular. Accordingly, *n vectors $\mathbf{v}_1, \ldots, \mathbf{v}_n$ of V^n are linearly independent if and only if A is nonsingular, where A is the matrix whose column vectors are $\mathbf{v}_1, \ldots, \mathbf{v}_n$.*

A set of k vectors $\mathbf{v}_1, \ldots, \mathbf{v}_k$ of V^n is said to be a *basis* for V^n if every vector $\mathbf{v}$ of V^n can be expressed in unique fashion as a linear combination of $\mathbf{v}_1, \ldots, \mathbf{v}_k$: that is, if

$$\mathbf{v} = c_1\mathbf{v}_1 + \cdots + c_k\mathbf{v}_k$$

for unique choices of the scalars $c_1, \ldots, c_k$.

We can now state a number of rules concerning linear independence and basis:

(a) The vectors $\mathbf{v}_1, \ldots, \mathbf{v}_k$ are linearly dependent if and only if one of these vectors is expressible as a linear combination of the others.

(b) If one of the vectors $\mathbf{v}_1, \ldots, \mathbf{v}_k$ is $\mathbf{0}$, then $\mathbf{v}_1, \ldots, \mathbf{v}_k$ are linearly dependent.

(c) If $\mathbf{v}_1, \ldots, \mathbf{v}_k$ are linearly independent, but $\mathbf{v}_1, \ldots, \mathbf{v}_k, \mathbf{v}_{k+1}$ are linearly dependent, then $\mathbf{v}_{k+1}$ is expressible as a linear combination of $\mathbf{v}_1, \ldots, \mathbf{v}_k$.

(d) If $\mathbf{v}_1, \ldots, \mathbf{v}_k$ are linearly independent and $h < k$, then $\mathbf{v}_1, \ldots, \mathbf{v}_h$ are linearly independent.

(e) (Rule for comparing coefficients). If $\mathbf{v}_1, \ldots, \mathbf{v}_k$ are linearly independent and

$$a_1\mathbf{v}_1 + \cdots + a_k\mathbf{v}_k = b_1\mathbf{v}_1 + \cdots + b_k\mathbf{v}_k,$$

then $a_1 = b_1, a_2 = b_2, \ldots, a_k = b_k$.

(f) There exist n linearly independent vectors in V^n: for example, the vectors

$$\mathbf{e}_1 = (1, 0, \ldots, 0), \quad \mathbf{e}_2 = (0, 1, 0, \ldots, 0), \quad \ldots, \quad \mathbf{e}_n = (0, \ldots, 0, 1). \tag{1–47}$$

(g) There do not exist $n + 1$ linearly independent vectors in V^n.

(h) If $\mathbf{v}_1, \ldots, \mathbf{v}_n$ are linearly independent vectors in V^n, then $\mathbf{v}_1, \ldots, \mathbf{v}_n$ form a basis for V^n; in particular, $\mathbf{e}_1, \ldots, \mathbf{e}_n$ form a basis for V^n.

(i) Every basis for V^n consists of n linearly independent vectors.

(j) If $k < n$ and $\mathbf{v}_1, \ldots, \mathbf{v}_k$ are linearly independent, then there exist $\mathbf{v}_{k+1}, \ldots, \mathbf{v}_n$ such that $\mathbf{v}_1, \ldots, \mathbf{v}_n$ form a basis for V^n.

(k) If $\mathbf{v}_1, \ldots, \mathbf{v}_k$ are linearly independent vectors in V^n, and $\mathbf{u}_1, \ldots, \mathbf{u}_{k+1}$ are all expressible as linear combinations of $\mathbf{v}_1, \ldots, \mathbf{v}_k$, then $\mathbf{u}_1, \ldots, \mathbf{u}_{k+1}$ are linearly dependent.

We here prove the rules (g) and (i) and leave the remaining proofs to Problem 10 below.

Proof of (g). Let $\mathbf{v}_1, \ldots, \mathbf{v}_{n+1}$ be linearly independent vectors in V^n. Then by rule (d), $\mathbf{v}_1, \ldots, \mathbf{v}_n$ are also linearly independent. Hence the matrix A, whose columns are $\mathbf{v}_1, \ldots, \mathbf{v}_n$, is nonsingular. Therefore the equation $A\mathbf{c} = \mathbf{v}_{n+1}$ has a unique solution for $\mathbf{c}$; that is,

$$\mathbf{v}_{n+1} = c_1\mathbf{v}_1 + \cdots + c_n\mathbf{v}_n.$$

By rule (a), $\mathbf{v}_1, \ldots, \mathbf{v}_{n+1}$ would then be linearly dependent, contrary to assumption. Therefore, there cannot be $n + 1$ linearly independent vectors in V^n.

Proof of (i). Let $\mathbf{v}_1, \ldots, \mathbf{v}_k$ be a basis for V^n. Then $\mathbf{v}_1, \ldots, \mathbf{v}_k$ are linearly independent. For if $c_1\mathbf{v}_1 + \cdots + c_k\mathbf{v}_k = \mathbf{0}$, then

$$c_1 = 0, \ldots, c_k = 0,$$

since, by definition of a basis, $\mathbf{0}$ can be represented in only one way as a linear combination of $\mathbf{v}_1, \ldots, \mathbf{v}_k$. Hence by rules (d) and (g), we must have $k \leqq n$. Let us suppose $k < n$. Then we can express $\mathbf{e}_1, \ldots, \mathbf{e}_n$ as linear combinations of $\mathbf{v}_1, \ldots, \mathbf{v}_k$: say, $A\mathbf{b}_1 = \mathbf{e}_1, \ldots, A\mathbf{b}_n = \mathbf{e}_n$, where A is the $n \times k$ matrix whose column vectors are $\mathbf{v}_1, \ldots, \mathbf{v}_k$. Accordingly, $AB = I$, where B is the $k \times n$ matrix whose column vectors are $\mathbf{b}_1, \ldots, \mathbf{b}_n$. But for $k < n$, a product of an $n \times k$ matrix and a $k \times n$ matrix is always singular (see the Remark at the close of Section 1–4). Hence we have a contradiction and we must have $k = n$. Thus rule (i) is proved.

The basis $\mathbf{e}_1, \ldots, \mathbf{e}_n$ defined in (1–47) is called the *standard basis* of V^n. For $n = 3$, this is the familar basis $\mathbf{i}, \mathbf{j}, \mathbf{k}$. In general,

$$\mathbf{v} = (v_1, \ldots, v_n) = v_1\mathbf{e}_1 + \cdots + v_n\mathbf{e}_n. \tag{1–48}$$

We now prove a theorem of much importance for the geometry of E^n:

THEOREM. *If* $\mathbf{u}$ *and* $\mathbf{v}$ *are vectors of* V^n, *then*

$$|\mathbf{u} \cdot \mathbf{v}| \leqq |\mathbf{u}|\,|\mathbf{v}| \quad \text{(Cauchy-Schwarz inequality).} \tag{1–49}$$

The equality holds precisely when $\mathbf{u}$ *and* $\mathbf{v}$ *are linearly dependent. Furthermore,*

$$|\mathbf{u} + \mathbf{v}| \leqq |\mathbf{u}| + |\mathbf{v}| \quad \text{(triangle inequality).} \tag{1–50}$$

The equals sign holds precisely when $\mathbf{u} = h\mathbf{v}$ *or* $\mathbf{v} = h\mathbf{u}$ *with* $h \geqq 0$.

Proof. If $\mathbf{v} \neq \mathbf{0}$ and $\mathbf{u}$ is not a scalar multiple of $\mathbf{v}$, then $\mathbf{u} + t\mathbf{v} \neq \mathbf{0}$ for every scalar t and hence $|\mathbf{u} + t\mathbf{v}|^2 > 0$ for all t, so that

$$(\mathbf{u} + t\mathbf{v}) \cdot (\mathbf{u} + t\mathbf{v}) > 0 \quad \text{or} \quad P(t) = |\mathbf{u}|^2 + 2t(\mathbf{u} \cdot \mathbf{v}) + t^2|\mathbf{v}|^2 > 0$$

for all t. Hence the discriminant of the quadratic function $P(t)$ must be negative; that is,

$$4(\mathbf{u} \cdot \mathbf{v})^2 - 4|\mathbf{u}|^2|\mathbf{v}|^2 < 0,$$

so that $|\mathbf{u} \cdot \mathbf{v}| < |\mathbf{u}|\,|\mathbf{v}|$. Similarly, $|\mathbf{u} \cdot \mathbf{v}| < |\mathbf{u}|\,|\mathbf{v}|$ if $\mathbf{u} \neq \mathbf{0}$ and $\mathbf{v}$ is not a scalar multiple of $\mathbf{u}$. If $\mathbf{u} = c\mathbf{v}$ (in particular, if $\mathbf{u} = \mathbf{0}$), then

$$\begin{aligned} |\mathbf{u} \cdot \mathbf{v}| &= |c\mathbf{v} \cdot \mathbf{v}| = |c(\mathbf{v} \cdot \mathbf{v})| \\ &= |c|\,|\mathbf{v}|^2 = |c\mathbf{v}|\,|\mathbf{v}| = |\mathbf{u}|\,|\mathbf{v}|, \end{aligned}$$

by rule X in (1–44) and the definition (1–43). Similarly, equality holds if $\mathbf{v} = c\mathbf{u}$ (in particular, if $\mathbf{v} = \mathbf{0}$). Thus (1–49) is proved, along with the conditions for equality.

To prove (1–50), we write:

$$\begin{aligned} |\mathbf{u} + \mathbf{v}|^2 &= (\mathbf{u} + \mathbf{v}) \cdot (\mathbf{u} + \mathbf{v}) = |\mathbf{u}|^2 + |\mathbf{v}|^2 + 2\mathbf{u} \cdot \mathbf{v} \\ &= (|\mathbf{u}| + |\mathbf{v}|)^2 + 2\mathbf{u} \cdot \mathbf{v} - 2|\mathbf{u}|\,|\mathbf{v}| \leqq (|\mathbf{u}| + |\mathbf{v}|)^2. \end{aligned} \tag{1–51}$$

The last inequality follows from (1–49). By taking square roots, we obtain (1–50). The investigation of the case of equality is left to Problem 8 below.

Now let P_1, P_2, P_3 be three points of E^n. By (1–37), (1–39) and (1–43), $|\overrightarrow{P_1P_2}|$ equals the distance d from P_1 to P_2. We write $d(P_1, P_2) = |\overrightarrow{P_1P_2}|$. It follows that $d(P_1, P_2) \geqq 0$ and that $d(P_1, P_2) = 0$ only for $P_1 = P_2$. Also, $d(P_1, P_2) = d(P_2, P_1)$, since $|\overrightarrow{P_2P_1}| = |-\overrightarrow{P_1P_2}| = |\overrightarrow{P_1P_2}|$. Finally, from (1–39), $\overrightarrow{P_1P_3} = \overrightarrow{P_1P_2} + \overrightarrow{P_2P_3}$, so that, by (1–50),

$$|\overrightarrow{P_1P_3}| \leqq |\overrightarrow{P_1P_2}| + |\overrightarrow{P_2P_3}| \quad \text{or} \quad d(P_1, P_3) \leqq d(P_1, P_2) + d(P_2, P_3);$$

that is, the length of one side of a triangle is less than or equal to the sum of the lengths of the other two sides. This explains the term "triangle inequality" in (1–50).

Because of the theorem, we can define the *angle* θ between two nonzero vectors $\mathbf{u}$, $\mathbf{v}$ by the conditions

$$\cos\theta = \frac{\mathbf{u} \cdot \mathbf{v}}{|\mathbf{u}|\,|\mathbf{v}|}, \quad 0 \leqq \theta \leqq \pi, \tag{1–52}$$

for by (1–49), $\mathbf{u} \cdot \mathbf{v}/(|\mathbf{u}|\,|\mathbf{v}|)$ lies between -1 and 1 (inclusive). With the aid of distance and angle, one can now develop geometry in n-dimensional space in familiar fashion.

We define two vectors $\mathbf{u}$ and $\mathbf{v}$ to be *orthogonal* (or *perpendicular*) if $\mathbf{u} \cdot \mathbf{v} = 0$; that is, by (1–52) if they form an angle of $\pi/2$ (or if one of the vectors is $\mathbf{0}$). We define an *orthogonal system* of vectors as a system of k nonzero vectors $\mathbf{v}_1, \ldots, \mathbf{v}_k$ such that each two vectors of the system are orthogonal; the set of vectors is necessarily linearly independent (Problem 9 below), so that by rule (g) above, $k \leqq n$. An orthogonal system of *unit* vectors is called an *orthonormal system*. There exists an orthonormal system of n vectors: namely, $\mathbf{e}_1, \ldots, \mathbf{e}_n$. Every orthogonal system of n vectors forms a basis for V^n, by rule (h) above; we call such a system an *orthogonal basis* (or, for a system of unit vectors, an *orthonormal basis*).

Given k linearly independent vectors $\mathbf{v}_1, \ldots, \mathbf{v}_k$, it is always possible to construct an orthogonal system of k vectors $\mathbf{u}_1, \ldots, \mathbf{u}_k$, each of which is a linear combination of $\mathbf{v}_1, \ldots, \mathbf{v}_k$ (Gram-Schmidt orthogonalization process). We illustrate this for $k = 3$; the construction is then easily pictured in 3-dimensional space:

$$\begin{aligned} \mathbf{u}_1 &= \mathbf{v}_1, \\ \mathbf{u}_2 &= \mathbf{v}_2 - (\mathbf{v}_2 \cdot \mathbf{u}_1)\frac{\mathbf{u}_1}{|\mathbf{u}_1|^2}, \\ \mathbf{u}_3 &= \mathbf{v}_3 - (\mathbf{v}_3 \cdot \mathbf{u}_1)\frac{\mathbf{u}_1}{|\mathbf{u}_1|^2} - (\mathbf{v}_3 \cdot \mathbf{u}_2)\frac{\mathbf{u}_2}{|\mathbf{u}_2|^2}. \end{aligned} \tag{1–53}$$

To obtain $\mathbf{u}_2$, we subtract from $\mathbf{v}_2$ the vector obtained by "projecting" $\mathbf{v}_2$ on $\mathbf{v}_1$; to obtain $\mathbf{u}_3$, we subtract from $\mathbf{v}_3$ its projections on $\mathbf{u}_1$ and $\mathbf{u}_2$. By (1–53), $\mathbf{u}_1 = \mathbf{v}_1$, $\mathbf{u}_2$ is a linear combination of $\mathbf{v}_1$ and $\mathbf{v}_2$, and $\mathbf{u}_3$ is a linear combination of $\mathbf{v}_1, \mathbf{v}_2, \mathbf{v}_3$; also $\mathbf{v}_1 = \mathbf{u}_1$, $\mathbf{v}_2$ is a linear combination of $\mathbf{u}_1$ and $\mathbf{u}_2$, and $\mathbf{v}_3$ is a linear combination of $\mathbf{u}_1$, $\mathbf{u}_2$ and $\mathbf{u}_3$. The proof of these assertions and of the orthogonality is left to Problem 11 below. The process clearly extends by induction to any finite number of linearly independent vectors. Furthermore, in general the process is such that, for each h, $\mathbf{u}_h$ is a linear combination of $\mathbf{v}_1, \ldots, \mathbf{v}_h$, and $\mathbf{v}_h$ is a linear combination of $\mathbf{u}_1, \ldots, \mathbf{u}_h$.

Remark. With the aid of the Gram-Schmidt process, we can extend rule (j) above as follows:

(l) If $k < n$ and $\mathbf{u}_1, \ldots, \mathbf{u}_k$ form an orthogonal system in V^n, then there exist vectors $\mathbf{u}_{k+1}, \ldots, \mathbf{u}_n$ such that $\mathbf{u}_1, \ldots, \mathbf{u}_n$ form an orthogonal basis.

The proof is left as an exercise (Problem 10(j) below).

***1–9 Axioms for V^n.** From the rule (h) of Section 1–8) we can state:

XIII. V^n has a basis consisting of n vectors.

It is of considerable interest that the properties I through XII of (1–44) plus rule XIII can be regarded as a set of axioms, from which one can deduce all other properties of the vectors in n-dimensional space, without reference to the sets of components. For example, $\mathbf{u} + \mathbf{0} = \mathbf{u}$, since

$$\mathbf{u} + \mathbf{0} = 1\mathbf{u} + 0\mathbf{u} = (1 + 0)\mathbf{u} = \mathbf{u}$$

by VI, VII and IV respectively. Also the equation $\mathbf{v} + \mathbf{u} = \mathbf{w}$ has a solution for $\mathbf{u}$: namely, the vector $\mathbf{w} + (-1)\mathbf{v}$, which we also denote by $\mathbf{w} - \mathbf{v}$; for

$$\begin{aligned}\mathbf{v} + [\mathbf{w} + (-1)\mathbf{v}] &= \mathbf{v} + [(-1)\mathbf{v} + \mathbf{w}] = [1\mathbf{v} + (-1)\mathbf{v}] + \mathbf{w} \\ &= 0\mathbf{v} + \mathbf{w} = \mathbf{0} + \mathbf{w} = \mathbf{w}\end{aligned}$$

by II, VI, IV, VII, I, and the rule $\mathbf{u} + \mathbf{0} = \mathbf{u}$ just proved. The solution is unique, for $\mathbf{v} + \mathbf{u} = \mathbf{w}$ implies $-\mathbf{v} + (\mathbf{v} + \mathbf{u}) = -\mathbf{v} + \mathbf{w} = \mathbf{w} - \mathbf{v}$ or $(-\mathbf{v} + \mathbf{v}) + \mathbf{u} = \mathbf{w} - \mathbf{v}$ or $\mathbf{u} = \mathbf{w} - \mathbf{v}$; here we wrote $-\mathbf{v}$ for $(-1)\mathbf{v}$, as is usual. Other algebraic properties can be deduced in similar fashion. Also, the norm $|\mathbf{u}|$ is defined in terms of the scalar product by Eq. (1–43); its properties derive from those of the scalar product. In particular, the Cauchy-Schwarz inequality and triangle inequality follow as above, without use of components.

One can introduce coordinates or components on the basis of the axioms themselves, for rule XIII guarantees existence of a basis $\mathbf{v}_1, \ldots, \mathbf{v}_n$. The Gram-Schmidt process can then be used to construct an orthogonal basis $\mathbf{u}_1, \ldots, \mathbf{u}_n$. We "normalize" these vectors by dividing by their norms to obtain an orthonormal basis:

$$\mathbf{e}_1^* = \frac{\mathbf{u}_1}{|\mathbf{u}_1|}, \quad \ldots, \quad \mathbf{e}_n^* = \frac{\mathbf{u}_n}{|\mathbf{u}_n|}. \tag{1–54}$$

For an arbitrary vector **v** we can now write, in unique fashion:

$$\mathbf{v} = v_1^* \mathbf{e}_1^* + \cdots + v_n^* \mathbf{e}_n^*, \tag{1-55}$$

and call $(v_1^*, \ldots, v_n^*)$ the components of **v** with respect to the basis $\mathbf{e}_1^*, \ldots, \mathbf{e}_n^*$. We note that

$$\mathbf{v} \cdot \mathbf{e}_1^* = v_1^* (\mathbf{e}_1^* \cdot \mathbf{e}_1^*) + v_2^* (\mathbf{e}_2^* \cdot \mathbf{e}_1^*) + \cdots + v_n^* (\mathbf{e}_n^* \cdot \mathbf{e}_1^*) = v_1^*,$$

because the basis is orthonormal. In general,

$$v_1^* = \mathbf{v} \cdot \mathbf{e}_1^*, \ldots, v_n^* = \mathbf{v} \cdot \mathbf{e}_n^*. \tag{1-56}$$

On the basis of rules I through XII, we now verify (see Problem 12 below) that

$$\begin{aligned} \mathbf{u} + \mathbf{v} &= (u_1^* + v_1^*)\mathbf{e}_1^* + \cdots + (u_n^* + v_n^*)\mathbf{e}_n^*, \\ h\mathbf{v} &= (hv_1^*)\mathbf{e}_1^* + \cdots + (hv_n^*)\mathbf{e}_n^*, \\ \mathbf{u} \cdot \mathbf{v} &= u_1^* v_1^* + \cdots + u_n^* v_n^*. \end{aligned} \tag{1-57}$$

This simply means that the vector operations can be defined in terms of components just as in (1–40), (1–41) and (1–42). Accordingly, while the new basis may be different from the original one (this is simply a choice of new axes), all properties described with the old components can be found just as well with the new ones. This means that *all* properties of the vectors can be found from (1–44) and XIII alone. This leads to the following definition:

Definition. A Euclidean n-dimensional vector space is a collection of objects **u**, **v**, . . . called vectors, including a zero vector **0**, for which the operations of addition, multiplication by scalars, and scalar product are defined and obey the rules I through XIII.

By virtue of our discussion, there is, except for notation, really only one Euclidean n-dimensional vector space, namely, V^n.

We have emphasized the axiomatic approach to vectors. There is a similar discussion for points. In fact, given a Euclidean n-dimensional vector space, we can assign components $(v_1^*, \ldots, v_n^*)$ to each vector as above. To the zero vector $(0, \ldots, 0)$, we then assign an origin O; to each vector $(v_1^*, \ldots, v_n^*)$ we assign a point P and interpret $v_1^*, \ldots, v_n^*$ as the coordinates of P. All the concepts of geometry can now be introduced as above, and we recover our n-dimensional Euclidean space, the same (except for notation) as E^n.

One can ask whether there is any value in the generalization of vectors to n-dimensional space, since we live in a three-dimensional world. The answer is that the generalization has proved to be exceedingly valuable. The equations of a mechanical system having "N degrees of freedom" are easier to describe and understand in terms of a vector space of dimension $2N$. In the kinetic theory of gases, an n-dimensional space (the "phase space") is of fundamental importance; here, n may be as large as 10^{23}.

In relativity a four-dimensional space is needed. In quantum mechanics, one is in fact forced to consider the limiting case $n = \infty$; surprisingly enough, this theory of a vector space of infinite dimension turns out to be closely related to the theory of Fourier series (Chapter 7 below).

Problems

1. In V^4 let $\mathbf{u} = (3, 2, 1, 0)$, $\mathbf{v} = (1, 0, 1, 2)$, $\mathbf{w} = (5, 4, 1, -2)$.
(a) Find $\mathbf{u} + \mathbf{v}$, $\mathbf{u} + \mathbf{w}$, $2\mathbf{u}$, $-3\mathbf{v}$, $0\mathbf{w}$.
(b) Find $3\mathbf{u} - 2\mathbf{v}$, $2\mathbf{v} + 3\mathbf{w}$, $\mathbf{u} - \mathbf{w}$, $\mathbf{u} + \mathbf{v} - 2\mathbf{w}$.
(c) Find $\mathbf{u} \cdot \mathbf{v}$, $\mathbf{u} \cdot \mathbf{w}$, $|\mathbf{u}|$, $|\mathbf{v}|$.
(d) Show that $\mathbf{w}$ can be expressed as a linear combination of $\mathbf{u}$ and $\mathbf{v}$ and hence that $\mathbf{u}$, $\mathbf{v}$, $\mathbf{w}$ are linearly dependent.

2. In E^n the line segment P_1P_2 joining points P_1, P_2 is formed of all points P such that $\overrightarrow{P_1P} = t\overrightarrow{P_1P_2}$, where $0 \leqq t \leqq 1$.
(a) In E^4, show that $(5, 10, -1, 8)$ is on the line segment from $(1, 2, 3, 4)$ to $(7, 14, -3, 10)$.
(b) In E^4, is $(2, 8, 1, 6)$ on the line segment joining $(1, 7, 2, 5)$ to $(9, 2, 0, 7)$?
(c) In E^5, find the *midpoint* of the line segment from P_1: $(2, 0, 1, 3, 7)$ to P_2: $(10, 4, -3, 3, 5)$; that is, find the point P on the segment such that $|\overrightarrow{P_1P}| = |\overrightarrow{P_2P}|$.
(d) In E^5, trisect the line segment joining $(7, 1, 7, 9, 6)$ to $(16, -2, 10, 3, 0)$.
(e) In E^n, show that if P is on the line segment P_1P_2, then $|\overrightarrow{P_1P}| + |\overrightarrow{P_2P}| = |\overrightarrow{P_1P_2}|$.

3. Prove the Pythagorean theorem in E^n: if $\overrightarrow{P_1P_2}$ is orthogonal to $\overrightarrow{P_2P_3}$, then $|\overrightarrow{P_1P_3}|^2 = |\overrightarrow{P_1P_2}|^2 + |\overrightarrow{P_2P_3}|^2$. [Hint: Write the left side as

$$(\overrightarrow{P_1P_2} + \overrightarrow{P_2P_3}) \cdot (\overrightarrow{P_1P_2} + \overrightarrow{P_2P_3}).]$$

4. Prove the Law of Cosines in E^n: if $\overrightarrow{P_1P_2}$ and $\overrightarrow{P_1P_3}$ are nonzero vectors, then

$$|\overrightarrow{P_2P_3}|^2 = |\overrightarrow{P_1P_2}|^2 + |\overrightarrow{P_1P_3}|^2 - 2|\overrightarrow{P_1P_2}|\,|\overrightarrow{P_1P_3}| \cos\theta,$$

where θ is the angle between $\overrightarrow{P_1P_2}$ and $\overrightarrow{P_1P_3}$.

5. (a) In E^5 find the sides and angles of the triangle with vertices

$$(1, 2, 3, 4, 5), \quad (5, 4, 2, 3, 1) \quad \text{and} \quad (2, 2, 2, 2, 2).$$

(b) Prove: in E^n, the sum of the angles of a triangle is π. [Hint: For a triangle with sides a, b, c in E^n, construct the triangle with the same sides in the plane E^2; why is this possible? Then show, with the aid of the Law of Cosines, that corresponding angles of the two triangles are equal.]

6. Prove the rules (1–44), with the aid of (1–38) through (1–43).

7. Prove Cauchy's inequality:

$$\left[\sum_{i=1}^{n} u_i v_i\right]^2 \leqq \sum_{i=1}^{n} u_i^2 \sum_{i=1}^{n} v_i^2.$$

[Hint: Use (1–49).]

8. Prove the rule for equality in (1–50) as stated in the theorem. [Hint: Use (1–51) to reduce the problem to that for $|\mathbf{u}\cdot\mathbf{v}| = |\mathbf{u}|\,|\mathbf{v}|$. Use the first part of the theorem and the meaning of linear dependence to prove the desired result.]

9. Prove: The vectors of an orthogonal system are linearly independent.

10. Prove the following rules for vectors in V^n, with the aid of the hints given.
(a) Rule (a) [Hint: If $c_1\mathbf{v}_1 + \cdots + c_k\mathbf{v}_k = \mathbf{0}$ and, for example, $c_k \neq 0$, then express $\mathbf{v}_k$ as a linear combination of $\mathbf{v}_1, \ldots, \mathbf{v}_{k-1}$. For the converse, suppose $\mathbf{v}_k = c_1\mathbf{v}_1 + \cdots + c_{k-1}\mathbf{v}_{k-1}$ and write this equation in the form $c_1\mathbf{v}_1 + \cdots + c_k\mathbf{v}_k = \mathbf{0}$.]
(b) Rule (b) [Hint: Suppose $\mathbf{v}_k = \mathbf{0}$ and choose $c_1, \ldots, c_k$ not all 0 so that $c_1\mathbf{v}_1 + \cdots + c_k\mathbf{v}_k = \mathbf{0}$.]
(c) Rule (c) [Hint: One has $c_1\mathbf{v}_1 + \cdots + c_{k+1}\mathbf{v}_{k+1} = \mathbf{0}$ with not all c's 0. Show that c_{k+1} can't be 0.]
(d) Rule (d) (e) Rule (e) (f) Rule (f)
(g) Rule (h) [Hint: Apply rules (g) and (e).]
(h) Rule (j) [Hint: By rule (i), $\mathbf{v}_1, \ldots, \mathbf{v}_k$ do not form a basis. Hence one can choose $\mathbf{v}_{k+1}$ so that $\mathbf{v}_1, \ldots, \mathbf{v}_{k+1}$ are linearly independent. If $k + 1 < n$, repeat the process.]
(i) Rule (k) [Hint: Apply rule (j) to obtain a basis $\mathbf{v}_1, \ldots, \mathbf{v}_n$. By rule (g), $\mathbf{u}_1, \ldots, \mathbf{u}_{k+1}, \mathbf{v}_{k+1}, \ldots, \mathbf{v}_n$ are linearly dependent, so that

$$c_1\mathbf{u}_1 + \cdots + c_{k+1}\mathbf{u}_{k+1} + c_{k+2}\mathbf{v}_{k+1} + \cdots + c_{n+1}\mathbf{v}_n = \mathbf{0}.$$

Show from the hypotheses that $c_{k+2} = 0, \ldots, c_{n+1} = 0$.]
(j) Rule (l) [Hint: Let $\mathbf{v}_1 = \mathbf{u}_1, \ldots, \mathbf{v}_k = \mathbf{u}_k$. Then choose $\mathbf{v}_{k+1}, \ldots, \mathbf{v}_n$ as in rule (j) so that $\mathbf{v}_1, \ldots, \mathbf{v}_n$ form a basis. Now apply the Gram-Schmidt process to $\mathbf{v}_1, \ldots, \mathbf{v}_n$.]

11. (a) Let $\mathbf{v}_1, \mathbf{v}_2, \mathbf{v}_3$ be linearly independent. Prove that the vectors (1–53) are nonzero and form an orthogonal system. Prove also that, for $h = 1, 2, 3$, $\mathbf{u}_h$ is a linear combination of $\mathbf{v}_1, \ldots, \mathbf{v}_h$ and $\mathbf{v}_h$ is a linear combination of $\mathbf{u}_1, \ldots, \mathbf{u}_h$.
(b) Carry out the process (1–53) in V^3 with $\mathbf{v}_1 = \mathbf{i}$, $\mathbf{v}_2 = \mathbf{i} + \mathbf{j} + \mathbf{k}$, $\mathbf{v}_3 = 2\mathbf{j} + \mathbf{k}$, and graph.

12. Prove that if $u_1^*, \ldots, u_n^*$ and $v_1^*, \ldots, v_n^*$ are the components of $\mathbf{u}$ and $\mathbf{v}$ with respect to the orthonormal basis $\mathbf{e}_1^*, \ldots, \mathbf{e}_n^*$, then (1–57) holds.

Answers

1. (a) $(4, 2, 2, 2)$, $(8, 6, 2, -2)$, $(6, 4, 2, 0)$, $(-3, 0, -3, -6)$, $(0, 0, 0, 0) = \mathbf{0}$. (b) $(7, 6, 1, -4)$, $(17, 12, 5, -2)$, $(-2, -2, 0, 2)$, $(-6, -6, 0, 6)$. (c) $4, 24, \sqrt{14}, \sqrt{6}$.
2. (b) No. (c) $(6, 2, -1, 3, 6)$, (d) $(10, 0, 8, 7, 4)$, $(13, -1, 9, 5, 2)$.
5. (a) $\sqrt{38}, \sqrt{15}, \sqrt{15}, \cos^{-1}(-\frac{4}{15}), \cos^{-1}(19/\sqrt{570}), \cos^{-1}(19/\sqrt{570})$.

1–10 Linear mappings. We now consider the simultaneous linear equations

$$\begin{aligned} a_{11}x_1 + \cdots + a_{1n}x_n &= y_1, \\ \vdots \qquad\qquad & \quad \vdots \\ a_{m1}x_1 + \cdots + a_{mn}x_n &= y_m \end{aligned} \tag{1–58}$$

or, in matrix form,

$$A\mathbf{x} = \mathbf{y}, \tag{1–58$'$}$$

where $\mathbf{x} = \text{col}\,(x_1, \ldots, x_n)$, $\mathbf{y} = \text{col}\,(y_1, \ldots, y_m)$. (In this section all vectors will be written as column vectors.) Through (1–58) or (1–58$'$) a vector $\mathbf{y}$ of V^m is assigned to each vector $\mathbf{x}$ of V^n. Thus we have a *function* or *mapping* (Section 0–7) whose domain is V^n and whose range is contained in V^m. Furthermore, if $\mathbf{x}_1$ and $\mathbf{x}_2$ are in V^n and c_1, c_2 are scalars, then

$$A(c_1\mathbf{x}_1 + c_2\mathbf{x}_2) = c_1(A\mathbf{x}_1) + c_2(A\mathbf{x}_2); \tag{1–59}$$

that is, *the mapping* (1–58$'$) *assigns to each linear combination* $c_1\mathbf{x}_1 + c_2\mathbf{x}_2$ *the corresponding linear combination of the values assigned to* $\mathbf{x}_1$ *and* $\mathbf{x}_2$. We call such a mapping linear:

Definition. Let T be a mapping of V^n into V^m. Then T is *linear* if for every choice of $\mathbf{x}_1$ and $\mathbf{x}_2$ in V^n and every pair of scalars c_1, c_2, one has

$$T(c_1\mathbf{x}_1 + c_2\mathbf{x}_2) = c_1T(\mathbf{x}_1) + c_2T(\mathbf{x}_2). \tag{1–60}$$

Thus for every $m \times n$ matrix A, the equation $\mathbf{y} = A\mathbf{x}$ defines a linear mapping T of V^n into V^m with $T(\mathbf{x}) = A\mathbf{x}$.

Conversely, if T *is a linear mapping of* V^n *into* V^m, *then there is an* $m \times n$ *matrix* A *such that* $T(\mathbf{x}) = A\mathbf{x}$ *for all* $\mathbf{x}$ *in* V^n. To prove this assertion, we first note that (1–60) implies the general rule

$$T(c_1\mathbf{x}_1 + \cdots + c_k\mathbf{x}_k) = c_1T(\mathbf{x}_1) + \cdots + c_kT(\mathbf{x}_k). \tag{1–61}$$

Now, let $\mathbf{e}_1, \ldots, \mathbf{e}_n$ be the standard basis in V^n and let

$$T(\mathbf{e}_j) = \mathbf{u}_j, \quad j = 1, \ldots, n. \tag{1–62}$$

For each vector $\mathbf{x} = (x_1, \ldots, x_n) = x_1\mathbf{e}_1 + \cdots + x_n\mathbf{e}_n$ in V^n, it then follows from (1–61) that

$$\begin{aligned} T(\mathbf{x}) &= x_1T(\mathbf{e}_1) + \cdots + x_nT(\mathbf{e}_n) = x_1\mathbf{u}_1 + \cdots + x_n\mathbf{u}_n \\ &= A\ \text{col}\,(x_1, \ldots, x_n) = A\mathbf{x}, \end{aligned}$$

where A is the $m \times n$ matrix whose column vectors are $\mathbf{u}_1, \ldots, \mathbf{u}_n$.

It follows that the study of linear mappings of V^n into V^m is equivalent to the study of mappings of the form $\mathbf{y} = A\mathbf{x}$, where A is an $m \times n$ matrix. For each linear mapping T, we call A *the matrix of* T, or *the matrix representing* T. For each matrix A, we call the corresponding linear mapping T the *linear mapping* T *determined by* A; we also write more concisely: the linear mapping $\mathbf{y} = A\mathbf{x}$, or even, the linear mapping A.

From the proof just given we see that for each linear mapping T, the column vectors $\mathbf{u}_1, \ldots, \mathbf{u}_n$ of the matrix A representing T are the vectors $T(\mathbf{e}_1), \ldots, T(\mathbf{e}_n)$ of V^m. Thus if $A = (a_{ij})$, then

$$A\mathbf{e}_j = \text{col}\,(a_{1j}, \ldots, a_{mj}) = \mathbf{u}_j. \tag{1–63}$$

Furthermore, as in the proof, we can write

$$\mathbf{y} = A\mathbf{x} = A\ \text{col}\,(x_1, \ldots, x_n) = x_1\mathbf{u}_1 + \cdots + x_n\mathbf{u}_n.$$

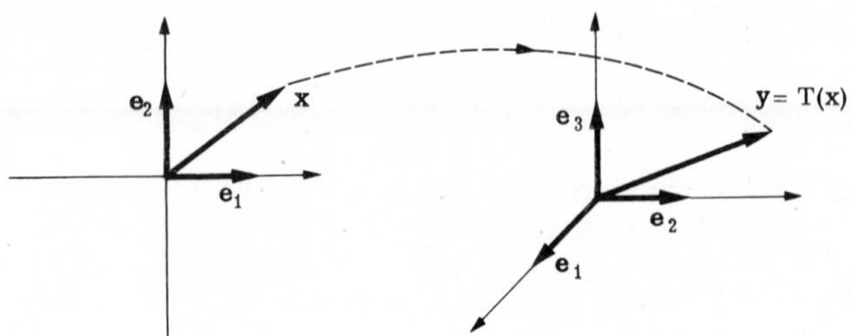

FIG. 1–2. Linear mapping of V^2 into V^3.

Thus, the linear mapping A assigns to each vector $\mathbf{x} = \text{col}\,(x_1, \ldots, x_n)$ the linear combination $x_1\mathbf{u}_1 + \cdots + x_n\mathbf{u}_n$ of the column vectors

$$\mathbf{u}_1, \ldots, \mathbf{u}_n \quad (\text{vectors of } V^m).$$

It is very helpful to think of a linear mapping as suggested in Fig. 1–2, in which the vectors are represented by directed segments from the origin; here $n = 2$, $m = 3$.

EXAMPLE 1. Let the linear mapping T have the matrix $A = \begin{bmatrix} 2 & 3 \\ 1 & 2 \end{bmatrix}$. Evaluate $T(\mathbf{x})$ for $\mathbf{x} = (1, 0) = \mathbf{x}_1$, $\mathbf{x} = (0, 1) = \mathbf{x}_2$, $\mathbf{x} = (2, -1) = \mathbf{x}_3$, $\mathbf{x} = (1, -1) = \mathbf{x}_4$ and graph.

Solution. $T(\mathbf{x}_1) = A \ \text{col}\,(1, 0) = \text{col}\,(2, 1)$, $T(\mathbf{x}_2) = A \ \text{col}\,(0, 1) = \text{col}\,(3, 2)$, $T(\mathbf{x}_3) = A \ \text{col}\,(2, -1) = \text{col}\,(1, 0)$, $T(\mathbf{x}_4) = A \ \text{col}\,(1, -1) = \text{col}\,(-1, -1)$. (See Fig. 1–3.)

We note that $\mathbf{x}_1 = \mathbf{e}_1$ and $\mathbf{x}_2 = \mathbf{e}_2$, so that

$$T(\mathbf{x}_1) = T(\mathbf{e}_1) = \mathbf{u}_1 = \text{col}\,(2, 1),$$

the first column vector of A. Similarly, $T(\mathbf{x}_2) = \mathbf{u}_2 = \text{col}\,(3, 2)$, the second column vector of A. Also $\mathbf{x}_3 = 2\mathbf{e}_1 - \mathbf{e}_2$, so that

$$T(\mathbf{x}_3) = 2\mathbf{u}_1 - \mathbf{u}_2 = 2(2, 1) - (3, 2) = (1, 0).$$

Similarly, $T(\mathbf{x}_4) = T(\mathbf{e}_1 - \mathbf{e}_2) = \mathbf{u}_1 - \mathbf{u}_2 = (-1, -1)$.

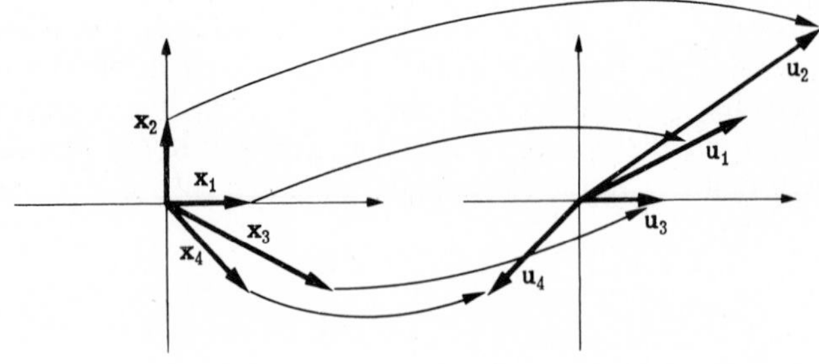

FIG. 1–3. Linear mapping of V^2 into V^2.

EXAMPLE 2. Find a linear mapping T of V^2 into V^3 such that

$$T((1, 0)) = (2, 1, 2) \quad \text{and} \quad T((0, 1)) = (5, 3, 7).$$

Solution. T has the matrix $A = \begin{bmatrix} 2 & 5 \\ 1 & 3 \\ 2 & 7 \end{bmatrix}$.

Notation. We shall write $T(x_1, \ldots, x_n)$ for $T((x_1, \ldots, x_n))$, since the extra parentheses are not necessary.

For each particular linear mapping T of V^n into V^m, one is concerned with such properties as the following:

(a) the *range* of T: the set of all $\mathbf{y}$ for which $T(\mathbf{x}) = \mathbf{y}$ for at least one $\mathbf{x}$;

(b) whether T maps V^n *onto* V^m: that is, whether the range of T is all of V^m;

(c) whether T is one-to-one: that is, whether, for every $\mathbf{y}$, $T(\mathbf{x}) = \mathbf{y}$ has at most one solution $\mathbf{x}$;

(d) the set of all $\mathbf{x}$ for which $T(\mathbf{x}) = \mathbf{0}$; this set is called the *kernel* of T;

(e) the set of all $\mathbf{x}$ for which $T(\mathbf{x}) = \mathbf{y}_0$, where $\mathbf{y}_0$ is a given element of V^m.

We remark that $T(\mathbf{0})$ must be $\mathbf{0}$. For $T(\mathbf{0}) = T(0\mathbf{0}) = 0T(\mathbf{0}) = \mathbf{0}$. Hence *the kernel of T always contains the zero vector of V^n.* With regard to (c) and (e) we have a useful rule:

THEOREM. *Let T be a linear mapping of V^n into V^m and let $T(\mathbf{x}_0) = \mathbf{y}_0$ for a particular $\mathbf{x}_0$ and $\mathbf{y}_0$. Then all solutions $\mathbf{x}$ of the equation $T(\mathbf{x}) = \mathbf{y}_0$ are given by*

$$\mathbf{x} = \mathbf{x}_0 + \mathbf{z},$$

where $\mathbf{z}$ is an arbitrary vector in the kernel of T. Hence T is one-to-one precisely when the kernel of T consists of $\mathbf{0}$ alone.

Proof. We are given that $T(\mathbf{x}_0) = \mathbf{y}_0$. If also $T(\mathbf{x}) = \mathbf{y}_0$, then

$$T(\mathbf{x} - \mathbf{x}_0) = T(\mathbf{x}) - T(\mathbf{x}_0) = \mathbf{y}_0 - \mathbf{y}_0 = \mathbf{0}.$$

Hence $\mathbf{x} - \mathbf{x}_0$ is an element $\mathbf{z}$ of the kernel of T:

$$\mathbf{x} - \mathbf{x}_0 = \mathbf{z}, \quad \text{or} \quad \mathbf{x} = \mathbf{x}_0 + \mathbf{z}.$$

Conversely, if $\mathbf{x} = \mathbf{x}_0 + \mathbf{z}$, where $\mathbf{z}$ is in the kernel of T, then

$$T(\mathbf{x}) = T(\mathbf{x}_0) + T(\mathbf{z}) = T(\mathbf{x}_0) + \mathbf{0} = \mathbf{y}_0.$$

The last sentence of the theorem follows from the previous result, since T is one-to-one precisely when the equation $T(\mathbf{x}) = \mathbf{y}_0$ has exactly one solution, for every $\mathbf{y}_0$ for which there is a solution.

EXAMPLE 3. Let T map V^2 into V^3 and have the matrix

$$A = \begin{bmatrix} 2 & 4 \\ 3 & 6 \\ 1 & 2 \end{bmatrix}.$$

Then $A\mathbf{x} = \mathbf{0}$ is equivalent to

$$2x_1 + 4x_2 = 0,\ 3x_1 + 6x_2 = 0,\ x_1 + 2x_2 = 0.$$

This is satisfied for $x_1 = -2x_2$; that is, by all vectors $t(2, -1)$ of V^2. We note that A col $(2, 3) = (16, 24, 8)$; that is, $T(\mathbf{x}) = (16, 24, 8)$ is satisfied for $\mathbf{x} = \mathbf{x}_0 = (2, 3)$. Hence, all vectors $\mathbf{x}$ such that $T(\mathbf{x}) = (16, 24, 8)$ are given by

$$\mathbf{x} = (2, 3) + t(2, -1),$$

where t is arbitrary.

EXAMPLE 4. T maps V^3 into V^2 and

$$T(x_1, x_2, x_3) = (y_1, y_2) = (x_1 - x_2, x_2 - x_3).$$

To find the kernel, we must solve the equations $x_1 - x_2 = 0$, $x_2 - x_3 = 0$. The solutions are given by all (x_1, x_2, x_3) such that $x_1 = x_2 = x_3$; hence, by all vectors $t(1, 1, 1)$, where t is an arbitrary scalar. Here T is not one-to-one. However, the range of T is all of V^2, for the equations

$$x_1 - x_2 = y_1, \quad x_2 - x_3 = y_2$$

can always be solved for x_1, x_2, x_3—for example, by taking $x_3 = 0$, $x_2 = y_2$, and $x_1 = y_1 + y_2$. Therefore, T maps V^3 *onto* V^2 but is not one-to-one.

EXAMPLE 5. T maps V^n into V^m and $T(\mathbf{x}) = \mathbf{0}$ for every $\mathbf{x}$. Thus T has the matrix $O = O_{mn}$. We call T the *zero mapping*, and sometimes also denote this mapping by O. T is clearly linear and T is neither one-to-one nor onto.

EXAMPLE 6. T maps V^n into V^n and $T(\mathbf{x}) = \mathbf{x}$ for every $\mathbf{x}$. Thus T has the matrix $I = I_n$. We call T the *identity mapping*, and sometimes also denote this mapping by I. T is clearly linear, is one-to-one and onto.

EXAMPLE 7. Let T map V^3 into V^3 and have the matrix

$$A = \begin{bmatrix} 2 & 3 & 1 \\ 1 & 0 & 2 \\ 1 & 2 & 0 \end{bmatrix}$$

Find the range of T.

Solution. The range of T consists of all $\mathbf{y} = x_1\mathbf{u}_1 + x_2\mathbf{u}_2 + x_3\mathbf{u}_3$, where $\mathbf{u}_1, \mathbf{u}_2, \mathbf{u}_3$ are the column vectors of A:

$$\mathbf{u}_1 = \text{col}\,(2, 1, 1), \quad \mathbf{u}_2 = \text{col}\,(3, 0, 2), \quad \mathbf{u}_3 = \text{col}\,(1, 2, 0).$$

Thus the range consists of all linear combinations of $\mathbf{u}_1, \mathbf{u}_2, \mathbf{u}_3$. If $\mathbf{u}_1, \mathbf{u}_2, \mathbf{u}_3$ are linearly independent, then they form a basis for V^3 and the range is V^3. However, we verify that $2\mathbf{u}_1 - \mathbf{u}_2 - \mathbf{u}_3 = \mathbf{0}$, so that these vectors are linearly dependent. Furthermore, we see that $\mathbf{u}_1, \mathbf{u}_2$ are linearly independent and that $\mathbf{u}_3$ is expressible as a linear combination of $\mathbf{u}_1, \mathbf{u}_2$. Hence the range is given by all linear combinations of $\mathbf{u}_1$ and $\mathbf{u}_2$, and this set is not all of V^3. Thus T does not map V^3 onto V^3 and the equation $T(\mathbf{x}) = \mathbf{y}_0$ has no solution for $\mathbf{x}$, for some choices of $\mathbf{y}_0$; for example, $T(\mathbf{x}) = (1, 0, 0)$ has no solution, as one can verify (the vector $(1, 0, 0)$ is not a linear combination of $\mathbf{u}_1$ and $\mathbf{u}_2$).

Problems

In these problems all vectors are written as column vectors. Also, the following matrices are referred to:

$$A = \begin{bmatrix} 2 & 1 \\ 3 & 5 \end{bmatrix}, \quad B = \begin{bmatrix} 2 & 3 \\ 4 & 6 \end{bmatrix}, \quad C = \begin{bmatrix} 1/\sqrt{2} & -1/\sqrt{2} \\ 1/\sqrt{2} & 1/\sqrt{2} \end{bmatrix}, \quad D = \begin{bmatrix} 1 & 0 \\ 0 & -1 \end{bmatrix},$$

$$E = \begin{bmatrix} -1 & 0 \\ 0 & -1 \end{bmatrix}, \quad F = \begin{bmatrix} 3 & 1 & 2 \\ 0 & 2 & 4 \end{bmatrix}, \quad G = \begin{bmatrix} 1 & 4 & 3 \\ 2 & 8 & 6 \end{bmatrix},$$

$$H = \begin{bmatrix} 2 & 1 \\ 4 & 2 \\ -6 & -3 \end{bmatrix}, \quad J = \begin{bmatrix} 2 & 1 \\ 1 & 2 \\ 1 & 2 \end{bmatrix}, \quad K = \begin{bmatrix} 3 & 1 & 2 \\ 1 & 0 & 1 \\ 5 & 2 & 3 \end{bmatrix}, \quad L = \begin{bmatrix} 1 & 0 & 1 \\ 0 & 1 & 0 \\ 0 & 1 & 1 \end{bmatrix},$$

$$M = \begin{bmatrix} 2 & 0 & 0 \\ 0 & 2 & 0 \\ 0 & 0 & 2 \end{bmatrix} = 2I, \quad N = \begin{bmatrix} 1 & 0 & 0 \\ 0 & 2 & 0 \\ 0 & 0 & 3 \end{bmatrix}.$$

1. Let the linear mapping T have the matrix A.
(a) Evaluate $T(1, 0)$, $T(0, 1)$, $T(2, -1)$, $T(-1, 1)$ and graph.
(b) Find the kernel of T, determine whether T is one-to-one, and find all $\mathbf{x}$ such that $T(\mathbf{x}) = (2, 3)$.
(c) Find the range of T and determine whether T maps V^2 onto V^2.

2. Let the linear mapping T have the matrix B.
(a) Evaluate $T(1, 0)$, $T(0, 1)$, $T(1, -1)$, $T(-1, -1)$ and graph.
(b) Find the kernel of T, determine whether T is one-to-one, and find all $\mathbf{x}$ such that $T(\mathbf{x}) = (2, 4)$.
(c) Find the range of T and determine whether T maps V^2 onto V^2.

3. Let T map V^n into V^m and have the matrix F.
(a) Find n and m.
(b) Find the kernel of T, and determine whether T is one-to-one.
(c) Find the range of T and determine whether T maps V^n onto V^m.

4. (a), (b), (c) Proceed as in Problem 3 with matrix G.

5. (a), (b), (c) Proceed as in Problem 3 with matrix H.

6. (a), (b), (c) Proceed as in Problem 3 with matrix J.

7. (a), (b), (c) Proceed as in Problem 3 with matrix K.

8. (a), (b), (c) Proceed as in Problem 3 with matrix L.

9. Let the linear mapping T have the equation $\mathbf{y} = C\mathbf{x}$. For general $\mathbf{x}$, find the angle between $\mathbf{x}$ and $T(\mathbf{x}) = \mathbf{y}$, as vectors in V^2, and also compare $|\mathbf{x}|$ and $|T(\mathbf{x})|$. From these results, interpret T geometrically. [Hint: Consider $\mathbf{x}$ as $\overrightarrow{OP}$ and $\mathbf{y}$ as $\overrightarrow{OQ}$, where O is the origin of E^2 and P and Q are points of E^2.]

10. Let the linear mapping T have the equation $\mathbf{y} = D\mathbf{x}$. Regard $\mathbf{x}$ as $\overrightarrow{OP}$, $\mathbf{y}$ as $\overrightarrow{OQ}$, as in Problem 9, and describe geometrically the relation between $\mathbf{x}$ and $\mathbf{y} = T(\mathbf{x})$.

11. Interpret each of the following linear mappings geometrically, as in Problems 9 and 10:

(a) $\mathbf{y} = E\mathbf{x}$. (b) $\mathbf{y} = M\mathbf{x}$.

12. If T maps V^n into V^n, then $\mathbf{x}$ and $\mathbf{y} = T(\mathbf{x})$ are vectors in the same n-dimensional vector space and can be compared (see Problems 9, 10 and 11). If $T(\mathbf{x})$ is a scalar multiple of $\mathbf{x}$, for some nonzero $\mathbf{x}$, say $T(\mathbf{x}) = \lambda\mathbf{x}$, then we say that x is an *eigenvector* of T, associated with the *eigenvalue* λ. If T has the matrix A, then $\mathbf{x}$ is also an eigenvector of A, associated with the eigenvalue λ, as in Section 1–5.

Prove: if $\mathbf{v}_1, \ldots, \mathbf{v}_k$ are eigenvectors of the linear mapping T, associated with distinct eigenvalues $\lambda_1, \ldots, \lambda_k$, then $\mathbf{v}_1, \ldots, \mathbf{v}_k$ are linearly independent. [Hint: Let T have the matrix A. Use induction. Verify that the assertion is true for $k = 1$. Assume that it is true for $k = m < n$ and prove it for $k = m + 1$ by assuming $c_1\mathbf{v}_1 + \cdots + c_{m+1}\mathbf{v}_{m+1} = \mathbf{0}$, multiplying by A on the left, and then eliminating $\mathbf{v}_{m+1}$ to obtain

$$c_1(\lambda_{m+1} - \lambda_1)\mathbf{v}_1 + \cdots + c_m(\lambda_{m+1} - \lambda_m)\mathbf{v}_m = \mathbf{0}.$$

Now use the induction hypothesis to conclude that $c_1 = 0, \ldots, c_m = 0$ and hence also $c_{m+1} = 0$.]

13. Let T be a linear mapping of V^n into V^m. Prove: If $\mathbf{v}_1, \ldots, \mathbf{v}_k$ are linearly dependent vectors of V^n, then $T(\mathbf{v}_1), \ldots, T(\mathbf{v}_k)$ are linearly dependent vectors of V^m. Is the converse true? Explain.

Answers

1. (a) $(2, 3)$, $(1, 5)$, $(3, 1)$, $(-1, 2)$. (b) Kernel is $\mathbf{0}$ alone; T is one-to-one, $T(\mathbf{x}) = (2, 3)$ only for $\mathbf{x} = (1, 0)$. (c) Range is V^2; T maps V^2 onto V^2.

2. (a) $(2, 4)$, $(3, 6)$, $(-1, -2)$, $(-5, -10)$. (b) Kernel is all $t(3, -2)$; T is not one-to-one; $T(\mathbf{x}) = (2, 4)$ for $\mathbf{x} = (1, 0) + t(3, -2)$, $-\infty < t < \infty$; (c) Range is all $t(1, 2)$; T does not map V^2 onto V^2.

3. (a) $n = 3$, $m = 2$; (b) all $t(0, 2, -1)$, not one-to-one. (c) Range is V^2; T maps V^3 onto V^2.

4. (a) $n = 3$, $m = 2$. (b) All $t_1(-4, 1, 0) + t_2(-3, 0, 1)$, not one-to-one. (c) Range is all $t(1, 2)$; T does not map V^3 onto V^2.

5. (a) $n = 2$, $m = 3$. (b) All $t(1, -2)$, not one-to-one. (c) Range is all $t(1, 2, -3)$; T does not map V^2 onto V^3.

6. (a) $n = 2$, $m = 3$. (b) $\mathbf{0}$ alone, one-to-one. (c) Range is all $t_1(2, 1, 1) + t_2(1, 2, 2)$; T does not map V^2 onto V^3.

7. (a) $n = 3$, $m = 3$. (b) All $t(1, -1, -1)$, not one-to-one. (c) Range is all $t_1(3, 1, 5) + t_2(1, 0, 2)$; T does not map V^3 onto V^3.

8. (a) $n = 3$, $m = 3$. (b) $\mathbf{0}$ alone, one-to-one. (c) Range is V^3; T maps V^3 onto V^3.

9. Angle is $\pi/4$, $|\mathbf{x}| = |T(\mathbf{x})|$. 10. Reflection in x-axis.

11. (a) Reflection in origin. (b) All vectors stretched in ratio 2 to 1.

***1–11 Linear manifolds in E^n.** We now want to define lines, planes and higher-dimensional analogues of these objects in E^n. We are motivated by the equation of a line in 3-dimensional space. As in Section 0–6,

$$\overrightarrow{OP} = \overrightarrow{OP_0} + t\mathbf{q}, \quad -\infty < t < \infty, \tag{1–64}$$

is the equation of a line through P_0 with velocity vector $\mathbf{q}$ (assumed nonzero). A plane in 3-dimensional space also has a vector equation:

$$\overrightarrow{OP} = \overrightarrow{OP_0} + t_1\mathbf{w}_1 + t_2\mathbf{w}_2, \quad -\infty < t_1 < \infty, \quad -\infty < t_2 < \infty \tag{1–65}$$

in terms of two parameters t_1 and t_2. Here $\mathbf{w}_1$ and $\mathbf{w}_2$ are two linearly independent vectors representable as $\overrightarrow{P_0P_1}$ and $\overrightarrow{P_0P_2}$ respectively, where P_0, P_1, P_2 are all in the plane (Fig. 1–4). The point P is in the plane precisely when $\overrightarrow{P_0P}$, $\overrightarrow{P_0P_1}$ and $\overrightarrow{P_0P_2}$ are coplanar—that is, when $\overrightarrow{P_0P}$ is expressible as a linear combination of $\mathbf{w}_1$ and $\mathbf{w}_2$: $\overrightarrow{P_0P} = t_1\mathbf{w}_1 + t_2\mathbf{w}_2$. If we replace $\overrightarrow{P_0P}$ by $\overrightarrow{OP} - \overrightarrow{OP_0}$ we obtain (1–65). As t_1 and t_2 vary over all real values, we obtain all points P in the plane; also each point P corresponds to a unique pair of parameter values t_1, t_2, since $\mathbf{w}_1$ and $\mathbf{w}_2$ are linearly independent.

We regard a line as a 1-dimensional linear manifold, a plane as a 2-dimensional linear manifold, and now make our general definition:

Definition. A k-dimensional linear manifold H in E^n is the set H of all points P such that

$$\begin{gathered}\overrightarrow{OP} = \overrightarrow{OP_0} + t_1\mathbf{w}_1 + \cdots + t_k\mathbf{w}_k, \\ -\infty < t_1 < \infty, \quad \ldots, \quad -\infty < t_k < \infty,\end{gathered} \tag{1–66}$$

where P_0 is a given point of E^n and $\mathbf{w}_1, \ldots, \mathbf{w}_k$ are k given linearly independent vectors of V^n. For $k = 1$, the linear manifold is called a *line;* for $k = 2$, the linear manifold is called a *plane*.

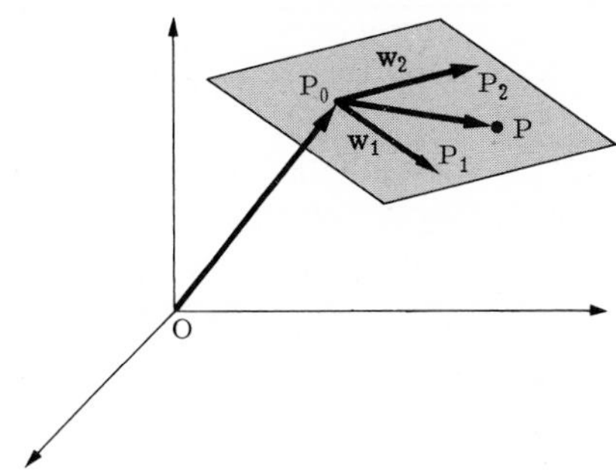

FIG. 1–4. Vector equation of a plane in space.

For $k = n$, $\mathbf{w}_1, \ldots, \mathbf{w}_n$ are a basis for V^n and hence the linear manifold coincides with the whole n-dimensional space E^n. For $k < n$, $\mathbf{w}_1, \ldots, \mathbf{w}_k$ cannot be a basis for V^n, by rule (i) of Section 1–8, so that the linear manifold is only part, not all, of E^n.

It is convenient to allow $k = 0$ and to consider the point P_0 itself (defined by the equation $\overrightarrow{OP} = \overrightarrow{OP_0}$) to be a *zero-dimensional linear manifold.*

In general, we call $\mathbf{w}_1, \ldots, \mathbf{w}_k$ in (1–66) a *basis* for the linear manifold H. There are many bases for a given linear manifold, but they all contain the same number, k, of vectors (see below). Hence we are justified in calling H k-dimensional.

The vector equation (1–66) can also be written in the form

$$\overrightarrow{P_0P} = t_1\mathbf{w}_1 + \cdots + t_k\mathbf{w}_k \tag{1–66$'$}$$

(as in the discussion of (1–65) above). If we equate components of vectors, (1–66′) becomes a set of n *parametric equations*

$$\begin{aligned} x_1 &= x_{10} + t_1w_{11} + \cdots + t_kw_{1k}, \\ &\vdots \\ x_n &= x_{n0} + t_1w_{n1} + \cdots + t_kw_{nk}. \end{aligned} \tag{1–66$''$}$$

These equations are in turn equivalent to the matrix equation

$$\mathbf{x} = \mathbf{x}_0 + W\mathbf{t}, \tag{1–66$'''$}$$

where $\mathbf{x} = \text{col}\,(x_1, \ldots, x_n)$, $\mathbf{x}_0 = \text{col}\,(x_{10}, \ldots, x_{n0})$, $W = (w_{ij})$, and $\mathbf{t} = \text{col}\,(t_1, \ldots, t_k)$.

EXAMPLE 1. The equations $x_1 = 3 + 2t$, $x_2 = 5 - t$, $x_3 = 4 + t$, $x_4 = 1 + 7t$ are parametric equations of a line in E^4. The equivalent vector equation is $\overrightarrow{OP} = (3, 5, 4, 1) + t(2, -1, 1, 7)$ or, more concisely,

$$\overrightarrow{P_0P} = t(2, -1, 1, 7),$$

where P_0 is $(3, 5, 4, 1)$.

EXAMPLE 2. In E^n, $\overrightarrow{OP} = t\mathbf{e}_1$ is the vector equation of a line L. The points of L are all points $(t, 0, \ldots, 0)$ for $-\infty < t < \infty$. We call L the x_1-axis of E^n. In the same way, $\overrightarrow{OP} = t\mathbf{e}_2$ represents the x_2-axis, and so on.

EXAMPLE 3. In E^n, $\overrightarrow{OP} = t_1\mathbf{e}_1 + t_2\mathbf{e}_2$ is the vector equation of a plane H through O: $(0, \ldots, 0)$. The points of H are just the points $(x_1, x_2, 0, \ldots, 0)$ of E^n. We call H the x_1x_2-coordinate plane of E^n. We observe that, as a geometric object, H is the same as the x_1x_2-plane of plane analytic geometry; that is, H is the same as E^2. For example, the distance between the points $(x_1', x_2', 0, \ldots, 0)$ and $(x_1'', x_2'', 0, \ldots, 0)$ of H is

$$[(x_1'' - x_1')^2 + (x_2'' - x_2')^2]^{\frac{1}{2}}.$$

The zero coordinates can be ignored. In a similar fashion, we define the x_hx_j-coordinate plane of E^n by the equation $\overrightarrow{OP} = t_1\mathbf{e}_h + t_2\mathbf{e}_j \quad (h < j)$.

Generalizing Examples 2 and 3, we see that the equation

$$\overrightarrow{OP} = t_1\mathbf{e}_1 + \cdots + t_k\mathbf{e}_k$$

is the equation of the $x_1 \ldots x_k$-coordinate linear manifold, formed of all points $(x_1, x_2, \ldots, x_k, 0, \ldots, 0)$ and that, as a geometric object, this k-dimensional linear manifold is the same as E^k.

Example 4. In E^4, the equations $x_1 = 1 + t_1$, $x_2 = 3 + t_2$, $x_3 = 5$, $x_4 = 2$ are parametric equations of a plane H. The equations can be written in the vector form

$$\overrightarrow{P_0P} = t_1(1, 0, 0, 0) + t_2(0, 1, 0, 0) = t_1\mathbf{e}_1 + t_2\mathbf{e}_2,$$

where P_0 is the point $(1, 3, 5, 2)$. We can take P_0 as a new origin and introduce new axes parallel to the old ones through P_0; that is, we introduce new coordinates (x_1', x_2', x_3', x_4') by writing

$$\overrightarrow{P_0P} = x_1'\mathbf{e}_1 + x_2'\mathbf{e}_2 + x_3'\mathbf{e}_3 + x_4'\mathbf{e}_4.$$

The equation $|\overrightarrow{P_1P_2}| = |\overrightarrow{P_0P_2} - \overrightarrow{P_0P_1}|$ then shows that distance can be calculated as usual in terms of the new coordinates, just as with the old ones. In the new coordinates H is the set of all points $(x_1', x_2', 0, 0)$; that is, it is the $x_1'x_2'$-plane.

In general, in E^n, $\overrightarrow{P_0P} = t_1\mathbf{e}_1 + \cdots + t_k\mathbf{e}_k$ is the vector equation of a k-dimensional linear manifold H. If we shift the origin to P_0 and define new coordinates by the equation $\overrightarrow{P_0P} = x_1'\mathbf{e}_1 + \cdots + x_n'\mathbf{e}_n$, then H is the set of points $(x_1', \ldots, x_k', 0, \ldots, 0)$. Hence H is a k-dimensional coordinate linear manifold.

Example 5. In E^4, the equation $\overrightarrow{OP} = t_1(1, 2, 1, 5) + t_2(3, -1, -6, 1)$ represents a plane H. The vectors

$$\mathbf{w}_1 = (1, 2, 1, 5) \quad \text{and} \quad \mathbf{w}_2 = (3, -1, -6, 1)$$

happen to be orthogonal: $\mathbf{w}_1 \cdot \mathbf{w}_2 = 3 - 2 - 6 + 5 = 0$. Now by rule (1) of Section 1–8 we can choose $\mathbf{w}_3$ and $\mathbf{w}_4$ so that $\mathbf{w}_1, \ldots, \mathbf{w}_4$ form an orthogonal system. We normalize these vectors by dividing by their lengths to obtain an orthonormal basis $\mathbf{e}_1', \ldots, \mathbf{e}_4'$. We then introduce new coordinates in E^4 by the equation $\overrightarrow{OP} = x_1'\mathbf{e}_1' + \cdots + x_4'\mathbf{e}_4'$, and again verify that that distance can be calculated in terms of the new coordinates in the usual way. In the new coordinates, H is formed all of points $(x_1', x_2', 0, 0)$ and is again a coordinate plane.

Generalizing all these examples, we see that every k-dimensional linear manifold H in E^n, for proper choice of coordinates, becomes a coordinate linear manifold and that, as a geometric object, H is the same as k-dimensional Euclidean space E^k. It follows that *every theorem of plane analytic geometry applies to every plane in* E^n and that *every theorem of solid analytic geometry applies to every 3-dimensional linear manifold in* E^n.

EXAMPLE 6. Represent the plane H: $2x_1 + x_2 - x_3 = 5$ in E^3 by parametric equations and by a vector equation.

Solution. We can write $x_1 = t_1$, $x_2 = t_2$ and then find $x_3 = 2t_1 + t_2 - 5$. These are parametric equations for H. The corresponding vector equation is $\overrightarrow{OP} = (0, 0, -5) + t_1(1, 0, 2) + t_2(0, 1, 1)$. We remark that

$$\mathbf{w}_1 = (1, 0, 2) \quad \text{and} \quad \mathbf{w}_2 = (0, 1, 1)$$

are linearly independent as required.

EXAMPLE 7. Show that $x_1 + x_2 + 2x_3 - x_4 = 2$ is the equation of a 3-dimensional linear manifold H in E_4.

Solution. We set $x_1 = t_1$, $x_2 = t_2$, $x_3 = t_3$ and find $x_4 = t_1 + t_2 + 2t_3 - 2$, so that we obtain the vector equation $\overrightarrow{P_0P} = t_1\mathbf{w}_1 + t_2\mathbf{w}_2 + t_3\mathbf{w}_3$, where P_0 is $(0, 0, 0, -2)$, $\mathbf{w}_1 = (1, 0, 0, 1)$, $\mathbf{w}_2 = (0, 1, 0, 1)$, $\mathbf{w}_3 = (0, 0, 1, 2)$. For every choice of t_1, t_2, t_3 the corresponding point P: (x_1, x_2, x_3, x_4) satisfies the given linear equation and every point satisfying the given equation corresponds to a unique set of parameter values t_1, t_2, t_3. Also the vectors $\mathbf{w}_1$, $\mathbf{w}_2$, $\mathbf{w}_3$ are linearly independent. For

$$c_1(1, 0, 0, 1) + c_2(0, 1, 0, 1) + c_3(0, 0, 1, 2) = \mathbf{0} = (0, 0, 0, 0)$$

implies $c_1 = 0$, $c_2 = 0$, $c_3 = 0$, $c_1 + c_2 + 2c_3 = 0$, so that

$$c_1 = c_2 = c_3 = 0.$$

Hence the given equation represents a 3-dimensional linear manifold H in E^4.

Generalizing Examples 6 and 7, we see that every linear equation

$$b_1x_1 + \cdots + b_nx_n = h \quad (\text{not all } b_i = 0)$$

describes an $(n - 1)$-dimensional linear manifold H in E^n.

We know that every line in 3-dimensional space can be represented as the intersection of two planes, hence as the set of common solutions of *two* linear equations. What can we say about the solutions of two such equations in E^n?

EXAMPLE 8. Show that the equations

$$2x_1 + 4x_2 - x_3 + x_4 = 2, \quad 4x_1 - 2x_2 + x_3 + x_4 = 0$$

represent a plane in E^4.

Solution. We set $x_1 = t_1$, $x_2 = t_2$ and solve for x_3, x_4. To this end, we first write our equations as follows:

$$\begin{aligned} -x_3 + x_4 &= 2 - 2t_1 - 4t_2, \\ x_3 + x_4 &= -4t_1 + 2t_2. \end{aligned}$$

Thus we have two simultaneous linear equations for x_3, x_4. Since the determinant of coefficients is not 0, we can solve. We find easily:

$$x_3 = -1 - t_1 + 3t_2, \quad x_4 = 1 - 3t_1 - t_2.$$

As for Example 7, we find that the points $(x_1, \ldots, x_4)$ satisfying the given equation are the points of the 2-dimensional linear manifold

$$\overrightarrow{OP} = (0, 0, -1, 1) + t_1(1, 0, -1, -3) + t_2(0, 1, 3, -1).$$

Generalizing the last example, we expect that "in general" two simultaneous linear equations in $x_1, \ldots, x_n$ should represent an $(n - 2)$-dimensional linear manifold in E^n and that "in general," for $k \leqq n$, k simultaneous linear equations in $x_1, \ldots, x_n$ should represent an $(n - k)$-dimensional linear manifold in E^n. These assertions will be considered in detail in Section 1–13 below. Also it will be shown in that section that every k-dimensional linear manifold in E^n can be represented as the set of common solutions of $n - k$ linear equations in $x_1, \ldots, x_n$.

Remarks. The vector equation of a k-dimensional linear manifold H can be written in many different ways. If

$$\overrightarrow{P_0P} = t_1\mathbf{w}_1 + \cdots + t_k\mathbf{w}_k$$

is one equation for H, and P_1 is any point of H, then one can replace P_0 by P_1. For since P_1 is in H, we have $\overrightarrow{P_0P_1} = c_1\mathbf{w}_1 + \cdots + c_k\mathbf{w}_k$ for some choice of $c_1, \ldots, c_k$. But $\overrightarrow{P_0P} = \overrightarrow{P_0P_1} + \overrightarrow{P_1P}$ and hence the given equation of H can be written:

$$\overrightarrow{P_1P} = (t_1 - c_1)\mathbf{w}_1 + \cdots + (t_k - c_k)\mathbf{w}_k$$

or, in terms of new parameters $\tau_1 = t_1 - c_1, \ldots, \tau_k = t_k - c_k$,

$$\overrightarrow{P_1P} = \tau_1\mathbf{w}_1 + \cdots + \tau_k\mathbf{w}_k, \quad -\infty < \tau_1 < \infty, \quad \ldots, \quad -\infty < \tau_k < \infty.$$

Thus we have replaced P_0 by P_1 in the equation.

We can also replace $\mathbf{w}_1, \ldots, \mathbf{w}_k$ by another basis for H. Each of the vectors $\mathbf{w}_1, \ldots, \mathbf{w}_k$ can be regarded as a vector "in H": that is, as a vector joining two points of H. For example, the point P_1 of H obtained by setting $t_1 = 1$, $t_2 = 0, \ldots, t_k = 0$ in the equation of H is such that $\overrightarrow{P_0P_1} = 1\mathbf{w}_1 + \mathbf{0} + \cdots + \mathbf{0} = \mathbf{w}_1$. Similarly, $\mathbf{w}_2 = \overrightarrow{P_0P_2}, \ldots, \mathbf{w}_k = \overrightarrow{P_0P_k}$. Thus changing the basis for H is equivalent to replacing $P_1, \ldots, P_k$ by new points $Q_1, Q_2, \ldots, Q_m$ such that every vector $\overrightarrow{P_0P}$ in H is expressible uniquely as a linear combination of $\mathbf{u}_1 = \overrightarrow{P_0Q_1}$, $\mathbf{u}_2 = \overrightarrow{P_0Q_2}, \ldots, \mathbf{u}_m = \overrightarrow{P_0Q_m}$. The number m must equal k. For the vectors $\mathbf{u}_1, \ldots, \mathbf{u}_m$ must be linearly independent and all are expressible as linear combinations of $\mathbf{w}_1, \ldots, \mathbf{w}_k$. By rule (k) of Section 1–8, $m \leqq k$. But $\mathbf{w}_1, \ldots, \mathbf{w}_k$ are also linearly independent and all are expressible as linear combinations of $\mathbf{u}_1, \ldots, \mathbf{u}_m$. Therefore for the same reason, $k \leqq m$, and we conclude that $k = m$. Thus we are fully justified in calling H k-dimensional.

Problems

1. Verify that each of the following is the vector equation of a k-dimensional linear manifold H in E^n, give the values of k and n, and determine whether the points P_1, P_2 are in H.

(a) $\overrightarrow{OP} = (3, 5, 4, 1) + t(2, -1, 1, 3)$, P_1: $(3, 5, 4, 1)$, P_2: $(5, 4, 5, 4)$.
(b) $\overrightarrow{OP} = (1, 0, 2, 2, 1) + t(3, 5, 7, 0, 4)$, P_1: $(4, 5, 9, 2, 5)$, P_2: $(0, \dots, 0)$.
(c) $\overrightarrow{OP} = (3, 1, 2, 1) + t_1(0, 0, 0, 1) + t_2(0, 0, 1, 1)$, P_1: $(3, 1, 3, 3)$, P_2: $(4, 1, 0, 1)$.
(d) $\overrightarrow{OP} = (1, 2, 3, 4, 5, 6) + t_1(1, 0, 1, 0, 1, 0) + t_2(0, 1, 0, 1, 0, 1) + t_3(1, 1, 1, 1, 1, 0)$, P_1: $(1, 2, 3, 4, 5, 6)$ and P_2: $(3, 2, 5, 4, 7, 7)$.
(e) $\overrightarrow{OP} = t_1(1, 0, 0, 0, 0) + t_2(0, 0, 1, 0, 0)$, P_1: $(0, 1, 0, 0, 0)$, P_2: $(2, 0, 3, 0, 0)$.
(f) $\overrightarrow{OP} = t_1(0, 0, 0, 0, 0, 1) + t_2(0, 0, 0, 0, 1, 0) + t_3(0, 0, 0, 1, 0, 0)$, P_1: $(0, 0, 0, 7, 4, 0)$, P_2: $(0, 0, 0, 0, 0, 0)$.

2. Show that each of the following equations is the equation of an $(n - 1)$-dimensional linear manifold H in E^n and obtain a vector equation for H.

(a) $x_1 + x_2 = 1$ in E^2.
(b) $2x_1 - 5x_2 - x_3 = 7$ in E^3.
(c) $x_1 + x_2 - x_3 - x_4 = 4$ in E^4.
(d) $x_1 + 2x_2 + x_3 + 2x_4 = 5$ in E^4.
(e) $2x_1 + x_2 - 3x_3 - x_4 + x_5 = 2$ in E^5.
(f) $x_1 - x_2 + x_3 - x_4 + x_5 - x_6 = 1$ in E^6.
(g) $x_1 + x_2 = 0$ in E^4.
(h) $x_1 - x_2 + x_3 = 0$ in E^5.
(i) $x_1 = 0$ in E^7.
(j) $x_2 = 0$ in E^{20}.

3. Show that the set of solutions of each of the following sets of simultaneous linear equations forms a k-dimensional linear manifold H in E^n and obtain a vector equation for H.

(a) $x_1 + x_2 + x_3 = 1$, $2x_1 + x_2 - x_3 = 1$ in E^3.
(b) $x_1 + 4x_2 + x_3 = 6$, $3x_1 - x_2 - x_3 = 1$ in E^3.
(c) $2x_1 + x_2 - x_3 + x_4 = 0$, $x_1 + x_2 + x_3 + 2x_4 = 0$ in E^4.
(d) $x_1 - x_2 + x_3 + x_4 = 1$, $2x_1 + x_2 + 2x_3 + x_4 = 0$ in E^4.
(e) $x_1 - x_2 = 1$, $x_2 - x_3 = 0$, $x_3 - x_4 = 1$ in E^4.
(f) $x_1 + x_2 + x_3 = 1$, $x_2 + x_3 + x_4 = 1$, $x_3 + x_4 + x_5 = 1$ in E^5.

4. (a) Show that the *three* equations

$$x_1 + x_2 + x_3 - x_4 = 2, \quad x_1 - x_2 - x_3 - x_4 = 0, \quad x_1 + 5x_2 + 5x_3 - x_4 = 6$$

represent a *plane* in E^4. This is an exception to the "general" rule: k linear equations in E^n describe an $(n - k)$-dimensional linear manifold. Explain.

(b) Give an example of three equations in E^6 representing a 4-dimensional linear manifold.

5. Let H be a k-dimensional linear manifold in E^n with basis $\mathbf{w}_1, \dots, \mathbf{w}_k$.

(a) Prove: if $\mathbf{u}_1, \dots, \mathbf{u}_k$ are linearly independent vectors, each of which is a linear combination of $\mathbf{w}_1, \dots, \mathbf{w}_k$, then $\mathbf{u}_1, \dots, \mathbf{u}_k$ is a basis for H.

Remark. The converse of the statement in (a) is proved in the text above.

(b) Prove: H has an orthogonal basis $\mathbf{u}_1, \ldots, \mathbf{u}_k$ and hence new coordinates can be chosen in E^n so that H becomes the $x'_1 \ldots x'_k$-coordinate linear manifold.

6. Prove the following geometric theorems:

(a) In E^n there is one and only one line through two given distinct points.

(b) In E^n, if H is a k-dimensional linear manifold, with $k = 2$, and P_1, P_2 are points of H, then the line through P_1 and P_2 lies in H.

(c) In E^n, with $n \geqq 2$, there is one and only one plane through three given points which are not collinear.

Answers

1. (a) $k = 1$, $n = 4$, P_1 and P_2 in H. (b) $k = 1$, $n = 5$, P_1 in H, P_2 not. (c) $k = 2$, $n = 4$, P_1 in H, P_2 not. (d) $k = 3$, $n = 6$, P_1 and P_2 in H. (e) $k = 2$, $n = 5$, P_1 not in H, P_2 in. (f) $k = 3$, $n = 6$, P_1 and P_2 in H.

2. (a) $\overrightarrow{OP} = (0, 1) + t(1, -1)$.
(b) $\overrightarrow{OP} = (0, 0, -7) + t_1(1, 0, 2) + t_2(0, 1, -5)$.
(c) $\overrightarrow{OP} = (0, 0, 0, -4) + t_1(1, 0, 0, 1) + t_2(0, 1, 0, 1) + t_3(0, 0, 1, -1)$.
(d) $\overrightarrow{OP} = 5\mathbf{e}_1 + t_1(\mathbf{e}_2 - 2\mathbf{e}_1) + t_2(\mathbf{e}_3 - \mathbf{e}_1) + t_3(\mathbf{e}_4 - 2\mathbf{e}_1)$.
(e) $\overrightarrow{OP} = 2\mathbf{e}_5 + t_1(\mathbf{e}_1 - 2\mathbf{e}_5) + t_2(\mathbf{e}_2 - \mathbf{e}_5) + t_3(\mathbf{e}_3 + 3\mathbf{e}_5) + t_4(\mathbf{e}_4 + \mathbf{e}_5)$.
(f) $\overrightarrow{OP} = -\mathbf{e}_6 + t_1(\mathbf{e}_1 + \mathbf{e}_6) + t_2(\mathbf{e}_2 - \mathbf{e}_6) + t_3(\mathbf{e}_3 + \mathbf{e}_6) + t_4(\mathbf{e}_4 - \mathbf{e}_6) + t_5(\mathbf{e}_5 + \mathbf{e}_6)$.
(g) $\overrightarrow{OP} = t_1(\mathbf{e}_2 - \mathbf{e}_1) + t_2\mathbf{e}_3 + t_3\mathbf{e}_4$.
(h) $\overrightarrow{OP} = t_1(\mathbf{e}_1 + \mathbf{e}_2) + t_2(\mathbf{e}_3 - \mathbf{e}_1) + t_3\mathbf{e}_4 + t_4\mathbf{e}_5$.
(i) $\overrightarrow{OP} = \sum_{i=2}^{7} t_i\mathbf{e}_i$. (j) $\overrightarrow{OP} = t_1\mathbf{e}_1 + \sum_{i=3}^{20} t_i\mathbf{e}_i$.

3. (a) $\overrightarrow{OP} = (0, 1, 0) + (t/2)(2, -3, 1)$.
(b) $\overrightarrow{OP} = (0, \frac{7}{3}, -\frac{10}{3}) + (t/3)(3, -4, 13)$.
(c) $\overrightarrow{OP} = t_1(1, 0, 1, -1) + (t_2/3)(0, 3, 1, -2)$.
(d) $\overrightarrow{OP} = (0, 0, -1, 2) + t_1(1, 0, -1, 0) + t_2(0, 1, -2, 3)$.
(e) $\overrightarrow{OP} = (0, -1, -1, -2) + t(1, 1, 1, 1)$.
(f) $\overrightarrow{OP} = \mathbf{e}_3 + t_1(\mathbf{e}_1 - \mathbf{e}_3 + \mathbf{e}_4) + t_2(\mathbf{e}_2 - \mathbf{e}_3 + \mathbf{e}_5)$.

***1–12 Rank of a matrix or linear mapping.** Now let A be an $m \times n$ matrix, whose column vectors are $\mathbf{u}_1, \ldots, \mathbf{u}_n$, so that

$$A\mathbf{e}_1 = \mathbf{u}_1, \ldots, A\mathbf{e}_n = \mathbf{u}_n.$$

The vectors $\mathbf{u}_1, \ldots, \mathbf{u}_n$ are n vectors in V^m and may or may not be linearly independent; they are surely dependent if $n > m$. However, in any case we can let r be the largest integer such that r of the vectors $\mathbf{u}_1, \ldots, \mathbf{u}_n$ are linearly independent; we call this integer r the *rank* of the matrix A, or of the linear mapping T whose matrix is A. If all of $\mathbf{u}_1, \ldots, \mathbf{u}_n$ are $\mathbf{0}$, then we set $r = 0$; in this case $A = O$. Otherwise, $1 \leqq r \leqq n$, and r cannot exceed m.

EXAMPLE 1. Consider the matrices

$$A = \begin{bmatrix} 1 & 3 \\ 2 & 5 \end{bmatrix}, \quad B = \begin{bmatrix} 2 & 4 \\ 6 & 12 \end{bmatrix}, \quad C = \begin{bmatrix} 2 & 0 \\ 1 & 0 \end{bmatrix}, \quad D = \begin{bmatrix} 2 & 3 & 5 \\ 1 & 2 & 3 \end{bmatrix},$$

$$E = \begin{bmatrix} 1 & 2 \\ 3 & 1 \\ 4 & 6 \end{bmatrix}, \quad F = \begin{bmatrix} 1 & 0 & 2 \\ 3 & 1 & 3 \\ 5 & 2 & 4 \end{bmatrix}.$$

For A we see that $\mathbf{u}_1 = (1, 2)$, $\mathbf{u}_2 = (3, 5)$ are linearly independent, so that A has rank 2. For B, $\mathbf{u}_1 = (2, 6)$, $\mathbf{u}_2 = (4, 12)$, so that $\mathbf{u}_2 = 2\mathbf{u}_1$, and B has rank 1. For C, $\mathbf{u}_2 = (0, 0) = 0\mathbf{u}_1$, so that C has rank 1. For D, $\mathbf{u}_1$ and $\mathbf{u}_2$ are linearly independent, so that without considering $\mathbf{u}_3$, we see that the rank must be 2, for in V^2 there can be no more than two linearly independent vectors; in fact, here $\mathbf{u}_3 = \mathbf{u}_1 + \mathbf{u}_2$. For E, we see that $\mathbf{u}_1 = (1, 3, 4)$ and $\mathbf{u}_2 = (2, 1, 6)$ are linearly independent, so that the rank is 2.

For F we have $\mathbf{u}_1 = (1, 3, 5)$, $\mathbf{u}_2 = (0, 1, 2)$, $\mathbf{u}_3 = (2, 3, 4)$, and hence $\mathbf{u}_1$, $\mathbf{u}_2$ are linearly independent. However, it is not immediately obvious whether $\mathbf{u}_1, \mathbf{u}_2, \mathbf{u}_3$ are linearly independent. To settle this, we seek scalars c_1, c_2 such that $\mathbf{u}_3 = c_1\mathbf{u}_1 + c_2\mathbf{u}_2$ and are led to the equations

$$2 = c_1 + 0c_2, \quad 3 = 3c_1 + c_2, \quad 4 = 5c_1 + 2c_1.$$

These are all satisfied for $c_1 = 2$ and $c_2 = -3$, so that $\mathbf{u}_3 = 2\mathbf{u}_1 - 3\mathbf{u}_2$. Therefore, $\mathbf{u}_1, \mathbf{u}_2, \mathbf{u}_3$ are linearly dependent and F has rank 2, not 3.

The methods of the example, especially that for F, can be used generally to find the rank of a matrix. Furthermore, one has a useful criterion based on determinants. For a given matrix A, we define a *minor* of A to be the *determinant* of a square array obtained from A by deleting certain rows and columns of A (or, if A is square, by possibly deleting nothing). For example, the matrix E of Example 1 has the following minors of order 2:

$$\begin{vmatrix} 1 & 2 \\ 3 & 1 \end{vmatrix} = -5, \quad \begin{vmatrix} 3 & 1 \\ 4 & 6 \end{vmatrix} = 14, \quad \begin{vmatrix} 1 & 2 \\ 4 & 6 \end{vmatrix} = -2$$

and E has 6 minors of order 1, consisting of the individual entries of E. We can now state:

The rank r of a nonzero matrix always equals the largest integer p such that A has a minor of order p not equal to zero.

For example, matrix E has nonzero minors of order 2 (and E has no minors of order greater than 2), so that E has rank 2. For the matrix F of Example 1, the minors are

$$\begin{vmatrix} 1 & 0 & 2 \\ 3 & 1 & 3 \\ 5 & 2 & 4 \end{vmatrix}, \quad \begin{vmatrix} 1 & 0 \\ 3 & 1 \end{vmatrix}, \quad \begin{vmatrix} 1 & 2 \\ 3 & 3 \end{vmatrix}, \quad \ldots$$

(There are 19 minors in all.) The first we find to be equal to 0, the second equal to 1. We conclude that F has rank 2.

For a proof of the rule on minors, see *CLA* Chapter 10, Section 10–14. From the rule, we deduce another useful way of finding the rank:

The rank r of a nonzero matrix A equals the largest integer ρ *such that A has* ρ *linearly independent row vectors.*

To prove the assertion we let $B = A'$, the transpose of A (Section 1–6), so that the row vectors of A are the column vectors of B. The rank of B can be computed by minors as above. But the minors of B equal the corresponding minors of A, since the value of a determinant is unaffected if we interchange rows and columns. Hence the rank of B equals the rank of A, and therefore the maximal number of linearly independent column vectors of B equals the rank of A, but these column vectors are the row vectors of A, and the assertion follows.

As an illustration, we consider the matrix E of Example 1. Here the row vectors are the vectors (1, 2), (3, 1) and (4, 6) of V^2. The first two are linearly independent. Therefore, the rank is 2, as above.

Now let A be an $m \times n$ matrix and let us consider the linear mapping $\mathbf{y} = A\mathbf{x}$. We consider $(x_1, \ldots, x_n)$ and $(y_1, \ldots, y_m)$ as *points* in E^n and E^m respectively, so that the mapping A assigns a point of E^m to each point of E^n. The *range* of the mapping A is then a set of points in E^m. We can also consider the *kernel* of A: this is the set of all $\mathbf{x}$ for which $A\mathbf{x} = \mathbf{0}$, considered as a set of points in E^n. We can now state some important properties of such a mapping:

Theorem. *Let A be an* $m \times n$ *matrix of rank r and consider A as defining a mapping* $\mathbf{y} = A\mathbf{x}$ *of* E^n *into* E^m. *Then the range of A is an r-dimensional linear manifold in* E^m, *containing the origin of* E^m. *The kernel of A is a k-dimensional linear manifold in* E^n, *containing the origin of* E^n. *Furthermore,*

$$k + r = n.$$

Proof. If A is the zero matrix, then $r = 0$, the range of A is the origin of E^m (a 0-dimensional linear manifold), and the kernel of A is all of E^n (an n-dimensional linear manifold); thus $r = 0$, $k = n$ and the assertions are correct for this case.

Now let A be a nonzero matrix and let $\mathbf{u}_1, \ldots, \mathbf{u}_n$ be the successive column vectors of A. Then we know that r, but no more than r, of these vectors are linearly independent. Let us suppose, for example, that $\mathbf{u}_1, \ldots, \mathbf{u}_r$ are linearly independent. Then, by rule (c) of Section 1–8, each of $\mathbf{u}_{r+1}, \ldots, \mathbf{u}_n$ must be expressible as a linear combination of $\mathbf{u}_1, \ldots, \mathbf{u}_r$. Now the range of A consists of all $\mathbf{y} = x_1\mathbf{u}_1 + \cdots + x_n\mathbf{u}_n$—that is, of all linear combinations of $\mathbf{u}_1, \ldots, \mathbf{u}_n$. But all such linear combinations can be expressed as linear combinations of $\mathbf{u}_1, \ldots, \mathbf{u}_r$ alone. Also every linear combination of $\mathbf{u}_1, \ldots, \mathbf{u}_r$ is in the range:

$$x_1\mathbf{u}_1 + \cdots + x_r\mathbf{u}_r = A\mathbf{x} \quad \text{for} \quad \mathbf{x} = \text{col}\,(x_1, \ldots, x_r, 0, \ldots, 0).$$

Therefore, the range consists precisely of all $\mathbf{y} = t_1\mathbf{u}_1 + \cdots + t_r\mathbf{u}_r$. If we now interpret $\mathbf{y}$ as $\overrightarrow{OP}$, the range is simply the r-dimensional linear manifold $\overrightarrow{OP} = t_1\mathbf{u}_1 + \cdots + t_r\mathbf{u}_r$ in E^m.

To analyze the kernel, we first express $\mathbf{u}_{r+1}, \ldots, \mathbf{u}_n$ in terms of $\mathbf{u}_1, \ldots, \mathbf{u}_r$:

$$\mathbf{u}_{r+1} = b_{11}\mathbf{u}_1 + \cdots + b_{1r}\mathbf{u}_r, \quad \ldots, \quad \mathbf{u}_n = b_{k1}\mathbf{u}_1 + \cdots + b_{kr}\mathbf{u}_r,$$

where $k = n - r$. We then set

$$\mathbf{z}_1 = b_{11}\mathbf{e}_1 + \cdots + b_{1r}\mathbf{e}_r - \mathbf{e}_{r+1}, \quad \ldots, \quad \mathbf{z}_k = b_{k1}\mathbf{e}_1 + \cdots + b_{kr}\mathbf{e}_r - \mathbf{e}_n, \tag{1–67}$$

where $\mathbf{e}_1, \ldots, \mathbf{e}_n$ is the standard basis of E^n. Then

$$\begin{aligned} A\mathbf{z}_1 &= b_{11}A\mathbf{e}_1 + \cdots + b_{1r}A\mathbf{e}_r - A\mathbf{e}_{r+1} \\ &= b_{11}\mathbf{u}_1 + \cdots + b_{1r}\,\mathbf{u}_r - \mathbf{u}_{r+1} = \mathbf{0} \end{aligned}$$

and similarly $A\mathbf{z}_2 = \mathbf{0}, \ldots, A\mathbf{z}_k = \mathbf{0}$. Thus $\mathbf{z}_1, \ldots, \mathbf{z}_k$ are all in the kernel of A. Furthermore, $\mathbf{e}_1, \ldots, \mathbf{e}_r, \mathbf{z}_1, \ldots, \mathbf{z}_k$ are together a set of n vectors which form a basis for V^n. For these vectors are linearly independent: if

$$c_1\mathbf{e}_1 + \cdots + c_r\mathbf{e}_r + c_{r+1}\mathbf{z}_1 + \cdots + c_n\mathbf{z}_k = \mathbf{0}, \tag{1–68}$$

then by (1–67) we would obtain a linear combination of $\mathbf{e}_1, \ldots, \mathbf{e}_n$ equal to $\mathbf{0}$; here the coefficient of $\mathbf{e}_{r+1}$ is $-c_{r+1}, \ldots,$ of $\mathbf{e}_n$ is $-c_n$. Hence these coefficients must be 0 and hence, by (1–68), $c_1, \ldots, c_r$ are also 0, and the linear independence is proved. Since we have n linearly independent vectors, we have a basis for V^n. Thus an arbitrary $\mathbf{x}$ of V^n can be written as

$$\mathbf{x} = d_1\mathbf{e}_1 + \cdots + d_r\mathbf{e}_r + t_1\mathbf{z}_1 + \cdots + t_k\mathbf{z}_k. \tag{1–69}$$

Then $A\mathbf{x} = d_1\mathbf{u}_1 + \cdots + d_r\mathbf{u}_r$, since $\mathbf{z}_1, \ldots, \mathbf{z}_k$ are in the kernel of A. Accordingly $A\mathbf{x} = \mathbf{0}$—that is, $\mathbf{x}$ is in the kernel of A—precisely when $d_1 = 0, \ldots, d_r = 0$; that is, precisely when

$$\mathbf{x} = \overrightarrow{OP} = t_1\mathbf{z}_1 + \cdots + t_k\mathbf{z}_k.$$

Therefore the kernel of A is a k-dimensional linear manifold through the origin, where $k = n - r$. The vectors $\mathbf{z}_1, \ldots, \mathbf{z}_k$ form a basis for this linear manifold.

Remark. If $r = n$, then (1–69) becomes $\mathbf{x} = x_1\mathbf{e}_1 + \cdots + x_n\mathbf{e}_n$ and $A\mathbf{x} = x_1\mathbf{u}_1 + \cdots + x_n\mathbf{u}_n = \mathbf{0}$ if and only if $\mathbf{x} = \mathbf{0}$. Thus $k = 0$ and the kernel of A is the 0-dimensional linear manifold consisting of the origin of E^n alone.

COROLLARY 1. *Let A be an $m \times n$ matrix. Let $(A \quad \mathbf{y})$ denote the matrix obtained from A by adjoining* col $(y_1, \ldots, y_m)$ *as an $(n+1)$-st column:*

$$(A \quad \mathbf{y}) = \begin{bmatrix} a_{11} & \cdots & a_{1n} & y_1 \\ & & \vdots & \\ a_{m1} & \cdots & a_{mn} & y_m \end{bmatrix}$$

Then $\mathbf{y}$ is in the range of A if and only if A and $(A \quad \mathbf{y})$ have the same rank.

For as in the proof of the theorem, $\mathbf{y}$ is in the range of A precisely when $\mathbf{y}$ is expressible as $c_1\mathbf{u}_1 + \cdots + c_r\mathbf{u}_r$; that is, precisely when $\mathbf{u}_1, \ldots, \mathbf{u}_r, \mathbf{y}$ are linearly dependent, or precisely when A and $(A \quad \mathbf{y})$ both have rank r.

COROLLARY 2. *There exists a one-to-one linear mapping of V^n onto V^m if and only if $m = n$.*

Proof. If $m = n$, then I is a one-to-one linear mapping of V^n onto V^m. Conversely, if a one-to-one linear mapping A of V^n onto V^m exists, then A must have a kernel consisting of $\mathbf{0}$ alone, so that $k = 0$ and hence, by the theorem r must equal n; but if the range is to be all of V^m, then r must also equal m. Hence $m = n$.

Nullity, geometric meaning of rank and nullity. The integer k in the theorem is called the *nullity* of A, or of the linear mapping T whose matrix is A. From the theorem, $k = n - r$, where n is the number of columns of A and r is the rank of A. We have defined the nullity and rank of the linear mapping T by referring to the matrix A representing T. However, the theorem shows that the rank and nullity have a geometric meaning, when we consider T as a mapping of E^n into E^m: the rank, r, of T is the dimension of the range of T; the nullity, k, of T is the dimension of the kernel of T. Since the range and kernel are linear manifolds, these dimensions are well defined. For example, the range of T might be a plane, the kernel of T a line; then $r = 2$ and $k = 1$.

Maximum rank. We have seen that the rank r of an $m \times n$ matrix A is at most equal to m and at most equal to n. Or we can say, r is at most equal to the smaller of m and n. When A has the largest rank allowed by these conditions, we say that A has *maximum rank*. Thus if A has maximum rank and $m \leqq n$, then $r = m$; if $n \leqq m$, then $r = n$. The matrix A of Example 1 has maximum rank, namely 2; but B has less than maximum rank, namely 1 (2 allowed).

*1–13 Applications of rank.

Let m linear equations in n unknowns $x_1, \ldots, x_n$ be given:

$$\begin{aligned} a_{11}x_1 + \cdots + a_{1n}x_n &= y_1, \\ &\vdots \\ a_{m1}x_1 + \cdots + a_{mn}x_n &= y_m. \end{aligned} \tag{1–70}$$

The properties of the solutions of such equations correspond to statements about linear mappings, matrices and linear manifolds. We explore these relationships for various cases. We let $A = (a_{ij})$ be the matrix of coefficients and let A have rank r and nullity $k = n - r$.

Case I. $n = m$. Here we have n equations in n unknowns and A is a square matrix. If A has maximum rank n, then $\det A \neq 0$ and A is nonsingular. Hence for each $\mathbf{y}$, the equation $A\mathbf{x} = \mathbf{y}$ has a unique solution for $\mathbf{x}$: $\mathbf{x} = A^{-1}\mathbf{y}$. Thus the linear mapping A is one-to-one and maps V^n onto V^n. The equation $\mathbf{x} = A^{-1}\mathbf{y}$ defines the *inverse mapping*, which is also linear. Let, for example, $n = 3$ and let us interpret $\mathbf{y} = A\mathbf{x} = x_1\mathbf{u}_1 + x_2\mathbf{u}_2 + x_3\mathbf{u}_3$ as describing a mapping of E^3 onto E^3.

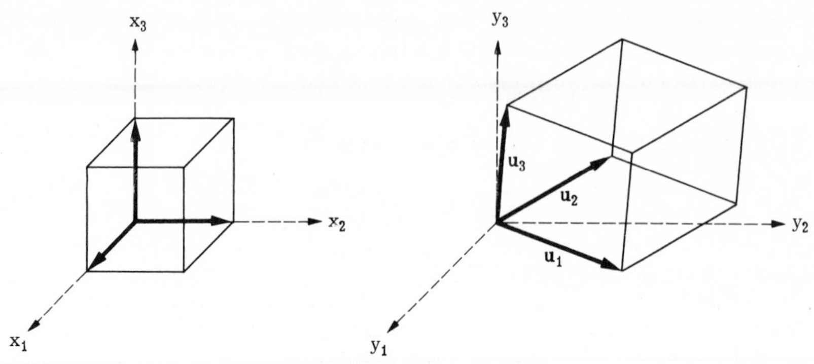

FIG. 1–5. Relationship between volumes for mapping $\mathbf{y} = A\mathbf{x}$.

The points $0 \leqq x_1 \leqq 1$, $0 \leqq x_2 \leqq 1$, $0 \leqq x_3 \leqq 1$ form a cube in the $\mathbf{x}$-space. The corresponding points in the $\mathbf{y}$-space:

$$\mathbf{y} = x_1\mathbf{u}_1 + x_2\mathbf{u}_2 + x_3\mathbf{u}_3, \quad 0 \leqq x_1 \leqq 1, \quad 0 \leqq x_2 \leqq 1, \quad 0 \leqq x_3 \leqq 1$$

fill a parallelepiped whose edge vectors are $\mathbf{u}_1$, $\mathbf{u}_2$, $\mathbf{u}_3$ (Fig. 1–5). The volume of the parallelepiped is

$$\pm\mathbf{u}_1 \cdot \mathbf{u}_2 \times \mathbf{u}_3 = \pm\det A$$

(see Section 0–6). Hence the cube of volume 1 becomes a parallelepiped of volume $|\det A|$, or we can say: the volume of the cube is multiplied by $|\det A|$ in going from the $\mathbf{x}$-space to the $\mathbf{y}$-space. It can be shown that all volumes are multiplied by the same factor $|\det A|$ in going from the $\mathbf{x}$-space to the $\mathbf{y}$-space; in fact there is a similar statement concerning n-dimensional volume for general n. For a discussion, see *CLA*, Section 11–18. This relationship is very important for the calculus of several variables, where $\det A$ becomes a *Jacobian determinant* (see Sections 2–7, 4–8, and 5–14).

If A has less than maximum rank, then $k = n - r > 0$ and the kernel of A is a k-dimensional linear manifold $K: \mathbf{x} = t_1\mathbf{z}_1 + \cdots + t_k\mathbf{z}_k$. Thus the homogeneous equations related to (1–70):

$$\begin{aligned} a_{11}x_1 + \quad \cdots \quad + a_{1n}x_n &= 0, \\ \vdots \qquad\qquad & \\ a_{n1}x_1 + \quad \cdots \quad + a_{nn}x_n &= 0, \end{aligned} \tag{1–71}$$

have infinitely many solutions, forming the k-dimensional linear manifold K. The range of A is an r-dimensional linear manifold H in E^n and, since $r < n$, H is not all of E^n. Thus for some choices of $y_1, \ldots, y_n$, the given linear equations (1–70) have no solution for $x_1, \ldots, x_n$. A solution exists precisely when $\mathbf{y}$ is in H—that is, when A and $(A \quad \mathbf{y})$ have the same rank, as in Corollary 1 of Section 1–12. Furthermore, since the kernel does not reduce to $\mathbf{0}$ alone, even when a solution exists, it is not unique. By the theorem of Section 1–10, for each $\mathbf{y}_0$ for which a solution $\mathbf{x}_0$ exists, all solutions are given by $\mathbf{x} = \mathbf{x}_0 + \mathbf{z}$, where $\mathbf{z}$ is an arbitrary vector in the kernel of A; that is, the solutions are given by

$$\mathbf{x} = \mathbf{x}_0 + t_1\mathbf{z}_1 + \cdots + t_k\mathbf{z}_k,$$

and they form a k-dimensional linear manifold.

EXAMPLE 1.

$$\begin{aligned} 3x_1 + 2x_2 - x_3 &= y_1, \\ x_1 - x_2 + x_3 &= y_2, \\ 5x_1 + 5x_2 - 3x_3 &= y_3. \end{aligned} \qquad A = \begin{bmatrix} 3 & 2 & -1 \\ 1 & -1 & 1 \\ 5 & 5 & -3 \end{bmatrix}$$

Here $n = 3$. We find that $\det A = 0$, so that A has less than maximum rank. We easily find nonzero second order minors of A and conclude that A has rank $r = 2$. Hence $k = 3 - 2 = 1$ and the solutions of the corresponding homogeneous equations

$$\begin{aligned} 3x_1 + 2x_2 - x_3 &= 0, \\ x_1 - x_2 + x_3 &= 0, \\ 5x_1 + 5x_2 - 3x_3 &= 0 \end{aligned}$$

(kernel of A) form a 1-dimensional linear manifold in E^3—that is, a straight line in 3-dimensional space. If we solve the first two equations for x_2 and x_3 in terms of x_1, we obtain $x_2 = -4x_1$, $x_3 = -5x_1$, and these values satisfy the third equation. Hence we can write

$$x_1 = t, \quad x_2 = -4t, \quad x_3 = -5t \quad \text{or} \quad \mathbf{x} = t(1, -4, -5)$$

to describe the line of solutions.

For the given nonhomogeneous equations we reason that the range of A is a 2-dimensional linear manifold H in E^3 (a plane in space), so for some choices of $\mathbf{y}$ there is no solution. We observe that the first two column vectors $\mathbf{u}_1$, $\mathbf{u}_2$ of A are linearly independent and hence form a basis for H. Thus $\mathbf{y}$ is in the range if and only if $\mathbf{y}$ is a linear combination of these two vectors. For example $(5, 0, 10) = \mathbf{u}_1 + \mathbf{u}_2$, and thus the equations

$$\begin{aligned} 3x_1 + 2x_2 - x_3 &= 5, \\ x_1 - x_2 + x_3 &= 0, \\ 5x_1 + 5x_2 - 3x_3 &= 10 \end{aligned}$$

have solutions. One solution is $\mathbf{e}_1 + \mathbf{e}_2 = (1, 1, 0)$, since

$$A(\mathbf{e}_1 + \mathbf{e}_2) = A\mathbf{e}_1 + A\mathbf{e}_2 = \mathbf{u}_1 + \mathbf{u}_2.$$

All solutions are given by $\mathbf{x} = (1, 1, 0) + t(1, -4, -5)$, since we found the kernel of A to consist of all $\mathbf{z} = t(1, -4, -5)$.

Case II. $m > n$ (more equations than unknowns). Let A have maximum rank, so that $r = n$. Then the linear equations (1–70) can be written

$$\mathbf{y} = A\mathbf{x} = x_1\mathbf{u}_1 + \cdots + x_n\mathbf{u}_n,$$

where $\mathbf{u}_1, \ldots, \mathbf{u}_n$ are linearly independent. Hence the range of A is an n-dimensional linear manifold H in E^m, and $\mathbf{u}_1, \ldots, \mathbf{u}_n$ are a basis for H.

Since $m > n$, H is not all of E^m. However, since $r = n$, $k = 0$, and A is a one-to-one mapping. Actually we have here simply the usual parametric representation of the linear manifold H, with $x_1, \ldots, x_n$ as parameters. As for the case $m = n$, one can consider the relationship between the volume of a portion of the $\mathbf{x}$-space and the volume of the corresponding portion of H. Here it can be shown that, in going from the $\mathbf{x}$-space to H, volumes are multiplied by $(\det B)^{\frac{1}{2}}$, where B is the $n \times n$ matrix $(\mathbf{u}_i \cdot \mathbf{u}_j)$. (See *CLA* Section 11–18).

Examples with A of less than maximum rank are left to the problems.

Case III. $m < n$ (more unknowns than equations). Let A have maximum rank, so that $r = m$. Hence $k = n - m > 0$. The range of A is an m-dimensional linear manifold in E^m, and hence the range is all of E^m. Therefore, for each $\mathbf{y}$ the equations (1–70) have a solution, and in fact have infinitely many solutions $\mathbf{x} = \mathbf{x}_0 + \mathbf{z}$, where $\mathbf{z}$ is an arbitrary vector in the kernel of A. The kernel itself is a k-dimensional linear manifold, formed of all $\overrightarrow{OP} = t_1\mathbf{z}_1 + \cdots + t_k\mathbf{z}_k$. This is illustrated in Section 1–11 above (Examples 6, 7, 8).

Again we leave to the exercises the discussion of the cases in which A has less than maximum rank.

For A of maximum rank, we have just seen that for each fixed choice of $y_1, \ldots, y_m$, Eqs. (1–70) describe an $(n - m)$-dimensional linear manifold in E^n. This answers one question raised in Section 1–11. In that section we also asked whether, for $1 \leqq m < n$, every $(n - m)$-dimensional linear manifold H in E^n can be represented as the set of solutions of m linear equations in $x_1, \ldots, x_n$. This we answer affirmatively as follows. Let H be given:

$$\mathbf{x} = \overrightarrow{OP} = \mathbf{x}_0 + t_1\mathbf{z}_1 + \cdots + t_k\mathbf{z}_k, \tag{1–72}$$

where $k = n - m > 0$. Since $\mathbf{z}_1, \ldots, \mathbf{z}_k$ are linearly independent, we can always choose $m = n - k$ of the vectors $\mathbf{e}_1, \ldots, \mathbf{e}_n$ so that $\mathbf{z}_1, \ldots, \mathbf{z}_k$ and these m vectors form a basis for V^n (Problem 12 below); for example, let $\mathbf{z}_1, \ldots, \mathbf{z}_k, \mathbf{e}_1, \ldots, \mathbf{e}_m$ be a basis, so that the $n \times n$ matrix C having these vectors as column vectors is nonsingular. We then seek a matrix A whose kernel has $\mathbf{z}_1, \ldots, \mathbf{z}_k$ as basis. If we make

$$\begin{aligned} A\mathbf{z}_i &= \mathbf{0} \quad \text{for} \quad i = 1, \ldots, k, \\ A\mathbf{e}_j &= \boldsymbol{\epsilon}_j \quad \text{for} \quad j = 1, \ldots, m, \end{aligned} \tag{1–73}$$

where $\boldsymbol{\epsilon}_1, \ldots, \boldsymbol{\epsilon}_m$ are the standard basis for V^m, then

$$A(c_1\mathbf{z}_1 + \cdots + c_k\mathbf{z}_k + c_{k+1}\mathbf{e}_1 + \cdots + c_n\mathbf{e}_m) = \mathbf{0}$$

precisely when $c_{k+1} = 0, \ldots, c_n = 0$. Thus A will be as desired. But (1–73) is equivalent to the matrix equation

$$AC = B,$$

where B is the $m \times n$ matrix whose column vectors are $\mathbf{0}$ (k times) and then $\boldsymbol{\epsilon}_1, \ldots, \boldsymbol{\epsilon}_m$. Hence we obtain

$$A = BC^{-1}.$$

If we let $\mathbf{y}_0 = A\mathbf{x}_0$, then the solutions of the equation

$$A\mathbf{x} = \mathbf{y}_0 \tag{1–74}$$

are given by (1–72) as desired. Since (1–74) is equivalent to m linear equations in $x_1, \ldots, x_n$, our problem is solved.

EXAMPLE 2. H is the plane

$$\mathbf{x}_1 = (1, 0, 3, 2, 1) + t_1(2, 1, 1, 1, 3) + t_2(1, 0, 3, 1, 4)$$

in E^5. Here $\mathbf{z}_1 = (2, 1, 1, 1, 3)$ and $\mathbf{z}_2 = (1, 0, 3, 1, 4)$, and $\mathbf{z}_1, \mathbf{z}_2, \mathbf{e}_1, \mathbf{e}_2, \mathbf{e}_3$ form a basis for V^5, since the matrix C having these vectors as column vectors is nonsingular:

$$\det C = \begin{vmatrix} 2 & 1 & 1 & 0 & 0 \\ 1 & 0 & 0 & 1 & 0 \\ 1 & 3 & 0 & 0 & 1 \\ 1 & 1 & 0 & 0 & 0 \\ 3 & 4 & 0 & 0 & 0 \end{vmatrix} = \begin{vmatrix} 1 & 1 \\ 3 & 4 \end{vmatrix} = 1 \neq 0.$$

Since $k = 2$, $m = 5 - 2 = 3$ and B is the 3×5 matrix

$$\begin{bmatrix} 0 & 0 & 1 & 0 & 0 \\ 0 & 0 & 0 & 1 & 0 \\ 0 & 0 & 0 & 0 & 1 \end{bmatrix}.$$

We readily find C^{-1} and then

$$A = BC^{-1} = \begin{bmatrix} 0 & 0 & 1 & 0 & 0 \\ 0 & 0 & 0 & 1 & 0 \\ 0 & 0 & 0 & 0 & 1 \end{bmatrix} \begin{bmatrix} 0 & 0 & 0 & 4 & -1 \\ 0 & 0 & 0 & -3 & 1 \\ 1 & 0 & 0 & -5 & 1 \\ 0 & 1 & 0 & -4 & 1 \\ 0 & 0 & 1 & 5 & -2 \end{bmatrix}$$

$$= \begin{bmatrix} 1 & 0 & 0 & -5 & 1 \\ 0 & 1 & 0 & -4 & 1 \\ 0 & 0 & 1 & 5 & -2 \end{bmatrix}.$$

Since $A\mathbf{x}_0 = A \operatorname{col}(1, 0, 3, 2, 1) = \operatorname{col}(-8, -7, 11)$, our three linear equations are

$$\begin{aligned} x_1 - 5x_4 + x_5 &= -8, \\ x_2 - 4x_4 + x_5 &= -7, \\ x_3 + 5x_4 - 2x_5 &= 11. \end{aligned}$$

Problems

1. Find the rank of each of the following matrices, as given at the beginning of the set of problems following Section 1–10: (a) A, (b) B, (c) E, (d) F, (e) G, (f) H, (g) J, (h) K, (i) L.

2. Give an example of an $m \times n$ matrix of rank r for the given values, if possible:

(a) $m = 2, \quad n = 2, \quad r = 1$
(b) $m = 3, \quad n = 2, \quad r = 1$
(c) $m = 5, \quad n = 3, \quad r = 4$
(d) $m = 4, \quad n = 3, \quad r = 2$
(e) $m = 1, \quad n = 3, \quad r = 0$
(f) $m = 3, \quad n = 4, \quad r = 4$

3. Let the following matrices be given:

$$A = \begin{bmatrix} 3 & 1 \\ 1 & 2 \\ 1 & 2 \end{bmatrix}, \quad B = \begin{bmatrix} 2 & 1 & 2 \\ 3 & 0 & 1 \\ 2 & 1 & 4 \end{bmatrix}, \quad C = \begin{bmatrix} 1 & -1 & 1 & 2 \\ 2 & 1 & 0 & 3 \\ 1 & 2 & 4 & 1 \end{bmatrix}, \quad D = \begin{bmatrix} 2 & 1 & 2 \\ 3 & 1 & 1 \\ 1 & 1 & 3 \end{bmatrix},$$

$$E = \begin{bmatrix} 1 & 0 \\ 1 & 1 \\ 3 & 1 \end{bmatrix}, \quad F = \begin{bmatrix} 1 & 2 & 3 & 1 & 1 \\ 2 & -1 & 0 & 2 & 1 \\ 3 & 1 & 3 & 3 & 2 \end{bmatrix}, \quad G = \begin{bmatrix} 1 & 0 & 0 & 0 \\ 1 & 1 & 0 & 0 \\ 1 & 1 & 1 & 3 \\ 5 & 1 & 2 & 6 \end{bmatrix}.$$

(a) Verify that A has rank 2 and give the values of p and q such that A has as range a p-dimensional linear manifold in E^q.

(b) Determine whether $\mathbf{y} = B\mathbf{x}$ describes a one-to-one mapping of E^3 onto E^3.

(c) Describe the nature of the set of solutions of $G\mathbf{x} = \mathbf{0}$.

(d) Give the values of p and q such that the equation $C\mathbf{x} = \text{col}\,(1, 0, 2)$ represents a p-dimensional linear manifold in E^q.

(e) Show that the equation $\mathbf{y} = D\mathbf{x}$ does not describe a one-to-one mapping of E^3 onto E^3.

(f) Find a vector equation for the p-dimensional linear manifold which is the range of E.

(g) Find the dimension of the linear manifold which is the kernel of F.

4. Determine whether a solution exists **x** and, if it does, find all solutions:

(a) $3x_1 + x_2 = y_1, \quad 5x_1 + 2x_2 = y_2 \quad (y_1, y_2 \text{ given})$.

(b) $6x_1 + 4x_2 = 1, \quad 9x_1 + 6x_2 = 2$.

(c) $x_1 + x_2 - x_3 = y_1, \quad 2x_1 + x_2 + x_3 = y_2, \quad 2x_1 + x_2 = y_3 \quad (y_1, y_2, y_3 \text{ given})$.

(d) $2x_1 + x_2 - x_3 = 3, \quad x_1 + x_2 + x_3 = 2, \quad 3x_1 + 2x_2 = 5$.

(e) $x_1 - x_2 + x_3 + x_4 = 0, \quad x_1 + x_2 + x_3 = 0, \quad x_2 + x_3 + x_4 = 0, \quad x_1 + x_3 + x_4 = 0$.

(f) $x_1 + x_2 - x_3 - x_4 = 0, \quad 2x_1 - x_2 + x_3 + x_4 = 0, \quad 7x_1 + x_2 - x_3 - x_4 = 0, \quad x_1 - 5x_2 + 5x_3 + 5x_4 = 0$.

5. For each of the following sets of linear equations determine whether (i) the equations can be considered as parametric equations of a p-dimensional linear manifold in E^q or (ii) for each fixed choice of $(y_1, y_2, \ldots)$ the *solutions* of the equations form a p-dimensional linear manifold in E^q. In each case give the values of p and q.

(a) $x_1 + x_2 - x_3 = y_1, \; x_1 - x_2 = y_2, \; x_1 + 2x_2 + x_3 = y_3, \; x_1 - x_2 + 3x_3 = y_4$.

(b) $x_1 - x_2 = y_1, \quad 2x_1 + x_2 = y_2, \quad x_1 - 2x_2 = y_3$.

(c) $2x_1 + x_2 - x_3 = y_1, \quad x_1 - x_2 + x_3 = y_2$.

(d) $3x_1 = y_1, \quad 2x_1 = y_2, \quad 5x_1 = y_3$.

(e) $x_1 + x_2 = y_1 + 1, \quad x_1 - x_2 = y_2 - 2, \quad 2x_1 + x_2 = y_3$.

(f) $2x_1 - x_2 + x_3 - x_4 = y_1, \quad x_1 + x_2 + x_3 + x_4 = y_2$.

(g) $x_1 + 3x_2 - x_3 + x_4 = y_1 + 2, \quad 2x_1 - 2x_2 + 5x_3 = y_2 + 5$.

6. Each of the following is the vector equation of a linear manifold. Find a set of simultaneous linear equations for the manifold:

(a) $\mathbf{x} = (3, 4, 0) + t_1(1, 1, 0) + t_2(1, 0, 1)$.

(b) $\mathbf{x} = (2, 5, 1) + t_1(5, 3, 2) + t_2(2, 1, 4)$.

(c) $\mathbf{x} = (1, 0, 5, 2) + t_1(1, 1, 1, 0) + t_2(1, 1, 0, 1)$.

(d) $\mathbf{x} = (1, 1, 5, 7) + t_1(2, 3, 1, 1) + t_2(3, 2, 1, 1)$.

(e) $\mathbf{x} = t_1(1, 1, 0, 1, -1) + t_2(0, 0, 1, 1, 2) + t_3(1, 0, 0, 0, 1)$.

(f) $\mathbf{x} = t_1(1, 2, -1, 0, 0) + t_2(3, 2, 1, 0, 0) + t_3(2, 0, 1, 0, 0) + t_4(2, 0, 0, 1, 1)$.

7. For the system (1–70) let $m > n$, so that one has Case II, but let A have rank r less than n. Show that the range of A is an r-dimensional linear manifold H in E^m and that the mapping A is not one-to-one. Show also that if, for example, the first r column vectors $\mathbf{u}_1, \ldots, \mathbf{u}_r$ of A are linearly independent, then $\mathbf{y} = x_1\mathbf{u}_1 + \cdots + x_r\mathbf{u}_r$ is a vector equation for H with $x_1, \ldots, x_r$ as parameters.

8. Show that, for each of the following systems of form (1–70), A has less than maximal rank and find a vector equation for the range of A (see Problem 7).

(a) $3x_1 + 6x_2 = y_1, \quad -x_1 - 2x_2 = y_2, \quad 2x_1 + 4x_2 = y_3.$
(b) $x_1 + 2x_2 = y_1, \quad 5x_1 + 10x_2 = y_2, \quad 0 = y_3, \quad 3x_1 + 6x_2 = y_4.$
(c) $-x_1 + 3x_2 + x_3 = y_1, \quad 2x_1 + x_2 + 5x_3 = y_2, \quad x_1 + 4x_2 + 6x_3 = y_3,$ $5x_1 - 2x_2 + 8x_3 = y_4.$

9. For the system (1–70) let $m < n$, so that one has Case III, but let A have rank $r < m$. (a) Show that for each fixed $\mathbf{y}_0$, the equation $A\mathbf{x} = \mathbf{y}_0$ has a solution $\mathbf{x}_0$ precisely when $(A \quad \mathbf{y}_0)$ has rank r, and that there are choices of $\mathbf{y}_0$ for which there is no solution. (b) Show that, if $\mathbf{y}_0$ is such that the equation $A\mathbf{x} = \mathbf{y}_0$ has a solution, then all solutions form an $(n - r)$-dimensional linear manifold H in E^n; show also that, if, for example, the first r *row* vectors of A are linearly independent, then the first r equations (1–70) are a set of equations for H.

10. Show that for each of the following systems of form (1–70) with given $\mathbf{y}$, the matrix A has less than maximal rank, determine whether there are solutions $\mathbf{x}$ and, if these solutions form an $(n - r)$-dimensional linear manifold H, find a set of r equations for H (see Problem 9).

(a) $2x_1 + 2x_2 - x_3 = 1, \quad 4x_1 + 4x_2 - 2x_3 = 3.$
(b) $x_1 - x_2 + x_3 - 3x_4 = 2, \quad 3x_1 - 3x_2 + 3x_3 - 9x_4 = 6.$
(c) $2x_1 + x_2 - x_3 - 3x_4 = 1, \quad x_1 + 5x_2 - x_3 + x_4 = 1,$ $5x_1 - 2x_2 - 2x_3 - 10x_4 = 2.$
(d) $x_1 + 2x_2 - x_3 + x_4 = 2, \quad 2x_1 - x_2 + x_3 - x_4 = 3, \quad 5x_1 + x_3 - x_4 = 5.$
(e) $2x_1 + x_2 - x_3 - x_4 = 0, \quad x_1 - x_2 + x_3 - 2x_4 = 0, \quad x_1 + x_2 - x_3 = 0.$
(f) $x_1 + x_2 - x_3 - x_5 = 0, \quad 3x_1 + 2x_2 + 4x_4 = 0, \quad x_2 - 3x_3 - 4x_4 - 3x_5 = 0, \quad 7x_1 + 5x_2 - x_3 + 8x_4 - x_5 = 0.$

11. (a) Show that two $(n - 1)$-dimensional linear manifolds in E^n:

$$H_1\colon a_{11}x_1 + \cdots + a_{1n}x_n = k_1, \quad H_2\colon a_{21}x_1 + \cdots + a_{2n}x_n = k_2$$

either do not intersect, or coincide, or intersect in an $(n - 2)$-dimensional linear manifold, and state in terms of matrices when each case occurs.

(b) Show generally that a p-dimensional linear manifold and a q-dimensional linear manifold in E^n intersect, if at all, in a k-dimensional linear manifold, where

$$k \geqq p + q - n.$$

Illustrate by lines and planes in E^3.

12. Let $1 \leqq k < n$ and let $\mathbf{z}_1, \ldots, \mathbf{z}_k$ be linearly independent vectors in V^n. Show that $n - k$ of the basis vectors $\mathbf{e}_1, \ldots, \mathbf{e}_n$ can be chosen so that these $n - k$ vectors and $\mathbf{z}_1, \ldots, \mathbf{z}_k$ together form a basis. [Hint: By rule (k) of Section 1–8, at least one of the vectors $\mathbf{e}_1, \ldots, \mathbf{e}_n$ is not expressible as a linear combination of $\mathbf{z}_1, \ldots, \mathbf{z}_k$. For example, let $\mathbf{e}_1$ be such a one. Then $\mathbf{z}_1, \ldots, \mathbf{z}_k, \mathbf{e}_1$ are linearly independent. If $k + 1 < n$, repeat the process, until n vectors are obtained as required.]

Answers

1. (a) 2, (b) 1, (c) 2, (d) 2, (e) 1, (f) 1, (g) 2, (h) 2, (i) 3.

3. (a) $p = 2$, $q = 3$. (b) It does. (c) $r = 3$, a 1-dimensional linear manifold in E^4. (d) $p = 1$, $q = 4$. (f) $\mathbf{y} = x_1(1, 1, 3) + x_2(0, 1, 1)$. (g) 3.

4. (a) $x_1 = 2y_1 - y_2$, $x_2 = -5y_1 + 3y_2$. (b) no solutions. (c) $x_1 = -y_1 - y_2 + 2y_3$, $x_2 = 2y_1 + 2y_2 - 3y_3$, $x_3 = y_2 - y_3$. (d) $\mathbf{x} = (1, 1, 0) + t(2, -3, 1)$. (e) $\mathbf{x} = \mathbf{0}$. (f) $\mathbf{x} = t_1(0, 1, 1, 0) + t_2(0, 1, 0, 1)$.

5. (a) as in (i), $p = 3$, $q = 4$. (b) as in (i), $p = 2$, $q = 3$. (c) as in (ii), $p = 1$, $q = 3$. (d) as in (i), $p = 1$, $q = 3$. (e) as in (i), $p = 2$, $q = 3$. (f) as in (ii), $p = 2$, $q = 4$. (g) as in (ii), $p = 2$, $q = 4$.

6. (a) $x_1 - x_2 - x_3 + 1 = 0$. (b) $10x_1 - 16x_2 - x_3 = -61$. (c) $x_1 - x_3 - x_4 = -6$, $x_2 - x_3 - x_4 = -7$. (d) $x_1 + x_2 - 5x_3 = -23$, $-x_3 + x_4 = 2$. (e) $x_1 + 4x_3 - 2x_4 - x_5 = 0$, $x_2 + x_3 - x_4 = 0$. (f) $x_4 - x_5 = 0$.

***1–14 Change of basis and coordinates.** We pointed out in Section 1–9 above that, in addition to the standard orthogonal basis $\mathbf{e}_1, \ldots, \mathbf{e}_n$ for V^n, there are other orthonormal bases, and that one can introduce components with respect to each such basis. Let $\mathbf{e}_1^*, \ldots, \mathbf{e}_n^*$ be a "new" orthonormal basis for V^n, so that each vector $\mathbf{v}$ has components $(v_1^*, \ldots, v_n^*)$ relative to this basis. We now ask how the old components $v_1, \ldots, v_n$ are related to the new ones $v_1^*, \ldots, v_n^*$. To this end we remark that

$$\mathbf{v} = v_1\mathbf{e}_1 + \cdots + v_n\mathbf{e}_n = v_1^*\mathbf{e}_1^* + \cdots + v_n^*\mathbf{e}_n^*. \tag{1–75}$$

Now we can express each $\mathbf{e}_j$ in terms of the new basis:

$$\mathbf{e}_j = a_{1j}\mathbf{e}_1^* + \cdots + a_{nj}\mathbf{e}_n^*, \quad j = 1, \ldots, n, \tag{1–76}$$

where by orthonormality,

$$a_{ij} = \mathbf{e}_j \cdot \mathbf{e}_i^*. \tag{1–77}$$

Hence by (1–75),

$$v_1(a_{11}\mathbf{e}_1^* + \cdots + a_{n1}\mathbf{e}_n^*) + \cdots + v_n(a_{1n}\mathbf{e}_1^* + \cdots + a_{nn}\mathbf{e}_n^*) = v_1^*\mathbf{e}_1^* + \cdots + v_n^*\mathbf{e}_n^*.$$

Since $\mathbf{e}_1^*, \ldots, \mathbf{e}_n^*$ form a basis, we can compare coefficients of $\mathbf{e}_1^*, \ldots, \mathbf{e}_n^*$. We conclude:

$$v_1^* = a_{11}v_1 + \cdots + a_{1n}v_n, \ldots, v_n^* = a_{n1}v_1 + \cdots + a_{nn}v_n. \tag{1–78}$$

Therefore, with $A = (a_{ij})$,

$$\begin{bmatrix} v_1^* \\ \vdots \\ v_n^* \end{bmatrix} = A \begin{bmatrix} v_1 \\ \vdots \\ v_n \end{bmatrix}. \tag{1–79}$$

This equation shows that we have in effect a linear mapping, with matrix A, of the vector space of n-tuples $(v_1, \ldots, v_n)$ into the vector space of n-tuples $(v_1^*, \ldots, v_n^*)$. From (1–79) or from (1–76) it follows that the *j-th column* of A is simply the n-tuple of new components of $\mathbf{e}_j$. Hence A is nonsingular. Furthermore, each column of A is a unit vector:

$$a_{1j}^2 + \cdots + a_{nj}^2 = 1, \tag{1–80}$$

and each pair of column vectors of A are orthogonal:

$$a_{1j}a_{1k} + \cdots + a_{nj}a_{nk} = 0 \quad \text{for} \quad j \neq k. \tag{1–81}$$

Furthermore, from (1–77), *the i-th row of A is simply the set of old components of* $\mathbf{e}_i^*$. Hence

$$a_{i1}^2 + \cdots + a_{in}^2 = 1 \tag{1–82}$$

and each two row vectors of A are orthogonal:

$$a_{i1}a_{h1} + \cdots + a_{in}a_{hn} = 0 \quad \text{for} \quad i \neq h. \tag{1–83}$$

From (1–80) and (1–81) or from (1–82) and (1–83) we see that A is an *orthogonal* matrix (Section 1–7). Accordingly,

$$AA' = I \quad \text{or} \quad A' = A^{-1}.$$

Therefore, from (1–79) we can obtain the old components in terms of the new ones as follows:

$$\begin{bmatrix} v_1 \\ \vdots \\ v_n \end{bmatrix} = A' \begin{bmatrix} v_1^* \\ \vdots \\ v_n^* \end{bmatrix} \tag{1–84}$$

From the results of Section 1–7 it follows that for *every* orthogonal matrix A of order n, the equations (1–79) and (1–84) describe the relationship between new and old components of vectors when a new orthonormal basis $\mathbf{e}_1^*, \ldots, \mathbf{e}_n^*$ is introduced in V^n. The columns of A are the vectors $\mathbf{e}_1, \ldots, \mathbf{e}_n$ in new components; the rows of A are the vectors $\mathbf{e}_1^*, \ldots, \mathbf{e}_n^*$ in old components.

EXAMPLE 1. In V^2, we start with the old basis $\mathbf{i}$, $\mathbf{j}$ and introduce a new one $\mathbf{i}^*$, $\mathbf{j}^*$, where

$$\mathbf{i}^* = \tfrac{3}{5}\mathbf{i} - \tfrac{4}{5}\mathbf{j}, \quad \mathbf{j}^* = \tfrac{4}{5}\mathbf{i} + \tfrac{3}{5}\mathbf{j}.$$

We verify at once that $\mathbf{i}^*, \mathbf{j}^*$ form an orthonormal system (they are orthogonal and each is a unit vector); see Fig. 1–6. Therefore we have the orthogonal matrix

$$A = \begin{bmatrix} \frac{3}{5} & -\frac{4}{5} \\ \frac{4}{5} & \frac{3}{5} \end{bmatrix}.$$

Each vector $\mathbf{v} = v_1\mathbf{i} + v_2\mathbf{j}$ has thus new components (v_1^*, v_2^*) given by

$$\begin{bmatrix} v_1^* \\ v_2^* \end{bmatrix} = \begin{bmatrix} \frac{3}{5} & -\frac{4}{5} \\ \frac{4}{5} & \frac{3}{5} \end{bmatrix} \begin{bmatrix} v_1 \\ v_2 \end{bmatrix}. \tag{1–85}$$

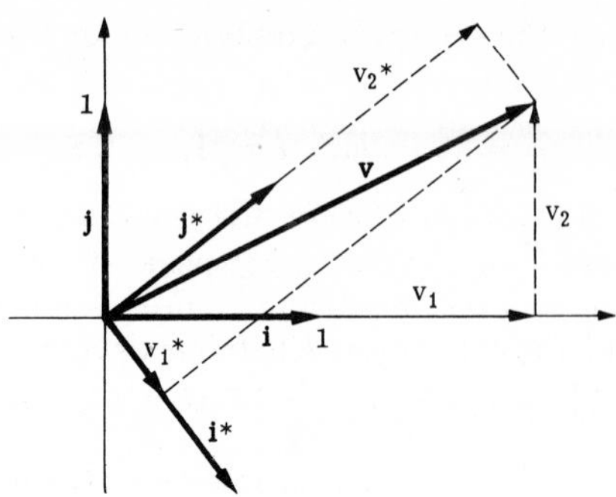

FIG. 1–6. Change of basis in the plane.

For example, the vector $\mathbf{v} = 2\mathbf{i} + \mathbf{j}$ has $v_1 = 2$, $v_2 = 1$ and by (1–85) we find $v_1^* = \frac{2}{5}$, $v_2^* = \frac{11}{5}$. The two sets of components are shown in Figure 1–6.

EXAMPLE 2. In V^3, we start with the old basis $\mathbf{i}, \mathbf{j}, \mathbf{k}$ and introduce a new orthonormal basis $\mathbf{i}^*, \mathbf{j}^*, \mathbf{k}^*$, where

$$\mathbf{i}^* = \tfrac{2}{3}\mathbf{i} + \tfrac{2}{3}\mathbf{j} + \tfrac{1}{3}\mathbf{k}, \quad \mathbf{j}^* = \tfrac{2}{3}\mathbf{i} - \tfrac{1}{3}\mathbf{j} - \tfrac{2}{3}\mathbf{k}, \quad \mathbf{k}^* = \tfrac{1}{3}\mathbf{i} - \tfrac{2}{3}\mathbf{j} + \tfrac{2}{3}\mathbf{k}.$$

Here, for example, the first row of A is $(\frac{2}{3}, \frac{2}{3}, \frac{1}{3})$ and in general

$$\begin{bmatrix} v_1^* \\ v_2^* \\ v_3^* \end{bmatrix} = \begin{bmatrix} \frac{2}{3} & \frac{2}{3} & \frac{1}{3} \\ \frac{2}{3} & -\frac{1}{3} & -\frac{2}{3} \\ \frac{1}{3} & -\frac{2}{3} & \frac{2}{3} \end{bmatrix} \begin{bmatrix} v_1 \\ v_2 \\ v_3 \end{bmatrix}.$$

As a check, we observe that the three *columns* of A also form an orthonormal system; these columns are the sets of components of $\mathbf{i}, \mathbf{j}, \mathbf{k}$ with respect to the basis $\mathbf{i}^*, \mathbf{j}^*, \mathbf{k}^*$. For example,

$$\mathbf{j} = \tfrac{2}{3}\mathbf{i}^* - \tfrac{1}{3}\mathbf{j}^* - \tfrac{2}{3}\mathbf{k}^*.$$

New coordinates in E^n. To introduce new coordinates in E^n, we choose a new origin O^* and a new orthonormal basis $\mathbf{e}_1^*, \ldots, \mathbf{e}_n^*$. For a general point P we can then write

$$\overrightarrow{O^*P} = x_1^*\mathbf{e}_1^* + \cdots + x_n^*\mathbf{e}_n^*$$

and call $(x_1^*, \ldots, x_n^*)$ the new coordinates of P. Hence $(x_1^*, \ldots, x_n^*)$ are simply the new components of the vector $\mathbf{v} = \overrightarrow{O^*P}$. If O^* has old coordinates $(h_1, \ldots, h_n)$ and P has old coordinates $(x_1, \ldots, x_n)$, then $\mathbf{v}$ has old components $(x_1 - h_1, \ldots, x_n - h_n)$.

Let A be the orthogonal matrix relating old and new components, as above. Then $A^{-1} = A'$ and we have

$$\begin{bmatrix} x_1^* \\ \vdots \\ x_n^* \end{bmatrix} = A \begin{bmatrix} x_1 - h_1 \\ \vdots \\ x_n - h_n \end{bmatrix} \quad \text{and} \quad \begin{bmatrix} x_1 - h_1 \\ \vdots \\ x_n - h_n \end{bmatrix} = A' \begin{bmatrix} x_1^* \\ \vdots \\ x_n^* \end{bmatrix}.$$

If we simply change the origin, but do not change the basis, then $A = I$ and

$$x_i^* = x_i - h_i, \quad i = 1, \ldots, n.$$

Here we speak of a *translation of axes*. If the origin is unchanged, but a new orthonormal basis is introduced, we speak of a rotation of axes only if there is no "reversal of orientation." For 3-dimensional space, this means that $\mathbf{e}_1^*$, $\mathbf{e}_2^*$, $\mathbf{e}_3^*$ must be a *positive triple* or $\det A > 0$ (Section 0–6). For n-dimensional space a similar reasoning leads again to the condition $\det A > 0$.

Nonorthogonal bases. For many problems concerning V^n the natural basis is not orthonormal or even orthogonal. We know that any set of n linearly independent vectors $\mathbf{u}_1^*, \ldots, \mathbf{u}_n^*$ can serve as a basis. If we refer the vector $\mathbf{v} = (v_1, \ldots, v_n)$ (old components) to the new basis $\mathbf{u}_1^*, \ldots, \mathbf{u}_n^*$, we again obtain new components $(v_1^*, \ldots, v_n^*)$ so that

$$\mathbf{v} = v_1^*\mathbf{u}_1^* + \cdots + v_n^*\mathbf{u}_n^*.$$

Again we ask how old components $(v_1, \ldots, v_n)$ and new components $(v_1^*, \ldots, v_n^*)$ are related. To this end we let $\mathbf{e}_j$ have components $(a_{1j}, \ldots, a_{nj})$ with respect to the new basis $\mathbf{u}_1^*, \ldots, \mathbf{u}_n^*$, so that

$$\mathbf{e}_j = a_{1j}\mathbf{u}_1^* + \cdots + a_{nj}\mathbf{u}_n^*.$$

Then we are again led to (1–79), where A is the n by n matrix whose j-th column is $\mathbf{e}_j$ in terms of the new basis. However, A need no longer be orthogonal. We do know that the correspondence between old and new components must be one-to-one, so that A is nonsingular. Thus we can solve for old components in terms of new:

$$\begin{bmatrix} v_1 \\ \vdots \\ v_n \end{bmatrix} = A^{-1}\begin{bmatrix} v_1^* \\ \vdots \\ v_n^* \end{bmatrix}. \tag{1–86}$$

The j-th column of A^{-1} gives the old components of $\mathbf{u}_j^*$.

Furthermore, given a nonsingular $n \times n$ matrix A, one can always interpret A as the matrix for a change of basis as in (1–79) and (1–86).

Oblique coordinates in E^n. Corresponding to nonorthogonal bases $\mathbf{u}_1^*, \ldots, \mathbf{u}_n^*$ in V^n, one has oblique coordinates in E^n. The same reasoning as above leads to an equation $\overrightarrow{O^*P} = x_1^*\mathbf{u}_1^* + \cdots + x_n^*\mathbf{u}_n^*$ assigning new coordinates $(x_1^*, \ldots, x_n^*)$ to P relative to a new origin O^*: $(h_1, \ldots, h_n)$. Old coordinates and new coordinates are then related by an equation

$$\operatorname{col}(x_1^*, \ldots, x_n^*) = A \operatorname{col}(x_1 - h_1, \ldots, x_n - h_n),$$

where A is nonsingular.

Effect of change of basis on matrix representing a linear mapping. Let T be a linear mapping of V^n into V^m. In assigning a matrix A to T we used the particular components of each vector in V^n or V^m with respect to the given bases: the standard basis in each case. If we change basis in V^n

and/or in V^m, we replace the old components by new ones and, accordingly, we change the matrix representing T. In old components in V^n and V^m, let T be given by

$$\text{col}\,(y_1, \ldots, y_m) = A\ \text{col}\,(x_1, \ldots, x_n). \tag{1–87}$$

If we introduce a new basis $\mathbf{u}_1^*, \ldots, \mathbf{u}_n^*$ in V^n, then the vector $\mathbf{x}$ obtains new components $(x_1^*, \ldots, x_n^*)$ given by an equation

$$\text{col}\,(x_1^*, \ldots, x_n^*) = C\ \text{col}\,(x_1, \ldots, x_n), \tag{1–88}$$

where C is nonsingular (and C is orthogonal if the new basis is orthonormal). If we also introduce a new basis $\mathbf{v}_1^*, \ldots, \mathbf{v}_m^*$ in V^m, then $\mathbf{y}$ also obtains new components $(y_1^*, \ldots, y_m^*)$, given by an equation

$$\text{col}\,(y_1^*, \ldots, y_m^*) = C_1\,\text{col}\,(y_1, \ldots, y_m), \tag{1–89}$$

where C_1 is an $m \times m$ nonsingular matrix. We now want to express $(y_1^*, \ldots, y_n^*)$ in terms of $(x_1^*, \ldots, x_n^*)$. From the last three equations we find easily:

$$\text{col}\,(y_1^*, \ldots, y_m^*) = C_1AC^{-1}\,\text{col}\,(x_1^*, \ldots, x_n^*). \tag{1–90}$$

Hence the new matrix representing T is

$$B = C_1AC^{-1}.$$

We remark that, for $(x_1^*, \ldots, x_n^*) = (0, \ldots, 0, 1, 0, \ldots, 0) = \mathbf{u}_j^*$ (1 in the j-th position), (1–90) gives $(y_1^*, \ldots, y_m^*)$ as the j-th column of B. Hence we can always obtain the matrix B as the matrix whose columns are the new m-tuples $(y_1^*, \ldots, y_m^*)$ assigned by T to the new basis vectors $\mathbf{u}_1^*, \ldots, \mathbf{u}_n^*$ in V^n. Equivalently, we have

$$T(\mathbf{u}_j^*) = b_{1j}\mathbf{v}_1^* + \cdots + b_{mj}\mathbf{v}_m^*, \quad j = 1, \ldots, n.$$

When T maps V^n into the same space V^n, one normally refers $\mathbf{x}$ and $\mathbf{y} = T(\mathbf{x})$ to the same basis. In this case $C_1 = C$ and (1–90) becomes

$$\text{col}\,(y_1^*, \ldots, y_n^*) = CAC^{-1}\,\text{col}\,(x_1^*, \ldots, x_n^*).$$

Thus by changing basis in V^n we replace the $n \times n$ matrix A representing T by the new matrix CAC^{-1}. One says that CAC^{-1} is *similar* to A (Section 1–5). If A has n real distinct eigenvalues $\lambda_1, \ldots, \lambda_n$, then as pointed out in Section 1–5, there is a diagonal matrix $B = \text{diag}\,(\lambda_1, \ldots, \lambda_n)$ similar to A. Hence in this case we can choose a new basis in V^n so that T has the equations

$$y_1^* = \lambda_1x_1^*, \ldots, y_n^* = \lambda_nx_n^*.$$

If the new basis in V^n is orthonormal, then C is orthogonal and $B = CAC^{-1}$ is *orthogonally congruent* to A (Section 1–7).

EXAMPLE 3. Let T map V^3 into V^2 and let T have the matrix

$$A = \begin{bmatrix} 1 & 2 & 3 \\ 4 & 1 & 2 \end{bmatrix}$$

(referred to standard bases). Let the new basis

$$\mathbf{u}_1^* = (1, 1, 0), \quad \mathbf{u}_2^* = (1, 2, 1), \quad \mathbf{u}_3^* = (0,1, 3)$$

be chosen in V^3 and let the new basis

$$\mathbf{v}_1^* = (1, 2), \quad \mathbf{v}_2^* = (2, 3)$$

be chosen in V^2. We seek the matrix for T referred to the new bases. To this end we remark that in V^3 the matrix C relating new and old components has inverse

$$C^{-1} = \begin{bmatrix} 1 & 1 & 0 \\ 1 & 2 & 1 \\ 0 & 1 & 3 \end{bmatrix}.$$

The j-th column of C^{-1} is $\mathbf{u}_j^*$ (in old components). Similarly in V^2 the matrix C_1 relating new and old components has inverse

$$C_1^{-1} = \begin{bmatrix} 1 & 2 \\ 2 & 3 \end{bmatrix}.$$

We now find C_1 as the inverse of C_1^{-1} and then

$$B = C_1AC^{-1} = \begin{bmatrix} -3 & 2 \\ 2 & -1 \end{bmatrix} \begin{bmatrix} 1 & 2 & 3 \\ 4 & 1 & 2 \end{bmatrix} \begin{bmatrix} 1 & 1 & 0 \\ 1 & 2 & 1 \\ 0 & 1 & 3 \end{bmatrix} = \begin{bmatrix} 1 & -8 & -19 \\ 1 & 8 & 15 \end{bmatrix}.$$

We check by verifying that

$$\begin{aligned} T(\mathbf{u}_1^*) &= A \text{ col } (1, 1, 0) = \text{col } (3, 5) = \mathbf{v}_1^* + \mathbf{v}_2^*, \\ T(\mathbf{u}_2^*) &= A \text{ col } (1, 2, 1) = \text{col } (8, 8) = -8\mathbf{v}_1^* + 8\mathbf{v}_2^*, \\ T(\mathbf{u}_3^*) &= A \text{ col } (0, 1, 3) = \text{col } (11, 7) = -19\mathbf{v}_1^* + 15\mathbf{v}_2^*. \end{aligned}$$

Problems

1. In V^2 let $\mathbf{i}^* = (\mathbf{i} - \mathbf{j})/\sqrt{2}$, $\mathbf{j}^* = (\mathbf{i} + \mathbf{j})/\sqrt{2}$. (a) Verify that $(\mathbf{i}^*, \mathbf{j}^*)$ is an orthonormal system. (b) Find the matrix A expressing new components in terms of old ones as in (1–79), if the basis $\mathbf{i}, \mathbf{j}$ is replaced by $\mathbf{i}^*, \mathbf{j}^*$. (c) Find the new components of $3\mathbf{i} + 2\mathbf{j}$ and verify graphically. (d) Find the old components of $2\mathbf{i}^* - \mathbf{j}^*$ and verify graphically.

2. In V^3, let $\mathbf{i}^* = (1/\sqrt{3})(\mathbf{i} + \mathbf{j} + \mathbf{k})$ and $\mathbf{j}^* = (1/\sqrt{6})(2\mathbf{i} - \mathbf{j} - \mathbf{k})$. (a) Find $\mathbf{k}^*$ so that $\mathbf{i}^*, \mathbf{j}^*, \mathbf{k}^*$ is a positive triple and forms an orthonormal basis. (b) Find the matrix A relating new components to old ones as in (1–79) if the basis $\mathbf{i}, \mathbf{j}, \mathbf{k}$ is replaced by $\mathbf{i}^*, \mathbf{j}^*, \mathbf{k}^*$. (c) Find the new components of $2\mathbf{i} + 3\mathbf{j} + \mathbf{k}$. (d) Find the old components of $\mathbf{i}^* - \mathbf{k}^*$.

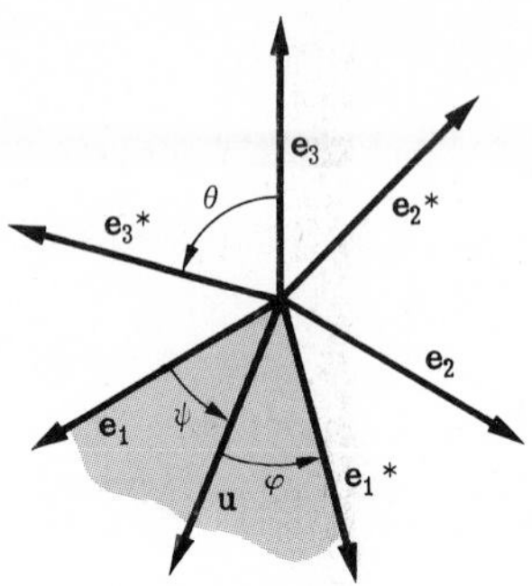

FIG. 1–7 Euler angles.

3. (a) In E^4 axes are translated to a new origin at $(3, 5, -1, 2)$. Find the new coordinates of the points P: $(0, 2, 1, 5)$, Q: $(1, 3, 4, -1)$ and O: $(0, 0, 0, 0)$ (old coordinates).

(b) Translate axes in E^3 so that the new equation of the sphere

$$x_1^2 + x_2^2 + x_3^2 - 4x_1 + 6x_2 - 10x_3 - 11 = 0$$

has the standard form $x_1^{*2} + x_2^{*2} + x_3^{*2} = a^2$ and find the center and radius.

4. (a) In E^2 new coordinates (x^*, y^*) are obtained by rotating axes through angle ω (measured as in trigonometry). Show that

$$\begin{bmatrix} x^* \\ y^* \end{bmatrix} = \begin{bmatrix} \cos\omega & \sin\omega \\ -\sin\omega & \cos\omega \end{bmatrix} \begin{bmatrix} x \\ y \end{bmatrix}.$$

(b) To obtain new axes in space, one can proceed as follows: One rotates axes through angle ψ about the z-axis, so that $\mathbf{e}_1$ moves to the position $\mathbf{u} = \cos\psi\mathbf{e}_1 + \sin\psi\mathbf{e}_2$. Then one rotates axes through angle θ about an axis through O having the direction of $\mathbf{u}$, so that $\mathbf{e}_3$ moves to the new position $\mathbf{e}_3^* = \cos\theta\mathbf{e}_3 + \sin\theta\mathbf{u}\times\mathbf{e}_3$. Finally one rotates axes through angle φ about an axis through O having the direction of $\mathbf{e}_3^*$, so that $\mathbf{u}$ is moved to the position $\mathbf{e}_1^* = \cos\varphi\mathbf{u} + \sin\varphi\mathbf{e}_3^*\times\mathbf{u}$, and takes $\mathbf{e}_2^* = \mathbf{e}_3^*\times\mathbf{e}_1^*$. Obtain the matrix $A = (\mathbf{e}_i^*\cdot\mathbf{e}_j)$. (See Fig. 1–7.)

Remark. The procedure described provides all new positively oriented coordinate systems with origin O. The angles ψ, θ, φ are called *Euler angles*. They are of importance in mechanics, in describing the motion of a solid body, and in astronomy, in describing the motion of planets.

5. Let T be a linear mapping of V^3 into V^2 with matrix $\begin{bmatrix} 5 & 1 & 1 \\ 7 & 2 & 0 \end{bmatrix}$. Find the matrix representing T for each of the following choices of new bases in V^3 and V^2:

(a) $\mathbf{i}+\mathbf{j}-\mathbf{k}$, $\mathbf{i}-\mathbf{j}+\mathbf{k}$, $-\mathbf{i}+\mathbf{j}+\mathbf{k}$ in V^3; $\mathbf{i}+2\mathbf{j}$, $2\mathbf{i}+3\mathbf{j}$ in V^2.
(b) $\mathbf{i}+\mathbf{j}$, $\mathbf{j}+\mathbf{k}$, $\mathbf{i}+\mathbf{k}$ in V^3; $\mathbf{i}$, $-\mathbf{i}+\mathbf{j}$ in V^2.

6. Let C_1 be an $m\times m$ matrix, A an $m\times n$ matrix, C an $n\times n$ matrix, and let C_1 and C be nonsingular. Show that A and C_1AC^{-1} have the same rank. [Hint: Consider A and C_1AC^{-1} as different matrices representing the same linear mapping. Interpret the rank geometrically.]

Answers

1. Denote $1/\sqrt{2}$ by c. (b) $\begin{bmatrix} c & -c \\ c & c \end{bmatrix}$, (c) col $(c, 5c)$, (d) col $(c, -3c)$.

2. Denote $1/\sqrt{3}$ by a, $1/\sqrt{2}$ by b. (a) $b\mathbf{j} - b\mathbf{k}$,
(b) $\begin{bmatrix} a & a & a \\ 2ab & -ab & -ab \\ 0 & b & -b \end{bmatrix}$, (c) $(2\sqrt{3}, 0, \sqrt{2})$, (d) $(a, a - b, a + b)$.

3. (a) $(-3, -3, 2, 3)$, $(-2, -2, 5, -3)$, $(-3, -5, 1, -2)$. (b) New origin at $(2, -3, 5)$ which is center of sphere, radius 7.

4. (b) A has row vectors
$(\cos\varphi\cos\psi - \sin\varphi\sin\psi\cos\theta,\ \cos\varphi\sin\psi + \sin\varphi\cos\psi\cos\theta,\ \sin\varphi\sin\theta)$,
$(-\sin\varphi\cos\psi - \cos\varphi\sin\psi\cos\theta,\ -\sin\varphi\sin\psi + \cos\varphi\cos\psi\cos\theta,\ \cos\varphi\sin\theta)$,
$(\sin\psi\sin\theta,\ -\cos\psi\sin\theta,\ \cos\theta)$.

5. (a) $\begin{bmatrix} 3 & -5 & -1 \\ 1 & 5 & -1 \end{bmatrix}$, (b) $\begin{bmatrix} 15 & 4 & 13 \\ 9 & 2 & 7 \end{bmatrix}$.

***1–15 Other vector spaces.** Thus far in this chapter we have considered only Euclidean n-dimensional vector spaces. One is often led to consider sets of objects which can be added and multiplied by scalars, in accordance with the familiar rules, but for which there may be no scalar product or norm. We call such a set, with the given operations, a *vector space*. The term "Euclidean" is reserved for a vector space having a scalar product for which *all* the rules (1–44) of Section 1–8 are satisfied. In each vector space one can define linear independence and dependence in the usual way. One then calls the vector space *n-dimensional* if it contains n but no more than n linearly independent vectors. For some vector spaces one can find n linearly independent vectors for *every* positive integer n; one then speaks of an *infinite-dimensional* vector space. For technical reasons, it is also useful to introduce a zero-dimensional vector space V^0, consisting of $\mathbf{0}$ alone; this vector space has no sets of linearly independent vectors.

For convenience we define a vector space formally:

Definition. A vector space V is a collection of objects $\mathbf{u}, \mathbf{v}, \ldots$ called vectors, including a zero vector $\mathbf{0}$, for which addition and multiplication by real scalars are defined in accordance with the following rules:

$$\begin{array}{ll} \text{I. } \mathbf{u} + \mathbf{v} = \mathbf{v} + \mathbf{u}. & \text{II. } (\mathbf{u} + \mathbf{v}) + \mathbf{w} = \mathbf{u} + (\mathbf{v} + \mathbf{w}). \\ \text{III. } h(\mathbf{u} + \mathbf{v}) = h\mathbf{u} + h\mathbf{v}. & \text{IV. } (a + b)\mathbf{u} = a\mathbf{u} + b\mathbf{u}. \\ \text{V. } (ab)\mathbf{u} = a(b\mathbf{u}), & \text{VI. } 1\mathbf{u} = \mathbf{u}. \\ \text{VII. } 0\mathbf{u} = \mathbf{0}. & \end{array} \qquad (1\text{–}91)$$

For many vector spaces the objects called vectors do not arise in a geometrical context and our standard boldface or arrow notations are not appropriate; see, for instance, Examples 1 and 2 below.

As pointed out in Section 1–9, other rules can be deduced from the rules I, ..., VII: for example, the rule stating that subtraction is always possible and is unique, and the rule $\mathbf{u} + \mathbf{0} = \mathbf{u}$. Furthermore, the discussion of basis and linear independence carries over to an arbitrary vector space. In particular, the rules (a), (b), (c), (d), (e) of Section 1–8 remain valid. Rules (f) and (g) are essentially the definition of an n-dimensional vector space. Rules (h), (i), (j), and (k) remain valid in the form:

(h') If $\mathbf{v}_1, \ldots, \mathbf{v}_n$ are linearly independent vectors in an n-dimensional vector space V, then $\mathbf{v}_1, \ldots, \mathbf{v}_n$ form a basis for V.

(i') Every basis for an n-dimensional vector space consists of n linearly independent vectors.

(j') If $k < n$ and $\mathbf{v}_1, \ldots, \mathbf{v}_k$ are linearly independent vectors in an n-dimensional vector space V, then there exist $\mathbf{v}_{k+1}, \ldots, \mathbf{v}_n$ so that $\mathbf{v}_1, \ldots, \mathbf{v}_n$ form a basis for V.

(k') If $\mathbf{v}_1, \ldots, \mathbf{v}_k$ are linearly independent vectors in a vector space, V and $\mathbf{u}_1, \ldots, \mathbf{u}_{k+1}$ are all expressible as linear combinations of $\mathbf{v}_1, \ldots, \mathbf{v}_k$, then $\mathbf{u}_1, \ldots, \mathbf{u}_{k+1}$ are linearly dependent.

The discussion of a general n-dimensional vector space V can in fact be referred back to the case of V^n by the following procedure. Let $\mathbf{v}_1, \ldots, \mathbf{v}_n$ be n linearly independent vectors in V. Since V does not contain $n + 1$ linearly independent vectors, it follows that every vector of V is expressible uniquely as a linear combination of $\mathbf{v}_1, \ldots, \mathbf{v}_n$, so that $\mathbf{v}_1, \ldots, \mathbf{v}_n$ form a basis. Hence for every vector $\mathbf{u}$ in V we can write

$$\mathbf{u} = u_1\mathbf{v}_1 + \cdots + u_n\mathbf{v}_n$$

and, as in Section 1–14, we call $u_1, \ldots, u_n$ the components of $\mathbf{u}$ with respect to the basis $\mathbf{v}_1, \ldots, \mathbf{v}_n$. We can now identify the vectors of V with the corresponding n-tuples, writing $\mathbf{u} = (\mathbf{u}_1, \ldots, \mathbf{u}_n)$. Then

$$c\mathbf{u} = (cu_1, \ldots, cu_n), \quad \mathbf{u} + \mathbf{w} = (u_1 + w_1, \ldots, u_n + w_n)$$

exactly as in V^n. Hence (apart from considerations related to the scalar product or norm), V is the same as V^n.

For a general, not necessarily finite-dimensional, vector space V, one can take advantage of the theory of V^n in the following way. Let $\mathbf{v}_1 \ldots, \mathbf{v}_n$ be linearly independent vectors in V. Then all the linear combinations of $\mathbf{v}_1, \ldots, \mathbf{v}_n$ themselves form a vector space V', as in the preceding paragraph, and we can identify V' with V^n by replacing each vector of V' by the corresponding n-tuple of components with respect to $\mathbf{v}_1, \ldots, \mathbf{v}_n$. Hence V' forms an n-dimensional vector space. We call V' an n-dimensional *subspace* of V. The term "subspace" is used more generally for any vector space contained in a given vector space V.

As a special case of the preceding paragraph, we conclude that if all vectors in a vector space V are expressible as linear combinations of n linearly independent vectors $\mathbf{v}_1, \ldots, \mathbf{v}_n$, then V is n-dimensional and $\mathbf{v}_1, \ldots, \mathbf{v}_n$ from a basis for V.

We proceed to give some examples of vector spaces:

EXAMPLE 1. Let V consist of *all* polynomials in x: $3 - x$, $5 + x + 2x^2$, $7 - 3x^2 + 9x^3, \ldots$ The sum of two polynomials is again a polynomial; a scalar times a polynomial is again a polynomial:

$$3(5 + x + 2x^2) = 15 + 3x + 6x^2.$$

There is a zero polynomial: 0. The rules I, . . . , VII are satisfied, as one sees at once. For example, if $p(x)$ and $q(x)$ are polynomials, then $p(x) + q(x) = q(x) + p(x)$ (equality meaning identity). Thus rule I holds. Hence *V is a vector space.* We observe that the n polynomials $1, x, x^2, \ldots, x^{n-1}$ are linearly independent. For suppose that

$$c_1 1 + c_2 x + \cdots + c_n x^{n-1} = 0;$$

that is, let $c_1 + c_2 x + \cdots + c_n x^{n-1}$ coincide with the 0 polynomial and hence have the value 0 for all x. Now a polynomial of degree k has at most k roots, and hence $c_1, \ldots, c_n$ must all be 0. Therefore, $1, x, \ldots, x^{n-1}$ are linearly independent. Accordingly, *V is an infinite-dimensional* vector space.

EXAMPLE 2. Let V consist of all polynomials of degree at most 3: that is, of all polynomials of the form $a_0 + a_1 x + a_2 x^2 + a_3 x^3$. The sum of two such polynomials is again a polynomial of degree at most 3; there is a similar statement for a scalar times such a polynomial. The rules I through VII all hold. Therefore, V is a vector space. The four polynomials $1, x, x^2, x^3$ are in V and are linearly independent. Thus we can consider V as the space V', as above, formed of all linear combinations of four linearly independent vectors in the vector space of Example 1. Accordingly, as above, V is a four-dimensional subspace of the vector space of all polynomials.

EXAMPLE 3. Let V consist of all 2×2 matrices. Then again addition and multiplication by scalars yield matrices of the same size and, by the rules of Section 1–2, all the rules I through VII are satisfied. The "zero vector" is $O = O_{22}$. In V the four matrices

$$E_{11} = \begin{bmatrix} 1 & 0 \\ 0 & 0 \end{bmatrix}, \quad E_{12} = \begin{bmatrix} 0 & 1 \\ 0 & 0 \end{bmatrix}, \quad E_{21} = \begin{bmatrix} 0 & 0 \\ 1 & 0 \end{bmatrix}, \quad E_{22} = \begin{bmatrix} 0 & 0 \\ 0 & 1 \end{bmatrix}$$

are linearly independent. For the equation

$$c_1 E_{11} + c_2 E_{12} + c_3 E_{21} + c_4 E_{22} = O$$

is equivalent to

$$\begin{bmatrix} c_1 & c_2 \\ c_3 & c_4 \end{bmatrix} = O = \begin{bmatrix} 0 & 0 \\ 0 & 0 \end{bmatrix}$$

and hence to $c_1 = 0, \ldots, c_4 = 0$. Furthermore, every 2×2 matrix A is expressible uniquely as a linear combination of $E_{11}, \ldots, E_{22}$: $A = (a_{ij}) =$

$a_{11}E_{11} + \cdots + a_{22}E_{22}$. It follows as above that V is 4-dimensional and that $E_{11}, \ldots, E_{22}$ form a basis for V.

By similar reasoning, we show that, for each fixed choice of m and n, the set V of all $m \times n$ matrices forms a vector space of dimension mn. The mn matrices E_{ij}, having 1 as the ij-entry and all other entries 0, form a basis for V.

Complex vector spaces. One can extend the concept of vector space by allowing the scalars to be complex numbers. One then obtains a *complex vector* space. The previous theory carries over with the obvious changes.

EXAMPLE 4. The set of all complex numbers forms a complex vector space, whose dimension is 1, with basis consisting of any one nonzero complex number.

EXAMPLE 5. The set of all polynomials with complex coefficients forms an infinite-dimensional complex vector space.

EXAMPLE 6. The set of all functions of form $ae^{2ix} + be^{-2ix}$, where a and b are complex constants, forms a complex vector space of dimension 2, with basis e^{2ix}, e^{-2ix}.

Problems

1. Show that each of the following sets of objects, with the usual operations of addition and multiplication by scalars, forms a vector space. Give the dimension in each case and, if the dimension is finite, give a basis.

(a) All polynomials of degree at most 2.

(b) All polynomials containing no term of odd degree: $3 + 5x^2 + x^4$, $x^2 - x^{10}, \ldots$

(c) all *trigonometric polynomials:*

$$a_0 + a_1 \cos x + b_1 \sin x + \cdots + a_n \cos nx + b_n \sin nx.$$

(d) All functions of the form $ae^x + be^{-x}$.

(e) All 3×3 diagonal matrices.

(f) All 4×4 symmetric matrices A; that is, all matrices A such that $A = A'$.

(g) All functions $y = f(x)$, $-\infty < x < \infty$, such that $y'' + y = 0$.

(h) All functions $y = f(x)$, $-\infty < x < \infty$, such that $y''' - y' = 0$.

(i) All functions $f(x)$ which are defined and continuous for $0 \leqq x \leqq 1$.

(j) All functions $f(x)$ which are defined and have a continuous derivative for $0 \leqq x \leqq 1$.

(k) All infinite sequences: $x_1, x_2, \ldots, x_n, \ldots$

(l) All convergent sequences.

2. Show that the four polynomials 1, $1 + x$, $1 + x + x^2$, $1 + x + x^3$ form a basis for the vector space of Example 2.

3. Show that the following matrices form a basis for the vector space of Example 3:

$$\begin{bmatrix} 1 & 2 \\ 1 & -3 \end{bmatrix}, \begin{bmatrix} 0 & 1 \\ 2 & -1 \end{bmatrix}, \begin{bmatrix} 5 & 2 \\ 7 & 1 \end{bmatrix}, \begin{bmatrix} 0 & 1 \\ 1 & 0 \end{bmatrix}.$$

4. Show that the functions $\cos 2x$, $\sin 2x$ form a basis for the complex vector space of Example 6.

Answers

1. (a) 3, basis: $1, x, x^2$. (b) infinite. (c) infinite. (d) 2, basis: e^x, e^{-x}. (e) 3, basis: E_{11}, E_{22}, E_{33}. (f) 10, basis: $E_{11}, E_{22}, E_{33}, E_{44}, E_{12} + E_{21}, E_{13} + E_{31}, E_{14} + E_{41}, E_{23} + E_{32}, E_{24} + E_{42}, E_{34} + E_{43}$. (g) 2, basis: $\cos x, \sin x$. (h) 3, basis: $1, e^x, e^{-x}$. (i) infinite. (j) infinite. (k) infinite (l) infinite.

Suggested References

BIRKHOFF, G., and MACLANE, S., *Survey of Modern Algebra*, 3rd ed. New York: Macmillan, 1965.

CURTIS, CHARLES W., *Linear Algebra: An Introductory Approach*, 2nd ed. Boston: Allyn and Bacon, 1968.

GANTMACHER, F. R., *The Theory of Matrices*, transl. by K. A. Hirsch, 2 vols. New York: Chelsea Publishing Co., 1960.

KAPLAN, WILFRED, and LEWIS, DONALD J., *Calculus and Linear Algebra*. New York: John Wiley and Sons, Inc., 1970–1971.

KUIPER, NICOLAAS H., *Linear Algebra and Geometry*, transl. by A. van der Sluis. New York: Interscience, 1962.

PERLIS, SAM, *Theory of Matrices*. Reading, Mass.: Addison-Wesley, 1952.

CHAPTER 2

Differential Calculus of Functions of Several Variables

2–1 Functions of several variables. If to each point (x, y) of a certain part of the xy plane is assigned a real number z, then z is said to be given as a function of the two real variables x and y. Thus

$$z = x^2 - y^2, \quad z = x \sin xy \quad [\text{all } (x, y)]$$

are such functions. Many such functions are considered, without special mention, in the theory of functions of one variable (Sections 0–6 to 0–9). For example, the function $y = a^x$ is a function of both a and x, as is the function $y = \log_a x$. The basic theorems on differentiation relate to the functions $y = u + v$, $y = u \cdot v$, $y = u/v$, i.e., to certain functions of u and v, where u and v are functions of x. In many cases a function of one variable can be considered as a function of two variables with one of the two variables held fixed; e.g., $y = x^3$ is obtained from $y = x^n$ (a function of x and n) by assigning to n the value 3.

Similar remarks apply to functions of three or more variables. Thus

$$u = xyz, \quad u = x^2 + y^2 + z^2 - t^2$$

give u as a function of three and four variables respectively.

It will be seen throughout the following that the theory of functions of three or more variables differs only slightly from that of functions of two variables. For this reason, most of the emphasis will be on functions of two variables. On the other hand, there are basic differences between the calculus of functions of one variable and that of functions of two variables.

Functions of two, three, four and even millions of variables occur in physics. The following are simple examples:

$$p = \frac{RT}{V} \text{ (ideal gas law)},$$

$$L = \frac{\pi r^4 \theta n}{2} \text{ (torque on a wire)},$$

$$E = \frac{m}{2}\sum_{i=1}^{N} (u_i^2 + v_i^2 + w_i^2) \text{ (energy of an ideal gas in terms of velocity components of the } N \text{ molecules)}.$$

2–2 Domains and regions. Most of the theory of functions of one variable is given in terms of a function defined over an interval: $a \leqq x \leqq b$. For functions of x and y a corresponding concept is needed. A natural one would be a rectangle: $a \leqq x \leqq b$, $c \leqq y \leqq d$. But many problems require more complicated sets; circles, ellipses, etc. In order to have sufficient generality to cover all practical cases, it is necessary to formulate the concept of a domain.

The general term *set of points* in the xy plane means any sort of collection of points, finite or infinite in number: the points $(0, 0)$ and $(1, 0)$; the points on the line $y = x$; the points inside the circle $x^2 + y^2 = 1$. (See Section 0–7).

A *neighborhood* of a point (x_1, y_1) will mean the set of points inside a circle having center (x_1, y_1) and radius δ; we can thus speak of a *neighborhood of radius* δ. Each point (x, y) of the neighborhood satisfies the inequality:

$$(x - x_1)^2 + (y - y_1)^2 < \delta^2. \tag{2–1}$$

A set of points is called *open* if every point (x_1, y_1) of the set has a neighborhood lying wholly within the set. The interior of a circle is open, as is the interior of an ellipse or a square; these open sets are defined by inequalities such as the following:

$$x^2 + y^2 < 1; \quad \frac{x^2}{2} + \frac{y^2}{3} < 1; \quad |x| < 1 \quad \text{and} \quad |y| < 1.$$

The entire xy plane is open, as is a half-plane such as the "right half-plane": $x > 0$. However, the interior of a circle plus the circumference is not open; for no neighborhood of a point on the circumference lies entirely in the set.

A set E is called *closed* if the points of the plane which are not in E form an open set. Thus the points on and outside the circle $x^2 + y^2 = 1$ form a closed set. The points of the circumference themselves form a closed set, as do the points on and interior to the circumference.

A set is called *bounded* if the whole set can be enclosed in a circle of sufficiently large radius. Thus the points of the square: $|x| \leqq 1$, $|y| \leqq 1$ form a bounded set; this set is also closed. The points interior to an ellipse: $x^2 + 2y^2 < 1$ form a bounded open set.

A nonempty open set is called a *connected open set* or a *domain* if, besides being open, it has the property that any two points P, Q of the set can be joined by a broken line lying wholly within the set. Thus the interior of a circle is a domain. (As in Section 0–7, the word "domain" is also used for the set on which a given function is defined. The context will make clear in which sense the word is to be understood. Furthermore, in many cases the domain of definition of a function is also a domain as defined here.)

We remark that a *domain D cannot be formed of two nonoverlapping open sets.* For example, the points for which $|x| > 0$ form an open set E composed of two parts: the set of points for which $x > 0$ and the set for which $x < 0$. The set E is not a domain, since the points $(-1, 0)$ and $(1, 0)$ lie in E but cannot be joined by a broken line in E.

A *boundary point* of a set is a point every neighborhood of which contains at least one point in the set and at least one point not in the set. Thus the boundary points of the circular domain: $x^2 + y^2 < 1$ are the points of the circumference: $x^2 + y^2 = 1$. No boundary point of an open set can belong to the set; however, every boundary point of a closed set belongs to the set.

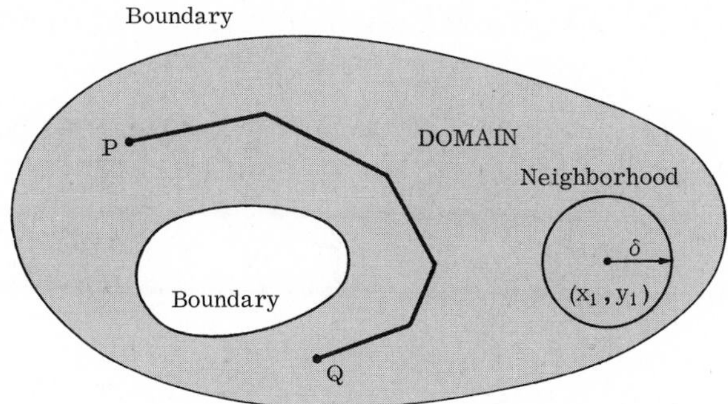

FIG. 2–1. Set concepts.

The term *region* will be used to describe a set consisting of a domain plus, perhaps, some or all of its boundary points. Thus a region may be a domain (if no boundary points are included). If all boundary points are included, the region is called a *closed region;* it then necessarily forms a *closed* set. Thus a circle plus interior: $x^2 + y^2 \leqq 1$ is a closed region. A domain is sometimes called an *open region.*

It will be found that for most practical problems a domain is defined by one or more inequalities, and the boundary of a domain is defined by one or more equations, while a closed region is given by a combination of the two; e.g.,

$$\begin{aligned} xy &< 1 \text{ is a domain,} \\ xy &= 1 \text{ is its boundary,} \\ xy &\leqq 1 \text{ is a closed region.} \end{aligned}$$

These concepts are illustrated in Fig. 2–1.

The extension of these ideas to three or more dimensions is not difficult; for four or more dimensions graphical representation is essentially hopeless. Thus a *neighborhood* of a point (x_1, y_1, z_1) in space is the set of points (x, y, z) inside a sphere:

$$(x - x_1)^2 + (y - y_1)^2 + (z - z_1)^2 < \delta^2,$$

and the other definitions can be repeated without change. In general, in n-dimensional space E^n (Section 1–8), a neighborhood of radius δ of a point A: $(a_1, \ldots, a_n)$ consists of all points P: $(x_1, \ldots, x_n)$ such that $d(A, P) < \delta$; that is, such that

$$|\overrightarrow{AP}|^2 = (x_1 - a_1)^2 + \cdots + (x_n - a_n)^2 < \delta^2.$$

The other concepts, such as open set and closed set, are defined in terms of neighborhoods as above.

The definitions can also be adapted to the case of one dimension. A *neighborhood* of a point x_1 of the x axis is an interval: $x_1 - \delta < x < x_1 + \delta$. A *domain* on the x axis is one of the following four types of sets: (1) an *open interval:* $a < x < b$; (2) an *infinite open interval:* $a < x$; (3) an *infinite open interval:* $x < b$; (4) the entire x axis. A *bounded closed region* on the x axis is a *closed interval:* $a \leqq x \leqq b$.

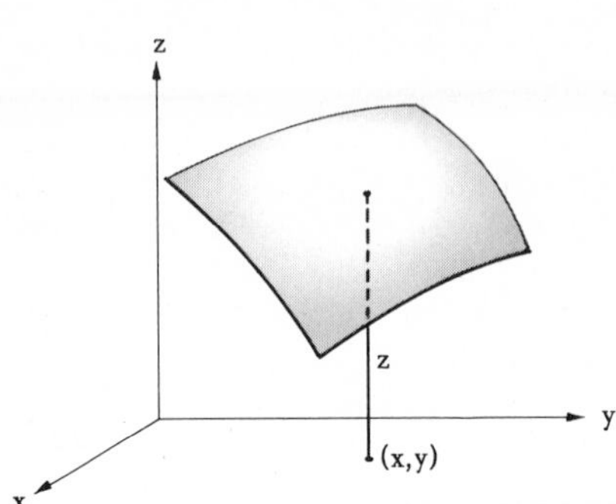

FIG. 2–2. Function of two variables.

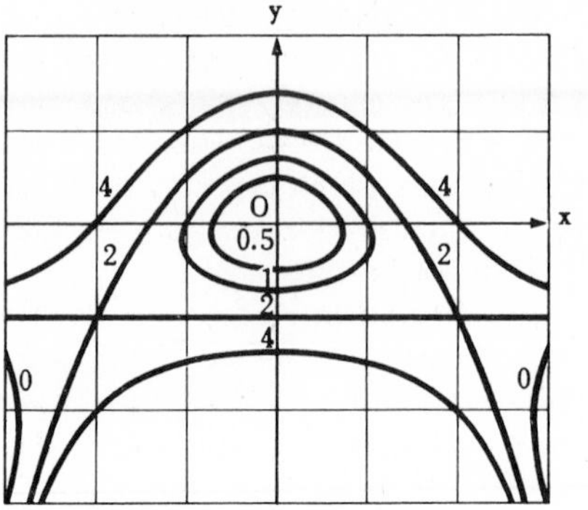

FIG. 2–3. Level curves.

2–3 Functional notation. Level curves and level surfaces. Most of the functions to be considered will be defined in a domain or, occasionally, in a closed region. The notation: "$z = f(x, y)$ in domain D" will mean that z is given as a function of x and y for all points in a domain D of the xy plane. The variables x and y are called *independent variables*, while z is *dependent*. Similarly one writes: "$u = f(x, y, z)$ in domain D" or "$w = f(x, y, z, u)$ in domain D" for functions of more than two variables. As with functions of one variable, the functional notation also serves to indicate corresponding values for a given function. Thus, if $z = f(x, y)$ is defined by the equation: $z = \sqrt{1 - x^2 - y^2}$ in the domain: $x^2 + y^2 < 1$, then $f(0, 0) = 1, f(\frac{1}{2}, \frac{1}{2}) = \sqrt{\frac{1}{2}}$, etc.

The functional relationship: $z = f(x, y)$ is sometimes written thus: $z = z(x, y)$. For functions of three or more variables one writes similarly: $u = u(x, y, z)$, $w = w(x, y, z, u)$.

A function of two variables can be represented graphically by a surface in three-dimensional space, as shown in Fig. 2–2. For functions of three or more variables, this representation is not available.

Another method for representing functions of two variables is that of *level curves* or *contour lines*. This is the method used in making a contour map or topographical map. One plots the loci:

$$f(x, y) = c_1, \quad f(x, y) = c_2, \ldots$$

for various choices of the constants $c_1, c_2, \ldots$; each locus $f(x, y) = c$ is called a *level curve* of $f(x, y)$; it may actually be formed of several distinct curves. This is illustrated in Fig. 2–3, in which $f(x, y) = x^2y + x^2 + 2y^2$; the value of c is shown on each curve. The level curves often provide a better understanding of the function than a sketch of the surface $z = f(x, y)$.

The method just described is available in principle for functions of three variables; here one draws the *level surfaces:* $f(x, y, z) = c_1$, $f(x, y, z) = c_2$, ... for appropriate choices of $c_1, c_2, \ldots$. The surfaces of constant gravitational potential (approximately spheres) about the earth illustrate this. The surfaces of constant temperature or pressure are of importance in meteorology.

One can also represent a function of three variables with the aid of level *curves;* for if, for example, z is fixed, then $f(x, y, z)$ becomes a function of x and y and can be represented by its level curves in an xy plane as above. If this is done for several values of z, one obtains a corresponding number of diagrams which together represent the function. This is common practice in meteorology, in which lines of constant pressure ("isobars") are plotted for various altitudes.

For functions of four or more variables the loci, $f = \text{const}$, are "hypersurfaces" in a space of four or more dimensions. These level hypersurfaces are mainly of theoretical interest. To represent the function graphically, one is forced to fix one or more of the variables and thereby obtain level surfaces in three-dimensional space or level curves in a plane.

A function $f(x, y)$ is said to be *bounded* when (x, y) is restricted to a set E, if there is a number M such that $|f(x, y)| < M$ when (x, y) is in E. For example, $z = x^2 + y^2$ is bounded, with $M = 2$, if $|x| < 1$ and $|y| < 1$. The function $z = \tan(x + y)$ is not bounded for $|x + y| < \frac{1}{2}\pi$.

2–4 Limits and continuity. Let $z = f(x, y)$ be given in a domain D, and let (x_1, y_1) be a point of D or a boundary point of D. Then the equation

$$\lim_{\substack{x \to x_1 \\ y \to y_1}} f(x, y) = c \tag{2–2}$$

means the following: given any $\epsilon > 0$, a $\delta > 0$ can be found such that for every (x, y) in D and within the neighborhood of (x_1, y_1) of radius δ, except possibly for (x_1, y_1) itself, one has

$$|f(x, y) - c| < \epsilon. \tag{2–3}$$

In other terms, if (x, y) is in D and

$$0 < (x - x_1)^2 + (y - y_1)^2 < \delta^2, \tag{2–4}$$

then (2–3) holds. Thus, if the variable point (x, y) is sufficiently close to (but not at) its limiting position (x_1, y_1), the value of the function is as close as desired to its limiting value c.

If the point (x_1, y_1) is in D and

$$\lim_{\substack{x \to x_1 \\ y \to y_1}} f(x, y) = f(x_1, y_1), \tag{2–5}$$

then $f(x, y)$ is said to be *continuous* at (x_1, y_1). If this holds for every point (x_1, y_1) of D, then $f(x, y)$ is said to be *continuous in* D.

The notions of limits and continuity can be extended to more complicated sets, for example, to closed regions. The above definitions can be repeated essentially without change. Thus, if $f(x, y)$ is defined in a closed region R and (x_1, y_1) is in R, then (2–2) is said to hold if, for given $\epsilon > 0$, a $\delta > 0$ can be found such that (2–3) holds whenever (x, y) *is in* R and is within distance δ of (x_1, y_1), but not at (x_1, y_1). If (2–5) holds, then $f(x, y)$ is continuous at (x_1, y_1). Similar definitions hold if $f(x, y)$ is defined only on a curve in the xy plane. The notions of limits and continuity must always be considered *relative to the set in which the function is defined.*

Continuity for functions of two variables is a more subtle requirement than for functions of one variable. Such a simple function as

$$z = \frac{x^2 - y^2}{x^2 + y^2}$$

is badly discontinuous at the origin, without becoming infinite there. For z has limit 0 if (x, y) approaches the origin on the line $x = y$, has limit 1 if (x, y) approaches the origin on the x axis, and has limit -1 if (x, y) approaches the origin on the y axis. Thus no limiting value can be assigned at $(0, 0)$. It should be noted that the level curves of this function are straight lines, all passing through $(0, 0)$: this alone shows that there is a discontinuity at the origin.

However, the fundamental theorem on limits and continuity (Section 0–7) holds without change:

THEOREM. *Let $u = f(x, y)$ and $v = g(x, y)$ both be defined in the domain D of the xy plane. Let*

$$\lim_{\substack{x \to x_1 \\ y \to y_1}} f(x, y) = u_1, \quad \lim_{\substack{x \to x_1 \\ y \to y_1}} g(x, y) = v_1. \tag{2–6}$$

Then

$$\lim_{\substack{x \to x_1 \\ y \to y_1}} [f(x, y) + g(x, y)] = u_1 + v_1, \tag{2–7}$$

$$\lim_{\substack{x \to x_1 \\ y \to y_1}} [f(x, y) \cdot g(x, y)] = u_1 \cdot v_1, \tag{2–8}$$

$$\lim_{\substack{x \to x_1 \\ y \to y_1}} \frac{f(x, y)}{g(x, y)} = \frac{u_1}{v_1} \quad (v_1 \neq 0). \tag{2–9}$$

If $f(x, y)$ and $g(x, y)$ are continuous at (x_1, y_1), then so also are the functions

$$f(x, y) + g(x, y), \quad f(x, y) \cdot g(x, y), \quad \frac{f(x, y)}{g(x, y)},$$

provided, in the last case, $g(x_1, y_1) \neq 0$.

Let $F(u, v)$ be defined and continuous in a domain D_0 of the uv plane and let $F[f(x, y), g(x, y)]$ be defined for (x, y) in D. Then, if (u_1, v_1) is in D_0,

$$\lim_{\substack{x \to x_1 \\ y \to y_1}} F[f(x, y), g(x, y)] = F(u_1, v_1). \tag{2–10}$$

If $f(x, y)$ and $g(x, y)$ are continuous at (x_1, y_1), then so also is $F[f(x, y), g(x, y)]$.

Proof. We first consider the composite function $F[f(x, y), g(x, y)]$, which is the most fundamental notion of the whole theorem. Since

$F[u, v]$ is assumed continuous in D_0, one has

$$\lim_{\substack{u \to u_1 \\ v \to v_1}} F[u, v] = F[u_1, v_1]; \tag{2–11}$$

by (2–6), as (x, y) approaches (x_1, y_1), (u, v) approaches (u_1, v_1), so that, by (2–11),

$$\lim_{\substack{x \to x_1 \\ y \to y_1}} F[f(x, y),\ g(x, y)] = F[\lim_{\substack{x \to x_1 \\ y \to y_1}} f(x, y),\ \lim_{\substack{x \to x_1 \\ y \to y_1}} g(x, y)] = F[u_1, v_1].$$

Thus (2–10) is established. If f and g are continuous at (x_1, y_1), then $f(x_1, y_1) = u_1$ and $g(x_1, y_1) = v_1$, so that, by (2–10),

$$\lim_{\substack{x \to x_1 \\ y \to y_1}} F[f(x, y), g(x, y)] = F[f(x_1, y_1), g(x_1, y_1)];$$

i.e., $F[f(x, y), g(x, y)]$ is continuous at (x_1, y_1).

Now one verifies easily that the particular $F[u, v] \equiv u + v$ is continuous for all values of u and v. If (2–10) is applied to this choice of F, one finds

$$\lim_{\substack{x \to x_1 \\ y \to y_1}} [f(x, y) + g(x, y)] = u_1 + v_1,$$

which is (2–7); by the same reasoning one concludes that if $f(x, y)$ and $g(x, y)$ are continuous at (x_1, y_1), then so is $f(x, y) + g(x, y)$.

The statements about products and quotients follow in the same way by consideration of the special functions $F \equiv u \cdot v$ and $F \equiv u/v$. One need only show that these functions are continuous (for $v \neq 0$ in the second case). This can be done directly by applying the ϵ, δ definition above, or as follows. One shows that the functions $u + v$ and $u - v$ are continuous functions of u and v, also that $\frac{1}{4}w$ and w^2 are continuous functions of w (theorems on functions of one variable). It follows from the function-of-function rule just proved that $(u + v)^2$ and $(u - v)^2$ are continuous and hence that

$$u \cdot v \equiv \tfrac{1}{4}[(u + v)^2 - (u - v)^2]$$

is continuous for all (u, v). Finally one shows that $1/v$ is a continuous function of v for $v \neq 0$ (function of one variable) and hence that

$$\frac{u}{v} \equiv u \cdot \frac{1}{v}$$

is a continuous function of u and v for $v \neq 0$.

The theorem above can be restated in analogous form for functions of three or more variables and, in the case of the composite function, for combinations of functions of one and two variables, one and three variables, etc.:

$$F[f(x, y)],\quad F[f(t), g(t)],\quad F[f(x, y, z)],\quad F[f(t), g(t), h(t)], \ldots.$$

By virtue of this theorem one can conclude that *polynomial* functions such as

$$w = x^3y + 3xz^2 - xyz$$

are continuous for all values of the variables, whereas *rational* functions, such as

$$w = \frac{x^2y - x}{1 - x^2 - y}$$

are continuous except where the denominator is 0.

Problems

1. Give several examples of functions of several variables occurring in geometry (area and volume formulas, law of cosines, etc.).

2. Represent the following functions (a) by sketching a surface, (b) by drawing level curves:

(i) $z = 3 - x - 3y$ (iii) $z = \sin(x + y)$
(ii) $z = x^2 + y^2 + 1$ (iv) $z = e^{xy}$

3. Analyze the following functions by describing their level surfaces in space:

(a) $u = x^2 + y^2 + z^2$ (c) $w = x^2 + y^2 - z$
(b) $u = x + y + z$ (d) $w = x^2 + y^2$

4. Determine the values of the following limits, wherever the limit exists:

(a) $\lim\limits_{\substack{x \to 0 \\ y \to 0}} \dfrac{x^2 - y^2}{1 + x^2 + y^2}$ (c) $\lim\limits_{\substack{x \to 0 \\ y \to 0}} \dfrac{(1 + y^2)\sin x}{x}$

(b) $\lim\limits_{\substack{x \to 0 \\ y \to 0}} \dfrac{x}{x^2 + y^2}$ (d) $\lim\limits_{\substack{x \to 0 \\ y \to 0}} \dfrac{1 + x - y}{x^2 + y^2}$

5. Show that the following functions are discontinuous at (0, 0) and graph the corresponding surfaces:

(a) $z = \dfrac{x}{x - y}$ (b) $z = \log(x^2 + y^2)$

6. Describe the sets in which the following functions are defined:

(a) $z = e^{x-y}$ (c) $z = \sqrt{1 - x^2 - y^2}$

(b) $z = \log(x^2 + y^2 - 1)$ (d) $u = \dfrac{xy}{z}$

7. Prove the theorem: *Let $f(x, y)$ be defined in domain D and continuous at the point (x_1, y_1) of D. If $f(x_1, y_1) > 0$, then there is a neighborhood of (x_1, y_1) in which $f(x, y) > \frac{1}{2}f(x_1, y_1) > 0$.* [Hint: use $\epsilon = \frac{1}{2}f(x_1, y_1)$ in the definition of continuity.

8. Prove the theorem: *Let $f(x, y)$ be continuous in domain D. Let $f(x, y)$ be positive for at least one point of D and negative for at least one point of D. Then $f(x, y) = 0$ for at least one point of D.* [Hint: use Prob. 7 to conclude that the set A where $f(x, y) > 0$ and the set B where $f(x, y) < 0$ are open. If $f(x, y) \neq 0$ in D, then D is formed of the two nonoverlapping open sets A and B; this is not possible by Section 2–2.]

Remark. The analogous proposition for functions of one variable is the *intermediate value theorem:* if $f(x)$ is continuous on an interval containing a and b, and $f(a) < 0, f(b) > 0$, then $f(x) = 0$ for some x between a and b.

Answers

4. (a) 0, (b) no limit, (c) 1, (d) ∞. 6. (a) all (x, y), a domain; (b) $x^2 + y^2 > 1$, a domain; (c) $x^2 + y^2 \leqq 1$, a closed region; (d) all (x, y, z) except the points of the xy plane, an open set, not a domain.

2–5 Partial derivatives. Let $z = f(x, y)$ be defined in a domain D of the xy plane and let (x_1, y_1) be a point of D. The function $f(x, y_1)$ then depends on x alone and is defined in an interval about x_1. Hence its derivative with respect to x at $x = x_1$ may exist. If it does, its value is called the *partial derivative of $f(x, y)$ with respect to x at* (x_1, y_1), and is denoted by

$$\frac{\partial f}{\partial x}(x_1, y_1) \quad \text{or by} \quad \left.\frac{\partial z}{\partial x}\right|_{(x_1, y_1)}$$

One has thus, by the definition of the derivative:

$$\frac{\partial f}{\partial x}(x_1, y_1) = \lim_{\Delta x \to 0} \frac{f(x_1 + \Delta x, y_1) - f(x_1, y_1)}{\Delta x}. \tag{2–12}$$

When the point (x_1, y_1) at which the derivative is being evaluated is evident, one can write simply $\partial z/\partial x$ or $\partial f/\partial x$ for the derivative. Other notations commonly used are z_x, f_x, f_1 or, more explicitly, $z_x(x_1, y_1)$, $f_x(x_1, y_1)$, $f_1(x_1, y_1)$. When subscripts are used, there can be confusion with the symbol for components of a vector; hence, *when vectors and partial derivatives are being used together, a notation such as $\partial z/\partial x$ or $\partial f/\partial x$ is preferable for partial derivatives.*

The function $z = f(x, y)$ can be represented by a surface in space. The equation $y = y_1$ then represents a plane cutting the surface in a curve.

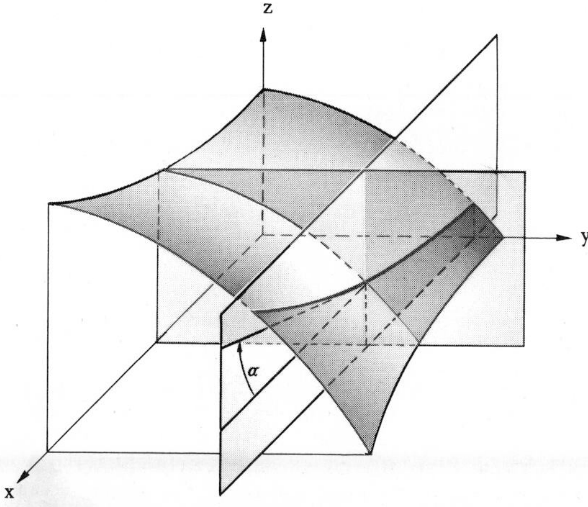

FIG. 2–4. Partial derivatives.

The partial derivative $\partial z/\partial x$ at (x_1, y_1) can then be interpreted as the slope of the tangent to the curve, i.e., as $\tan \alpha$, where α is the angle shown in Fig. 2–4. In this figure $z = 5 + x^2 - y^2$ and the derivative is being computed at the point $x = 1, y = 2$. For $y = 2, z = 1 + x^2$ so that the derivative along the curve is $2x$; for $x = 1$, one finds $f_x(1, 2)$ to be 2.

The partial derivative $\left.\dfrac{\partial z}{\partial y}\right|_{(x_1, y_1)}$ is defined similarly; one now holds x constant, equal to x_1, and differentiates $f(x_1, y)$ with respect to y. One has thus

$$\frac{\partial f}{\partial y}(x_1, y_1) = \left.\frac{\partial z}{\partial y}\right|_{(x_1, y_1)} = \lim_{\Delta y \to 0} \frac{f(x_1, y_1 + \Delta y) - f(x_1, y_1)}{\Delta y}. \tag{2–13}$$

This can also be interpreted as the slope of the tangent to the curve in which the plane $x = x_1$ cuts the surface $z = f(x, y)$. One also writes $f_y(x_1, y_1)$, $f_2(x_1, y_1)$ for this derivative.

If the point (x_1, y_1) is now varied, one obtains (wherever the derivative exists) a new function of two variables: the function $f_x(x, y)$. Similarly the derivative $\partial z/\partial y$ at a variable point (x, y) is a function $f_y(x, y)$. For explicit functions $z = f(x, y)$, evaluation of these derivatives is carried out as in ordinary calculus, for one is always differentiating a function of one variable, the *other being treated as a constant.* For example, if $z = x^2 - y^2$, then

$$\frac{\partial z}{\partial x} = 2x, \quad \frac{\partial z}{\partial y} = -2y.$$

The above definitions extend at once to functions of three or more variables. If $w = g(x, y, u, v)$, then one partial derivative at (x, y, u, v) is

$$\frac{\partial w}{\partial u} = \lim_{\Delta u \to 0} \frac{g(x, y, u + \Delta u, v) - g(x, y, u, v)}{\Delta u}. \tag{2–14}$$

When only three variables x, y, z are involved in a discussion, the notation $\partial z/\partial x$ is self-explanatory: x and y are independent and y is held constant. However, when four or more variables are involved, the partial derivative symbol by itself is ambiguous. Thus, if x, y, u, v are involved, then $\partial u/\partial x$ may be interpreted as $f_x(x, y)$, where $u = f(x, y)$, $v = g(x, y)$; one may also interpret $\partial u/\partial x$ as $h_x(x, y, v)$, where $u = h(x, y, v)$. For this reason, when four or more variables are involved, it is advisable to supplement the partial derivative symbol by indicating the variables held constant. For example,

$$\left(\frac{\partial z}{\partial x}\right)_y \text{ means } f_x(x, y), \text{ where } z = f(x, y),$$

$$\left(\frac{\partial u}{\partial x}\right)_{yv} \text{ means } h_x(x, y, v), \text{ where } u = h(x, y, v).$$

The independent variables consist of the variable with respect to which the differentiation is being made, plus all variables appearing as subscripts.

Example 1. If $w = xuv + u - 2v$, then

$$\frac{\partial w}{\partial x} = uv, \quad \frac{\partial w}{\partial u} = xv + 1, \quad \frac{\partial w}{\partial v} = xu - 2.$$

Example 2. If u, v, x, y are related by the equations

$$u = x - y, \quad v = x + y,$$

then

$$\left(\frac{\partial u}{\partial x}\right)_y = 1, \quad \left(\frac{\partial v}{\partial x}\right)_y = 1,$$

whereas

$$\left(\frac{\partial u}{\partial x}\right)_v = 2, \quad \left(\frac{\partial v}{\partial x}\right)_u = 2,$$

since u can be expressed in terms of x and v by the equation

$$u = 2x - v,$$

from which one also obtains

$$v = 2x - u.$$

Example 3. If $x^2 + y^2 - z^2 = 1$, then

$$2x - 2z\frac{\partial z}{\partial x} = 0, \quad 2y - 2z\frac{\partial z}{\partial y} = 0,$$

whence

$$\frac{\partial z}{\partial x} = \frac{x}{z}, \quad \frac{\partial z}{\partial y} = \frac{y}{z} \quad (z \neq 0).$$

2–6 Total differential. Fundamental lemma. In forming the above partial derivatives $\partial z/\partial x$ and $\partial z/\partial y$, changes Δx and Δy in x and y were considered separately; we now consider the effect of changing x and y together. Let (x, y) be a fixed point of D and let $(x + \Delta x, y + \Delta y)$ be a second point of D. Then the function $z = f(x, y)$ changes by an amount Δz in going from (x, y) to $(x + \Delta x, y + \Delta y)$:

$$\Delta z = f(x + \Delta x, y + \Delta y) - f(x, y). \tag{2–15}$$

This defines Δz as a function of Δx and Δy (x and y being considered as constants), with the special property:

$$\Delta z = 0 \quad \text{when} \quad \Delta x = 0 \quad \text{and} \quad \Delta y = 0.$$

For example, if $z = x^2 + xy + xy^2$, then

$$\begin{aligned}\Delta z &= (x + \Delta x)^2 + (x + \Delta x)(y + \Delta y) + (x + \Delta x)(y + \Delta y)^2 - x^2 - xy - xy^2 \\ &= \Delta x(2x + y + y^2) + \Delta y(x + 2xy) + \overline{\Delta x}^2 + \Delta x\, \Delta y(1 + 2y) \\ &\quad + \overline{\Delta y}^2 x + \Delta x\, \overline{\Delta y}^2.\end{aligned}$$

Here Δz is of the form

$$\Delta z = a\,\Delta x + b\,\Delta y + c\,\overline{\Delta x}^2 + d\,\Delta x\,\Delta y + e\,\overline{\Delta y}^2 + f\,\Delta x\,\overline{\Delta y}^2,$$

i.e., *a linear function of* Δx *and* Δy *plus terms of higher degree.*

In general, the function $z = f(x, y)$ is said to have a *total differential* at the point (x, y) if at this point

$$\Delta z = a\,\Delta x + b\,\Delta y + \epsilon_1 \cdot \Delta x + \epsilon_2 \cdot \Delta y, \tag{2–16}$$

where a and b are independent of Δx, Δy, and ϵ_1 and ϵ_2 are functions of Δx and Δy such that

$$\lim_{\substack{\Delta x \to 0 \\ \Delta y \to 0}} \epsilon_1 = 0, \quad \lim_{\substack{\Delta x \to 0 \\ \Delta y \to 0}} \epsilon_2 = 0; \tag{2–17}$$

the linear function of Δx and Δy:

$$a\,\Delta x + b\,\Delta y$$

is then termed the *total differential* of z at the point (x, y) and is denoted by dz:

$$dz = a\,\Delta x + b\,\Delta y. \tag{2–18}$$

If Δx *and* Δy *are sufficiently small,* dz *gives a close approximation to* Δz. More precisely, one can write

$$\Delta z = \Delta x(a + \epsilon_1) + \Delta y(b + \epsilon_2),$$

where a and b are constants; by (2–17) the percentage error in each term caused by replacing ϵ_1 and ϵ_2 by 0 can be made as small as desired by choosing Δx and Δy sufficiently small. (This argument fails if a or b is 0.)

In the above example Δz has a total differential at each point (x, y), with

$$a = 2x + y + y^2, \quad b = x + 2xy,$$

and

$$\epsilon_1 = \Delta x + \Delta y(1 + 2y), \quad \epsilon_2 = x\,\Delta y + \Delta x\,\Delta y.$$

THEOREM. *If* $z = f(x, y)$ *has a total differential* (2–18) *at the point* (x, y), *then*

$$a = \frac{\partial z}{\partial x}, \quad b = \frac{\partial z}{\partial y}; \tag{2–19}$$

that is, the two partial derivatives exist at (x, y) *and have the given values.*

Proof. Set $\Delta y = 0$. Then, by (2–16) and (2–17),

$$\frac{\partial z}{\partial x} = \lim_{\Delta x \to 0} \frac{\Delta z}{\Delta x} = \lim_{\Delta x \to 0} \frac{\Delta x(a + \epsilon_1)}{\Delta x} = \lim_{\Delta x \to 0} (a + \epsilon_1) = a.$$

Similarly one shows that $\partial z/\partial y = b$.

The existence of the partial derivatives at the point (x, y) is not sufficient to guarantee existence of the total differential; however, their continuity near the point is sufficient for this.

Fundamental Lemma. *If $z = f(x, y)$ has continuous first partial derivatives in D, then z has a differential*

$$dz = \frac{\partial z}{\partial x}\,\Delta x + \frac{\partial z}{\partial y}\,\Delta y \tag{2–20}$$

at every point (x, y) of D.

Proof. Let (x, y) be a fixed point of D. If x alone changes, one obtains a change Δz in z:

$$\Delta z = f(x + \Delta x, y) - f(x, y);$$

this difference can be evaluated by the law of the mean [equation (0–145)]; for, with y held fixed, z is a function of x having a continuous derivative $f_x(x, y)$. Thus one concludes:

$$f(x + \Delta x, y) - f(x, y) = f_x(x_1, y)\,\Delta x,$$

where x_1 is between x and $x + \Delta x$. Since $f_x(x, y)$ is continuous, the difference

$$\epsilon_1 = f_x(x_1, y) - f_x(x, y)$$

approaches zero as Δx approaches 0. Thus

$$f(x + \Delta x, y) - f(x, y) = f_x(x, y)\,\Delta x + \epsilon_1\,\Delta x. \tag{2–21}$$

Now if both x and y change, one obtains a change Δz in z:

$$\Delta z = f(x + \Delta x, y + \Delta y) - f(x, y).$$

This can be written as the sum of terms representing the effect of a change in x alone and a subsequent change in y alone:

$$\Delta z = [f(x + \Delta x, y) - f(x, y)] + [f(x + \Delta x, y + \Delta y) - f(x + \Delta x, y)]. \tag{2–22}$$

The first term can be evaluated by (2–21). The second is evaluated similarly, with z a function of y alone:

$$f(x + \Delta x, y + \Delta y) - f(x + \Delta x, y) = f_y(x + \Delta x, y_1)\,\Delta y,$$

where y_1 is between y and $y + \Delta y$. It follows from the continuity of $f_y(x, y)$ that the difference

$$\epsilon_2 = f_y(x + \Delta x, y_1) - f_y(x, y)$$

approaches 0 as *both Δx and Δy* approach 0. One has now

$$f(x + \Delta x, y + \Delta y) - f(x + \Delta x, y) = f_y(x, y)\,\Delta y + \epsilon_2\,\Delta y. \tag{2–23}$$

Equations (2–21), (2–22) and (2–23) now give

$$\Delta z = f_x(x, y)\,\Delta x + f_y(x, y)\,\Delta y + \epsilon_1\,\Delta x + \epsilon_2\,\Delta y,$$

where ϵ_1 and ϵ_2 satisfy (2–17). Thus z has a differential dz as stated in (2–20) and the Fundamental Lemma is proved.

For reasons to be explained below, Δx and Δy can be replaced by dx and dy in (2–20). Thus one has

$$dz = \frac{\partial z}{\partial x}\,dx + \frac{\partial z}{\partial y}\,dy, \tag{2–24}$$

which is the customary way of writing the differential.

The preceding analysis extends at once to functions of three or more variables. For example, if $w = f(x, y, u, v)$, then

$$dw = \frac{\partial w}{\partial x}\,dx + \frac{\partial w}{\partial y}\,dy + \frac{\partial w}{\partial u}\,du + \frac{\partial w}{\partial v}\,dv. \tag{2–25}$$

EXAMPLE 1. If $z = x^2 - y^2$, then $dz = 2x\,dx - 2y\,dy$.

EXAMPLE 2. If $w = \dfrac{xy}{z}$, then $dw = \dfrac{y}{z}\,dx + \dfrac{x}{z}\,dy - \dfrac{xy}{z^2}\,dz$.

Problems

1. Evaluate $\dfrac{\partial z}{\partial x}$ and $\dfrac{\partial z}{\partial y}$ if

(a) $z = \dfrac{y}{x^2 + y^2}$

(b) $z = y \sin xy$

(c) $x^3 + x^2y - x^2z + z^3 - 2 = 0$

(d) $z = \sqrt{e^{x+2y} - y^2}$

2. A certain function $f(x, y)$ is known to have the following values: $f(0, 0) = 0$, $f(1, 0) = 1$, $f(2, 0) = 4$, $f(0, 1) = -2$, $f(1, 1) = -1$, $f(2, 1) = 2$, $f(0, 2) = -4$, $f(1, 2) = -3$, $f(2, 2) = 0$. Compute approximately the derivatives $f_x(1, 1)$ and $f_y(1, 1)$.

3. Evaluate the indicated partial derivatives:

(a) $\left(\dfrac{\partial u}{\partial x}\right)_y$ and $\left(\dfrac{\partial v}{\partial y}\right)_x$ if $u = x^2 - y^2$, $v = x - 2y$,

(b) $\left(\dfrac{\partial x}{\partial u}\right)_y$ and $\left(\dfrac{\partial y}{\partial v}\right)_u$ if $u = x - 2y$, $v = u - 2y$.

4. Find the differentials of the following functions:

(a) $z = \dfrac{x}{y}$

(b) $z = \log \sqrt{x^2 + y^2}$

(c) $z = \arctan \dfrac{y}{x}$

(d) $u = \dfrac{1}{\sqrt{x^2 + y^2 + z^2}}$

5. If $z = x^2 + 2xy$, find Δz in terms of Δx and Δy for $x = 1$, $y = 1$. Plot Δz as a function of Δx and Δy and compare with the graph of dz in terms of Δx and Δy.

6. A certain function $z = f(x, y)$ is known to have the value $f(1, 2) = 3$ and derivatives $f_x(1, 2) = 2$, $f_y(1, 2) = 5$. Make "reasonable" estimates of $f(1.1, 1.8)$, $f(1.2, 1.8)$, $f(1.3, 1.8)$.

Answers

1. (a) $\dfrac{\partial z}{\partial x} = \dfrac{-2xy}{(x^2 + y^2)^2}$, $\dfrac{\partial z}{\partial y} = \dfrac{x^2 - y^2}{(x^2 + y^2)^2}$;

(b) $\dfrac{\partial z}{\partial x} = y^2 \cos xy$, $\dfrac{\partial z}{\partial y} = \sin xy + xy \cos xy$;

(c) $\dfrac{\partial z}{\partial x} = \dfrac{3x^2 + 2xy - 2xz}{x^2 - 3z^2}$, $\dfrac{\partial z}{\partial y} = \dfrac{x^2}{x^2 - 3z^2}$;

(d) $\dfrac{\partial z}{\partial x} = \dfrac{e^{x+2y}}{2\sqrt{e^{x+2y} - y^2}}$, $\dfrac{\partial z}{\partial y} = \dfrac{e^{x+2y} - y}{\sqrt{e^{x+2y} - y^2}}$.

2. Approximately $f_x(1, 1) = \dfrac{f(2, 1) - f(1, 1)}{1} = 3$,

or $\quad f_x(1, 1) = \dfrac{f(1, 1) - f(0, 1)}{1} = 1$, or $f_x(1, 1) = \dfrac{f(2, 1) - f(0, 1)}{2} = 2$.

It can be shown that the last value is "in general" the best estimate. For $f_y(1, 1)$ the analogous formulas give the estimates -2, -2, -2. This topic is discussed in Section 2–23.

3. (a) $\left(\dfrac{\partial u}{\partial x}\right)_y = 2x$, $\left(\dfrac{\partial v}{\partial y}\right)_x = -2$; (b) $\left(\dfrac{\partial x}{\partial u}\right)_y = 1$, $\left(\dfrac{\partial y}{\partial v}\right)_u = -\frac{1}{2}$.

4. (a) $\dfrac{y\,dx - x\,dy}{y^2}$, (b) $\dfrac{x\,dx + y\,dy}{x^2 + y^2}$, (c) $\dfrac{-y\,dx + x\,dy}{x^2 + y^2}$,

(d) $\dfrac{-(x\,dx + y\,dy + z\,dz)}{(x^2 + y^2 + z^2)^{\frac{3}{2}}}$.

5. $\Delta z = 4\Delta x + 2\Delta y + \overline{\Delta x}^2 + 2\Delta x\Delta y$, $dz = 4\Delta x + 2\Delta y$.

6. 2.2, 2.4, 2.6.

2–7 Differential of functions of *n* variables and of vector functions. The Jacobian matrix. For a function of n variables

$$y = f(x_1, \ldots, x_n) \tag{2–26}$$

the differential is obtained as in Section 2–6:

$$dy = f_{x_1}dx_1 + \cdots + f_{x_n}dx_n. \tag{2–27}$$

Thus it is a linear function of $dx_1, \ldots, dx_n$, whose coefficients $f_{x_1}, \ldots, f_{x_n}$ are the partial derivatives of f at the point considered. This linear function is a close approximation to the increment Δy in the sense described in Section 2–6:

$$\begin{aligned} \Delta y &= f(x_1 + dx_1, \ldots, x_n + dx_n) - f(x_1, \ldots, x_n) \\ &= f_{x_1}\,dx_1 + \cdots + f_{x_n}\,dx_n + \epsilon_1\,dx_1 + \cdots + \epsilon_n dx_n, \end{aligned} \tag{2–28}$$

where

$$\epsilon_1 \to 0, \ldots, \quad \epsilon_n \to 0 \quad \text{as} \quad dx_1 \to 0, \ldots, dx_n \to 0.$$

On occasion one has to deal with several functions of n variables:

$$\begin{cases} y_1 = f_1(x_1, \ldots, x_n), \\ \vdots \\ y_m = f_m(x_1, \ldots, x_n). \end{cases} \tag{2–29}$$

If these functions have continuous partial derivatives in a domain D of E^n, then all have differentials:

$$\begin{cases} dy_1 = \dfrac{\partial f_1}{\partial x_1} dx_1 + \cdots + \dfrac{\partial f_1}{\partial x_n} dx_n, \\ \vdots \\ dy_m = \dfrac{\partial f_m}{\partial x_1} dx_1 + \cdots + \dfrac{\partial f_m}{\partial x_n} dx_n. \end{cases} \tag{2–30}$$

These equations can be written in matrix form:

$$\begin{bmatrix} dy_1 \\ \vdots \\ dy_m \end{bmatrix} = \begin{bmatrix} \dfrac{\partial f_1}{\partial x_1} & \cdots & \dfrac{\partial f_1}{\partial x_n} \\ \vdots & & \vdots \\ \dfrac{\partial f_m}{\partial x_1} & \cdots & \dfrac{\partial f_m}{\partial x_n} \end{bmatrix} \begin{bmatrix} dx_1 \\ \vdots \\ dx_n \end{bmatrix}. \tag{2–31}$$

Thus the *vector* col $(dy_1, \ldots, dy_m)$ is obtained from the *vector*

$$\text{col}\,(dx_1, \ldots, dx_n)$$

by multiplication by the *matrix*

$$\left(\frac{\partial f_i}{\partial x_j}\right) = \begin{bmatrix} \dfrac{\partial f_1}{\partial x_1} & \cdots & \dfrac{\partial f_1}{\partial x_n} \\ \vdots & & \vdots \\ \dfrac{\partial f_m}{\partial x_1} & \cdots & \dfrac{\partial f_m}{\partial x_n} \end{bmatrix}. \tag{2–32}$$

This matrix is called the *Jacobian matrix* of the set of functions (2–29); its entries are partial derivatives of the functions (2–29), evaluated at a chosen point of D.

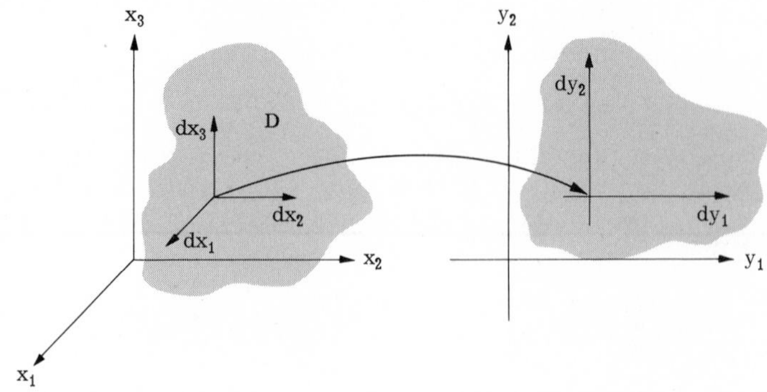

Fig. 2–5. The differential as a linear mapping approximating a given mapping.

The equations (2–29) assign a point $(y_1, \ldots, y_m)$ in E^m to each point $(x_1, \ldots, x_n)$ of D. Thus they describe a *mapping* of D into E^m (see Fig. 2–5, in which $n = 3$, $m = 2$). The linear equations (2–30) or (2–31) describe a *linear mapping* (Section 1–10) which approximates the given mapping near a chosen point; the linear mapping is expressed in terms of coordinates $(dx_1, \ldots, dx_n)$, with origin at the chosen point in D and axes parallel to the given axes, and coordinates $(dy_1, \ldots, dy_m)$ related similarly to the corresponding point in E^m.

We can simplify further by regarding (2–29) as a vector function of the vector $\mathbf{x} = (x_1, \ldots, x_n)$:

$$\mathbf{y} = \mathbf{f}(\mathbf{x}). \tag{2–29'}$$

Here $\mathbf{y} = (y_1, \ldots, y_m)$ and $\mathbf{f}$ is a *vector function* $(f_1, \ldots, f_m)$. We then write (2–31) in the concise form:

$$d\mathbf{y} = \mathbf{f_x}\, d\mathbf{x}. \tag{2–31'}$$

Here $d\mathbf{x} = \text{col}\,(dx_1, \ldots, dx_n)$, $d\mathbf{y} = \text{col}\,(dy_1, \ldots, dy_m)$ and $\mathbf{f_x}$ is an abbreviation for the Jacobian matrix $(\partial f_i/\partial x_j)$. We can also write $\partial y_i/\partial x_j$ for $\partial f_i/\partial x_j$ and are then led to write (2–31′) in the form

$$d\mathbf{y} = \mathbf{y_x}\, d\mathbf{x},$$

which is much like the formula $dy = y'dx$ for functions of one variable.

Example 1. The function $\mathbf{f}$ is defined by the equations

$$\begin{cases} y_1 = x_1^2 + x_2^2 - x_3^2, \\ y_2 = x_1^2 - x_2^2 + x_3^2, \\ y_3 = -x_1^2 + x_2^2 + x_3^2. \end{cases}$$

Hence

$$d\mathbf{y} = \begin{bmatrix} dy_1 \\ dy_2 \\ dy_3 \end{bmatrix} = \begin{bmatrix} 2x_1\,dx_1 + 2x_2\,dx_2 - 2x_3\,dx_3 \\ 2x_1\,dx_1 - 2x_2\,dx_2 + 2x_3\,dx_3 \\ -2x_1\,dx_1 + 2x_2\,dx_2 + 2x_3\,dx_3 \end{bmatrix}$$

$$= \begin{bmatrix} 2x_1 & 2x_2 & -2x_3 \\ 2x_1 & -2x_2 & 2x_3 \\ -2x_1 & 2x_2 & 2x_3 \end{bmatrix} \begin{bmatrix} dx_1 \\ dx_2 \\ dx_3 \end{bmatrix}.$$

At the point $(x_1, x_2, x_3) = (2, 1, 1)$, we find $(y_1, y_2, y_3) = (4, 4, -2)$ and

$$d\mathbf{y} = \begin{bmatrix} 4 & 2 & -2 \\ 4 & -2 & 2 \\ -4 & 2 & 2 \end{bmatrix} d\mathbf{x}.$$

If $\mathbf{x} = (2.01, 1.03, 1.02)$, then $d\mathbf{x} = (0.01, 0.03, 0.02)$ and the last equation gives

$$d\mathbf{y} = \begin{bmatrix} 4 & 2 & -2 \\ 4 & -2 & 2 \\ -4 & 2 & 2 \end{bmatrix} \begin{bmatrix} 0.01 \\ 0.03 \\ 0.02 \end{bmatrix} = \begin{bmatrix} 0.06 \\ 0.02 \\ 0.06 \end{bmatrix},$$

so that, approximately, $\mathbf{y} = (4.06, 4.02, -1.94)$; the exact value is $(4.0606, 4.0196, -1.9388)$.

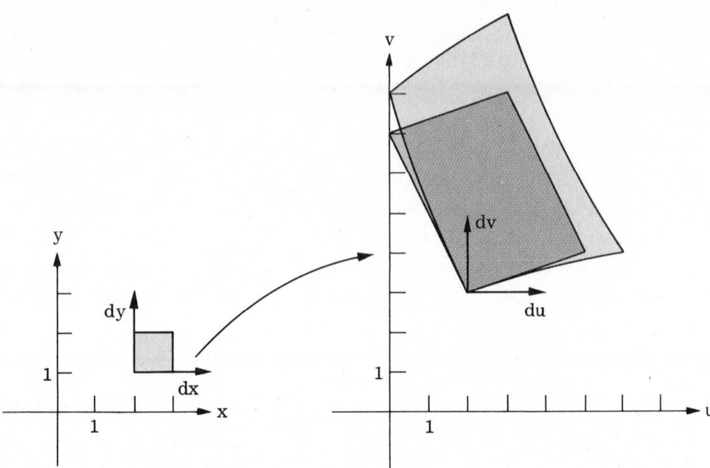

FIG. 2–6. Mapping and approximating linear mapping for Example 2.

EXAMPLE 2. $u = x^2 - xy, v = xy + y^2$. Here the independent variable vector is (x, y), the dependent variable vector is (u, v). We have

$$\begin{bmatrix} du \\ dv \end{bmatrix} = \begin{bmatrix} (2x - y)\,dx - xdy \\ ydx + (x + 2y)\,dy \end{bmatrix} = \begin{bmatrix} 2x - y & -x \\ y & x + 2y \end{bmatrix} \begin{bmatrix} dx \\ dy \end{bmatrix}.$$

At $(x, y) = (2, 1)$, $(u, v) = (2, 3)$ and the approximating linear mapping is

$$\begin{bmatrix} du \\ dv \end{bmatrix} = \begin{bmatrix} 3 & -2 \\ 1 & 4 \end{bmatrix} \begin{bmatrix} dx \\ dy \end{bmatrix}$$

(see Fig. 2–6). We study this linear mapping in more detail. For $dy = 0$, we have $du = 3\,dx$ and $dv = dx$, so that (du, dv) follows a line of slope $\frac{1}{3}$; for $dx = 0$, we have $du = -2\,dy$ and $dv = 4\,dy$, so that (du, dv) follows a line of slope -2. Similarly, the linear mapping can be studied along other lines. In particular, we verify that the points of the square

$$0 \leq dx \leq 1, \quad 0 \leq dy \leq 1$$

correspond to the points of the shaded parallelogram in the (du, dv) diagram. The area of the square is 1 and the area of the parallelogram is

$$|(3\mathbf{i} + \mathbf{j}) \times (-2\mathbf{i} + 4\mathbf{j})| = \begin{vmatrix} 3 & -2 \\ 1 & 4 \end{vmatrix} = 14 \text{ sq units.}$$

For the given nonlinear mapping $u = x^2 - xy$, $v = xy + y^2$, the lines $y = 1$ and $x = 2$ (on which $dy = 0$ and $dx = 0$, respectively) correspond to parabolas through (2, 3) in the uv-plane, as shown. The square $2 \leqq x \leqq 3$, $1 \leqq y \leqq 2$ (the same square as above) corresponds to a "curved parallelogram" as in the figure. Thus we see in a geometric way how the linear mapping approximates the given mapping. We observe that the linear mapping takes the line $dy = 0$ or the line $dx = 0$ to a line *tangent* to the curve obtained from the nonlinear mapping.

EXAMPLE 3. The mapping

$$x = \cos u \cos v, \quad y = \cos u \sin v, \quad z = \sin u$$

has the Jacobian matrix

$$\begin{bmatrix} \dfrac{\partial x}{\partial u} & \dfrac{\partial x}{\partial v} \\ \dfrac{\partial y}{\partial u} & \dfrac{\partial y}{\partial v} \\ \dfrac{\partial z}{\partial u} & \dfrac{\partial z}{\partial v} \end{bmatrix} = \begin{bmatrix} -\sin u \cos v & -\cos u \sin v \\ -\sin u \sin v & \cos u \cos v \\ \cos u & 0 \end{bmatrix}.$$

EXAMPLE 4. Let $w = F(x, y, z)$. Then the Jacobian matrix of F is the *row vector* $(\partial F/\partial x, \partial F/\partial y, \partial F/\partial z)$. We call this vector the *gradient vector* of F and denote it by ∇F or grad F. This is discussed in Section 2–12 below. Similarly, $F(x_1, \ldots, x_n)$ has as Jacobian matrix the row vector $(F_{x_1}, \ldots, F_{x_n})$, called the gradient vector of F.

We return to the general mapping $\mathbf{y} = \mathbf{f}(\mathbf{x})$ and assume $m = n$, so that the mapping is given by equations

$$\begin{cases} y_1 = f_1(x_1, \ldots, x_n), \\ \vdots \\ y_n = f_n(x_1, \ldots, x_n). \end{cases}$$

In this case the Jacobian matrix $\mathbf{y}_\mathbf{x} = (\partial f_i/\partial x_j)$ is *square* and we can form its determinant:

$$J = \det\left(\frac{\partial y_i}{\partial x_j}\right) = \begin{vmatrix} \dfrac{\partial y_1}{\partial x_1} & \cdots & \dfrac{\partial y_1}{\partial x_n} \\ \vdots & & \vdots \\ \dfrac{\partial y_n}{\partial x_1} & \cdots & \dfrac{\partial y_n}{\partial x_n} \end{vmatrix}.$$

We call this determinant the *Jacobian determinant* (or simply, the *Jacobian*) of the mapping. For the corresponding linear mapping $d\mathbf{y} = \mathbf{y}_\mathbf{x}\, d\mathbf{x}$, J is the determinant of the matrix $\mathbf{y}_\mathbf{x}$ of the mapping; this determinant measures the ratio of n-dimensional volumes (see Chapter 1, Section 1–13). Symbolically,

$$\Delta V_y = |\det \mathbf{y}_\mathbf{x}| \Delta V_x.$$

Since the linear mapping approximates the nonlinear one, we can say that the absolute value of the Jacobian determinant J measures the ratio of corresponding volumes for small regions near the chosen point. This relationship is studied further in Sections 2–9, 4–8 and 5–14 below. An illustration is provided by Example 2 above, in which $n = 2$ and $J = 14$ at the point; this is precisely the ratio of the *area* of the parallelogram to the area of the square to which the parallelogram corresponds. For $n = 1$, J becomes the derivative dy/dx and its absolute value does measure the ratio of corresponding *lengths* Δy, Δx: $|dy/dx| \sim |\Delta y|/|\Delta x|$ for small Δx, since $dy/dx = \lim\, (\Delta y/\Delta x)$ as $\Delta x \to 0$.

The Jacobian determinant is also denoted as follows:

$$J = \frac{\partial(y_1, \ldots, y_n)}{\partial(x_1, \ldots, x_n)} = \frac{\partial(f_1, \ldots, f_n)}{\partial(x_1, \ldots, x_n)}.$$

The concept of Jacobian determinant and these notations can also be applied to n functions of more than n variables. One simply forms the indicated partial derivatives, holding all other variables constant. For example, for $f(u, v, w)$, $g(u, v, w)$, one has

$$\frac{\partial(f, g)}{\partial(u, v)} = \begin{vmatrix} f_u & f_v \\ g_u & g_v \end{vmatrix}, \quad \frac{\partial(f, g)}{\partial(u, w)} = \begin{vmatrix} f_u & f_w \\ g_u & g_w \end{vmatrix}.$$

Problems

1. Obtain the Jacobian matrix for each of the following mappings:
(a) $y_1 = 2x_1 - 3x_2$, $y_2 = x_1 + 2x_2$. (b) $y_1 = x_1^2 - x_2^2$, $y_2 = 2x_1x_2$.
(c) $y_1 = x_1x_2x_3$, $y_2 = x_1^2x_3$.
(d) $u = x \cos y$, $v = x \sin y$, $w = x^2$.
(e) $w = x^2yz$. (f) $w = x^2 + y^2 - z^2$. (g) $x = t^2$, $y = t^3$, $z = t^4$.

2. Obtain the linear mapping $d\mathbf{y} = \mathbf{f}_\mathbf{x}\, d\mathbf{x}$ approximating the given mapping $\mathbf{y} = \mathbf{f}(\mathbf{x})$ near the specified point, and use the linear mapping to obtain an approximation to the value $\mathbf{f}(\mathbf{x})$ specified.
(a) $y_1 = x_1^2 + x_2^2$, $y_2 = x_1x_2$ at $(1, 1)$, approximate $\mathbf{f}(1.02, 1.03)$.
(b) $y_1 = x_1x_2 - x_3^2$, $y_2 = x_1x_2 + x_1x_3$ at $(2, 1, 1)$, approximate $\mathbf{f}(2.01, 1.02, 0.99)$.
(c) $u = e^x \cos y$, $v = e^x \sin y$, $w = 2e^x$ at $(0, \pi/2)$, approximate value of (u, v, w) for $(x, y) = (0.1, 1.6)$.
(d) $y_1 = x_2^2 + \cdots + x_n^2$, $y_2 = x_1^2 + x_3^2 + \cdots + x_n^2, \ldots, y_n = x_1^2 + \cdots + x_{n-1}^2$ at $(1, 0, \ldots, 0)$, approximate $\mathbf{f}(1, 0.1, \ldots, 0.1)$.

3. Obtain the Jacobian determinant requested:
(a) $\dfrac{\partial(u, v)}{\partial(x, y)}$ for $u = x^3 - 3xy^2$, $v = 3x^2y - y^3$.
(b) $\dfrac{\partial(u, v, w)}{\partial(x, y, z)}$ for $u = xe^y \cos z$, $v = xe^y \sin z$, $w = xe^y$.
(c) $\dfrac{\partial(f, g)}{\partial(u, v)}$ for $f(u, v, w) = u^2vw$, $g(u, v, w) = u^2v^2 - w^4$.
(d) $\dfrac{\partial(f, g, h)}{\partial(x, y, z)}$ for $f(x, y, z, t) = x^2 + 2y + z^2 - t^2$, $g(x, y, z, t) = xyz + t^2$, $h(x, y, z, t) = z^2 - t^2$.

4. For the mapping $u = e^x \cos y$, $v = e^x \sin y$ from the xy-plane to the uv-plane carry out the following steps:
(a) Evaluate the Jacobian determinant at $(1, 0)$.
(b) Show that the square R_{xy}: $0.9 \leqq x \leqq 1.1$, $-0.1 \leqq y \leqq 0.1$ corresponds to the region R_{uv} bounded by arcs of the circles $u^2 + v^2 = e^{1.8}$, $u^2 + v^2 = e^{2.2}$ and the rays $v = \pm\tan 0.1u$, $u \geqq 0$, and find the ratio of the area of R_{uv} to that of R_{xy}. Compare with the result of (a).
(c) Obtain the approximating linear mapping at $(1, 0)$ and find the region R'_{uv} corresponding to the square R_{xy} of part (b) under this linear mapping. Find the ratio of the area of R'_{uv} to that of R_{xy} and compare with the results of parts (a) and (b).

Answers

1. (a) $\begin{bmatrix} 2 & -3 \\ 1 & 2 \end{bmatrix}$, (b) $\begin{bmatrix} 2x_1 & -2x_2 \\ 2x_2 & 2x_1 \end{bmatrix}$, (c) $\begin{bmatrix} x_2x_3 & x_1x_3 & x_1x_2 \\ 2x_1x_3 & 0 & x_1^2 \end{bmatrix}$,

(d) $\begin{bmatrix} \cos y & -x \sin y \\ \sin y & x \cos y \\ 2x & 0 \end{bmatrix}$, (e) $(2xyz, x^2z, x^2y)$, (f) $(2x, 2y, -2z)$,
(g) col $(2t, 3t^2, 4t^3)$.

2. (a) $\begin{bmatrix} dy_1 \\ dy_2 \end{bmatrix} = \begin{bmatrix} 2 & 2 \\ 1 & 1 \end{bmatrix} \begin{bmatrix} dx_1 \\ dx_2 \end{bmatrix}$, (2.1, 1.05),

(b) $\begin{bmatrix} dy_1 \\ dy_2 \end{bmatrix} = \begin{bmatrix} 1 & 2 & -2 \\ 2 & 2 & 2 \end{bmatrix} \begin{bmatrix} dx_1 \\ dx_2 \\ dx_3 \end{bmatrix}$, (1.07, 4.04),

(c) $\begin{bmatrix} du \\ dv \\ dw \end{bmatrix} = \begin{bmatrix} 0 & -1 \\ 1 & 0 \\ 2 & 0 \end{bmatrix} \begin{bmatrix} dx \\ dy \end{bmatrix}$, (−0.03, 1.1, 2.2),

(d) $d\mathbf{y} = \begin{bmatrix} 0 & 0 & \cdots & 0 \\ 2 & 0 & \cdots & 0 \\ 2 & 0 & \cdots & 0 \\ \vdots & \vdots & & \vdots \\ 2 & 0 & \cdots & 0 \end{bmatrix} d\mathbf{x}$, (0, 1, . . . , 1).

3. (a) $9(x^2 + y^2)^2$, (b) 0, (c) $2u^3v^2w$, (d) $4z^2(x^2 - y)$.

4. (a) $e^2 = 7.39$, (b) 7.44, (c) $du = e\,dx$, $dv = e\,dy$, $e^2 = 7.39$.

2–8 Derivatives and differentials of composite functions. The functions to be considered in the following will be assumed to be defined in appropriate domains and to have continuous first partial derivatives, so that the corresponding differentials can be formed.

THEOREM. *If $z = f(x, y)$ and $x = g(t)$, $y = h(t)$, then*

$$\frac{dz}{dt} = \frac{\partial z}{\partial x}\frac{dx}{dt} + \frac{\partial z}{\partial y}\frac{dy}{dt}. \tag{2–33}$$

If $z = f(x, y)$ and $x = g(u, v)$, $y = h(u, v)$, then

$$\frac{\partial z}{\partial u} = \frac{\partial z}{\partial x}\frac{\partial x}{\partial u} + \frac{\partial z}{\partial y}\frac{\partial y}{\partial u}, \quad \frac{\partial z}{\partial v} = \frac{\partial z}{\partial x}\frac{\partial x}{\partial v} + \frac{\partial z}{\partial y}\frac{\partial y}{\partial v}. \tag{2–34}$$

In general, if $z = f(x, y, t, \ldots)$ and $x = g(u, v, w, \ldots)$, $y = h(u, v, w, \ldots)$, $t = p(u, v, w, \ldots)$, . . ., then

$$\begin{aligned} \frac{\partial z}{\partial u} &= \frac{\partial z}{\partial x}\frac{\partial x}{\partial u} + \frac{\partial z}{\partial y}\frac{\partial y}{\partial u} + \frac{\partial z}{\partial t}\frac{\partial t}{\partial u} + \cdots, \\ \frac{\partial z}{\partial v} &= \frac{\partial z}{\partial x}\frac{\partial x}{\partial v} + \frac{\partial z}{\partial y}\frac{\partial y}{\partial v} + \frac{\partial z}{\partial t}\frac{\partial t}{\partial v} + \cdots, \\ \frac{\partial z}{\partial w} &= \frac{\partial z}{\partial x}\frac{\partial x}{\partial w} + \frac{\partial z}{\partial y}\frac{\partial y}{\partial w} + \frac{\partial z}{\partial t}\frac{\partial t}{\partial w} + \cdots. \end{aligned} \tag{2–35}$$

These rules, known as "chain rules," are basic for computation of derivatives of composite functions. The equations (2–33), (2–34), (2–35) are concise statements of the relations between the derivatives involved. Thus in (2–33)

$$z = f[g(t), h(t)]$$

is the function of t whose derivative is denoted by dz/dt, while dx/dt and dy/dt stand for $g'(t)$ and $h'(t)$ respectively. The derivatives $\partial z/\partial x$ and $\partial z/\partial y$, which could be written $(\partial z/\partial x)_y$ and $(\partial z/\partial y)_x$, stand for $f_x(x, y)$ and $f_y(x, y)$. In (2–34).

$$z = f[g(u, v), h(u, v)]$$

is the function whose derivative with respect to u is denoted by $\partial z/\partial u$, which should be understood as $(\partial z/\partial u)_v$. A more precise statement of the first equation in (2–34) would be as follows:

$$\left(\frac{\partial z}{\partial u}\right)_v = \left(\frac{\partial z}{\partial x}\right)_y \left(\frac{\partial x}{\partial u}\right)_v + \left(\frac{\partial z}{\partial y}\right)_x \left(\frac{\partial y}{\partial u}\right)_v,$$

and similar remarks apply to the other equations.

The proof of (2–33) will be given as a sample; the other rules are proved in the same way. Let t be a fixed value and let x, y, z be the corresponding values of the functions g, h, and f. Then, for given Δt, Δx and Δy are determined as

$$\Delta x = g(t + \Delta t) - g(t), \quad \Delta y = h(t + \Delta t) - h(t),$$

while Δz is then determined as

$$\Delta z = f(x + \Delta x, y + \Delta y) - f(x, y).$$

By the Fundamental Lemma, one has

$$\Delta z = \frac{\partial z}{\partial x}\Delta x + \frac{\partial z}{\partial y}\Delta y + \epsilon_1 \Delta x + \epsilon_2 \Delta y.$$

Hence

$$\frac{\Delta z}{\Delta t} = \frac{\partial z}{\partial x}\frac{\Delta x}{\Delta t} + \frac{\partial z}{\partial y}\frac{\Delta y}{\Delta t} + \epsilon_1 \frac{\Delta x}{\Delta t} + \epsilon_2 \frac{\Delta y}{\Delta t}.$$

As Δt approaches 0, $\Delta x/\Delta t$ and $\Delta y/\Delta t$ approach the derivatives dx/dt, dy/dt respectively, while ϵ_1 and ϵ_2 approach 0, since Δx and Δy approach 0. Hence

$$\lim_{\Delta t \to 0} \frac{\Delta z}{\Delta t} = \frac{\partial z}{\partial x} \cdot \frac{dx}{dt} + \frac{\partial z}{\partial y} \cdot \frac{dy}{dt} + 0 \cdot \frac{dx}{dt} + 0 \cdot \frac{dy}{dt};$$

that is,

$$\frac{dz}{dt} = \frac{\partial z}{\partial x}\frac{dx}{dt} + \frac{\partial z}{\partial y}\frac{dy}{dt},$$

as was to be proved.

The three functions of t considered here: $x = g(t)$, $y = h(t)$, $z = f[g(t), h(t)]$ have differentials, as in Section 0–9:

$$dx = \frac{dx}{dt}\Delta t, \quad dy = \frac{dy}{dt}\Delta t, \quad dz = \frac{dz}{dt}\Delta t.$$

From (2–33) one concludes that

$$\frac{dz}{dt}\Delta t = \frac{\partial z}{\partial x}\left(\frac{dx}{dt}\Delta t\right) + \frac{\partial z}{\partial y}\left(\frac{dy}{dt}\Delta t\right),$$

i.e., that

$$dz = \frac{\partial z}{\partial x}dx + \frac{\partial z}{\partial y}dy.$$

But this is the same as (2–24), in which dx and dy are Δx and Δy, arbitrary increments of independent variables. Thus (2–24) holds whether x and y are independent and dz is the corresponding differential or whether x and y, and hence z, depend on t, so that dx, dy, dz are the differentials of these variables in terms of t.

A similar reasoning applies to (2–34). Here u and v are the independent variables on which x, y, and z depend. The corresponding differentials are

$$dx = \frac{\partial x}{\partial u}\Delta u + \frac{\partial x}{\partial v}\Delta v,\quad dy = \frac{\partial y}{\partial u}\Delta u + \frac{\partial y}{\partial v}\Delta v,\quad dz = \frac{\partial z}{\partial u}\Delta u + \frac{\partial z}{\partial v}\Delta v.$$

But (2–34) gives

$$\begin{aligned} dz &= \left(\frac{\partial z}{\partial x}\frac{\partial x}{\partial u} + \frac{\partial z}{\partial y}\frac{\partial y}{\partial u}\right)\Delta u + \left(\frac{\partial z}{\partial x}\frac{\partial x}{\partial v} + \frac{\partial z}{\partial y}\frac{\partial y}{\partial v}\right)\Delta v \\ &= \frac{\partial z}{\partial x}\left(\frac{\partial x}{\partial u}\Delta u + \frac{\partial x}{\partial v}\Delta v\right) + \frac{\partial z}{\partial y}\left(\frac{\partial y}{\partial u}\Delta u + \frac{\partial y}{\partial v}\Delta v\right) \\ &= \frac{\partial z}{\partial x}dx + \frac{\partial z}{\partial y}dy. \end{aligned}$$

Again (2–24) holds. Generalization of this to (2–35) permits one to conclude:

Theorem. *The differential formula*

$$dz = \frac{\partial z}{\partial x}dx + \frac{\partial z}{\partial y}dy + \frac{\partial z}{\partial t}dt + \cdots \tag{2–36}$$

which holds when $z = f(x, y, t, \ldots)$ and $dx = \Delta x$, $dy = \Delta y$, $dt = \Delta t, \ldots,$ remains true when $x, y, t, \ldots,$ and hence z, are all functions of other independent variables and $dx, dy, dt, \ldots, dz$ are the corresponding differentials.

As a consequence of this theorem, one can conclude: *Any equation in differentials which is correct for one choice of independent and dependent variables remains true for any other choice.* Another way of saying this is that any equation in differentials treats all variables on an equal basis. Thus, if

$$dz = 2dx - 3dy$$

at a given point, then

$$dx = \tfrac{1}{2}dz + \tfrac{3}{2}dy$$

is the corresponding differential of x in terms of y and z.

An important practical application of the theorem is that, in order to compute partial derivatives, one can first compute differentials, pretending that all variables are functions of a hypothetical single variable (e.g., t), so that *all the rules of ordinary differential calculus* (Section 0–9) *apply.* From the resulting differential formula, one can at once obtain all partial derivatives desired.

Example 1. If $z = \dfrac{x^2 - 1}{y}$, then

$$dz = \frac{2xy\,dx - (x^2 - 1)\,dy}{y^2}$$

by the quotient rule [(0–129) of Section 0–9]. Hence

$$\frac{\partial z}{\partial x} = \frac{2x}{y}, \quad \frac{\partial z}{\partial y} = \frac{1 - x^2}{y^2}.$$

Example 2. If $r^2 = x^2 + y^2$, then $r\,dr = x\,dx + y\,dy$, whence

$$\left(\frac{\partial r}{\partial x}\right)_y = \frac{x}{r}, \quad \left(\frac{\partial r}{\partial y}\right)_x = \frac{y}{r}, \quad \left(\frac{\partial x}{\partial r}\right)_y = \frac{r}{x}, \text{ etc.}$$

Example 3. If $z = \operatorname{arc\,tan} y/x$ $(x \neq 0)$, then

$$dz = \frac{1}{1 + \left(\dfrac{y}{x}\right)^2}\, d\left(\frac{y}{x}\right) = \frac{x\,dy - y\,dx}{x^2 + y^2}$$

and hence

$$\frac{\partial z}{\partial x} = -\frac{y}{x^2 + y^2}, \quad \frac{\partial z}{\partial y} = \frac{x}{x^2 + y^2}.$$

Problems

1. If (a) $y = u + v$, (b) $y = u \cdot v$, (c) $y = u/v$, where u and v are functions of x, then apply (2–33) to find dy/dx. Compare the results with (0–129) of Section 0–9.

2. If $y = u^v$, where u and v are functions of x, then find dy/dx by (2–33). [Hint: use (0–131) and (0–140) of Section 0–9].

3. If $y = \log_u v$, where u and v are functions of x, then find dy/dx by (2–33). [Hint: use (0–144) of Section 0–9.]

4. If $z = e^x \cos y$, while x and y are implicit functions of t defined by the equations

$$x^3 + e^x - t^2 - t = 1, \quad yt^2 + y^2t - t + y = 0,$$

then find dz/dt for $t = 0$. [Note that $x = 0$ and $y = 0$ for $t = 0$.]

5. If $u = f(x, y)$ and $x = r\cos\theta$, $y = r\sin\theta$, then show that

$$\left(\frac{\partial u}{\partial x}\right)^2 + \left(\frac{\partial u}{\partial y}\right)^2 = \left(\frac{\partial u}{\partial r}\right)^2 + \frac{1}{r^2}\left(\frac{\partial u}{\partial \theta}\right)^2.$$

[Hint: use the chain rules to evaluate the derivatives on the *right*-hand side.]

6. If $w = f(x, y)$ and $x = u \cosh v$, $y = u \sinh v$, then show that

$$\left(\frac{\partial w}{\partial x}\right)^2 - \left(\frac{\partial w}{\partial y}\right)^2 = \left(\frac{\partial w}{\partial u}\right)^2 - \frac{1}{u^2}\left(\frac{\partial w}{\partial v}\right)^2.$$

[Cf. hint for Prob. 5.]

7. If $z = f(ax + by)$, show that

$$b\frac{\partial z}{\partial x} - a\frac{\partial z}{\partial y} = 0.$$

8. Find $\partial z/\partial x$ and $\partial z/\partial y$ by first obtaining dz:

(a) $z = \log \sin (x^2y^2 - 1)$

(b) $z = x^2y^2\sqrt{1 - x^2 - y^2}$

(c) $x^2 + 2y^2 - z^2 = 1$

9. If $f(x, y)$ satisfies the identity

$$f(tx, ty) = t^n f(x, y)$$

for a fixed n, f is called *homogeneous* of degree n. Show that one then has the relation

$$x\frac{\partial f}{\partial x} + y\frac{\partial f}{\partial y} = nf(x, y).$$

This is *Euler's theorem on homogeneous functions.* [Hint: differentiate both sides of the identity with respect to t and then set $t = 1$.]

10. *The Stokes total time derivative in hydrodynamics.* Let $w = F(x, y, z, t)$, where $x = f(t)$, $y = g(t)$, $z = h(t)$, so that w can be expressed in terms of t alone. Show that

$$\frac{dw}{dt} = \frac{\partial w}{\partial x}\frac{dx}{dt} + \frac{\partial w}{\partial y}\frac{dy}{dt} + \frac{\partial w}{\partial z}\frac{dz}{dt} + \frac{\partial w}{\partial t}.$$

Here both dw/dt and $\partial w/\partial t = F_t(x, y, z, t)$ have meaning and are in general unequal. In hydrodynamics dx/dt, dy/dt, dz/dt are the velocity components of a moving fluid particle and dw/dt describes the variation of w "following the motion of the fluid." It is customary, following Stokes, to write Dw/Dt for dw/dt. [See H. Lamb, *Hydrodynamics,* 6th Edition (Cambridge University Press, 1932), p. 3.]

Answers

2. $\dfrac{dy}{dx} = vu^{v-1}\dfrac{du}{dx} + u^v \log u \dfrac{dv}{dx}.$

3. $\dfrac{dy}{dx} = -\dfrac{\log v}{u \log^2 u}\dfrac{du}{dx} + \dfrac{1}{v \log u}\dfrac{dv}{dx}.$

4. 1.

8. In all cases $dz = \dfrac{\partial z}{\partial x}dx + \dfrac{\partial z}{\partial y}dy$:

(a) $dz = 2 \cot (x^2y^2 - 1)(xy^2\,dx + x^2y\,dy)$,

(b) $dz = \dfrac{(2xy^2 - 3x^3y^2 - 2xy^4)\,dx + (2x^2y - 3x^2y^3 - 2x^4y)\,dy}{\sqrt{1 - x^2 - y^2}}$,

(c) $dz = \dfrac{x\,dx + 2y\,dy}{z}.$

2–9 The general chain rule. On occasion one deals with two sets of functions:

$$\begin{aligned} y_1 &= f_1(u_1, \dots, u_p), \\ &\vdots \\ y_m &= f_m(u_1, \dots, u_p), \end{aligned} \tag{2–37}$$

and

$$\begin{aligned} u_1 &= g_1(x_1, \dots, x_n), \\ &\vdots \\ u_p &= g_p(x_1, \dots, x_n). \end{aligned} \tag{2–38}$$

If one substitutes the functions (2–38) in the functions (2–37), one obtains composite functions

$$\begin{aligned} y_1 &= f_1(g_1(x_1, \dots, x_n), \quad \dots, \quad g_p(x_1, \dots, x_n)) = F_1(x_1, \dots, x_n), \\ &\vdots \\ y_m &= f_m(g_1(x_1, \dots, x_n), \quad \dots, \quad g_p(x_1, \dots, x_n)) = F_m(x_1, \dots, x_n). \end{aligned} \tag{2–39}$$

Under the appropriate hypotheses, one can obtain the partial derivatives of these composite functions by chain rules, as in the previous section:

$$\frac{\partial y_i}{\partial x_j} = \frac{\partial y_i}{\partial u_1}\frac{\partial u_1}{\partial x_j} + \cdots + \frac{\partial y_i}{\partial u_p}\frac{\partial u_p}{\partial x_j} \quad (i = 1, \dots, m, \ j = 1, \dots, n). \tag{2–40}$$

The formulas (2–40) can be expressed concisely in matrix language. The partial derivatives $\partial y_i/\partial x_j$ are the entries in the $m \times n$ matrix

$$\left(\frac{\partial y_i}{\partial x_j}\right) = \begin{bmatrix} \frac{\partial y_1}{\partial x_1} & \frac{\partial y_1}{\partial x_2} & \cdots & \frac{\partial y_1}{\partial x_n} \\ \vdots & \vdots & & \vdots \\ \frac{\partial y_m}{\partial x_1} & \frac{\partial y_m}{\partial x_2} & \cdots & \frac{\partial y_m}{\partial x_n} \end{bmatrix}. \tag{2–41}$$

This is the Jacobian matrix of the mapping (2–39) (see Section 2–7). The formulas (2–40) involve two other Jacobian matrices:

$$\left(\frac{\partial y_i}{\partial u_j}\right) = \begin{bmatrix} \frac{\partial y_1}{\partial u_1} & \cdots & \frac{\partial y_1}{\partial u_p} \\ \vdots & & \vdots \\ \frac{\partial y_m}{\partial u_1} & \cdots & \frac{\partial y_m}{\partial u_p} \end{bmatrix}, \quad \left(\frac{\partial u_i}{\partial x_j}\right) = \begin{bmatrix} \frac{\partial u_1}{\partial x_1} & \cdots & \frac{\partial u_1}{\partial x_n} \\ \vdots & & \vdots \\ \frac{\partial u_p}{\partial x_1} & \cdots & \frac{\partial u_p}{\partial x_n} \end{bmatrix}. \tag{2–42}$$

The chain rules (2–40) state that the product of the last two Jacobian matrices equals the previous one:

$$\left(\frac{\partial y_i}{\partial x_j}\right) = \left(\frac{\partial y_i}{\partial u_j}\right)\left(\frac{\partial u_i}{\partial x_j}\right). \tag{2–43}$$

This equation is called the *general chain rule.* It includes all the rules of the previous section.

EXAMPLE 1. Let $y_1 = u_1u_2 - u_1u_3$, $y_2 = u_1u_3 + u_2^2$, $u_1 = x_1 \cos x_2 + (x_1 - x_2)^2$, $u_2 = x_1 \sin x_2 + x_1x_2$, $u_3 = x_1^2 - x_1x_2 + x_2^2$. Then by (2–43)

$$\left(\frac{\partial y_i}{\partial x_j}\right) = \begin{bmatrix} u_2 - u_3 & u_1 & -u_1 \\ u_3 & 2u_2 & u_1 \end{bmatrix}.$$

$$\begin{bmatrix} \cos x_2 + 2(x_1 - x_2) & -x_1 \sin x_2 - 2(x_1 - x_2) \\ \sin x_2 + x_2 & x_1 \cos x_2 + x_1 \\ 2x_1 - x_2 & 2x_2 - x_1 \end{bmatrix}.$$

On the right u_1, u_2, u_3 can be expressed in terms of x_1, x_2 and the two matrices can be multiplied. However, for many purposes it is sufficient to leave the result in indicated form. In particular, to obtain numerical values, one can substitute the appropriate values and multiply the matrices only as a last step. For example, for $x_1 = 1$, $x_2 = 0$, we obtain $u_1 = 2$, $u_2 = 0$, $u_3 = 1$ and hence

$$\left(\frac{\partial y_i}{\partial x_j}\right) = \begin{bmatrix} -1 & 2 & -2 \\ 1 & 0 & 2 \end{bmatrix} \begin{bmatrix} 3 & -2 \\ 0 & 2 \\ 2 & -1 \end{bmatrix} = \begin{bmatrix} -7 & 6 \\ 7 & -4 \end{bmatrix};$$

that is to say, $\partial y_1/\partial x_1 = -7$, $\partial y_1/\partial x_2 = 6$, $\partial y_2/\partial x_1 = 7$, $\partial y_2/\partial x_1 = 7$, and $\partial y_2/\partial x_2 = -4$.

Differentials and the chain rule. If we take differentials in Eqs. (2–37) and (2–38), we obtain the equations

$$\begin{aligned} dy_1 &= \frac{\partial y_1}{\partial u_1} du_1 + \cdots + \frac{\partial y_1}{\partial u_p} du_p, \\ &\vdots \\ dy_m &= \frac{\partial y_m}{\partial u_1} du_1 + \cdots + \frac{\partial y_m}{\partial u_p} du_p, \end{aligned} \tag{2–44}$$

and

$$\begin{aligned} du_1 &= \frac{\partial u_1}{\partial x_1} dx_1 + \cdots + \frac{\partial u_1}{\partial x_n} dx_n, \\ &\vdots \\ du_p &= \frac{\partial u_p}{\partial x_1} dx_1 + \cdots + \frac{\partial u_p}{\partial x_n} dx_n. \end{aligned} \tag{2–45}$$

In (2–44), $du_1, \ldots, du_p$ are arbitrary increments $\Delta u_1, \ldots, \Delta u_p$, whereas in (2–45) they are functions of the arbitrary increments $dx_1(= \Delta x_1), \ldots, dx_n(= \Delta x_n)$. However, we know from Section 2–8 that the relationships are the same no matter how we interpret the differentials. We can write these equations in matrix form:

$$\begin{bmatrix} dy_1 \\ \vdots \\ dy_m \end{bmatrix} = \left(\frac{\partial y_i}{\partial u_j}\right) \begin{bmatrix} du_1 \\ \vdots \\ du_p \end{bmatrix}, \quad \begin{bmatrix} du_1 \\ \vdots \\ du_p \end{bmatrix} = \left(\frac{\partial u_i}{\partial x_j}\right) \begin{bmatrix} dx_1 \\ \vdots \\ dx_n \end{bmatrix}. \tag{2–46}$$

If we eliminate the vector col $(du_1, \ldots, du_p)$ in these equations, we obtain

$$\begin{bmatrix} dy_1 \\ \vdots \\ dy_m \end{bmatrix} = \left(\frac{\partial y_i}{\partial u_j}\right)\left(\frac{\partial u_i}{\partial x_j}\right)\begin{bmatrix} dx_1 \\ \vdots \\ dx_n \end{bmatrix} \tag{2–47}$$

or

$$dy_1 = \left(\frac{\partial y_1}{\partial u_1}\frac{\partial u_1}{\partial x_1} + \cdots + \frac{\partial y_1}{\partial u_p}\frac{\partial u_p}{\partial x_1}\right) dx_1 + \left(\frac{\partial y_1}{\partial u_1}\frac{\partial u_1}{\partial x_2} + \cdots\right) dx_2 + \cdots$$

and so on. From these equations we can read off $\partial y_1/\partial x_1$, $\partial y_1/\partial x_2, \ldots$ Clearly the results are the same as (2–40) or (2–43) above. Thus (2–47) can be termed the *general chain rule in differential form.*

The preceding development can be carried out even more concisely in terms of the notations of Section 2–7: The given functions are really vector functions:

$$\mathbf{y} = \mathbf{f}(\mathbf{u}), \quad \mathbf{u} = \mathbf{g}(\mathbf{x}).$$

If we take differentials, we obtain

$$d\mathbf{y} = \mathbf{y_u}\, d\mathbf{u}, \quad d\mathbf{u} = \mathbf{u_x}\, d\mathbf{x}$$

and hence

$$d\mathbf{y} = \mathbf{y_u}\,\mathbf{u_x}\, d\mathbf{x}, \tag{2–48}$$

so that

$$\mathbf{y_x} = \mathbf{y_u}\mathbf{u_x}. \tag{2–49}$$

This last equation is the same as (2–43); the previous one is the same as (2–47).

We saw in Section 2–7 that a Jacobian matrix such as $\mathbf{y_x}$ is the matrix of the linear transformation approximating the given, in general nonlinear, mapping $\mathbf{y} = \mathbf{f}(\mathbf{x})$. Thus Eq. (2–43) asserts that if we have successive mappings $\mathbf{u} = \mathbf{g}(\mathbf{x})$, $\mathbf{y} = \mathbf{f}(\mathbf{u})$ and hence obtain a composite mapping $\mathbf{y} = \mathbf{f}(\mathbf{g}(\mathbf{x}))$, then the matrix of the linear approximation of the composite mapping is obtained by *multiplying* the approximating matrices of the two stages. In the special case when $\mathbf{f}$ and $\mathbf{g}$ are linear, then we have $\mathbf{u} = B\mathbf{x}$, $\mathbf{y} = A\mathbf{u}$ for appropriate matrices A and B ($\mathbf{u_x} = B$, $\mathbf{y_u} = A$), and the composite mapping is $\mathbf{y} = A(B\mathbf{x}) = AB\mathbf{x}$, as in Section 1–3; thus $\mathbf{y_x} = AB = \mathbf{y_u}\mathbf{u_x}$.

Case of square matrices. In the preceding analysis, let $m = n = p$, so that all the Jacobian matrices appearing are *square*, and each has a determinant—the *Jacobian determinant* of the corresponding mapping, as in Section 2–7. For example,

$$\det \mathbf{y_u} = \begin{vmatrix} \dfrac{\partial y_1}{\partial u_1} & \cdots & \dfrac{\partial y_1}{\partial u_n} \\ \vdots & & \vdots \\ \dfrac{\partial y_n}{\partial u_1} & \cdots & \dfrac{\partial y_n}{\partial u_n} \end{vmatrix} = \frac{\partial(y_1, \ldots, y_n)}{\partial(u_1, \ldots, u_n)}.$$

To the equation (2–49) we can apply the rule det AB = det A det A ((1–13) in Section 1–4), to obtain the following very useful rule:

$$\det \mathbf{y_x} = \det \mathbf{y_u} \det \mathbf{u_x}; \tag{2–50}$$

that is,

$$\frac{\partial(y_1, \ldots, y_n)}{\partial(x_1, \ldots, x_n)} = \frac{\partial(y_1, \ldots, y_n)}{\partial(u_1, \ldots, u_n)} \frac{\partial(u_1, \ldots, u_n)}{\partial(x_1, \ldots, x_n)}. \tag{2–51}$$

If, for example, $n = 2$, then each determinant here can be interpreted as in Section 2–7 as plus or minus the ratio of small corresponding areas, and (2–51) states roughly that

$$\frac{\Delta A_y}{\Delta A_x} = \frac{\Delta A_y}{\Delta A_u} \frac{\Delta A_u}{\Delta A_x},$$

where we have written ΔA_x for an "area element" in the x_1x_2-plane, and similarly for ΔA_y, ΔA_u. There is a similar interpretation for $n = 3$, in terms of volumes, and for higher n in terms of higher-dimensional volume.

Problems

1. Find the Jacobian matrix $(\partial y_i/\partial x_j)$ in the form of a product of two matrices and evaluate the matrix for the given values of $x_1, x_2, \ldots$

(a) $y_1 = u_1u_2 - 3u_1$, $y_2 = u_2^2 + 2u_1u_2 + 2u_1 - u_2$; $u_1 = x_1 \cos 3x_2$, $u_2 = x_1 \sin 3x_2$; $x_1 = 0$, $x_2 = 0$.

(b) $y_1 = u_1^2 + u_2^2 - 3u_1 + u_3$, $y_2 = u_1^2 - u_2^2 + 2u_1 - 3u_3$; $u_1 = x_1x_2^2x_3^2$, $u_2 = x_1x_2^2x_3$, $u_3 = x_1^2x_2x_3$; $x_1 = 1$, $x_2 = 1$, $x_3 = 1$.

(c) $y_1 = u_1e^{u_2}$, $y_2 = u_1e^{-u_2}$, $y_3 = u_1^2$; $u_1 = x_1^2 + x_2$, $u_2 = 2x_1^2 - x_2$; $x_1 = 1$, $x_2 = 0$.

(d) $y_1 = u_1^2 + \cdots + u_n^2 - u_1^2$, $y_2 = u_1^2 + \cdots + u_n^2 - u_2^2, \ldots, y_n = u_1^2 + \cdots + u_n^2 - u_n^2$; $u_1 = x_1^2 + x_1x_2$, $u_2 = x_1^2 + 2x_1x_2, \ldots, u_n = x_1^2 + nx_1x_2$; $x_1 = 1$, $x_2 = 0$.

2. (a) Find $\partial(z, w)/\partial(x, y)$ for $x = 1, y = 0$ if $z = u^3 + 3u^2v - v^3 + u^2 - v^2$, $w = u^3 + v^3 - 2u^2$; $u = x \cos xy$, $v = x \sin xy + x^2 - y^2$.

(b) Find $\partial(x, y)/\partial(s, t)$ for $s = 0, t = 0$ if $x = (z^2 + w^2)^{1/2}$, $y = w(z^2 + w^2)^{-1/2}$; $z = (s + t + 1)^{-1}$, $w = (2s - t + 1)^{-1}$.

3. Justify the rules, under appropriate hypotheses:

(a) If $\mathbf{y} = \mathbf{f}(\mathbf{u})$, $\mathbf{u} = \mathbf{g}(\mathbf{v})$, $\mathbf{v} = \mathbf{h}(\mathbf{x})$, then $\mathbf{y_x} = \mathbf{y_u u_v v_x}$.

(b) $\displaystyle \frac{\partial(z, w)}{\partial(x, y)} = \frac{\partial(z, w)}{\partial(u, v)} \frac{\partial(u, v)}{\partial(s, t)} \frac{\partial(s, t)}{\partial(x, y)}.$

4. For certain functions $f(x, y)$, $g(x, y)$, $p(u, v)$, $q(u, v)$ it is known that $f(x_0, y_0) = u_0$, $g(x_0, y_0) = v_0$ and that $f_x(x_0, y_0) = 2$, $f_y(x_0, y_0) = 3$, $g_x(x_0, y_0) = -1$, $g_y(x_0, y_0) = 5$, $p_u(u_0, v_0) = 7$, $p_v(u_0, v_0) = 1$, $q_u(u_0, v_0) = -3$, $q_v(u_0, v_0) = 2$. Let $z = F(x, y) = p(f(x, y), g(x, y))$, $w = G(x, y) = q(f(x, y), g(x, y))$ and find the Jacobian matrix of $z(x, y)$, $w(x, y)$ at (x_0, y_0).

Answers

1. (a) $\begin{bmatrix} u_2 - 3 & u_1 \\ 2u_2 + 2 & 2u_2 + 2u_1 - 1 \end{bmatrix} \begin{bmatrix} \cos 3x_2 & -3x_1 \sin 3x_2 \\ \sin 3x_2 & 3x_1 \cos 3x_2 \end{bmatrix}, \begin{bmatrix} -3 & 0 \\ 2 & 0 \end{bmatrix}$

(b) $\begin{bmatrix} 2u_1 - 3 & 2u_2 & 1 \\ 2u_1 + 2 & -2u_2 & -3 \end{bmatrix} \begin{bmatrix} x_2x_3^2 & x_1x_3^2 & 2x_1x_2x_3 \\ x_2^2x_3 & 2x_1x_2x_3 & x_1x_2^2 \\ 2x_1x_2x_3 & x_1^2x_3 & x_1^2x_2 \end{bmatrix}, \quad \begin{bmatrix} 3 & 4 & 1 \\ -4 & -3 & 3 \end{bmatrix}.$

(c) $\begin{bmatrix} e^{u_2} & u_1e^{u_2} \\ e^{-u_2} & -u_1e^{-u_2} \\ 2u_1 & 0 \end{bmatrix} \begin{bmatrix} 2x_1 & 1 \\ 4x_1 & -1 \end{bmatrix}, \quad \begin{bmatrix} 6e^2 & 0 \\ -2e^{-2} & 2e^{-2} \\ 4 & 2 \end{bmatrix}.$

(d) $\begin{bmatrix} 0 & 2u_2 & \dots & 2u_n \\ 2u_1 & 0 & \dots & 2u_n \\ \vdots & \vdots & & \vdots \\ 2u_1 & & \dots & 0 \end{bmatrix} \begin{bmatrix} 2x_1 + x_2 & x_1 \\ 2x_1 + 2x_2 & 2x_1 \\ \vdots & \vdots \\ 2x_1 + nx_2 & nx_1 \end{bmatrix}, \quad \begin{bmatrix} 4(n-1) & n^2 + n - 2 \\ 4(n-1) & n^2 + n - 4 \\ \vdots & \vdots \\ 4(n-1) & n^2 + n - 2n \end{bmatrix}.$

2. (a) 31, (b) $-\frac{3}{2}$.

4. $\begin{bmatrix} 13 & 26 \\ -8 & 1 \end{bmatrix}.$

2–10 Implicit functions. If $F(x, y, z)$ is a given function of x, y, and z, then the equation

$$F(x, y, z) = 0 \tag{2–52}$$

is a relation which may describe one or several functions z of x and y. Thus, if $x^2 + y^2 + z^2 - 1 = 0$, then

$$z = \sqrt{1 - x^2 - y^2} \quad \text{or} \quad z = -\sqrt{1 - x^2 - y^2},$$

both functions being defined for $x^2 + y^2 \leqq 1$. Either function is said to be *implicitly defined* by the equation $x^2 + y^2 + z^2 - 1 = 0$.

Similarly, an equation

$$F(x, y, z, w) = 0 \tag{2–53}$$

may define one or more implicit functions w of x, y, z. If two such equations are given:

$$F(x, y, z, w) = 0, \quad G(x, y, z, w) = 0, \tag{2–54}$$

it is in general possible (at least in theory) to reduce the equations by elimination to the form

$$w = f(x, y), \quad z = g(x, y), \tag{2–55}$$

i.e., to obtain two functions of two variables. In general, if m equations in n unknowns are given ($m < n$), it is possible to solve for m of the variables in terms of the remaining $n - m$ variables; the *number of dependent variables equals the number of equations.*

The main question to be considered here is of the following type. Suppose a particular solution of the m simultaneous equations is known, e.g., a quadruple (x_1, y_1, z_1, w_1) of values satisfying (2–54); then one seeks to determine the behavior of the m dependent variables as functions of the independent variables near the given point; e.g., for (2–54), one wishes to study $f(x, y)$ and $g(x, y)$ near (x_1, y_1), given that $f(x_1, y_1) = w_1$ and

$g(x_1, y_1) = z_1$. Determining behavior of a function near a point will consist here of finding the first partial derivatives at the point; when the first derivatives are known, the total differential can be found and hence a *linear approximation* to the function is known.

It will in fact be seen that the essential step to be taken consists of a *linearization* of the given simultaneous equations. This has its counterpart in the formulation of laws of physics; one seeks to describe natural phenomena by means of the simplest possible equations. These are usually linear in form and are usually valid only when the variables are restricted to a narrow range. The complete description of the phenomena in general involves simultaneous nonlinear equations, which are far more difficult to grasp. A typical example is Hooke's Law for a spring: $F = k^2x$; the linear dependence of force F on displacement x is valid only as a first approximation.

To analyze an equation of form (2–52), we assume that $z = f(x, y)$ is a differentiable function which satisfies the equation, so that

$$F(x, y, f(x, y)) = 0. \tag{2–56}$$

We assume that this relation holds for (x, y) in a domain D and that the points (x, y, z), for (x, y) in D and $z = f(x, y)$, all lie in a domain in which F is differentiable. Then from (2–52) or (2–56) we obtain:

$$F_x\,dx + F_y\,dy + F_z dz = 0. \tag{2–57}$$

Here z is considered to be the function $f(x, y)$, so that $dz = f_x\,dx + f_y\,dy$. From (2–57) we deduce:

$$dz = -\frac{F_x}{F_z}\,dx - \frac{F_y}{F_z}\,dy, \tag{2–58}$$

so that (provided $F_z \neq 0$ at the points considered)

$$f_x = \frac{\partial z}{\partial x} = -\frac{F_x}{F_z}, \quad f_y = \frac{\partial z}{\partial y} = -\frac{F_y}{F_z}. \tag{2–59}$$

These are the desired expressions for the derivatives.

Example 1. $x^2 + y^2 + z^2 = 1$ or $x^2 + y^2 + z^2 - 1 = 0$. We imitate the procedure of the preceding paragraph:

$$2x\,dx + 2y\,dy + 2z\,dz = 0,$$

$$dz = -\frac{x}{z}\,dx - \frac{y}{z}\,dy,$$

$$\frac{\partial z}{\partial x} = -\frac{x}{z}, \quad \frac{\partial z}{\partial y} = -\frac{y}{z} \quad (z \neq 0).$$

These equations apply to each differentiable function satisfying the given equation—in particular to the function $z = (1 - x^2 - y^2)^{\frac{1}{2}}$. At $x = \frac{1}{2}$, $y = \frac{1}{2}$, we find $z = 1/\sqrt{2}$, and hence

$$\frac{\partial z}{\partial x} = -\frac{x}{z} = -\frac{\sqrt{2}}{2}, \quad \frac{\partial z}{\partial y} = -\frac{y}{z} = -\frac{\sqrt{2}}{2}.$$

We remark that the same results can be obtained by taking *partial derivatives* instead of differentials in the given equation. From (2–52), with z considered as a function of x and y, we differentiate with respect to x and then with respect to y to obtain

$$F_x + F_z \frac{\partial z}{\partial x} = 0, \quad F_y + F_z \frac{\partial z}{\partial y} = 0,$$

from which (2–59) again follows. In the case of Ex. 1, we obtain

$$2x + 2z \frac{\partial z}{\partial x} = 0, \quad 2y + 2z \frac{\partial z}{\partial y} = 0.$$

For the pair of equations (2–54), we assume that differentiable functions $w = f(x, y)$, $z = g(x, y)$ satisfy the equations and can then take differentials:

$$\begin{aligned} F_x\,dx + F_y\,dy + F_z\,dz + F_w\,dw &= 0, \\ G_x\,dx + G_y\,dy + G_z\,dz + G_w\,dw &= 0. \end{aligned} \tag{2–60}$$

We consider these equations as simultaneous *linear* equations for dz and dw and solve by elimination or by determinants. Cramer's rule yields

$$dz = \frac{\begin{vmatrix} -F_x\,dx - F_y\,dy & F_w \\ -G_x\,dx - G_y\,dy & G_w \end{vmatrix}}{\begin{vmatrix} F_z & F_w \\ G_z & G_w \end{vmatrix}}, \quad dw = \frac{\begin{vmatrix} F_z & -F_x\,dx - F_y\,dy \\ G_z & -G_x\,dx - G_y\,dy \end{vmatrix}}{\begin{vmatrix} F_z & F_w \\ G_z & G_w \end{vmatrix}},$$

and hence

$$dz = -\frac{\begin{vmatrix} F_x & F_w \\ G_x & G_w \end{vmatrix}}{\begin{vmatrix} F_z & F_w \\ G_z & G_w \end{vmatrix}}\,dx - \frac{\begin{vmatrix} F_y & F_w \\ G_y & G_w \end{vmatrix}}{\begin{vmatrix} F_z & F_w \\ G_z & G_w \end{vmatrix}}\,dy,$$

$$dw = -\frac{\begin{vmatrix} F_z & F_x \\ G_z & G_x \end{vmatrix}}{\begin{vmatrix} F_z & F_w \\ G_z & G_w \end{vmatrix}}\,dx - \frac{\begin{vmatrix} F_z & F_y \\ G_z & G_y \end{vmatrix}}{\begin{vmatrix} F_z & F_w \\ G_z & G_w \end{vmatrix}}\,dy.$$

Thus we can read off partial derivatives. Since the determinants appearing are Jacobian determinants, we can write the derivatives in terms of Jacobians:

$$\begin{aligned} \frac{\partial z}{\partial x} &= -\frac{\dfrac{\partial(F, G)}{\partial(x, w)}}{\dfrac{\partial(F, G)}{\partial(z, w)}}, \quad & \frac{\partial z}{\partial y} &= -\frac{\dfrac{\partial(F, G)}{\partial(y, w)}}{\dfrac{\partial(F, G)}{\partial(z, w)}}, \\ \frac{\partial w}{\partial x} &= -\frac{\dfrac{\partial(F, G)}{\partial(z, x)}}{\dfrac{\partial(F, G)}{\partial(z, w)}}, \quad & \frac{\partial w}{\partial y} &= -\frac{\dfrac{\partial(F, G)}{\partial(z, y)}}{\dfrac{\partial(F, G)}{\partial(z, w)}}. \end{aligned} \tag{2–61}$$

Here we must assume that the Jacobian $\partial(F, G)/\partial(z, w)$ in the denominator is different from 0 at the points considered.

EXAMPLE 2. $2x^2 + y^2 + z^2 - zw = 0,$
$x^2 + y^2 + 2z^2 + zw - 8 = 0.$

We observe that the equations are satisfied for $x = 1$, $y = 1$, $z = 1$, $w = 4$, and seek differentiable functions $z(x, y)$, $w(x, y)$ satisfying the equations near this point. If such functions exist, then

$$4x\,dx + 2y\,dy + (2z - w)\,dz - z\,dw = 0,$$
$$2x\,dx + 2y\,dy + (4z + w)\,dz + z\,dw = 0.$$

By elimination we find

$$6x\,dx + 4y\,dy + 6z\,dz = 0,$$
$$6x(2z + w)\,dx + 4y(z + w)\,dy - 6z^2\,dw = 0$$

and hence

$$dz = -\frac{x}{z}\,dx - \frac{2y}{3z}\,dy,$$

$$dw = \frac{x(2z + w)}{z^2}\,dx + \frac{2y(z + w)}{3z^2}\,dy,$$

from which we can read off partial derivatives: $\partial z/\partial x = -x/z$ and so on. We could have applied (2–61) directly. For example,

$$\frac{\partial z}{\partial x} = -\frac{\begin{vmatrix} 4x & -z \\ 2x & z \end{vmatrix}}{\begin{vmatrix} 2z - w & -z \\ 4z + w & z \end{vmatrix}} = \frac{-6xz}{6z^2} = -\frac{x}{z}.$$

We observe that the determinant in the denominator equals $6z^2$, and this is different from 0 near $z = 1$.

We could also have taken partial derivatives. By differentiating with respect to x, we obtain

$$4x + (2z - w)\frac{\partial z}{\partial x} - z\frac{\partial w}{\partial x} = 0,$$

$$2x + (4z + w)\frac{\partial z}{\partial x} + z\frac{\partial w}{\partial x} = 0.$$

Elimination gives $\partial z/\partial x = -x/z$, $\partial w/\partial x = x(2z + w)/z^2$. Taking differentials saves time, since all partial derivatives are obtained at once.

The reasoning generalizes to an arbitrary set of m equations in $m + n$ unknowns, say

$$\begin{aligned} F_1(y_1, \ldots, y_m, x_1, \ldots, x_n) &= 0, \\ &\vdots \\ F_m(y_1, \ldots, y_m, x_1, \ldots, x_n) &= 0. \end{aligned} \tag{2–62}$$

We seek m differentiable functions

$$\begin{aligned} y_1 &= f_1(x_1, \ldots, x_n), \\ &\vdots \\ y_m &= f_m(x_1, \ldots, x_n) \end{aligned} \tag{2–63}$$

satisfying the equations. Assuming differentiability as above, we obtain from (2–62)

$$\begin{aligned} F_{1y_1}\,dy_1 + \cdots + F_{1y_m}\,dy_m + F_{1x_1}\,dx_1 + \cdots + F_{1x_n}\,dx_n &= 0, \\ &\vdots \\ F_{my_1}\,dy_1 + \cdots + F_{my_m}\,dy_m + F_{mx_1}\,dx_1 + \cdots + F_{mx_n}\,dx_n &= 0, \end{aligned} \tag{2–64}$$

where $dy_1, \ldots, dy_m$ are the differentials of the functions sought. Equations (2–64) are m *linear* equations in the m unknowns $dy_1, \ldots, dy_m$ and we have in effect *linearized* our problem. If the appropriate determinant is not 0, we can solve Eqs. (2–64) for $dy_1, \ldots, dy_m$. That determinant is

$$\begin{vmatrix} F_{1y_1} & \cdots & F_{1y_m} \\ \vdots & & \vdots \\ F_{my_1} & \cdots & F_{my_m} \end{vmatrix} = \frac{\partial(F_1, \ldots, F_m)}{\partial(y_1, \ldots, y_m)}. \tag{2–65}$$

The equations can be solved by elimination or by determinants.

We can also use matrices here. Equations (2–64) can be written:

$$\mathbf{F_y}\,d\mathbf{y} + \mathbf{F_x}\,d\mathbf{x} = \mathbf{0}, \tag{2–66}$$

where $\mathbf{F_y}$ is the Jacobian matrix of the coefficients of $dy_1, \ldots, dy_m$ in (2–64); that is, it is the matrix whose determinant is given in (2–65). Similarly,

$$\mathbf{F_x} = \begin{bmatrix} F_{1x_1} & \cdots & F_{1x_n} \\ \vdots & & \vdots \\ F_{mx_1} & \cdots & F_{mx_n} \end{bmatrix}. \tag{2–67}$$

Finally, $d\mathbf{y} = \text{col}\,(dy_1, \ldots, dy_m)$ and $d\mathbf{x} = \text{col}\,(dx_1, \ldots, dx_n)$. If $\mathbf{F_y}$ is not singular at the point considered—that is, if the Jacobian determinant (2–65) is not 0, then $\mathbf{F_y}$ has an inverse and we can write:

$$\begin{aligned} \mathbf{F_y}\,d\mathbf{y} &= -\mathbf{F_x}\,d\mathbf{x}, \\ d\mathbf{y} &= -(\mathbf{F_y})^{-1}\mathbf{F_x}\,d\mathbf{x}. \end{aligned} \tag{2–68}$$

This equation gives $dy_1, \ldots, dy_m$ and hence all partial derivatives of the unknown functions $f_1, \ldots, f_m$.

We can also take partial derivatives in (2–62). By differentiation with respect to x_j, we obtain:

$$\begin{aligned} F_{1y_1}\frac{\partial y_1}{\partial x_j} + \cdots + F_{1y_m}\frac{\partial y_m}{\partial x_j} + F_{1x_j} &= 0, \\ &\vdots \\ F_{my_1}\frac{\partial y_1}{\partial x_j} + \cdots + F_{my_m}\frac{\partial y_m}{\partial x_j} + F_{mx_j} &= 0. \end{aligned}$$

These are m linear equations for the m unknowns $\partial y_1/\partial x_j, \ldots, \partial y_m/\partial x_j$.

If the determinant (2–65) is not 0, we can solve for the unknowns. Cramer's rule gives, for example,

$$\frac{\partial y_1}{\partial x_j} = -\begin{vmatrix} F_{1x_j} & F_{1y_2} & \dots & F_{1y_m} \\ \vdots & \vdots & & \vdots \\ F_{mx_j} & F_{my_2} & \dots & F_{my_m} \end{vmatrix} \div \begin{vmatrix} F_{1y_1} & F_{1y_2} & \dots & F_{1y_m} \\ \vdots & \vdots & & \vdots \\ F_{my_1} & F_{my_2} & \dots & F_{my_m} \end{vmatrix}$$

$$= -\frac{\dfrac{\partial(F_1, \dots, F_m)}{\partial(x_j, y_2, \dots, y_m)}}{\dfrac{\partial(F_1, \dots, F_m)}{\partial(y_1, \dots, y_m)}}.$$

Similar formulas are obtained for $\partial y_2/\partial x_j, \dots, \partial y_m/\partial x_j$. In general,

$$\frac{\partial y_i}{\partial x_j} = -\frac{\dfrac{\partial(F_1, \dots, F_m)}{\partial(y_1, \dots, y_{i-1}, x_j, y_{i+1}, \dots, y_m)}}{\dfrac{\partial(F_1, \dots, F_m)}{\partial(y_1, \dots, y_m)}} \tag{2–69}$$

$$(i = 1, \dots, m, \quad j = 1, \dots, n).$$

The denominator here is the Jacobian (2–65). The numerator is obtained from the denominator by replacing the i-th column by col $(\partial F_1/\partial x_j, \dots, \partial F_m/\partial x_j)$. An illustration is given by Eqs. (2–61) above, where $m = 2$ and $n = 2$ (2 equations in 4 unknowns). The following are additional illustrations:

1 equation in 2 unknowns: $F(x, y) = 0$. Here

$$\frac{dy}{dx} = -\frac{F_x}{F_y} \quad (F_y \neq 0). \tag{2–70}$$

1 equation in 3 unknowns: $F(x, y, z) = 0$. Here, as in (2–59),

$$\frac{\partial z}{\partial x} = -\frac{F_x}{F_z}, \quad \frac{\partial z}{\partial y} = -\frac{F_y}{F_z} \quad (F_z \neq 0). \tag{2–71}$$

2 equations in 3 unknowns: $F(x, y, z) = 0$, $G(x, y, z) = 0$. Here

$$\frac{dz}{dx} = -\frac{\dfrac{\partial(F, G)}{\partial(y, x)}}{\dfrac{\partial(F, G)}{\partial(y, z)}}, \quad \frac{dy}{dx} = -\frac{\dfrac{\partial(F, G)}{\partial(x, z)}}{\dfrac{\partial(F, G)}{\partial(y, z)}}, \tag{2–72}$$

where

$$\frac{\partial(F, G)}{\partial(y, z)} = \begin{vmatrix} F_y & F_z \\ G_y & G_z \end{vmatrix} \neq 0.$$

3 equations in 5 unknowns: $F(x, y, z, u, v) = 0$, $G(x, y, z, u, v) = 0$, $H(x, y, z, u, v) = 0$.

Here

$$\frac{\partial x}{\partial u} = -\frac{\dfrac{\partial(F, G, H)}{\partial(u, y, z)}}{\dfrac{\partial(F, G, H)}{\partial(x, y, z)}}, \quad \frac{\partial x}{\partial v} = -\frac{\dfrac{\partial(F, G, H)}{\partial(v, y, z)}}{\dfrac{\partial(F, G, H)}{\partial(x, y, z)}},$$

$$\frac{\partial y}{\partial u} = -\frac{\dfrac{\partial(F, G, H)}{\partial(x, u, z)}}{\dfrac{\partial(F, G, H)}{\partial(x, y, z)}}, \quad \frac{\partial y}{\partial v} = -\frac{\dfrac{\partial(F, G, H)}{\partial(x, v, z)}}{\dfrac{\partial(F, G, H)}{\partial(x, y, z)}}, \tag{2–73}$$

$$\frac{\partial z}{\partial u} = -\frac{\dfrac{\partial(F, G, H)}{\partial(x, y, u)}}{\dfrac{\partial(F, G, H)}{\partial(x, y, z)}}, \quad \frac{\partial z}{\partial v} = -\frac{\dfrac{\partial(F, G, H)}{\partial(x, y, v)}}{\dfrac{\partial(F, G, H)}{\partial(x, y, z)}}.$$

For greater precision the partial derivatives appearing should have subscripts attached. For example, in (2–73), $\partial x/\partial u$ should be $(\partial x/\partial u)_v$.

The entire preceding discussion has been based on the assumption that implicit functions are in fact defined by the equations. This assumption is not always fulfilled. For example, the equations

$$x^2 + y^2 + u^2 + v^2 + 1 = 0, \quad x^2 - y^2 + 2uv = 0$$

define no functions at all. It can be shown that if

(a) Eqs. (2–62) are all satisfied at one point $(y_1^0, \ldots, y_m^0, x_1^0, \ldots, x_n^0)$, in a neighborhood of which all functions $F_1, \ldots, F_m$ are differentiable, and

(b) the Jacobian determinant (2–65) is not 0 at this point,

then in an appropriate neighborhood of $(x_1^0, \ldots, x_n^0)$ Eqs. (2–62) do define a unique set of differentiable functions (2–63) with $f_i(x_1^0, \ldots, x_n^0) = y_i^0$ for $i = 1, \ldots, m$. This is the Implicit Function Theorem. For a proof of a typical case, see *CLA* Section 12–15.

If condition (b) fails, one may be able to choose a different set of m dependent variables (say, $x_1, y_2, \ldots, y_m$), consider these as functions of the remaining n variables and then find that the new form of condition (b) holds. If no such choice succeeds, one is dealing with a degenerate case requiring special study. (See also Section 2–22.)

2–11 Inverse functions. Curvilinear coordinates. A pair of functions

$$x = f(u, v), \quad y = g(u, v) \tag{2–74}$$

can be regarded as a mapping from the uv-plane to the xy-plane (see Section 2–7 above). Under appropriate conditions, this mapping is a one-to-one correspondence between a domain D_{uv} in the uv-plane and a domain D_{xy} in the xy-plane (Figure 2–7). One can then consider the *inverse mapping* which takes each point (x, y) in D_{xy} to the unique point (u, v) such that (2–74) holds. The inverse mapping is given by functions

$$u = \varphi(x, y), \quad v = \psi(x, y). \tag{2–75}$$

However, we may have difficulty in solving (2–74) to obtain these functions explicitly. We may, nevertheless, consider Eqs. (2–74), in the form

$$f(u, v) - x = 0, \quad g(u, v) - y = 0$$

as implicit equations for the functions (2–75). We can then seek partial derivatives as in the preceding section. With $F(x, y, u, v) = f(u, v) - x$ and $G(x, y, u, v) = g(u, v) - y$, we have, for example,

$$\frac{\partial u}{\partial x} = -\frac{\dfrac{\partial(F, G)}{\partial(x, v)}}{\dfrac{\partial(F, G)}{\partial(u, v)}} = -\frac{\begin{vmatrix} -1 & f_v \\ 0 & g_v \end{vmatrix}}{\begin{vmatrix} f_u & f_v \\ g_u & g_v \end{vmatrix}} = \frac{g_v}{\begin{vmatrix} f_u & f_v \\ g_u & g_v \end{vmatrix}}.$$

The Jacobian determinant in the denominator is simply the Jacobian of the mapping (2–74). As above, we assume it is not zero at the points considered.

The fact that the functions (2–75) are solutions of (2–74) means that

$$f[\varphi(x, y), \psi(x, y)] \equiv x, \quad g[\varphi(x, y), \psi(x, y)] \equiv y.$$

(If we apply a mapping and then the inverse mapping, the composite is the identity mapping.) Hence by the rule (2–51)

$$\frac{\partial(x, y)}{\partial(u, v)} \frac{\partial(u, v)}{\partial(x, y)} = \begin{vmatrix} 1 & 0 \\ 0 & 1 \end{vmatrix} = 1. \tag{2–76}$$

Thus *the Jacobian of the inverse mapping is the reciprocal of the Jacobian of the mapping.*

Curvilinear coordinates. We can also regard (2–74) as equations describing a transformation from rectangular to curvilinear coordinates in the xy-plane. For fixed v, (x, y) traces a curve with u as parameter; for fixed u, (x, y) traces a curve with v as parameter. The curves mentioned serve as lines $v = \text{const}$, $u = \text{const}$ in the curvilinear coordinate system (Fig. 2–7).

The most common example of curvilinear coordinates is provided by polar coordinates (Fig. 2–8). Here $x = r \cos \theta$, $y = r \sin \theta$.

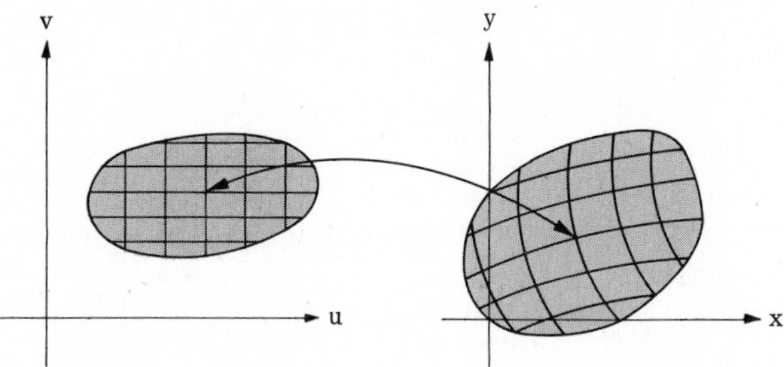

FIG. 2–7. Mapping, inverse mapping and curvilinear coordinates.

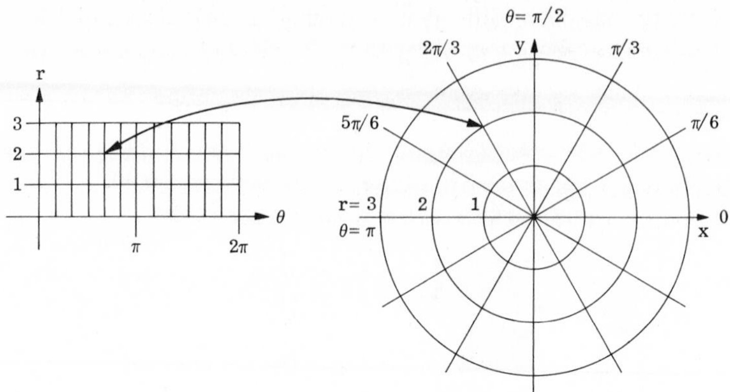

FIG. 2–8. Polar coordinates as curvilinear coordinates.

The inverse functions (2–75) express the curvilinear coordinates in terms of rectangular coordinates. For polar coordinates, the inverse functions are

$$r = \sqrt{x^2 + y^2}, \quad \theta = \tan^{-1}\frac{y}{x},$$

for an appropriate interpretation of the inverse tangent.

The preceding discussion extends to 3-dimensional space and generally to E^n. One considers a mapping

$$x_1 = f_1(u_1, \ldots, u_n), \ldots, x_n = f_n(u_1, \ldots, u_n) \tag{2–77}$$

or, in vector language, a vector function

$$\mathbf{x} = \mathbf{f}(\mathbf{u}).$$

Under appropriate conditions there is an inverse mapping

$$u_1 = g_1(x_1, \ldots, x_n), \ldots, u_n = g_n(x_1, \ldots, x_n) \tag{2–78}$$

or $\mathbf{u} = \mathbf{g}(\mathbf{x})$. The partial derivatives of the inverse functions (2–78) can be obtained as above, by the procedures for implicit functions. One can also reason that the differentials satisfy a matrix equation

$$d\mathbf{x} = \mathbf{f_u}d\mathbf{u}$$

and hence

$$d\mathbf{u} = (\mathbf{f_u})^{-1}\,d\mathbf{x}.$$

This shows that the Jacobian matrix of the inverse mapping is simply the inverse of the Jacobian matrix of the mapping (assumed to be nonsingular). In concise notation,

$$\mathbf{u_x} = (\mathbf{x_u})^{-1},$$

so that

$$\det \mathbf{u_x} = \frac{1}{\det \mathbf{x_u}} \quad \text{and} \quad \det \mathbf{u_x} \det \mathbf{x_u} = 1 \tag{2–79}$$

as in (2–76). The concept of curvilinear coordinates also applies to (2–77). Common examples, for $n = 3$, are the two systems

$$x = r\cos\theta, \quad y = r\sin\theta, \quad z = z, \tag{2–80}$$

$$x = \rho\sin\phi\cos\theta, \quad y = \rho\sin\phi\sin\theta, \quad z = \rho\cos\phi, \tag{2–81}$$

(cf. Section 0–5), which are the equations of transformation from rectangular coordinates to cylindrical and spherical coordinates respectively.

Problems

1. Given that

$$2x + y - 3z - 2u = 0, \quad x + 2y + z + u = 0,$$

find the following partial derivatives:

$$\left(\frac{\partial x}{\partial y}\right)_z, \quad \left(\frac{\partial y}{\partial x}\right)_u, \quad \left(\frac{\partial z}{\partial u}\right)_x, \quad \left(\frac{\partial y}{\partial z}\right)_x.$$

2. Given that

$$x^2 + y^2 + z^2 - u^2 + v^2 = 1, \quad x^2 - y^2 + z^2 + u^2 + 2v^2 = 21,$$

(a) find du and dv in terms of dx, dy, and dz at the point $x = 1$, $y = 1$, $z = 2$, $u = 3$, $v = 2$;
(b) find $(\partial u/\partial x)_{y,\,z}$ and $(\partial v/\partial y)_{x,\,z}$ at this point;
(c) find approximately the values of u and v for $x = 1.1$, $y = 1.2$, $z = 1.8$.

3. For the transformation: $x = r\cos\theta$, $y = r\sin\theta$ from rectangular to polar coordinates, verify the relations:

(a) $dx = \cos\theta\,dr - r\sin\theta\,d\theta, \quad dy = \sin\theta\,dr + r\cos\theta\,d\theta.$

(b) $dr = \cos\theta\,dx + \sin\theta\,dy, \quad d\theta = -\dfrac{\sin\theta}{r}\,dx + \dfrac{\cos\theta}{r}\,dy.$

(c) $\left(\dfrac{\partial x}{\partial r}\right)_\theta = \cos\theta, \quad \left(\dfrac{\partial x}{\partial r}\right)_y = \sec\theta, \quad \left(\dfrac{\partial r}{\partial x}\right)_y = \cos\theta, \quad \left(\dfrac{\partial r}{\partial x}\right)_\theta = \sec\theta.$

(d) $\dfrac{\partial(x, y)}{\partial(r, \theta)} = r, \quad \dfrac{\partial(r, \theta)}{\partial(x, y)} = \dfrac{1}{r}.$

4. Given that

$$\begin{aligned} x^2 - y\cos(uv) + z^2 &= 0,\\ x^2 + y^2 - \sin(uv) + 2z^2 &= 2,\\ xy - \sin u\cos v + z &= 0, \end{aligned}$$

find $(\partial x/\partial u)_v$ and $(\partial x/\partial v)_u$ at the point $x = 1$, $y = 1$, $u = \dfrac{\pi}{2}$, $v = 0$, $z = 0$.

5. Find $(\partial u/\partial x)_y$ if

$$x^2 - y^2 + u^2 + 2v^2 = 1, \quad x^2 + y^2 - u^2 - v^2 = 2.$$

6. Given the mapping

$$x = u - 2v, \quad y = 2u + v,$$

(a) write the equations of the inverse mapping, (b) evaluate the Jacobian of the mapping and that of the inverse mapping.

7. Given the mapping

$$x = u^2 - v^2, \quad y = 2uv,$$

(a) compute its Jacobian, (b) evaluate $\left(\dfrac{\partial u}{\partial x}\right)_y$ and $\left(\dfrac{\partial v}{\partial x}\right)_y$.

8. Given the mapping

$$x = f(u, v), \quad y = g(u, v),$$

with Jacobian $J = \dfrac{\partial(x, y)}{\partial(u, v)}$, show that for the inverse functions one has

$$\frac{\partial u}{\partial x} = \frac{1}{J}\frac{\partial y}{\partial v}, \quad \frac{\partial u}{\partial y} = -\frac{1}{J}\frac{\partial x}{\partial v}, \quad \frac{\partial v}{\partial x} = -\frac{1}{J}\frac{\partial y}{\partial u}, \quad \frac{\partial v}{\partial y} = \frac{1}{J}\frac{\partial x}{\partial u}.$$

Use these results to check Prob. 3(b) above.

9. Given the mapping

$$x = f(u, v, w), \quad y = g(u, v, w), \quad z = h(u, v, w),$$

with Jacobian $J = \dfrac{\partial(x, y, z)}{\partial(u, v, w)}$, show that for the inverse functions one has

$$\frac{\partial u}{\partial x} = \frac{1}{J}\frac{\partial(y, z)}{\partial(v, w)}, \quad \frac{\partial u}{\partial y} = \frac{1}{J}\frac{\partial(z, x)}{\partial(v, w)}, \quad \frac{\partial u}{\partial z} = \frac{1}{J}\frac{\partial(x, y)}{\partial(v, w)},$$

$$\frac{\partial v}{\partial x} = \frac{1}{J}\frac{\partial(y, z)}{\partial(w, u)}, \quad \frac{\partial v}{\partial y} = \frac{1}{J}\frac{\partial(z, x)}{\partial(w, u)}, \quad \frac{\partial v}{\partial z} = \frac{1}{J}\frac{\partial(x, y)}{\partial(w, u)},$$

$$\frac{\partial w}{\partial x} = \frac{1}{J}\frac{\partial(y, z)}{\partial(u, v)}, \quad \frac{\partial w}{\partial y} = \frac{1}{J}\frac{\partial(z, x)}{\partial(u, v)}, \quad \frac{\partial w}{\partial z} = \frac{1}{J}\frac{\partial(x, y)}{\partial(u, v)}.$$

10. For the transformation (2–81) from rectangular to spherical coordinates, (a) compute the Jacobian $\dfrac{\partial(x, y, z)}{\partial(\rho, \phi, \theta)}$, (b) evaluate $\partial\rho/\partial y$, $\partial\phi/\partial z$, $\partial\theta/\partial x$ for the inverse transformation (cf. Prob. 9).

11. Let equations $F_1(x_1, x_2, x_3, x_4) = 0$, $F_2(x_1, x_2, x_3, x_4) = 0$ be given. At a certain point where the equations are satisfied, it is known that

$$\frac{\partial F_i}{\partial x_j} = \begin{bmatrix} 3 & 1 & 0 & 2 \\ 5 & 1 & -1 & 4 \end{bmatrix}.$$

(a) Evaluate $(\partial x_1/\partial x_3)_{x_4}$ and $(\partial x_1/\partial x_4)_{x_3}$ at the point.
(b) Evaluate $(\partial x_1/\partial x_3)_{x_2}$ and $(\partial x_4/\partial x_3)_{x_2}$ at the point.
(c) Evaluate $\partial(x_1, x_2)/\partial(x_3, x_4)$ and $\partial(x_3, x_4)/\partial(x_1, x_2)$ at the point.

12. Prove: If $F(x, y, z) = 0$, then $\left(\dfrac{\partial z}{\partial x}\right)_y\left(\dfrac{\partial x}{\partial y}\right)_z\left(\dfrac{\partial y}{\partial z}\right)_x = -1.$

13. Prove that, if $x = f(u, v)$, $y = g(u, v)$, then

$$\left(\frac{\partial x}{\partial u}\right)_v\left(\frac{\partial u}{\partial x}\right)_y = \left(\frac{\partial y}{\partial v}\right)_u\left(\frac{\partial v}{\partial y}\right)_x$$

and

$$\left(\frac{\partial x}{\partial v}\right)_u\left(\frac{\partial v}{\partial x}\right)_y = \left(\frac{\partial u}{\partial y}\right)_x\left(\frac{\partial y}{\partial u}\right)_v,$$

also that

$$\left(\frac{\partial x}{\partial y}\right)_u\left(\frac{\partial y}{\partial x}\right)_u = 1.$$

14. In *thermodynamics* the variables p (pressure), T (temperature), U (internal energy), and V (volume) occur. For each substance these are related by two equations, so that any two of the four variables can be chosen as independent, the other two then being dependent. In addition, the second law of thermodynamics implies the relation

(a) $\dfrac{\partial U}{\partial V} - T\dfrac{\partial p}{\partial T} + p = 0,$

when V and T are independent. Show that this relation can be written in each of the following forms:

(b) $\dfrac{\partial T}{\partial V} + T\dfrac{\partial p}{\partial U} - p\dfrac{\partial T}{\partial U} = 0 \qquad (U, V \text{ indep.}),$

(c) $T - p\dfrac{\partial T}{\partial p} + \dfrac{\partial(T, U)}{\partial(V, p)} = 0 \qquad (V, p \text{ indep.}),$

(d) $\dfrac{\partial U}{\partial p} + T\dfrac{\partial V}{\partial T} + p\dfrac{\partial V}{\partial p} = 0 \qquad (p, T \text{ indep.}),$

(e) $\dfrac{\partial T}{\partial p} - T\dfrac{\partial V}{\partial U} + p\dfrac{\partial(V, T)}{\partial(U, p)} = 0 \quad (U, p \text{ indep.}),$

(f) $T\dfrac{\partial(p, V)}{\partial(T, U)} - p\dfrac{\partial V}{\partial U} - 1 = 0 \qquad (T, U \text{ indep.}).$

[Hint: the relation (a) implies that, if

$$dU = a\,dV + b\,dT, \quad dp = c\,dV + e\,dT$$

are the expressions for dU and dp in terms of dV and dT, then $a - Te + p = 0$. To prove (b), for example, one assumes relations

$$dT = \alpha\,dV + \beta\,dU, \quad dp = \gamma\,dV + \delta\,dU.$$

If these are solved for dU and dp in terms of dV and dT, then one obtains expressions for a and e in terms of $\alpha, \beta, \gamma, \delta$. If these expressions are substituted in the equation $a - Te + p = 0$, one has an equation in $\alpha, \beta, \gamma, \delta$. Since $\alpha = \partial T/\partial V$, etc., the relation of form (b) is obtained. The others are proved in the same way.]

Answers

1. $\left(\dfrac{\partial x}{\partial y}\right)_z = -\dfrac{5}{4}, \quad \left(\dfrac{\partial y}{\partial x}\right)_u = -\dfrac{5}{7}, \quad \left(\dfrac{\partial z}{\partial u}\right)_x = -\dfrac{5}{7}, \quad \left(\dfrac{\partial y}{\partial z}\right)_x = \dfrac{1}{5}.$

2. (a) $du = \frac{1}{9}(dx + 3\,dy + 2\,dz), \quad dv = -\frac{1}{3}(dx + 2\,dz);$

 (b) $\left(\dfrac{\partial u}{\partial x}\right)_{y,z} = \frac{1}{9}, \quad \left(\dfrac{\partial v}{\partial y}\right)_{x,z} = 0;$ (c) $u = 3.033, \quad v = 2.1.$

4. $\left(\dfrac{\partial x}{\partial u}\right)_v = 0, \quad \left(\dfrac{\partial x}{\partial v}\right)_u = \dfrac{\pi}{12}.$ 5. $\left(\dfrac{\partial u}{\partial x}\right)_y = \dfrac{3x}{u}.$

6. (a) $u = \frac{1}{5}(x + 2y), \quad v = \frac{1}{5}(y - 2x);$ (b) $J = 5, \quad J$ for inverse $= \frac{1}{5}.$

7. (a) $J = 4(u^2 + v^2),$ (b) $\left(\dfrac{\partial u}{\partial x}\right)_y = \dfrac{u}{2(u^2 + v^2)}, \quad \left(\dfrac{\partial v}{\partial x}\right)_y = -\dfrac{v}{2(u^2 + v^2)}.$

10. (a) $J = \rho^2 \sin\phi;$ (b) $\dfrac{\partial \rho}{\partial y} = \sin\phi\sin\theta, \quad \dfrac{\partial \phi}{\partial z} = -\dfrac{\sin\phi}{\rho}, \quad \dfrac{\partial \theta}{\partial x} = -\dfrac{\sin\theta}{\rho\sin\phi}.$

11. (a) $\frac{1}{2}, -1;$ (b) $-1, \frac{3}{2};$ (c) $-1, -1.$

2–12 Geometrical applications. Let the functions $f(t)$, $g(t)$, $h(t)$ be defined and have continuous derivatives for $\alpha \leqq t \leqq \beta$. Then, as in Sections 0–7 and 0–9, the equations

$$x = f(t), \quad y = g(t), \quad z = h(t) \tag{2–82}$$

are parametric equations of a curve in space. A tangent line to such a curve is defined as the limiting position of a secant through two points (x_1, y_1, z_1) and $(x_1 + \Delta x, y_1 + \Delta y, z_1 + \Delta z)$ on the curve, corresponding to parameter values t_1 and $t_1 + \Delta t$, as Δt approaches 0. This is illustrated in Fig. 2–9. The three ratios

$$\frac{\Delta x}{\Delta t}, \quad \frac{\Delta y}{\Delta t}, \quad \frac{\Delta z}{\Delta t},$$

form a set of direction components of the secant; when Δt approaches 0, these ratios approach as limits the derivatives

$$f'(t_1), \quad g'(t_1), \quad h'(t_1),$$

which can accordingly be interpreted as direction components of the tangent line. If all three derivatives are 0 for $t = t_1$, this reasoning fails; in such a case it may be possible to obtain the tangent line by repeated differentiation (Prob. 14 below).

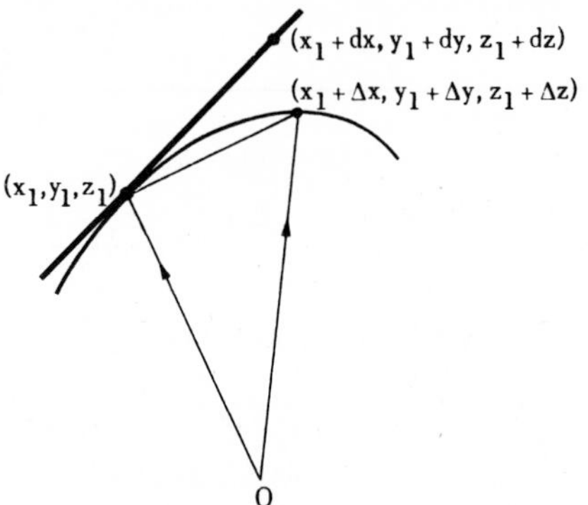

FIG. 2–9. Tangent to a curve.

We assume now that not all three derivatives are 0. Then the parametric equations of the tangent line are as follows (Section 0–5):

$$x - x_1 = f'(t_1)(t - t_1), \quad y - y_1 = g'(t_1)(t - t_1), \quad z - z_1 = h'(t_1)(t - t_1). \tag{2–83}$$

If we take differentials in (2–82), we obtain the equations

$$dx = f'(t)\,dt, \quad dy = g'(t)\,dt, \quad dz = h'(t)\,dt. \tag{2–84}$$

These are the same as (2–83), if we let $t = t_1$ and identify the differentials dx, dy, dz with changes in x, y, z, t along the tangent line; that is, we write

$$dx = x - x_1, \quad dy = y - y_1, \quad dz = z - z_1, \quad dt = t - t_1, \tag{2–85}$$

where (x, y, z) is a point on the tangent line at (x_1, y_1, z_1). This illustrates a basic principle, namely, that *taking differentials corresponds to forming tangents.*

These results can be put in vector form. We introduce the position vector

$$\mathbf{r} = x\mathbf{i} + y\mathbf{j} + z\mathbf{k} = f(t)\mathbf{i} + g(t)\mathbf{j} + h(t)\mathbf{k}$$

of a variable point (x, y, z) on the given curve. Then, as in Section 0–9, the "velocity vector"

$$\frac{d\mathbf{r}}{dt} = f'(t)\mathbf{i} + g'(t)\mathbf{j} + h'(t)\mathbf{k} \tag{2–86}$$

is a tangent vector to the curve. If we multiply this equation by dt, we obtain an equation

$$d\mathbf{r} = f'(t)\,dt\mathbf{i} + g'(t)\,dt\mathbf{j} + h'(t)\,dt\mathbf{k}, \tag{2–87}$$

which is the vector form of (2–84). The vector $d\mathbf{r}$ is the "differential displacement vector":

$$d\mathbf{r} = dx\mathbf{i} + dy\mathbf{j} + dz\mathbf{k} = (x - x_1)\mathbf{i} + (y - y_1)\mathbf{j} + (z - z_1)\mathbf{k}, \tag{2–88}$$

where (x, y, z) is a point on the tangent line. The magnitude of $d\mathbf{r}$ is the scalar

$$|d\mathbf{r}| = \sqrt{dx^2 + dy^2 + dz^2} = ds, \tag{2–89}$$

where s is arc length on the curve, as in Section 0–9. The magnitude of the vector $d\mathbf{r}/dt$ is the "speed"

$$\left|\frac{d\mathbf{r}}{dt}\right| = \frac{|d\mathbf{r}|}{dt} = \frac{ds}{dt},$$

if we assume s increases with t. The vector

$$\mathbf{T} = \frac{d\mathbf{r}}{ds} = \frac{d\mathbf{r}}{dt}\bigg/\frac{ds}{dt}$$

is a unit tangent vector to the curve. If we use arc length s as parameter on the curve, the vector (2–86) becomes precisely the unit tangent vector $\mathbf{T}$:

$$\mathbf{T} = \frac{d\mathbf{r}}{ds} = \frac{dx}{ds}\mathbf{i} + \frac{dy}{ds}\mathbf{j} + \frac{dz}{ds}\mathbf{k}. \tag{2–90}$$

Now let an equation

$$F(x, y, z) = 0 \tag{2–91}$$

be given; in general, this represents a surface in space. Let (x_1, y_1, z_1) be a point on this surface, and let

$$x = f(t), \quad y = g(t), \quad z = h(t) \tag{2–92}$$

be a curve *in the surface*, passing through (x_1, y_1, z_1) when $t = t_1$. One has thus

$$F[f(t), g(t), h(t)] \equiv 0. \tag{2–93}$$

Taking differentials, one concludes:

$$\frac{\partial F}{\partial x}\,dx + \frac{\partial F}{\partial y}\,dy + \frac{\partial F}{\partial z}\,dz = 0, \tag{2–94}$$

where $\partial F/\partial x$, $\partial F/\partial y$, $\partial F/\partial z$ are to be evaluated at (x_1, y_1, z_1) and $dx = f'(t_1)\,dt$, $dy = g'(t_1)\,dt$, $dz = h'(t_1)\,dt$; by (2–85) these differentials can be replaced by $x - x_1$, $y - y_1$, $z - z_1$, where (x, y, z) is a point on the tangent to the given curve at (x_1, y_1, z_1). Accordingly, (2–94) can be written:

$$\left.\frac{\partial F}{\partial x}\right|_{(x_1, y_1, z_1)}(x - x_1) + \left.\frac{\partial F}{\partial y}\right|_{(x_1, y_1, z_1)}(y - y_1) + \left.\frac{\partial F}{\partial z}\right|_{(x_1, y_1, z_1)}(z - z_1) = 0. \tag{2–95}$$

This is the equation of a plane containing the tangent line to the chosen curve. However, (2–95) no longer depends on the particular curve chosen; all tangent lines to curves in the surface through (x_1, y_1, z_1) lie in the one plane (2–95), which is termed the *tangent plane to the surface at* (x_1, y_1, z_1). [If all three partial derivatives of F are 0 at the chosen point, equation (2–95) fails to determine a plane and the definition breaks down.]

It should be remarked that (2–94) is equivalent to (2–95), i.e., that again *the operation of taking differentials yields the tangent.* As a further illustration of this, consider a curve determined by two intersecting surfaces:

$$F(x, y, z) = 0, \quad G(x, y, z) = 0. \tag{2–96}$$

The corresponding differential relations:

$$\frac{\partial F}{\partial x} dx + \frac{\partial F}{\partial y} dy + \frac{\partial F}{\partial z} dz = 0, \quad \frac{\partial G}{\partial x} dx + \frac{\partial G}{\partial y} dy + \frac{\partial G}{\partial z} dz = 0 \tag{2–97}$$

represent two intersecting *tangent planes* at the point (x_1, y_1, z_1) at the point considered; the intersection of these planes is the *tangent line* to the curve (2–96). To obtain the equations in the usual form, the partial derivatives must be evaluated at (x_1, y_1, z_1) and the differentials dx, dy,dz must be replaced by $x - x_1$, $y - y_1$, $z - z_1$.

These results can also be put in vector form. First of all, (2–95) shows that the vector $(\partial F/\partial x)\mathbf{i} + (\partial F/\partial y)\mathbf{j} + (\partial F/\partial z)\mathbf{k}$ is *normal* to the tangent plane at (x_1, y_1, z_1). This vector is known as the *gradient vector* of the function $F(x, y, z)$ (see Fig. 2–10). We write:

$$\text{grad } F = \frac{\partial F}{\partial x}\mathbf{i} + \frac{\partial F}{\partial y}\mathbf{j} + \frac{\partial F}{\partial z}\mathbf{k}. \tag{2–98}$$

The notation ∇F (read "del F") for grad F will also be used; this is discussed in the following section. There is a gradient vector of F at each point of the domain of definition at which the partial derivatives exist; in particular, there is a gradient vector at each point of the surface $F(x, y, z) = 0$ considered. Thus the gradient vector has a "point of application" and should be thought of as a *bound vector*.

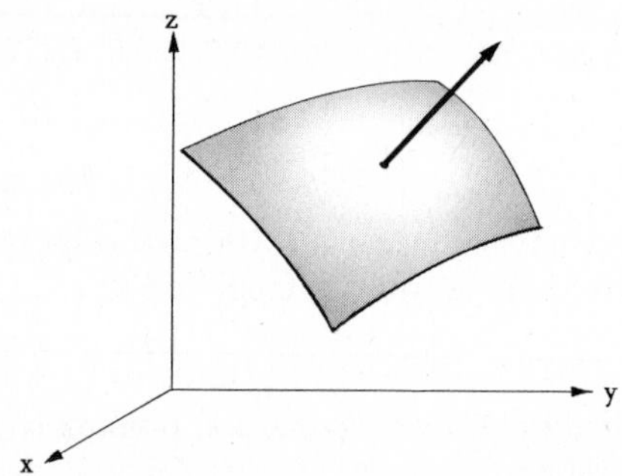

FIG. 2–10. Gradient vector of F and surface $F(x, y, z) = \text{const}$.

The equation (2–94) for the tangent plane can now be written in the vector form:

$$\text{grad } F \cdot d\mathbf{r} = 0 \quad (\mathbf{r} = x\mathbf{i} + y\mathbf{j} + z\mathbf{k}), \tag{2–99}$$

while the two equations (2–97) for the tangent line become

$$\text{grad } F \cdot d\mathbf{r} = 0, \quad \text{grad } G \cdot d\mathbf{r} = 0. \tag{2–100}$$

Since (2–100) expresses the fact that $d\mathbf{r} = dx\mathbf{i} + dy\mathbf{j} + dz\mathbf{k}$ is perpendicular to both grad F and grad G, one concludes that

$$d\mathbf{r} \times (\operatorname{grad} F \times \operatorname{grad} G) = \mathbf{0}; \tag{2–101}$$

this equation again represents the tangent line. The vector grad $F \times$ grad G has components

$$\begin{vmatrix} \frac{\partial F}{\partial y} & \frac{\partial F}{\partial z} \\ \frac{\partial G}{\partial y} & \frac{\partial G}{\partial z} \end{vmatrix}, \quad \begin{vmatrix} \frac{\partial F}{\partial z} & \frac{\partial F}{\partial x} \\ \frac{\partial G}{\partial z} & \frac{\partial G}{\partial x} \end{vmatrix}, \quad \begin{vmatrix} \frac{\partial F}{\partial x} & \frac{\partial F}{\partial y} \\ \frac{\partial G}{\partial x} & \frac{\partial G}{\partial y} \end{vmatrix}.$$

Hence the tangent line can be written in the symmetric form

$$\frac{x - x_1}{\begin{vmatrix} \frac{\partial F}{\partial y} & \frac{\partial F}{\partial z} \\ \frac{\partial G}{\partial y} & \frac{\partial G}{\partial z} \end{vmatrix}} = \frac{y - y_1}{\begin{vmatrix} \frac{\partial F}{\partial z} & \frac{\partial F}{\partial x} \\ \frac{\partial G}{\partial z} & \frac{\partial G}{\partial x} \end{vmatrix}} = \frac{z - z_1}{\begin{vmatrix} \frac{\partial F}{\partial x} & \frac{\partial F}{\partial y} \\ \frac{\partial G}{\partial x} & \frac{\partial G}{\partial y} \end{vmatrix}} \tag{2–102}$$

or, in terms of Jacobians, in the form

$$\frac{x - x_1}{\frac{\partial(F, G)}{\partial(y, z)}} = \frac{y - y_1}{\frac{\partial(F, G)}{\partial(z, x)}} = \frac{z - z_1}{\frac{\partial(F, G)}{\partial(x, y)}}. \tag{2–103}$$

This discussion and that of the preceding section show the significance of the differential. Taking differentials in an equation or system of equations corresponds on the one hand to replacement of the equations by *linear* equations in the variables $dx, dy, \ldots$, and on the other hand to replacement of curves and surfaces by *tangent* lines and planes.

***2–13 Manifolds in n-dimensional space.** When more than three variables are involved, the geometrical interpretations of the preceding section appear to lose meaning. However, one can very easily define the notions of lines, planes and k-dimensional linear manifolds in n-dimensional space, as in Section 1–11. With the aid of these concepts, one can then discuss tangents and normals.

In particular, if we have n equations

$$x_i = F_i(t_1, \ldots, t_k), \quad i = 1, \ldots, \quad n, \quad k < n, \tag{2–104}$$

then we can interpret $t_1, \ldots, t_k$ as parameters and the equations themselves as parametric equations of a *k-dimensional manifold* in E^n. If we take differentials, we obtain (in vector notation) the equation

$$d\mathbf{x} = \mathbf{F_t}\, d\mathbf{t}, \tag{2–105}$$

which describes the corresponding *linear* mapping, whose range is the k-dimensional linear manifold tangent to the k-dimensional manifold (2–104) at the chosen point. To ensure that (2–105) actually represents such a linear manifold, we must require that the Jacobian matrix $\mathbf{F_t}$ has rank k at the point (Sections 1–12 and 1–13).

Similarly, if we have m simultaneous equations

$$F_i(x_1, \ldots, x_n) = 0, \quad i = 1, \ldots, \quad m, \quad m < n, \tag{2–106}$$

then "in general" these equations represent an $(n-m)$-dimensional manifold in E^n. We obtain the tangent linear manifold at a chosen point by taking differentials to obtain, in vector notation,

$$\mathbf{F_x}\, d\mathbf{x} = \mathbf{0}. \tag{2–107}$$

This equation is equivalent to m linear equations in n unknowns. It describes the kernel of the Jacobian matrix $\mathbf{F_x}$, evaluated at the chosen point. To ensure that the equations actually describe an $(n - m)$-dimensional linear manifold, one must require that $\mathbf{F_x}$ has maximum rank, m, at the chosen point.

In the case of a single equation

$$F(x_1, \ldots, x_n) = 0, \tag{2–108}$$

the tangent equation becomes

$$F_{x_1}\, dx_1 + \cdots + F_{x_n}\, dx_n = 0, \tag{2–109}$$

that is,

$$F_{x_1}(x_1 - x_1^0) + \cdots + F_{x_n}(x_n - x_n^0) = 0, \tag{2–110}$$

where P_0: $(x_1^0, \ldots, x_n^0)$ is the point of the manifold at which the tangent linear manifold is sought and $F_{x_1} = \partial F/\partial x_1, \ldots, F_{x_n} = \partial F/\partial x_n$ are evaluated at this point. We can interpret the vector $(F_{x_1}, \ldots, F_{x_n})$, which is a $1 \times n$ Jacobian matrix, as the *gradient* of F, as in the preceding section:

$$\nabla F = \operatorname{grad} F = \left(\frac{\partial F}{\partial x_1}, \ldots, \frac{\partial F}{\partial x_n}\right). \tag{2–111}$$

The equation (2–110) then states that, at the point P_0 chosen, ∇F is orthogonal to each vector $\overrightarrow{P_0P}$ joining P_0 to an arbitrary point P of the tangent linear manifold. Hence ∇F provides a *normal vector* to this linear manifold, and we can interpret it as a normal vector $\mathbf{n}$ for the given $(n-1)$-dimensional manifold (2–108) at P_0.

Problems

1. Given the equations $x = \sin t$, $y = \sin t$, $z = \cos 2t$ of a space curve,

(a) sketch the curve,

(b) find the equations of the tangent line at the point P for which $t = \dfrac{\pi}{4}$,

(c) find the equation of a plane cutting the curve at right angles at P.

2. (a) Show that the curve of Prob. 1 lies in the surface $x^2 + y^2 + z = 1$.

(b) Find the equation of the plane tangent to the surface at the point P of Prob. 1(b).

(c) Show that the tangent line to the curve at P lies in the tangent plane to the surface.

3. For each of the following surfaces find the tangent plane and normal line at the point indicated, verifying that the point is in the surface:

(a) $x^2 + y^2 + z^2 = 9$ at $(2, 2, 1)$
(b) $e^{x^2+y^2} - z^2 = 0$ at $(0, 0, 1)$
(c) $x^3 - xy^2 + yz^2 - z^3 = 0$ at $(1, 1, 1)$
(d) $x^2 + y^2 - z^2 = 0$ at $(0, 0, 0)$

Why does the procedure break down in (d)? Show by graphing that a solution is impossible.

4. Show that the tangent plane at (x_1, y_1, z_1) to a surface given by an equation $z = f(x, y)$ is as follows:

$$z - z_1 = \frac{\partial f}{\partial x}(x - x_1) + \frac{\partial f}{\partial y}(y - y_1).$$

Obtain the equations of the normal line.

5. Find the tangent plane and normal line to the following surfaces at the points shown:

(a) $z = x^2 + y^2$ at $(1, 1, 2)$, (b) $z = \sqrt{1 - x^2 - y^2}$ at $(\frac{2}{3}, \frac{2}{3}, \frac{1}{3})$. (See Prob. 4.)

6. For each of the following curves (represented by intersecting surfaces), find the equations of the tangent line at the point indicated, verifying that the point is on the curve:

(a) $x^2 + y^2 + z^2 = 9,\quad x^2 + y^2 - 8z^2 = 0$ at $(2, 2, 1)$;
(b) $x^2 + y^2 = 1,\quad x + y + z = 0$ at $(1, 0, -1)$;
(c) $x^2 + y^2 + z^2 = 9,\quad x^2 + 2y^2 + 3z^2 = 9$ at $(3, 0, 0)$.

Why does the procedure break down in (c)? Show that solution is impossible.

7. Show that the curve

$$x^2 - y^2 + z^2 = 1, \quad xy + xz = 2$$

is tangent to the surface

$$xyz - x^2 - 6y = -6$$

at the point $(1, 1, 1)$.

8. Show that the equation of the plane normal to the curve

$$F(x, y, z) = 0, \quad G(x, y, z) = 0$$

at the point (x_1, y_1, z_1) can be written in the vector form

$$d\mathbf{r} \cdot \operatorname{grad} F \times \operatorname{grad} G = 0,$$

and write out the equation in rectangular coordinates. Use the results to find the normal planes to the curves of Prob. 6(a) and (b) at the points given.

9. Determine a plane normal to the curve: $x = t^2$, $y = t$, $z = 2t$ and passing through the point $(1, 0, 0)$.

10. Find the gradient vectors of the following functions:

(a) $F = x^2 + y^2 + z^2$, (b) $F = 2x^2 + y^2$.

Plot a level surface of each function and verify that the gradient vector is always normal to the level surface.

11. Three equations of form $x = f(u, v)$, $y = g(u, v)$, $z = h(u, v)$ can be considered as parametric equations of a surface, for elimination of u and v leads in general to a single equation $F(x, y, z) = 0$.

(a) Show that the vector

$$\left(\frac{\partial f}{\partial u}\mathbf{i}+\frac{\partial g}{\partial u}\mathbf{j}+\frac{\partial h}{\partial u}\mathbf{k}\right)\times\left(\frac{\partial f}{\partial v}\mathbf{i}+\frac{\partial g}{\partial v}\mathbf{j}+\frac{\partial h}{\partial v}\mathbf{k}\right)\equiv\frac{\partial(g,h)}{\partial(u,v)}\mathbf{i}+\frac{\partial(h,f)}{\partial(u,v)}\mathbf{j}+\frac{\partial(f,g)}{\partial(u,v)}\mathbf{k}$$

is normal to the surface at the point (x_1, y_1, z_1), where $x_1 = f(u_1, v_1)$, $y_1 = g(u_1, v_1)$, $z_1 = h(u_1, v_1)$ and the derivatives are evaluated at this point.

(b) Write the equation of the tangent plane.

(c) Apply the results to find the tangent plane to the surface

$$x = \cos u \cos v, \quad y = \cos u \sin v, \quad z = \sin u$$

at the point for which $u = \pi/4$, $v = \pi/4$.

(d) Show that the surface (c) is a sphere.

12. Find the equations of a tangent line to a curve given by equations:

$$z = f(x, y), \quad z = g(x, y).$$

13. Three equations

$$F(x, y, z, t) = 0, \quad G(x, y, z, t) = 0, \quad H(x, y, z, t) = 0$$

can be regarded as implicit parametric equations of a curve in terms of the parameter t. Find the equations of the tangent line.

14. Let a curve in space be given in parametric form by an equation: $\mathbf{r} = \mathbf{r}(t)$, $t_1 \leqq t \leqq t_2$, where $\mathbf{r} = \overrightarrow{OP}$ and O is fixed.

(a) Show that if $d\mathbf{r}/dt \equiv \mathbf{0}$, then the curve degenerates into a single point.

(b) Let $d\mathbf{r}/dt = \mathbf{0}$ at a point P_0 on the curve, at which $t = t_0$. Show that if $d^2\mathbf{r}/dt^2 \neq \mathbf{0}$ for $t = t_0$, then $d^2\mathbf{r}/dt^2$ represents a tangent vector to the curve at P_0 and, in general, that if

$$\frac{d\mathbf{r}}{dt} = 0, \quad \frac{d^2\mathbf{r}}{dt^2} = 0, \quad \ldots, \quad \frac{d^n\mathbf{r}}{dt^n} = 0, \quad \frac{d^{n+1}\mathbf{r}}{dt^{n+1}} \neq 0$$

for $t = t_0$, then

$$\left.\frac{d^{n+1}\mathbf{r}}{dt^{n+1}}\right|_{t=t_0}$$

represents a tangent vector to the curve at P_0. [Hint: for $t = t_0$, the vector

$$\frac{\mathbf{r}(t)-\mathbf{r}(t_0)}{(t-t_0)^{n+1}} = \frac{f(t)-f(t_0)}{(t-t_0)^{n+1}}\mathbf{i} + \frac{g(t)-g(t_0)}{(t-t_0)^{n+1}}\mathbf{j} + \frac{h(t)-h(t_0)}{(t-t_0)^{n+1}}\mathbf{k}$$

represents a secant to the curve, as in Fig. 2–9. Now let $t \to t_0$ and evaluate the limit of this vector with the aid of de l'Hôpital's rule (Prob. 25 at end of Introduction), noting that $f'(t_0) = 0$, $g'(t_0) = 0$, $h'(t_0) = 0$, but at least one of the three $(n + 1)$-st derivatives is not 0.]

(c) Show that if the parameter t is the arc length s, then $d\mathbf{u}/ds$ has magnitude 1 and hence is always a tangent vector.

Answers

1. (b) $\dfrac{x-\dfrac{\sqrt{2}}{2}}{\sqrt{2}} = \dfrac{y-\dfrac{\sqrt{2}}{2}}{\sqrt{2}} = \dfrac{z}{-4}$, (c) $\sqrt{2}x + \sqrt{2}y - 4z = 2$.

2. (b) $\sqrt{2}x + \sqrt{2}y + z = 2$.

3. (a) $2x + 2y + z = 9$, $\dfrac{x-2}{2} = \dfrac{y-2}{2} = \dfrac{z-1}{1}$; (b) $z = 1$, $x = 0$, $y = 0$; (c) $2x - y - z = 0$, $\dfrac{x-1}{2} = \dfrac{y-1}{-1} = \dfrac{z-1}{-1}$.

5. (a) $z - 2 = 2(x-1) + 2(y-1)$, $\dfrac{x-1}{2} = \dfrac{y-1}{2} = \dfrac{z-2}{-1}$;

(b) $z - \frac{1}{3} = -2(x - \frac{2}{3}) - 2(y - \frac{2}{3})$, $\dfrac{x-\frac{2}{3}}{2} = \dfrac{y-\frac{2}{3}}{2} = \dfrac{z-\frac{1}{3}}{1}$.

6. (a) $x + y = 4$, $z = 1$; (b) $x = 1$, $y + z + 1 = 0$.

8. $\begin{vmatrix} x - x_1 & y - y_1 & z - z_1 \\ \dfrac{\partial F}{\partial x} & \dfrac{\partial F}{\partial y} & \dfrac{\partial F}{\partial z} \\ \dfrac{\partial G}{\partial x} & \dfrac{\partial G}{\partial y} & \dfrac{\partial G}{\partial z} \end{vmatrix} = 0$, $\begin{vmatrix} x-2 & y-2 & z-1 \\ 4 & 4 & 2 \\ 4 & 4 & -16 \end{vmatrix} = 0$,

$$\begin{vmatrix} x-1 & y & z+1 \\ 2 & 0 & 0 \\ 1 & 1 & 1 \end{vmatrix} = 0.$$

9. $y + 2z = 0$. 10. (a) $2x\mathbf{i} + 2y\mathbf{j} + 2z\mathbf{k}$, (b) $4x\mathbf{i} + 2y\mathbf{j}$. 11. $x + y + \sqrt{2}z = 2$.

12. $\dfrac{x - x_1}{\dfrac{\partial g}{\partial y} - \dfrac{\partial f}{\partial y}} = \dfrac{y - y_1}{\dfrac{\partial f}{\partial x} - \dfrac{\partial g}{\partial x}} = \dfrac{z - z_1}{\dfrac{\partial(f, g)}{\partial(x, y)}}$. 13. $\dfrac{x - x_1}{\dfrac{\partial(F, G, H)}{\partial(t, y, z)}} = \dfrac{y - y_1}{\dfrac{\partial(F, G, H)}{\partial(x, t, z)}} = \dfrac{z - z_1}{\dfrac{\partial(F, G, H)}{\partial(x, y, t)}}$.

2–14 The directional derivative. Let $F(x, y, z)$ be given in a domain D of space. To compute the partial derivative $\partial F/\partial x$ at a point (x, y, z) of this domain, one considers the ratio of the change ΔF in the function F from (x, y, z) to $(x + \Delta x, y, z)$ to the change Δx in x. Thus only the values of F along a line parallel to the x axis are considered. Similarly $\partial F/\partial y$ and $\partial F/\partial z$ involve a consideration of how F changes along parallels to the y and z axes respectively. It appears unnatural to restrict attention to these three directions. Accordingly, one defines the *directional derivative* of F in a given direction as the limit of the ratio

$$\frac{\Delta F}{\Delta s}$$

of the change in F to the distance Δs moved in the given direction, as Δs approaches 0.

Let the direction in question be given by a nonzero vector $\mathbf{v}$. The directional derivative of F in direction $\mathbf{v}$ at the point (x, y, z) is then denoted by $\nabla_v F(x, y, z)$ or, more concisely, by $\nabla_v F$. A displacement from (x, y, z) in direction $\mathbf{v}$ corresponds to changes Δx, Δy, Δz proportional to the components v_x, v_y, v_z; that is,

$$\Delta x = hv_x, \quad \Delta y = hv_y, \quad \Delta z = hv_z,$$

where h is a positive scalar. The displacement is thus simply the vector

$h\mathbf{v}$, and its magnitude Δs is $h|\mathbf{v}|$. The directional derivative is now by definition the limit:

$$\nabla_v F = \lim_{h\to 0} \frac{F(x + hv_x, y + hv_y, z + hv_z) - F(x, y, z)}{h|\mathbf{v}|}. \tag{2-112}$$

If F has a total differential at (x, y, z), then, as in Section 2–6 above,

$$\begin{aligned}\Delta F &= F(x + hv_x, y + hv_y, z + hv_z) - F(x, y, z)\\ &= \frac{\partial F}{\partial x} hv_x + \frac{\partial F}{\partial y} hv_y + \frac{\partial F}{\partial z} hv_z + \epsilon_1 hv_x + \epsilon_2 hv_y + \epsilon_3 hv_z.\end{aligned}$$

Thus

$$\frac{\Delta F}{h|\mathbf{v}|} = \frac{\partial F}{\partial x}\frac{v_x}{|\mathbf{v}|} + \frac{\partial F}{\partial y}\frac{v_y}{|\mathbf{v}|} + \frac{\partial F}{\partial z}\frac{v_z}{|\mathbf{v}|} + \epsilon_1 \frac{v_x}{|\mathbf{v}|} + \epsilon_2 \frac{v_y}{|\mathbf{v}|} + \epsilon_3 \frac{v_z}{|\mathbf{v}|}.$$

If h approaches zero, the last three terms approach 0, while the left-hand side approaches $\nabla_v F$. One has thus the equation

$$\nabla_v F = \frac{\partial F}{\partial x}\frac{v_x}{|\mathbf{v}|} + \frac{\partial F}{\partial y}\frac{v_y}{|\mathbf{v}|} + \frac{\partial F}{\partial z}\frac{v_z}{|\mathbf{v}|}. \tag{2-113}$$

Now $v_x/|\mathbf{v}|$, $v_y/|\mathbf{v}|$, $v_z/|\mathbf{v}|$ are simply the components of a unit vector $\mathbf{u}$ in the direction of $\mathbf{v}$ and are thus, by Section 0–6, the direction cosines of $\mathbf{v}$:

$$\mathbf{u} = \frac{1}{|\mathbf{v}|}(v_x\mathbf{i} + v_y\mathbf{j} + v_z\mathbf{k}) = \cos\alpha\mathbf{i} + \cos\beta\mathbf{j} + \cos\gamma\mathbf{k}.$$

Accordingly, (2–113) can be written in the following form:

$$\nabla_v F = \frac{\partial F}{\partial x}\cos\alpha + \frac{\partial F}{\partial y}\cos\beta + \frac{\partial F}{\partial z}\cos\gamma. \tag{2-114}$$

One has thus the fundamental rule:

The directional derivative of a function $F(x, y, z)$ is given by

$$\frac{\partial F}{\partial x}\cos\alpha + \frac{\partial F}{\partial y}\cos\beta + \frac{\partial F}{\partial z}\cos\gamma,$$

where α, β, γ are the direction angles of the direction chosen.

The right-hand side of (2–114) can be interpreted as the scalar product of the vector grad $F = \nabla F = (\partial F/\partial x)\mathbf{i} + (\partial F/\partial y)\mathbf{j} + (\partial F/\partial z)\mathbf{k}$ and the unit vector $\mathbf{u}$. Thus by Section 0–6 the *directional derivative equals the component of grad F in the direction of* $\mathbf{v}$:

$$\nabla_v F = \nabla F \cdot \frac{\mathbf{v}}{|\mathbf{v}|} = \text{comp}_v\, \nabla F. \tag{2-115}$$

It is for this reason that the notation $\nabla_v F$ is used for the directional derivative. In the special case when $\mathbf{v} = \mathbf{i}$, so that one is computing the directional derivative in the x direction, one writes

$$\nabla_v F = \nabla_x F = \text{comp}_x\, \nabla F = \frac{\partial F}{\partial x}; \tag{2-116}$$

similarly the directional derivatives in the y and z directions are

$$\nabla_y F = \frac{\partial F}{\partial y}, \quad \nabla_z F = \frac{\partial F}{\partial z}.$$

There are other situations in which a partial derivative notation is used for the directional derivative. A common one is that in which one is computing the directional derivative at a point (x, y, z) of a surface S along a given direction normal to S. If $\mathbf{n}$ is a unit normal vector in the chosen direction, one writes

$$\nabla_n F = \nabla F \cdot \mathbf{n} = \frac{\partial F}{\partial n}. \tag{2–117}$$

Equation (2–115) gives further information about the vector $\nabla F =$ grad F. For from (2–115) one concludes that the directional derivative at a given point is a maximum when it is in the direction of ∇F; this maximum value is simply

$$|\nabla F| = \sqrt{\left(\frac{\partial F}{\partial x}\right)^2 + \left(\frac{\partial F}{\partial y}\right)^2 + \left(\frac{\partial F}{\partial z}\right)^2}. \tag{2–118}$$

Thus *the gradient vector points in the direction in which F increases most rapidly and its length is the rate of increase in that direction.* If $\mathbf{v}$ makes an angle θ with ∇F, then the directional derivative in the direction of $\mathbf{v}$ is

$$\nabla_v F = |\nabla F| \cos\theta. \tag{2–119}$$

Thus, if $\mathbf{v}$ is tangent to a level surface:

$$F(x, y, z) = \text{constant},$$

at (x, y, z), then $\nabla_v F = 0$; for ∇F is normal to such a level surface.

One sometimes refers to a "directional derivative along a curve." By this is meant the directional derivative along a direction tangent to the curve. Let the curve be given in terms of arc length as parameter:

$$x = f(s), \quad y = g(s), \quad z = h(s),$$

and let the direction chosen be that of increasing s. The vector

$$\mathbf{u} = \frac{dx}{ds}\mathbf{i} + \frac{dy}{ds}\mathbf{j} + \frac{dz}{ds}\mathbf{k}$$

is then tangent to the curve and has length 1 (cf. Section 2–12). Hence the directional derivative along the curve is

$$\nabla_u F = \frac{\partial F}{\partial x}\frac{dx}{ds} + \frac{\partial F}{\partial y}\frac{dy}{ds} + \frac{\partial F}{\partial z}\frac{dz}{ds} = \frac{dF}{ds}; \tag{2–120}$$

that is, it is the rate of change of F with respect to arc length on the curve.

The meaning of the directional derivative can be clarified by visualizing the following experiment. A traveler in a balloon carries with him a thermometer. At intervals he records the temperature. If his reading at position A is 42° and at the position B is 44°, he would estimate that the directional derivative of the temperature in the direction $\overrightarrow{AB}$ is positive and has a value given approximately by $2° \div d$, where d is the distance $|\overrightarrow{AB}|$; if $|\overrightarrow{AB}| = 5000$ ft, the estimate would be .0004° per foot. If he continued traveling past B in the same direction he would expect the temperature to rise at about the same rate. If the balloon moves on a curved path at a known constant speed, the traveler can compute the directional derivative *along the path* without looking out of the balloon.

The foregoing discussion has been given for three dimensions. It can be specialized to two dimensions without difficulty. Thus one considers a function $F(x, y)$ and its rate of change along a given direction $\mathbf{v}$ in the xy plane. If $\mathbf{v}$ makes an angle α with the positive x axis, one can write

$$\nabla_v F = \nabla_\alpha F = \frac{\partial F}{\partial x}\cos\alpha + \frac{\partial F}{\partial y}\sin\alpha; \tag{2-121}$$

for $\mathbf{v}$ has direction cosines $\cos\alpha$ and $\cos\beta = \cos(\frac{1}{2}\pi - \alpha) = \sin\alpha$. Again the directional derivative is the component of grad $F = (\partial F/\partial x)\mathbf{i} + (\partial F/\partial y)\mathbf{j}$ in the given direction; the directional derivative at a given point has its maximum in the direction of grad F, the value being

$$|\nabla F| = \sqrt{\left(\frac{\partial F}{\partial x}\right)^2 + \left(\frac{\partial F}{\partial y}\right)^2}. \tag{2-122}$$

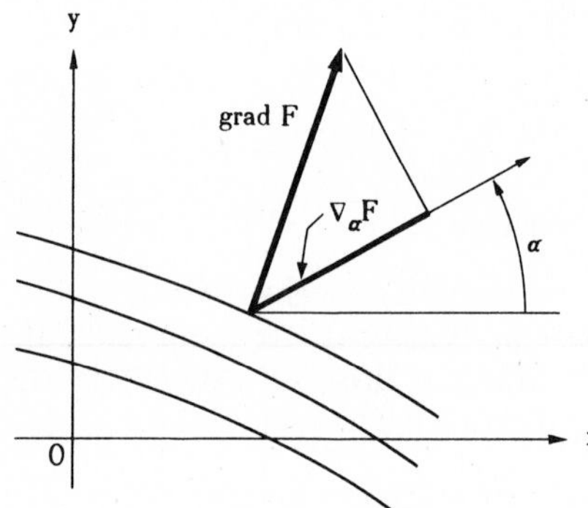

FIG. 2–11. Gradient of $F(x, y)$ and curves $F(x, y) = \text{const.}$

The directional derivative is zero along a level curve of F, as suggested in the accompanying Fig. 2–11.

If the level curves are interpreted as contour lines of a landscape, i.e., of the surface $z = F(x, y)$, then the directional derivative means simply the rate of climb in the given direction. The rate of climb in the direction of steepest ascent is the "gradient," precisely the term in common use. The bicyclist zigzagging up a hill is taking advantage of the component rule to reduce the directional derivative.

Problems

1. Evaluate the directional derivatives of the following functions for the points and directions given:

(a) $F(x, y, z) = 2x^2 - y^2 + z^2$ at $(1, 2, 3)$ in the direction of the line from $(1, 2, 3)$ to $(3, 5, 0)$;

(b) $F(x, y, z) = x^2 + y^2$ at $(0, 0, 0)$ in the direction of the vector $\mathbf{u} = a\mathbf{i} + b\mathbf{j} + c\mathbf{k}$; discuss the significance of the result;

(c) $F(x, y) = e^x \cos y$ at $(0, 0)$ in a direction making an angle of 60° with the x axis;

(d) $F(x, y) = 2x - 3y$ at $(1, 1)$ along the curve $y = x^2$ in the direction of increasing x;

(e) $F(x, y, z) = 3x - 5y + 2z$ at $(2, 2, 1)$ in the direction of the outer normal of the surface $x^2 + y^2 + z^2 = 9$;

(f) $F(x, y, z) = x^2 + y^2 - z^2$ at $(3, 4, 5)$ along the curve $x^2 + y^2 - z^2 = 0$, $2x^2 + 2y^2 - z^2 = 25$ in the direction of increasing x; explain the answer.

2. Evaluate $\partial F/\partial n$, where $\mathbf{n}$ is the outer normal to the surface given, at a general point (x, y, z) of the surface given:

(a) $F = x^2 - y^2$, surface $x^2 + y^2 + z^2 = 4$;
(b) $F = xyz$, surface $x^2 + 2y^2 + 4z^2 = 8$.

3. Prove that, if $u = f(x, y)$ and $v = g(x, y)$ are functions such that

$$\frac{\partial u}{\partial x} = \frac{\partial v}{\partial y}, \quad \frac{\partial u}{\partial y} = -\frac{\partial v}{\partial x},$$

then

$$\nabla_\alpha u = \nabla_{\alpha + \frac{1}{2}\pi}\, v$$

for every angle α.

4. Show that, if $u = f(x, y)$, then the directional derivatives of u along the line $\theta = \text{const}$, and circles $r = \text{const}$ (polar coordinates) are given respectively by

$$\nabla_\theta u = \frac{\partial u}{\partial r}, \quad \nabla_{\theta + \frac{1}{2}\pi}\, u = \frac{1}{r}\frac{\partial u}{\partial \theta}.$$

5. Show that under the hypotheses of Prob. 3 one has

$$\frac{\partial u}{\partial r} = \frac{1}{r}\frac{\partial v}{\partial \theta}, \quad \frac{1}{r}\frac{\partial u}{\partial \theta} = -\frac{\partial v}{\partial r}.$$

[Hint: use Prob. 4.]

6. Let arc length s be measured from the point $(2, 0)$ of the circle $x^2 + y^2 = 4$, starting in the direction of increasing y. If $u = x^2 - y^2$, evaluate du/ds on this circle. Check the result by using both the directional derivative and the explicit expression for u in terms of s. At what point of the circle does u have its smallest value?

7. Under the hypotheses of Prob. 3, show that $\partial u/\partial s = \partial v/\partial n$ along each curve C of the domain in which u and v are given, for appropriate direction of the normal $\mathbf{n}$.

8. Determine the points (x, y) and directions for which the directional derivative of $u = 3x^2 + y^2$ has its largest value, if (x, y) is restricted to lie on the circle $x^2 + y^2 = 1$.

9. A function $F(x, y, z)$ is known to have the following values: $F(1, 1, 1) = 1$, $F(2, 1, 1) = 4$, $F(2, 2, 1) = 8$, $F(2, 2, 2) = 16$. Compute approximately the directional derivatives:

$$\nabla_i F, \quad \nabla_{i+j} F, \quad \nabla_{i+j+k} F$$

at the point $(1, 1, 1)$.

Answers

1. (a) $-\sqrt{22}$, (b) 0, (c) $\frac{1}{2}$, (d) $-\dfrac{4}{\sqrt{5}}$, (e) $-\frac{2}{3}$, (f) 0.

2. (a) $x^2 - y^2$, (b) $\dfrac{7xyz}{\sqrt{x^2 + 4y^2 + 16z^2}}$.

6. $du/ds = -2xy$. Minimum is -4 at $(0, \pm 2)$.

8. Maximum is 6 in direction $\mathbf{i}$ at $(1, 0)$, in direction $-\mathbf{i}$ at $(-1, 0)$.

9. 3, $\dfrac{7}{\sqrt{2}}$, $\dfrac{15}{\sqrt{3}}$.

2–15 Partial derivatives of higher order. Let a function $z = F(x, y)$ be given; its two partial derivatives $\partial z/\partial x$ and $\partial z/\partial y$ are themselves functions of x and y:

$$\frac{\partial z}{\partial x} = F_x(x, y), \quad \frac{\partial z}{\partial y} = F_y(x, y).$$

Hence each can be differentiated with respect to x and y; one thus obtains the four *second partial derivatives:*

$$\frac{\partial^2 z}{\partial x^2} = F_{xx}(x, y), \quad \frac{\partial^2 z}{\partial y\, \partial x} = F_{xy}(x, y), \quad \frac{\partial^2 z}{\partial x\, \partial y} = F_{yx}(x, y), \quad \frac{\partial^2 z}{\partial y^2} = F_{yy}(x, y). \tag{2–123}$$

Thus $\partial^2 z/\partial x^2$ is the result of differentiating $\partial z/\partial x$ with respect to x, while $\partial^2 z/\partial y\, \partial x$ is the result of differentiation of $\partial z/\partial x$ with respect to y. Here a simplification is possible, if all derivatives concerned are continuous in the domain considered, for one can prove that

$$\frac{\partial^2 z}{\partial y\, \partial x} = \frac{\partial^2 z}{\partial x\, \partial y}; \tag{2–124}$$

i.e., the order of differentiation is immaterial.

Third and higher order partial derivatives are defined in a similar fashion and again, under appropriate assumptions of continuity, the order of differentiation does not matter. Thus one obtains four third partial derivatives:

$$\frac{\partial^3 z}{\partial x^3}, \quad \frac{\partial^3 z}{\partial x^2\, \partial y} = \frac{\partial^3 z}{\partial x\, \partial y\, \partial x} = \frac{\partial^3 z}{\partial y\, \partial x^2}, \quad \frac{\partial^3 z}{\partial x\, \partial y^2} = \frac{\partial^3 z}{\partial y\, \partial x\, \partial y} = \frac{\partial^3 z}{\partial y^2\, \partial x}, \quad \frac{\partial^3 z}{\partial y^3}.$$

A clue as to why the order of differentiation does not matter can be obtained by considering the mixed partial derivatives of $z = x^n y^m$. Here one has

$$\frac{\partial z}{\partial x} = nx^{n-1}y^m, \quad \frac{\partial^2 z}{\partial y\, \partial x} = nmx^{n-1}y^{m-1},$$

$$\frac{\partial z}{\partial y} = mx^n y^{m-1}, \quad \frac{\partial^2 z}{\partial x\, \partial y} = nmx^{n-1}y^{m-1}.$$

Thus the order does not matter in this case. A similar reasoning applies to a sum of such terms with constant factors, i.e., a polynomial in x and y·

$$z = a_0 + a_1 x + a_2 y + a_3 x^2 + a_4 xy + a_5 y^2 + \cdots + a_s x^p y^q.$$

It is essentially because an "arbitrary" function can be approximated by such a polynomial near a given point that the rule holds. [For a proof one is referred to *CLA*, Section 12–20.]

Other notations for higher derivatives are illustrated by the following examples:

$$\frac{\partial^2 z}{\partial x^2} = z_{xx} = f_{11}(x, y), \quad \frac{\partial^2 z}{\partial x\,\partial y} = z_{yx} = f_{21}(x, y),$$

$$\frac{\partial^3 w}{\partial x\,\partial y\,\partial z} = w_{zyx} = f_{321}(x, y, z), \quad \frac{\partial^4 w}{\partial x^2\,\partial z^2} = w_{zzxx} = f_{3311}(x, y, z).$$

If $z = f(x, y)$, the *Laplacian* of z, denoted by Δz or $\nabla^2 z$, is the expression

$$\Delta z = \nabla^2 z = \frac{\partial^2 z}{\partial x^2} + \frac{\partial^2 z}{\partial y^2}. \tag{2–125}$$

The Δ symbol here must not be confused with the symbol for increments; for this reason the $\nabla^2 z$ notation is preferable. If $w = f(x, y, z)$, the Laplacian of w is defined analogously:

$$\Delta w = \nabla^2 w = \frac{\partial^2 w}{\partial x^2} + \frac{\partial^2 w}{\partial y^2} + \frac{\partial^2 w}{\partial z^2}.$$

The origin of the ∇^2 symbol lies in the interpretation of ∇ as a "vector differential operator":

$$\nabla = \frac{\partial}{\partial x}\mathbf{i} + \frac{\partial}{\partial y}\mathbf{j} + \frac{\partial}{\partial z}\mathbf{k}.$$

One then has symbolically:

$$\nabla^2 = \nabla \cdot \nabla = \frac{\partial^2}{\partial x^2} + \frac{\partial^2}{\partial y^2} + \frac{\partial^2}{\partial z^2}.$$

This point of view will be discussed further in Chapter 3.

If $z = f(x, y)$ has continuous second derivatives in a domain D and

$$\nabla^2 z = 0 \tag{2–126}$$

in D, then z is said to be *harmonic* in D. The same term is used for a function of three variables which has continuous second derivatives in a domain D in space and whose Laplacian is 0 in D. The two equations for harmonic functions:

$$\frac{\partial^2 z}{\partial x^2} + \frac{\partial^2 z}{\partial y^2} = 0, \quad \frac{\partial^2 w}{\partial x^2} + \frac{\partial^2 w}{\partial y^2} + \frac{\partial^2 w}{\partial z^2} = 0, \tag{2–127}$$

are known as the *Laplace equations* in two and three dimensions respectively.

Another important combination of derivatives occurs in the *biharmonic equation:*

$$\frac{\partial^4 z}{\partial x^4} + 2\frac{\partial^4 z}{\partial x^2\,\partial y^2} + \frac{\partial^4 z}{\partial y^4} = 0, \tag{2–128}$$

which arises in the theory of elasticity. The combination which appears here can be expressed in terms of the Laplacian, for one has

$$\nabla^2(\nabla^2 z) = \frac{\partial^4 z}{\partial x^4} + 2\frac{\partial^4 z}{\partial x^2\,\partial y^2} + \frac{\partial^4 z}{\partial y^4}.$$

If we write: $\nabla^4 z = \nabla^2(\nabla^2 z)$, then the biharmonic equation can be written:

$$\nabla^4 z = 0. \tag{2–129}$$

Its solutions are termed *biharmonic* functions. This can again be generalized to functions of three variables, (2–129) suggesting the definition to be used.

Harmonic functions arise in the theory of electromagnetic fields, in fluid dynamics, in the theory of heat conduction, and many other parts of physics; applications will be discussed in Chapters 5, 9, and 10. Biharmonic functions are used mainly in elasticity.

2–16 Higher derivatives of composite functions. Let $z = f(x, y)$ and $x = g(t)$, $y = h(t)$, so that z can be expressed in terms of t alone. The derivative dz/dt can then be evaluated by the chain rule (2–33) of Section 2–8 above:

$$\frac{dz}{dt} = \frac{\partial z}{\partial x}\frac{dx}{dt} + \frac{\partial z}{\partial y}\frac{dy}{dt}. \tag{2–130}$$

By applying the product rule one obtains the following expression for the second derivative:

$$\frac{d^2z}{dt^2} = \frac{d}{dt}\left(\frac{dz}{dt}\right) = \frac{\partial z}{\partial x}\frac{d^2x}{dt^2} + \frac{dx}{dt}\frac{d}{dt}\left(\frac{\partial z}{\partial x}\right) + \frac{\partial z}{\partial y}\frac{d^2y}{dt^2} + \frac{dy}{dt}\frac{d}{dt}\left(\frac{\partial z}{\partial y}\right).$$

To evaluate the expressions $(d/dt)(\partial z/\partial x)$ and $(d/dt)(\partial z/\partial y)$, one uses (2–130) again, this time applied to $\partial z/\partial x$ and $\partial z/\partial y$ rather than to z.

$$\frac{d}{dt}\left(\frac{\partial z}{\partial x}\right) = \frac{\partial^2 z}{\partial x^2}\frac{dx}{dt} + \frac{\partial^2 z}{\partial y\,\partial x}\frac{dy}{dt},$$

$$\frac{d}{dt}\left(\frac{\partial z}{\partial y}\right) = \frac{\partial^2 z}{\partial x\,\partial y}\frac{dx}{dt} + \frac{\partial^2 z}{\partial y^2}\frac{dy}{dt}.$$

One thus finds the rule:

$$\frac{d^2z}{dt^2} = \frac{\partial z}{\partial x}\frac{d^2x}{dt^2} + \frac{\partial^2 z}{\partial x^2}\left(\frac{dx}{dt}\right)^2 + 2\frac{\partial^2 z}{\partial x\,\partial y}\frac{dx}{dt}\frac{dy}{dt} + \frac{\partial^2 z}{\partial y^2}\left(\frac{dy}{dt}\right)^2 + \frac{\partial z}{\partial y}\frac{d^2y}{dt^2}. \tag{2–131}$$

This is a new chain rule.

Similarly, if $z = f(x, y)$, $x = g(u, v)$, $y = h(u, v)$, so that (2–34) holds, one has

$$\frac{\partial z}{\partial u} = \frac{\partial z}{\partial x}\frac{\partial x}{\partial u} + \frac{\partial z}{\partial y}\frac{\partial y}{\partial u}, \quad \frac{\partial^2 z}{\partial u^2} = \frac{\partial z}{\partial x}\frac{\partial^2 x}{\partial u^2} + \frac{\partial}{\partial u}\left(\frac{\partial z}{\partial x}\right)\frac{\partial x}{\partial u} + \frac{\partial z}{\partial y}\frac{\partial^2 y}{\partial u^2} + \frac{\partial}{\partial u}\left(\frac{\partial z}{\partial y}\right)\frac{\partial y}{\partial u}. \tag{2–132}$$

Applying (2–132) again, one finds

$$\frac{\partial}{\partial u}\left(\frac{\partial z}{\partial x}\right) = \frac{\partial^2 z}{\partial x^2}\frac{\partial x}{\partial u} + \frac{\partial^2 z}{\partial y\,\partial x}\frac{\partial y}{\partial u}, \quad \frac{\partial}{\partial u}\left(\frac{\partial z}{\partial y}\right) = \frac{\partial^2 z}{\partial x\,\partial y}\frac{\partial x}{\partial u} + \frac{\partial^2 z}{\partial y^2}\frac{\partial y}{\partial u},$$

so that

$$\frac{\partial^2 z}{\partial u^2} = \frac{\partial z}{\partial x}\frac{\partial^2 x}{\partial u^2} + \left(\frac{\partial^2 z}{\partial x^2}\right)\left(\frac{\partial x}{\partial u}\right)^2 + 2\frac{\partial^2 z}{\partial x\,\partial y}\frac{\partial x}{\partial u}\frac{\partial y}{\partial u} + \frac{\partial^2 z}{\partial y^2}\left(\frac{\partial y}{\partial u}\right)^2 + \frac{\partial z}{\partial y}\frac{\partial^2 y}{\partial u^2}. \quad (2\text{–}133)$$

It should be remarked that (2–133) is a special case of (2–131), since v is treated as a constant throughout.

Rules for $\partial^2 z/\partial u\,\partial v$, $\partial^2 z/\partial v^2$ and for higher derivatives can be formed, analogous to (2–131) and (2–133). These rules are of importance, but in most practical cases it is better to use only the chain rules (2–33), (2–34), (2–35), applying repeatedly if necessary. One reason for this is that simplifications are obtained if the derivatives occurring are expressed in terms of the right variables, and a complete description of all possible cases would be too involved to be useful.

The variations possible can be illustrated by the following example, which concerns only functions of one variable.

Let $y = f(x)$ and $x = e^t$. Then

$$\frac{dy}{dt} = \frac{dy}{dx}\frac{dx}{dt} = \frac{dy}{dx}e^t.$$

Hence

$$\begin{aligned}\frac{d^2y}{dt^2} &= \frac{d}{dt}\left(\frac{dy}{dx}\right)e^t + \frac{dy}{dx}e^t\\ &= \frac{d^2y}{dx^2}\frac{dx}{dt}e^t + \frac{dy}{dx}e^t\\ &= \frac{d^2y}{dx^2}e^{2t} + \frac{dy}{dx}e^t.\end{aligned}$$

One could also write

$$\frac{dy}{dt} = \frac{dy}{dx}e^t = x\frac{dy}{dx}$$

and then

$$\begin{aligned}\frac{d^2y}{dt^2} &= \frac{d}{dt}\left(x\frac{dy}{dx}\right) = \frac{d}{dx}\left(x\frac{dy}{dx}\right)\frac{dx}{dt} = x\frac{d}{dx}\left(x\frac{dy}{dx}\right)\\ &= \frac{d^2y}{dx^2}x^2 + \frac{dy}{dx}x.\end{aligned}$$

The second method is clearly simpler than the first; the answers obtained are equivalent because of the equation $x = e^t$.

2–17 The Laplacian in polar, cylindrical, and spherical coordinates. An important application of the method of the preceding section is the transformation of the Laplacian to its expression for other coordinate systems.

We consider first the two-dimensional Laplacian

$$\nabla^2 w = \frac{\partial^2 w}{\partial x^2} + \frac{\partial^2 w}{\partial y^2}$$

and its expression in terms of polar coordinates r, θ. Thus we are given $w = f(x, y)$ and $x = r\cos\theta$, $y = r\sin\theta$ and we wish to express $\nabla^2 w$ in terms of r, θ, and derivatives of w with respect to r and θ. The solution is as follows. One has

$$\frac{\partial w}{\partial x} = \frac{\partial w}{\partial r}\frac{\partial r}{\partial x} + \frac{\partial w}{\partial \theta}\frac{\partial \theta}{\partial x}, \quad \frac{\partial w}{\partial y} = \frac{\partial w}{\partial r}\frac{\partial r}{\partial y} + \frac{\partial w}{\partial \theta}\frac{\partial \theta}{\partial y}, \tag{2–134}$$

by the chain rule. To evaluate $\partial r/\partial x$, $\partial \theta/\partial x$, $\partial r/\partial y$, $\partial \theta/\partial y$ we use the equations

$$dx = \cos\theta\, dr - r\sin\theta\, d\theta, \quad dy = \sin\theta\, dr + r\cos\theta\, d\theta.$$

These can be solved for dr and $d\theta$ by determinants or by elimination to give

$$dr = \cos\theta\, dx + \sin\theta\, dy, \quad d\theta = -\frac{\sin\theta}{r}\,dx + \frac{\cos\theta}{r}\,dy.$$

Hence

$$\frac{\partial r}{\partial x} = \cos\theta, \quad \frac{\partial r}{\partial y} = \sin\theta, \quad \frac{\partial \theta}{\partial x} = -\frac{\sin\theta}{r}, \quad \frac{\partial \theta}{\partial y} = \frac{\cos\theta}{r}.$$

Thus (2–134) can be written as follows:

$$\frac{\partial w}{\partial x} = \cos\theta\,\frac{\partial w}{\partial r} - \frac{\sin\theta}{r}\frac{\partial w}{\partial \theta}, \quad \frac{\partial w}{\partial y} = \sin\theta\,\frac{\partial w}{\partial r} + \frac{\cos\theta}{r}\frac{\partial w}{\partial \theta}. \tag{2–135}$$

These equations provide general rules for expressing derivatives with respect to x or y in terms of derivatives with respect to r and θ. By applying the first equation to the function $\partial w/\partial x$, one finds that

$$\frac{\partial^2 w}{\partial x^2} = \frac{\partial}{\partial x}\left(\frac{\partial w}{\partial x}\right) = \cos\theta\,\frac{\partial}{\partial r}\left(\frac{\partial w}{\partial x}\right) - \frac{\sin\theta}{r}\frac{\partial}{\partial \theta}\left(\frac{\partial w}{\partial x}\right);$$

by (2–135) this can be written as follows:

$$\frac{\partial^2 w}{\partial x^2} = \cos\theta\,\frac{\partial}{\partial r}\left(\cos\theta\,\frac{\partial w}{\partial r} - \frac{\sin\theta}{r}\frac{\partial w}{\partial \theta}\right) - \frac{\sin\theta}{r}\frac{\partial}{\partial \theta}\left(\cos\theta\,\frac{\partial w}{\partial r} - \frac{\sin\theta}{r}\frac{\partial w}{\partial \theta}\right).$$

The rule for differentiation of a product gives finally

$$\frac{\partial^2 w}{\partial x^2} = \cos^2\theta\,\frac{\partial^2 w}{\partial r^2} - \frac{2\sin\theta\cos\theta}{r}\frac{\partial^2 w}{\partial r\,\partial \theta} + \frac{\sin^2\theta}{r^2}\frac{\partial^2 w}{\partial \theta^2} + \frac{\sin^2\theta}{r}\frac{\partial w}{\partial r} + \frac{2\sin\theta\cos\theta}{r^2}\frac{\partial w}{\partial \theta}. \tag{2–136}$$

In the same manner one finds

$$\frac{\partial^2 w}{\partial y^2} = \frac{\partial}{\partial y}\left(\frac{\partial w}{\partial y}\right) = \sin\theta\,\frac{\partial}{\partial r}\left(\sin\theta\,\frac{\partial w}{\partial r} + \frac{\cos\theta}{r}\frac{\partial w}{\partial \theta}\right) + \frac{\cos\theta}{r}\frac{\partial}{\partial \theta}\left(\sin\theta\,\frac{\partial w}{\partial r} + \frac{\cos\theta}{r}\frac{\partial w}{\partial \theta}\right),$$

$$\frac{\partial^2 w}{\partial y^2} = \sin^2\theta\,\frac{\partial^2 w}{\partial r^2} + \frac{2\sin\theta\cos\theta}{r}\,\frac{\partial^2 w}{\partial r\,\partial\theta} + \frac{\cos^2\theta}{r^2}\,\frac{\partial^2 w}{\partial\theta^2} + \frac{\cos^2\theta}{r}\,\frac{\partial w}{\partial r} - \frac{2\sin\theta\cos\theta}{r^2}\,\frac{\partial w}{\partial\theta}. \tag{2–137}$$

Adding (2–136) and (2–137), we conclude:

$$\nabla^2 w = \frac{\partial^2 w}{\partial x^2} + \frac{\partial^2 w}{\partial y^2} = \frac{\partial^2 w}{\partial r^2} + \frac{1}{r^2}\frac{\partial^2 w}{\partial\theta^2} + \frac{1}{r}\frac{\partial w}{\partial r}; \tag{2–138}$$

this is the desired result.

Equation (2–138) at once permits one to write the expression for the three-dimensional Laplacian in cylindrical coordinates; for the transformation of coordinates:

$$x = r\cos\theta,\quad y = r\sin\theta,\quad z = z$$

involves only x and y and in the same way as above. One finds:

$$\nabla^2 w = \frac{\partial^2 w}{\partial x^2} + \frac{\partial^2 w}{\partial y^2} + \frac{\partial^2 w}{\partial z^2} = \frac{\partial^2 w}{\partial r^2} + \frac{1}{r^2}\frac{\partial^2 w}{\partial\theta^2} + \frac{1}{r}\frac{\partial w}{\partial r} + \frac{\partial^2 w}{\partial z^2}. \tag{2–139}$$

A procedure similar to the above gives the three-dimensional Laplacian in spherical coordinates (Section 0–5):

$$\nabla^2 w = \frac{\partial^2 w}{\partial x^2} + \frac{\partial^2 w}{\partial y^2} + \frac{\partial^2 w}{\partial z^2} = \frac{\partial^2 w}{\partial\rho^2} + \frac{1}{\rho^2}\frac{\partial^2 w}{\partial\phi^2} + \frac{1}{\rho^2\sin^2\phi}\frac{\partial^2 w}{\partial\theta^2} + \frac{2}{\rho}\frac{\partial w}{\partial\rho} + \frac{\cot\phi}{\rho^2}\frac{\partial w}{\partial\phi}. \tag{2–140}$$

(See Prob. 8 below.)

2–18 Higher derivatives of implicit functions. In Section 2–10 above procedures are given for obtaining differentials or first partial derivatives of functions implicitly defined by simultaneous equations. Since the results are in the form of *explicit* expressions for the first partial derivatives [cf. (2–71), (2–72), (2–73), etc.], these derivatives can be differentiated explicitly. An example will illustrate the situation. Let x and y be defined as functions of u and v by the implicit equations:

$$x^2 + y^2 + u^2 + v^2 = 1,\quad x^2 + 2y^2 - u^2 + v^2 = 1.$$

Then with $F(x, y, u, v) = x^2 + y^2 + u^2 + v^2 - 1$, $G(x, \ldots) = x^2 + 2y^2 - u^2 + v^2 - 1$,

$$\frac{\partial x}{\partial u} = -\frac{\dfrac{\partial(F, G)}{\partial(u, y)}}{\dfrac{\partial(F, G)}{\partial(x, y)}} = -\frac{\begin{vmatrix} 2u & 2y \\ -2u & 4y \end{vmatrix}}{\begin{vmatrix} 2x & 2y \\ 2x & 4y \end{vmatrix}} = -\frac{3u}{x},$$

$$\frac{\partial y}{\partial u} = -\frac{\dfrac{\partial(F, G)}{\partial(x, u)}}{\dfrac{\partial(F, G)}{\partial(x, y)}} = -\frac{\begin{vmatrix} 2x & 2u \\ 2x & -2u \end{vmatrix}}{\begin{vmatrix} 2x & 2y \\ 2x & 4y \end{vmatrix}} = \frac{2u}{y}.$$

Hence

$$\frac{\partial^2 x}{\partial u^2} = \frac{3u}{x^2}\frac{\partial x}{\partial u} - \frac{3}{x} = \frac{3u}{x^2}\left(\frac{-3u}{x}\right) - \frac{3}{x} = -\frac{9u^2}{x^3} - \frac{3}{x},$$

$$\frac{\partial^2 y}{\partial u^2} = -\frac{2u}{y^2}\frac{\partial y}{\partial u} + \frac{2}{y} = -\frac{2u}{y^2}\left(\frac{2u}{y}\right) + \frac{2}{y} = -\frac{4u^2}{y^3} + \frac{2}{y}.$$

Problems

1. Find the indicated partial derivatives:

(a) $\dfrac{\partial^2 w}{\partial x^2}$ and $\dfrac{\partial^2 w}{\partial y^2}$ if $w = \dfrac{1}{\sqrt{x^2+y^2}}$,

(b) $\dfrac{\partial^2 w}{\partial x^2}$ and $\dfrac{\partial^2 w}{\partial y^2}$ if $w = \arctan\dfrac{y}{x}$,

(c) $\dfrac{\partial^3 w}{\partial x\,\partial y^2}$ and $\dfrac{\partial^3 w}{\partial x^2\,\partial y}$ if $w = e^{x^2-y^2}$.

Formula (1–138) can be used as a check for (a) and (b).

2. Verify that the mixed derivatives are identical for the following cases:

(a) $\dfrac{\partial^2 z}{\partial x\,\partial y}$ and $\dfrac{\partial^2 z}{\partial y\,\partial x}$ for $z = \dfrac{x}{x^2+y^2}$,

(b) $\dfrac{\partial^3 w}{\partial x\,\partial y\,\partial z}$, $\dfrac{\partial^3 w}{\partial z\,\partial y\,\partial x}$ and $\dfrac{\partial^3 w}{\partial y\,\partial z\,\partial x}$ for $w = \sqrt{x^2+y^2+z^2}$.

3. Show that the following functions are harmonic in x and y:
(a) $e^x \cos y$, (b) $x^3 - 3xy^2$, (c) $\log\sqrt{x^2+y^2}$.

4. (a) Show that every harmonic function is biharmonic.
(b) Show that the following functions are biharmonic in x and y:

$$xe^x \cos y, \quad x^4 - 3x^2y^2.$$

5. (a) Prove the identity:

$$\nabla^2(uv) = u\,\nabla^2 v + v\,\nabla^2 u + 2\,\nabla u \cdot \nabla v$$

for functions u and v of x and y.
(b) Prove the identity of (a) for functions of x, y and z.
(c) Prove that if u and v are harmonic in 2 or 3 dimensions, then

$$w = xu + v$$

is biharmonic. [Hint: use the identity of (a) and (b).]
(d) Prove that if u and v are harmonic in 2 or 3 dimensions, then

$$w = r^2 u + v$$

is biharmonic, where $r^2 = x^2 + y^2$ for 2 dimensions and $r^2 = x^2 + y^2 + z^2$ for 3 dimensions.

6. Establish a chain rule analogous to (2–133) for $\partial^2 z/\partial u\,\partial v$.

7. Use the rule (2–133), applied to $\partial^2 w/\partial x^2$ and $\partial^2 w/\partial y^2$, to prove (2–138).

8. Prove (2–140). [Hint: use (2–139) to express $\nabla^2 w$ in cylindrical coordinates; then note that the equations of transformation from (z, r) to (ρ, ϕ) are the same as those from (x, y) to (r, θ).]

9. Prove that the biharmonic equation in x and y becomes

$$w_{rrrr} + \frac{2}{r^2} w_{rr\theta\theta} + \frac{1}{r^4} w_{\theta\theta\theta\theta} + \frac{2}{r} w_{rrr} - \frac{2}{r^3} w_{r\theta\theta} - \frac{1}{r^2} w_{rr} + \frac{4}{r^4} w_{\theta\theta} + \frac{1}{r^3} w_r = 0$$

in polar coordinates (r, θ). [Hint: use (2–138).]

10. If u and v are functions of x and y defined by the equations

$$xy + uv = 1, \quad xu + yv = 1,$$

find $\partial^2 u/\partial x^2$.

11. If u and v are inverse functions of the system

$$x = u^2 - v^2, \quad y = 2uv,$$

show that u is harmonic.

12. A *differential equation* is an equation relating one or more variables and their derivatives; the Laplace equation and biharmonic equation above are illustrations (see Chapters 8 and 10). A basic tool in the solution of differential equations, i.e., in determining the functions which satisfy the equations, is that of introducing new variables by appropriate substitution formulas. The introduction of polar coordinates in the Laplace equation in Section 2–17 illustrates this. The substitutions can involve independent or dependent variables or both; in each case one must indicate which of the new variables are to be treated as independent and which as dependent. Make the indicated substitutions in the following differential equations:

(a) $\dfrac{dy}{dx} = \dfrac{2x}{y + x^2}$; new variables y (dep.) and $u = x^2$ (indep.);

(b) $\dfrac{dy}{dx} = \dfrac{2x - y + 1}{x + y - 4}$; new variables $v = y - 3$ (dep.) and $u = x - 1$ (indep.);

(c) $x^2 \dfrac{d^3y}{dx^3} + 3x \dfrac{d^2y}{dx^2} + \dfrac{dy}{dx} = 0$; new variables y (dep.) and $t = \log x$ (indep.);

(d) $\dfrac{d^2y}{dx^2} + \left(\dfrac{dy}{dx}\right)^3 = 0$; new variables x (dep.) and y (indep.);

(e) $\dfrac{d^2y}{dx^2} - 4x \dfrac{dy}{dx} + y(3x^2 - 2) = 0$; new variables $v = e^{-x^2}y$ (dep.) and x (indep.);

(f) $a \dfrac{\partial u}{\partial x} + b \dfrac{\partial u}{\partial y} = 0$ (a, b constants); new variables u (dep.), $z = bx - ay$ (indep.), and $w = ax + by$ (indep.);

(g) $\dfrac{\partial^2 u}{\partial x^2} - \dfrac{\partial^2 u}{\partial y^2} = 0$; new variables u (dep.), $z = x + y$ (indep.), and $w = x - y$ (indep.).

Answers

1. (a) $\dfrac{2x^2 - y^2}{(x^2 + y^2)^{\frac{5}{2}}}, \quad \dfrac{2y^2 - x^2}{(x^2 + y^2)^{\frac{5}{2}}}$; (b) $\dfrac{2xy}{(x^2 + y^2)^2}, \quad \dfrac{-2xy}{(x^2 + y^2)^2}$;
(c) $4xe^{x^2-y^2}(2y^2 - 1), \quad -4ye^{x^2-y^2}(2x^2 + 1)$.

6. $\dfrac{\partial z}{\partial x}\dfrac{\partial^2 x}{\partial u\,\partial v} + \dfrac{\partial z}{\partial y}\dfrac{\partial^2 y}{\partial u\,\partial v} + \dfrac{\partial^2 z}{\partial x^2}\dfrac{\partial x}{\partial u}\dfrac{\partial x}{\partial v} + \dfrac{\partial^2 z}{\partial x\,\partial y}\left(\dfrac{\partial x}{\partial u}\dfrac{\partial y}{\partial v} + \dfrac{\partial x}{\partial v}\dfrac{\partial y}{\partial u}\right) + \dfrac{\partial^2 z}{\partial y^2}\dfrac{\partial y}{\partial u}\dfrac{\partial y}{\partial v}.$

10. $\dfrac{2(u^2 - y^2)}{(1 - 2ux)^3}(2u - 3u^2x - xy^2).$

12. (a) $\dfrac{dy}{du} = \dfrac{1}{y + u}$, (b) $\dfrac{dv}{du} = \dfrac{2u - v}{u + v}$, (c) $\dfrac{d^3y}{dt^3} = 0$, (d) $\dfrac{d^2x}{dy^2} - 1 = 0$,
(e) $\dfrac{d^2v}{dx^2} - x^2v = 0$, (f) $\dfrac{\partial u}{\partial w} = 0$, (g) $\dfrac{\partial^2 u}{\partial z\,\partial w} = 0$.

2–19 Maxima and minima of functions of several variables. We first recall the basic facts concerning maxima and minima of functions of one variable. Let $y = f(x)$ be defined and differentiable in a closed interval: $a \leqq x \leqq b$, and let x_0 be a number between a and b: $a < x_0 < b$. The function $f(x)$ is said to have a *relative maximum* at x_0 if $f(x) \leqq f(x_0)$ for x sufficiently close to x_0. It follows from the very definition of the derivative that if $f'(x_0) > 0$, then $f(x) > f(x_0)$ for all $x > x_0$ and sufficiently close to x_0; similarly, if $f'(x_0) < 0$, then $f(x) > f(x_0)$ for all $x < x_0$ and sufficiently close to x_0. Hence *at a relative maximum necessarily* $f'(x_0) = 0$. A *relative minimum* of $f(x)$ is defined by the condition: $f(x) \geqq f(x_0)$ for all x sufficiently close to x_0. A reasoning such as the preceding enables one to conclude that *at a relative minimum* of $f(x)$ *necessarily* $f'(x_0) = 0$.

The points x_0 at which $f'(x_0) = 0$ are termed the *critical points* of $f(x)$. While every relative maximum and minimum occurs at a critical point, a critical point need not give either maximum or minimum. This is illustrated by the function $y = x^3$ at $x = 0$. This function has a critical point at $x = 0$, but the point is neither maximum nor minimum, and is an example of what is termed a *horizontal inflection point*. This is illustrated in Fig. 2–12.

Let x be a critical point, so that $f'(x_0) = 0$, and let it be assumed that $f''(x_0) > 0$. Then $f(x)$ *has a relative minimum at* x_0. For by the law of the mean, when $x > x_0$, $f(x) - f(x_0) = f'(x_1)(x - x_0)$, where $x_0 < x_1 < x$. Since $f''(x_0) > 0$, $f'(x_1) > 0$ for $x_1 > x_0$ and x_1 sufficiently close to x_0. Hence $f(x) - f(x_0) > 0$ for $x > x_0$ and x sufficiently close to x_0. Similarly, $f(x) - f(x_0) > 0$ for $x < x_0$ and x sufficiently close to x_0. Therefore $f(x)$ has a relative minimum at x_0. An analogous reasoning applies when $f''(x_0) < 0$. Accordingly, one can state the rules:

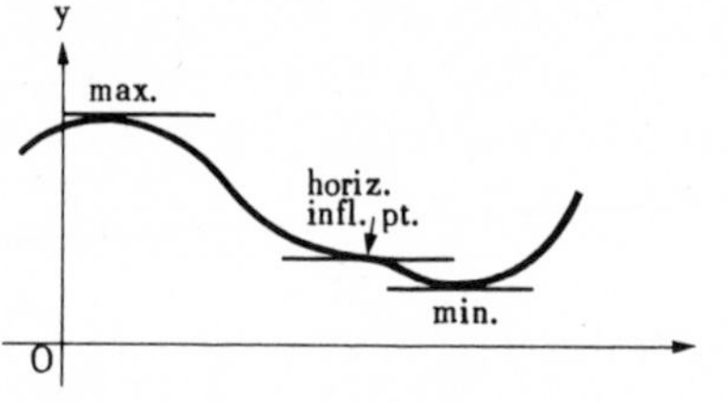

FIG. 2–12. Critical points of function $y = f(x)$.

If $f'(x_0) = 0$ *and* $f''(x_0) > 0$, *then* $f(x)$ *has a relative minimum at* x_0; *if* $f'(x_0) = 0$ *and* $f''(x_0) < 0$, *then* $f(x)$ *has a relative maximum at* x_0.

These two rules cover most cases of interest. It can of course happen that $f''(x_0) = 0$ at a critical point, in which case one must consider the higher derivatives. A repeated application of the reasoning above gives the following general rule:

Let $f'(x_0) = 0, f''(x_0) = 0, \ldots, f^{(n)}(x_0) = 0$, *but* $f^{(n+1)}(x_0) \neq 0$; *then* $f(x)$ *has a relative maximum at* x_0 *if* n *is odd and* $f^{(n+1)}(x_0) < 0$; $f(x)$ *has a relative minimum at* x_0 *if* n *is odd and* $f^{(n+1)}(x_0) > 0$; $f(x)$ *has neither relative maximum nor relative minimum at* x_0 *but a horizontal inflection point at* x_0 *if* n *is even.*

The above discussion has been restricted to points x_0 *within* the interval of definition of $f(x)$, and to *relative* maxima and minima. The notion of relative maximum and minimum can readily be extended to the end-points a and b and rules can be formulated in terms of derivatives. However, the interest in these points arises mainly in connection with the *absolute* maximum and minimum of $f(x)$ on the interval $a \leqq x \leqq b$. A function $f(x)$ is said to have an *absolute maximum* M for a certain range of values of x if $f(x_0) = M$ for some x_0 of the range given and $f(x) \leqq M$ for all x of the range; an absolute minimum is defined similarly, with the condition $f(x) \geqq M$ replacing the condition $f(x) \leqq M$. The following theorem is then fundamental.

THEOREM. *If* $f(x)$ *is continuous in the closed interval:* $a \leqq x \leqq b$, *then* $f(x)$ *has an absolute minimum* M_1 *and an absolute maximum* M_2 *on this interval.*

The proof of this theorem requires a more profound analysis of the real number system; the reader is referred to *CLA*, Section 2–14.

It should be remarked that the inclusion of the end-values is essential for this theorem, as the simple example: $y = x$ for $0 < x < 1$ illustrates; here the function has no minimum or maximum *on the range given.* For the range $0 \leqq x \leqq 1$, the absolute minimum is 0, for $x = 0$, and the absolute maximum is 1, for $x = 1$. The example: $y = \tan x$ for $-\pi/2 < x < \pi/2$ illustrates a function with no absolute minimum or maximum; in this case adjoining values at the end-points $x = \pm\pi/2$ does not help.

A function $f(x)$ is said to be *bounded* for a given range of x if there is a constant K such that $|f(x)| \leqq K$ on this range. The theorem above implies that, *if* $f(x)$ *is continuous in the closed interval:* $a \leqq x \leqq b$, *then* $f(x)$ *is bounded in this interval;* for K can be chosen as the larger of $|M_1|$, $|M_2|$. For example, let $y = \sin x$ for $0 \leqq x \leqq \pi$. This function has the absolute minimum $M_1 = 0$, the absolute maximum $M_2 = 1$ and is bounded, with $K = 1$. The example: $y = x$ for $0 < x < 1$ illustrates a bounded function $(K = 1)$ having neither absolute minimum nor absolute maximum; the example: $y = \tan x$ for $-\frac{1}{2}\pi < x < \frac{1}{2}\pi$ illustrates an unbounded function.

To find the absolute maximum M of a function $y = f(x)$, differentiable for $a \leqq x \leqq b$, one can reason as follows: if $f(x_0) = M$ and $a < x_0 < b$, then $f(x)$ has necessarily a relative maximum at x_0; thus the absolute maximum occurs either at a critical point within the interval or at $x = a$ or $x = b$. One therefore first locates all critical points within the interval and compares the values of y at these points with the values at $x = a$ and

$x = b$; the largest value of y is the maximum sought. It is thus not necessary to use the second derivative test described above. The absolute minimum can be found in the same way.

The determination of critical points and their classification into maxima, minima, or neither are important also for graphing functions; from a knowledge of the critical points and the corresponding values of y, a very good first approximation to the graph of $y = f(x)$ can be obtained.

After these preliminaries, we can consider the analogous questions for *functions of two or more variables.* Let $z = f(x, y)$ be defined and continuous in a domain D. This function is said to have a *relative maximum* at (x_0, y_0) if $f(x, y) \leqq f(x_0, y_0)$ for (x, y) sufficiently close to (x_0, y_0) and to have a *relative minimum* at (x_0, y_0) if $f(x, y) \geqq f(x_0, y_0)$ for (x, y) sufficiently close to (x_0, y_0). Let (x_0, y_0) give a relative maximum of $f(x, y)$; then the function $f(x, y_0)$, which depends on x alone, has a relative maximum at x_0, as illustrated in Fig. 2–13. Hence, if $f_x(x_0, y_0)$ exists, then $f_x(x_0, y_0) = 0$; similarly, if $f_y(x_0, y_0)$ exists, then $f_y(x_0, y_0) = 0$. Points (x, y) at which both partial derivatives are 0 are termed *critical points* of f. As before, one concludes that every relative maximum and every relative minimum occurs at a critical point of f, if f_x and f_y exist in D.

One might expect that the nature of a critical point could be determined by studying the functions $f(x, y_0)$ and $f(x_0, y)$, with the aid of the second derivatives as above. First of all, it should be remarked that one of these functions can have a maximum at (x_0, y_0), while the other has a minimum. This is illustrated by the function $z = 1 + x^2 - y^2$ at $(0, 0)$; this function has a minimum with respect to x at $x = 0$ for $y = 0$ and a maximum with respect to y at $y = 0$ for $x = 0$, as shown in Fig. 2–14. This critical point is an example of what is termed a *saddle point;* this will be discussed further below. The level curves of z are shown in Fig. 2–14; the configuration shown is typical of that at a saddle point.

A further complication is the following: it can happen that $z = f(x, y_0)$ has a relative maximum for $x = x_0$ and $z = f(x_0, y)$ has also a relative maximum for $y = y_0$, but $z = f(x, y)$ does *not* have a relative maximum at (x_0, y_0). This is illustrated by the function $z = 1 - x^2 + 4xy - y^2$ at $(0, 0)$. For $y = 0$, $z = 1 - x^2$, with maximum for $x = 0$; for $x = 0$, $z = 1 - y^2$, with maximum for $y = 0$. On the other hand, for $y = x$, $z = 1 + 2x^2$, so that the section of the surface by the plane $y = x$ has a *minimum* at $x = 0$. This is shown in Fig. 2–15. The level curves, shown in the same figure, again reveal the presence of a saddle point.

A better understanding of this example and a clue to a general procedure can be obtained by introducing cylindrical coordinates (r, θ). One has then $z = 1 - r^2(1 - 2\sin 2\theta)$. If one sets $\theta = \text{const} = \alpha$, one obtains z as a function of r. For $\alpha = 0$, one obtains the xz trace: $z = 1 - r^2$; for $\alpha = \pi/2$, one obtains the yz trace: $z = 1 - r^2$; for $\alpha = \pi/4$, one obtains the trace in the plane $y = x$: $z = 1 + r^2$. For a general α one has the trace: $z = 1 - r^2(1 - 2\sin 2\alpha)$; r can here be allowed both positive and negative values. On each trace, $\partial z/\partial r = -2r(1 - 2\sin 2\alpha)$; thus z, as a function of r, has a critical point for $r = 0$. However, $\partial^2 z/\partial r^2 =$

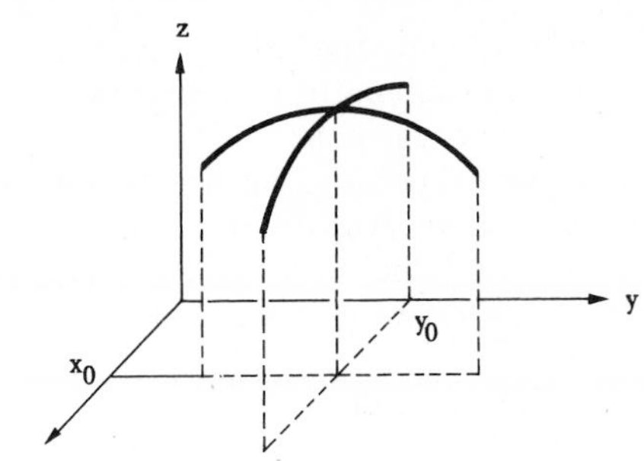

Fig. 2–13. Maximum of $z = f(x, y)$.

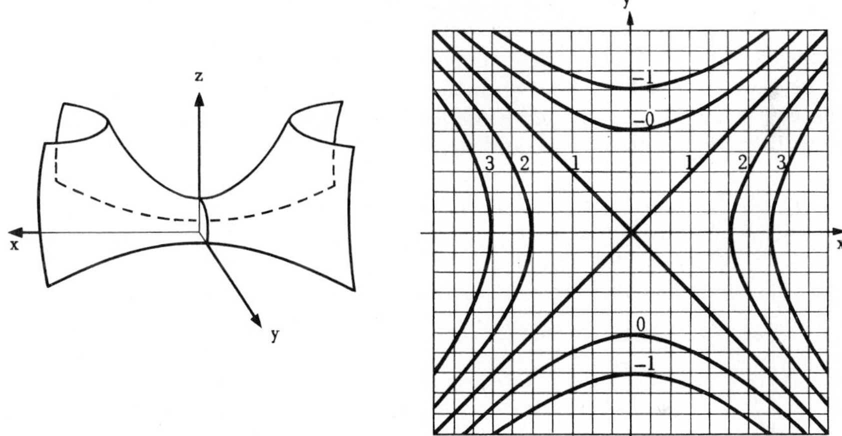

Fig. 2–14. Saddle-point ($z = 1 + x^2 - y^2$).

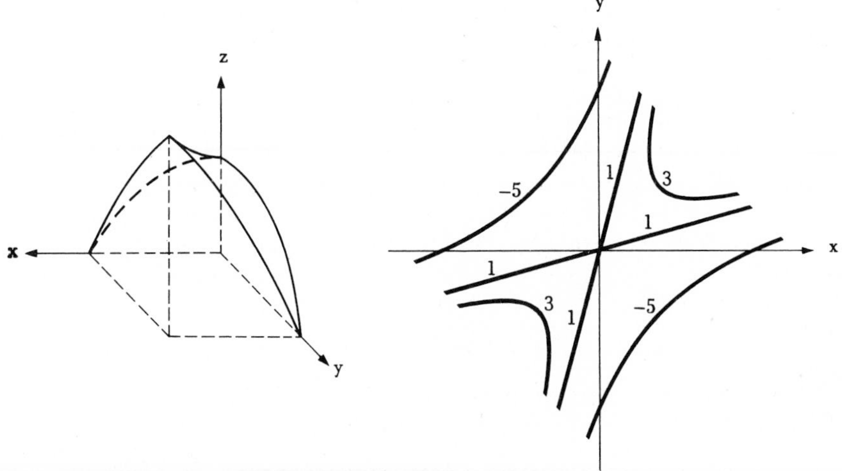

Fig. 2–15. $z = 1 - x^2 + 4xy - y^2$.

$-2(1 - 2\sin 2\alpha)$. This second derivative takes on both positive and negative values, as shown graphically in Fig. 2–16. The second derivative test for functions of one variable at once shows that both relative maxima and relative minima occur in corresponding directions α.

In order to generalize this analysis to an arbitrary function $z = f(x, y)$, we make use of the directional derivative in direction α, defined in Section 2–14 above. This we again denote by $\nabla_\alpha z$ and recall the formula:

$$\nabla_\alpha z = \frac{\partial z}{\partial x}\cos\alpha + \frac{\partial z}{\partial y}\sin\alpha.$$

At a critical point (x_0, y_0) one has necessarily $\nabla_\alpha z = 0$; this means that, in each plane $(x - x_0)\sin\alpha - (y - y_0)\cos\alpha = 0$, z has a critical point at (x_0, y_0), when z is regarded as a function of "directed distance" s from (x_0, y_0), in the xy plane; this is illustrated in Fig. 2–17. As in the example above, the type of critical point may vary with the direction chosen. In order to analyze the type, one introduces the second derivative of z with respect to s, i.e., the *second directional derivative in direction* α. This is simply the quantity $\nabla_\alpha \nabla_\alpha z$, and one has

$$\begin{aligned}\nabla_\alpha \nabla_\alpha z &= \nabla_\alpha\left(\frac{\partial z}{\partial x}\cos\alpha + \frac{\partial z}{\partial y}\sin\alpha\right)\\ &= \frac{\partial^2 z}{\partial x^2}\cos^2\alpha + 2\frac{\partial^2 z}{\partial x\,\partial y}\sin\alpha\cos\alpha + \frac{\partial^2 z}{\partial y^2}\sin^2\alpha. \qquad (2\text{–}141)\end{aligned}$$

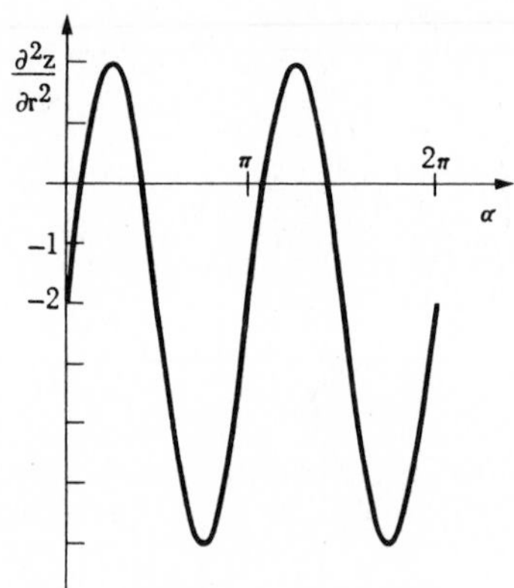

Fig. 2–16. $\partial^2 z/\partial r^2$ for function of Fig. 2–15.

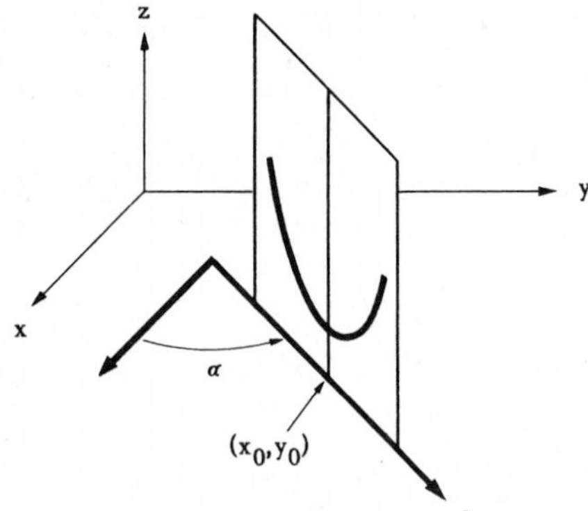

Fig. 2–17. Critical point of $z = f(x, y)$ analyzed by vertical sections.

This is precisely the quantity $\partial^2 z/\partial r^2$ of the above example. In order to guarantee a relative maximum for z at (x_0, y_0), one can require that this second derivative be negative for all α, $0 \leqq \alpha \leqq 2\pi$; for this ensures that z has a relative maximum as a function of s in each plane in direction α, and in fact (all second derivatives being assumed continuous) that z has a relative maximum at the point; cf. Prob. 10 below. A similar reasoning applies to relative minima. One therefore obtains the rule:

If $\partial z/\partial x = 0$ and $\partial z/\partial y = 0$ at (x_0, y_0) and

$$\frac{\partial^2 z}{\partial x^2}\cos^2\alpha + 2\frac{\partial^2 z}{\partial x\,\partial y}\sin\alpha\cos\alpha + \frac{\partial^2 z}{\partial y^2}\sin^2\alpha < 0 \tag{2–142}$$

for $x = x_0$, $y = y_0$ and all α: $0 \leqq \alpha \leqq 2\pi$, then $z = f(x, y)$ has a relative maximum at (x_0, y_0); if $\partial z/\partial x = 0$ and $\partial z/\partial y = 0$ and

$$\frac{\partial^2 z}{\partial x^2}\cos^2\alpha + 2\frac{\partial^2 z}{\partial x\,\partial y}\sin\alpha\cos\alpha + \frac{\partial^2 z}{\partial y^2}\sin^2\alpha > 0 \tag{2–143}$$

for $x = x_0$, $y = y_0$ and all α: $0 \leqq \alpha \leqq 2\pi$, then $z = f(x, y)$ has a relative minimum at (x_0, y_0).

It thus appears that the study of the critical points is reduced to the analysis of the expression

$$A\cos^2\alpha + 2B\sin\alpha\cos\alpha + C\sin^2\alpha, \tag{2–144}$$

where the abbreviations

$$A = \frac{\partial^2 z}{\partial x^2}(x_0, y_0),\quad B = \frac{\partial^2 z}{\partial x\,\partial y}(x_0, y_0),\quad C = \frac{\partial^2 z}{\partial y^2}(x_0, y_0) \tag{2–145}$$

are used. Here an algebraic analysis reduces the question to a simpler one, for one has the theorem:

If $B^2 - AC < 0$ and $A + C < 0$, then the expression (2–144) is negative for all α; if $B^2 - AC < 0$ and $A + C > 0$, then the expression (2–145) is positive for all α.

Proof. Let the expression (2–144) be denoted by $P(\alpha)$. Let $B^2 - AC < 0$ and $A + C < 0$. Then $P(\pm\pi/2) = C < 0$, for if $C \geqq 0$, then $A + C < 0$ implies: $A < 0$, so that $AC \leqq 0$; this contradicts the condition $B^2 - AC < 0$. Similarly one shows: $P(0) = A < 0$. For $\alpha \neq \pm\pi/2$ one has

$$P(\alpha) = \cos^2\alpha(A + 2B\tan\alpha + C\tan^2\alpha).$$

Thus $P(\alpha)$ is positive, negative, or 0, according as the quadratic expression

$$Q(u) = Cu^2 + 2Bu + A \quad (u = \tan\alpha)$$

is positive, negative, or 0. Since $B^2 - AC < 0$, $Q(u)$ has no real roots (Section 0–3); thus $Q(u)$ is always positive or always negative. For $u = 0, Q = A < 0$. Hence $Q(u)$ is always negative and the same holds for $P(\alpha)$. Accordingly, the first assertion is proved. The second is proved in the same way.

If $B^2 - AC > 0$, the proof just given shows that $P(\alpha)$ will be positive for some values of α and negative for others, as in the example of Fig. 2–15 above. In this case the critical point is termed a *saddle point.* If $B^2 - AC = 0$, there will be some directions in which $P(\alpha) = 0$, and one must introduce higher derivatives in order to determine the nature of the critical point.

We summarize the results obtained:

THEOREM. *Let $z = f(x, y)$ be defined and have continuous first and second partial derivatives in a domain D. Let (x_0, y_0) be a point of D for which $\partial z/\partial x$ and $\partial z/\partial y$ are 0. Let*

$$A = \frac{\partial^2 z}{\partial x^2}(x_0, y_0), \quad B = \frac{\partial^2 z}{\partial x\, \partial y}(x_0, y_0), \quad C = \frac{\partial^2 z}{\partial y^2}(x_0, y_0).$$

Then one has the following cases:

$$B^2 - AC < 0 \text{ and } A + C < 0, \textit{ relative maximum at } (x_0, y_0); \tag{2–146}$$
$$B^2 - AC < 0 \text{ and } A + C > 0, \textit{ relative minimum at } (x_0, y_0); \tag{2–147}$$
$$B^2 - AC > 0, \textit{ saddle point at } (x_0, y_0); \tag{2–148}$$
$$B^2 - AC = 0, \textit{ nature of critical point undetermined.} \tag{2–149}$$

If A, B, and C are all 0, so that $P(\alpha) \equiv 0$, one can study the critical point with the aid of the third derivative: $\nabla_\alpha\nabla_\alpha\nabla_\alpha z$. This is illustrated by the function: $z = x^3 - 3xy^2$, whose level curves are graphed in Fig. 2–18. If the third derivatives are all 0, one can go on to higher derivatives as for functions of one variable. In essence, the whole problem is reduced to one for functions of one variable, and the directional derivative is the chief tool in this reduction.

In the accompanying Fig. 2–18, examples are given of various types of critical points; in each case the level curves of the function are shown.

The preceding discussion has concerned only relative maxima and minima. Just as with functions of one variable, one can define the notions of absolute maximum and absolute minimum; again their investigation calls for a study of the function at the "ends" of the domain D of definition, i.e., on the boundary. In order to ensure existence of absolute maximum and minimum, one must demand that the domain D itself be *bounded* (cf. Section 2–2). One has then as for functions of one variable:

THEOREM. *Let D be a bounded domain of the xy plane. Let $f(x, y)$ be defined and continuous in the closed region E formed of D plus its boundary. Then $f(x, y)$ has an absolute maximum and an absolute minimum in E.*

The proof resembles that for functions of one variable; see *CLA*, Section 12–25. The fact that D is a domain is actually inessential for this theorem; the theorem remains valid for any bounded closed set E (Section 2–2).

Again one has the corollary that a function $f(x, y)$ which is continuous in a bounded closed region (or set) E is necessarily bounded on E; i.e., $|f(x, y)| \leqq M$ for (x, y) in E and for a suitable choice of M.

To determine the absolute maximum and minimum of a function $f(x, y)$ defined on a bounded closed region E, one proceeds as for functions of one variable. One determines the critical points of f inside E and the values of f at the critical points; one then determines the maximum and minimum of f on the boundary. Among these values the desired maximum and minimum will be found.

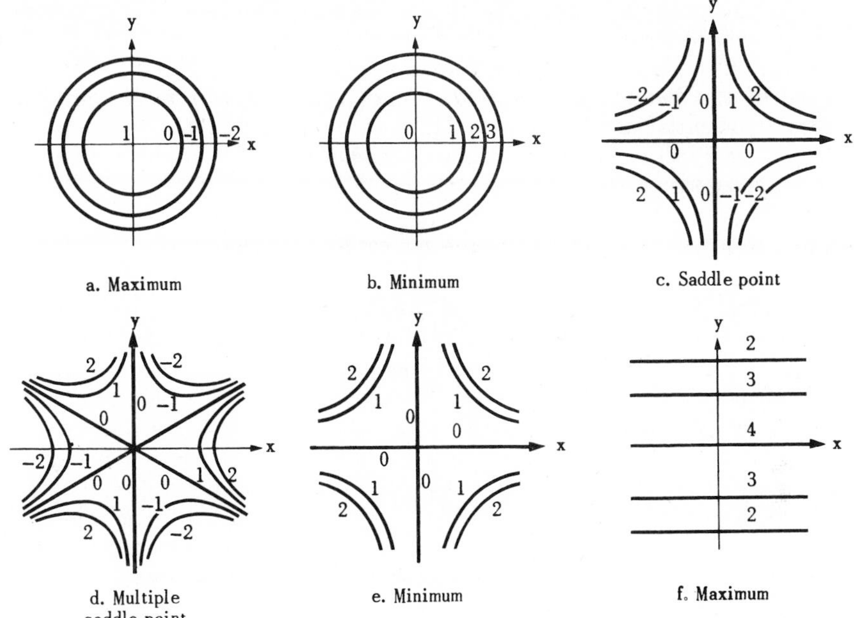

FIG. 2–18. Examples of critical points of $f(x, y)$ at $(0, 0)$. The corresponding functions are as follows: (a) $z = 1 - x^2 - y^2$, (b) $z = x^2 + y^2$, (c) $z = xy$, (d) $z = x^3 - 3xy^2$, (e) $z = x^2y^2$, (*f*) $z = 4 - y^2$.

EXAMPLE. Find the maximum and minimum of $z = x^2 + 2y^2 - x$ on the set $x^2 + y^2 \leqq 1$.

One has

$$\frac{\partial z}{\partial x} = 2x - 1, \quad \frac{\partial z}{\partial y} = 4y.$$

Thus the critical point is at $(\frac{1}{2}, 0)$, at which $z = -\frac{1}{4}$. On the boundary of E one has $x^2 + y^2 = 1$, so that $z = 2 - x - x^2$, $-1 \leqq x \leqq 1$. For this function we find the absolute maximum to be $2\frac{1}{4}$, at the critical point $x = -\frac{1}{2}$; the absolute minimum is 0, at the end $x = 1$. Thus the absolute maximum is $2\frac{1}{4}$, occurring at $\left(-\frac{1}{2}, \pm\frac{\sqrt{3}}{2}\right)$, and the absolute minimum is $-\frac{1}{4}$, occurring at $(\frac{1}{2}, 0)$. Since the minimum occurs *inside* E, it is also a relative minimum, as can be checked by (2–146 . . . (2–149). One has here

$$A = \frac{\partial^2 z}{\partial x^2} = 2, \quad B = \frac{\partial^2 z}{\partial x\, \partial y} = 0, \quad C = \frac{\partial^2 z}{\partial y^2} = 4,$$

so that (2–147) is satisfied.

The maximum and minimum on the boundary can also be analyzed by the method of Lagrange multipliers, to be explained in the following section.

All of the foregoing extends to functions of three or more variables without essential modification. Thus at a critical point (x_0, y_0, z_0) of $w = f(x, y, z)$ all three derivatives $\partial w/\partial x$, $\partial w/\partial y$, $\partial w/\partial z$ are 0, so that the directional derivative:

$$\nabla_u w = \frac{\partial w}{\partial x}\cos\alpha + \frac{\partial w}{\partial y}\cos\beta + \frac{\partial w}{\partial z}\cos\gamma$$

in the direction of an arbitrary unit vector $\mathbf{u} = \cos\alpha\mathbf{i} + \cos\beta\mathbf{j} + \cos\gamma\mathbf{k}$ is zero at the point (x_0, y_0, z_0). This directional derivative is the derivative dw/ds of a function of one variable s, serving as coordinate on a line through (x_0, y_0, z_0) in direction $\mathbf{u}$. To analyze the critical point, one uses the second derivative $d^2w/ds^2 = \nabla_u \nabla_u w$:

$$\begin{aligned}\nabla_u \nabla_u w &= \frac{\partial}{\partial x}(\nabla_u w)\cos\alpha + \frac{\partial}{\partial y}(\nabla_u w)\cos\beta + \frac{\partial}{\partial z}(\nabla_u w)\cos\gamma \\ &= \frac{\partial^2 w}{\partial x^2}\cos^2\alpha + 2\frac{\partial^2 w}{\partial x\,\partial y}\cos\alpha\cos\beta + 2\frac{\partial^2 w}{\partial x\,\partial z}\cos\alpha\cos\gamma \qquad (2\text{–}150) \\ &\quad + \frac{\partial^2 w}{\partial y^2}\cos^2\beta + 2\frac{\partial^2 w}{\partial y\,\partial z}\cos\beta\cos\gamma + \frac{\partial^2 w}{\partial z^2}\cos^2\gamma.\end{aligned}$$

If this expression is positive for all $\mathbf{u}$, then w has a relative minimum at (x_0, y_0, z_0). Algebraic criteria for the positiveness of this "quadratic form" can be obtained; see Section 2–21 below.

***2–20 Maxima and minima for functions with side conditions. Lagrange multipliers.** A problem of considerable importance for applications is that of maximizing or minimizing a function of several variables, where the variables are related by one or more equations, termed "side conditions." Thus the problem of finding the radius of the largest sphere inscribable in the ellipsoid $x^2 + 2y^2 + 3z^2 = 6$ is equivalent to minimizing the function $w = x^2 + y^2 + z^2$, with the side condition: $x^2 + 2y^2 + 3z^2 = 6$.

To handle such problems one can, if possible, eliminate some of the variables by using the side conditions and eventually reduce the problem to an ordinary maximum and minimum problem such as that considered in the preceding section. This procedure is not always feasible and the following procedure is often more convenient; it also treats the variables in a more symmetrical way, so that various simplifications may be possible.

To illustrate the method, we consider the problem of maximizing $w = f(x, y, z)$, where equations $g(x, y, z) = 0$ and $h(x, y, z) = 0$ are given. The equations $g = 0$ and $h = 0$ describe two surfaces in space and the problem is thus one of maximizing $f(x, y, z)$ as (x, y, z) varies on the curve of intersection of these surfaces. At a maximum point, the derivative of f along the curve, i.e., the directional derivative along the tangent to the curve, must be 0. This directional derivative is the component of the vector ∇f along the tangent. It follows that ∇f must lie in a plane normal to the curve at the point. This plane also contains the vectors ∇g and ∇h (Section 2–12 above); i.e., the vectors ∇f, ∇g, and ∇h are coplanar at the point. Hence (Section 0–6) there must exist scalars λ_1 and λ_2 such that

$$\nabla f + \lambda_1 \nabla g + \lambda_2 \nabla h = \mathbf{0} \tag{2–151}$$

at the critical point. This is equivalent to three scalar equations:

$$\begin{gathered}\frac{\partial f}{\partial x} + \lambda_1 \frac{\partial g}{\partial x} + \lambda_2 \frac{\partial h}{\partial x} = 0, \quad \frac{\partial f}{\partial y} + \lambda_1 \frac{\partial g}{\partial y} + \lambda_2 \frac{\partial h}{\partial y} = 0, \\ \frac{\partial f}{\partial z} + \lambda_1 \frac{\partial g}{\partial z} + \lambda_2 \frac{\partial h}{\partial z} = 0.\end{gathered} \tag{2–152}$$

These three equations, together with the equations $g(x, y, z) = 0$, $h(x, y, z) = 0$, serve as five equations in the five unknowns $x, y, z, \lambda_1, \lambda_2$. By solving them for x, y, z, one locates the critical points on the curve; the critical points can be tested further by using the "second directional derivative," as in the preceding section.

It has been tacitly assumed here that the surfaces $g = 0, h = 0$ do actually intersect in a curve and that ∇g and ∇h are linearly independent. Cases in which these conditions fail are degenerate and require further investigation (cf. Section 2–22).

The method described applies quite generally. To find the critical points of $w = f(x, y, z, u, \ldots)$, where the variables $x, y, z, \ldots$ are related by equations: $g(x, y, z, u, \ldots) = 0$, $h(x, y, z, u, \ldots) = 0, \ldots$, one solves the system of equations

$$\begin{gathered}\frac{\partial f}{\partial x} + \lambda_1 \frac{\partial g}{\partial x} + \lambda_2 \frac{\partial h}{\partial x} + \cdots = 0, \quad \frac{\partial f}{\partial y} + \lambda_1 \frac{\partial g}{\partial y} + \lambda_2 \frac{\partial h}{\partial y} + \cdots = 0, \ldots, \\ g(x, y, z, u, \ldots) = 0, \quad h(x, y, z, u, \ldots) = 0, \ldots\end{gathered} \tag{2–153}$$

for the unknowns $x, y, z, u, \ldots, \lambda_1, \lambda_2, \ldots$. The parameters $\lambda_1, \lambda_2, \ldots$ are known as *Lagrange multipliers*.

Example. To find the critical points of $w = xyz$, subject to the condi tion $x^2 + y^2 + z^2 = 1$, one forms the function

$$f + \lambda g = xyz + \lambda(x^2 + y^2 + z^2 - 1)$$

and then obtains four equations:

$$yz + 2\lambda x = 0, \quad xz + 2\lambda y = 0, \quad xy + 2\lambda z = 0, \quad x^2 + y^2 + z^2 = 1.$$

Multiplying the first three by x, y, z, respectively, adding, and using the fourth equation, one finds $\lambda = -\frac{1}{2}(3xyz)$. Using this relation, one easily finds the 14 critical points $(0, 0, \pm 1)$, $(0, \pm 1, 0)$, $(\pm 1, 0, 0)$, $(\pm\sqrt{3}/3, \pm\sqrt{3}/3, \pm\sqrt{3}/3)$. The first six listed are saddle points, while the remaining eight furnish four minima and four maxima, as simple considerations of signs show.

***2–21 Maxima and minima of quadratic forms on the unit sphere.** One of the most important applications of the method of Lagrange multipliers is to the problem of maximizing or minimizing a quadratic form in n variables (Section 1–6)

$$w = f(x_1, \ldots, x_n) = \sum_{i=1}^{n} \sum_{j=1}^{n} a_{ij}x_ix_j \tag{2–154}$$

subject to the side condition

$$x_1^2 + \cdots + x_n^2 = 1. \tag{2–155}$$

We can interpret (2–155) as the equation of the *unit sphere* in E^n, so that our problem is that of maximizing or minimizing f on the unit sphere. As in Section 1–6, we let $A = (a_{ij})$ be the $n \times n$ matrix of coefficients of the quadratic form and can always assume that f is written in a form which makes A *symmetric:* $a_{ij} = a_{ji}$ for all i and j, or $A = A'$. As in Section 1–6, we can then also write

$$f(x_1, \ldots, x_n) = \mathbf{x}'A\mathbf{x}, \quad (\mathbf{x} = \text{col}\,(x_1, \ldots, x_n)).$$

We proceed as in the preceding section and are led to the $n + 1$ equations:

$$\begin{gathered} \frac{\partial f}{\partial x_1} + \lambda \frac{\partial g}{\partial x_1} = 0, \quad \frac{\partial f}{\partial x_2} + \lambda \frac{\partial g}{\partial x_2} = 0, \ldots, \quad \frac{\partial f}{\partial x_n} + \lambda \frac{\partial g}{\partial x_n} = 0, \\ g(x_1, \ldots, x_n) = 0. \end{gathered} \tag{2–156}$$

Here f is as above and $g(x_1, \ldots, x_n) = 1 - x_1^2 - \cdots - x_n^2$. Since A is symmetric,

$$\begin{aligned} f(x_1, \ldots, x_n) &= a_{11}x_1^2 + a_{12}x_1x_2 + a_{12}x_2x_1 + a_{13}x_1x_3 + a_{13}x_3x_1 + \cdots \\ &= a_{11}x_1^2 + 2a_{12}x_1x_2 + 2a_{13}x_1x_3 + \cdots \end{aligned}$$

(we are suggesting only the terms in x_1), and accordingly

$$\frac{\partial f}{\partial x_1} = 2a_{11}x_1 + 2a_{12}x_2 + 2a_{13}x_3 + \cdots.$$

Thus the equations (2–156) become

$$\begin{aligned} 2a_{11}x_1 + 2a_{12}x_2 + \cdots + 2a_{1n}x_n - 2\lambda x_1 &= 0, \\ &\vdots \\ 2a_{n1}x_1 + 2a_{n2}x_2 + \cdots + 2a_{nn}x_n - 2\lambda x_n &= 0, \\ x_1^2 + \cdots + x_n^2 &= 1. \end{aligned}$$

Here the first n equations are equivalent to the equation

$$A\mathbf{x} = \lambda\mathbf{x}.$$

Therefore, they state that $\mathbf{x}$ (if nonzero) is to be an *eigenvector* of A, associated with the *eigenvalue* λ (Section 1–5). The last equation states that $|\mathbf{x}| = 1$, so that $\mathbf{x}$ is a unit vector and cannot be $\mathbf{0}$. Thus we are seeking the unit eigenvectors of A. If $\mathbf{x}$ is such an eigenvector, then

$$\begin{aligned} f(x_1, \ldots, x_n) &= \mathbf{x}'A\mathbf{x} = \mathbf{x}'\lambda\mathbf{x} = \lambda\mathbf{x}'\mathbf{x} \\ &= \lambda(x_1^2 + \cdots + x_n^2) = \lambda. \end{aligned}$$

Hence the eigenvalues of A provide the values of f at its critical points on the unit sphere. In particular, the absolute maximum of f on the unit sphere equals the largest eigenvalue of A and the absolute minimum of f on the unit sphere equals the smallest eigenvalue of A.

EXAMPLE 1. Let $n = 2$ and let

$$f(x, y) = ax^2 + 2bxy + cy^2,$$

so that $A = \begin{bmatrix} a & b \\ b & c \end{bmatrix}$. Then the eigenvalues of A are the solutions of

$$\begin{vmatrix} a - \lambda & b \\ b & c - \lambda \end{vmatrix} = 0 \quad \text{or} \quad \lambda^2 - (a + c)\lambda + ac - b^2 = 0.$$

Hence

$$\lambda = \frac{a + c \pm \sqrt{(a + c)^2 - 4(ac - b^2)}}{2} = \frac{a + c \pm \sqrt{(a - c)^2 + 4b^2}}{2}.$$

The last form shows that the roots are always real (see Problem 12 following Section 1–7). If $a = c$ and $b = 0$, the roots are equal and equal to $(a + c)/2$. Otherwise the larger root is obtained from the plus sign, the smaller from the minus sign. Also, if $ac - b^2 > 0$, both roots have the same sign as $a + c$. Thus if $ac - b^2 > 0$ and $a + c > 0$, then both roots are positive; if $ac - b^2 > 0$ and $a + c < 0$, then both roots are negative.

Since our side condition is $x^2 + y^2 = 1$, we can write $x = \cos\alpha$, $y = \sin\alpha$, and f becomes $a\cos^2\alpha + 2b\sin\alpha\cos\alpha + c\sin^2\alpha$. When both roots are positive, f has a positive minimum and remains positive for all α; when both roots are negative, f has a negative maximum and remains negative for all α. Thus the conditions found provide a new proof of the rules given in Section 2–19 for the function

$$A\cos^2\alpha + 2B\sin\alpha\cos\alpha + C\sin^2\alpha.$$

In this example, we have thus far considered only the nature and significance of the eigenvalues. To find the eigenvectors, and hence to find (x, y) which maximizes or minimizes f, one proceeds as in Section 1–5.

For a general quadratic form f, one can imitate the example and ask for the conditions under which f has a positive minimum or a negative maximum on the unit sphere. We remark that every nonzero vector $\mathbf{x} = (x_1, \ldots, x_n)$ can be written as $k(u_1, \ldots, u_n)$, where $k = |\mathbf{x}|$ and $\mathbf{u} = (u_1, \ldots, u_n)$ is a unit vector. Hence

$$f(x_1, \ldots, x_n) = \sum\sum a_{ij}x_ix_j = k^2\sum\sum a_{ij}u_iu_j = k^2 f(u_1, \ldots, u_n).$$

Accordingly, if f has a positive minimum on the unit sphere, then f is positive for all nonzero $\mathbf{x}$. Conversely, if f is positive for all nonzero $\mathbf{x}$, then f is positive on the unit sphere, in particular, and hence f has a positive minimum on the unit sphere. Similarly, f has a negative maximum on the unit sphere, if and only if f is negative for all nonzero $\mathbf{x}$. A quadratic form which is positive (negative) for all nonzero $\mathbf{x}$ is said to be *positive definite* (*negative definite*). Hence we can say, for example: *f is positive definite if and only if all eigenvalues of A are positive.*

Another way to obtain these results is to choose new Cartesian coordinates $(y_1, \ldots, y_n)$ in E^n (Section 1–14). As pointed out in Section 1–7, for proper choice of the new coordinates, f becomes simply

$$\lambda_1 y_1^2 + \cdots + \lambda_n y_n^2,$$

where $\lambda_1, \ldots, \lambda_n$ are the (*necessarily real*) eigenvalues of A. It is thus clear that f is positive definite precisely when all the λ's are positive.

For further information on this topic, see Chapter 10 of Vol. 1 of the book by Gantmacher listed at the end of the chapter.

For a general function $F(x_1, \ldots, x_n)$, with critical point at $(x_1^0, \ldots, x_n^0)$ the method of Section 2–19 leads us to the quadratic form

$$\sum_{i,j} \frac{\partial^2 F}{\partial x_i \, \partial x_j} u_i u_j,$$

where $(u_1, \ldots, u_n)$ is a unit vector and all derivatives are evaluated at the critical point. If this form is positive definite, then F has a minimum at the critical point. Hence *if all eigenvalues of the matrix* $(\partial^2 F/\partial x_i \partial x_j)$ *are positive, there is a minimum. Similarly, if all eigenvalues are negative, then there is a maximum.*

Problems

1. Locate the critical points of the following functions, classify them, and graph the functions:

(a) $y = x^3 - 3x$, (b) $y = 2 \sin x + \sin 2x$, (c) $y = e^{-x} - e^{-2x}$.

2. Determine the nature of the critical point of $y = x^n$ ($n = 2, 3, \ldots$) at $x = 0$.

3. Determine the absolute maximum and absolute minimum, if they exist, of the following functions:

(a) $y = \cos x, \quad -\dfrac{\pi}{2} \leqq x \leqq \dfrac{\pi}{2}$ (c) $y = \tanh x$, all x

(b) $y = \log x, \quad 0 < x \leqq 1$ (d) $y = \dfrac{x}{1 + x^2}$, all x

4. Find the critical points of the following functions and test for maxima and minima:

(a) $z = \sqrt{1 - x^2 - y^2}$ (d) $z = x^2 - 5xy - y^2$
(b) $z = 1 + x^2 + y^2$ (e) $z = x^2 - 2xy + y^2$
(c) $z = 2x^2 - xy - 3y^2 - 3x + 7y$ (f) $z = x^3 - 3xy^2 + y^3$

5. Find the critical points of the following functions, classify, and graph the level curves of the functions:

(a) $z = e^{-x^2-y^2}$

(b) $z = x^4 - y^4$

(c) $z = \sin x \cosh y$

(d) $z = \dfrac{x}{x^2 + y^2}$

6. Find the critical points of the following functions with given side conditions and test for maxima and minima:

(a) $z = x^2 + 24xy + 8y^2$, where $x^2 + y^2 = 25$,
(b) $w = x + z$, where $x^2 + y^2 + z^2 = 1$,
(c) $w = xyz$, where $x^2 + y^2 = 1$ and $x - z = 0$.

7. Find the point of the curve

$$x^2 - xy + y^2 - z^2 = 1, \quad x^2 + y^2 = 1$$

nearest to the origin: (0, 0, 0).

8. Find the absolute minimum and maximum, if they exist, of the following functions:

(a) $z = \dfrac{1}{1 + x^2 + y^2}$, all (x, y), (b) $z = xy$, $x^2 + y^2 \leqq 1$,
(c) $w = x + y + z$, $x^2 + y^2 + z^2 \leqq 1$.

9. Determine whether the given quadratic form is positive definite:
(a) $3x^2 + 2xy + y^2$,
(b) $x^2 - xy - 2y^2$,
(c) $\frac{5}{3}x_1^2 + \frac{4}{3}x_1x_2 + 2x_2^2 + \frac{4}{3}x_2x_3 + \frac{7}{3}x_3^2$.

10. Prove the validity of the criterion (2–143) for a minimum, under the conditions stated. [Hint: the function $\nabla_\alpha \nabla_\alpha f(x_0, y_0)$ is continuous in α for $0 \leqq \alpha \leqq 2\pi$ and has a minimum M_1 in this interval; by (2–143), $M_1 > 0$. By the Fundamental Lemma of Section 2–6, $\partial z/\partial x$ and $\partial z/\partial y$ have differentials at (x_0, y_0). Show that this implies that

$$\nabla_\alpha f(x, y) = \nabla_\alpha f(x_0, y_0) + s\nabla_\alpha \nabla_\alpha f(x_0, y_0) + \epsilon s = s\nabla_\alpha \nabla_\alpha f(x_0, y_0) + \epsilon s,$$

where $x = x_0 + s \cos \alpha$, $y = y_0 + s \sin \alpha$ $(s > 0)$ and $|\epsilon|$ can be made as small as desired by choosing s sufficiently small. Choose δ so that $|\epsilon| < \frac{1}{2}M_1$ for $0 < s < \delta$ and show that

$$\nabla_\alpha f(x, y) = s[\nabla_\alpha \nabla_\alpha f(x_0, y_0) + \epsilon] > 0 \quad \text{for } 0 < s < \delta.$$

Accordingly, f increases steadily, as one recedes from (x_0, y_0) in the neighborhood of radius δ of (x_0, y_0).]

11. *The method of least squares.* Let 5 numbers e_1, e_2, e_3, e_4, e_5 be given. It is in general impossible to find a quadratic expression $f(x) = ax^2 + bx + c$ such that $f(-2) = e_1$, $f(-1) = e_2$, $f(0) = e_3$, $f(1) = e_4$, $f(2) = e_5$. However, one can try to make the "total square-error"

$$E = (f(-2) - e_1)^2 + (f(-1) - e_2)^2 + (f(0) - e_3)^2 + (f(1) - e_4)^2 + (f(2) - e_5)^2$$

as small as possible. Determine the values of a, b, and c which make E a minimum. This is the method of *least squares,* which is basic in the theory of statistics and in curve-fitting (see Chapter 7).

Answers

1. (a) max. at -1, min. at 1, (b) max. at $\pi/3 + 2n\pi$, min. at $-\pi/3 + 2n\pi$, horiz. infl. at $\pi + 2n\pi$ ($n = 0, \pm 1, \pm 2, \ldots$), (c) max. at log 2.

2. min. for $n = 2, 4, 6, \ldots$, horiz. infl. for $n = 3, 5, 7, \ldots$.

3. (a) max. $= 1$, min. $= 0$, (b) max. $= 0$, (c) no max. or min., (d) max. $= \frac{1}{2}$, min. $= -\frac{1}{2}$.

4. (a) max. at $(0, 0)$, (b) min. at $(0, 0)$, (c) saddle point at $(1, 1)$, (d) saddle point at $(0, 0)$, (e) critical point at every point of the line $y = x$, each point giving a relative min., (f) triple saddle point at $(0, 0)$.

5. (a) max. at $(0, 0)$, (b) saddle point at $(0, 0)$, (c) saddle points at $\pi/2 + 2n\pi$ ($n = 0, \pm 1, \pm 2, \ldots$), (d) no critical points [discontinuity at $(0, 0)$].

6. (a) max. at $(\pm 3, \pm 4)$, min. at $(\pm 4, \mp 3)$, (b) max. at $\left(\frac{1}{\sqrt{2}}, 0, \frac{1}{\sqrt{2}}\right)$, min. at $\left(-\frac{1}{\sqrt{2}}, 0, -\frac{1}{\sqrt{2}}\right)$, (c) max. at $(\pm\sqrt{\frac{2}{3}}, \sqrt{\frac{1}{3}}, \pm\sqrt{\frac{2}{3}})$ and $(0, -1, 0)$, min. at $(\pm\sqrt{\frac{2}{3}}, -\sqrt{\frac{1}{3}}, \pm\sqrt{\frac{2}{3}})$ and $(0, 1, 0)$. 7. $(0, \pm 1, 0)$ and $(\pm 1, 0, 0)$.

8. (a) max. $= 1$, (b) max. $= \frac{1}{2}$, min. $= -\frac{1}{2}$, (c) max. $= \sqrt{3}$, min. $= -\sqrt{3}$.

9. (a) and (c) are positive definite.

11. $a = \frac{1}{14}(2e_1 - e_2 - 2e_3 - e_4 + 2e_5)$, $b = \frac{1}{10}(-2e_1 - e_2 + e_4 + 2e_5)$, $c = \frac{1}{35}(-3e_1 + 12e_2 + 17e_3 + 12e_4 - 3e_5)$.

* **2–22 Functional dependence.** Throughout the present chapter the condition that a derivative or Jacobian be zero has played an important role. Thus in Section 2–10 the condition that the Jacobian *not* be zero was needed to obtain the derivatives of implicit functions; in Section 2–19 the condition that all partial derivatives be zero was used to locate critical points. In the present section we consider these questions from a more general point of view, with emphasis on certain extreme cases which are of importance.

Let $w = f(x, y)$ be given in a domain D. If $\nabla w = (\partial f/\partial x)\,\mathbf{i} + (\partial f/\partial y)\,\mathbf{j}$ is not $\mathbf{0}$ in D, then the level curves: $f(x, y) = \text{const}$ of w are well-defined curves, one through each point of D. This follows from the theorem on implicit functions referred to after (2–73); at each point (x_1, y_1) either $\partial f/\partial x \neq 0$ or $\partial f/\partial y \neq 0$, so that (2–70) gives a definite derivative for the implicit function. The family of level curves is thus without singularity, as shown in Fig. 2–19.

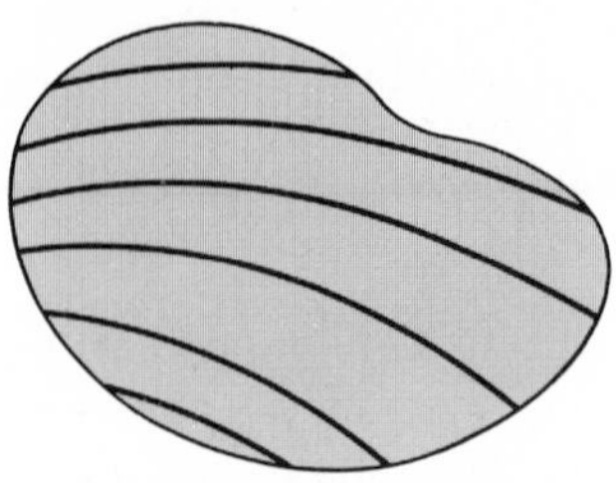

FIG. 2–19. Level curves of $f(x, y)$ with $\nabla f \neq \mathbf{0}$.

Singularities will be introduced if $\nabla w = \mathbf{0}$ at certain points of D. These are precisely the critical points considered in Section 2–19. Figure 2–18 illustrates some of the possible complications.

The extreme case is that in which $\nabla w \equiv \mathbf{0}$ in D, i.e., every point of D is a critical point. Here $\partial f/\partial x \equiv 0$ and $\partial f/\partial y \equiv 0$ in D and one concludes that f is constant in D:

THEOREM. *Let $f(x, y)$ be defined in domain D and let*

$$\frac{\partial f}{\partial x} \equiv 0, \quad \frac{\partial f}{\partial y} \equiv 0 \tag{2–157}$$

in D. Then there is a constant c such that

$$f \equiv c \tag{2–158}$$

in D.

Proof. Let $P_1\colon(x_1, y_1)$ and $P_2\colon(x_2, y_2)$ be two points of D such that the line segment P_1P_2 lies in D. Let $P\colon(x, y)$ vary on this segment and let s be the distance P_1P. The directional derivative of f at P in the direction $\overrightarrow{P_1P_2}$ is then equal to df/ds; since $\partial f/\partial x \equiv 0$, $\partial f/\partial y \equiv 0$, one has $df/ds \equiv 0$. Hence, by the familiar theorem for functions of one variable (Section 0–9) f, as a function of s, is constant on P_1P_2. Thus $f(x_1, y_1) = f(x_2, y_2) = c$ for some c. Every point of D can be joined to P_1 by a broken line (cf. Section 2–2 and Fig. 2–1); by repetition of the argument just given, one concludes that $f(x, y) = f(x_1, y_1) = c$ for every point (x, y) of D. The theorem is now proved.

Now let two functions $u = f(x, y)$, $v = g(x, y)$ be given in D. Suppose that $\nabla f \neq \mathbf{0}$ in D and $\nabla g \neq \mathbf{0}$ in D, so that both functions have well-defined level curves, as in Fig. 2–19. "In general" the two families of level curves determine curvilinear coordinates in D and a mapping of D onto a *domain* in the uv plane; this will be the case if the Jacobian $\partial(f, g)/\partial(x, y) \neq 0$ in D, as a study of the theorem on implicit functions shows. It should be noted that the condition $\partial(f, g)/\partial(x, y) = 0$ is the same as the condition $\nabla f \times \nabla g = \mathbf{0}$, i.e., that ∇f and ∇g are collinear vectors. For

$$\nabla f \times \nabla g = \begin{vmatrix} \mathbf{i} & \mathbf{j} & \mathbf{k} \\ \dfrac{\partial f}{\partial x} & \dfrac{\partial f}{\partial y} & 0 \\ \dfrac{\partial g}{\partial x} & \dfrac{\partial g}{\partial y} & 0 \end{vmatrix} = \mathbf{k}\,\frac{\partial(f, g)}{\partial(x, y)}.$$

When these vectors are collinear, the level curves $f = \text{const}$ and $g = \text{const}$ are tangent, the curvilinear coordinates are interfered with, and the mapping to the uv plane may degenerate.

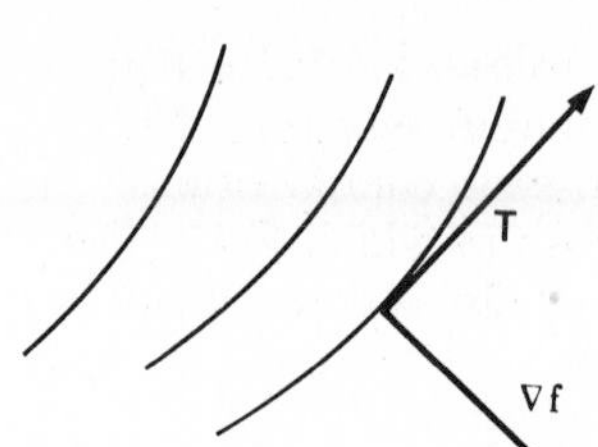

FIG. 2–20. Tangent to level curves.

The extreme case here is that for which $\nabla f \times \nabla g \equiv \mathbf{0}$ in D, or, equivalently, that $\partial(f, g)/\partial(x, y) \equiv 0$ in D. In this case *each level curve of f is a level curve of g* and vice versa. For a tangent vector to a level curve: $f = \text{const}$ is a vector perpendicular to the normal vector $(\partial f/\partial x)\mathbf{i} + (\partial f/\partial y)\mathbf{j}$; thus $\mathbf{T} = -(\partial f/\partial y)\mathbf{i} + (\partial f/\partial x)\mathbf{j}$ is such a tangent vector, as shown in Fig. 2–20. The component of ∇g in the direction of this tangent is 0, since

$$\nabla g \cdot \mathbf{T} = \frac{\partial g}{\partial x}\left(-\frac{\partial f}{\partial y}\right) + \frac{\partial g}{\partial y}\frac{\partial f}{\partial x} = \frac{\partial(f, g)}{\partial(x, y)} \tag{2–159}$$

and the Jacobian is 0 by assumption. Thus the directional derivative of g along the curve is 0 and g must be constant. Therefore, *if $\partial(f, g)/\partial(x, y) \equiv 0$ in D, then the level curves of f and g coincide. Conversely, if f and g have the same level curves, then $\partial(f, g)/\partial(x, y) \equiv 0$ in D*, for the argument just given can be reversed.

EXAMPLE. Let $f(x, y) = e^x \sin y$ and $g(x, y) = x + \log \sin y$ for $0 < x < 1$ and $0 < y < \pi$. Then

$$\frac{\partial(f, g)}{\partial(x, y)} = \begin{vmatrix} e^x \sin y & e^x \cos y \\ 1 & \cot y \end{vmatrix} = e^x \cos y - e^x \cos y \equiv 0.$$

In this example the functions f and g are related by an identity:

$$\log f(x, y) - g(x, y) \equiv 0$$

in the domain considered; that is, g is simply the function $\log f$, a "function of the function." From this fact it is clear that on a level curve $f = \text{const} = c$, one must also have $g = \text{const} = \log c$. In general, two functions f and g related by an identity:

$$F[f(x, y), g(x, y)] \equiv 0 \tag{2–160}$$

in a given domain D, are termed *functionally dependent* in D. Here $F[u, v]$ is a function of the variables u, v such that $F[f(x, y), g(x, y)]$ is defined in D and, in order to rule out degenerate cases, we assume $\nabla F \neq \mathbf{0}$ for the range of u, v involved.

THEOREM. *If $f(x, y)$ and $g(x, y)$ are differentiable in the domain D and are functionally dependent in D, then*

$$\frac{\partial(f, g)}{\partial(x, y)} \equiv 0, \tag{2–161}$$

so that the level curves of f and g coincide. Conversely, if (2–161) holds and $\nabla f \neq \mathbf{0}, \nabla g \neq \mathbf{0}$, then, in some neighborhood of each point of D, f and g are functionally dependent.

Proof. Let f and g be functionally dependent in D, so that (2–160) holds for a suitable $F[u, v]$. On differentiating (2–160) with respect to x and y and using the chain rules, one obtains the identities:

$$\frac{\partial F}{\partial u}\frac{\partial f}{\partial x} + \frac{\partial F}{\partial v}\frac{\partial g}{\partial x} \equiv 0, \quad \frac{\partial F}{\partial u}\frac{\partial f}{\partial y} + \frac{\partial F}{\partial v}\frac{\partial g}{\partial y} \equiv 0. \tag{2–162}$$

Since $\partial F/\partial u$ and $\partial F/\partial v$ are not both zero, these equations are consistent only if the "determinant of the coefficients" is zero (Section 0–3); i.e., only if

$$\begin{vmatrix} \dfrac{\partial f}{\partial x} & \dfrac{\partial g}{\partial x} \\ \dfrac{\partial f}{\partial y} & \dfrac{\partial g}{\partial y} \end{vmatrix} \equiv 0. \tag{2–163}$$

f = a_4, g = b_4

f = a_3, g = b_3

f = a_2, g = b_2

f = a_1, g = b_1

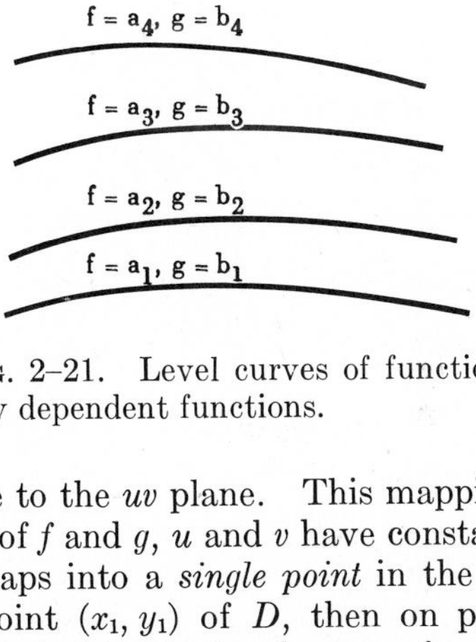

FIG. 2–21. Level curves of functionally dependent functions.

Thus (2–161) holds.

Conversely, let (2–161) hold, so that f and g have the same level curves (Fig. 2–21). The equations

$$u = f(x, y), \quad v = g(x, y) \tag{2–164}$$

then define a mapping from the xy plane to the uv plane. This mapping is degenerate, for along each level curve of f and g, u and v have constant values, so that the entire level curve maps into a *single point* in the uv plane. If one considers a particular point (x_1, y_1) of D, then on proceeding from this point in a direction normal to the level curve, f must either increase or decrease, since $\nabla f \neq 0$; similarly g must either increase or decrease. Thus a sufficiently small neighborhood in the xy plane is mapped by (2–164) on a curve in the uv plane expressible either as $u = \phi(v)$ or $v = \phi(u)$. Thus $f(x, y) - \phi[g(x, y)] \equiv 0$ in the neighborhood and f and g are functionally dependent.

The proof just given brings out the significance of the condition $\partial(f, g)/\partial(x, y) \equiv 0$ for the mapping (2–164): this mapping maps D not onto a domain but onto a *curve* or several curves. For the functions $f = e^x \sin y$, $g = x + \log \sin y$, the corresponding curve is given by part of the graph of $\log u - v = 0$.

The results obtained can be generalized to the case of 3 functions of 3 variables or, in general, to n functions of n variables. Thus, for 3 functions of 3 variables, functional dependence:

$$F[f(x, y, z), g(x, y, z), h(x, y, z)] \equiv 0$$

is equivalent as above to the condition

$$\frac{\partial(f, g, h)}{\partial(x, y, z)} \equiv 0.$$

This is in turn equivalent to the statement that the three vectors $\nabla f, \nabla g, \nabla h$ are coplanar at each point, so that the three families of level surfaces have a common tangent direction at each point.

One can also consider the case of n functions of m variables:

$$f_1(x_1, \ldots, x_m), \ldots, \quad f_n(x_1, \ldots, x_m),$$

where $n \leqq m$. Functional dependence:

$$F[f_1(x_1, \ldots, x_m), \ldots, \quad f_n(x_1, \ldots, x_m)] \equiv 0 \tag{2–165}$$

leads to equations generalizing (2–162):

$$\sum_{j=1}^{m} \frac{\partial F}{\partial u_j} \frac{\partial f_j}{\partial x_i} = 0, \quad i = 1, \ldots, m. \tag{2–166}$$

We assume that $F(u_1, \ldots, u_n)$ is such that its gradient vector

$$\nabla F = (\partial F/\partial u_1, \ldots, \partial F/\partial u_n)$$

is not $\mathbf{0}$ in the domain considered. Now (2–166) can be considered as m homogeneous linear equations in n unknowns, with matrix of coefficients

$$A = \begin{bmatrix} \dfrac{\partial f_1}{\partial x_1} & \cdots & \dfrac{\partial f_n}{\partial x_1} \\ \vdots & & \vdots \\ \dfrac{\partial f_1}{\partial x_m} & \cdots & \dfrac{\partial f_n}{\partial x_m} \end{bmatrix}.$$

This matrix is the *transpose* of the Jacobian matrix $\mathbf{f}_\mathbf{x}$. Since (2–166) holds, with $\nabla F \neq \mathbf{0}$, the kernel of A has dimension k at least 1. By the theorem of Section 1–12, $k + r = n$, where r is the rank of A. Hence $r = n - k < n$. Therefore every minor of A of order equal to n is 0. This is equivalent to the condition:

$$\frac{\partial(f_1, \ldots, f_n)}{\partial(x_{i1}, x_{i2}, \ldots, x_{in})} \equiv 0 \tag{2–167}$$

for all choices of n distinct indices $i_1, i_2, \ldots, i_n$ among the m numbers $1, \ldots, m$. Thus for two functions $f(x, y, z)$, $g(x, y, z)$ of three variables the condition is as follows:

$$\frac{\partial(f, g)}{\partial(x, y)} \equiv 0, \quad \frac{\partial(f, g)}{\partial(y, z)} \equiv 0, \quad \frac{\partial(f, g)}{\partial(x, z)} \equiv 0. \tag{2–168}$$

For $m = 1$, the condition reduces simply to

$$\frac{\partial f}{\partial x} = 0, \quad \frac{\partial f}{\partial y} = 0, \ldots \tag{2-169}$$

and hence, as in the first theorem of this section, to the identity

$$f(x, y, \ldots) \equiv \text{const}, \tag{2-170}$$

which can properly be interpreted as a kind of "functional dependence."

One can prove that if the rank of the matrix A is equal to $n - 1$ throughout D, then in some neighborhood of each point of D the functions $f_1, \ldots, f_n$ are functionally dependent, so that a relation of form (2-165) holds. For more information, see pages 287-295 of the book by Buck listed at the end of the chapter.

If m is greater than n, the question loses much of its interest, for here there is always some form of functional dependence. Thus, given three functions of two variables:

$$u = f(x, y), \quad v = g(x, y), \quad w = h(x, y),$$

one can "in general" eliminate x and y and obtain a single equation:

$$F(u, v, w) = 0.$$

This is equivalent to the statement that the three vectors ∇f, ∇g, ∇h in the xy plane are necessarily coplanar.

* 2-23 Derivatives and differences.

The differential calculus is based on taking limits of expressions of form $\Delta y/\Delta x$ as the increment Δx approaches 0. In some practical problems it may not be feasible to obtain the limit; this is the case, in particular, when the function being differentiated is known only in tabular form for selected values of the independent variables. It is important to be able to compute derivatives approximately in such a case. In this section we consider briefly some special cases of this problem; a more extended discussion is given in Chapter 10.

Let $y = f(x)$ be given for $x = x_1$, $x = x_2$, $x = x_3, \ldots$, as shown in Fig. 2-22. To compute the derivative at x_1, one can use the definition of the derivative thus:

$$f'(x_1) \sim \frac{f(x_2) - f(x_1)}{x_2 - x_1}. \tag{2-171}$$

A $\sim$ sign will be used throughout for an approximate formula. This is equivalent to using the slope of the chord joining the points $[x_1, f(x_1)]$, $[x_2, f(x_2)]$. At the point x_2 a similar formula can be used; however, another possibility suggests itself: namely, to use the slope of

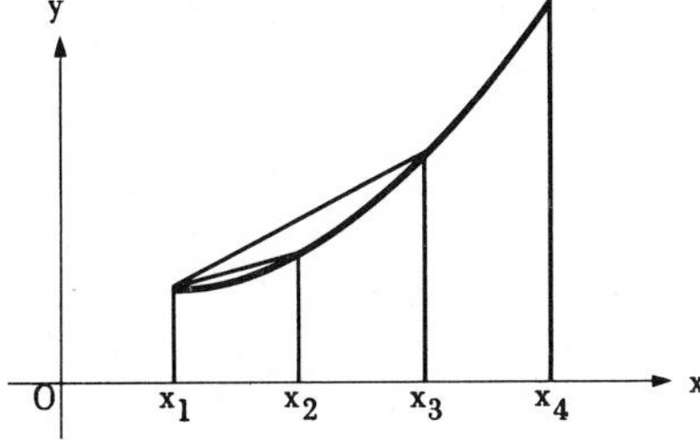

FIG. 2-22. Approximation of derivative.

the chord joining the points to the left and right. One could thus write:

$$f'(x_2) \sim \frac{f(x_3) - f(x_1)}{x_3 - x_1}. \tag{2–172}$$

The reliability of this as an improvement over (2–171) depends on how irregularly the points $x_1, x_2, x_3, \ldots$ are spaced. For the case of equal spacing, at intervals $h = x_2 - x_1 = x_3 - x_2 = \ldots$, (2–172) reduces to the following:

$$f'(x_2) \sim \frac{f(x_2 + h) - f(x_2 - h)}{2h}. \tag{2–173}$$

It can be shown, under appropriate assumptions, that this formula is "in general" considerably better than (2–171). For a convex curve such as that of Fig. 2–22 graphical analysis makes this evident.

In order to evaluate the second derivatives of $f(x)$, one can first of all use the computed values of $f'(x)$, however they were obtained. In the case of equal spacing, the following reasoning can be made. The quotients

$$\frac{f(x_2 + h) - f(x_2)}{h}, \quad \frac{f(x_2) - f(x_2 - h)}{h}$$

give good approximations to $f'(x)$ at the respective *halfway* points $x_2 + \frac{1}{2}h$, $x_2 - \frac{1}{2}h$. Using the principle of (2–172) again, one would thus evaluate $f''(x_2)$ by the formula

$$f''(x_2) \sim \frac{\dfrac{f(x_2 + h) - f(x_2)}{h} - \dfrac{f(x_2) - f(x_2 - h)}{h}}{h},$$

which reduces to

$$f''(x_2) \sim \frac{f(x_2 + h) - 2f(x_2) + f(x_2 - h)}{h^2}. \tag{2–174}$$

It should be remarked that the numerator here is simply the "second difference" of f at x_2, i.e., the difference of the "first differences":

$$f(x_2 + h) - f(x_2), \quad f(x_2) - f(x_2 - h).$$

Higher derivatives can be obtained in similar fashion. The formulas already given can also be used for partial derivatives involving only one independent variable. Thus, if $f(x, y)$ is tabulated at intervals of h in x, one has

$$\frac{\partial^2 f}{\partial x^2} \sim \frac{f(x + h, y) - 2f(x, y) + f(x - h, y)}{h^2} \tag{2–175}$$

and, if f is tabulated at the same interval for y, one obtains for the Laplacian of f the expression

$$\nabla^2 f = \frac{\partial^2 f}{\partial x^2} + \frac{\partial^2 f}{\partial y^2} \sim \frac{f(x+h, y) + f(x, y+h) + f(x-h, y) + f(x, y-h) - 4f(x, y)}{h^2}. \tag{2–176}$$

The Laplace equation in difference form is thus the equation:

$$f(x, y) = \tfrac{1}{4}[f(x + h, y) + f(x, y + h) + f(x - h, y) + f(x, y - h)]. \quad (2\text{–}177)$$

This equation is the basis of most numerical methods for solving the Laplace equation; see Chapter 10. It should be noted that (2–177) states that the value of f at (x, y) is equal to the average of its values at the corners of a square with center at (x, y); the orientation of the square can be shown to be irrelevant here.

To compute a mixed derivative such as $\partial^2 f/\partial x\, \partial y$ one can follow its definition, i.e., first compute $\partial f/\partial y$ and then $\partial/\partial x\, (\partial f/\partial y)$. For $\partial f/\partial y$ at Q and P of Fig. 2–23 one obtains the expressions

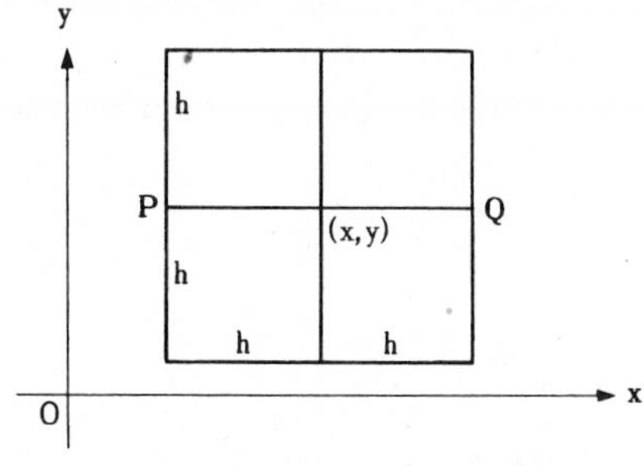

Fig. 2–23. Approximation of second partial derivatives.

$$\frac{f(x + h, y + h) - f(x + h, y - h)}{2h}, \quad \frac{f(x - h, y + h) - f(x - h, y - h)}{2h}.$$

Thus one has the formula

$$\frac{\partial^2 f}{\partial x\, \partial y} \sim \frac{f(x+h, y+h) - f(x+h, y-h) - f(x-h, y+h) + f(x-h, y-h)}{4h^2}. \quad (2\text{–}178)$$

Problems

1. A function $f(x, y)$, defined for all (x, y), satisfies the conditions

$$f(x, 0) = \sin x, \quad \frac{\partial f}{\partial y} \equiv 0.$$

Evaluate $f(\pi/2, 2)$, $f(\pi, 3)$, $f(x, 1)$.

2. Two functions $f(x, y)$ and $g(x, y)$ are such that

$$\nabla f \equiv \nabla g$$

in a domain D. Show that

$$f \equiv g + c$$

for some constant c.

3. Determine all functions $f(x, y)$ whose second partial derivatives are identically 0.

4. A function $f(x, y)$, defined for all (x, y), is such that $\dfrac{\partial f}{\partial y} \equiv 0$. Show that there is a function $g(x)$ such that

$$f(x, y) \equiv g(x).$$

5. Determine all functions $f(x, y)$ such that $\dfrac{\partial^2 f}{\partial x\, \partial y} \equiv 0$ for all (x, y). [Hint: cf. Prob. 4.]

6. Show that the following sets of functions are functionally dependent:

(a) $f = \dfrac{y}{x}, \quad g = \dfrac{x-y}{x+y}$;
(b) $f = x^2 + 2xy + y^2 + 2x + 2y, \quad g = e^x e^y$;
(c) $f = x^2y - xy^2 + xyz, \quad g = xy + x - y + z, \quad h = x^2 + y^2 + z^2 - 2yz + 2xz$;
(d) $f = u + v - x, \quad g = x - y + u, \quad h = u - 2v + 5x - 3y$.

7. Find an identity relating each of the sets of functions of Prob. 6.

8. Plot the level curves of the functions f and g of Prob. 6(a) and (b).

9. Let $f(x, y)$ and $g(x, y, u)$ be such that

$$\frac{\partial f}{\partial x}\frac{\partial g}{\partial y} - \frac{\partial f}{\partial y}\frac{\partial g}{\partial x} = 0$$

when $u = f(x, y)$. Then show that

$$f(x, y) \quad \text{and} \quad g[x, y, f(x, y)]$$

are functionally dependent.

10. Let $u(x, y)$ and $v(x, y)$ be harmonic in a domain D and have no critical points in D. Show that if u and v are functionally dependent, then they are "linearly" dependent: $u = av + b$ for suitable constants a and b. [Hint: assume a relation of form $u = f(v)$ and take the Laplacian of both sides.]

11. Tabulate $f(x) = e^x - x$ for $x = 0, .1, .2, .3, .4, .5$. Compute $f'(x)$ from the tabulated values for $x = 0, .1, .2, .3, .4, .5$. Compute $f''(x)$ from the tabulated values for $x = .1, .2, .3, .4$. Compare with the exact values: $f'(x) = e^x - 1$, $f''(x) = e^x$.

12. Compare the accuracy of (2–145) and (2–147) for $f(x) = ax^2 + bx + c$ and for $f(x) = ax^3 + bx^2 + cx + d$.

13. Graph the function $f(x)$ having the following values: $f(1) = 1$, $f(1.1) = 1.2$, $f(1.2) = 1.1$, $f(1.3) = 1.4$, $f(1.4) = 1.4$, $f(1.5) = 1.6$. Note that the graph is approximately that of $y = x$. Compute $f'(x)$ from the tabulated values. The irregularity obtained illustrates the difficulty of differentiating an empirical function. It is usually preferable to smooth the data before differentiation.

14. Tabulate the function $f(x, y) = x^3y - y^3$ for $x = 0, 1, 2, 3, 4$, $y = 0, 1, 2, 3, 4$ and compute $f_x(2, 2), f_y(2, 2), f_{xx}(2, 2), f_{xy}(2, 2), f_{xxx}(2, 2)$ from the tabulated values. Compare with the exact values.

15. Show that the harmonic function $u = x^2 - y^2$ satisfies the "Laplace equation in difference form" (2–177).

16. Obtain a difference equation for the biharmonic equation (2–128) in (x, y) analogous to (2–177) for the Laplace equation.

Answers

1. 1, 0, $\sin x$. 3. $ax + by + c$, a, b, c arbitrary constants.

5. $f(x, y) = g(x) + h(y)$, $g(x)$ and $h(y)$ being "arbitrary functions."

16. $$\begin{aligned} f(x, y) = \tfrac{1}{20}\{&-2[f(x + h, y + h) + f(x - h, y + h) + f(x - h, y - h) \\ &+ f(x + h, y - h)] + 8[f(x + h, y) + f(x, y + h) \\ &+ f(x - h, y) + f(x, y - h)] - [f(x + 2h, y) + f(x, y + 2h) \\ &+ f(x - 2h, y) + f(x, y - 2h)]\}. \end{aligned}$$

Suggested References

Apostol, Tom M., *Mathematical Analysis: A Modern Approach to Advanced Calculus*. Reading, Mass.: Addison-Wesley, 1957.

Buck, R. C., *Advanced Calculus*, 2nd ed. New York: McGraw-Hill, 1965.

Courant, Richard J., *Differential and Integral Calculus*, transl. by E. J. McShane, 2 vols. New York: Interscience, 1947.

Gantmacher, F. R., *Theory of Matrices*, transl. by K. A. Hirsch, 2 vols. New York: Chelsea Publishing Co., 1960.

CHAPTER 3

Vector Differential Calculus

3–1 Introduction. In Sections 0–7 and 0–9, the path of a moving point P is described by giving its position vector $\mathbf{r} = \overrightarrow{OP}$ as a function of time t. It is then shown how this vector function can be differentiated to give the velocity vector of the moving point and again to give the acceleration vector. These operations can properly be termed a part of "vector differential calculus."

In the present chapter there is again a natural physical model — a fluid in motion. At each point of the fluid one has a velocity vector $\mathbf{v}$, the velocity of the "fluid particle" located at the given point. Such vectors are defined for all points of the fluid and together form what is termed a *vector field*. This is illustrated in Fig. 3–1. The field may change with time or remain the same (*stationary* flow).

One can again trace the paths of the individual particles, the "stream lines," and determine the acceleration vector $\mathbf{a} = d\mathbf{v}/dt$ for each. However, one can also consider the velocity field at a given time as describing a vector $\mathbf{v}$ which is a function of x, y, and z, i.e., of position in space. The vector function $\mathbf{v}(x, y, z)$ can then be differentiated with respect to x, y, and z; that is, one can consider the rate and manner in which $\mathbf{v}$ varies from point to point in space.

The description of the variation of $\mathbf{v}$ turns out to require not merely partial derivatives, but special combinations of these, the *divergence* and *curl*. At each point of the field a scalar: div $\mathbf{v}$, the divergence of $\mathbf{v}$, and a vector: curl $\mathbf{v}$ will be defined. The divergence measures the net rate at which matter is being transported away from the neighborhood of each point and the condition:

$$\text{div } \mathbf{v} \equiv 0$$

describes the *incompressible* flow of a fluid. The curl is essentially a measure of the *angular velocity* of the motion; in the case when the fluid is rotating as a rigid body with angular velocity ω about the z axis, the curl of $\mathbf{v}$ is everywhere equal to $2\omega\mathbf{k}$ (Problem 16 following Section 3–6 below).

There are other important physical examples of vector fields, such as force fields arising from gravitational attraction or from electromagnetic sources; the familiar experiment showing the effect of a magnet on iron filings illustrates the latter.

In many cases, the vectors of the field are parallel to a fixed plane and form the same pattern in each plane parallel to this plane. In this case, the study of the field can be reduced to a two-dimensional problem. One is thus led to study *vector fields* in the plane, as illustrated in Fig. 3–2.

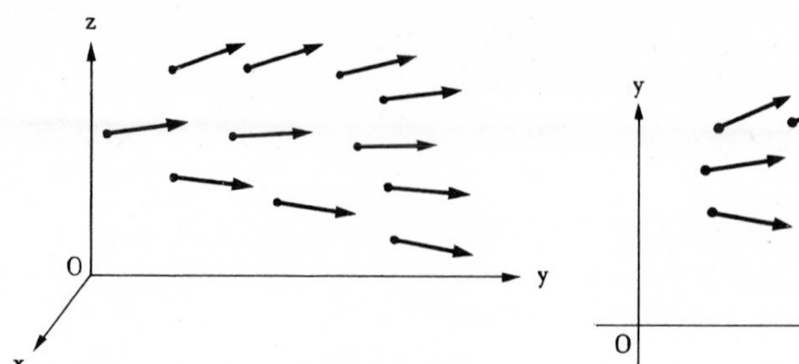

FIG. 3–1. Vector field in space.

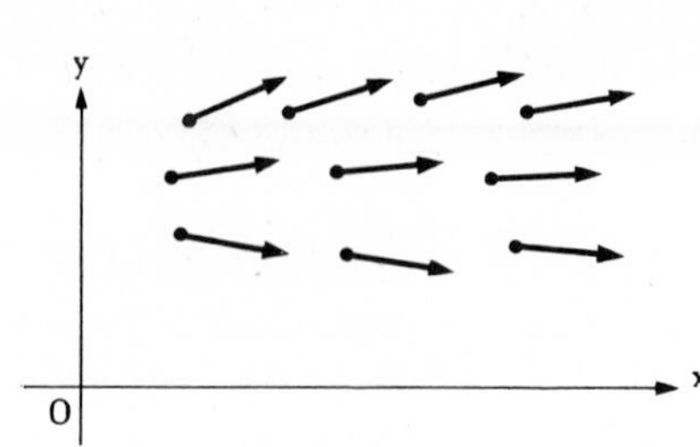

FIG. 3–2. Vector field in the plane.

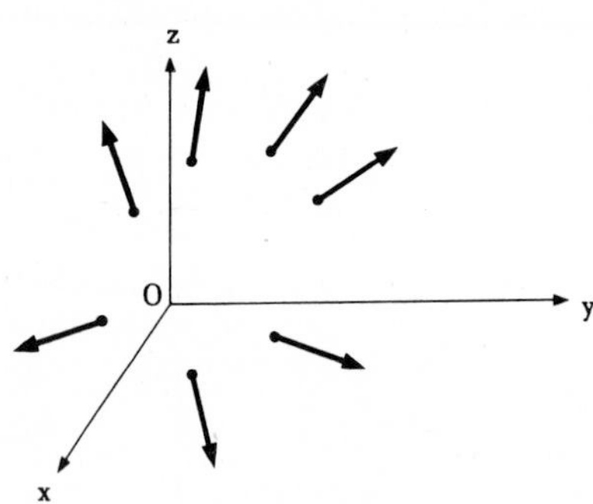

FIG. 3–3.

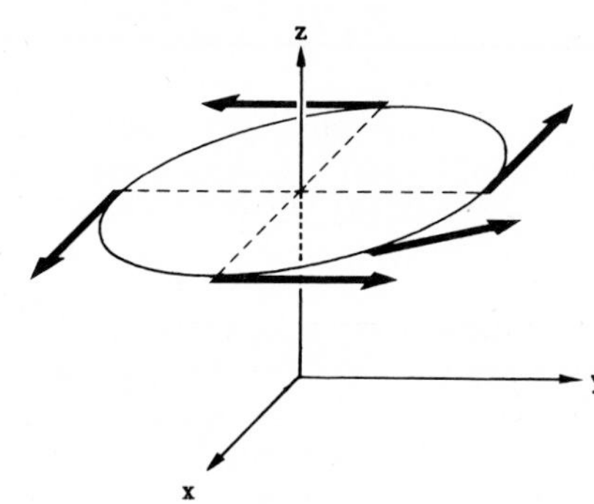

FIG. 3–4.

3–2 Vector fields and scalar fields. If to each point (x, y, z) of a domain D in space a vector $\mathbf{v} = \mathbf{v}(x, y, z)$ is assigned, then a *vector field* is said to be given in D. Each vector $\mathbf{v}$ of the field will be regarded as a *bound vector* (Section 1–2) attached to the corresponding point (x, y, z). If $\mathbf{v}$ is expressed in terms of components: $\mathbf{v} = v_x\mathbf{i} + v_y\mathbf{j} + v_z\mathbf{k}$, then these components will also vary from point to point, so that one has

$$\mathbf{v} = v_x(x, y, z)\mathbf{i} + v_y(x, y, z)\mathbf{j} + v_z(x, y, z)\mathbf{k}, \tag{3–1}$$

that is, each vector field is equivalent to a triple of scalar functions of the three variables x, y, z.

EXAMPLE 1. Let $\mathbf{v} = x\mathbf{i} + y\mathbf{j} + z\mathbf{k}$. Here $\mathbf{v} = \mathbf{0}$ at the origin; at other points $\mathbf{v}$ is a vector pointing away from the origin, as illustrated in Fig. 3–3.

EXAMPLE 2. Let $\mathbf{v} = -y\mathbf{i} + x\mathbf{j}$. Here the vectors can be interpreted as the velocity vectors of a rigid rotation about the z axis. This is illustrated in Fig. 3–4.

EXAMPLE 3. Let $\mathbf{F}$ denote the gravitational force exerted on a particle of mass m at P: (x, y, z) by a mass M concentrated at the origin. Newton's law of gravitation gives

$$\mathbf{F} = -k\frac{Mm}{r^2}\frac{\mathbf{r}}{r}, \tag{3–2}$$

where $\mathbf{r} = \overrightarrow{OP}$ and k is a universal constant. Here $\mathbf{r}/r$ is a unit vector so that $\mathbf{F}$ has magnitude

$$F = \frac{kMm}{r^2}$$

and the force is inversely proportional to the square of the distance.

The notion of vector field can be specialized to two dimensions. Thus a vector field $\mathbf{v}$ in a domain D of the xy plane is given by

$$\mathbf{v} = v_x(x, y)\mathbf{i} + v_y(x, y)\mathbf{j}, \tag{3–3}$$

where $v_x(x, y)$ and $v_y(x, y)$ are two scalar functions of x and y defined in D. Two-dimensional fields arise in applications mainly in connection with *planar problems*, i.e., problems concerning a vector field $\mathbf{v}$ in space such that $\mathbf{v}$ is always parallel to the xy plane and $\mathbf{v}$ is independent of z; that is, $v_z = 0$ and v_x and v_y depend only on x and y, as in (3–3). From (3–3) one can construct the vectors $\mathbf{v}$ first in the xy plane and then, using exactly the same vectors, in any plane parallel to the xy plane.

Example 4. Let $\mathbf{F}$ be the field

$$\mathbf{F} = \frac{1}{[(x+1)^2 + y^2][(x-1)^2 + y^2]}\,[2(x^2 - y^2 - 1)\mathbf{i} + 4xy\mathbf{j}]. \tag{3–4}$$

This is illustrated in Fig. 3–5 and can be interpreted as the electric force field due to two infinite straight wires, perpendicular to the xy plane at $(1, 0)$ and $(-1, 0)$, homogeneously and oppositely charged with electricity.

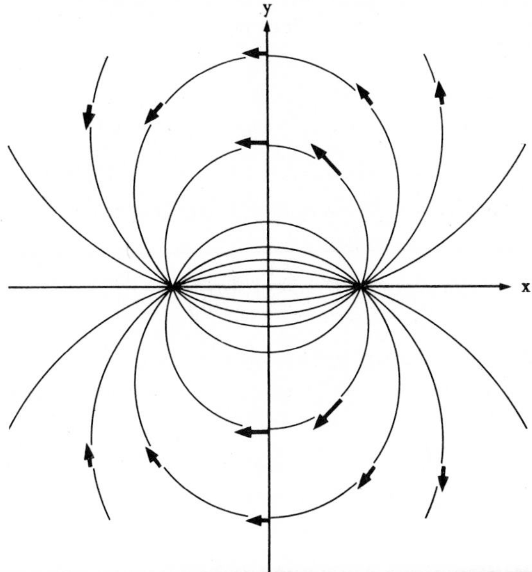

Fig. 3–5.

If we assign to each point of a domain D in space a scalar, rather than a vector, we obtain a *scalar field* in D. For example, the temperature at each point in a room defines a scalar field. If coordinates x, y, z are introduced, the scalar field determines a function $f(x, y, z)$ in D. One can also consider scalar fields in the plane; each such scalar field is described by a function $f(x, y)$.

It will be seen that scalar fields give rise to vector fields (e.g., the field of the vector grad f) and that vector fields give rise to scalar fields (e.g., the field of the scalar $|\mathbf{v}|$).

3–3 The gradient field. Let a scalar field f be given in space and a coordinate system chosen, so that $f = f(x, y, z)$ and is defined in a certain domain of space. If the first partial derivatives of f exist in this domain, then they form the components of the vector grad f, the *gradient* of the scalar f. Thus one has

$$\operatorname{grad} f = \frac{\partial f}{\partial x}\mathbf{i} + \frac{\partial f}{\partial y}\mathbf{j} + \frac{\partial f}{\partial z}\mathbf{k}. \tag{3–5}$$

For example, if $f = x^2y - z^2$, then

$$\operatorname{grad} f = 2xy\mathbf{i} + x^2\mathbf{j} - 2z\mathbf{k}.$$

The formula (3–5) can be written in the following symbolic form:

$$\operatorname{grad} f = \left(\frac{\partial}{\partial x}\mathbf{i} + \frac{\partial}{\partial y}\mathbf{j} + \frac{\partial}{\partial z}\mathbf{k}\right)f, \tag{3–6}$$

where the suggested multiplication actually leads to a differentiation. The expression in parentheses is denoted by the symbol ∇ and is called "del" or "nabla." Thus

$$\nabla \equiv \frac{\partial}{\partial x}\mathbf{i} + \frac{\partial}{\partial y}\mathbf{j} + \frac{\partial}{\partial z}\mathbf{k}; \tag{3–7}$$

∇ is a "vector differential operator." By itself, the ∇ has no numerical significance; it takes on such significance when applied to a function, i.e., in forming

$$\nabla f \equiv \operatorname{grad} f \equiv \frac{\partial f}{\partial x}\mathbf{i} + \frac{\partial f}{\partial y}\mathbf{j} + \frac{\partial f}{\partial z}\mathbf{k}. \tag{3–8}$$

The operator ∇ will be shown to be exceedingly useful.

It was shown in Section 2–10 that the directional derivative of the scalar f in the direction of the unit vector $\mathbf{u} = \cos\alpha\mathbf{i} + \cos\beta\mathbf{j} + \cos\gamma\mathbf{k}$ is given by

$$\nabla_u f = \nabla f \cdot \mathbf{u} = \frac{\partial f}{\partial x}\cos\alpha + \frac{\partial f}{\partial y}\cos\beta + \frac{\partial f}{\partial z}\cos\gamma. \tag{3–9}$$

This shows that grad f has a meaning independent of the coordinate system chosen: *its component in a given direction represents the rate of change* *of f in that direction.* In particular, grad f points in the direction of maximum increase of f.

The gradient obeys the following laws:

$$\text{grad}\,(f+g) = \text{grad}\,f + \text{grad}\,g, \tag{3–10}$$
$$\text{grad}\,(fg) = f\,\text{grad}\,g + g\,\text{grad}\,f; \tag{3–11}$$

i.e., with the ∇ symbol,

$$\nabla(f+g) = \nabla f + \nabla g, \quad \nabla(fg) = f\,\nabla g + g\,\nabla f. \tag{3–12}$$

These hold, provided grad f and grad g exist in the domain considered. The proofs are left for the problems.

If f is a constant c, (3–11) reduces to the simpler condition:

$$\text{grad}\,(cg) = c\,\text{grad}\,g \quad (c = \text{constant}). \tag{3–13}$$

If the terms in z are dropped, the preceding discussion specializes at once to two dimensions. Thus for $f = f(x, y)$ one has

$$\begin{aligned} \text{grad}\,f \equiv \nabla f &\equiv \frac{\partial f}{\partial x}\mathbf{i} + \frac{\partial f}{\partial y}\mathbf{j}, \\ \nabla &\equiv \frac{\partial}{\partial x}\mathbf{i} + \frac{\partial}{\partial y}\mathbf{j}. \end{aligned} \tag{3–14}$$

Problems

1. Sketch the following vector fields:
(a) $\mathbf{v} = (x^2 - y^2)\mathbf{i} + 2xy\mathbf{j}$ (c) $\mathbf{v} = -y\mathbf{i} + x\mathbf{j} + \mathbf{k}$
(b) $\mathbf{u} = (x - y)\mathbf{i} + (x + y)\mathbf{j}$

2. Sketch the level curves or surfaces of the following scalar fields:
(a) $f = xy$, (b) $f = x^2 + y^2 - z^2$.

3. Determine grad f for the scalar fields of Prob. 2 and sketch several of the corresponding vectors.

4. Show that the gravitational field (3–2) is the gradient of the scalar

$$f = \frac{kMm}{r}.$$

5. Show that the force field (3–4) is the gradient of the scalar

$$f = \log\frac{\sqrt{(x-1)^2 + y^2}}{\sqrt{(x+1)^2 + y^2}}.$$

6. Prove (3–10) and (3–11).

3–4 The divergence of a vector field. Given a vector field $\mathbf{v}$ in a domain D of space, one has (for a given coordinate system) three scalar functions v_x, v_y, v_z. If these possess first partial derivatives in D, we can form in all nine partial derivatives, which we arrange to form a matrix:

$$\begin{matrix} \frac{\partial v_x}{\partial x} & \frac{\partial v_x}{\partial y} & \frac{\partial v_x}{\partial z} \\ \frac{\partial v_y}{\partial x} & \frac{\partial v_y}{\partial y} & \frac{\partial v_y}{\partial z} \\ \frac{\partial v_z}{\partial x} & \frac{\partial v_z}{\partial y} & \frac{\partial v_z}{\partial z}. \end{matrix}$$

From three of these the scalar div **v**, the *divergence* of **v**, is constructed by the formula:

$$\operatorname{div} \mathbf{v} = \frac{\partial v_x}{\partial x} + \frac{\partial v_y}{\partial y} + \frac{\partial v_z}{\partial z}. \tag{3–15}$$

It will be noted that the derivatives used form a diagonal (*principal* diagonal) of the matrix.

For example, if $\mathbf{v} = x^2\mathbf{i} - xy\mathbf{j} + xyz\mathbf{k}$, then

$$\operatorname{div} \mathbf{v} = 2x - x + xy = x + xy.$$

Formula (3–15) can be written in the symbolic form:

$$\operatorname{div} \mathbf{v} = \nabla \cdot \mathbf{v}; \tag{3–16}$$

for, treating ∇ as a vector, one has

$$\begin{aligned} \nabla \cdot \mathbf{v} &= \left(\frac{\partial}{\partial x}\mathbf{i} + \frac{\partial}{\partial y}\mathbf{j} + \frac{\partial}{\partial z}\mathbf{k}\right) \cdot (v_x\mathbf{i} + v_y\mathbf{j} + v_z\mathbf{k}) \\ &= \frac{\partial v_x}{\partial x} + \frac{\partial v_y}{\partial y} + \frac{\partial v_z}{\partial z} = \operatorname{div} \mathbf{v}. \end{aligned}$$

The definition of the divergence at first appears to be quite arbitrary and to depend on the choice of axes in space. It will be seen in Section 3–8 below that this is not the case. The divergence has in fact a definite physical significance. In fluid dynamics it appears as a measure of the rate of decrease of density at a point. More precisely, let $\mathbf{u} = \mathbf{u}(x, y, z, t)$ denote the velocity vector of a fluid motion and let $\rho = \rho(x, y, z, t)$ denote the density. Then $\mathbf{v} = \rho\mathbf{u}$ is a vector whose divergence satisfies the equation

$$\operatorname{div} \mathbf{v} = -\frac{\partial \rho}{\partial t}. \tag{3–17}$$

This is in fact the "continuity equation" of fluid mechanics. If the fluid is incompressible, this reduces to the simpler equation

$$\operatorname{div} \mathbf{u} = 0. \tag{3–18}$$

The law (3–17) will be established in Chapter 5; the derivation of (3–18) from (3–17) is considered in Prob. 2 below.

The divergence also plays an important part in the theory of electromagnetic fields. Here the divergence of the electric force vector **E** satisfies the equation

$$\operatorname{div} \mathbf{E} = 4\pi\rho, \tag{3–19}$$

where ρ is the charge density. Thus, where there is no charge, one has

$$\operatorname{div} \mathbf{E} = 0. \tag{3–20}$$

The divergence has the basic properties:

$$\operatorname{div}(\mathbf{u} + \mathbf{v}) = \operatorname{div} \mathbf{u} + \operatorname{div} \mathbf{v}, \tag{3–21}$$

$$\operatorname{div}(f\mathbf{u}) = f \operatorname{div} \mathbf{u} + \operatorname{grad} f \cdot \mathbf{u}; \tag{3–22}$$

i.e., with the nabla symbol:

$$\nabla \cdot (\mathbf{u} + \mathbf{v}) = \nabla \cdot \mathbf{u} + \nabla \cdot \mathbf{v}, \quad \nabla \cdot (f\mathbf{u}) = f(\nabla \cdot \mathbf{u}) + (\nabla f \cdot \mathbf{u}).$$

The proofs are left to the problems.

3–5 The curl of a vector field. From the remaining six partial derivatives of the square array of Section 3–4, one constructs a new vector field, curl **v**, by the definition:

$$\operatorname{curl} \mathbf{v} = \left(\frac{\partial v_z}{\partial y} - \frac{\partial v_y}{\partial z}\right)\mathbf{i} + \left(\frac{\partial v_x}{\partial z} - \frac{\partial v_z}{\partial x}\right)\mathbf{j} + \left(\frac{\partial v_y}{\partial x} - \frac{\partial v_x}{\partial y}\right)\mathbf{k}. \tag{3–23}$$

It will be noted that each component is formed of elements symmetrically placed relative to the principal diagonal. The curl can be expressed in terms of ∇, for one has

$$\operatorname{curl} \mathbf{v} = \nabla \times \mathbf{v} = \begin{vmatrix} \mathbf{i} & \mathbf{j} & \mathbf{k} \\ \dfrac{\partial}{\partial x} & \dfrac{\partial}{\partial y} & \dfrac{\partial}{\partial z} \\ v_x & v_y & v_z \end{vmatrix}. \tag{3–24}$$

The determinant must be expanded by minors of the first row, i.e., so as to yield (3–23).

The fact that this vector field has a meaning independent of the choice of axes will also be shown in Section 3–8. The curl is important in the analysis of the velocity fields of fluid dynamics and in the analysis of electromagnetic force fields. The curl can be interpreted as measuring angular motion of a fluid (see Prob. 16 below) and the condition

$$\operatorname{curl} \mathbf{v} = \mathbf{0} \tag{3–25}$$

for a velocity field **v** characterizes what are termed *irrotational flows*. The analogous equation

$$\operatorname{curl} \mathbf{E} = \mathbf{0} \tag{3–26}$$

for the electric force vector **E** holds when only electrostatic forces are present.

The curl satisfies the basic laws:

$$\text{curl}\,(\mathbf{u} + \mathbf{v}) = \text{curl}\,\mathbf{u} + \text{curl}\,\mathbf{v}, \tag{3–27}$$
$$\text{curl}\,(f\mathbf{u}) = f\,\text{curl}\,\mathbf{u} + \text{grad}\,f \times \mathbf{u}. \tag{3–28}$$

The proofs are left to the problems.

3–6 Combined operations. As a result of the new definitions, we now have at our disposal the operations listed in Table 3–1.

TABLE 3–1

	OPERATION	SYMBOLS	WHERE DISCUSSED (SECTION NO.)
Algebraic Operations	(a) sum of scalars	$f + g$	0–1
	(b) product of scalars	fg	0–1
	(c) sum of vectors	$\mathbf{u} + \mathbf{v}$	0–6
	(d) scalar times vector	$f\mathbf{u}$	0–6
	(e) scalar product	$\mathbf{u} \cdot \mathbf{v}$	0–6
	(f) vector product	$\mathbf{u} \times \mathbf{v}$	0–6
Differential Operations	(g) derivative of scalar	$\frac{df}{dt}, \frac{\partial f}{\partial x}$	0–9, 2–5
	(h) derivative of vector	$\frac{d\mathbf{v}}{dt}$	0–9
	(i) gradient of scalar	$\nabla f \equiv \text{grad}\,f$	2–12, 3–3
	(j) divergence of vector	$\nabla \cdot \mathbf{v} \equiv \text{div}\,\mathbf{v}$	3–4
	(k) curl of vector	$\nabla \times \mathbf{v} \equiv \text{curl}\,\mathbf{v}$	3–5

The theory of vector algebra, discussed in Section 0–6, concerns the properties of the algebraic operations and their combinations. The theory of vector differential calculus concerns the theory of the differential operations (g) to (k) and their combinations with each other and with the algebraic operations (a) to (f).

The combinations of (g) and (h) with the algebraic operations are discussed in Section 0–9.

The combinations of (i), (j), and (k) with (a) and (c) are discussed in Sections 3–3, 3–4, and 3–5. The results can be summarized in the one rule:

operator on sum = sum of operators on terms. (3–29)

The combinations of (i), (j), (k) with (b) and (d) are also discussed in Sections 3–3 to 3–5. The results include the important case of scalar constant times scalar or vector. Here we have the general rule:

operator on scalar constant factor = scalar factor times operator; (3–30)

i.e., a scalar constant can be factored out. Thus $\nabla(cf) = c\,\nabla f$, $\nabla \cdot (c\mathbf{u}) = c\nabla \cdot \mathbf{u}$, etc. The rules (3–29) and (3–30) characterize what are called *linear operators;* thus grad, div, and curl are linear operators.

If one considers the other possible combinations, one obtains a long list of identities, some of which will be considered here. The proofs are left to the problems. All derivatives occurring are assumed continuous.

Curl of a gradient. Here one has the rule:

$$\text{curl grad } f = \mathbf{0}. \tag{3-31}$$

This relation is suggested by the fact that curl grad $f = \nabla \times (\nabla f)$, i.e., has the appearance of the vector product of collinear vectors. There is an important converse:

$$\text{if curl } \mathbf{v} = \mathbf{0}, \quad \text{then} \quad \mathbf{v} = \text{grad } f \text{ for some } f; \tag{3-32}$$

further assumptions are needed here and the rule (3-32) must be used with caution. A proof and full discussion are given in Chapter 5. A vector field $\mathbf{v}$ such that curl $\mathbf{v} = \mathbf{0}$ is often termed *irrotational.*

Divergence of a curl. Here one concludes that

$$\text{div curl } \mathbf{v} = 0; \tag{3-33}$$

this relation is again suggested by a vector identity, for div curl $\mathbf{v} = \nabla \cdot (\nabla \times \mathbf{v})$, so that one has an expression resembling a scalar triple product of coplanar vectors (Section 0-6). Again there is a converse:

$$\text{if div } \mathbf{u} = 0, \quad \text{then} \quad \mathbf{u} = \text{curl } \mathbf{v} \text{ for some } \mathbf{v}; \tag{3-34}$$

as with (3-32) these are restrictions on the use of (3-34) and one is again referred to Chapter 5. A vector field $\mathbf{u}$ such that div $\mathbf{u} = 0$ is often termed *solenoidal*.

Divergence of a vector product. Here one has

$$\text{div } (\mathbf{u} \times \mathbf{v}) = \mathbf{v} \cdot \text{curl } \mathbf{u} - \mathbf{u} \cdot \text{curl } \mathbf{v}. \tag{3-35}$$

Divergence of a gradient. If one expands in terms of components, one finds that

$$\text{div grad } f = \frac{\partial^2 f}{\partial x^2} + \frac{\partial^2 f}{\partial y^2} + \frac{\partial^2 f}{\partial z^2}. \tag{3-36}$$

The expression on the right is known as the Laplacian of f and is also denoted by Δf or by $\nabla^2 f$, since div grad $f = \nabla \cdot (\nabla f)$. A function f (having continuous second derivatives) such that div grad $f = 0$ in a domain is called *harmonic* in that domain. The equation satisfied by f:

$$\frac{\partial^2 f}{\partial x^2} + \frac{\partial^2 f}{\partial y^2} + \frac{\partial^2 f}{\partial z^2} = 0 \tag{3-37}$$

is called *Laplace's equation* (see Sections 2-15 and 2-17).

Curl of a curl. Here an expansion into components yields the relation:

$$\text{curl curl } \mathbf{u} = \text{grad div } \mathbf{u} - (\nabla^2 u_x \mathbf{i} + \nabla^2 u_y \mathbf{j} + \nabla^2 u_z \mathbf{k}). \tag{3-38}$$

If one defines the Laplacian of a vector $\mathbf{u}$ to be the vector

$$\nabla^2 \mathbf{u} = \nabla^2 u_x \mathbf{i} + \nabla^2 u_y \mathbf{j} + \nabla^2 u_z \mathbf{k}, \tag{3-39}$$

then (3–38) becomes

$$\text{curl curl } \mathbf{u} = \text{grad div } \mathbf{u} - \nabla^2\mathbf{u}. \tag{3–40}$$

This identity can be written as an expression for the *gradient of a divergence:*

$$\text{grad div } \mathbf{u} = \text{curl curl } \mathbf{u} + \nabla^2\mathbf{u}. \tag{3–41}$$

The identities listed here, together with those previously obtained, cover all of interest except for those for *gradient of a scalar product* and *curl of a vector product;* these two are considered in Probs. 13 and 14 below.

Problems

1. Prove (3–21) and (3–22).

2. Prove that the continuity equation (3–17) can be written in the form

$$\frac{\partial \rho}{\partial t} + \text{grad } \rho \cdot \mathbf{u} + \rho \text{ div } \mathbf{u} = 0$$

or, in terms of the Stokes derivative (Prob. 10 following Section 2–8), thus:

$$\frac{D\rho}{Dt} + \rho \text{ div } \mathbf{u} = 0.$$

Prove that (3–17) reduces to (3–18) when $\rho \equiv \text{const.}$ It will be shown in Chapter 5 that the same simplification can be made when ρ is variable, provided the fluid is incompressible. This follows from the fact that $D\rho/Dt$ measures the variation in density at a point moving with the fluid; for an incompressible fluid, this local density cannot vary.

3. Prove (3–27) and (3–28).

4. Prove (3–31). Verify by applying to $f = \dfrac{1}{\sqrt{x^2 + y^2 + z^2}}$.

5. Given the vector field $\mathbf{v} = 2xyz\mathbf{i} + x^2z\mathbf{j} + x^2y\mathbf{k}$, verify that curl $\mathbf{v} = \mathbf{0}$. Find all functions f such that grad $f = \mathbf{v}$.

6. Prove (3–33). Verify by applying to $\mathbf{v} = x^2yz\mathbf{i} - x^3y^3\mathbf{j} + xyz^3\mathbf{k}$.

7. Given the vector field $\mathbf{v} = 2x\mathbf{i} + y\mathbf{j} - 3z\mathbf{k}$, verify that div $\mathbf{v} = 0$. Find all vectors $\mathbf{u}$ such that curl $\mathbf{u} = \mathbf{v}$. [Hint: first remark that, on the basis of (3–32), all solutions of the equation curl $\mathbf{u} = \mathbf{v}$ are given by $\mathbf{u} = \mathbf{u}_0 + \text{grad} f$, where f is an arbitrary scalar and $\mathbf{u}_0$ is any one vector whose curl is $\mathbf{v}$. To find $\mathbf{u}_0$, assume $\mathbf{u}_0 \cdot \mathbf{k} = 0$.]

8. Prove (3–36). Verify that the function f of Prob. 4 is harmonic in space (except at the origin). [This function, which represents the electrostatic potential of a charge of $+1$ at the origin, is in a sense the fundamental harmonic function in space, for every harmonic function in space can be represented as a sum, or limit of a sum, of such functions.]

9. Prove (3–35).

10. Prove (3–38).

11. Prove the following identities:

(a) div $[\mathbf{u} \times (\mathbf{v} \times \mathbf{w})] = (\mathbf{u} \cdot \mathbf{w})$ div $\mathbf{v} - (\mathbf{u} \cdot \mathbf{v})$ div $\mathbf{w}$ + grad $(\mathbf{u} \cdot \mathbf{w}) \cdot \mathbf{v}$ − grad $(\mathbf{u} \cdot \mathbf{v}) \cdot \mathbf{w}$,

(b) div (grad $f \times f$ grad g) $= 0$,

(c) curl (curl $\mathbf{v}$ + grad f) = curl curl $\mathbf{v}$, (d) $\nabla^2 f$ = div (curl $\mathbf{v}$ + grad f).

These should be established by means of the identities already found in this chapter and not by expanding into components.

12. One defines the scalar product $\mathbf{u} \cdot \nabla$, with $\mathbf{u}$ on the *left* of the operator ∇, as the operator

$$\mathbf{u} \cdot \nabla = u_x \frac{\partial}{\partial x} + u_y \frac{\partial}{\partial y} + u_z \frac{\partial}{\partial z}.$$

This is thus quite unrelated to $\nabla \cdot \mathbf{u} = \text{div } \mathbf{u}$. The operator $\mathbf{u} \cdot \nabla$ can be applied to a scalar f:

$$(\mathbf{u} \cdot \nabla)f = u_x \frac{\partial f}{\partial x} + u_y \frac{\partial f}{\partial y} + u_z \frac{\partial f}{\partial z} = \mathbf{u} \cdot (\nabla f);$$

thus an associative law holds. The operator $\mathbf{u} \cdot \nabla$ can also be applied to a vector $\mathbf{v}$:

$$(\mathbf{u} \cdot \nabla)\mathbf{v} = u_x \frac{\partial \mathbf{v}}{\partial x} + u_y \frac{\partial \mathbf{v}}{\partial y} + u_z \frac{\partial \mathbf{v}}{\partial z},$$

where the partial derivatives $\partial \mathbf{v}/dt$, . . . are defined just as is $d\mathbf{v}/dt$ in Section 0–9; thus one has

$$\frac{\partial \mathbf{v}}{\partial x} = \frac{\partial v_x}{\partial x}\mathbf{i} + \frac{\partial v_y}{\partial x}\mathbf{j} + \frac{\partial v_z}{\partial x}\mathbf{k}.$$

(a) Show that, if $\mathbf{u}$ is a unit vector, then $(\mathbf{u} \cdot \nabla)f = \nabla_{\mathbf{u}} f$.
(b) Evaluate $[(\mathbf{i} - \mathbf{j}) \cdot \nabla]f$.
(c) Evaluate $[(x\mathbf{i} - y\mathbf{j}) \cdot \nabla](x^2\mathbf{i} - y^2\mathbf{j} + z^2\mathbf{k})$.

13. Prove the identity (cf. Prob. 12):

$$\text{grad } (\mathbf{u} \cdot \mathbf{v}) = (\mathbf{u} \cdot \nabla)\mathbf{v} + (\mathbf{v} \cdot \nabla)\mathbf{u} + (\mathbf{u} \times \text{curl } \mathbf{v}) + (\mathbf{v} \times \text{curl } \mathbf{u}).$$

14. Prove the identity (cf. Prob. 12):

$$\text{curl } (\mathbf{u} \times \mathbf{v}) = \mathbf{u} \text{ div } \mathbf{v} - \mathbf{v} \text{ div } \mathbf{u} + (\mathbf{v} \cdot \nabla)\mathbf{u} - (\mathbf{u} \cdot \nabla)\mathbf{v}.$$

15. Let $\mathbf{n}$ be the unit outer normal vector to the sphere $x^2 + y^2 + z^2 = 9$ and let $\mathbf{u}$ be the vector $(x^2 - z^2)(\mathbf{i} - \mathbf{j} + 3\mathbf{k})$. Evaluate $\partial/\partial n$ (div $\mathbf{u}$) at (2, 2, 1).

16. A rigid body is rotating about the z-axis with angular velocity ω. Show that a typical particle of the body follows a path

$$\overrightarrow{OP} = r \cos (\omega t + \alpha)\mathbf{i} + r \sin (\omega t + \alpha)\mathbf{j} + z\mathbf{k},$$

where α, r and z are constant, and that at each instant the velocity is

$$\mathbf{v} = \boldsymbol{\omega} \times \overrightarrow{OP},$$

where $\boldsymbol{\omega} = \omega\mathbf{k}$ (the *angular velocity vector* of the motion). Evaluate div $\mathbf{v}$ and curl $\mathbf{v}$.

17. A steady fluid motion has velocity $\mathbf{u} = y\mathbf{i}$. Show that all points which move do so on straight lines and that the flow is incompressible. Determine the volume occupied at time $t = 1$ by the points which at time $t = 0$ fill the cube bounded by the coordinate planes and the planes $x = 1$, $y = 1$, $z = 1$.

18. A steady fluid motion has velocity $\mathbf{u} = x\mathbf{i}$. Show that all points either do not move or else move on straight lines. Determine the volume occupied at time $t = 1$ by the points which at time $t = 0$ fill the cube of Prob. 17. [Hint: show that the paths of the individual points are given by $x = c_1e^t$, $y = c_2$, $z = c_3$, where c_1, c_2, c_3 are constants.] Is the flow incompressible?

Answers

5. $x^2yz + \text{const.}$ 7. $yz\mathbf{i} - 2xz\mathbf{j} + \operatorname{grad} f$, f arbitrary.
12. (b) $\dfrac{\partial f}{\partial x} - \dfrac{\partial f}{\partial y}$, (c) $2x^2\mathbf{i} + 2y^2\mathbf{j}$. 15. $-\frac{2}{3}$. 16. $\operatorname{div} \mathbf{v} = 0$, $\operatorname{curl} \mathbf{v} = 2\boldsymbol{\omega}$. 17. Vol. = 1. 8. Vol. = e.

*** 3–7 Curvilinear coordinates in space. Orthogonal coordinates.** The discussion of the preceding sections has been confined to a fixed rectangular coordinate system in space. We now consider the extension of the theory to curvilinear coordinates, such as cylindrical or spherical coordinates.

New coordinates u, v, w can be introduced in space by equations

$$x = f(u, v, w), \quad y = g(u, v, w), \quad z = h(u, v, w). \tag{3–42}$$

It will be assumed that f, g, and h are defined and have continuous first partial derivatives in a domain D_1 of uvw space and that equations (3–42) can be solved uniquely for u, v, and w:

$$u = F(x, y, z), \quad v = G(x, y, z), \quad w = H(x, y, z). \tag{3–43}$$

The inverse functions will be defined in a domain D of xyz space and we term u, v, w *curvilinear coordinates* in D. We assume that the Jacobian

$$J = \frac{\partial(x, y, z)}{\partial(u, v, w)} \tag{3–44}$$

is positive in D_1.

If v and w are assigned constant values v_0 and w_0, while u is allowed to vary, equations (3–42) define a curve in D, on which u is parameter. As the numbers v_0 and w_0 are varied, one obtains a family of curves in D, like the parallels to one of the axes in rectangular coordinates. For example, in spherical coordinates ρ, ϕ, θ, the curves $\phi = \phi_0$, $\theta = \theta_0$ (ρ variable) are rays through the origin.

If we write $\mathbf{r} = x\mathbf{i} + y\mathbf{j} + z\mathbf{k}$ for the position vector of a point (x, y, z), the equations (3–42) can be interpreted as defining a vector function $\mathbf{r} = \mathbf{r}(u, v, w)$. When $v = v_0$, $w = w_0$, this is the vector representation $\mathbf{r} = \mathbf{r}(u, v_0, w_0)$ of the curve of the preceding paragraph. The tangent vector to this curve is defined as in Sections 0–9 and 2–12 to be the derivative of $\mathbf{r}$ with respect to the parameter u. Here we must write $\partial\mathbf{r}/\partial u$ for the derivative, to indicate that v and w are held constant. Similarly, $\partial\mathbf{r}/\partial v$ is tangent to a curve $u = \text{const}$, $w = \text{const}$ and $\partial\mathbf{r}/\partial w$ is tangent to a curve $u = \text{const}$, $v = \text{const}$. This is illustrated in Fig. 3–6.

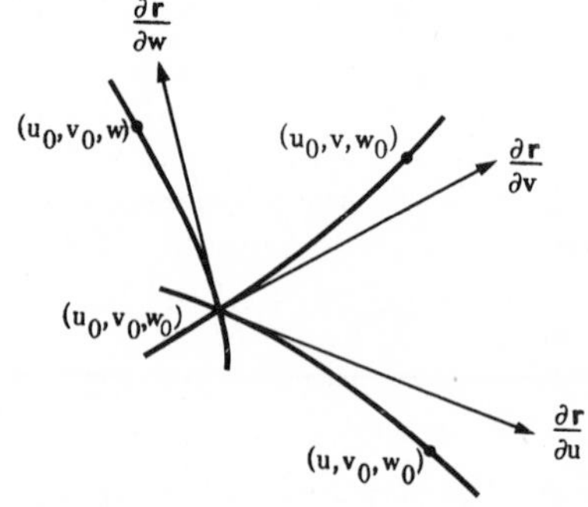

FIG. 3–6. Curvilinear coordinates in space.

We write:

$$\alpha = \left|\frac{\partial \mathbf{r}}{\partial u}\right|, \quad \beta = \left|\frac{\partial \mathbf{r}}{\partial v}\right|, \quad \gamma = \left|\frac{\partial \mathbf{r}}{\partial w}\right|. \tag{3–45}$$

Accordingly,

$$\alpha = \sqrt{\left(\frac{\partial x}{\partial u}\right)^2 + \left(\frac{\partial y}{\partial u}\right)^2 + \left(\frac{\partial z}{\partial u}\right)^2}$$

is the "speed," in terms of time u, with which a curve $v = v_0$, $w = w_0$ is traced, and $ds = \alpha\, du$ is the element of distance.

The tangent vectors $\partial \mathbf{r}/\partial u$, $\partial \mathbf{r}/\partial v$, $\partial \mathbf{r}/\partial w$ and Jacobian J can be expressed in terms of x, y, and z by the equations

$$\frac{\partial \mathbf{r}}{\partial u} = J(\nabla G \times \nabla H), \quad \frac{\partial \mathbf{r}}{\partial v} = J(\nabla H \times \nabla F), \quad \frac{\partial \mathbf{r}}{\partial w} = J(\nabla F \times \nabla G), \tag{3–46}$$

$$J = \frac{1}{\dfrac{\partial(u, v, w)}{\partial(x, y, z)}} = \frac{1}{\nabla F \cdot \nabla G \times \nabla H}. \tag{3–47}$$

The gradients ∇F, ∇G, ∇H can be expressed in terms of u, v, w by the equations

$$\nabla F = \frac{\dfrac{\partial \mathbf{r}}{\partial v} \times \dfrac{\partial \mathbf{r}}{\partial w}}{J}, \quad \nabla G = \frac{\dfrac{\partial \mathbf{r}}{\partial w} \times \dfrac{\partial \mathbf{r}}{\partial u}}{J}, \quad \nabla H = \frac{\dfrac{\partial \mathbf{r}}{\partial u} \times \dfrac{\partial \mathbf{r}}{\partial v}}{J}. \tag{3–48}$$

The proofs are left to Probs. 1 to 3 below. Because of the assumption: $J > 0$, the vectors $\partial \mathbf{r}/\partial u$, $\partial \mathbf{r}/\partial v$, $\partial \mathbf{r}/\partial w$ form a positive triple; because of (3–47), ∇F, ∇G, ∇H also form a positive triple.

We note two further sets of identities:

$$\frac{1}{J}\frac{\partial \mathbf{r}}{\partial u} = \operatorname{curl}\,(G\,\nabla H), \quad \frac{1}{J}\frac{\partial \mathbf{r}}{\partial v} = \operatorname{curl}\,(H\,\nabla F), \quad \frac{1}{J}\frac{\partial \mathbf{r}}{\partial w} = \operatorname{curl}\,(F\,\nabla G); \tag{3–49}$$

$$\operatorname{div}\left(\frac{1}{J}\frac{\partial \mathbf{r}}{\partial u}\right) = 0, \quad \operatorname{div}\left(\frac{1}{J}\frac{\partial \mathbf{r}}{\partial v}\right) = 0, \quad \operatorname{div}\left(\frac{1}{J}\frac{\partial \mathbf{r}}{\partial w}\right) = 0. \tag{3–50}$$

The first reduces to (3–46) on application of the identity (3–28). The second then follows from (3–33).

The curvilinear coordinate system defined by (3–42) and (3–43) is said to be *orthogonal* if the tangent vectors $\partial \mathbf{r}/\partial u$, $\partial \mathbf{r}/\partial v$, $\partial \mathbf{r}/\partial w$ at each point of D form a triple of mutually perpendicular vectors. It will be seen that, when this is the case, important simplifications in the formulas are possible. The most commonly used curvilinear coordinates are orthogonal and for this reason we shall confine attention to this case. *In this and the following section the coordinates will be assumed to be orthogonal.*

As a first consequence of orthogonality, we remark that $(1/\alpha)\, \partial \mathbf{r}/\partial u$, $(1/\beta)\, \partial \mathbf{r}/\partial v$, $(1/\gamma)\, \partial \mathbf{r}/\partial w$ are a positive triple of mutually perpendicular unit vectors. Hence

$$\left(\frac{1}{\alpha}\frac{\partial \mathbf{r}}{\partial u}\right)\cdot\left(\frac{1}{\beta}\frac{\partial \mathbf{r}}{\partial v}\right)\times\left(\frac{1}{\gamma}\frac{\partial \mathbf{r}}{\partial w}\right) = 1.$$

Accordingly, by (3–44),

$$J = \frac{\partial \mathbf{r}}{\partial u}\cdot\frac{\partial \mathbf{r}}{\partial v}\times\frac{\partial \mathbf{r}}{\partial w} = \alpha\beta\gamma. \tag{3–51}$$

By (3–48),

$$\alpha\,\nabla F = \frac{\alpha}{J}\left(\frac{\partial \mathbf{r}}{\partial v}\times\frac{\partial \mathbf{r}}{\partial w}\right) = \left(\frac{1}{\beta}\frac{\partial \mathbf{r}}{\partial v}\right)\times\left(\frac{1}{\gamma}\frac{\partial \mathbf{r}}{\partial w}\right) = \frac{1}{\alpha}\frac{\partial \mathbf{r}}{\partial u}.$$

A similar reasoning applies to ∇G and ∇H; we conclude that $\alpha\,\nabla F$, $\beta\,\nabla G$, $\gamma\,\nabla H$ are also mutually perpendicular unit vectors and

$$\alpha\,\nabla F = \frac{1}{\alpha}\frac{\partial \mathbf{r}}{\partial u},\quad \beta\,\nabla G = \frac{1}{\beta}\frac{\partial \mathbf{r}}{\partial v},\quad \gamma\,\nabla H = \frac{1}{\gamma}\frac{\partial \mathbf{r}}{\partial w}. \tag{3–52}$$

The surfaces $F = \text{const}$, $G = \text{const}$, $H = \text{const}$ must hence meet at right angles; they form what is called a *triply orthogonal* family of surfaces. Conversely, when the vectors ∇F, ∇G, ∇H are mutually perpendicular throughout D, the coordinates must be orthogonal (Prob. 4 below).

A curve in D can be described by equations: $x = x(t)$, $y = y(t)$, $z = z(t)$, or, by (3–43), in terms of the curvilinear coordinates by equations: $u = u(t)$, $v = v(t)$, $w = w(t)$. The element of arc ds on such a curve is defined by the equation

$$ds^2 = dx^2 + dy^2 + dz^2. \tag{3–53}$$

Hence

$$\begin{aligned} ds^2 &= \left(\frac{\partial x}{\partial u}\,du + \frac{\partial x}{\partial v}\,dv + \frac{\partial x}{\partial w}\,dw\right)^2 + \left(\frac{\partial y}{\partial u}\,du + \cdots\right)^2 + \left(\frac{\partial z}{\partial u}\,du + \cdots\right)^2 \\ &= \left|\frac{\partial \mathbf{r}}{\partial u}\right|^2 du^2 + \left|\frac{\partial \mathbf{r}}{\partial v}\right|^2 dv^2 + \left|\frac{\partial \mathbf{r}}{\partial w}\right|^2 dw^2 \\ &\quad + 2\left(\frac{\partial \mathbf{r}}{\partial u}\cdot\frac{\partial \mathbf{r}}{\partial v}\right)du\,dv + 2\left(\frac{\partial \mathbf{r}}{\partial v}\cdot\frac{\partial \mathbf{r}}{\partial w}\right)dv\,dw + 2\left(\frac{\partial \mathbf{r}}{\partial w}\cdot\frac{\partial \mathbf{r}}{\partial u}\right)dw\,du. \end{aligned}$$

Since the coordinates are orthogonal, we conclude:

$$ds^2 = \alpha^2\,du^2 + \beta^2\,dv^2 + \gamma^2\,dw^2. \tag{3–54}$$

Now $\alpha\,du$, $\beta\,dv$, $\gamma\,dw$ are the elements of arc on the u curves, v curves, w curves respectively. The expression (3–54) is built up as a sum of squares of the elements in three coordinate directions just as in (3–53). This is one of the basic properties of orthogonal coordinates. The above derivation shows that (3–54) can hold only when $\partial\mathbf{r}/\partial u$, $\partial\mathbf{r}/\partial v$, $\partial\mathbf{r}/\partial w$ are mutually perpendicular, so that (3–54) itself can be used to define what is meant by orthogonal coordinates.

We note also that $\alpha\beta\gamma\,du\,dv\,dw$ can be interpreted as the volume dV of an "elementary rectangular parallelepiped." Hence by (3–44) and (3–51)

$$dV = \alpha\beta\gamma\, du\, dv\, dw = J\, du\, dv\, dw = \frac{\partial(x, y, z)}{\partial(u, v, w)}\, du\, dv\, dw.$$

This formula will be discussed in Chapter 4.

From equations (3–52) and the identity (3–31) we deduce the important rule:

$$\operatorname{curl}\left(\frac{1}{\alpha^2}\frac{\partial \mathbf{r}}{\partial u}\right) = \mathbf{0}, \quad \operatorname{curl}\left(\frac{1}{\beta^2}\frac{\partial \mathbf{r}}{\partial v}\right) = \mathbf{0}, \quad \operatorname{curl}\left(\frac{1}{\gamma^2}\frac{\partial \mathbf{r}}{\partial w}\right) = \mathbf{0}. \tag{3–55}$$

*** 3–8 Vector operations in orthogonal curvilinear coordinates.** Now let a vector field $\mathbf{p}$ be given in D. The vector $\mathbf{p}$ can be described by its components p_x, p_y, p_z in terms of the given rectangular system. However, at each point of D, the vectors $(1/\alpha)\, \partial\mathbf{r}/\partial u$, $(1/\beta)\, \partial\mathbf{r}/\partial v$, $(1/\gamma)\, \partial\mathbf{r}/\partial w$ are a triple of mutually perpendicular unit vectors. Hence we can write

$$\mathbf{p} = p_u \frac{1}{\alpha}\frac{\partial \mathbf{r}}{\partial u} + p_v \frac{1}{\beta}\frac{\partial \mathbf{r}}{\partial v} + p_w \frac{1}{\gamma}\frac{\partial \mathbf{r}}{\partial w} \tag{3–56}$$

in terms of the components p_u, p_v, p_w in the directions of the three unit vectors. It should be emphasized that the triple of unit vectors in general *varies from point to point* in D. By (3–52), we can also write

$$\mathbf{p} = p_u\alpha\, \nabla F + p_v\beta\, \nabla G + p_w\gamma\, \nabla H. \tag{3–56$'$}$$

The components p_u, p_v, p_w can be computed from the components p_x, p_y, p_z in the rectangular system. For example,

$$p_u = \mathbf{p} \cdot \frac{1}{\alpha}\frac{\partial \mathbf{r}}{\partial u} = \frac{1}{\alpha}\left(p_x \frac{\partial x}{\partial u} + p_y \frac{\partial y}{\partial u} + p_z \frac{\partial z}{\partial u}\right). \tag{3–57}$$

Similarly, the components p_x, p_y, p_z can be computed from p_u, p_v, p_w:

$$p_x = \mathbf{p} \cdot \mathbf{i} = p_u \frac{1}{\alpha}\frac{\partial \mathbf{r}}{\partial u} \cdot \mathbf{i} + p_v \frac{1}{\beta}\frac{\partial \mathbf{r}}{\partial v} \cdot \mathbf{i} + p_w \frac{1}{\gamma}\frac{\partial \mathbf{r}}{\partial w} \cdot \mathbf{i},$$

$$p_x = \frac{1}{\alpha} p_u \frac{\partial x}{\partial u} + \frac{1}{\beta} p_v \frac{\partial x}{\partial v} + \frac{1}{\gamma} p_w \frac{\partial x}{\partial w}. \tag{3–58}$$

If the representation (3–56$'$) is used, one finds:

$$p_u = \alpha\left(p_x \frac{\partial u}{\partial x} + p_y \frac{\partial u}{\partial y} + p_z \frac{\partial u}{\partial z}\right), \tag{3–57$'$}$$

$$p_x = \alpha p_u \frac{\partial u}{\partial x} + \beta p_v \frac{\partial v}{\partial x} + \gamma p_w \frac{\partial w}{\partial x}. \tag{3–58$'$}$$

Since the vectors $(1/\alpha)\, \partial\mathbf{r}/\partial u$, $(1/\beta)\, \partial\mathbf{r}/\partial v$, $(1/\gamma)\, \partial\mathbf{r}/\partial w$ form a positive triple of mutually perpendicular unit vectors, the operations: scalar times vector, sum of vectors, scalar product, and vector product can be carried out in terms of components in the directions of these unit vectors just as in terms of x, y, and z components. In particular,

$$\begin{aligned}
[\mathbf{p}+\mathbf{q}]_u &= p_u + q_u, \quad [\mathbf{p}+\mathbf{q}]_v = p_v + q_v, \ldots, \\
[\phi\mathbf{p}]_u &= \phi p_u, \quad [\phi\mathbf{p}]_v = \phi p_v, \ldots, \\
\mathbf{p}\cdot\mathbf{q} &= p_u q_u + p_v q_v + p_w q_w, \\
[\mathbf{p}\times\mathbf{q}]_u &= p_v q_w - p_w q_v, \quad [\mathbf{p}\times\mathbf{q}]_v = p_w q_u - p_u q_w, \ldots.
\end{aligned} \tag{3–59}$$

On the other hand, because the base vectors vary from point to point, the differential operations become more complicated:

$$[\operatorname{grad}\phi]_u = \frac{1}{\alpha}\frac{\partial\phi}{\partial u}, \quad [\operatorname{grad}\phi]_v = \frac{1}{\beta}\frac{\partial\phi}{\partial v}, \quad [\operatorname{grad}\phi]_w = \frac{1}{\gamma}\frac{\partial\phi}{\partial w}, \tag{3–60}$$

$$\operatorname{div}\mathbf{p} = \frac{1}{\alpha\beta\gamma}\left[\frac{\partial}{\partial u}(\beta\gamma p_u) + \frac{\partial}{\partial v}(\gamma\alpha p_v) + \frac{\partial}{\partial w}(\alpha\beta p_w)\right], \tag{3–61}$$

$$\begin{aligned}
[\operatorname{curl}\mathbf{p}]_u &= \frac{1}{\beta\gamma}\left[\frac{\partial}{\partial v}(\gamma p_w) - \frac{\partial}{\partial w}(\beta p_v)\right], \\
[\operatorname{curl}\mathbf{p}]_v &= \frac{1}{\gamma\alpha}\left[\frac{\partial}{\partial w}(\alpha p_u) - \frac{\partial}{\partial u}(\gamma p_w)\right], \\
[\operatorname{curl}\mathbf{p}]_w &= \frac{1}{\alpha\beta}\left[\frac{\partial}{\partial u}(\beta p_v) - \frac{\partial}{\partial v}(\alpha p_u)\right].
\end{aligned} \tag{3–62}$$

To prove (3–60), we use (3–57):

$$[\operatorname{grad}\phi]_u = \frac{1}{\alpha}\left(\frac{\partial\phi}{\partial x}\frac{\partial x}{\partial u} + \frac{\partial\phi}{\partial y}\frac{\partial y}{\partial u} + \frac{\partial\phi}{\partial z}\frac{\partial z}{\partial u}\right) = \frac{1}{\alpha}\frac{\partial\phi}{\partial u}.$$

The other components are found in the same way. We remark that $[\operatorname{grad}\phi]_u$ is the directional derivative of ϕ along a u curve, i.e., $d\phi/ds$ in terms of arc length s on the curve. Since $ds = \alpha\, du$, the result (3–60) follows at once.

To prove (3–61), we use (3–51) and (3–56) to write:

$$\mathbf{p} = (\beta\gamma p_u)\left(\frac{1}{J}\frac{\partial\mathbf{r}}{\partial u}\right) + (\gamma\alpha p_v)\left(\frac{1}{J}\frac{\partial\mathbf{r}}{\partial v}\right) + (\alpha\beta p_w)\left(\frac{1}{J}\frac{\partial\mathbf{r}}{\partial w}\right).$$

The divergence of $\mathbf{p}$ is the sum of the divergences of the terms on the right. By (3–22) and (3–50), the divergence of the first term is

$$\operatorname{grad}(\beta\gamma p_u)\cdot\left(\frac{1}{J}\frac{\partial\mathbf{r}}{\partial u}\right) + \beta\gamma p_u \operatorname{div}\left(\frac{1}{J}\frac{\partial\mathbf{r}}{\partial u}\right) = (\operatorname{grad}\beta\gamma p_u)\cdot\left(\frac{1}{J}\frac{\partial\mathbf{r}}{\partial u}\right).$$

By (3–60) this can be written as

$$\frac{\alpha}{J}(\operatorname{grad}\beta\gamma p_u)\cdot\frac{1}{\alpha}\frac{\partial\mathbf{r}}{\partial u} = \frac{\alpha}{J}[\operatorname{grad}\beta\gamma p_u]_u = \frac{1}{J}\frac{\partial}{\partial u}(\beta\gamma p_u) = \frac{1}{\alpha\beta\gamma}\frac{\partial}{\partial u}(\beta\gamma p_u).$$

This gives the first term on the right of (3–61); the others are found in the same way.

The proof of (3–62) is left to Prob. 5 below.

From (3–60) and (3–61) we obtain an expression for the Laplacian in orthogonal curvilinear coordinates:

$$\nabla^2\phi = \operatorname{div}\operatorname{grad}\phi = \frac{1}{\alpha\beta\gamma}\left[\frac{\partial}{\partial u}\left(\frac{\beta\gamma}{\alpha}\frac{\partial\phi}{\partial u}\right) + \frac{\partial}{\partial v}\left(\frac{\gamma\alpha}{\beta}\frac{\partial\phi}{\partial v}\right) + \frac{\partial}{\partial w}\left(\frac{\alpha\beta}{\gamma}\frac{\partial\phi}{\partial w}\right)\right]. \tag{3–63}$$

Remark. If the curvilinear coordinates are not orthogonal, the very notion of vector components must be generalized. This leads to *tensor analysis.* See Section 3–9.

The formulas (3–60), (3–61), (3–62) can be applied to the special case in which the new coordinates are obtained simply by choosing new rectangular coordinates u, v, w in space. This can be accomplished by choosing a positive triple $\mathbf{i}_1, \mathbf{j}_1, \mathbf{k}_1$ of unit vectors and then choosing axes u, v, w through an origin O_1: (x_1, y_1, z_1) having the directions $\mathbf{i}_1, \mathbf{j}_1, \mathbf{k}_1$ respectively (See Section 1–14). The new coordinates (u, v, w) of a point P: (x, y, z) are defined by the equation:

$$\overrightarrow{O_1P} = u\mathbf{i}_1 + v\mathbf{j}_1 + w\mathbf{k}_1.$$

One then finds

$$\begin{aligned} u &= \overrightarrow{O_1P}\cdot\mathbf{i}_1 = [(x - x_1)\mathbf{i} + (y - y_1)\mathbf{j} + (z - z_1)\mathbf{k}]\cdot\mathbf{i}_1 \\ &= (x - x_1)(\mathbf{i}\cdot\mathbf{i}_1) + (y - y_1)(\mathbf{j}\cdot\mathbf{i}_1) + (z - z_1)(\mathbf{k}\cdot\mathbf{i}_1); \end{aligned}$$

similar expressions are found for v and w. Also

$$x = \overrightarrow{OP}\cdot\mathbf{i} = (\overrightarrow{OO_1} + \overrightarrow{O_1P})\cdot\mathbf{i} = x_1 + u(\mathbf{i}_1\cdot\mathbf{i}) + v(\mathbf{j}_1\cdot\mathbf{i}) + w(\mathbf{k}_1\cdot\mathbf{i}),$$

and similar expressions are found for y and z. These two sets of equations correspond to (3–43) and (3–42). We note that in all cases the functions involved are *linear.*

The quantities α, β, γ can be evaluated for this case without computation. For, since the new coordinates are rectangular and *no change of scale is made,* one must have

$$ds^2 = du^2 + dv^2 + dw^2 \tag{3–64}$$

for element of arc on a general curve. Hence

$$\alpha = \beta = \gamma = 1. \tag{3–65}$$

If we substitute these values in (3–60), (3–61), (3–62), we find that the basic formulas (3–5), (3–15), (3–23) reappear, with x replaced by u, y by v, z by w. For example,

$$\operatorname{div}\mathbf{p} = \frac{\partial p_u}{\partial u} + \frac{\partial p_v}{\partial v} + \frac{\partial p_w}{\partial w}.$$

This shows that *the fundamental definitions* (3–5), (3–15), (3–23) *of grad, div, curl are not dependent on the particular coordinate system chosen.*

If a change of scale is made, the formulas will be altered. Physically this corresponds to a change in unit of length (e.g., from inches to feet) and one could hardly expect a temperature gradient, for example, to have the same value in degrees per inch as in degrees per foot. For this reason, in practice one must always specify the units used.

It has been assumed throughout that the Jacobian J is positive; this implies, in particular, that in the case just considered the vectors $\mathbf{i}_1$, $\mathbf{j}_1$, $\mathbf{k}_1$ form a positive triple. If J is negative, one finds that the only change in (3–60), (3–61), (3–62) is a reversal of sign in the components of the curl. In particular, if new rectangular coordinates are chosen, based on a negative triple $\mathbf{i}_1$, $\mathbf{j}_1$, $\mathbf{k}_1$, the curl as defined by (3–23) relative to these axes is the negative of the curl relative to the original x, y, and z axes. *If the orientation of the axes is reversed, the curl reverses direction.* This is not surprising if one thinks of the curl as an angular velocity vector. On the other hand, the gradient vector and divergence do not depend on the orientation. The vector product components in (3–59) also change sign if the orientation is reversed. This could be predicted from the very definition of the vector product in Section 0–6.

For further discussion of change of coordinates in space, see Chapter 4 of *Classical Mechanics*, by H. Goldstein (Cambridge: Addison-Wesley Press, 1950).

If the Jacobian J is 0 at a point P, it is in general impossible to use the curvilinear coordinates at P; in particular, the vectors $\partial\mathbf{r}/\partial u$, $\partial\mathbf{r}/\partial v$, $\partial\mathbf{r}/\partial w$ are coplanar, so that one cannot express an arbitrary vector $\mathbf{p}$ in terms of components p_u, p_v, p_w. Furthermore, the point P will in general correspond to many values of the coordinates (u, v, w); that is, the inverse functions (3–43) become ambiguous at P.

However, let us assume that the functions (3–42) remain continuous and differentiable in the domain D_1 of the coordinates (u, v, w) and have values in a domain D; then the chain rules (2–28) imply that every function $U(x, y, z)$ which is differentiable in D becomes a function

$$U[f(u, v, w), \quad g(u, v, w), \quad h(u, v, w)]$$

which is defined and differentiable in D_1. Accordingly, even though the inverse transformation (3–43) may fail to exist, *scalar* functions of (x, y, z) can be transformed into functions of (u, v, w) without difficulty.

Precisely this situation arises in cylindrical and spherical coordinates. For example, in cylindrical coordinates the equations (3–42) are as follows:

$$x = r\cos\theta, \quad y = r\sin\theta, \quad z = z;$$

these functions are defined and differentiable as often as desired for all real values of r, θ, z; as (r, θ, z) ranges over all possible combinations, (x, y, z) ranges over all of space. An inverse transformation:

$$r = \sqrt{x^2 + y^2}, \quad \theta = \arctan\frac{y}{x}, \quad z = z$$

can be defined by suitably restricting θ; but in no way can this be defined as a triple of continuous functions when (x, y, z) ranges over a domain D including points of the z axis. It is precisely when $r = 0$ that the Jacobian $J = r$ (Prob. 6 below) is zero.

It is shown in Prob. 6 that the Laplacian in cylindrical coordinates has the expression:

$$\nabla^2 U = \frac{1}{r^2}\left[r\frac{\partial}{\partial r}\left(r\frac{\partial U}{\partial r}\right) + \frac{\partial^2 U}{\partial \theta^2} + r^2\frac{\partial^2 U}{\partial z^2}\right]. \qquad (3\text{–}66)$$

This expression becomes meaningless when $r = 0$. However, if we know that U, when expressed in rectangular coordinates, has continuous first and second derivatives on the z axis, then $\nabla^2 U$ must also be a continuous function of (r, θ, z) for $r = 0$. Under these assumptions, $\nabla^2 U$ can be obtained from (3–66) for $r = 0$ by a limit process. For example, if

$$U = x^2 + x^2 y = r^2 \cos^2\theta + r^3 \cos^2\theta \sin\theta,$$

then

$$\nabla^2 U = 2 + 2y = 2 + 2r\sin\theta;$$

(3–66) gives the indeterminate expression

$$\nabla^2 U = \frac{1}{r^2}[2r^2 + 2r^3 \sin\theta].$$

While this function is indeterminate for $r = 0$, it has a definite limit as $r \to 0$; the limit is 2 and is independent of θ. By cancelling r^2 from numerator and denominator, we automatically remove the indeterminacy and assign the correct limiting value for $r = 0$:

$$\nabla^2 U = 2 + 2r\sin\theta.$$

A similar discussion applies to spherical coordinates ρ, ϕ, θ. In Prob. 7 below it is shown that $J = \rho^2 \sin\phi$, so that $J = 0$ on the z axis. The Laplacian is found to be

$$\nabla^2 U = \frac{1}{\rho^2 \sin^2\phi}\left[\sin^2\phi\frac{\partial}{\partial\rho}\left(\rho^2\frac{\partial U}{\partial\rho}\right) + \sin\phi\frac{\partial}{\partial\phi}\left(\sin\phi\frac{\partial U}{\partial\phi}\right) + \frac{\partial^2 U}{\partial\theta^2}\right]. \qquad (3\text{–}67)$$

This is indeterminate on the z axis ($\rho = 0$ or $\sin\phi = 0$); when $\nabla^2 U$ is known to be continuous on the z axis, the values of $\nabla^2 U$ on this line can be found from (3–67) by a limit process.

It is also possible to obtain the values of the Laplacian at the troublesome points directly in terms of derivatives. For example, $\nabla^2 U$ can be computed at the origin ($\rho = 0$) by the formula:

$$\nabla^2 U|_{\rho=0} = \frac{\partial^2 U}{\partial\rho^2}(0, \tfrac{1}{2}\pi, 0) + \frac{\partial^2 U}{\partial\rho^2}(0, \tfrac{1}{2}\pi, \tfrac{1}{2}\pi) + \frac{\partial^2 U}{\partial\rho^2}(0, 0, 0). \qquad (3\text{–}68)$$

The three terms on the right are simply the three terms of the Laplacian

$$\frac{\partial^2 U}{\partial x^2} + \frac{\partial^2 U}{\partial y^2} + \frac{\partial^2 U}{\partial z^2}$$

at the origin.

Problems

1. Prove (3–48). [Hint: use the result of Prob. 9 following Section 2–11.]

2. Prove (3–47). [Hint: use (3–48) and the identity (0–86) of Section 0–6.

3. Prove (3–46). [Hint: use (3–48) and the identity (0–86) of Section 0–6.

4. Prove that when the vectors ∇F, ∇G, ∇H are mutually perpendicular in D, the coordinates are orthogonal.

5. Prove (3–62). [Hint: use (3–56) to write:

$$\mathbf{p} = (\alpha p_u)\left(\frac{1}{\alpha^2}\frac{\partial \mathbf{r}}{\partial u}\right) + \beta p_v\left(\frac{1}{\beta^2}\frac{\partial \mathbf{r}}{\partial v}\right) + (\gamma p_w)\left(\frac{1}{\gamma^2}\frac{\partial \mathbf{r}}{\partial w}\right).$$

Use (3–28) and (3–55) to show that

$$\text{curl } \mathbf{p} = \text{grad } (\alpha p_u) \times \left(\frac{1}{\alpha^2}\frac{\partial \mathbf{r}}{\partial u}\right) + \text{grad } (\beta p_v) \times \frac{1}{\beta^2}\left(\frac{\partial \mathbf{r}}{\partial v}\right) + \cdots.$$

To compute $[\text{curl } \mathbf{p}]_u$, take the scalar product of both sides with $\dfrac{1}{\alpha}\dfrac{\partial \mathbf{r}}{\partial u}$ and use (3–59) and (3–60) to compute the scalar triple products.]

6. Verify the following relations for cylindrical coordinates $u = r$, $v = \theta$, $w = z$:

(a) the surfaces $r = \text{const}$, $\theta = \text{const}$, $z = \text{const}$ form a triply orthogonal family and

$$J = \frac{\partial(x, y, z)}{\partial(r, \theta, z)} = r$$

(b) the element of arc length is given by

$$ds^2 = dr^2 + r^2\, d\theta^2 + dz^2,$$

(c) the components of a vector $\mathbf{p}$ are given by

$$p_r = p_x \cos\theta + p_y \sin\theta, \quad p_\theta = -p_x \sin\theta + p_y \cos\theta, \quad p_z = p_z;$$

(d) grad U has components: $\dfrac{\partial U}{\partial r}, \quad \dfrac{1}{r}\dfrac{\partial U}{\partial \theta}, \quad \dfrac{\partial U}{\partial z};$

(e) $\text{div } \mathbf{p} = \dfrac{1}{r}\left[\dfrac{\partial}{\partial r}(rp_r) + \dfrac{\partial p_\theta}{\partial \theta} + r\dfrac{\partial p_z}{\partial z}\right];$

(f) curl $\mathbf{p}$ has components:

$$\frac{1}{r}\left[\frac{\partial p_z}{\partial \theta} - r\frac{\partial p_\theta}{\partial z}\right], \quad \left[\frac{\partial p_r}{\partial z} - \frac{\partial p_z}{\partial r}\right], \quad \frac{1}{r}\left[\frac{\partial}{\partial r}(rp_\theta) - \frac{\partial p_r}{\partial \theta}\right];$$

(g) $\nabla^2 U$ is given by (3–66).

7. Verify the following relations for spherical coordinates $u = \rho$, $v = \phi$, $w = \theta$:

(a) the surfaces $\rho = \text{const}$, $\phi = \text{const}$, $\theta = \text{const}$ form a triply orthogonal family and

$$J = \frac{\partial(x, y, z)}{\partial(\rho, \phi, \theta)} = \rho^2 \sin\phi;$$

(b) the element of arc length is given by

$$ds^2 = d\rho^2 + \rho^2\, d\phi^2 + \rho^2 \sin^2\phi\, d\theta^2;$$

(c) the components of a vector $\mathbf{p}$ are given by

$$\begin{aligned} p_\rho &= p_x \sin\phi\cos\theta + p_y \sin\phi\sin\theta + p_z\cos\phi, \\ p_\phi &= p_x \cos\phi\cos\theta + p_y \cos\phi\sin\theta - p_z \sin\phi, \\ p_\theta &= -p_x \sin\theta + q_y \cos\theta; \end{aligned}$$

(d) grad U has components $\dfrac{\partial U}{\partial \rho}, \quad \dfrac{1}{\rho}\dfrac{\partial U}{\partial \phi}, \quad \dfrac{1}{\rho\sin\phi}\dfrac{\partial U}{\partial \theta};$

(e) div $\mathbf{p} = \dfrac{1}{\rho^2 \sin\phi}\left[\sin\phi \dfrac{\partial}{\partial\rho}(\rho^2 p_\rho) + \rho \dfrac{\partial}{\partial\phi}(p_\phi \sin\phi) + \rho \dfrac{\partial p_\theta}{\partial\theta}\right]$;

(f) curl $\mathbf{p}$ has components:

$$\frac{1}{\rho\sin\phi}\left[\frac{\partial}{\partial\phi}(p_\theta \sin\phi) - \frac{\partial p_\phi}{\partial\theta}\right], \quad \frac{1}{\rho\sin\phi}\left[\frac{\partial p_\rho}{\partial\theta} - \sin\phi\frac{\partial}{\partial\rho}(\rho p_\theta)\right], \quad \frac{1}{\rho}\left[\frac{\partial}{\partial\rho}(\rho p_\phi) - \frac{\partial p_\rho}{\partial\phi}\right];$$

(g) $\nabla^2 U$ is given by (3–67).

8. *Curvilinear coordinates on a surface.* Equations

$$x = f(u, v), \quad y = g(u, v), \quad z = h(u, v)$$

can be interpreted as parametric equations of a surface S in space. They can be considered as a special case of (3–42), in which w is restricted to a constant value, while (u, v) varies over a domain D_0 of the uv plane; the surface S then corresponds to a surface $w =$ const for (3–42). We consider u, v as *curvilinear coordinates* on S. The two sets of curves $u =$ const and $v =$ const on S form families like the parallels to the axes in the xy plane. Graph the surface and the lines $u =$ const, $v =$ const, for the following cases:

(a) sphere: $x = \sin u \cos v, \quad y = \sin u \sin v, \quad z = \cos u$;
(b) cylinder: $x = \cos u, \quad y = \sin u, \quad z = v$;
(c) cone: $x = \sinh u \sin v, \quad y = \sinh u \cos v, \quad z = \sinh u$.

9. Let a surface S be given as in Prob. 8 and let $\mathbf{r} = x\mathbf{i} + y\mathbf{j} + z\mathbf{k}$.

(a) Show that $\partial\mathbf{r}/\partial u$ and $\partial\mathbf{r}/\partial v$ are vectors tangent to the lines $v =$ const, $u =$ const on the surface.

(b) Show that the curves $v =$ const, $u =$ const intersect at right angles, so that the coordinates are *orthogonal*, if and only if

$$\frac{\partial x}{\partial u}\frac{\partial x}{\partial v} + \frac{\partial y}{\partial u}\frac{\partial y}{\partial v} + \frac{\partial z}{\partial u}\frac{\partial z}{\partial v} = 0.$$

(c) Show that the element of arc on a curve $u = u(t)$, $v = v(t)$ on S is given by

$$ds^2 = E\,du^2 + 2F\,du\,dv + G\,dv^2,$$

$$E = \left|\frac{\partial\mathbf{r}}{\partial u}\right|^2 = \left(\frac{\partial x}{\partial u}\right)^2 + \left(\frac{\partial y}{\partial u}\right)^2 + \left(\frac{\partial z}{\partial u}\right)^2,$$

$$G = \left|\frac{\partial\mathbf{r}}{\partial v}\right|^2 = \left(\frac{\partial x}{\partial v}\right)^2 + \left(\frac{\partial y}{\partial v}\right)^2 + \left(\frac{\partial z}{\partial v}\right)^2,$$

$$F = \frac{\partial\mathbf{r}}{\partial u}\cdot\frac{\partial\mathbf{r}}{\partial v} = \frac{\partial x}{\partial u}\frac{\partial x}{\partial v} + \frac{\partial y}{\partial u}\frac{\partial y}{\partial v} + \frac{\partial z}{\partial u}\frac{\partial z}{\partial v}.$$

(d) Show that the coordinates are *orthogonal* precisely when

$$ds^2 = E\,du^2 + G\,dv^2.$$

For further theory of surfaces see the book by Struik listed at the end of this chapter.

***3–9 Tensors.** When nonorthogonal curvilinear coordinates are introduced, the methods of the preceding section are no longer adequate for the analysis of the basic vector operations. The desired analysis can be carried out with the aid of tensors, which we proceed to develop briefly here.

We first consider the very simple case of a change of scale in space. Let (x, y, z) be given Cartesian coordinates and let $(\bar{x}, \bar{y}, \bar{z})$ be new coordinates, where

$$\bar{x} = \lambda x, \quad \bar{y} = \lambda y, \quad \bar{z} = \lambda z, \tag{3-69}$$

λ being a positive constant scalar. Thus we have changed scale in the ratio $\lambda:1$, and $1/\lambda$ is our new unit of distance. For a point moving on a path, we have hitherto assigned a velocity vector $\mathbf{v} = (v_x, v_y, v_z)$ and thought of this as a definite geometric object, represented by a directed line segment. But the components of $\mathbf{v}$, as a velocity vector, depend on the coordinate system; that is, here, where we are considering a change of scale, these components depend on the unit of distance chosen. In the (x, y, z) coordinates we would assign

$$v_x = \frac{dx}{dt}, \quad v_y = \frac{dy}{dt}, \quad v_z = \frac{dz}{dt},$$

but in the $(\bar{x}, \bar{y}, \bar{z})$ coordinates we would assign

$$\bar{v}_x = \frac{d\bar{x}}{dt}, \quad \bar{v}_y = \frac{d\bar{y}}{dt}, \quad \bar{v}_z = \frac{d\bar{z}}{dt}.$$

By virtue of Eqs. (3–69), $d\bar{x}/dt = \lambda\, dx/dt$, so that $\bar{v}_x = \lambda v_x$ and in general

$$\bar{v}_x = \lambda v_x, \quad \bar{v}_y = \lambda v_y, \quad \bar{v}_z = \lambda v_z. \tag{3-70}$$

Thus in the new coordinates $(\bar{x}, \bar{y}, \bar{z})$ we assign to $\mathbf{v}$ the new components $(\bar{v}_x, \bar{v}_y, \bar{v}_z)$ which are λ times the previous components.

Now a vector $\mathbf{v}$ can also be obtained as the gradient vector of a function f: $\mathbf{v} = \operatorname{grad} f$, so that

$$v_x = \frac{\partial f}{\partial x}, \quad v_y = \frac{\partial f}{\partial y}, \quad v_z = \frac{\partial f}{\partial z}.$$

If we change scale by (3–69), we would still like $\mathbf{v}$ to be the gradient of f. But in the new coordinates f becomes $f(\bar{x}/\lambda, \bar{y}/\lambda, \bar{z}/\lambda) = \bar{f}(\bar{x}, \bar{y}, \bar{z})$ and this has gradient

$$\bar{v}_x = \frac{\partial \bar{f}}{\partial \bar{x}} = \frac{1}{\lambda} f_x, \quad \bar{v}_y = \frac{\partial \bar{f}}{\partial \bar{y}} = \frac{1}{\lambda} f_y, \quad \bar{v}_z = \frac{\partial \bar{f}}{\partial \bar{z}} = \frac{1}{\lambda} f_z.$$

Hence now

$$\bar{v}_x = \frac{1}{\lambda} v_x, \quad \bar{v}_y = \frac{1}{\lambda} v_y, \quad \bar{v}_z = \frac{1}{\lambda} v_z. \tag{3-71}$$

Thus when we change scale by (3–69), the three components of the gradient vector are *divided* by λ. This result is to be expected, since the gradient vector measures the rate of change of f with respect to distance (in various directions), and we are changing the unit of distance. If for example, $\lambda = 2$, then the new unit of distance is half the old one and the amount of change in f per unit of distance is half as much as before, whereas velocity components are twice as large as before, since one covers twice as many units of distance per unit of time.

The two different rules (3–70) and (3–71) show that we really have two different ways of assigning components to vectors. In the case of (3–70) one is dealing with a "contravariant vector," and in the case of (3–71) with a "covariant vector." In obtaining the two types of components, we have emphasized the change in unit of distance. However, in changing coordinates, we can always keep in mind the original unit of distance (as a standard of reference), so that all distances can ultimately be restated in the original units. This concept of a standard unit of distance will be important in the development to follow.

In tensor analysis one allows changes of coordinates much more general than changes of scale. It is more convenient to number our coordinates as (x^1, x^2, x^3) and simply to refer to the (x^i) coordinates. We can in fact reason generally about n-dimensional space, and allow i to go from 1 to n. The use of superscripts rather than subscripts is required by a certain consistency of all the tensor notations. If we introduce new coordinates $(\bar{x}^i)$, then we have equations

$$x^i = x^i(\bar{x}^1, \ldots, \bar{x}^n), \quad i = 1, \ldots, n, \tag{3–72}$$

relating new and old coordinates. We consider these equations only in a neighborhood D of a certain point and assume that they define a one-to-one mapping with inverse

$$\bar{x}^i = \bar{x}^i(x^1, \ldots, x^n), \quad i = 1, \ldots, n, \tag{3–73}$$

and that all functions in (3–72) and (3–73) are differentiable, with continuous second partial derivatives and nonzero Jacobian

$$\frac{\partial(x^1, \ldots, x^n)}{\partial(\bar{x}^1, \ldots, \bar{x}^n)}.$$

It is essential for the theory of tensors that we allow *all* such changes of coordinates.

In the following discussion it will be convenient to denote by $(\xi^1, \ldots, \xi^n)$ a *fixed* Cartesian coordinate system in E^n, in terms of which distance and angle are measured as usual. We denote by $(x^1, \ldots, x^n)$ and $(\bar{x}^1, \ldots, \bar{x}^n)$ two other, generally curvilinear, coordinate systems introduced as above in the neighborhood D of a point, and related by Eqs. (3–72) and (3–73) to each other and related in similar fashion to the (ξ^i). We shall refer to the (ξ^i) as *standard coordinates.*

Now let a vector field be given in the chosen neighborhood. Then the vectors all have sets of components *in each coordinate system.* For a *contravariant vector field* (or *briefly, contravariant vector*), the components in the (x^i) coordinates will be denoted by $(u^1, \ldots, u^n)$ (upper indices), and in the $(\bar{x}^i)$ coordinates by $(\bar{u}^1, \ldots, \bar{u}^n)$. Furthermore, for any two such coordinate systems, the components are to be related by the rule

$$\bar{u}^i = \sum_{j=1}^{n} \frac{\partial \bar{x}^i}{\partial x^j} u^j. \tag{3–74}$$

This rule is chosen to fit the case when the vector in question is a velocity vector:

$$u^i = \frac{dx^i}{dt}, \quad i = 1, \ldots, n.$$

For then

$$\overline{u}^i = \frac{d\overline{x}^i}{dt} = \sum_{j=1}^n \frac{\partial \overline{x}^i}{\partial x^j}\frac{dx^j}{dt} = \sum_{j=1}^n \frac{\partial \overline{x}^i}{\partial x^j} u^j.$$

For a *covariant vector*, we denote the components in the two systems of coordinates by $(u_1, \ldots, u_n)$ and $(\overline{u}_1, \ldots, \overline{u}_n)$ respectively and require that

$$\overline{u}_i = \sum_{j=1}^n \frac{\partial x^j}{\partial \overline{x}^i} u_j. \tag{3–75}$$

This rule is chosen to fit the case when the vector in question is a gradient vector:

$$u_i = \frac{\partial f}{\partial x^i}, \quad i = 1, \ldots, n.$$

For then

$$\overline{u}_i = \frac{\partial}{\partial \overline{x}^i} f(x^1(\overline{x}^1, \ldots), \ldots, x^n(\overline{x}^1, \ldots))$$

$$= \sum_{j=1}^n \frac{\partial f}{\partial x^j}\frac{\partial x^j}{\partial \overline{x}^i} = \sum_{j=1}^n \frac{\partial x^j}{\partial \overline{x}^i} u_j.$$

Each contravariant vector field in the chosen neighborhood D can be obtained in the following way. One first selects a vector field U^i in the (ξ^i) coordinates—that is, n functions $U^1(\xi^1, \ldots, \xi^n), \ldots, U^n(\xi^1, \ldots, \xi^n)$ defined in D. Then to each other coordinate system, say (x^i), one assigns components by the equations analogous to (3–74):

$$u^i = \sum_{j=1}^n \frac{\partial x^i}{\partial \xi^j} U^j. \tag{3–76}$$

Thus in $(\overline{x}^i)$ one has similarly

$$\overline{u}^i = \sum_{j=1}^n \frac{\partial \overline{x}^i}{\partial \xi^j} U^j. \tag{3–77}$$

Now the inverse of the matrix $(\partial x^i/\partial \xi^j)$ is the matrix $(\partial \xi^i/\partial x^j)$. Hence we can solve (3–76) to obtain

$$U^j = \sum_{k=1}^n \frac{\partial \xi^j}{\partial x^k} u^k.$$

If we substitute in (3–77) we obtain

$$\overline{u}^i = \sum_{j=1}^n \sum_{k=1}^n \frac{\partial \overline{x}^i}{\partial \xi^j}\frac{\partial \xi^j}{\partial x^k} u^k = \sum_{k=1}^n \sum_{j=1}^n \frac{\partial \overline{x}^i}{\partial \xi^j}\frac{\partial \xi^j}{\partial x^k} u^k = \sum_{k=1}^n \frac{\partial \overline{x}^i}{\partial x^k} u^k,$$

so that (3–74) holds. Accordingly, once we have assigned components in the standard coordinates, we automatically obtain components in all other coordinate systems, related by (3–74), and a contravariant vector is obtained.

A similar reasoning applies to covariant vector fields (Problem 5 below).

From this discussion it follows that, starting with n functions $f_i(\xi^1, \ldots, \xi^n)$, defined in D, we can generate either a contravariant vector field u^i or a covariant field u_i. For the contravariant vector, we assign components $U^i = f_i(\xi^1, \ldots, \xi^n)$ in standard coordinates and hence obtain the u^i as above in all other allowed coordinates (x^i); for the covariant vector, we assign components $U_i = f_i(\xi^1, \ldots, \xi^n)$ in standard coordinates and hence obtain the u_i in all other coordinate systems (x^i). The contravariant and covariant vector fields thus obtained are related in a special way: namely, in that in standard coordinates, corresponding components are equal:

$$U^i(\xi^1, \ldots, \xi^n) = f_i(\xi^1, \ldots, \xi^n) = U_i(\xi^1, \ldots, \xi^n).$$

In such a case we say that the contravariant vector u^i and covariant vector u_i are *associated*. We regard the u^i and u_i as different aspects of one underlying geometric object $\mathbf{u}$, a vector field abstracted from its components. Thus we speak of u^i as contravariant components of $\mathbf{u}$, u_i as covariant components of $\mathbf{u}$. In standard coordinates, the two types of components coincide and can be identified with our usual vector components.

The contravariant and covariant vectors are the tensors of *first order*. We also introduce tensors of *order zero*, as scalar functions whose values are not changed by coordinate transformations. Thus in the two coordinate systems (x^i) and $(\bar{x}^i)$ such a function is given by $f(x^1, \ldots, x^n)$ and $\bar{f}(\bar{x}^1, \ldots, \bar{x}^n)$ respectively, where

$$f(x^1, \ldots, x^n) = \bar{f}(\bar{x}^1, \ldots, \bar{x}^n)$$

whenever (x^i) and $(\bar{x}^i)$ refer to the same point; that is,

$$f(x^1(\bar{x}^1, \ldots), \ldots, x^n(\bar{x}^1, \ldots,)) \equiv \bar{f}(\bar{x}^1, \ldots, \bar{x}^n). \tag{3–78}$$

We also call such a tensor an *invariant*.

We can also introduce tensors of higher order. A tensor of order two requires two indices and hence can be arranged as a matrix. For example, we denote by v_{ij} a tensor of order two which is *covariant* in both indices; here v_{ij} $(i = 1, \ldots, n, j = 1, \ldots, n)$ are the n^2 components of the tensor in the (x^i) coordinate system. In the second coordinate system $(\bar{x}^i)$ the tensor has components $\bar{v}_{ij}$ and we require that

$$\bar{v}_{ij} = \sum_{k=1}^{n} \sum_{l=1}^{n} \frac{\partial x^k}{\partial \bar{x}^i} \frac{\partial x^l}{\partial \bar{x}^j} v_{kl}. \tag{3–79}$$

Because of the many such sums appearing in tensor analysis, one agrees to drop the sigma signs and to write (3–79) simply as

$$\bar{v}_{ij} = \frac{\partial x^k}{\partial \bar{x}^i} \frac{\partial x^l}{\partial \bar{x}^j} v_{kl},$$

with the understanding that we sum over each index which appears more than once (k and l in this case). This notational rule is called the *summation convention.*

By v_j^i we denote a second order tensor, called *mixed*, which is contravariant in the index i and covariant in the index j, and require that, for a change of coordinates,

$$\bar{v}_j^i = \frac{\partial \bar{x}^i}{\partial x^k}\frac{\partial x^l}{\partial \bar{x}^j} v_l^k. \tag{3-80}$$

By v^{ij} we denote a second order tensor which is contravariant in both indices and require that

$$\bar{v}^{ij} = \frac{\partial \bar{x}^i}{\partial x^k}\frac{\partial \bar{x}^j}{\partial x^l} v^{kl}. \tag{3-81}$$

Similar definitions are given for tensors of third, fourth and higher orders. For example, w_k^{ij} denotes a tensor contravariant in i and j, and covariant in k, whereby

$$\bar{w}_k^{ij} = \frac{\partial \bar{x}^i}{\partial x^p}\frac{\partial \bar{x}^j}{\partial x^r}\frac{\partial x^s}{\partial \bar{x}^k} w_s^{pr}. \tag{3-82}$$

The tensors of higher order appear in many geometrical and physical theories. In particular, they are needed for the basic operations on vector fields, as we shall illustrate.

Each tensor of higher order can also be obtained by first assigning components in standard coordinates (ξ^i) and then using the appropriate rule to obtain components in an arbitrary coordinate system (x^i). This is proved just as for contravariant and covariant vectors above. In particular, starting with a set of functions $f_{ijklm}(\xi^1, \ldots, \xi^n)$, where each index $i, j, \ldots$ runs from 1 to n, one can choose some indices as contravariant, others as covariant , in some chosen order in each case, and then introduce corresponding tensor components in D: for example,

$$W_{klj}^{im}(\xi^1, \ldots, \xi^n) = f_{ijklm}(\xi^1, \ldots, \xi^n).$$

Then the rules tell us how to define the components w_{klj}^{im} in (x^i), to obtain a tensor of the type indicated. All these tensors, arising from the same set of functions in standard coordinates, are said to be *associated*, and we regard them as different aspects of one geometric object **w**. From $f_{ij}(\xi^1, \ldots, \xi^n)$ we obtain in this way six associated tensors

$$u_{ij}, \quad u_{ji}, \quad u_j^i, \quad u_i^j, \quad u^{ij}, \quad u^{ji}.$$

A very important tensor of second order arises in discussing distance in curvilinear coordinates. We start with standard coordinates (ξ^i) in D. For a point moving along a path in D the usual reasoning leads us to the formula

$$\frac{ds}{dt} = \left(\sum_{i=1}^{n}\left(\frac{d\xi^i}{dt}\right)^2\right)^{\frac{1}{2}}$$

for the speed. Correspondingly, we write

$$ds^2 = d\xi^i\, d\xi^i$$

(summation convention!) for the square of the "element of distance" ds. If we now introduce curvilinear coordinates (x^i), but require that *distance be unchanged*, then we find

$$ds^2 = d\xi^i\, d\xi^i = \left(\frac{\partial \xi^i}{\partial x^k}\, dx^k\right)\left(\frac{\partial \xi^i}{\partial x^l}\, dx^l\right) = \frac{\partial \xi^i}{\partial x^k}\frac{\partial \xi^i}{\partial x^l}\, dx^k\, dx^l.$$

Hence in curvilinear coordinates ds^2 appears as a "quadratic differential form"

$$ds^2 = g_{kl}\, dx^k\, dx^l, \tag{3–83}$$

where

$$g_{kl} = \frac{\partial \xi^i}{\partial x^k}\frac{\partial \xi^i}{\partial x^l} = g_{lk} \quad (k = 1, \ldots, n,\ l = 1, \ldots, n). \tag{3–84}$$

(Thus the matrix (g_{kl}) is symmetric.) By this procedure we assign sets of components g_{kl} in every curvilinear coordinate system. For two sets of coordinates (x^i) and $(\bar{x}^i)$ we have

$$g_{ij} = \frac{\partial \xi^r}{\partial x^i}\frac{\partial \xi^r}{\partial x^j}, \quad \bar{g}_{ij} = \frac{\partial \xi^r}{\partial \bar{x}^i}\frac{\partial \xi^r}{\partial \bar{x}^j},$$

and hence

$$\bar{g}_{ij} = \left(\frac{\partial \xi^r}{\partial x^k}\frac{\partial x^k}{\partial \bar{x}^i}\right)\left(\frac{\partial \xi^r}{\partial x^l}\frac{\partial x^l}{\partial \bar{x}^j}\right) = \frac{\partial x^k}{\partial \bar{x}^i}\frac{\partial x^l}{\partial \bar{x}^j}\frac{\partial \xi^r}{\partial x^k}\frac{\partial \xi^r}{\partial x^l}$$

$$= \frac{\partial x^k}{\partial \bar{x}^i}\frac{\partial x^l}{\partial \bar{x}^j}\, g_{kl}.$$

Therefore, g_{ij} is indeed a second order tensor, covariant in both indices. We call g_{ij} the *fundamental metric tensor*.

We observe that the g_{ij} reduce to δ_{ij} (Kronecker delta) in standard coordinates, since in the (ξ^i)

$$ds^2 = d\xi^i\, d\xi^i = \delta_{ij}\, d\xi^i\, d\xi^j.$$

Hence we can regard the g_{ij} as the covariant tensor obtained from δ_{ij} (constant functions) in standard coordinates. From these functions we obtain six associated tensors as above. However, since $\delta_{ji} = \delta_{ij}$, $g_{ji} = g_{ij}$ (as noted above), so that these two associated tensors are the same. By setting $G_j^i = \delta_{ij}$ in standard coordinates (ξ^i), we obtain a mixed second order tensor g_j^i. In the (x^i) coordinates we obtain

$$g_j^i = \frac{\partial x^i}{\partial \xi^k}\frac{\partial \xi^l}{\partial x^j}\,\delta_{kl} = \frac{\partial x^i}{\partial \xi^k}\frac{\partial \xi^k}{\partial x^j} = \delta_{ij}$$

(since the matrices $(\partial x^i/\partial \xi^j)$ and $(\partial \xi^i/\partial x^j)$ are inverses of each other).

Hence $$g^i_j = g^j_i = \delta_{ij} \quad \text{(constant functions)} \tag{3–85}$$

in all coordinates. We often write δ^i_j for this tensor. Next, by setting $G^{ij} = \delta_{ij}$ in standard coordinates, we obtain a contravariant second order tensor g^{ij}; as above,

$$g^{ij} = \frac{\partial x^i}{\partial \xi^k}\frac{\partial x^j}{\partial \xi^l}\delta_{kl} = \frac{\partial x^i}{\partial \xi^k}\frac{\partial x^j}{\partial \xi^k} = g^{ji}. \tag{3–86}$$

Comparison with (3–84) suggests that (g^{ij}) is the inverse matrix of (g_{ij}): that is, that

$$g_{i\alpha}g^{\alpha j} = \delta_{ij}. \tag{3–87}$$

This can be directly verified (Problem 6 below). It follows that both matrices are nonsingular. We denote by g the determinant of (g_{ij}):

$$g = \det (g_{ij}). \tag{3–88}$$

Thus g is a scalar function, depending on the coordinate system; it is not an invariant. One can show that *g must be positive* (Problem 7 below).

We thus see that our fundamental metric tensor has three aspects:

$$g_{ij} = g_{ji}, \quad g^i_j = g^j_i = \delta^j_i, \quad g^{ij} = g^{ji} \quad \text{(inverse matrix of the } g_{ij}). \tag{3–89}$$

Tensor algebra. Tensors can be combined in certain ways to yield new tensors. We can *add* two tensors of the same type, to obtain another of the same type (same number of contravariant and covariant indices): for example,

$$v^i_j + w^i_j = u^i_j \tag{3–90}$$

defines the sum of the tensors v^i_j and w^i_j. Here we are giving the rule for a typical (x^i) coordinate system, and assume that v^i_j and w^i_j are both given in the neighborhood of a point. Thus we understand that, in an $(\bar{x}^i)$ coordinate system,

$$\bar{v}^i_j + \bar{w}^i_j = \bar{u}^i_j.$$

We can now verify that the sum is indeed a tensor (Problem 8 below).

We can *multiply* a tensor by an *invariant*; for example, from u_{ij} and f we form

$$fu_{ij} = v_{ij} \tag{3–91}$$

and verify that this again leads to a tensor v_{ij} of the type indicated. (Problem 9 below).

We can multiply two tensors to form their *tensor product:* for example, from first order tensors u_i, v_j, w^k, z^l we can form

$$s_{ij} = u_i v_j, \quad t^k_i = u_i w^k, \quad q^{kl} = w^k z^l$$

and can verify that these are tensors of second order of the types indicated (Problem 10 below).

We can *contract* a tensor by summing over a pair of indices, one contravariant, one covariant. For example, from u^{ij}_{kl} we can form

$$w_k^j = u_{ki}^{ij} \quad \text{and} \quad v_k^i = u_{kj}^{ij}$$

(summed over i and j respectively). Each such contraction lowers the order by two. We can verify that the procedure described always leads to a tensor (Problem 11 below). In particular, if a_{ij}, u^i and v^j are tensors, then

$$a_{ij}u^iv^j = f$$

is an invariant, for $a_{ij}u^kv^l$ is a product of tensors and hence is a tensor of order 4; f is obtained from this tensor by contracting twice.

We can *raise or lower indices* of a given tensor $u_{kl\dots}^{ij\dots}$ by the following procedure. We multiply by the metric tensor g_{pq} or g^{pq} to first increase the order by two. Then we use contraction, based on q and one selected index of $u_{kl\dots}^{ij\dots}$, to lower the order by two. For example,

$$g_{i\alpha}u_{kl}^{\alpha j} \text{ is a new tensor } u_{kli}^{j}, \quad g^{k\alpha}u_{\alpha l}^{ij} \text{ is a new tensor } u_l^{ijk}.$$

We have some freedom here in where we locate the new index. However, this freedom corresponds to a choice among several associated tensors. In fact, all tensors obtained by this process are associated with the initial tensor $u_{kl\dots}^{ij\dots}$. To see this, we carry out the first process in standard coordinates (ξ^i). In these coordinates, $g_{ij} = \delta_{ij}$, $g^{ij} = \delta_{ij}$, $u_{kl\dots}^{ij\dots}$ becomes $U_{kl\dots}^{ij\dots}$ and $g_{i\alpha}u_{kl\dots}^{\alpha i\dots}$ becomes

$$\delta_{i\alpha}U_{kl\dots}^{\alpha j\dots} = U_{kl\dots i}^{j\dots}.$$

But the left side equals $U_{kl\dots}^{ij\dots}$. Hence in standard coordinates we have merely moved an index to a new position. This shows that we are dealing with associated tensors.

Covariant and contravariant derivatives. We seek a process, analogous to forming partial derivatives, which we can apply to tensor components to yield a new tensor. In doing so we shall add one index. We can make this covariant or contravariant, and obtain corresponding covariant and contravariant derivatives. We make the process definite by simply requiring that, in standard coordinates (ξ^i), we are in fact computing partial derivatives.

For example, for a (differentiable) contravariant vector u^i, in (x^i), we obtain the corresponding U^i in (ξ^i), then form $W_j^i = \partial U^i/\partial \xi^j$. The W_j^i determine a mixed second order tensor, becoming w_j^i in (x^i) and, since we have added a covariant index, we call w_j^i the *covariant derivative* of u^i and denote it by $\Delta_j u^i$. It remains to calculate the w_j^i. We have

$$U^i = \frac{\partial \xi^i}{\partial x^l}u^l,$$

$$W_j^i = \frac{\partial U^i}{\partial \xi^j} = \frac{\partial U^i}{\partial x^k}\frac{\partial x^k}{\partial \xi^j} = \left(\frac{\partial^2 \xi^i}{\partial x^k\,\partial x^l}u^l + \frac{\partial \xi^i}{\partial x^l}\frac{\partial u^l}{\partial x^k}\right)\frac{\partial x^k}{\partial \xi^j},$$

$$w_j^i = W_q^p\frac{\partial x^i}{\partial \xi^p}\frac{\partial \xi^q}{\partial x^j} = \frac{\partial^2 \xi^p}{\partial x^k\,\partial x^l}\frac{\partial x^k}{\partial \xi^q}\frac{\partial x^i}{\partial \xi^p}\frac{\partial \xi^q}{\partial x^j}u^l + \frac{\partial \xi^p}{\partial x^l}\frac{\partial x^k}{\partial \xi^q}\frac{\partial x^i}{\partial \xi^p}\frac{\partial \xi^q}{\partial x^j}\frac{\partial u^l}{\partial x^k}.$$

Here we can simplify by noting various products of inverse matrices appearing and can write our result as

$$\Delta_j u^i = w^i_j = \frac{\partial u^i}{\partial x^j} + \Gamma^i_{jl} u^l, \tag{3–92}$$

where

$$\Gamma^i_{jl} = \frac{\partial^2 \xi^p}{\partial x^j\, \partial x^l} \frac{\partial x^i}{\partial \xi^p} = \Gamma^i_{lj}. \tag{3–93}$$

The Γ^i_{jl} are called *Christoffel symbols;* they are not the components of a tensor. One can express them in terms of the components of the fundamental metric tensor (Problem 12 below):

$$\Gamma^i_{jl} = \tfrac{1}{2} g^{is} \left(\frac{\partial g_{js}}{\partial x^l} + \frac{\partial g_{ls}}{\partial x^j} - \frac{\partial g_{jl}}{\partial x^s} \right). \tag{3–94}$$

By proceeding in this way we obtain covariant derivatives of other types of tensors. For example, we find

$$\Delta_j u_i = \frac{\partial u_i}{\partial x^j} - \Gamma^l_{ij} u_l, \tag{3–95}$$

$$\Delta_k u^i_j = \frac{\partial u^i_j}{\partial x^k} + \Gamma^i_{kl} u^l_j - \Gamma^l_{kj} u^i_l, \tag{3–96}$$

$$\Delta_k u^i_{jh} = \frac{\partial u^i_{jh}}{\partial x^k} + \Gamma^i_{kl} u^l_{jh} - \Gamma^l_{kj} u^i_{lh} - \Gamma^l_{kh} u^i_{jl}. \tag{3–97}$$

From these results the general pattern should be clear.

Contravariant derivatives are found in similar fashion. In fact one has the simple rule: for every tensor A,

$$\Delta^k A = g^{k\alpha}\, \Delta_\alpha A. \tag{3–98}$$

To justify this we observe that the right side is a tensor of the type desired and that in standard coordinates (ξ^i) it reduces to the set of partial derivatives of the components of A with respect to x^k. Equation (3–98) expresses contravariant derivatives in terms of covariant derivatives and shows that $\Delta_k A$ and $\Delta^k A$ are associated tensors.

For an invariant f, one finds

$$\Delta_k f = \frac{\partial f}{\partial x^k}, \quad \Delta^k f = g^{k\alpha} \frac{\partial f}{\partial x^k}. \tag{3–99}$$

Thus the covariant derivative of an invariant f is simply the *gradient* of f:

$$\Delta_k f = \operatorname{grad} f = \frac{\partial f}{\partial x^k}. \tag{3–100}$$

The contravariant derivative of f is the associated contravariant vector.

We can now define the *divergence* of a vector field. We want this to reduce to the usual divergence in standard coordinates. If $\mathbf{u}$ has contravariant components u^i, then we define

$$\operatorname{div} \mathbf{u} = \operatorname{div} u^i = \Delta_\alpha u^\alpha, \tag{3–101}$$

for this reduces to the expected expression

$$\frac{\partial U^1}{\partial \xi^1} + \cdots + \frac{\partial U^n}{\partial \xi^n} \tag{3–102}$$

in standard coordinates. Since $\Delta_\alpha u^\alpha$ is obtained by contraction of the tensor $\Delta_i u^j$ it is a tensor of order 0, an invariant. We can obtain the same invariant from the covariant components of $\mathbf{u}$:

$$\operatorname{div} \mathbf{u} = \operatorname{div} u_i = \Delta^\alpha u_\alpha, \tag{3–103}$$

for in standard coordinates the components become $U_i = U^i$ and again (3–102) is obtained.

One can also obtain (Problem 13 below) the following useful expression for the divergence:

$$\operatorname{div} \mathbf{u} = \frac{1}{\sqrt{g}} \frac{\partial}{\partial x^\lambda} (\sqrt{g}\, g^{\alpha\lambda} u_\alpha). \tag{3–104}$$

From the divergence and gradient we obtain the *Laplacian* of an invariant:

$$\nabla^2 f = \operatorname{div} \operatorname{grad} f = \Delta^\alpha \Delta_\alpha f. \tag{3–105}$$

We can also obtain this from the contravariant components of grad f:

$$\nabla^2 f = \Delta_\alpha \Delta^\alpha f. \tag{3–106}$$

Corresponding to (3–104) one has also the useful expression:

$$\nabla^2 f = \frac{1}{\sqrt{g}} \frac{\partial}{\partial x^\lambda} \left(\sqrt{g}\, g^{\alpha\lambda} \frac{\partial f}{\partial x^\alpha}\right). \tag{3–107}$$

For the curl of a vector field our usual definition is specially adapted to 3-dimensional space. The appropriate generalization turns out to be a *second order* tensor. We define the *curl* of $\mathbf{u}$ to be the second order tensor

$$b_{ij} = \Delta_i u_j - \Delta_j u_i. \tag{3–108}$$

However, we verify (Problem 14 below) that this is equivalent to

$$\operatorname{curl} \mathbf{u} = b_{ij} = \frac{\partial u_j}{\partial x^i} - \frac{\partial u_i}{\partial x^j}. \tag{3–109}$$

Thus the form which (3–108) takes in standard coordinates is the form which it takes in arbitrary coordinates. (The same is true for the gradient operation.) By (3–109) we have

$$b_{ij} = -b_{ji}, \quad b_{ij} = 0 \text{ if } i = j. \tag{3–110}$$

In 3-dimensional space, the curl has nine components. By (3–110) three of these are 0 and the other six come in pairs of opposite sign:

$$b_{23} = -b_{32} = \frac{\partial u_3}{\partial x^2} - \frac{\partial u_2}{\partial x^3}, \quad b_{31} = -b_{13} = \frac{\partial u_1}{\partial x^3} - \frac{\partial u_3}{\partial x^1},$$
$$b_{12} = -b_{21} = \frac{\partial u_2}{\partial x^1} - \frac{\partial u_1}{\partial x^2}. \tag{3–111}$$

These three values are the familiar components of the curl vector in space: $\partial u_z/\partial y - \partial u_y/\partial z, \ldots$ It is because these *three* quantities determine the curl that the curl can be interpreted as a vector in 3-dimensional space. In 4-dimensional space, six components are needed and one cannot associate these with a vector. For $n = 2$ one has essentially one component

$$b_{12} = -b_{21} = \frac{\partial u_2}{\partial x^1} - \frac{\partial u_1}{\partial x^2}$$

(which is the $\partial Q/\partial x - \partial P/\partial y$ of Green's theorem—see Chapter 5).

Case of orthogonal curvilinear coordinates. We consider here only the case of 3-dimensional space. In Section 3–7 it is shown that for orthogonal coordinates one has $ds^2 = \alpha^2\,dx^2 + \beta^2\,dy^2 + \gamma^2\,dz^2$—that is, in the present notations,

$$ds^2 = g_{11}\,dx^1\,dx^1 + g_{22}\,dx^2\,dx^2 + g_{33}\,dx^3\,dx^3.$$

Thus (g_{ij}) is a diagonal matrix and its inverse is the diagonal matrix diag $(1/g_{11}, 1/g_{22}, 1/g_{33})$, its determinant is $g = g_{11}g_{22}g_{33}$. From (3–107) the Laplacian of f is

$$\frac{1}{\sqrt{g}}\frac{\partial}{\partial x^1}\left(\frac{\sqrt{g}}{g_{11}}\frac{\partial f}{\partial x^1}\right) + \frac{1}{\sqrt{g}}\frac{\partial}{\partial x^2}\left(\frac{\sqrt{g}}{g_{22}}\frac{\partial f}{\partial x^2}\right) + \frac{1}{\sqrt{g}}\frac{\partial}{\partial x^3}\left(\frac{\sqrt{g}}{g_{33}}\frac{\partial f}{\partial x^3}\right).$$

With $g_{11} = \alpha^2$, $g_{22} = \beta^2$, $g_{33} = \gamma^2$, $g = \alpha^2\beta^2\gamma^2$, this reduces to (3–63) above.

One can discuss div $\mathbf{u}$ in the same way for the case of orthogonal coordinates. One appears to obtain a formula differing from (3–61). However, as (3–57′) shows, the components p_u, p_v, p_w of Section 3–8 are neither contravariant nor covariant components of the vector $\mathbf{p}$; these components are obtained geometrically by perpendicular projection. From (3–57′) we see that p_u/α, p_v/β, p_w/γ are contravariant components of $\mathbf{p}$ and can now verify that (3–61) is in agreement with (3–104) for the case of orthogonal coordinates.

For further information on tensors, one is referred to the books by Brand, Coburn, Hay, Levi-Civita, Rainich and Weyl listed at the end of this chapter.

Problems

1. In E^2 let (ξ^1, ξ^2) be standard coordinates and let (x^1, x^2) be new coordinates given by $x^1 = 3\xi^1 + 2\xi^2$, $x^2 = 4\xi^1 + 3\xi^2$. Find the (x^i) components of the following tensors, for which the components in standard coordinates are given.
(a) u_i, where $U_1 = \xi^1\xi^2$, $U_2 = \xi^1 - \xi^2$.
(b) v^i, where $V^1 = \xi^1 \cos \xi^2$, $V^2 = \xi^1 \sin \xi^2$.

(c) w_{ij}, where $W_{11} = 0$, $W_{12} = \xi^1\xi^2$, $W_{21} = -\xi^1\xi^2$, $W_{22} = 0$.
(d) z^i_j, where $Z^1_1 = \xi^1 + \xi^2$, $Z^1_2 = Z^2_1 = 3\xi^1 + 2\xi^2$, $Z^2_2 = \xi^1 - \xi^2$.

2. For the coordinates of Problem 1 find (a) the components of the fundamental metric tensor g_{ij} in standard coordinates and in the (x^i), and (b) the g^i_j and g^{ij} in both sets of coordinates.

3. With reference to the coordinates and tensors of Problem 1 and with the aid of the results of Problem 2, find the following tensor components in (x^i): (a) u^i, associated to u_i; (b) v_i, associated to v^i; (c) w^{ij}, associated to w_{ij}.

4. (a) Show that if a tensor has all components 0 at a point in one coordinate system, then all components are 0 at the point in every allowed coordinate system.
(b) Show that if two tensors of the same type have corresponding components equal (identically) in one coordinate system, then corresponding components are equal in every allowed coordinate system. (One then says that the tensors are equal.)
(c) Show that if a tensor u_{ij} is such that $u_{ij} = u_{ji}$ in one coordinate system (x^i), then $\bar{u}_{ij} = \bar{u}_{ji}$ in every other coordinate system $(\bar{x}^i)$. (One then calls the tensor *symmetric*.)
(d) Show that if a tensor u_{ij} is such that $u_{ij} = -u_{ji}$ in one coordinate system (x^i), then $\bar{u}_{ij} = -\bar{u}_{ji}$ in every other coordinate system $(\bar{x}^i)$.
(One calls such a tensor *antisymmetric*.)

5. Prove: if components U_i are defined in (ξ^i), and corresponding components u_i are defined in every other allowed coordinate system (x^i) by the equation $u_i = (\partial \xi^j/\partial x^i)U_j$, then the components $\bar{u}_i$ in $(\bar{x}^i)$ satisfy (3–75), so that a covariant vector has been defined.

6. Prove (3–87) from (3–84) and (3–86).

7. Prove that $g = \det(g_{ij})$ is positive. [Hint: Show from (3–84) that (g_{ij}) is the product of two matrices and then take determinants.]

8. (a) Show that, the tensors v^i_j and w^i_j being given in a neighborhood D, Eq. (3–90) defines u^i_j as components of a tensor.
(b) Generalize the result of part (a) to addition of two tensors of arbitrary type.

9. Show that multiplication of all tensor components by the same invariant, as in (3–91), defines a new tensor.

10. Let u_i, v_j, w^k, z^l, p_{mh} be tensors. Show that each of the following tensor products is a tensor:
(a) $u_iv_j = s_{ij}$, (b) $u_iw^k = t^k_i$, (c) $w^kz^l = q^{kl}$, (d) $w^kp_{mh} = d^k_{mh}$.

11. (a) Let u^{ij}_{kl} be a tensor of the type indicated. Show that each of the following is a tensor: $w^j_k = u^{ij}_{ki}$, $v^i_k = u^{ij}_{kj}$.
(b) Prove generally that contraction of a tensor yields a tensor.

12. Show from (3–93) that the Γ^i_{jl} can be expressed as in (3–94). [Hint: Show first that $\Gamma^i_{jl}g_{is} = (\partial^2\xi^\alpha/\partial x^j\partial x^l)(\partial\xi^\alpha/\partial x^s)$. Next, with the aid of (3–84), compute the quantity in parentheses in (3–94) and show that it equals $2\Gamma^i_{jl}g_{is}$. Now multiply both sides by g^{st}.]

13. Derive the formula (3–104). [Hint: First show that

$$\operatorname{div} \mathbf{u} = (\partial u^\alpha/\partial x^\alpha) + \Gamma^\alpha_{\alpha l}u^l.$$

Next put $i = j = \alpha$ in (3–94), multiply out and show that the last two terms cancel, so that $\Gamma^{\alpha}_{\alpha l} = \frac{1}{2}g^{\alpha s}(\partial g_{\alpha s}/\partial x^l)$. Now obtain $\partial g/\partial x^l$, from the determinant g, as a sum of n determinants, the α'th of which is obtained from the given determinant by differentiating each element in the α-th row with respect to x^l. Expand each of these determinants by minors of the α-th row to obtain

$$\partial g/\partial x^l = A_{\alpha s}\partial g_{\alpha s}/\partial x^l,$$

where $A_{\alpha s}$ is the cofactor of $g_{\alpha s}$ in the determinant g. Now $g^{\alpha s} = g^{s\alpha} = A_{\alpha s}/g$ (see Problem 7 following Section 1–4). Thus conclude that $\Gamma^{\alpha}_{\alpha l} = (2g)^{-1}(\partial g/\partial x^l)$ and show that this permits one to write div **u** in the form asserted.]

14. Derive (3–109) from (3–108), using (3–95).

15. We define the *norm* of a vector **u** to be its norm or length in standard coordinates. Thus for components U^i or U_i the length is

$$|\mathbf{u}| = (U^iU^i)^{\frac{1}{2}} = (U_iU_i)^{\frac{1}{2}}.$$

Show that $|\mathbf{u}|$ can be written in the forms:

$$|\mathbf{u}| = (u_iu^i)^{\frac{1}{2}}, \quad |\mathbf{u}| = (g_{ij}u^iu^j)^{\frac{1}{2}}, \quad |\mathbf{u}| = (g^{ij}u_iu_j)^{\frac{1}{2}},$$

each of which shows that the length is an invariant.

16. The *inner product* of two vectors **u, v** is defined to be the invariant

$$(\mathbf{u}, \mathbf{v}) = u_iv^i.$$

Show that this can be written in the forms:

$$(\mathbf{u}, \mathbf{v}) = g_{ij}u^iv^j, \quad (\mathbf{u}, \mathbf{v}) = g^{ij}u_iv_j$$

and that

$$(\mathbf{u}, \mathbf{u}) = |\mathbf{u}|^2.$$

17. (a) Prove: if u^i, v^j are contravariant vectors and $b_{ij}u^iv^j$ is an invariant, then b_{ij} is a covariant tensor.

(b) Prove generally: if $u_i, v_j, \ldots, w^k, z^l, \ldots$ are covariant and contravariant vectors and

$$b^{ij\ldots}_{kl\ldots}u_iv_j\ldots w^k z^l \ldots$$

is an invariant, then $b^{ij\ldots}_{kl\ldots}$ is a tensor of the type indicated.

Answers

1. (a) $u_1 = -36x^1x^1 + 51x^1x^2 - 18x^2x^2 - 28x^1 + 20x^2$, $u_2 = 24x^1x^1 - 34x^1x^2 + 12x^2x^2 + 21x^1 - 15x^2$. (b) $v^1 = (3x^1 - 2x^2)\,[3\cos(-4x^1 + 3x^2) + 2\sin(-4x^1 + 3x^2)]$, $v^2 = (3x^1 - 2x^2)\,[(4\cos(-4x^1 + 3x^2) + 3\sin(-4x^1 + 3x^2)]$. (c) $w_{11} = w_{22} = 0$, $w_{12} = -w_{21} = -12x^1x^1 + 17x^1x^2 - 6x^2x^2$. (d) $z^1_1 = -71x^1 + 49x^2$, $z^1_2 = 53x^1 - 36x^2$, $z^2_1 = -103x^1 + 72x^2$, $z^2_2 = 77x^1 - 53x^2$.

2. In standard coordinates all reduce to δ_{ij}. In (x^i), $g_{11} = g^{22} = 25$, $g_{12} = g_{21} = -g^{12} = -g^{21} = -18$, $g_{22} = g^{11} = 13$, $g^i_j = \delta_{ij}$ always.

3. (a) $u^1 = 13u_1 + 18u_2$, $u^2 = 18u_1 + 25u_2$, where u_1, u_2 are as in the answer to Prob. 1 (a). (b) $v_1 = 25v^1 - 18v^2$, $v_2 = -18v^1 + 13v^2$, where v^1, v^2 are as in the answer to Prob. 1 (b). (c) $w^{11} = w^{22} = 0$, $w^{12} = -w^{21} = w_{12}$, where w_{12} is as in the answer to Problem 1 (c).

Suggested References

Brand, Louis, *Vector and Tensor Analysis.* New York: John Wiley and Sons, Inc., 1947.

Coburn, N., *Vector and Tensor Analysis.* New York: Macmillan, 1955.

Gibbs, J. Willard, *Vector Analysis.* New Haven: Yale University Press, 1913.

Hay, G. E., *Vector and Tensor Analysis.* New York: Dover, 1954.

Levi-Civita, T., *The Absolute Differential Calculus,* transl. by M. Long. London: Blackie and Son, 1927.

Phillips, H. B., *Vector Analysis.* New York: John Wiley and Sons, Inc., 1933.

Rainich, G. Y., *Mathematics of Relativity.* New York: John Wiley and Sons, Inc., 1950. Especially Chapters 1, 2, and 4.

Struik, Dirk J., *Lectures on Classical Differential Geometry.* Cambridge: Addison-Wesley Press, Inc., 1950.

Weyl, Hermann, *Space, Time, Matter,* transl. by H. L. Brose. London Methuen, 1922.

CHAPTER 4

Integral Calculus of Functions of Several Variables

4–1 Introduction. The purpose of this chapter is to study integration processes which involve functions of two or more variables. In particular, multiple integrals:

$$\iint\limits_R f(x, y)\, dx\, dy, \quad \iiint\limits_R f(x, y, z)\, dx\, dy\, dz, \ldots,$$

as well as single and multiple integrals depending on a parameter:

$$\int_a^b f(x, t)\, dx, \quad \int_a^{t^2} f(x)\, dx, \quad \iint\limits_R f(x, y, t)\, dx\, dy, \ldots,$$

will be investigated.

Basic for all these topics are the notions of *definite integral* of a function of one variable:

$$\int_a^b f(x)\, dx = \lim_{\substack{n\to\infty \\ \max \Delta_i x \to 0}} \sum_{i=1}^{n} f(x_i^*)\, \Delta_i x \tag{4–1}$$

and of *indefinite integral* of a function of one variable:

$$\int f(x)\, dx = F(x) + C. \tag{4–2}$$

The reader is referred to Section 0–10 for a statement of the definitions and basic properties of the definite and indefinite integral.

In order to ensure that the definite and indefinite integral and the relations between them are fully understood, the first two sections will be devoted to a consideration of the *numerical evaluation* of both definite and indefinite integrals. The procedures described will be useful themselves for applications, but it is hoped that they will serve the additional purpose of clarifying the concepts involved.

4–2 Numerical evaluation of definite integrals. If a definite integral such as

$$\int_0^{\frac{1}{2}\pi} x \sin x\, dx$$

is given, then its value can be found by the formula

$$\int_a^b f(x)\, dx = F(b) - F(a), \quad F'(x) = f(x), \tag{4–3}$$

provided an indefinite integral $F(x)$ of $f(x)$ can be found for the given interval. Thus in this example,

$$\int_0^{\frac{1}{2}\pi} x \sin x \, dx = (-x \cos x + \sin x)\Big|_0^{\frac{1}{2}\pi} = 1,$$

where the indefinite integral was found by integration by parts [Eq. (0–178)].

While the methods for finding indefinite integrals are extensive and long tables for this purpose do exist [cf. the references following Eq. (0–178)], the methods are far from complete. The integrals

$$\int e^{-x^2}\, dx, \quad \int \frac{\sin x}{x}\, dx, \quad \int \frac{dx}{\sqrt{2 - \cos^2 x}}$$

will be found unassailable by any of the familiar methods; it can, in fact, be shown that these integrals cannot be expressed in terms of elementary functions.

Accordingly, the numerical evaluation of a definite integral such as

$$\int_0^1 e^{-x^2}\, dx$$

cannot be carried out by (4–3) in terms of elementary functions. To evaluate this integral, two basic methods are available: use of power series and use of numerical formulas such as the trapezoidal rule and Simpson's rule. Because of the connection of the integral with area, the integral can also be computed by plotting the function $y = f(x) = e^{-x^2}$ and measuring the area bounded by the x axis, the plotted curve and the lines $x = 0$, $x = 1$. This area can be measured by counting squares on a graph paper or by using a *planimeter*, a mechanical instrument devised for this purpose. One can even cut out the area being measured and weigh the piece of paper cut out; if the density of the paper (gm per sq cm) is known, the weight of the piece divided by the density will yield the desired area. A variety of special definite integrals can also be evaluated by the method of *residues*, described in Chapter 9.

Here we shall confine attention to the simpler numerical formulas, but use the concept of area to help in understanding these formulas. Thus we shall think in terms of the area shaded in Fig. 4–1. Here $f(x)$ is continuous for $a \leqq x \leqq b$ and the integral

$$\int_a^b f(x)\, dx$$

is precisely equal to the area. When $f(x)$ is negative, areas are counted negatively.

Rectangular formula.

$$\int_a^b f(x)\, dx \sim f(x_1^*)\, \Delta_1 x + \cdots + f(x_n^*)\, \Delta_n x. \tag{4–4}$$

Here a subdivision of the interval $a \leqq x \leqq b$ is assumed: $a = x_0 < x_1 < \cdots < x_n = b$ and $\Delta_1 x, \ldots, \Delta_n x$ are the lengths of the successive subintervals: $\Delta_i x = x_i - x_{i-1}$. The numbers $x_1^*, \ldots, x_n^*$ are chosen arbitrarily within the respective subintervals, as shown in Fig. 4–1. The $\sim$ sign in (4–4) indicates that only an approximation is intended. However, the continuity of $f(x)$ implies the existence of the integral as limiting value of the right-hand side when the number n of subdivisions becomes infinite, while the maximum $\Delta_i x$ approaches 0; in other words, for a fine enough subdivision, the right-hand side differs from the left-hand side by less than a prescribed number ϵ, no matter how $x_1^*, \ldots, x_n^*$ are chosen. Actually, the law of the mean (0–172) implies that for each subdivision $x_1^*, \ldots, x_n^*$ can be chosen so that $f(x_1^*)\,\Delta_1 x$ equals the integral from a to x_1, $f(x_2^*)\,\Delta_2 x$ equals the integral from x_1 to $x_2, \ldots$; with these choices the right-hand side of (4–4) *equals* the left. However, in general we do not know how to choose $x_1^*, \ldots, x_n^*$ to achieve this equality unless we know the value of the integral over each subinterval.

As Fig. 4–1 shows, the right-hand side represents the sum of the areas of the rectangles with bases $\Delta_1 x, \ldots, \Delta_n x$ shown, hence the term "rectangular formula."

In practice, the points x_i^* will be chosen either at the left of the subinterval, at the right, or in the middle. It is also customary to choose the $\Delta_i x$ to be all equal, with common value

$$\Delta x = \frac{b-a}{n}. \tag{4–5}$$

With the x_i^* all chosen at the right, the formula (4–4) then reduces to

$$\int_a^b f(x)\,dx \sim \frac{b-a}{n}\,[f(x_1) + f(x_2) + \cdots + f(x_n)]. \tag{4–6}$$

This formula shows most clearly that integration is basically a process of addition. This is illustrated in Table 4–1 in which the integral $\int_0^{10} x^2\,dx$ is computed, with $n = 10$, so that $\Delta x = 1$.

Table 4–1

x	$y = x^2$
1	1
2	4
3	9
4	16
5	25
6	36
7	49
8	64
9	81
10	100
	385

$$\int_0^{10} x^2\,dx \sim 385.$$

If one divides both sides of (4–6) by $(b - a)$, one obtains another interpretation of the integral:

$$\frac{1}{b-a}\int_a^b f(x)\,dx \sim \frac{f(x_1) + f(x_2) + \cdots + f(x_n)}{n}. \tag{4–7}$$

The right-hand side is simply the *average* or *arithmetic mean* of the values $f(x_1), \ldots, f(x_n)$. Since the right-hand side approaches the left as $n \to \infty$, the left-hand side can be thought of as the *average* or *arithmetic mean* of the function $f(x)$ over the interval $a \leqq x \leqq b$:

$$\frac{1}{b-a}\int_a^b y\,dx = \bar{y} = [\text{average of } y = f(x)]. \tag{4–8}$$

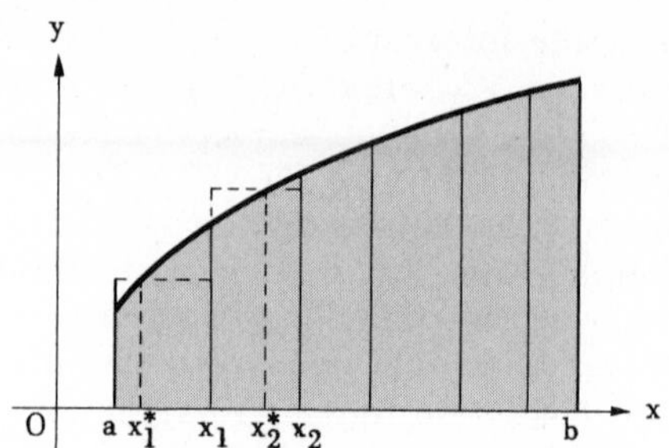

FIG. 4–1. Rectangular formula for $\int_a^b f(x)\,dx$.

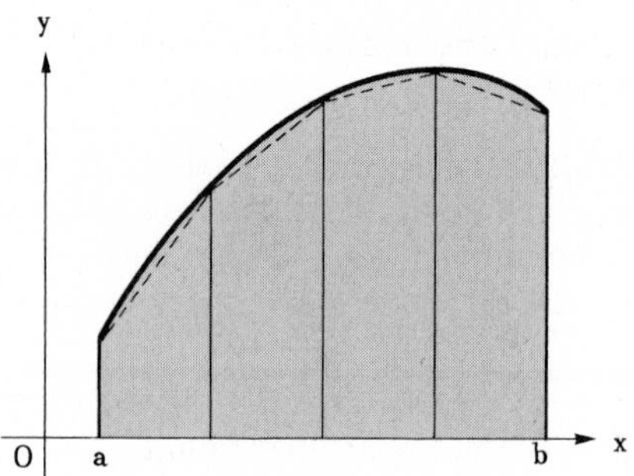

FIG. 4–2. Trapezoidal rule for $\int_a^b f(x)\,dx$.

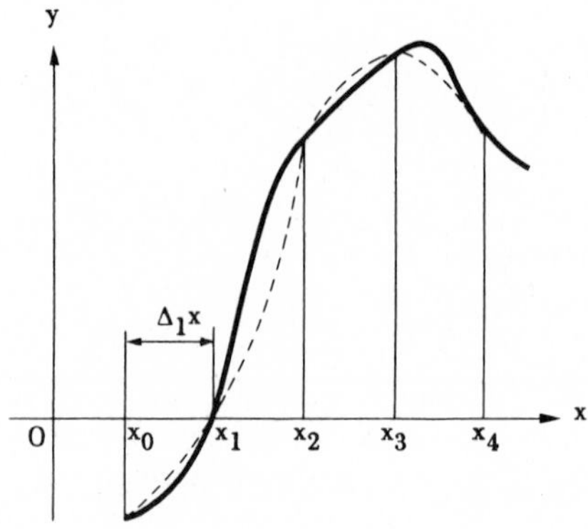

FIG. 4–3. Simpson's rule for $\int_a^b f(x)\,dx$.

Trapezoidal rule.

$$\int_a^b f(x)\,dx \sim \Delta_1 x\,\frac{f(a)+f(x_1)}{2}+\cdots+\Delta_n x\,\frac{f(x_{n-1})+f(b)}{2}. \tag{4–9}$$

Here one replaces the rectangles by the trapezoids shown in Fig. 4–2. The continuity of $f(x)$ implies that

$$f(x_1^*) = \frac{f(a)+f(x_1)}{2}$$

for some choice of x_1^* between a and x_1, so that

$$\Delta_1 x\,\frac{f(a)+f(x_1)}{2} = f(x_1^*)\,\Delta_1 x.$$

In other words, the trapezoidal rule is simply one special way of choosing the points $x_1^*, x_2^*, \ldots, x_n^*$ in (4–4). For a simple curve, such as that shown, it is clear that (4–9) will give a more accurate result than (4–6), for example.

If all the subintervals are chosen equal to Δx, so that (4–5) holds, the trapezoidal rule becomes

$$\int_a^b f(x)\,dx \sim \frac{b-a}{2n}\,[f(a)+2f(x_1)+2f(x_2)+\cdots+2f(x_{n-1})+f(b)]. \tag{4–10}$$

If this is applied to $\int_0^{10} x^2\,dx$, with $\Delta x = 1$ as above, one finds

$$\int_0^{10} x^2\,dx \sim \tfrac{10}{20}\,[0+2+8+\cdots+162+100] = 335$$

This is much closer to the true value, $333\frac{1}{3}$.

Simpson's rule.

$$\int_a^b f(x)\,dx \sim \Delta_1 x\,\frac{f(a)+4f(x_1)+f(x_2)}{3}+\Delta_3 x\,\frac{f(x_2)+4f(x_3)+f(x_4)}{3}$$
$$+\cdots+\Delta_{2n-1}\,x\,\frac{f(x_{2n-2})+4f(x_{2n-1})+f(b)}{3}. \tag{4–11}$$

Here the interval from a to b is subdivided into an *even* number, $2n$, of subintervals by the points $a = x_0, x_1, \ldots, x_{2n} = b$ and moreover

$$\Delta_1 x = x_1 - x_0 = x_2 - x_1, \quad \Delta_3 x = x_3 - x_2 = x_4 - x_3, \ldots; \tag{4–12}$$

i.e., the first two subintervals are equal, the next two are equal, and so on. This is illustrated in Fig. 4–3.

Simpson's rule is based on the idea of replacing the curve $y = f(x)$ between x_0 and x_2 by the parabola

$$y = Ax^2 + Bx + C,$$

which coincides with $y = f(x)$ for $x = x_0$, $x = x_1$, $x = x_2$. The area under the parabola is shown (Prob. 8 below) to depend only on the values of $f(x)$ at these three points and to equal

$$\Delta_1 x \frac{f(x_0) + 4f(x_1) + f(x_2)}{3}. \tag{4-13}$$

This is the first term of (4–11); the other terms are obtained in the same way. The parabolic arcs are shown in Fig. 4–3; it is clear that for a "smooth" function $f(x)$, the area under the parabola will give a good approximation to the area under the given curve.

If the subdivisions are all chosen equal, so that $\Delta_1 x = \Delta_3 x = \cdots = (b - a)/2n$, (4–11) takes the simpler form

$$\int_a^b f(x)\,dx \sim \frac{b-a}{6n}[f(a) + 4f(x_1) + 2f(x_2) + 4f(x_3) + \cdots + 2f(x_{2n-2}) + 4f(x_{2n-1}) + f(b)]. \tag{4-14}$$

Simpson's rule can also be interpreted as a special case of the rectangular rule. The expression (4–13) can be written:

$$2\,\Delta_1 x \frac{f(x_0) + 4f(x_1) + f(x_2)}{6},$$

i.e., as the width Δx of the interval from x_0 to x_2 times a weighted average of the values of $f(x)$ at the ends and center of this interval. It is easily shown that the continuity of $f(x)$ implies existence of a value x^* between x_0 and x_2 such that this weighted average is precisely $f(x^*)$. Accordingly, (4–13) can be written as $\Delta x \cdot f(x^*)$, i.e., as the area of a rectangle with base Δx. If this is done with the other terms of (4–11), this takes on the form of the rectangular formula (4–4). Accordingly, one is assured that, as the number of subdivisions is increased, while the maximum $\Delta_i x$ approaches 0, the expression on the right of (4–11) does approach the integral as limit.

If $f(x)$ is itself a polynomial of second degree in x, Simpson's rule must, of course, give the precise value of the integral. It is of interest to note that this remains true if $f(x)$ is a *third degree* polynomial: $Ax^3 + Bx^2 + Cx + D$ (see Prob. 9 below).

It should be remarked that the methods described apply equally well to functions $f(x)$ given only in tabulated form. In effect, one is interpolating between the tabulated values by a straight line or a parabola in computing the integrals by the above formulas.

Estimate of error. The following theorem is basic for estimates of error in integrals:

$$\textit{If } |f(x)| \leqq M \textit{ for } a \leqq x \leqq b, \textit{ then } \left|\int_a^b f(x)\,dx\right| \leqq M(b-a). \tag{4-15}$$

The function $f(x)$ need not be continuous here, but need only be such that the integral exists. For example, $f(x)$ can be "piecewise continuous," with jump discontinuities, as suggested in Fig. 4–4 (see Section 4–4).

The inequality (4–15) can be proved from (0–173) of the Introduction or directly as follows. By repeated application of the inequality $|a + b| \leqq |a| + |b|$ [cf. (0–4)], one concludes that

$$\left|\sum_{i=1}^{n} f(x_i^*)\,\Delta_i x\right| \leqq \sum_{i=1}^{n} |f(x_i^*)|\,\Delta_i x.$$

Since $|f(x)| \leqq M$, the sum on the right is at most equal to

$$\sum_{i=1}^{n} M\,\Delta_i x = M\sum_{i=1}^{n} \Delta_i x = M(b-a).$$

Thus

$$\left|\sum_{i=1}^{n} f(x_i^*)\,\Delta_i x\right| \leqq M(b-a),$$

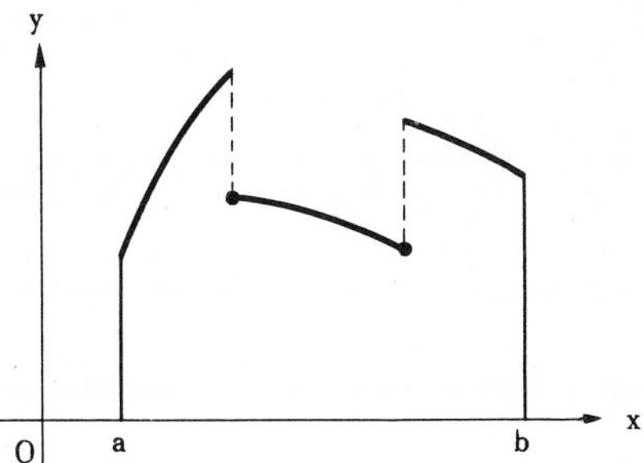

FIG. 4–4. Piecewise continuous function.

and it follows that the integral which is the limit of the left-hand side must satisfy the inequality

$$\left|\int_a^b f(x)\,dx\right| \leqq M(b-a).$$

We now apply (4–15) to estimating the error in using the rectangular, trapezoidal or Simpson's rule. In each case, the function $f(x)$ is replaced by another function $g(x)$, whose integral is then computed. For the rectangular rule, $g(x)$ is a "step-function"; for the trapezoidal rule, $g(x)$ is piecewise linear. The error caused by the replacement of g by f is

$$E = \int_a^b g(x)\,dx - \int_a^b f(x)\,dx = \int_a^b [g(x) - f(x)]\,dx.$$

Now suppose it is known that $|g(x) - f(x)| \leqq M$ for $a \leqq x \leqq b$, i.e., that g differs from f by at most M. Then by (4–15) the absolute value of the error satisfies the inequality:

$$|E| = \left|\int_a^b [g(x) - f(x)]\,dx\right| \leqq M(b-a). \tag{4–16}$$

Thus in each particular application of an approximate formula for integration, one need only estimate the maximum of the difference between the function f and the approximating function; if this difference is known to be at most M, the error is in absolute value at most M times the length of the interval of integration. Usually, the error will be much less than this, partly because the worst difference between f and g is used in the estimate and partly because positive and negative errors may cancel each other.

Remark. If, for example, $f(x)$ is continuous for $a \leqq x \leqq b$, then $|f(x)|$ can also be shown to be continuous and the discussion above shows that

$$\left|\int_a^b f(x)\,dx\right| \leqq \int_a^b |f(x)|\,dx. \tag{4–17}$$

This inequality tells more than (4–15), but (4–15) is used more often. The inequality (0–113):

$$M_1(b-a) \leqq \int_a^b f(x)\,dx \leqq M_2(b-a), \tag{4–18}$$

where $M_1 \leqq f(x) \leqq M_2$ for $a \leqq x \leqq b$, can be used to give both upper and lower estimates of the integral.

For further information on approximate methods of integration, the reader is referred to the books by Willers, Kaplan and Lewis (Sections 7–21 through 7–23), Henrici, and Ralston that are listed at the end of this chapter.

Problems

1. Evaluate the integral $\int_0^{10} (x^3 - 10x^2)\,dx$

(a) by graphing on cross-section paper and counting squares,

(b) by the rectangular rule, using left-hand end points, with 10 equal subintervals,

(c) by the rectangular rule, using right-hand end points, with 10 equal subintervals,

(d) by the trapezoidal rule, with 10 equal subintervals,

(e) by Simpson's rule (4–14), with $n = 5$,

(f) by finding an indefinite integral.

2. Graph the curve $y = \sin x$, $0 \leqq x \leqq \dfrac{\pi}{2}$, with equal scales on the two axes. Then evaluate the integral $\int_0^{\pi/2} \sin x\,dx$ by the rectangular rule (4–4), with 4 equal subintervals and the numbers $x_1^*, \ldots, x_4^*$ chosen so that each of the 4 rectangles appears, from the graph, to give a good approximation to the corresponding area. Compare the result with the precise value of the integral.

3. Use the known result

$$\int_0^1 \frac{1}{1+x^2}\,dx = \frac{\pi}{4}$$

to evaluate π by computing the integral on the left by Simpson's rule (4–14), with $n = 5$.

4. Study the convergence of the rectangular sum on the right of (4–6) to the integral on the left by computing $\int_0^1 x^2\,dx$ with $n = 2, 5, 10, 20$.

5. Compute $\int_0^1 e^{-x^2}\,dx$ by Simpson's rule (4–14) with $n = 1$ and $n = 2$.

6. Show that $\int_0^{\pi/2} e^{-x^2} \sin x\,dx < 1$.

7. The function $\sin(x^2)$ can be accurately approximated by the polynomial $x^2 - \frac{1}{6}x^6$ for small x. Verify graphically that this approximation becomes poorer as x increases and estimate the error in using this approximation to evaluate $\int_0^1 \sin(x^2)\,dx$.

8. Show that, if $f(x) = Ax^2 + Bx + C$, where A, B, C are constants, then

$$\int_{x_0}^{x_0+2h} f(x)\,dx = \frac{h}{3}\,[f(x_0) + 4f(x_0 + h) + f(x_0 + 2h)]. \tag{a}$$

This is the basis of Simpson's rule.

9. Show that Eq. (a) continues to hold if $f(x) = Ax^3 + Bx^2 + Cx + D$.

10. Show that, if $f(x)$ is continuous for $0 \leqq x \leqq 1$, then

$$\lim_{n\to\infty} \frac{1}{n}\left[f\left(\frac{1}{n}\right) + f\left(\frac{2}{n}\right) + \cdots + f\left(\frac{n-1}{n}\right) + f\left(\frac{n}{n}\right)\right] = \int_0^1 f(x)\,dx.$$

11. Using the result of Prob. 10, show that

(a) $\displaystyle\lim_{n\to\infty} \frac{1 + 2 + \cdots + n}{n^2} = \int_0^1 x\,dx = \frac{1}{2},$

(b) $\displaystyle\lim_{n\to\infty} \frac{1^2 + 2^2 + \cdots + n^2}{n^3} = \frac{1}{3},$

(c) $\displaystyle\lim_{n\to\infty} \frac{1^P + 2^P + \cdots + n^P}{n^{P+1}} = \frac{1}{P+1},\quad P \geqq 0,$

(d) $\displaystyle\lim_{n\to\infty} \frac{1}{n}\{(n+1)(n+2)\cdots(2n)\}^{\frac{1}{n}} = \frac{4}{e}.$ [Hint: use $f(x) = \log(1 + x)$.]

12. Find the average value of the function given over the range indicated:

(a) $f(x) = \sin x,\quad 0 \leqq x \leqq \dfrac{\pi}{2};$

(b) $f(x) = \sin x,\quad -\dfrac{\pi}{2} \leqq x \leqq 0;$

(c) $f(x) = \sin^2 x,\quad 0 \leqq x \leqq \dfrac{\pi}{2};$

(d) $f(x) = ax + b,\quad x_1 \leqq x \leqq x_2.$

13. Prove: If $f(x)$ is continuous for $a \leqq x \leqq b$ and

$$\int_a^b [f(x)]^2\,dx = 0,$$

then $f(x) \equiv 0$. [Hint: let $g(x) = [f(x)]^2$. If $g(x)$ is not identically 0, then $g(x_1) > 0$ for some x_1, $a < x_1 < b$. Hence, as in Prob. 7 following Section 2–4, there is a δ such that $g(x) > \frac{1}{2}g(x_1) > 0$ for $x_1 - \delta \leqq x_1 \leqq x_1 + \delta$. Use (4–18) to show that the integral of $g(x)$ from $x_1 - \delta$ to $x_1 + \delta$ must be greater than 0 and hence that the integral from a to b would have to be greater than 0, contrary to assumption.]

Answers

1. (b) -825, (c) -825, (d) -825, (e) $-833\frac{1}{3}$, (f) $-833\frac{1}{3}$.

4. 0.625, 0.440, 0.385, 0.35875.

5. 0.747, 0.746. 7. 0.0082 (worst error at $x = 1$).

12. (a) $\dfrac{2}{\pi}$, (b) $-\dfrac{2}{\pi}$, (c) $\dfrac{1}{2}$, (d) $b + \frac{1}{2}a(x_1 + x_2)$.

4–3 Numerical evaluation of indefinite integrals. Elliptic integrals. As remarked in the preceding section, the indefinite integral

$$\int e^{-x^2}\,dx$$

cannot be expressed in terms of elementary functions. It is, however, possible to evaluate the definite integral

$$\int_a^b e^{-x^2}\,dx$$

as accurately as desired, by the methods of Section 4–2. We shall now see that a similar statement holds for the indefinite integral.

The procedure to be given is an application of the fundamental theorem

$$\frac{d}{dx}\int_a^x f(x)\,dx = f(x) \tag{4–19}$$

referred to in Section 0–9. We first review the significance of (4–19) and its proof.

The integral

$$\int_a^x f(x)\,dx$$

is understood as the definite integral of $f(x)$ between the limits a and x. The "x" in the integral sign has actually nothing to do with the x in the upper limit, and one can make this clear by writing the integral as

$$\int_a^x f(t)\,dt.$$

The re-labelling of the variable of integration has no effect on the value of the integral, for this variable is only a "dummy variable," which is "integrated out."

If $f(x)$ is positive for the range of x considered, the integral in question can be interpreted as the area A under $y = f(x)$ between a and x, as shown in Fig. 4–5. This area A depends on x, and one has

$$A(x) = \int_a^x f(x)\,dx = \int_a^x f(t)\,dt.$$

The content of (4–19) is then the statement

$$\frac{dA}{dx} = f(x). \tag{4–20}$$

This can be made plausible by writing

$$\frac{dA}{dx} = \lim_{\Delta x \to 0} \frac{\Delta A}{\Delta x}$$

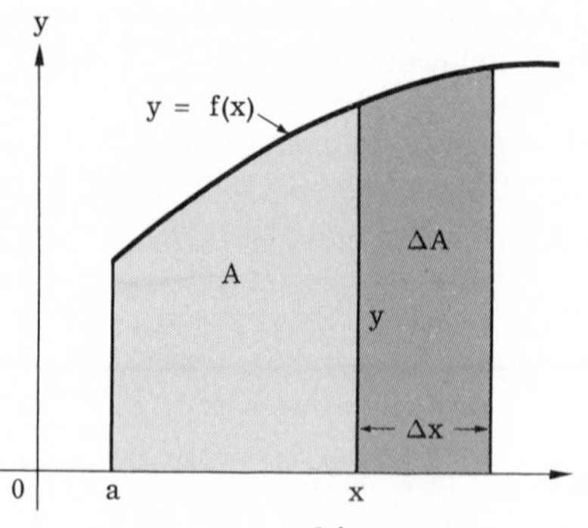

FIG. 4–5. $\dfrac{dA}{dx} = f(x)$.

and noting that, ΔA being approximately equal to the area of the rectangle with sides Δx and y of Fig. 4–5,

$$\lim_{\Delta x \to 0} \frac{\Delta A}{\Delta x} = \lim_{\Delta x \to 0} \frac{y\,\Delta x}{\Delta x} = y = f(x).$$

The argument just given is not a precise one, but it is very useful as an aid in understanding (4–19). A precise proof of (4–19) is as follows. Let $f(x)$ be defined and continuous for $a \leqq x \leqq b$. Let

$$F(x) = \int_a^x f(t)\,dt, \quad a \leqq x \leqq b.$$

$F(x)$ is the same as $A(x)$, but, if f changes sign, the area interpretation is less simple. Let x be fixed, $a < x < b$. Then

$$\begin{aligned} F(x + \Delta x) - F(x) &= \int_a^{x+\Delta x} f(t)\,dt - \int_a^x f(t)\,dt \\ &= \int_x^{x+\Delta x} f(t)\,dt. \end{aligned}$$

By the mean-value theorem (0–112), one has

$$\int_x^{x+\Delta x} f(t)\,dt = f(x_1)\,\Delta x,$$

where x_1 is between x and $x + \Delta x$. Hence

$$\frac{F(x + \Delta x) - F(x)}{\Delta x} = f(x_1).$$

As Δx approaches 0, x_1 must approach x and, since $f(x)$ is continuous, $f(x_1)$ must approach $f(x)$. Hence

$$F'(x) = \lim_{\Delta x \to 0} \frac{F(x + \Delta x) - F(x)}{\Delta x} = f(x).$$

A similar proof applies for $x = a$ or $x = b$ if one considers derivatives to the right or left. It should be noted that, since $F(x)$ has a derivative, it is necessarily continuous for $a \leqq x \leqq b$ (Section 0–9).

The significance of (4–19) for our numerical problem is as follows. The definite integral

$$F(x) = \int_a^x f(t)\,dt$$

defines a function $F(x)$ such that $F'(x) = f(x)$; i.e., $F(x)$ is an indefinite integral of $f(x)$; the complete indefinite integral is then given by

$$\int f(x)\,dx = F(x) + C.$$

The function $F(x)$ can be numerically evaluated, for each *fixed* x, by the procedures of the preceding section. If this is done for a sequence of values of x covering the range $a \leqq x \leqq b$, $F(x)$ will be known in *tabular* form and can be graphed. It is in this sense that the indefinite integral is being numerically evaluated.

EXAMPLE. To evaluate

$$\int e^{\sin x}\, dx \tag{4-21}$$

for $0 \leqq x \leqq 1$, one computes

$$F(x) = \int_0^x e^{\sin t}\, dt$$

for $x = 0, 0.1, 0.2, \ldots, 1$. The work is shown in Table 4–2. The integrals are computed by the trapezoidal rule. The fourth column, which shows $F(x)$, is the "cumulative sum" of the third column. Thus the third column gives the differences ΔF and, if these are divided by $\Delta x = 0.1$, one should obtain $f(x)$ again approximately. For example,

$$\frac{F(0.6) - F(0.5)}{0.1} = \frac{0.816 - 0.647}{0.1} = \frac{0.169}{0.1} = 1.69.$$

This gives $f(x)$ for x about equal to 0.55.

The complete indefinite integral (4–21) is now given by $F(x) + C$, where $F(x)$ is given by Table 4–2, for $0 \leqq x \leqq 1$. The range of x can clearly be

TABLE 4–2

x	$e^{\sin x}$	$\int_x^{x+0.1} e^{\sin t}\, dt$	$\int_0^x e^{\sin t}\, dt = F(x)$
0	1.00	0.105	0
0.1	1.11	0.116	0.105
0.2	1.22	0.129	0.221
0.3	1.35	0.142	0.350
0.4	1.48	0.155	0.492
0.5	1.62	0.169	0.647
0.6	1.75	0.182	0.816
0.7	1.90	0.197	0.998
0.8	2.05	0.212	1.195
0.9	2.18	0.225	1.407
1	2.32		1.632

extended as far as desired and the accuracy improved as required, so that the method should meet all practical needs.

Elliptic integrals. An integral of the form

$$\int \frac{dx}{\sqrt{1 - k^2 \sin^2 x}}, \quad 0 < k^2 < 1 \tag{4–22}$$

is an example of an *elliptic* integral. It can be shown that this integral cannot be evaluated in terms of elementary functions. Accordingly, some numerical method is called for. Because of the importance of this integral in applications, elaborate tables have been computed of the function

$$y = F(x) = \int_0^x \frac{dt}{\sqrt{1 - k^2 \sin^2 t}} \tag{4–23}$$

for various values of the constant k. Such tables are given in *A Short Table of Integrals*, 3rd ed., by B. O. Peirce (Boston: Ginn, 1929) and *Tables of Functions* by Jahnke and Emde (New York: Stechert, 1938).

The integral (4–23) is called an elliptic integral of first kind. Integrals of the second and third kinds are illustrated by

$$y = E(x) = \int_0^x \sqrt{1 - k^2 \sin^2 t}\, dt, \tag{4–24}$$

$$y = \int_0^x \frac{dt}{\sqrt{1 - k^2 \sin^2 t}(1 + a^2 \sin^2 t)}, \tag{4–25}$$

where $0 < k^2 < 1, a \neq 0, a^2 \neq k^2$. The integral (4–24) arises in computing the length of an arc of an ellipse (Prob. 4 below) and this is the basis of the term *elliptic* integral.

It can be shown that, if $R(x, y)$ is a rational function of x and y (Section 2–4) and $g(x)$ is a polynomial in x of degree 3 or 4, then the integral

$$\int R[x, \sqrt{g(x)}]\, dx$$

can be expressed as an elementary function plus elliptic integrals of the first, second, or third kinds. Accordingly, numerical evaluation of a few integrals permits precise evaluation of a broad class of integrals. This puts additional emphasis on the fact that every indefinite integral determines a function of x. For each new indefinite integral that we study and evaluate, a broad class of new functions becomes both numerically useful and open to a thorough analysis. Indeed, the trigonometric functions could have been *defined* in such a manner:

$$y = \int_0^x \frac{dt}{\sqrt{1 - t^2}}$$

defines $y = \arcsin x$ and hence $x = \sin y$. This is more awkward than the usual geometric definition, but all properties can be deduced from such a definition (cf. Prob. 3 below).

For further information on elliptic integrals we refer to the books of von Kármán and Biot (Chapter 4) and Whittaker and Watson (Chapters 20–22) listed at the end of the chapter. Some properties are obtained in Probs. 5 to 7 below.

Problems

1. Evaluate numerically:

(a) $\int x\,dx$ for $0 \leqq x \leqq 10$, using $x = 0, 1, 2, \ldots$;
(b) $\int e^{-x^2}\,dx$ for $0 \leqq x \leqq 1$, using $x = 0, .1, .2, \ldots$.

The answers should be graphed and checked by differentiation.

2. Let $f(x)$ be continuous for $a \leqq x \leqq b$.

(a) Prove that

$$\frac{d}{dx}\int_x^b f(t)\,dt = -f(x), \quad a \leqq x \leqq b.$$

(b) Prove that

$$\frac{d}{dx}\int_a^{x^2} f(t)\,dt = 2xf(x^2), \quad a \leqq x^2 \leqq b.$$

[Hint: let $u = x^2$, so that the integral becomes a function $F(u)$. Then, by the function-of-function rule, $\dfrac{d}{dx}F(u) = F'(u)\dfrac{du}{dx}\cdot$]

(c) Prove that

$$\frac{d}{dx}\int_{x^2}^b f(t)\,dt = -2xf(x^2), \quad a \leqq x^2 \leqq b.$$

(d) Prove that

$$\frac{d}{dx}\int_{x^2}^{x^3} f(t)\,dt = 3x^2f(x^3) - 2xf(x^2), \quad a \leqq x^2 \leqq b, \quad a \leqq x^3 \leqq b.$$

[Hint: let $u = x^2$, $v = x^3$. Then the integral is $F(u, v)$. By the chain rule,

$$\frac{d}{dx}F(u, v) = \frac{\partial F}{\partial u}\frac{du}{dx} + \frac{\partial F}{\partial v}\frac{dv}{dx}\cdot\Big]$$

3. The function $\log x$ (base e always understood) can be *defined* by the equation

$$\log x = \int_1^x \frac{dt}{t}, \quad x > 0. \qquad \text{(a)}$$

(a) Use this equation to evaluate $\log 1$, $\log 2$, $\log 0.5$ approximately.

(b) Prove, from Eq. (a) above, that $\log x$ is defined and continuous for $0 < x < \infty$ and that

$$\frac{d \log x}{dx} = \frac{1}{x}\cdot$$

(c) Prove that $\log(ax) = \log a + \log x$ for $a > 0$ and $x > 0$. [Hint: let $F(x) = \log(ax) - \log x$. Use the result of (b) to show that $F'(x) \equiv 0$, so that $F(x) \equiv$ const. Take $x = 1$ to evaluate the constant.]

4. Let an ellipse be given by parametric equations: $x = a\cos\phi$, $y = b\sin\phi$, $b > a > 0$. Show that the length of arc from $\phi = 0$ to $\phi = \alpha$ is given by

$$s = b\int_0^\alpha \sqrt{1 - k^2\sin^2\phi}\,d\phi, \quad k^2 = \frac{b^2 - a^2}{b^2}.$$

5. Show that the function $F(x)$ defined by (4–23) has the properties:

(a) $F(x)$ is defined and continuous for all x;
(b) as x increases, $F(x)$ increases;
(c) $F(x + \pi) - F(x) = 2K$, where K is a constant;
(d) $\lim_{x\to\infty} F(x) = \infty$, $\lim_{x\to-\infty} F(x) = -\infty$.

6. The function $x = am(y)$ is defined as the inverse of the function $y = F(x)$ of (4–23). Using the results of Prob. 5, show that $am(y)$ is defined and continuous for all y and has the properties:

(a) as y increases, $am(y)$ increases:
(b) $am(y + 2K) = am(y) + \pi$;
(c) $\dfrac{dx}{dy} = \sqrt{1 - k^2\sin^2 x}$.

7. The functions $sn(y)$, $cn(y)$, $dn(y)$ are defined in terms of the function of Prob. 6 by the equations:

$$sn(y) = \sin[am(y)], \quad cn(y) = \cos[am(y)], \quad dn(y) = \sqrt{1 - k^2 sn^2(y)}.$$

Prove the following identities:

(a) $sn^2(y) + cn^2(y) = 1$;

(b) $\dfrac{d}{dy}\,sn(y) = cn(y)\,dn(y)$; (c) $\dfrac{d}{dy}\,cn(y) = -sn(y)\,dn(y)$;

(d) $sn(y + 4K) = sn(y)$; (e) $cn(y + 4K) = cn(y)$; (f) $dn(y + 2K) = dn(y)$.

The functions $sn(y)$, $cn(y)$, $dn(y)$ are called *elliptic functions.* It should be emphasized that they depend on k, in addition to y.

8. The Error Function: $y = erf(x)$ is defined by the equation:

$$y = erf(x) = \int_0^x e^{-t^2}\,dt.$$

This function is of great importance in probability and statistics and is tabulated in the books mentioned after (4–23). Establish the following properties:

(a) $erf(x)$ is defined and continuous for all x;
(b) $erf(-x) = -erf(x)$;
(c) $-1 < erf(x) < 1$ for all x.

4–4 Improper integrals. A function $f(x)$ is said to be *bounded,* for a certain range of x, if there exists a constant M such that $|f(x)| \leqq M$ for all values of x in the given range. Thus $y = x$ is bounded for $-1 \leqq x \leqq 1$ ($M = 1$), $y = e^x$ is bounded for $x \leqq 0$ ($M = 1$), $y = \log x$ is bounded for $0.1 \leqq x \leqq 2$ ($M = \log 10$); however, $y = x$ is not bounded for $0 \leqq x < \infty$, $y = e^x$ is not bounded for $0 \leqq x < \infty$, $y = \log x$ is not bounded for $0 < x \leqq 1$.

As pointed out in Section 2–15, a continuous function over a *closed* interval $a \leqq x \leqq b$ has a maximum and minimum and is therefore bounded on this interval. If the function is continuous only on an *open* interval, $a < x < b$, it need not be bounded, as the examples $y = 1/x$, $y = \log x$ for $0 < x < 1$ illustrate.

A function $f(x)$ can be bounded and well-defined in a closed interval $a \leqq x \leqq b$ and yet have discontinuities there. Typical discontinuities are the following:

(a) *jump discontinuities*, illustrated by

$$y = 1, \quad 0 \leqq x \leqq 1; \qquad y = \tfrac{1}{2}, \quad 1 < x \leqq 2;$$

(b) *oscillatory discontinuities*, illustrated by

$$y = \sin\frac{1}{x}, \quad 0 < x \leqq 1; \qquad y = 0, \quad x = 0.$$

These are shown in Fig. 4–6. More complicated types can occur.

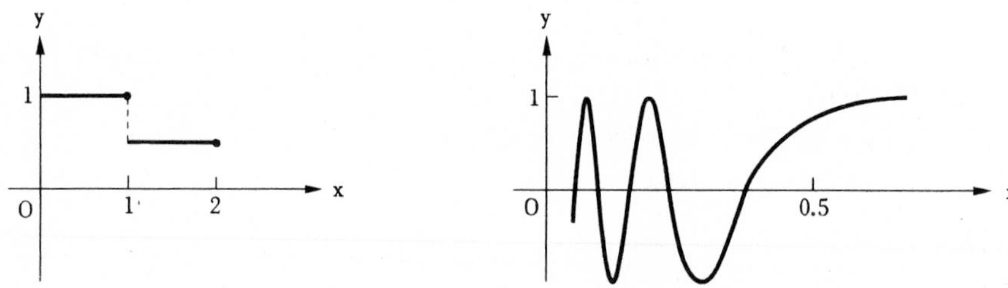

FIG. 4–6. Jump discontinuities and oscillatory discontinuities.

It may seem artificial to define a function $f(x)$ by two different formulas, valid for different parts of the interval of x. However, such functions appear naturally both in applied problems and in problems of geometry. For example, the force of gravity on a particle of mass m moving on an x axis, with $x = 0$ at the center of the earth, is given by

$$F = mg\frac{x}{R}, \quad 0 \leqq x \leqq R; \qquad F = mg\frac{R^2}{x^2}, \quad x \geqq R.$$

Also, the angle subtended by the circumference of a circle of radius a at a point P at distance r from the center has the constant value 2π when $0 \leqq r < a$ and is equal to $2 \arcsin (a/r)$ for $r \geqq a$. This function has a jump discontinuity for $r = a$.

If $f(x)$ is defined and bounded for $a \leqq x \leqq b$ and is continuous except at a finite number of points, then the definite integral, as defined by (4–1), continues to exist. This follows from the fact that the contribution to the sum $\Sigma f(x)\,\Delta x$ from a subinterval containing a discontinuity is in absolute value at most $M\,\Delta x$, where Δx is the length of that subinterval. As Δx approaches 0, this contribution must also approach 0. The convergence of the terms not involving discontinuities follows from the continuity of $f(x)$ on the rest of the interval. Thus the sum has a limit.

The most important case of such discontinuities is that of jumps. If $f(x)$ has jumps as in Fig. 4–6 at $x_1, x_2, \ldots, x_n$, then it is simplest to think of the integral as a sum of integrals:

$$\int_a^b f(x)\,dx = \int_a^{x_1} f(x)\,dx + \int_{x_1}^{x_2} f(x)\,dx + \cdots + \int_{x_n}^b f(x)\,dx.$$

In each term on the right, the integral is the same as that of a continuous function which coincides with $f(x)$ inside the interval.

If $f(x)$ is not bounded for $a \leqq x \leqq b$, then the integral, as defined by (4–1), loses its meaning. For the contribution $f(x)\,\Delta x$ to the sum $\Sigma f(x)\,\Delta x$ of a term corresponding to an interval within which $|f|$ takes arbitrarily large values can be made as large as desired; there will always be such a term and hence the limit cannot exist.

However, if $f(x)$ has only a finite number of discontinuities, it may be possible to assign the integral a meaning by a limiting process. Thus let $f(x)$ be defined for $a \leqq x \leqq b$ with discontinuities at $x = a$, $x = x_1$ and $x = b$; it is not even necessary that f be given values at these points. Then the integral is defined as follows (cf. Section 0–7):

$$\int_a^b f(x)\,dx = \lim_{a_1\to a+} \int_{a_1}^{a_2} f(x)\,dx + \lim_{a_3\to x_1-} \int_{a_2}^{a_3} f(x)\,dx + \lim_{a_4\to x_1+} \int_{a_4}^{a_5} f(x)\,dx + \lim_{a_6\to b-} \int_{a_5}^{a_6} f(x)\,dx,$$

provided *all four limits exist*. The auxiliary numbers a_2 and a_5 are chosen as suggested in Fig. 4–7. The value obtained is clearly independent of the choice of a_2 and a_5. The integral of such an unbounded function is termed an *improper integral* and is said to *converge* or to *diverge* according as the integral has a meaning or none by the limit process.

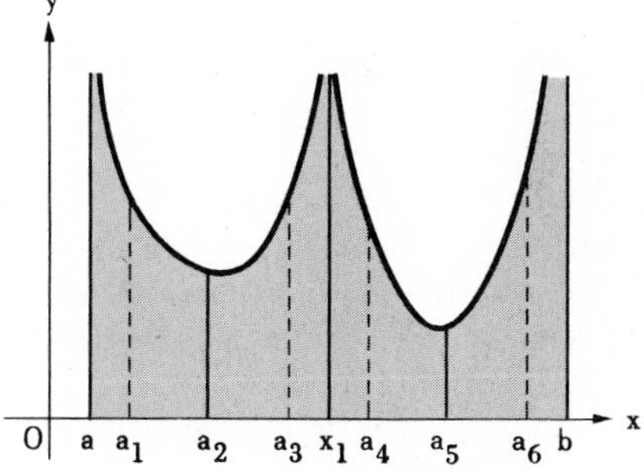

Fig. 4–7. Improper integral $\int_a^b f(x)\,dx$.

Example. For the integral

$$\int_0^2 \frac{x^2 - 3x + 1}{x(x-1)^2}\,dx,$$

one has infinite discontinuities at $x = 0$ and $x = 1$. Thus one has

$$\int_0^2 \frac{x^2 - 3x + 1}{x(x-1)^2}\,dx = \lim_{b\to 0+} \int_b^{0.5} \cdots dx + \lim_{c\to 1-} \int_{0.5}^{c} \cdots dx + \lim_{h\to 1+} \int_h^2 \cdots dx$$

$$= \lim_{b\to 0+} \left(\log x + \frac{1}{x-1}\right)\Bigg|_b^{0.5} + \lim_{c\to 1-} \left(\log x + \frac{1}{x-1}\right)\Bigg|_{0.5}^{c} + \lim_{h\to 1+} \left(\log x + \frac{1}{x-1}\right)\Bigg|_h^2,$$

since one has

$$\frac{x^2 - 3x + 1}{x(x-1)^2} = \frac{1}{x} - \frac{1}{(x-1)^2} = \frac{d}{dx}\left(\log x + \frac{1}{x-1}\right).$$

Now the first limit is $+\infty$, the second is $-\infty$, while the third is $-\infty$. Hence the integral is divergent.

Other types of improper integrals are obtained by allowing one or both of the limits of integration to become infinite. Thus the integrals

$$\int_0^\infty e^{-x}\,dx, \quad \int_{-\infty}^0 \sin x\,dx, \quad \int_{-\infty}^\infty \frac{1}{1+x^2}\,dx$$

are improper. The functions concerned here are continuous over each finite interval contained in the interval of integration, and the integrals can thus be assigned values by limit processes. For example,

$$\int_0^\infty e^{-x}\,dx = \lim_{b\to\infty}\int_0^b e^{-x}\,dx = \lim_{b\to\infty}(-e^{-b}+1) = 1,$$

$$\int_{-\infty}^0 \sin x\,dx = \lim_{a\to-\infty}\int_a^0 \sin x\,dx = \lim_{a\to-\infty}(-1+\cos a),$$

so that the integral is divergent;

$$\begin{aligned}\int_{-\infty}^\infty \frac{1}{1+x^2}\,dx &= \lim_{a\to-\infty}\int_a^0 \frac{1}{1+x^2}\,dx + \lim_{b\to\infty}\int_0^b \frac{1}{1+x^2}\,dx\\ &= \lim_{a\to-\infty}(-\operatorname{arc}\tan a) + \lim_{b\to\infty}(\operatorname{arc}\tan b)\\ &= -\left(-\frac{\pi}{2}\right) + \frac{\pi}{2} = \pi.\end{aligned}$$

If the function being integrated has a finite number of discontinuities within the interval of integration, these can be treated as before. The basic rule is that each discontinuity (near which the function is unbounded) and each infinite limit of integration calls for a separate limit process. The following example illustrates this:

$$\begin{aligned}\int_0^\infty \frac{1}{x^{\frac{1}{3}}}\,dx &= \lim_{a\to0+}\int_a^1 \frac{1}{x^{\frac{1}{3}}}\,dx + \lim_{b\to\infty}\int_1^b \frac{1}{x^{\frac{1}{3}}}\,dx\\ &= \lim_{a\to0+}\left(\tfrac{3}{2} - \tfrac{3}{2}a^{\frac{2}{3}}\right) + \lim_{b\to\infty}\left(\tfrac{3}{2}b^{\frac{2}{3}} - \tfrac{3}{2}\right)\\ &= \tfrac{3}{2} + \left(\infty - \tfrac{3}{2}\right) = \infty.\end{aligned}$$

Thus the integral is meaningless.

It should be remarked that the value of the integral $\int_a^b f(x)\,dx$ (if it exists) is not affected if one changes the value of $f(x)$ at a *finite* number of values of x; for, as in the case of a bounded function, the effect of these changes on the rectangular sums (4–4) must approach 0 as $n \to \infty$. One can even leave $f(x)$ wholly undefined at a finite number of points; the integral is then technically improper. However, if $f(x)$ is bounded, the integral can be obtained as a limit of rectangular sums (4–4), provided only that the points $x_1^*, \ldots, x_n^*$ are always chosen to be points at which $f(x)$ is defined. One can achieve the same effect by assigning arbitrary values to $f(x)$ at the points where the function was not defined; the value of the integral cannot depend on the values chosen.

***4–5 Tests for convergence of improper integrals. Numerical evaluation.** The examples of improper integrals in the preceding section could all be explicitly evaluated with the aid of indefinite integrals. For such integrals as

$$\int_0^\infty \frac{\sin x}{1+x^2}\,dx, \quad \int_0^1 \frac{\log x}{1+x^2}\,dx$$

this is not the case. If one is assured that the integral converges, then the limit process itself indicates how the integral can be evaluated numerically. Thus

$$\int_0^\infty \frac{\sin x}{1+x^2}\,dx = \lim_{b\to\infty}\int_0^b \frac{\sin x}{1+x^2}\,dx.$$

If the limit on the right exists, then, for a sufficiently large value of b, the integral from 0 to b gives the value sought to desired accuracy. The integral from 0 to b can then be found, again approximately, by the methods of Section 4–2.

It thus appears necessary to provide tests for the existence of improper integrals. The tests themselves will give information about the error made in stopping short of the limit of integration.

Since the theory of the tests parallels that for infinite series, we state them here mainly without proofs and refer the reader to Chapter 6 for a full discussion.

THEOREM I(a). *(Comparison test for convergence). Let $f(x)$ and $g(x)$ be continuous for $a < x \leqq b$, with*

$$0 \leqq |f(x)| \leqq g(x).$$

If

$$\int_a^b g(x)\,dx$$

converges, then

$$\int_a^b f(x)\,dx$$

converges and

$$0 \leqq \left|\int_a^b f(x)\,dx\right| \leqq \int_a^b g(x)\,dx. \tag{4–26}$$

Thus the integral

$$\int_0^\pi \frac{1}{\sqrt{x}}\,dx = \lim_{a\to 0+} 2\sqrt{x}\Big|_a^\pi = 2\sqrt{\pi}$$

and hence

$$\int_0^\pi \frac{\cos x}{\sqrt{x}}\,dx$$

is convergent, for

$$\left|\frac{\cos x}{\sqrt{x}}\right| \leqq \frac{1}{\sqrt{x}} \quad \text{for} \quad 0 < x \leqq \pi.$$

THEOREM I(b). *(Comparison test for divergence.) Let $f(x)$ and $g(x)$ be continuous for $a < x \leqq b$, with*

$$0 \leqq g(x) \leqq f(x).$$

If $\int_a^b g(x)\,dx$ diverges, then $\int_a^b f(x)\,dx$ diverges.

Thus the integral

$$\int_0^1 \frac{dx}{x} = \lim_{a\to 0+} \log x\Big|_a^1 = \infty$$

and hence

$$\int_0^1 \frac{\sqrt{1+x^2}}{x}\,dx$$

is divergent, for

$$\frac{\sqrt{1+x^2}}{x} > \frac{1}{x} \quad \text{for} \quad 0 < x \leqq 1.$$

It should be remarked that in the comparison test for convergence only the absolute value of $f(x)$ is considered, and the integral can accordingly be termed *absolutely convergent.* It is thus possible to apply this test even when $f(x)$ changes sign infinitely often. For example,

$$\int_0^1 \frac{\sin\frac{1}{x}}{\sqrt{x}}\,dx$$

can be shown to converge by comparison with

$$\int_0^1 \frac{dx}{\sqrt{x}}.$$

On the other hand, the test for divergence applies only to a positive $f(x)$. If $f(x)$ changes sign only a finite number of times, one can restrict attention to an interval $a < x \leqq c$, $c < b$, in which $f(x)$ has only one sign. In general, the convergence or divergence of the integral is determined by the behavior of f near the point of discontinuity a. Multiplying f by a constant will have no effect on convergence or divergence.

Theorems I(a) and I(b) are stated in terms of a discontinuity at $x = a$. The same tests apply to discontinuities at $x = b$, for a simple change of variable interchanges left and right limits.

In order to make the tests given really effective, one needs to know particular functions $g(x)$ for which the corresponding integrals are convergent or divergent. Such functions are provided by the following theorem.

THEOREM II. *The improper integral*

$$\int_a^b \frac{dx}{(x-a)^p} \quad (a < b) \tag{4–27}$$

converges for $p < 1$ and diverges for $p \geqq 1$.

For, when $p \neq 1$,

$$\int_a^b \frac{dx}{(x-a)^p} = \lim_{c \to a+} \frac{(x-a)^{1-p}}{1-p}\bigg|_c^b = \lim_{c \to a+} \frac{(b-a)^{1-p} - (c-a)^{1-p}}{1-p}.$$

If $p < 1$, the limit is $(b-a)^{1-p}/(1-p)$, and if $p > 1$, the limit is infinite. If $p = 1$, one has

$$\int_a^b \frac{dx}{x-a} = \lim_{c \to a+} \log(x-a)\bigg|_c^b = \lim_{c \to a+} [\log(b-a) - \log(c-a)] = \infty.$$

The condition $a < b$ is clearly unessential here, as the substitution $u = -x$ reduces the case $a > b$ to the case $a < b$. Accordingly, the integral

$$\int_a^b \frac{dx}{(b-x)^p} \quad (a < b)$$

must also converge for $p < 1$ and diverge for $p \geqq 1$.

EXAMPLE. The integral

$$\int_0^1 \frac{5x^2}{\sqrt{1-x^2}\,(x^3+1)}\,dx$$

converges, for the function being integrated can be written as

$$\frac{1}{\sqrt{1-x}}\,h(x), \quad h(x) = \frac{5x^2}{\sqrt{1+x}\,(x^3+1)}.$$

The function $h(x)$ is continuous for $0 \leqq x \leqq 1$, and is therefore bounded:

$$|h(x)| \leqq M.$$

It is not necessary to compute the number M [which would require finding absolute maximum and minimum of $h(x)$], for one has in any case

$$\left|\frac{h(x)}{\sqrt{1-x}}\right| \leqq \frac{M}{\sqrt{1-x}} \quad \text{for} \quad 0 \leqq x \leqq 1.$$

The integral

$$\int_0^1 \frac{M}{\sqrt{1-x}}\,dx$$

exists by Theorem II, the constant factor M having no effect on convergence. Hence

$$\int_0^1 \frac{5x^2}{\sqrt{1-x^2}\,(x^3+1)}\,dx$$

converges.

In the examples given, only the question of convergence or divergence has been considered. In the case of the convergent integrals, (4–26) provides a method of estimating the error in computing the integral numerically. Thus the integral $\int_0^\pi \frac{\cos x}{\sqrt{x}}\,dx$ can be computed by evaluating the integral from a to π, where a is close to 0. The error made is then

$$\int_0^a \frac{\cos x}{\sqrt{x}}\,dx \leqq \int_0^a \frac{1}{\sqrt{x}}\,dx = 2\sqrt{a}.$$

Thus, for $a = 0.01$, the error is at most 0.2. If the integral from 0.01 to π is computed by Simpson's rule with an error less than 0.2, the whole integral is known with an error of at most 0.4.

The comparison tests can be extended to integrals with infinite limits.

THEOREM III(a). *(Comparison test for convergence.) Let $f(x)$ and $g(x)$ be continuous for $x \geqq a$. If*

$$0 \leqq |f(x)| \leqq g(x)$$

and $\int_a^\infty g(x)$ *converges, then* $\int_a^\infty f(x)\,dx$ *converges and*

$$\left|\int_a^\infty f(x)\,dx\right| \leqq \int_a^\infty g(x)\,dx. \qquad (4\text{–}28)$$

THEOREM III(b). *(Comparison test for divergence.) Let $f(x)$ and $g(x)$ be continuous for $x \geqq a$. If*

$$0 \leqq g(x) \leqq f(x)$$

and $\int_a^\infty g(x)\,dx$ *diverges, then* $\int_a^\infty f(x)\,dx$ *diverges.*

There is a parallel to Theorem II, providing comparison functions $g(x)$:

THEOREM IV. *The improper integral*

$$\int_1^\infty \frac{dx}{x^p}$$

converges for $p > 1$ and diverges for $p \leqq 1$.

The proof is left to the exercises (Prob. 1 below). It should be noted that the convergence in Theorem II was for p *less* than 1, while in Theorem IV it holds for p *greater* than 1.

Integrals from $-\infty$ to a finite value are treated in the same way; they can be reduced to those from a to $+\infty$ by a substitution $u = -x$.

The inequality (4–28) again provides an estimate for the error in computing the integral numerically. For example,

$$\int_0^\infty e^{-x^2}\,dx$$

exists, since

$$e^{-x^2} < \frac{1}{x^2}, \quad x \geqq 1,$$

as follows from a simple discussion of derivatives. Hence one can evaluate the integral by computing numerically the integral from 0 to $x = a$. The error is then

$$\int_a^\infty e^{-x^2}\,dx \leqq \int_a^\infty \frac{dx}{x^2} = \frac{1}{a}. \tag{4–29}$$

Thus for $a = 10$, the error is at most 0.1 (see Prob. 8 below).

Problems

1. Prove Theorem IV above and evaluate the integral for $p > 1$.

2. Show that the following integrals converge by evaluating the corresponding limits:

(a) $\displaystyle\int_0^1 \frac{dx}{\sqrt{1-x^2}}$ (c) $\displaystyle\int_0^1 \log x\,dx$ (e) $\displaystyle\int_0^\infty x^2 e^{-x}\,dx$

(b) $\displaystyle\int_0^\infty e^{-x}\,dx$ (d) $\displaystyle\int_1^\infty \frac{dx}{x\sqrt{1+x^2}}$ (f) $\displaystyle\int_1^\infty \frac{\log x}{x^2}\,dx.$

3. Test for convergence or divergence:

(a) $\displaystyle\int_1^2 \frac{dx}{x^2-1}$ (c) $\displaystyle\int_0^{\frac{\pi}{2}} \tan x\,dx$ (e) $\displaystyle\int_1^\infty \frac{\sin x}{x^2}\,dx$

(b) $\displaystyle\int_0^1 \frac{\sin x}{x^{\frac{3}{2}}}\,dx$ (d) $\displaystyle\int_0^\infty \frac{x^2-1}{x^4+1}\,dx$ (f) $\displaystyle\int_0^\infty \frac{x}{\sqrt{1+x^3}}\,dx.$

4. Evaluate where possible:

(a) $\displaystyle\int_{-1}^1 \frac{dx}{x^{\frac{1}{3}}}$ (c) $\displaystyle\int_0^\infty \frac{dx}{1+x^2}$ (e) $\displaystyle\int_0^\infty \sin x\,dx$

(b) $\displaystyle\int_{-1}^1 \frac{dx}{x^3}$ (d) $\displaystyle\int_0^\infty \frac{x^2-x-1}{x(x^3+1)}\,dx$ (f) $\displaystyle\int_0^\infty (1-\tanh x)\,dx.$

5. Let $F(x)$ be defined as follows:

$$F(x) = 0 \quad \text{for} \quad x < 0, \quad F(x) = 1 \quad \text{for} \quad x \geqq 0.$$

Evaluate the following integrals:

(a) $\displaystyle\int_0^{\frac{\pi}{2}} F(x) \sin x \, dx$

(b) $\displaystyle\int_{\frac{-\pi}{2}}^{\frac{\pi}{2}} F(x) \cos x \, dx$

(c) $\displaystyle\int_{-\infty}^{\infty} F(x) e^{-x} \, dx$

(d) $\displaystyle\int_{-\pi}^{\pi} F(x) \sin nx \, dx \ (n > 0)$.

6. Evaluate

$$\int_0^{\frac{\pi}{2}} \frac{\cos x}{\sqrt{x}} \, dx$$

by computing the integral from $\pi/6$ to $\pi/2$ by the trapezoidal rule, with $\Delta x = \pi/6$. Estimate the error obtained.

7. Verify numerically (as accurately as desired) that

(a) $\displaystyle\int_0^{\infty} e^{-x^2} \, dx = \frac{\sqrt{\pi}}{2}$

(b) $\displaystyle\int_0^{\infty} \frac{dx}{(x^2 + 1)^3} = \frac{3\pi}{16}$

(c) $\displaystyle\int_0^{\infty} \frac{\sin^2 x}{x^2} \, dx = \frac{\pi}{2}$

(d) $\displaystyle\int_0^1 \frac{\log x}{1 - x} \, dx = -\frac{\pi^2}{6}$.

8. Show that

$$e^{-x^2} \leqq e^{-2ax + a^2} \quad \text{for} \quad x \geqq a$$

and hence improve the estimate (4–29) above.

9. Let $F(x) = x$ for $0 \leqq x \leqq 1$, $F(x) = 2x$ for $1 < x \leqq 2$. Show that

$$\int_0^2 F'(x) \, dx$$

exists. Is this equal to $F(2) - F(0)$? State conditions under which

$$\int_a^b F'(x) \, dx = F(b) - F(a),$$

when $F'(x)$ is defined and continuous except at a finite number of points and is bounded in the interval $a \leqq x \leqq b$.

10. Show that, if $f(x)$ is continuous for $a < x \leqq b$ and

$$\int_a^b f(x) \, dx$$

converges, then

$$\frac{d}{dx} \int_a^x f(t) \, dt = f(x), \quad a < x \leqq b.$$

11. Justify the proof of convergence of

$$\int_0^{\infty} \frac{dx}{1 + x^2}$$

by the method of setting $x = \tan \theta$ and thus reducing the integral to one which is no longer improper.

Answers

2. (a) $\frac{\pi}{2}$, (b) 1, (c) -1, (d) $\log(1+\sqrt{2})$, (e) 2, (f) 1.

3. (a) div, (b) conv, (c) div, (d) conv, (e) conv, (f) div.

4. (a) 0, (b) div, (c) $\frac{\pi}{2}$, (d) div, (e) div, (f) log 2.

5. (a) 1, (b) 1, (c) 1, (d) $(1 - \cos n\pi)/n$. 6. Trapezoidal rule gives 0.6, with an error (positive) of at most 0.1. Error due to stopping short of 0 is at most 1.4 (negative).

4–6 Double integrals. The definite integral $\int_a^b f(x)\,dx$ is defined with respect to a function $f(x)$ defined over an interval $a \leqq x \leqq b$. The double integral

$$\iint\limits_R f(x, y)\,dx\,dy$$

will be defined with reference to a function $f(x, y)$ defined over a closed region R of the xy plane. It is necessary further to assume that R is *bounded*, i.e., that R can be enclosed in a circle of sufficiently large radius; otherwise, just as in the case when a or b is infinite, the integral is improper.

The definition of the double integral parallels that of the definite integral. One subdivides the region R by drawing parallels to the x and y axes, as in Fig. 4–8. One considers only those rectangles which are within R and numbers them from 1 to n, denoting by $\Delta_i A$ the area of the ith rectangle. One then forms the sum

$$\sum_{i=1}^{n} f(x_i^*, y_i^*)\,\Delta_i A,$$

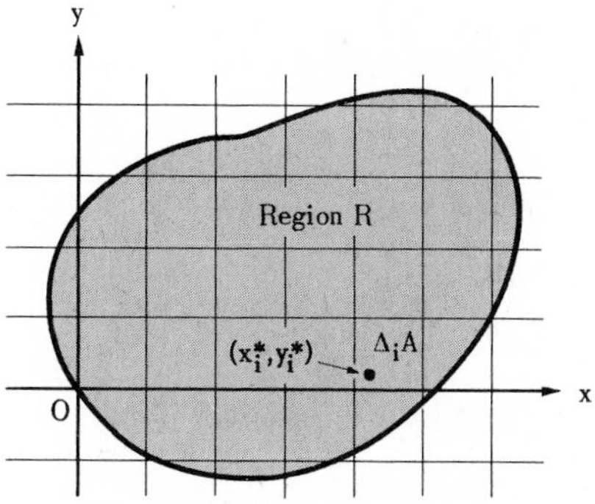

Fig. 4–8. The double integral.

where (x_i^*, y_i^*) is chosen arbitrarily in the ith rectangle. If this sum approaches a unique limit as n becomes infinite and the maximum diagonal of the rectangles approaches 0, then the double integral is defined as that limit:

$$\iint\limits_R f(x, y)\,dx\,dy = \lim \sum_{i=1}^{n} f(x_i^*, y_i^*)\,\Delta_i A. \tag{4–30}$$

The existence of the double integral can be demonstrated when f is continuous and R satisfies simple conditions, in particular when R can be split into a finite number of pieces, each of which is described by inequalities of the form

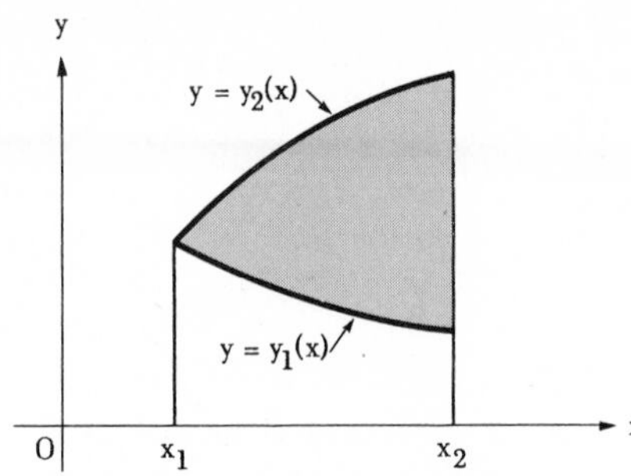

Fig. 4–9. Reduction of double integral to iterated integral.

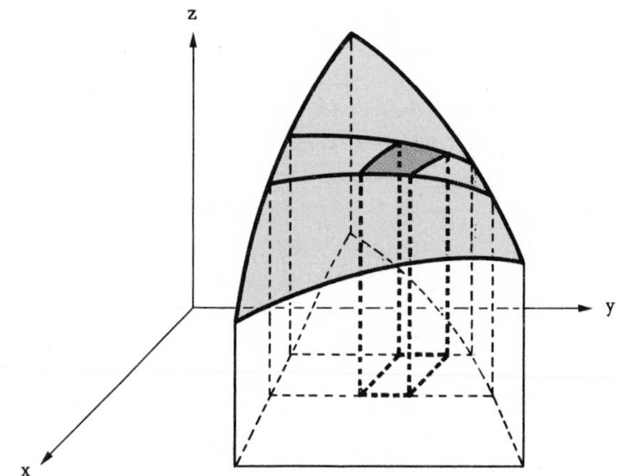

Fig. 4–10. The double integral as volume.

$$y_1(x) \leqq y \leqq y_2(x), \quad x_1 \leqq x \leqq x_2 \tag{4–31}$$

or of the form

$$x_1(y) \leqq x \leqq x_2(y), \quad y_1 \leqq y \leqq y_2, \tag{4–32}$$

where the functions $y_1(x)$, $y_2(x)$, $x_1(y)$, $x_2(y)$ are continuous. Fig. 4–9 illustrates the first form.

Just as the definite integral of a function $f(x)$ can be interpreted in terms of area, so can the double integral be interpreted in terms of volume, as suggested in Fig. 4–10. Here $f(x, y)$ is assumed to be positive over the region R (the same type as in Fig. 4–9) and the volume being computed is that beneath the surface $z = f(x, y)$ and above the region R in the xy plane. Each term $f(x_i^*, y_i^*) \Delta_i A$ in the sum (4–30) represents the volume of a rectangular parallelepiped with base $\Delta_i A$ and altitude $f(x_i^*, y_i^*)$, as shown in Fig. 4–10. It is clear that the sum of these volumes can be regarded as an approximation to the volume beneath the surface and that the double integral (4–30) can be used as definition of the volume.

The evaluation of double integrals for a region like that of Fig. 4–9 is facilitated by reduction to an *iterated integral*:

$$\int_{x_1}^{x_2}\left[\int_{y_1(x)}^{y_2(x)} f(x, y)\, dy\right] dx.$$

For each fixed value of x, the inner integral

$$\int_{y_1(x)}^{y_2(x)} f(x, y)\, dy$$

is simply a definite integral with respect to y of the continuous function $f(x, y)$. This integral can be interpreted as the area of a cross section, perpendicular to the x axis, of the volume being computed; this is suggested in Fig. 4–10. If the area of the cross section is denoted by $A(x)$, then the iterated integral gives

$$\int_{x_1}^{x_2} A(x)\, dx.$$

If this integral is interpreted as a limit of a sum of terms $\Sigma A(x)\, \Delta x$, then it is geometrically evident that the integral does represent the volume.

THEOREM. *If $f(x, y)$ is continuous in a closed region R described by the inequalities* (4–31), *then, for $x_1 \leqq x \leqq x_2$,*

$$\int_{y_1(x)}^{y_2(x)} f(x, y)\, dy$$

is a continuous function of x and

$$\iint_R f(x, y)\, dx\, dy = \int_{x_1}^{x_2}\int_{y_1(x)}^{y_2(x)} f(x, y)\, dy\, dx. \tag{4–33}$$

Similarly, if R is described by (4–32),

$$\iint_R f(x, y)\, dx\, dy = \int_{y_1}^{y_2}\int_{x_1(y)}^{x_2(y)} f(x, y)\, dx\, dy. \tag{4–34}$$

For a proof of existence of the double integral of a continuous function and of the reduction to an iterated integral, the reader is referred to *CLA*, Section 13–3.

EXAMPLE. Let R be the quarter-circle, $0 \leqq y \leqq \sqrt{1 - x^2}$, $0 \leqq x \leqq 1$, and let $f(x, y) = x^2 + y^2$. Then one has

$$\begin{aligned}\iint_R (x^2 + y^2)\, dx\, dy &= \int_0^1 \int_0^{\sqrt{1-x^2}} (x^2 + y^2)\, dy\, dx \\ &= \int_0^1 \left(x^2\sqrt{1 - x^2} + \tfrac{1}{3}(1 - x^2)^{\frac{3}{2}}\right) dx \\ &= \int_0^{\frac{\pi}{2}} \left(\sin^2\theta \cos^2\theta + \tfrac{1}{3}\cos^4\theta\right) d\theta = \frac{\pi}{8}.\end{aligned}$$

The double integral can be interpreted and applied in a variety of ways. The following are illustrations:

I. *Volume.* If $z = f(x, y)$ is the equation of a surface, then

$$V = \iint_R f(x, y)\, dx\, dy \tag{4–35}$$

gives the volume between the surface and the xy plane, volumes above the xy plane being counted positively and those below the xy plane being counted negatively.

II. *Area.* If one takes $f(x, y) \equiv 1$, one obtains

$$A = \text{area of } R = \iint_R dx\, dy. \tag{4–36}$$

III. *Mass.* If f is interpreted as density, i.e., as mass per unit area, then

$$M = \text{mass of } R = \iint_R f(x, y)\, dx\, dy. \tag{4–37}$$

IV. *Center of mass.* If f is density, then the center of mass $(\bar{x}, \bar{y})$ of the "thin plate" represented by R is located by the equations

$$M\bar{x} = \iint_R xf(x, y)\, dx\, dy, \quad M\bar{y} = \iint_R yf(x, y)\, dx\, dy, \tag{4–38}$$

where M is given by (4–37).

V. *Moment of inertia.* The moments of inertia of the thin plate about the x and y axes are given by the equations

$$I_x = \iint_R y^2 f(x, y)\, dx\, dy, \quad I_y = \iint_R x^2 f(x, y)\, dx\, dy, \tag{4–39}$$

while the polar moment of inertia about the origin O is given by

$$I_O = I_x + I_y = \iint_R (x^2 + y^2) f(x, y)\, dx\, dy. \tag{4–40}$$

The basic properties of the double integral are essentially the same as those for the definite integral (Section 0–10).

$$\iint_R [f(x, y) + g(x, y)]\, dx\, dy = \iint_R f(x, y)\, dx\, dy + \iint_R g(x, y)\, dx\, dy; \tag{4–41}$$

$$\iint_R cf(x, y)\, dx\, dy = c \iint_R f(x, y)\, dx\, dy \quad (c = \text{constant}); \tag{4–42}$$

$$\iint\limits_{R} f(x, y)\, dx\, dy = \iint\limits_{R_1} f(x, y)\, dx\, dy + \iint\limits_{R_2} f(x, y)\, dx\, dy, \tag{4–43}$$

where R is composed of two pieces R_1 and R_2 overlapping only at boundary points;

$$\iint\limits_{R} f(x, y)\, dx\, dy = f(x_1, y_1) \cdot A, \tag{4–44}$$

where A is the area of R, as in (4–36) above, and (x_1, y_1) is a suitably chosen point of R; if $M_1 \leqq f(x, y) \leqq M_2$ for (x, y) in R, then

$$M_1 A \leqq \iint\limits_{R} f(x, y)\, dx\, dy \leqq M_2 A, \tag{4–45}$$

where A is the area of R. The functions $f(x, y)$ and $g(x, y)$ are here assumed to be continuous in R. The proofs of these properties are essentially the same as for functions of one variable and will not be reviewed here.

From (4–45) or by the same method as that used to prove (4–15), one can prove the inequality:

$$\left| \iint\limits_{R} f(x, y)\, dx\, dy \right| \leqq M \cdot A, \tag{4–46}$$

where $|f(x, y)| \leqq M$ in R. This can be applied as in Section 4–2 for estimating errors.

Equation (4–44) is the *law of the mean for double integrals.* If both sides are divided by A, one obtains

$$f(x_1, y_1) = \frac{1}{A} \iint\limits_{R} f(x, y)\, dx\, dy. \tag{4–47}$$

The right-hand side can properly be termed the *average value of f in R* [cf. Eq. (4–8)]. The law of the mean thus asserts that a function continuous in R takes on its average value somewhere in R.

If R is a circular region R_r of radius r with center at (x_0, y_0), then one can consider the effect of letting r approach 0 in (4–47). The region R_r is thus being shrunk down to the point (x_0, y_0). Since (x_1, y_1) is always in R_r, it must approach (x_0, y_0); since f is assumed continuous, it must approach $f(x_0, y_0)$. Accordingly,

$$\lim_{r \to 0} \frac{1}{A_r} \iint\limits_{R_r} f(x, y)\, dx\, dy = f(x_0, y_0), \tag{4–48}$$

where $A_r = \pi r^2 =$ area of R. The operation on the left suggests what might be called "differentiation of the double integral with respect to area." Thus the derivative, in this sense, of the integral reproduces the function being integrated. Accordingly, (4–48) parallels the theorem

$$\frac{d}{dx}\int_a^x f(t)\,dt = f(x)$$

for functions of one variable. It should be remarked that the "incremental area" R_r of (4–48) could also be chosen as a rectangle, ellipse, etc.

Equation (4–48) has a physical interpretation, namely, the determination of density from mass. Thus, if a thin plate of variable density is given, one can measure the mass of the plate or any portion of it directly, simply by weighing the portion concerned. To determine the density at a particular point of the plate, one chooses a small area about this point, measures its mass and divides by its area. One is thus carrying out experimentally a stage of the limit process (4–48) in the form

$$\lim \frac{\text{mass}}{\text{area}} = \text{density}.$$

The following property of double integrals is established as for single integrals (Prob. 13 following Section 4–2): *If $f(x, y)$ is continuous in the bounded closed region R and*

$$\iint\limits_R [f(x, y)]^2\,dx\,dy = 0,$$

then $f(x, y) \equiv 0$ in R.

4–7 Triple integrals and multiple integrals in general. The notion of double integral generalizes to integrals of functions of three, four, ... variables:

$$\iiint\limits_R f(x, y, z)\,dx\,dy\,dz, \quad \iiiint\limits_R f(x, y, z, w)\,dx\,dy\,dz\,dw, \ldots.$$

These are called *triple, quadruple,* ... integrals. In general, one terms them *multiple* integrals, the double integral being the simplest multiple integral.

For the triple integral, for example, one considers a function $f(x, y, z)$ defined in a bounded closed region R of space. One subdivides R into rectangular parallelepipeds by planes parallel to the coordinate planes, numbers the parallelepipeds inside R from 1 to n, and denotes the ith volume by $\Delta_i V$. The triple integral is then obtained as the limit of a sum:

$$\iiint\limits_R f(x, y, z)\,dx\,dy\,dz = \lim \sum_{i=1}^{n} f(x_i^*, y_i^*, z_i^*)\,\Delta_i V \tag{4–49}$$

as the number n approaches ∞, while the maximum diagonal of the $\Delta_i V$ approaches 0. The point (x_i^*, y_i^*, z_i^*) is arbitrarily chosen in the ith parallelepiped. The existence of a unique limit can be shown if $f(x, y, z)$ is continuous in R.

The simplest theory holds for a region R described by inequalities such as the following:

$$x_1 \leqq x \leqq x_2, \quad y_1(x) \leqq y \leqq y_2(x), \quad z_1(x, y) \leqq z \leqq z_2(x, y). \tag{4–50}$$

For this region one can reduce the triple integral to an iterated integral by the equation

$$\iiint\limits_R f(x, y, z)\, dx\, dy\, dz = \int_{x_1}^{x_2} \int_{y_1(x)}^{y_2(x)} \int_{z_1(x,y)}^{z_2(x,y)} f(x, y, z)\, dz\, dy\, dx, \tag{4–51}$$

as in the theorem of Section 4–6.

EXAMPLE. If R is described by the inequalities

$$0 \leqq x \leqq 1, \quad 0 \leqq y \leqq x^2, \quad 0 \leqq z \leqq x + y$$

and $f = 2x - y - z$, one has

$$\begin{aligned} \iiint\limits_R f\, dx\, dy\, dz &= \int_0^1 \int_0^{x^2} \int_0^{x+y} (2x - y - z)\, dz\, dy\, dx \\ &= \frac{3}{2} \int_0^1 \int_0^{x^2} (x^2 - y^2)\, dy\, dx \\ &= \frac{3}{2} \int_0^1 \left(x^4 - \frac{x^6}{3}\right) dx = \frac{8}{35}. \end{aligned}$$

The limits of integration can be determined by the following procedure. Let the order chosen be $dz\, dy\, dx$ as in the above example. Then we determine the x limits x_1, x_2 first as the smallest and largest values of x on the figure R. We then consider a cross section: x = const of the figure, taken between x_1 and x_2. The y limits are now determined as the minimum and maximum values of y in the cross section; these depend on the x chosen, are hence $y_1(x)$ and $y_2(x)$. Finally, within the cross section we find the minimum and maximum values of z for each fixed y; these are the z limits; they depend on the constant values of x and y chosen, are hence $z_1(x, y)$ and $z_2(x, y)$. While the minima and maxima referred to can often be read off a sketch of R as a three-dimensional figure, it is usually easier to plot only typical cross sections as two-dimensional figures.

Since the definite integral can be interpreted as area, the double integral as volume, one would expect the triple integral to be interpreted as "hypervolume," or volume in a four-dimensional space. While such an interpretation has some value, it is simpler to think of mass; for example,

$$\iiint\limits_R f(x, y, z)\, dx\, dy\, dz = \text{mass of solid of density } f. \tag{4–52}$$

The other interpretations of the double integral also generalize:

$$\iiint\limits_R dx\, dy\, dz = \text{volume of } R; \tag{4–53}$$

$$M\bar{x} = \iiint\limits_R x f(x, y, z)\, dx\, dy\, dz, \ldots, \tag{4–54}$$

where $(\bar{x}, \bar{y}, \bar{z})$ is the center of mass;

$$I_x = \iiint\limits_R (y^2 + z^2) f(x, y, z)\, dx\, dy\, dz, \tag{4–55}$$

the moment of inertia about Ox.

The basic properties (4–41) to (4–45) also generalize to all multiple integrals. Thus, in particular, one has the law of the mean:

$$\iiint\limits_R f(x, y, z)\, dx\, dy\, dz = f(x_1, y_1, z_1) \cdot V, \tag{4–56}$$

where V is the volume of R. This can also be made the basis for "differentiating a triple integral with respect to volume," as in (4–48).

Problems

1. Evaluate the following integrals:

(a) $\iint\limits_R (x^2 + y^2)\, dx\, dy$, where R is the triangle with vertices (0, 0), (1, 0), (1, 1);

(b) $\iiint\limits_R u^2v^2w\, du\, dv\, dw$, where R is the region: $u^2 + v^2 \leqq 1$, $0 \leqq w \leqq 1$;

(c) $\iint\limits_R r^3 \cos\theta\, dr\, d\theta$, where R is the region: $1 \leqq r \leqq 2$, $\frac{\pi}{4} \leqq \theta \leqq \pi$.

2. Express the following in terms of multiple integrals and reduce to iterated integrals, but do not evaluate:

(a) the mass of a sphere whose density is proportional to the distance from one diametral plane;

(b) the coordinates of the center of mass of the sphere of part (a);

(c) the moment of inertia about the x axis of the solid filling the region $0 \leqq z \leqq 1 - x^2 - y^2$, $0 \leqq x \leqq 1$, $0 \leqq y \leqq 1 - x$ and having density proportional to xy.

3. The moment of inertia of a solid about an arbitrary line L is defined as

$$I_L = \iiint\limits_R d^2 f(x, y, z)\, dx\, dy\, dz,$$

where f is density and d is the distance from a general point (x, y, z) of the solid to the line L. Prove the *parallel axis theorem*:

$$I_L = I_{\bar{L}} + Mh^2,$$

where $\bar{L}$ is a line parallel to L through the center of mass, M is the mass, and h is the distance between L and $\bar{L}$. (Hint: take $\bar{L}$ to be the z axis.)

4. Let L be a line through the origin O with direction cosines l, m, n. Prove that

$$I_L = I_x l^2 + I_y m^2 + I_z n^2 - 2I_{xy} lm - 2I_{yz} mn - 2I_{zx} ln,$$

where

$$I_{xy} = \iiint\limits_R xyf(x, y, z)\, dx\, dy\, dz, \quad I_{yz} = \iiint\limits_R yzf \ldots .$$

The new integrals are called *products of inertia.* The locus

$$I_x x^2 + I_y y^2 + I_z z^2 - 2(I_{xy}xy + I_{yz}yz + I_{zx}zx) = 1$$

is an ellipsoid, called the *ellipsoid of inertia.*

Answers

1. (a) $1/3$, (b) $\pi/48$, (c) $(-15\sqrt{2})/8$.

4–8 Change of variables in integrals. For functions of one variable, the chain rule

$$\frac{dF}{du} = \frac{dF}{dx}\frac{dx}{du} \tag{4–57}$$

at once gives the rule for change of variable in a definite integral:

$$\int_{x_1}^{x_2} f(x)\, dx = \int_{u_1}^{u_2} f[x(u)] \frac{dx}{du}\, du. \tag{4–58}$$

Here $f(x)$ is assumed continuous at least for $x_1 \leqq x \leqq x_2$, $x = x(u)$ is defined for $u_1 \leqq u \leqq u_2$ and has a continuous derivative, with $x_1 = x(u_1)$, $x_2 = x(u_2)$, and $f[x(u)]$ is continuous for $u_1 \leqq u \leqq u_2$.

Proof. If $F(x)$ is an indefinite integral of $f(x)$, then

$$\int_{x_1}^{x_2} f(x)\, dx = F(x_2) - F(x_1).$$

But $F[x(u)]$ is then an indefinite integral of $f[x(u)]\dfrac{dx}{du}$, for (4–57) gives

$$\frac{dF}{du} = \frac{dF}{dx}\frac{dx}{du} = f(x)\frac{dx}{du} = f[x(u)]\frac{dx}{du},$$

when x is expressed in terms of u. Thus the integral on the right of (4–58) is

$$F[x(u_2)] - F[x(u_1)] = F(x_2) - F(x_1).$$

Since this is the same as the value of the left-hand member of (4–58), the rule is established.

It is worth noting that the emphasis in (4–58) is on the function $x(u)$ rather than on its inverse $u = u(x)$. Such an inverse will exist only when x is a steadily increasing function of u or a steadily decreasing function of u. This is not required for (4–58). In fact, the function $x(u)$ can take on values outside the interval $x_1 \leqq x \leqq x_2$, as illustrated in Fig. 4–11. However, $f[x(u)]$ must remain continuous for $u_1 \leqq u \leqq u_2$.

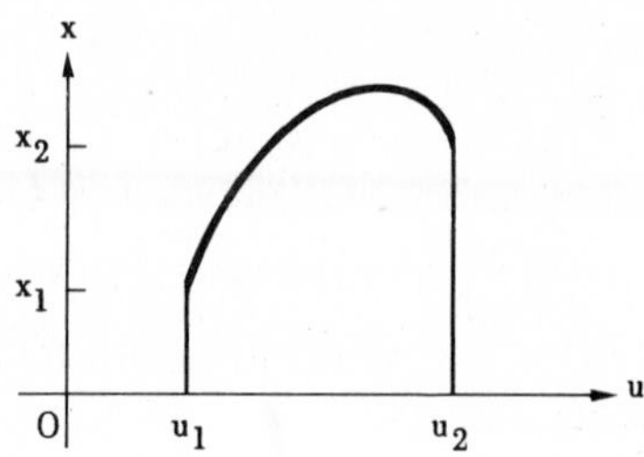

FIG. 4–11. The substitution $x = x(u)$ in a definite integral.

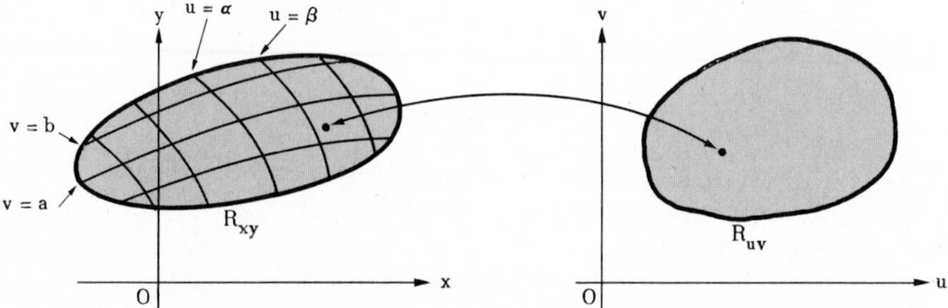

FIG. 4–12. Curvilinear coordinates for change of variables in a double integral.

There is a formula analogous to (4–58) for double integrals:

$$\iint\limits_{R_{xy}} f(x, y)\, dx\, dy = \iint\limits_{R_{uv}} f[x(u, v), y(u, v)] \left| \frac{\partial(x, y)}{\partial(u, v)} \right| du\, dv. \tag{4–59}$$

Here the functions

$$x = x(u, v), \quad y = y(u, v) \tag{4–60}$$

are assumed to be defined and have continuous derivatives in a region R_{uv} of the uv plane. The corresponding points (x, y) lie in the region R_{xy} of the xy plane, and it is assumed that the inverse functions

$$u = u(x, y), \quad v = v(x, y) \tag{4–61}$$

are defined and continuous in R_{xy}, so that the correspondence between R_{xy} and R_{uv} is *one-to-one*, as suggested in Fig. 4–12. The function $f(x, y)$ is assumed continuous in R_{xy}, so that $f[x(u, v), y(u, v)]$ is continuous in R_{uv}. Finally it is assumed that the Jacobian

$$J = \frac{\partial(x, y)}{\partial(u, v)}$$

is either positive throughout R_{uv} or negative throughout R_{uv}. It should be noted that it is the *absolute value* of J which is used in (4–59).

A proof of (4–59) will be given in the next chapter, with the aid of line integrals. Here we discuss the significance of (4–59) and its applications.

The equations (4–60) can be interpreted as an introduction of *curvilinear* coordinates in the xy plane, as suggested in Fig. 4–12. The lines $u =$ const and $v =$ const in R_{xy} form a system of curves like the parallels to the axes. It is natural to use them to cut the region R_{xy} into elements of area ΔA for formation of the double integral. With such curvilinear elements, the volume "beneath the surface $z = f(x, y)$" will still be approximated by $f(x, y)\,\Delta A$, where ΔA denotes the area of one of the curvilinear elements. If ΔA can be expressed as a multiple k of $\Delta u\,\Delta v$ and f is expressed in terms of u and v, one obtains a sum

$$\sum f[x(u, v), y(u, v)]k\,\Delta u\,\Delta v$$

which approaches a double integral

$$\iint\limits_{R_{uv}} f[x(u, v), y(u, v)]k\,du\,dv$$

as limit. The crucial question is thus the evaluation of the factor k. As (4–59) shows, one must prove

$$k = \left|\frac{\partial(x, y)}{\partial(u, v)}\right|.$$

The number k can also be interpreted as the ratio of an element of area ΔA_{xy} in the xy plane to the element $\Delta A_{uv} = \Delta u\,\Delta v$ in the uv plane. Thus one must show that

$$\left|\frac{\partial(x, y)}{\partial(u, v)}\right| = \lim \frac{\Delta A_{xy}}{\Delta A_{uv}}.$$

Example 1. Let

$$x = r\cos\theta, \quad y = r\sin\theta,$$

so that the curvilinear coordinates are polar coordinates. The element of area is approximately a rectangle with sides $r\,\Delta\theta$ and Δr, as shown in Fig. 4–13. Thus

$$\Delta A \sim r\,\Delta\theta\,\Delta r$$

and one expects the formula

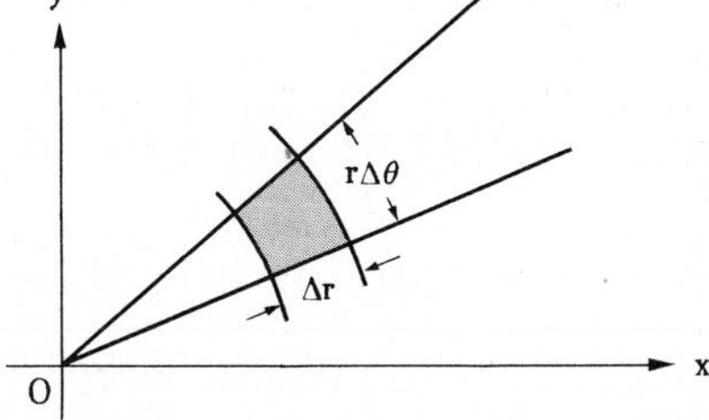

Fig. 4–13. Element of area in polar coordinates.

$$\iint\limits_{R_{xy}} f(x, y)\,dx\,dy = \iint\limits_{R_{r\theta}} f(r\cos\theta, r\sin\theta)r\,d\theta\,dr. \tag{4–62}$$

Now the Jacobian J in this case is

$$J = \frac{\partial(x, y)}{\partial(r, \theta)} = \begin{vmatrix} \cos\theta & -r\sin\theta \\ \sin\theta & r\cos\theta \end{vmatrix} = r.$$

Thus (4–62) is correct. The region $R_{r\theta}$ can be pictured in an $r\theta$ plane or, more simply, can be described by inequalities such as the following:

$$\alpha \leqq \theta \leqq \beta, \quad r_1(\theta) \leqq r \leqq r_2(\theta), \tag{4–63}$$

which can be read off the figure in the xy plane. From (4–63) one finds

$$\iint\limits_{R_{r\theta}} f(r\cos\theta, r\sin\theta) r\, d\theta\, dr = \int_\alpha^\beta \int_{r_1(\theta)}^{r_2(\theta)} f(r\cos\theta, r\sin\theta) r\, dr\, d\theta,$$

so that the integral has been reduced to an iterated integral in r and θ. It may be necessary to decompose the region $R_{r\theta}$ into several parts and obtain the integral as a sum of iterated integrals of the form given. For some problems it is simpler to integrate in the order $d\theta$, dr; the region $R_{r\theta}$ must then be described by inequalities

$$a \leqq r \leqq b, \quad \theta_1(r) \leqq \theta \leqq \theta_2(r), \tag{4–63'}$$

and the integral becomes

$$\int_a^b \int_{\theta_1(r)}^{\theta_2(r)} f(r\cos\theta, r\sin\theta) r\, d\theta\, dr.$$

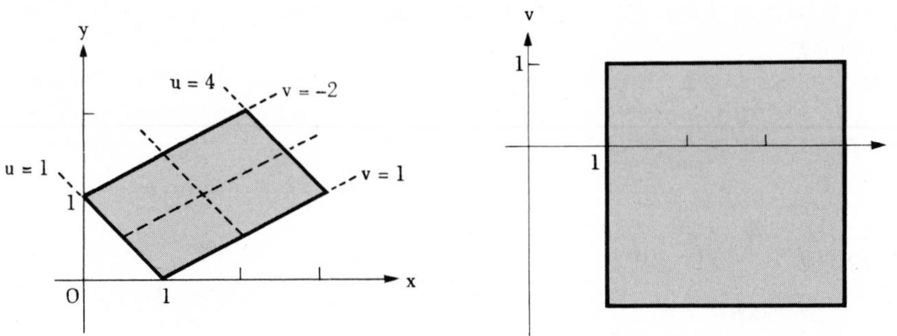

FIG. 4–14. Curvilinear coordinates $u = x + y$, $v = x - 2y$.

EXAMPLE 2. Let it be required to evaluate

$$\iint\limits_{R_{xy}} (x + y)^3\, dx\, dy,$$

where R_{xy} is the parallelogram shown in Fig. 4–14. The sides of R_{xy} are straight lines having equations of form

$$x + y = c_1, \quad x - 2y = c_2$$

for appropriate choices of c_1, c_2. It is therefore natural to introduce as new coordinates

$$u = x + y, \quad v = x - 2y.$$

The region R_{xy} then corresponds to the rectangle: $1 \leqq u \leqq 4$, $-2 \leqq v \leqq 1$.

The correspondence is clearly one-to-one. The Jacobian is

$$\frac{\partial(x, y)}{\partial(u, v)} = \frac{1}{\dfrac{\partial(u, v)}{\partial(x, y)}} = \frac{1}{\begin{vmatrix} 1 & 1 \\ 1 & -2 \end{vmatrix}} = -\frac{1}{3}.$$

Hence

$$\iint\limits_{R_{xy}} (x + y)^3 \, dx \, dy = \iint\limits_{R_{uv}} \frac{u^3}{3} \, du \, dv = \int_{-2}^{1} \int_{1}^{4} \frac{u^3}{3} \, du \, dv = 63\tfrac{3}{4}.$$

It should be noted that the limits of integration for the uv integral are determined from the figure and are not directly related to the limits which would be assigned to the corresponding iterated integral in the xy plane.

The fundamental formula (4–59) generalizes to triple integrals and multiple integrals of any order. Thus

$$\iiint\limits_{R_{xyz}} f(x, y, z) \, dx \, dy \, dz = \iiint\limits_{R_{uvw}} F(u, v, w) \left| \frac{\partial(x, y, z)}{\partial(u, v, w)} \right| du \, dv \, dw, \qquad (4\text{–}64)$$

where $F(u, v, w) = f[x(u, v, w), y(u, v, w), z(u, v, w)]$, under corresponding assumptions. Two important special cases are those of *cylindrical coordinates*:

$$\iiint\limits_{R_{xyz}} f(x, y, z) \, dx \, dy \, dz = \iiint\limits_{R_{r\theta z}} F(r, \theta, z) r \, dr \, d\theta \, dz, \qquad (4\text{–}65)$$

$$F(r, \theta, z) = f(r \cos \theta, r \sin \theta, z),$$

and *spherical coordinates*:

$$\iiint\limits_{R_{xyz}} f(x, y, z) \, dx \, dy \, dz = \iiint\limits_{R_{\rho\phi\theta}} F(\rho, \phi, \theta) \rho^2 \sin \phi \, d\rho \, d\phi \, d\theta, \qquad (4\text{–}66)$$

$$F(\rho, \phi, \theta) = f(\rho \sin \phi \cos \theta, \rho \sin \phi \sin \theta, \rho \cos \phi).$$

These are discussed in Probs. 7 and 8 below.

Remark. In order to determine the region R_{uv} for (4–59) and, in particular, to verify that the correspondence between R_{uv} and R_{xy} is one-to-one, a variety of techniques are available. In (4–60) or (4–61) one can set $u = \text{const}$ and plot the resulting level curves of u in R_{xy}. The same can be done for v. If these level curves have the property that a curve $u = c_1$ meets a curve $v = c_2$ in at most one point in R_{xy}, then the correspondence must be one-to-one. From the level curves one can follow the variation of u and v on the boundary of R_{xy} and thereby determine the boundary of R_{uv}. It can be shown that, if R_{xy} and R_{uv} are each bounded by a single closed curve, as in Fig. 4–12, if the correspondence between (x, y) and (u, v) is one-to-one on these boundary curves, and $J \neq 0$ in R_{uv}, then the correspondence is necessarily one-to-one in all of R_{xy} and R_{uv}. For a further discussion of this point one is referred to Section 5–14. Actually, the conditions that the correspondence be one-to-one and that

$J \neq 0$ are not vital for the theorem. It is shown in Section 5–14 that (4–59) can be written in a different form which covers the more general cases.

Problems

1. Evaluate with the aid of the substitution indicated:

(a) $\displaystyle\int_0^1 (1 - x^2)^{\frac{3}{2}}\, dx,\ x = \sin\theta;$

(b) $\displaystyle\int_0^1 \frac{1}{1 + \sqrt{1+x}}\, dx,\ x = u^2 - 1.$

2. Prove the formula

$$\int_{u_1}^{u_2} \phi'(u)\, du = \phi(u_2) - \phi(u_1)$$

as a special case of (4–58).

3. (a) Prove that (4–58) remains valid for improper integrals, i.e., if $f(x)$ is continuous for $x_1 \leqq x < x_2$, $x(u)$ is defined and has a continuous derivative for $u_1 \leqq u < u_2$, with $x(u_1) = x_1$, $\lim\limits_{u \to u_2} x(u) = x_2$, and $f[x(u)]$ is continuous for $u_1 \leqq u < u_2$. [Hint: use the fact that (4–58) holds with u_2 and x_2 replaced by u_0 and $x_0 = x(u_0)$, $u_1 < u_0 < u_2$. Then let u_0 approach u_2. One concludes that if either side of the equation has a limit, then the other side has a limit also and the limits are equal. Note that x_2 or u_2 or both may be ∞.]

(b) Evaluate $\displaystyle\int_1^\infty \frac{1}{x^2}\sinh\frac{1}{x}\, dx$ by setting $u = \dfrac{1}{x}\cdot$

(c) Evaluate $\displaystyle\int_0^\infty (1 - \tanh x)\, dx$ by setting $u = \tanh x$.

4. Evaluate the following integrals with the aid of the substitution suggested:

(a) $\displaystyle\iint\limits_{R_{xy}} (1 - x^2 - y^2)\, dx\, dy$, where R_{xy} is the region $x^2 + y^2 \leqq 1$, using $x = r\cos\theta$, $y = r\sin\theta$;

(b) $\displaystyle\iint\limits_{R_{xy}} (x - y)^2 \sin^2(x + y)\, dx\, dy$, where R_{xy} is the parallelogram with successive vertices $(\pi, 0)$, $(2\pi, \pi)$, $(\pi, 2\pi)$, $(0, \pi)$, using $u = x - y$, $v = x + y$.

5. Verify that the transformation

$$u = 2xy, \quad v = x^2 - y^2$$

defines a one-to-one mapping of the square $0 \leqq x \leqq 1$, $0 \leqq y \leqq 1$ onto a region of the uv plane. Express the integral

$$\iint\limits_{R_{xy}} \sqrt[3]{x^4 - 6x^2y^2 + y^4}\, dx\, dy$$

over the square as an iterated integral in u and v.

6. Transform the integrals given, using the substitutions indicated:

(a) $\int_0^1 \int_0^x \log(1 + x^2 + y^2)\, dy\, dx, \quad x = u + v, \quad y = u - v;$

(b) $\int_0^1 \int_{1-x}^{1+x} \sqrt{1 + x^2y^2}\, dy\, dx, \quad x = u, \quad y = u + v.$

7. Verify the correctness of (4–65) as a special case of (4–64). Show the geometric meaning of the volume element $r\,\Delta r\,\Delta\theta\,\Delta z$.

8. Verify the correctness of (4–66) as a special case of (4–64). Show the geometric meaning of the volume element $\rho^2 \sin\phi\,\Delta\rho\,\Delta\phi\,\Delta\theta$.

9. Transform to cylindrical coordinates but do not evaluate:

(a) $\iiint\limits_{R_{xyz}} x^2y\, dx\, dy\, dz$, where R_{xyz} is the region $x^2 + y^2 \leqq 1,\ 0 \leqq z \leqq 1$;

(b) $\int_0^1 \int_0^{\sqrt{1-x^2}} \int_0^{1+x+y} (x^2 - y^2)\, dz\, dy\, dx.$

10. Transform to spherical coordinates but do not evaluate:

(a) $\iiint\limits_{R_{xyz}} x^2y\, dx\, dy\, dz$, where R_{xyz} is the sphere: $x^2 + y^2 + z^2 \leqq a^2$;

(b) $\int_{-1}^1 \int_{-\sqrt{1-x^2}}^{\sqrt{1-x^2}} \int_{\sqrt{x^2+y^2}}^1 (x^2 + y^2 + z^2)\, dz\, dy\, dx.$

Answers

1. (a) $\frac{3}{16}\pi$, (b) $2\sqrt{2} - 2 + 2\log(2\sqrt{2} - 2)$. 3. (b) $\cosh 1 - 1$,
   (c) $\log 2$.

4. (a) $\frac{\pi}{2}$, (b) $\frac{\pi^4}{3}$. 5. $\int_0^2 \int_{\frac{u^2}{4}-1}^{1-\frac{u^2}{4}} \frac{\sqrt[3]{v^2 - u^2}}{4\sqrt{u^2 + v^2}}\, dv\, du.$

6. (a) $2\int_0^{\frac{1}{2}} \int_v^{1-v} \log(1 + 2u^2 + 2v^2)\, du\, dv,$

   (b) $\int_0^1 \int_{1-2u}^1 \sqrt{1 + u^2(u + v)^2}\, dv\, du.$

9. (a) $\int_0^{2\pi} \int_0^1 \int_0^1 r^4 \cos^2\theta \sin\theta\, dz\, dr\, d\theta,$

   (b) $\int_0^{\frac{\pi}{2}} \int_0^1 \int_0^{1+r(\cos\theta+\sin\theta)} r^3 \cos 2\theta\, dz\, dr\, d\theta.$

10. (a) $\int_0^{2\pi} \int_0^{\pi} \int_0^a \rho^5 \sin^4\phi \cos^2\theta \sin\theta\, d\rho\, d\phi\, d\theta,$

   (b) $\int_0^{2\pi} \int_0^{\frac{\pi}{4}} \int_0^{\sec\phi} \rho^4 \sin\phi\, d\rho\, d\phi\, d\theta.$

4–9 Arc length and surface area. In elementary calculus it is shown that a curve $y = f(x)$, $a \leqq x \leqq b$, has length

$$s = \int_a^b \sqrt{1 + \left(\frac{dy}{dx}\right)^2}\, dx \tag{4–67}$$

and that, if the curve is given parametrically by equations $x = x(t)$, $y = y(t)$ for $t_1 \leqq t \leqq t_2$, then it has length

$$s = \int_{t_1}^{t_2} \sqrt{\left(\frac{dx}{dt}\right)^2 + \left(\frac{dy}{dt}\right)^2}\, dt. \tag{4–68}$$

Furthermore, a curve $x = x(t)$, $y = y(t)$, $z = z(t)$ in space has length

$$s = \int_{t_1}^{t_2} \sqrt{\left(\frac{dx}{dt}\right)^2 + \left(\frac{dy}{dt}\right)^2 + \left(\frac{dz}{dt}\right)^2}\, dt. \tag{4–69}$$

The length is here defined as a limit of the lengths of inscribed polygons; the functions are assumed to have continuous derivatives over the intervals concerned.

The main question here is the generalization of (4–67) and (4–69) to the area of surfaces in space. For a surface $z = f(x, y)$ it will be seen that the area is given by

$$S = \iint\limits_{R_{xy}} \sqrt{1 + \left(\frac{\partial z}{\partial x}\right)^2 + \left(\frac{\partial z}{\partial y}\right)^2}\, dx\, dy. \tag{4–70}$$

This parallels (4–67). A surface in space can be represented parametrically by equations

$$x = x(u, v), \quad y = y(u, v), \quad z = z(u, v), \tag{4–71}$$

where u and v vary in a region R_{uv} of the uv plane. The area of the surface (4–71) is given by

$$S = \iint\limits_{R_{uv}} \sqrt{EG - F^2}\, du\, dv, \tag{4–72}$$

where

$$\begin{aligned} E &= \left(\frac{\partial x}{\partial u}\right)^2 + \left(\frac{\partial y}{\partial u}\right)^2 + \left(\frac{\partial z}{\partial u}\right)^2, \\ F &= \frac{\partial x}{\partial u}\frac{\partial x}{\partial v} + \frac{\partial y}{\partial u}\frac{\partial y}{\partial v} + \frac{\partial z}{\partial u}\frac{\partial z}{\partial v}, \\ G &= \left(\frac{\partial x}{\partial v}\right)^2 + \left(\frac{\partial y}{\partial v}\right)^2 + \left(\frac{\partial z}{\partial v}\right)^2. \end{aligned} \tag{4–73}$$

A complete justification of (4–70) and (4–72) turns out to involve a very delicate analysis, much more difficult than that of arc length. In particular, one cannot define surface area S simply as the limit of the areas of inscribed polyhedra. For a full discussion, the reader is referred to P. Franklin's *Treatise on Advanced Calculus*, pages 371–378 (New York: Wiley, 1940). Here we give an intuitive discussion of why (4–70) and (4–72) are to be expected.

We first note that the arc length s can be defined in a somewhat different way, using tangents rather than an inscribed polygon, as suggested in Fig. 4–15. Thus, for a curve $y = f(x)$, one subdivides the interval $a \leqq x \leqq b$ as for integrating $f(x)$. At a point x_i^* between x_{i-1} and x_i, one draws the corresponding tangent line to the curve:

$$y - y_i^* = f'(x_i^*)(x - x_i^*),$$

and denotes by $\Delta_i T$ the length of the segment of this line between x_{i-1} and x_i. It is natural to expect that

$$s = \lim \sum_{i=1}^{n} \Delta_i T$$

as n becomes infinite and max $\Delta_i x$ approaches 0. Now if α_i^* is the angle of inclination of $\Delta_i T$, so that $f'(x_i^*) = \tan \alpha_i^*$, one has

$$\Delta_i x = \Delta_i T \cos \alpha_i^*$$

or

$$\Delta_i T = \Delta_i x \sec \alpha_i^*. \tag{4–74}$$

The sum $\Sigma \Delta_i T$ thus becomes

$$\sum \sec \alpha_i^* \, \Delta_i x = \sum \sqrt{1 + f'(x_i^*)^2} \, \Delta_i x.$$

If $f'(x)$ is continuous for $a \leqq x \leqq b$, this sum approaches as limit the desired expression

$$\int_a^b \sec \alpha \, dx = \int_a^b \sqrt{1 + f'(x)^2} \, dx = s.$$

Now let a surface $z = f(x, y)$ be given, where $f(x, y)$ is defined and has continuous partial derivatives in a domain D. To find the area of the part

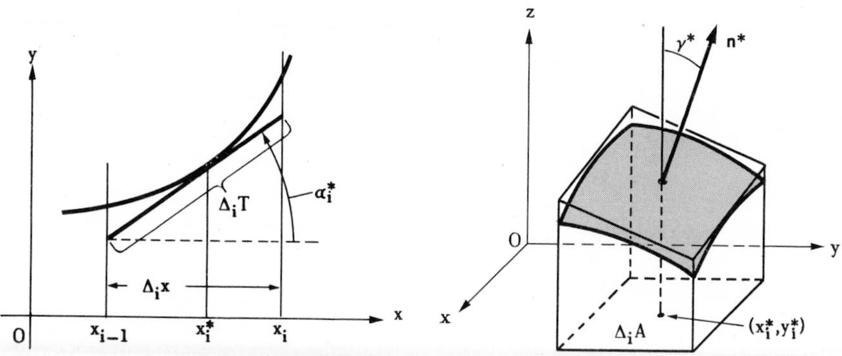

FIG. 4–15. Definition of arc length. FIG. 4–16. Definition of surface area.

of the surface above a bounded closed region R_{xy}, contained in D, we subdivide R_{xy} as in Fig. 4–8. Let (x_i^*, y_i^*) be a point of the ith rectangle; we then construct the tangent plane to the surface at the corresponding point:

$$z - z_i^* = f_x(x_i^*, y_i^*)(x - x_i^*) + f_y(x_i^*, y_i^*)(y - y_i^*). \tag{4–75}$$

Let $\Delta_i S^*$ be the area of the part of this tangent plane above the ith rectangle in the xy plane. Thus $\Delta_i S^*$ is the area of a certain parallelogram whose projection on the xy plane is a rectangle of area $\Delta_i A$. Now let $\mathbf{n}^*$ be the normal vector to the surface at the point of tangency:

$$\mathbf{n}^* = -\frac{\partial z}{\partial x}\mathbf{i} - \frac{\partial z}{\partial y}\mathbf{j} + \mathbf{k}. \tag{4–76}$$

Then it follows readily from geometry that

$$\Delta_i S^* = \sec \gamma_i^* \, \Delta_i A, \tag{4–77}$$

where γ_i^* is the angle between $\mathbf{n}^*$ and $\mathbf{k}$. This is analogous to (4–72) (cf. Prob. 4 below). Now from (4–76)

$$\cos \gamma_i^* = \frac{1}{\sqrt{1 + \left(\frac{\partial z}{\partial x}\right)^2 + \left(\frac{\partial z}{\partial y}\right)^2}} = \frac{\mathbf{n}^* \cdot \mathbf{k}}{|\mathbf{n}^*|},$$

so that

$$\sec \gamma_i^* = \sqrt{1 + \left(\frac{\partial z}{\partial x}\right)^2 + \left(\frac{\partial z}{\partial y}\right)^2},$$

all derivatives being evaluated at (x_i^*, y_i^*). Paralleling the procedure for arc length given above, it is natural to expect the surface area S to be obtained as the limit of a sum

$$\begin{aligned} \sum_{i=1}^{n} \Delta_i S^* &= \sum_{i=1}^{n} \sec \gamma_i^* \, \Delta_i A \\ &= \sum_{i=1}^{n} \sqrt{1 + \left(\frac{\partial z}{\partial x}\right)^2 + \left(\frac{\partial z}{\partial y}\right)^2} \, \Delta_i A, \end{aligned}$$

as the number n of subdivisions approaches ∞, while the maximum diagonal approaches 0. This limit is precisely the double integral

$$S = \iint_{R_{xy}} \sec \gamma \, dx \, dy = \iint_{R_{xy}} \sqrt{1 + \left(\frac{\partial z}{\partial x}\right)^2 + \left(\frac{\partial z}{\partial y}\right)^2} \, dx \, dy$$

as desired.

Surfaces in parametric form. The parametric equations

$$x = x(u, v), \quad y = y(u, v), \quad z = z(u, v), \quad (u, v) \text{ in } R_{uv} \tag{4–78}$$

of a surface can be regarded as a mapping of R_{uv} onto a "curved region" in space, as suggested in Fig. 4–17. The lines $u =$ const, $v =$ const on the surface determine *curvilinear coordinates on the surface.* When $v =$

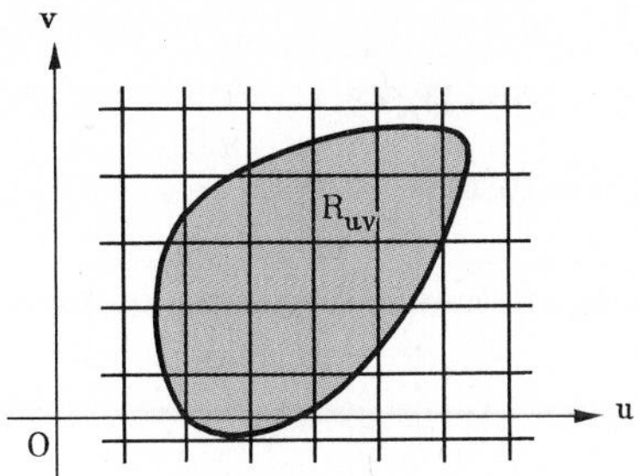

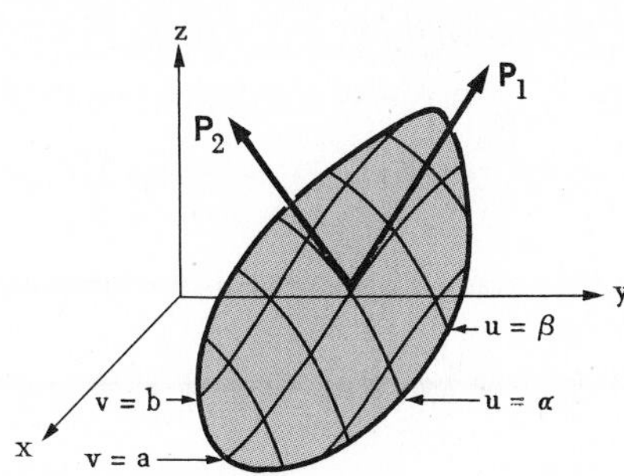

FIG. 4–17. Curvilinear coordinates on a surface.

const, (4–78) can be regarded as parametric equations of a curve; the tangent vector to such a curve is the vector

$$\mathbf{P}_1 = \frac{\partial x}{\partial u}\mathbf{i} + \frac{\partial y}{\partial u}\mathbf{j} + \frac{\partial z}{\partial u}\mathbf{k}, \tag{4–79}$$

in accordance with Section 1–16. Similarly, a curve u = const has tangent vector

$$\mathbf{P}_2 = \frac{\partial x}{\partial v}\mathbf{i} + \frac{\partial y}{\partial v}\mathbf{j} + \frac{\partial z}{\partial v}\mathbf{k}. \tag{4–80}$$

The lines u = const, v = const in R_{uv} can be used to subdivide R_{uv} as in the formation of the double integral. Each rectangle, of area $\Delta A = \Delta u\, \Delta v$, of this subdivision corresponds to a curved area element on the surface. To first approximation, this area element is a parallelogram with sides $|\mathbf{P}_1|\,\Delta u$ and $|\mathbf{P}_2|\,\Delta v$, for $\mathbf{P}_1$ can be interpreted as a velocity vector in terms of time u, so that

$$\frac{ds_1}{du} = |\mathbf{P}_1|,$$

where s_1 is distance along the line v = const chosen. For a small change Δu in the time u, the distance moved is approximately $ds_1 = |\mathbf{P}_1|\,\Delta u$. Similarly the other side of the parallelogram is approximately $|\mathbf{P}_2|\,\Delta v$. The vectors $\mathbf{P}_1$ and $\mathbf{P}_2$ are here evaluated at one corner of the parallelogram, as suggested in Fig. 4–17. The parallelogram has then as sides segments representing the vectors

$$\mathbf{P}_1\,\Delta u \quad \text{and} \quad \mathbf{P}_2\,\Delta v.$$

Its area is thus given by

$$|(\mathbf{P}_1\,\Delta u) \times (\mathbf{P}_2\,\Delta v)| = |\mathbf{P}_1 \times \mathbf{P}_2|\,\Delta u\,\Delta v.$$

If we grant that the sum of the areas of these parallelograms over the surface approaches the area of the surface as limit as the number of subdivisions of R_{uv} is increased, as in forming the double integral, then we conclude that the surface area is given by

$$S = \iint\limits_{R_{uv}} |\mathbf{P}_1 \times \mathbf{P}_2|\,du\,dv. \tag{4–81}$$

By means of the vector identity

$$|\mathbf{P}_1 \times \mathbf{P}_2|^2 = |\mathbf{P}_1|^2\,|\mathbf{P}_2|^2 - (\mathbf{P}_1 \cdot \mathbf{P}_2)^2 \tag{4–82}$$

(cf. Prob. 9 below), this can be written in the form

$$S = \iint\limits_{R_{uv}} \sqrt{|\mathbf{P}_1|^2\,|\mathbf{P}_2|^2 - (\mathbf{P}_1 \cdot \mathbf{P}_2)^2}\,du\,dv. \tag{4–83}$$

Comparison of (4–79), (4–80), and (4–73) shows that one has

$$|\mathbf{P}_1|^2 = E, \quad \mathbf{P}_1 \cdot \mathbf{P}_2 = F, \quad |\mathbf{P}_2|^2 = G. \tag{4–84}$$

Thus (4–83) reduces at once to the formula desired:

$$S = \iint\limits_{R_{uv}} \sqrt{EG - F^2}\,du\,dv. \tag{4–85}$$

It should be noted that, if the point (x, y, z) traces a general curve on the surface, the differential

$$d\mathbf{r} = dx\mathbf{i} + dy\mathbf{j} + dz\mathbf{k}$$

can be expressed as follows:

$$\begin{aligned} d\mathbf{r} &= \left(\frac{\partial x}{\partial u}\,du + \frac{\partial x}{\partial v}\,dv\right)\mathbf{i} + \left(\frac{\partial y}{\partial u}\,du + \frac{\partial y}{\partial v}\,dv\right)\mathbf{j} + \left(\frac{\partial z}{\partial u}\,du + \frac{\partial z}{\partial v}\,dv\right)\mathbf{k} \\ &= \mathbf{P}_1\,du + \mathbf{P}_2\,dv = \frac{\partial \mathbf{r}}{\partial u}\,du + \frac{\partial \mathbf{r}}{\partial v}\,dv. \end{aligned}$$

The element of arc on such a curve is defined by

$$ds^2 = dx^2 + dy^2 + dz^2 = |d\mathbf{r}|^2 = (\mathbf{P}_1\,du + \mathbf{P}_2\,dv) \cdot (\mathbf{P}_1\,du + \mathbf{P}_2\,dv).$$

Expanding the last product, one finds

$$ds^2 = |\mathbf{P}_1|^2\,du^2 + 2(\mathbf{P}_1 \cdot \mathbf{P}_2)\,du\,dv + |\mathbf{P}_2|^2\,dv^2. \tag{4–86}$$

Accordingly, by (4–84),

$$ds^2 = E\,du^2 + 2F\,du\,dv + G\,dv^2. \tag{4–87}$$

This shows the significance of E, F, G for the geometry on the surface.

Problems

1. Find the length of the circumference of a circle
(a) using the parametric representation

$$x = a\cos\theta, \quad y = a\sin\theta,$$

(b) using the parametric representation

$$x = a\,\frac{1 - t^2}{1 + t^2}, \quad y = a\,\frac{2t}{1 + t^2}.$$

2. Find the area of the surface of a sphere
(a) using the equation

$$z = \pm\sqrt{a^2 - x^2 - y^2},$$

(b) using the parametric equations

$$x = a \sin\phi \cos\theta, \quad y = a \sin\phi \sin\theta, \quad z = a \cos\phi.$$

3. Using the parametric equations of Prob. 2(b), set up a double integral for the area of a portion of the earth's surface bounded by two parallels of latitude and two meridians of longitude. Apply this to find the area of the United States of America, approximating this by the "rectangle" between parallels 30° N. and 47° N. and meridians 75° W. and 122° W. Take the radius of the earth to be 4000 miles.

4. Let a parallelogram be given in space whose sides represent the vectors **a** and **b**. Let **c** be a unit vector perpendicular to a plane C. (a) Show that $\mathbf{a} \times \mathbf{b} \cdot \mathbf{c}$ equals plus or minus the area of the projection of the parallelogram on C. (b) Show that this can also be written as $S \cos\gamma$, where S is the area of the parallelogram and γ is the angle between $\mathbf{a} \times \mathbf{b}$ and $\mathbf{c}$. (c) Show that one has

$$S = \sqrt{S_{yz}^2 + S_{zx}^2 + S_{xy}^2},$$

where S_{yz}, S_{zx}, S_{xy} are the areas of the projections of the parallelogram on the yz plane, zx plane, xy plane.

5. A surface of revolution is obtained by rotating a curve $z = f(x)$, $y = 0$ in the xz plane about the z axis. (a) Show that this surface has the equation $z = f(r)$ in cylindrical coordinates. (b) Show that the area of the surface is

$$S = \int_0^{2\pi}\int_a^b \sqrt{1 + f'(r)^2}\, r\, dr\, d\theta = 2\pi \int_a^b \sqrt{1 + f'(r)^2}\, r\, dr.$$

6. Prove (4–72) by assuming the surface in question can also be represented as $z = f(x, y)$ and has hence an area given by (4–70). The parametric equations (4–71) are to be regarded as a transformation of variables in the double integral (4–70), as in Section 4–8.

7. Show that (4–72) reduces to

$$S = \iint\limits_{R_{uv}} \left|\frac{\partial(x, y)}{\partial(u, v)}\right| du\, dv = \iint\limits_{R_{xy}} dx\, dy,$$

when u, v are curvilinear coordinates in a plane area R_{xy}.

8. Prove that, if a surface $z = f(x, y)$ is given in implicit form: $F(x, y, z) = 0$, then the surface area (4–70) becomes

$$\iint\limits_{R_{xy}} \frac{\sqrt{F_x^2 + F_y^2 + F_z^2}}{|F_z|}\, dx\, dy.$$

9. Prove the identity (4–82). [Hint: write the left side as $\mathbf{P}_1 \times \mathbf{P}_2 \cdot \mathbf{P}_1 \times \mathbf{P}_2$ and interchange the dot with one cross. Then use the identity (0–86) or (0–87).]

***4–10 Improper multiple integrals.** Since the discontinuities of functions of several variables can be much more complicated than those of functions of one variable, the discussion of improper multiple integrals is not as simple as for ordinary definite integrals. However, in principle, the analyses of Sections 4–4 and 4–5 carry over to multiple integrals.

The definition of boundedness of a function of several variables is the same as that of a function of one variable. Thus the function $\log(x^2 + y^2)$ is bounded for $1 \leqq x \leqq 2$, $1 \leqq y \leqq 2$ but is not bounded for $0 < x^2 + y^2 \leqq 1$. If a function $f(x, y)$ is defined and continuous throughout a bounded closed region R such as (4–31), except for a finite number of points of discontinuity, and if f is bounded, then the double integral

$$\iint_R f(x, y)\, dx\, dy$$

continues to exist as a limit of a sum, for the same reason as for functions of one variable. The discontinuities can even take the form of whole curves, finite in number and composed of "smooth curves," for the similar reason that these together form a set of *zero area*. An important case of this is when $f(x, y)$ is continuous and bounded only in a bounded domain D, nothing being known about values of f on the boundary of D. In such a case the integral

$$\iint_D f(x, y)\, dx\, dy$$

continues to exist as a limit of a sum, provided one evaluates f only where it is given, i.e., inside D. The same result is obtained if one assigns f arbitrary values, e.g., 0, on the boundary of D, provided this boundary is composed of smooth curves as above.

Example. The integral

$$\iint_D \sin\frac{y}{x}\, dx\, dy,$$

where D is the square domain: $0 < x < 1$, $0 < y < 1$, exists, even though the function is badly discontinuous on the y axis, since $\left|\sin\frac{y}{x}\right| \leqq 1$ in D.

For any such bounded function f one can approximate the integral over the whole region arbitrarily closely by the integral over a smaller region, avoiding the discontinuities. For if R is split into two regions R_1, R_2, overlapping only on boundary points, then

$$\iint_R f(x, y)\, dx\, dy = \iint_{R_1} f(x, y)\, dx\, dy + \iint_{R_2} f(x, y)\, dx\, dy.$$

Further, if $|f| \leqq M$ in R, then

$$\left|\iint\limits_{R_2} f(x, y)\,dx\,dy\right| \leqq M \cdot A_2, \tag{4–88}$$

where A_2 is the area of R_2, as follows from (4–46). If A_2 is sufficiently small, the integral over R_1 will approximate the integral over R as closely as desired.

True improper integrals arise, as for functions of one variable, when $f(x, y)$ is unbounded in R. The most important case of this is when $f(x, y)$ is defined and continuous but unbounded in a bounded domain D, with no information given about values of f on the boundary of D. In this case the limit of the sum $\Sigma f(x, y)\,\Delta A$ will fail to exist because f is unbounded. The integral of f over D is termed improper and is assigned a value by the limit process

$$\iint\limits_{D} f(x, y)\,dx\,dy = \lim_{R\to D} \iint\limits_{R} f(x, y)\,dx\,dy, \tag{4–89}$$

provided the limit exists. Here R denotes a closed region contained, with its boundary, in D, as suggested in Fig. 4–18. The limit process is understood as follows: the limit exists and has value K if, given $\epsilon > 0$, a particular region R_1 can be found, such that

$$\left|\iint\limits_{R} f(x, y)\,dx\,dy - K\right| < \epsilon$$

for all regions R containing R_1 and lying in D.

What amounts to a special case of the above is the case of a function having a *point discontinuity*, i.e., a function continuous in a domain D except at a single point P of D. It would be natural in this case to isolate the trouble at P by integrating up to a small circle of radius h about P and then letting h approach 0. This definition of the improper integral is equivalent to the preceding, provided $f(x, y)$ has the same sign (+ or −) near P. This is the most frequent case. Thus the integral

$$\iint\limits_{R} \frac{1}{r^p}\,dx\,dy, \quad p > 0,$$

where R is the circle $x^2 + y^2 \leqq 1$, is improper because of a point discontinuity at the origin. The limit process gives, in polar coordinates,

$$\begin{aligned}\iint\limits_{R} \frac{1}{r^p}\,dx\,dy &= \lim_{h\to 0} \int_0^{2\pi}\int_h^1 \frac{1}{r^p}\,r\,dr\,d\theta \\ &= \lim_{h\to 0} 2\pi\left(\frac{1}{2-p} - \frac{h^{2-p}}{2-p}\right), \quad (p \neq 2).\end{aligned} \tag{4–90}$$

Thus if $p < 2$, the integral converges to the value $2\pi/(2 - p)$ and if $p > 2$ the integral diverges. For $p = 2$, one obtains a logarithm and the integral again diverges. This result can be made the basis of a comparison test,

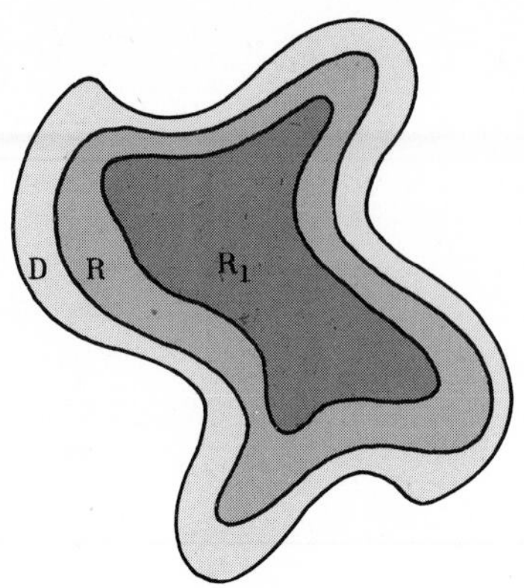

FIG. 4–18. The limit process $R \to D$.

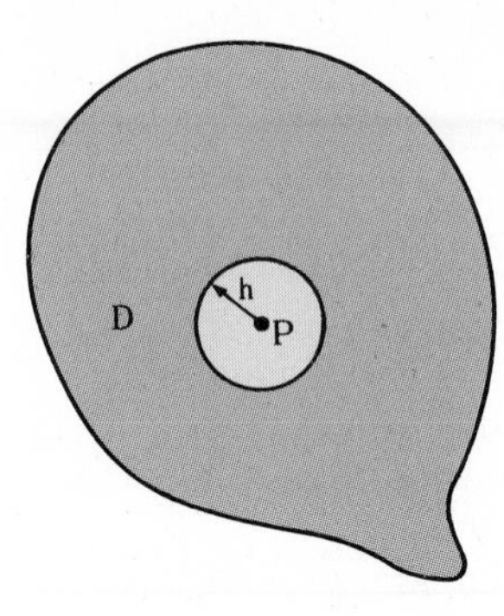

FIG. 4–19. Integral improper at a point P.

as with Theorems I(a), I(b), and II of Section 4–5. It should be noted that the critical value here is $p = 2$, as compared with $p = 1$ for single integrals.

A second type of improper integral, generalizing the definite integrals with infinite limits, is an integral

$$\iint_R f(x, y)\, dx\, dy$$

where R is an *unbounded* closed region. Here one obtains a value by a limit process just like that of (4–89) above. The most important case of this is that of a function continuous outside and on a circle $x^2 + y^2 = a^2$. If $f(x, y)$ is of one sign, the integral over this region R can be defined as the limit

$$\lim_{k \to \infty} \iint_{R_k} f(x, y)\, dx\, dy$$

where R_k is the region $a^2 \leqq x^2 + y^2 \leqq k^2$. Thus the improper integral

$$\iint_R \frac{1}{r^p}\, dx\, dy$$

has the value

$$\lim_{k \to \infty} \int_a^k \int_0^{2\pi} \frac{1}{r^p}\, d\theta\, r\, dr = \lim_{k \to \infty} 2\pi \frac{k^{2-p} - a^{2-p}}{2 - p},$$

which equals $2\pi a^{2-p}/(p - 2)$, for $p > 2$. For $p \leqq 2$, the integral diverges. Again one can set up comparison tests paralleling the theorems of Section 4–5.

While the emphasis here has been on double integrals, the statements hold with minor changes [affecting in particular the critical value of p for the integral (4–90)] for triple and other multiple integrals.

Problems

1. One way of evaluating the *error integral*

$$\int_0^\infty e^{-x^2}\,dx$$

is to use the equations

$$\left(\int_0^\infty e^{-x^2}dx\right)^2 = \int_0^\infty e^{-x^2}\,dx \int_0^\infty e^{-y^2}\,dy = \int_0^\infty\int_0^\infty e^{-x^2-y^2}\,dx\,dy$$

and to evaluate the double integral by polar coordinates. Carry out this evaluation, showing that the integral equals $\frac{1}{2}\sqrt{\pi}$; also discuss the significance of the above equations in terms of the limit definitions of the improper integrals.

2. Show that the integral

$$\iint\limits_R \log\sqrt{x^2+y^2}\,dx\,dy$$

converges, where R is the region $x^2 + y^2 \leqq 1$, and find its value. This can be interpreted as the *logarithmic potential,* at the origin, of a uniform mass distribution over the circle.

3. Show that the integral

$$\iiint\limits_R \frac{1}{r^p}\,dx\,dy\,dz, \quad r = \sqrt{x^2+y^2+z^2},$$

over the spherical region $x^2 + y^2 + z^2 \leqq 1$ converges for $p < 3$, and find its value. For $p = 1$, this is the *Newtonian potential* of a uniform mass distribution over the solid sphere.

4. Test for convergence or divergence:

(a) $\displaystyle\iint\limits_R \frac{x-y}{x^2+y^2}\,dx\,dy$, over the square: $|x| < 1$, $|y| < 1$;

(b) $\displaystyle\iint\limits_R \frac{\log(x^2+y^2)}{\sqrt{x^2+y^2}}\,dx\,dy$ over the circle: $x^2 + y^2 \leqq 1$;

(c) $\displaystyle\iint\limits_R \log(x^2+y^2)\,dx\,dy$ over the region: $x^2 + y^2 \geqq 1$;

(d) $\displaystyle\iiint\limits_R \log(x^2+y^2+z^2)\,dx\,dy\,dz$ over the solid: $x^2 + y^2 + z^2 \leqq 1$.

Answers

4. (a) conv, (b) conv, (c) div, (d) conv.

4–11 Integrals depending on a parameter — Leibnitz's rule. A definite integral

$$\int_a^b f(x, t)\,dx$$

of a continuous function $f(x, t)$ has a value which depends on the choice of t, so that one can write

$$\int_a^b f(x, t)\,dx = F(t). \tag{4-91}$$

One calls such an expression an *integral depending on a parameter* and t is termed the parameter. Thus

$$\int_0^{\frac{\pi}{2}} \frac{dx}{\sqrt{1 - k^2 \sin^2 x}}$$

is an integral depending on the parameter k; this example happens to be a *complete elliptic integral* (Section 4–3 above).

If an integral depending on a parameter can be evaluated in terms of familiar functions, it becomes simply an explicit function of one variable. Thus, for example,

$$\int_0^{\pi} \sin(xt)\,dx = \frac{1}{t} - \frac{\cos(\pi t)}{t} \quad (t \neq 0).$$

However, it can easily happen, as the above elliptic integral illustrates, that the integral cannot be expressed in terms of elementary functions. In such a case, the function of the parameter is nevertheless well-defined. It can be evaluated as accurately as desired for each particular parameter value and then tabulated; precisely this has been done for the above elliptic integral and tables thereof are easily accessible (Section 4–3).

The question to be studied here is the evaluation of the *derivative* of a function $F(t)$ defined by an integral as in Eq. (4–91):

Leibnitz's rule. Let $f(x, t)$ be continuous and have a continuous derivative $\partial f/\partial t$ in a domain of the xt plane which includes the rectangle $a \leqq x \leqq b$, $t_1 \leqq t \leqq t_2$. Then for $t_1 < t < t_2$

$$\frac{d}{dt}\int_a^b f(x, t)\,dx = \int_a^b \frac{\partial f}{\partial t}(x, t)\,dx. \tag{4-92}$$

In other words, differentiation and integration can be interchanged; e.g.,

$$\frac{d}{dt}\int_0^{\pi} \sin(xt)\,dx = \int_0^{\pi} x\cos(xt)\,dx.$$

Here both sides can be fully evaluated, so that the result can be checked.

Proof of Leibnitz's rule. Let

$$g(t) = \int_a^b \frac{\partial f}{\partial t}(x, t)\,dx \quad (t_1 \leqq t \leqq t_2).$$

Since $\partial f/\partial t$ is continuous, one concludes from the theorem of Section 4–6 that $g(t)$ is continuous for $t_1 \leqq t \leqq t_2$. Now for $t_1 < t_3 < t_2$

$$\int_{t_1}^{t_3} g(t)\, dt = \int_{t_1}^{t_3}\int_a^b \frac{\partial f}{\partial t}(x, t)\, dx\, dt;$$

by the theorem referred to one can interchange the order of integration:

$$\int_{t_1}^{t_3} g(t)\, dt = \int_a^b \int_{t_1}^{t_3} \frac{\partial f}{\partial t}(x, t)\, dt\, dx = \int_a^b [f(x, t_3) - f(x, t_1)]\, dx$$

$$= \int_a^b f(x, t_3)\, dx - \int_a^b f(x, t_1)\, dx = F(t_3) - F(t_1),$$

where $F(t)$ is defined by (4–91). If we now let t_3 be simply a variable t, we have

$$F(t) - F(t_1) = \int_{t_1}^{t} g(t)\, dt.$$

Both sides can now be differentiated with respect to t. By the fundamental theorem (4–19), one obtains

$$F'(t) = g(t) = \int_a^b \frac{\partial f}{\partial t}(x, t)\, dx.$$

Thus the rule is proved.

The notion of integral depending on a parameter extends at once to multiple integrals, and Leibnitz's rule also generalizes. Thus

$$\frac{d}{d\alpha}\int_1^2\int_1^2 \sqrt{x^\alpha + y^\alpha}\, dx\, dy = \int_1^2\int_1^2 \frac{x^\alpha \log x + y^\alpha \log y}{2\sqrt{x^\alpha + y^\alpha}}\, dx\, dy.$$

It can be extended to improper integrals, but with complications; this is discussed in Chapter 6.

One has at times to consider expressions like (4–91) in which the limits of integration a and b themselves depend on the parameter t. For example, one might have

$$F(t) = \int_{t^2}^{t^3} e^{-x^2 t}\, dx.$$

Here again one has a method for finding the derivative in terms of an integral.

THEOREM. *Let $f(x, t)$ satisfy the condition stated above for Leibnitz's rule. In addition, let $a(t)$ and $b(t)$ be defined and have continuous derivatives for $t_1 < t < t_2$. Then for $t_1 < t < t_2$*

$$\frac{d}{dt}\int_{a(t)}^{b(t)} f(x, t)\, dx = f[b(t), t]b'(t) - f[a(t), t]\, a'(t) + \int_{a(t)}^{b(t)} \frac{\partial f}{\partial t}(x, t)\, dx. \quad (4\text{–}93)$$

Thus, for example,

$$\frac{d}{dt}\int_{t^2}^{t^3} e^{-x^2 t}\, dx = e^{-t^7}(3t^2) - e^{-t^5}(2t) + \int_{t^2}^{t^3} e^{-x^2 t}(-x^2)\, dx.$$

Proof. Let $u = b(t)$, $v = a(t)$, $w = t$, so that the integral $F(t)$ can be written as follows:

$$F(t) = \int_v^u f(x, w)\, dx = G(u, v, w),$$

where u, v, w all depend on t. Hence by the chain rule

$$\frac{dF}{dt} = \frac{\partial G}{\partial u}\frac{du}{dt} + \frac{\partial G}{\partial v}\frac{dv}{dt} + \frac{\partial G}{\partial w}\frac{dw}{dt}.$$

It will be seen that the three terms here correspond to the three terms on the right of (4–93). Indeed one has

$$\frac{\partial G}{\partial u} = \frac{\partial}{\partial u}\int_v^u f(x, w)\, dx = f(u, w),$$

by the fundamental theorem (4–19). Since $u = b(t)$, $du/dt = b'(t)$ and

$$\frac{\partial G}{\partial u}\frac{du}{dt} = f[b(t), t]b'(t).$$

The second term is accounted for similarly, the minus sign appearing because

$$\frac{\partial}{\partial v}\int_v^u f(x, w)\, dx = \frac{\partial}{\partial v}\left\{-\int_u^v f(x, w)\, dx\right\} = -f(v, w).$$

Finally

$$\frac{\partial G}{\partial w} = \frac{\partial}{\partial w}\int_v^u f(x, w)\, dx = \int_v^u \frac{\partial f}{\partial w}(x, w)\, dx$$

by Leibnitz's rule. Since $w = t$, $dw/dt = 1$ and the third term is accounted for.

Problems

1. Obtain the indicated derivatives in the form of integrals:

(a) $\dfrac{d}{dt}\displaystyle\int_{\frac{\pi}{2}}^{\pi} \frac{\cos(xt)}{x}\, dx$ (c) $\dfrac{d}{du}\displaystyle\int_1^2 \log(xu)\, dx$

(b) $\dfrac{d}{dt}\displaystyle\int_1^2 \frac{x^2}{(1 - tx)^2}\, dx$ (d) $\dfrac{d^n}{dy^n}\displaystyle\int_1^2 \frac{\sin x}{x - y}\, dx$

2. Obtain the indicated derivatives:

(a) $\dfrac{d}{dx}\displaystyle\int_1^x t^2\, dt$ (c) $\dfrac{d}{dt}\displaystyle\int_{t^3}^2 \log(1 + x^2)\, dx$

(b) $\dfrac{d}{dt}\displaystyle\int_1^{t^2} \sin(x^2)\, dx$ (d) $\dfrac{d}{dx}\displaystyle\int_x^{\tan x} e^{-t^2}\, dt$

3. Prove the following:

(a) $\dfrac{d}{d\alpha}\displaystyle\int_{\sin\alpha}^{\cos\alpha} \log(x+\alpha)\,dx = \log\frac{\cos\alpha+\alpha}{\sin\alpha+\alpha} - [\sin\alpha \log(\cos\alpha+\alpha) + \cos\alpha\log(\sin\alpha+\alpha)];$

(b) $\dfrac{d}{du}\displaystyle\int_0^{\frac{\pi}{2u}} u \sin ux\,dx = 0;$

(c) $\dfrac{d}{dy}\displaystyle\int_y^{y^2} e^{-x^2y^2}\,dx = 2ye^{-y^6} - e^{-y^4} - 2y\int_y^{y^2} x^2e^{-x^2y^2}\,dx.$

4. (a) Evaluate $\displaystyle\int_0^1 x^n \log x\,dx$ by differentiating both sides of the equation $\displaystyle\int_0^1 x^n\,dx = \frac{1}{n+1}$ with respect to n $(n > -1)$.

(b) Evaluate $\displaystyle\int_0^\infty x^n e^{-ax}\,dx$ by repeated differentiation of $\displaystyle\int_0^\infty e^{-ax}\,dx$ $(a > 0)$.

(c) Evaluate $\displaystyle\int_0^\infty \frac{dy}{(x^2+y^2)^n}$ by repeated differentiation of $\displaystyle\int_0^\infty \frac{dy}{x^2+y^2}$.

[In (b) and (c) the improper integrals are of a type to which Leibnitz's rule is applicable, as is shown in Chapter 6. The result of (a) can be explicitly verified.]

5. Leibnitz's rule extends to indefinite integrals in the form:

$$\frac{\partial}{\partial t}\int f(x,t)\,dx + C = \int \frac{\partial}{\partial t} f(x,t)\,dx. \tag{a}$$

There is still an arbitrary constant in the equation, because we are evaluating an *indefinite* integral. Thus from the equation

$$\int e^{tx}\,dx = \frac{e^{tx}}{t} + C,$$

one deduces that

$$\int xe^{tx}\,dx = e^{tx}\left(\frac{x}{t} - \frac{1}{t^2}\right) + C_1.$$

(a) By differentiating n times, prove that

$$\int \frac{dx}{(x^2+a)^n} = \frac{(-1)^{n-1}}{(n-1)!}\frac{\partial^{n-1}}{\partial a^{n-1}}\left(\frac{1}{\sqrt{a}}\arctan\frac{x}{\sqrt{a}}\right) + C \ (a > 0).$$

(b) Prove: $\int x^n \cos ax\,dx = \dfrac{\partial^n}{\partial a^n}\left(\dfrac{\sin ax}{a}\right) + C,\ n = 4, 8, 12, \ldots.$

(c) Let $\int f(x,t)\,dx = F(x,t) + C$, so that $\partial F/\partial x = f(x,t)$. Show that Eq. (a) is equivalent to the statement

$$\frac{\partial^2 F}{\partial x\,\partial t} = \frac{\partial^2 F}{\partial t\,\partial x}.$$

6. Consider a one-dimensional fluid motion, the flow taking place along the x axis. Let $v = v(x, t)$ be the velocity at position x at time t, so that, if x is the coordinate of a fluid particle at time t, one has $dx/dt = v$. If $f(x, t)$ is any scalar associated with the flow (velocity, acceleration, density, ...), one can study the variation of f following the flow with the aid of the Stokes derivative:

$$\frac{Df}{Dt} = \frac{\partial f}{\partial x}\frac{dx}{dt} + \frac{\partial f}{\partial t}$$

[see Prob. 10 following Section 2–7]. A piece of the fluid occupying an interval $a_0 \leqq x \leqq b_0$ when $t = 0$ will occupy an interval $a(t) \leqq x \leqq b(t)$ at time t, where $\dfrac{da}{dt} = v(a, t)$, $\dfrac{db}{dt} = v(b, t)$. The integral

$$F(t) = \int_{a(t)}^{b(t)} f(x, t)\,dx$$

is then an integral of f over a definite piece of the fluid, whose position varies with time; if f is density, this is the mass of the piece. Show that

$$\frac{dF}{dt} = \int_{a(t)}^{b(t)} \left[\frac{\partial f}{\partial t}(x, t) + \frac{\partial}{\partial x}(fv)\right] dx = \int_{a(t)}^{b(t)} \left(\frac{Df}{Dt} + f\frac{dv}{dx}\right) dx.$$

This is generalized to arbitrary three-dimensional flows in Section 5–15 below.

7. Let $f(\alpha)$ be continuous for $0 \leqq \alpha \leqq 2\pi$. Let

$$u(r, \theta) = \frac{1}{2\pi}\int_0^{2\pi} f(\alpha)\,\frac{1 - r^2}{1 + r^2 - 2r\cos(\theta - \alpha)}\,d\alpha$$

for $r < 1$, r and θ being polar coordinates. Show that u is harmonic for $r < 1$. This is the *Poisson integral formula*.

Answers

1. (a) $-\displaystyle\int_{\frac{\pi}{2}}^{\pi} \sin(xt)\,dx$, (b) $\displaystyle\int_1^2 \frac{2x^3}{(1 - tx)^3}\,dx$, (c) $\displaystyle\int_1^2 \frac{1}{u}\,dx$,

(d) $n!\displaystyle\int_1^2 \frac{\sin x}{(x - y)^{n+1}}\,dx$.

2. (a) x^2, (b) $2t \sin t^4$, (c) $-3t^2 \log(1 + t^6)$, (d) $\sec^2 x e^{-\tan^2 x} - e^{-x^2}$.

4. (a) $\dfrac{-1}{(n + 1)^2}$, (b) $\dfrac{n!}{a^{n+1}}$, (c) $\dfrac{\pi}{2}\dfrac{1 \cdot 3 \cdots (2n - 3)}{2 \cdot 4 \cdots (2n - 2)}\dfrac{1}{x^{2n-1}}$, $x > 0$.

Suggested References

Courant, Richard J., *Differential and Integral Calculus*, transl. by E. J. McShane, 2 vols. New York: Interscience, 1947.

Franklin, Philip, *A Treatise on Advanced Calculus*. New York: John Wiley and Sons, Inc., 1940.

Goursat, Édouard, *A Course in Mathematical Analysis*, Vol. 1, transl. by E. R. Hedrick. New York: Ginn, 1904.

HENRICI, PETER K., *Elements of Numerical Analysis.* New York: John Wiley and Sons, Inc., 1964.

KAPLAN, WILFRED, and LEWIS, DONALD J., *Calculus and Linear Algebra.* New York: John Wiley and Sons, Inc., 1970.

RALSTON, ANTHONY, *First Course in Numerical Analysis.* New York: McGraw-Hill, 1965.

SCARBOROUGH, JAMES B., *Numerical Mathematical Analysis.* Baltimore: Johns Hopkins Press, 1950.

VON KÁRMÁN, THEODORE, and BIOT, MAURICE A., *Mathematical Methods in Engineering.* New York: McGraw-Hill, 1940.

WHITTAKER, E. T., and WATSON, G. N., *Modern Analysis,* 4th ed. Cambridge: Cambridge University Press, 1940.

WIDDER, DAVID V., *Advanced Calculus.* New York: Prentice-Hall, Inc., 1947.

WILLERS, F. A., *Practical Analysis,* transl. by R. T. Beyer. New York: Dover, 1948.

CHAPTER 5

Vector Integral Calculus

PART I. TWO-DIMENSIONAL THEORY

5–1 Introduction. The topic of this chapter is *line and surface integrals.* It will be seen that these can both be regarded as integrals of vectors and that the principal theorems can be most simply stated in terms of vectors; hence the title "vector integral calculus."

A familiar line integral is that of arc length: $\int_C ds$. The subscript C indicates that one is measuring the length of a curve C, as in Fig. 5–1. If C is given in parametric form: $x = x(t)$, $y = y(t)$, the line integral reduces to an ordinary definite integral:

$$\int_C ds = \int_{t_1}^{t_2} \sqrt{\left(\frac{dx}{dt}\right)^2 + \left(\frac{dy}{dt}\right)^2}\, dt.$$

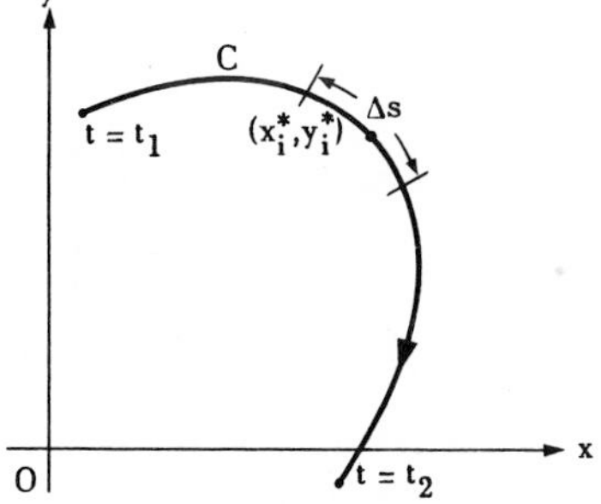

FIG. 5–1. Line integral.

If the curve C represents a wire whose density (mass per unit length) varies along C, then the wire has a total mass

$$M = \int_C f(x, y)\, ds,$$

where $f(x, y)$ is the density at the point (x, y) of the wire. The new integral can be expressed in terms of a parameter as above or can be thought of simply as a limit of a sum

$$\int_C f(x, y)\, ds = \lim \sum_{i=1}^{n} f(x_i^*, y_i^*)\, \Delta_i s.$$

Here the curve has been subdivided into n pieces of lengths $\Delta_1 s, \Delta_2 s, \ldots, \Delta_n s$, and the point (x_i^*, y_i^*) lies on the ith piece. The limit is taken as n becomes infinite, while the maximum $\Delta_i s$ approaches 0.

A third example of a line integral is that of *work.* If a particle moves from one end of C to the other under the influence of a force $\mathbf{F}$, the work done by this force is defined as

$$\int_C F_T\, ds,$$

where F_T denotes the component of $\mathbf{F}$ on the tangent $\mathbf{T}$ in the direction of motion. This integral can be thought of as a limit of a sum as above. However, another interpretation is possible. We first recall (Section 0–6) that the work done by a constant force $\mathbf{F}$ in moving a particle from A to B on the line segment AB is $\mathbf{F} \cdot \overrightarrow{AB}$; for this scalar product is equal to $|\mathbf{F}| \cdot \cos \alpha \cdot |\overrightarrow{AB}|$, α being the angle between $\mathbf{F}$ and $\overrightarrow{AB}$, and hence to the product of force component in direction of motion by the distance moved. Now the motion of the particle along C can be thought of as the sum of many small displacements along line segments, as suggested in Fig. 5–2. If these displacements are denoted by $\Delta_1\mathbf{r}, \Delta_2\mathbf{r}, \ldots, \Delta_n\mathbf{r}$, the work done would be approximated by a sum of form

$$\sum_{i=1}^{n} \mathbf{F}_i \cdot \Delta_i\mathbf{r},$$

where $\mathbf{F}_i$ is the force acting for the ith displacement. The limiting form of this is again equal to the line integral $\int F_T\, ds$, but because of the way the limit is obtained, we can also write it as

$$\int_C \mathbf{F} \cdot d\mathbf{r}.$$

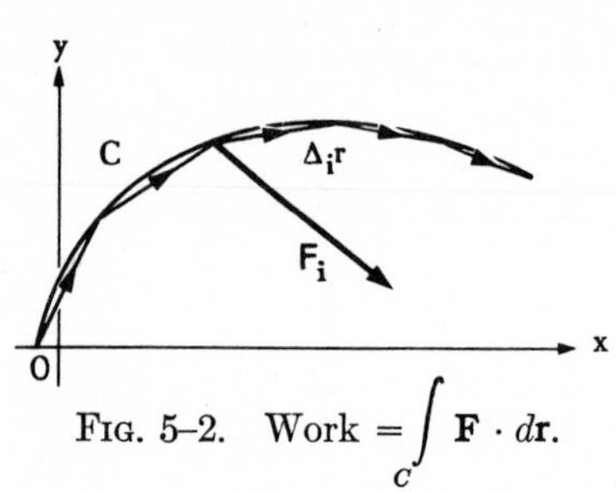

Fig. 5–2. Work $= \int_C \mathbf{F} \cdot d\mathbf{r}$.

One can thus write

$$\text{Work} = \int_C F_T\, ds = \int_C \mathbf{F} \cdot d\mathbf{r}.$$

If the displacement vector $\Delta\mathbf{r}$ and force $\mathbf{F}$ are expressed in components,

$$\mathbf{F} = F_x\mathbf{i} + F_y\mathbf{j}, \quad \Delta\mathbf{r} = \Delta x\mathbf{i} + \Delta y\mathbf{j},$$

the element of work $\mathbf{F} \cdot \Delta\mathbf{r}$ becomes

$$\mathbf{F} \cdot \Delta\mathbf{r} = F_x\, \Delta x + F_y\, \Delta y.$$

The total amount of work done is then approximated by a sum of form

$$\sum (F_x\, \Delta x + F_y\, \Delta y) = \sum F_x\, \Delta x + \sum F_y\, \Delta y.$$

The limiting form of this is a sum of two integrals:

$$\int_C F_x\, dx + \int_C F_y\, dy.$$

The first integral represents the work done by the x component of the force, the second integral represents the work done by the y component of the force.

It thus appears that one has three types of line integrals to consider, namely the types

$$\int_C f(x, y)\, ds, \quad \int_C P(x, y)\, dx, \quad \int_C Q(x, y)\, dy,$$

which are limits of sums

$$\sum f(x, y)\, \Delta s, \quad \sum P(x, y)\, \Delta x, \quad \sum Q(x, y)\, \Delta y.$$

The foregoing gives the basis for the theory of line integrals in the plane. A very slight extension of these ideas leads to line integrals in space:

$$\int_C f(x, y, z)\, ds, \quad \int_C f(x, y, z)\, dx, \quad \ldots.$$

Surface integrals appear as a natural generalization, with the surface area element $d\sigma$ replacing the arc element ds:

$$\iint_S f(x, y, z)\, d\sigma = \lim \sum f(x, y, z)\, \Delta\sigma.$$

There are corresponding component integrals

$$\iint_S f(x, y, z)\, dx\, dy, \quad \iint_S f(x, y, z)\, dy\, dz, \ldots$$

and a vector surface integral

$$\iint_S \mathbf{F} \cdot d\boldsymbol{\sigma} = \iint_S (\mathbf{F} \cdot \mathbf{n})\, d\sigma,$$

where $d\boldsymbol{\sigma} = \mathbf{n}\, d\sigma$ is the "area element vector," $\mathbf{n}$ being a unit normal vector to the surface.

It will be seen that the basic theorems, those of Green, Gauss, and Stokes, concern the relations between line, surface, and volume (triple) integrals. These correspond to fundamental physical relations between such quantities as flux, circulation, divergence, and curl. The applications will be considered at the end of the chapter.

5–2 Line integrals in the plane. We now state in precise form the definitions outlined in the preceding section.

By a *smooth curve* C in the xy plane will be meant a curve representable in the form:

$$x = \phi(t), \quad y = \psi(t), \quad h \leqq t \leqq k, \tag{5–1}$$

where x and y are continuous and have continuous derivatives for $h \leqq t \leqq k$. The curve C can be assigned a direction, which will usually be that of increasing t. If A denotes the point $[\phi(h), \psi(h)]$, and B the point

$[\phi(k), \psi(k)]$, then C can be thought of as the path of a point moving continuously from A to B. This path may cross itself, as for the curve C_1 of Fig. 5–3. If the initial point A and terminal point B coincide, C is termed a *closed* curve; if, in addition, (x, y) moves from A to $B = A$ without retracing any other point, C is called a *simple closed* curve (curve C_2 of Fig. 5–3).

Let C be a smooth curve as above, with positive direction that of increasing t. Let $f(x, y)$ be a function defined at least when (x, y) is on C. The line integral $\int_C f(x, y)\, dx$ is defined as a limit:

$$\int_C f(x, y)\, dx = \lim \sum_{i=1}^{n} f(x_i^*, y_i^*)\, \Delta_i x. \tag{5–2}$$

The limit refers to a subdivision of C as indicated in Fig. 5–4. The successive subdivision points are A: (x_0, y_0), (x_1, y_1), ..., B: (x_n, y_n). These correspond to parameter values: $h = t_0 < t_1 < \cdots < t_n = k$. The point (x_i^*, y_i^*) is some point of C between (x_{i-1}, y_{i-1}) and (x_i, y_i); i.e., (x_i^*, y_i^*) corresponds to a parameter value t_i^*, where $t_{i-1} \leqq t_i^* \leqq t_i$. $\Delta_i x$ denotes the difference $x_i - x_{i-1}$. The limit is taken as n becomes infinite and the largest $\Delta_i t$ approaches 0, where $\Delta_i t = t_i - t_{i-1}$. Similarly,

$$\int_C f(x, y)\, dy = \lim \sum f(x_i^*, y_i^*)\, \Delta_i y, \tag{5–3}$$

where $\Delta_i y = y_i - y_{i-1}$.

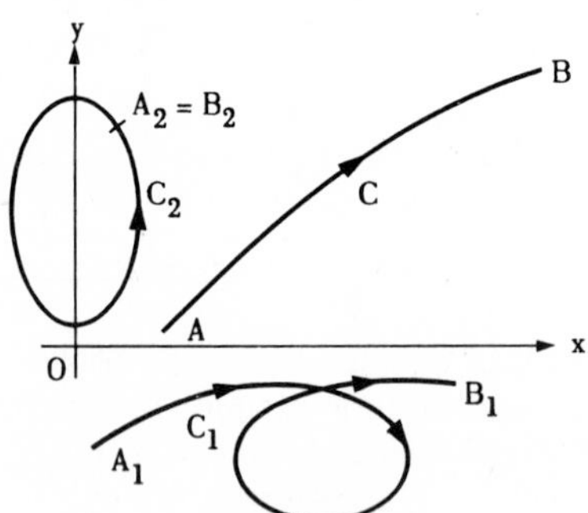

FIG. 5–3. Paths of integration.

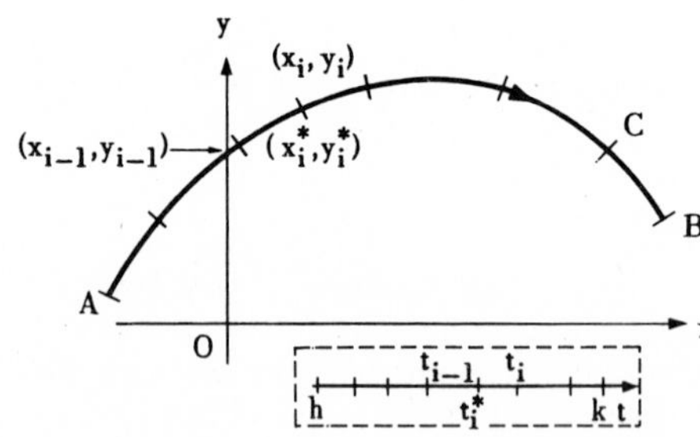

FIG. 5–4. Definition of line integral.

The significance of these definitions is guaranteed by the following basic theorems:

I. *If $f(x, y)$ is continuous on C, then*

$$\int_C f(x, y)\, dx \quad \textit{and} \quad \int_C f(x, y)\, dy \ \textit{exist.}$$

II. *If $f(x, y)$ is continuous on C, then*

$$\int\limits_C f(x, y)\,dx = \int_h^k f[\phi(t), \psi(t)]\phi'(t)\,dt, \tag{5–4}$$

$$\int\limits_C f(x, y)\,dy = \int_h^k f[\phi(t), \psi(t)]\psi'(t)\,dt. \tag{5–5}$$

The formulas (5–4) and (5–5) reduce the integrals to ordinary definite integrals and are thus essential for computation of particular integrals. Thus, let C be the path: $x = 1 + t$, $y = t^2$, $0 \leqq t \leqq 1$, directed with increasing t. Then

$$\int\limits_C (x^2 - y^2)\,dx = \int_0^1 [(1 + t)^2 - t^4]\,dt = \tfrac{32}{15},$$

$$\int\limits_C (x^2 - y^2)\,dy = \int_0^1 [(1 + t)^2 - t^4]2t\,dt = 2\tfrac{1}{2}.$$

It is logically easier to prove II first, for I is an immediate consequence of II. To prove II, one notes that the sum $\sum f(x_i^*, y_i^*)\,\Delta_i x$ can be written as

$$\sum_{i=1}^n f[\phi(t_i^*), \psi(t_i^*)]\,\frac{\Delta_i x}{\Delta_i t}\,\Delta_i t.$$

Now $\Delta_i x = x_i - x_{i-1} = \phi'(t_i^{**})\,\Delta_i t$ by the Law of the Mean [Eq. (0–145)]. Hence the sum can be written as

$$\sum_{i=1}^n F(t_i^*)\phi'(t_i^{**})\,\Delta_i t,$$

where $F(t) = f[\phi(t), \psi(t)]$ and t_i^* and t_i^{**} are both between t_{i-1} and t_i. It is easily shown [see *CLA*, Section 12–25] that this sum approaches as limit the integral

$$\int_h^k F(t)\phi'(t)\,dt = \int_h^k f[\phi(t), \psi(t)]\phi'(t)\,dt$$

as required. Formula (5–5) is proved in the same way.

We remark that the value of a line integral on C does not depend on the particular parametrization of C, but only on the order in which the points of C are traced. (See Prob. 5 below.)

In many applications the path C is not itself smooth, but is composed of a finite number of arcs, each of which is smooth. Thus C might be a broken line. In this case C is termed *piecewise* smooth. The line integral along C is simply, by definition, the sum of the integrals along the pieces. One verifies at once that (5–2), (5–3), and the theorems I and II continue to hold. In (5–4) and (5–5) the functions $\phi'(t)$ and $\psi'(t)$ will have jump discontinuities, which will not interfere with the existence of the integral (cf. Section 4–4). *Throughout this book all paths of integration for line integrals will be piecewise smooth, unless otherwise specified.*

If the curve C is represented in the form

$$y = g(x), \quad a \leqq x \leqq b,$$

then one can regard x itself as parameter, replacing t; i.e., C is given by the equations

$$x = x, \quad y = g(x), \quad a \leqq x \leqq b$$

in terms of the parameter x. If the direction of C is that of increasing x, (5–4) and (5–5) become

$$\int_C f(x, y)\, dx = \int_a^b f[x, g(x)]\, dx, \tag{5-6}$$

$$\int_C f(x, y)\, dy = \int_a^b f[x, g(x)]g'(x)\, dx. \tag{5-7}$$

The ordinary definite integral $\int_a^b y\, dx$, where $y = g(x)$, is a special case of (5–6).

Similarly, if C is represented in the form:

$$x = F(y), \quad c \leqq y \leqq d$$

and the direction of C is that of increasing y, then

$$\int_C f(x, y)\, dx = \int_c^d f[F(y), y]F'(y)\, dy, \tag{5-8}$$

$$\int_C f(x, y)\, dy = \int_c^d f[F(y), y]\, dy. \tag{5-9}$$

In most applications, the line integrals appear as a combination,

$$\int_C P(x, y)\, dx + \int_C Q(x, y)\, dy,$$

which is abbreviated as follows:

$$\int_C [P(x, y)\, dx + Q(x, y)\, dy] \quad \text{or} \quad \int_C P(x, y)\, dx + Q(x, y)\, dy,$$

the brackets being used only when necessary.

In the formulas thus far the direction of C has been that of increasing parameter. If the opposite direction is chosen, upper and lower limits are reversed on all integrals. Thus (5–4) becomes

$$\int_C f(x, y)\, dx = \int_k^h f[\phi(t), \psi(t)]\, \phi'(t)\, dt. \tag{5-4'}$$

The line integral is therefore multiplied by -1. Often it is most convenient to specify the path by its equations in some form and to indicate the direction by using the initial and terminal points as lower and upper limits:

$$\int_{C\ A}^{B} P\,dx + Q\,dy \quad \text{or} \quad \int_{C\ (x_1,\,y_1)}^{(x_2,\,y_2)} P\,dx + Q\,dy.$$

It will be seen later that, under certain conditions, one need only prescribe initial and terminal points:

$$\int_A^B P\,dx + Q\,dy.$$

EXAMPLE 1. To evaluate

$$\int_{C\ (1,0)}^{(-1,0)} (x^3 - y^3)\,dy,$$

where C is the semicircle $y = \sqrt{1 - x^2}$ shown in Fig. 5–5, one can represent C parametrically:

$$x = \cos t, \quad y = \sin t, \quad 0 \leqq t \leqq \pi,$$

and the integral becomes

$$\int_0^{\pi} (\cos^3 t - \sin^3 t) \cos t\,dt = \frac{3\pi}{8}.$$

One can use x as parameter and the integral becomes

$$\int_1^{-1} [x^3 - (1 - x^2)^{\frac{3}{2}}] \frac{-x}{\sqrt{1 - x^2}}\,dx;$$

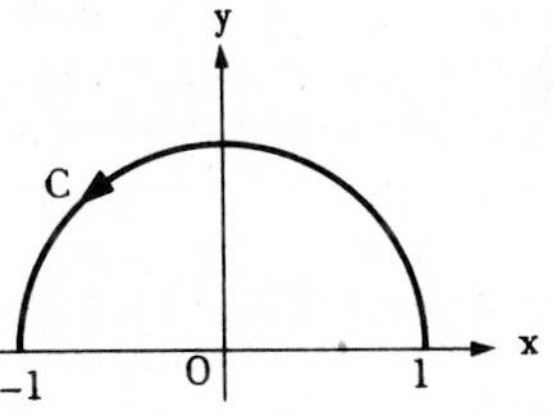

FIG. 5–5.

this is clearly in a more awkward form for integration. The substitution $x = \cos t$ brings one back to the parametric form. One can use y as parameter, but has then to split the integral into two parts, from $(1, 0)$ to $(0, 1)$ and from $(0, 1)$ to $(-1, 0)$:

$$\int_0^1 [(1 - y^2)^{\frac{3}{2}} - y^3]\,dy + \int_1^0 [-(1 - y^2)^{\frac{3}{2}} - y^3]\,dy = 2\int_0^1 (1 - y^2)^{\frac{3}{2}}\,dy.$$

Note that $x = \sqrt{1 - y^2}$ on the first part of the path and $x = -\sqrt{1 - y^2}$ on the second part.

EXAMPLE 2. Let C be the parabolic arc: $y = x^2$ from $(0, 0)$ to $(-1, 1)$. Then

$$\int_C xy^2\,dx + x^2y\,dy = \int_0^{-1} \left(xy^2 + x^2y\frac{dy}{dx}\right) dx = \int_0^{-1} (x^5 + 2x^5)\,dx = \tfrac{1}{2}.$$

If C is a *closed* curve, then there is no need to specify initial and terminal point, though the direction must be indicated. If C is a simple closed curve (traced just once), then one need only specify which of the two possible directions is chosen. The notations

$$\text{(a)} \oint P\,dx + Q\,dy, \quad \text{(b)} \oint P\,dx + Q\,dy$$

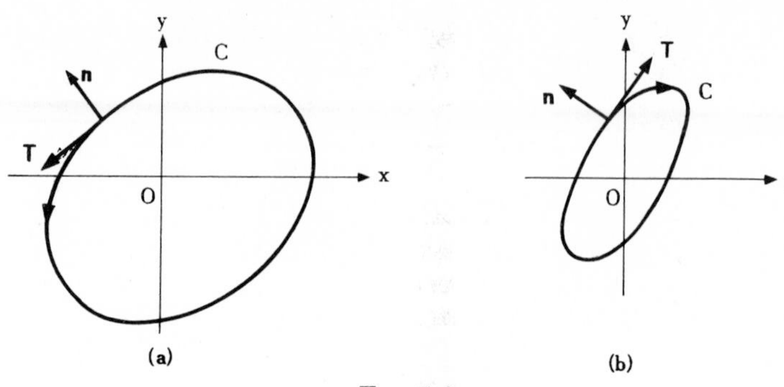

FIG. 5–6.

refer to the two cases of Figs. 5–6(a) and 5–6(b). The counterclockwise arrow refers to what is roughly a counterclockwise direction on C; this will be termed the *positive* direction (as for angular measure); the clockwise direction will be called the *negative* direction. It should be noted that the direction can be specified by reference to the unit tangent vector $\mathbf{T}$ in the direction of integration and the unit normal vector $\mathbf{n}$ which points to the outside of the region bounded by C; for the positive direction, $\mathbf{n}$ is 90° behind $\mathbf{T}$, as in Fig. 5–6(a); for the negative direction $\mathbf{n}$ is 90° ahead of $\mathbf{T}$ as in Fig. 5–6(b).

EXAMPLE 3. To evaluate

$$\oint_C y^2\,dx + x^2\,dy,$$

where C is the triangle with vertices (1, 0), (1, 1), (0, 0), shown in Fig. 5–7, one has to compute three integrals. The first is the integral from (0, 0) to (1, 0); along this path, $y = 0$ and, if x is the parameter, $dy = 0$. Hence the first integral is 0. The second integral is that from (1, 0) to (1, 1); if y is used as parameter, this reduces to

$$\int_0^1 dy = 1,$$

since $dx = 0$. For the third integral, from (1, 1) to (0, 0), x can be used as parameter, so that the integral is

$$\int_1^0 2x^2\,dx = -\tfrac{2}{3},$$

since $dy = dx$. Thus finally

$$\oint_C y^2\,dx + x^2\,dy = 0 + 1 - \tfrac{2}{3} = \tfrac{1}{3}.$$

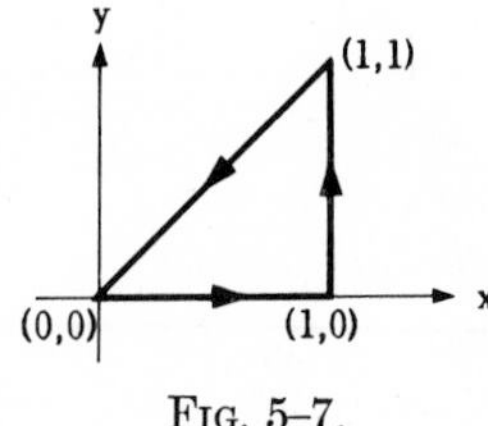

FIG. 5–7.

5–3 Integrals with respect to arc length. Basic properties of line integrals. For a smooth or piecewise smooth path C, as in the preceding section, arc length s is well-defined. Thus s can be defined as the distance traversed from the initial point ($t = h$) up to a general t:

$$s = \int_h^t \sqrt{\left(\frac{dx}{dt}\right)^2 + \left(\frac{dy}{dt}\right)^2}\, dt. \tag{5–10}$$

If the curve C is directed with increasing t, then s also increases in the direction of motion, going from 0 up to the length L of C. Let C be subdivided as in Fig. 5–4 and let $\Delta_i s$ denote the increment in s from t_{i-1} to t_i, i.e., the distance moved in this interval. One then makes the definition

$$\int_C f(x, y)\, ds = \lim_{\substack{n\to\infty \\ \max \Delta_i s \to 0}} \sum_{i=1}^{n} f(x_i^*, y_i^*)\, \Delta_i s. \tag{5–11}$$

If f is continuous on C, this integral will exist and can be evaluated as follows:

$$\int_C f(x, y)\, ds = \int_h^k f[\phi(t), \psi(t)]\, \sqrt{\phi'(t)^2 + \psi'(t)^2}\, dt. \tag{5–12}$$

This is proved in the same way as (5–4) and (5–5), with the aid of the formula

$$\frac{ds}{dt} = \sqrt{\left(\frac{dx}{dt}\right)^2 + \left(\frac{dy}{dt}\right)^2} = \sqrt{\phi'(t)^2 + \psi'(t)^2}.$$

One can in principle use s itself as the parameter on the curve C; if this is done, x and y become functions of s: $x = x(s)$, $y = y(s)$. The point $[x(s), y(s)]$ is then the position of the moving point after a distance s has been traversed. In this case, (5–11) reduces to a definite integral with respect to s:

$$\int_C f(x, y)\, ds = \int_0^L f[x(s), y(s)]\, ds. \tag{5–13}$$

If x is used as parameter, one has

$$\int_C f(x, y)\, ds = \int_a^b f[x, y(x)]\, \sqrt{1 + \left(\frac{dy}{dx}\right)^2}\, dx; \tag{5–14}$$

there is an analogous formula for y.

The basic combination

$$\int_C P\, dx + Q\, dy$$

referred to above can be written as an integral with respect to s as follows:

$$\int_C P\, dx + Q\, dy = \int_C (P \cos\alpha + Q \sin\alpha)\, ds, \tag{5–15}$$

where α is the angle between the positive x axis and a tangent vector in direction of increasing s. For, as shown in Fig. 5–8 (cf. Section 2–12), one has

$$\frac{dx}{ds} = \cos \alpha, \quad \frac{dy}{ds} = \sin \alpha \quad (5\text{–}16)$$

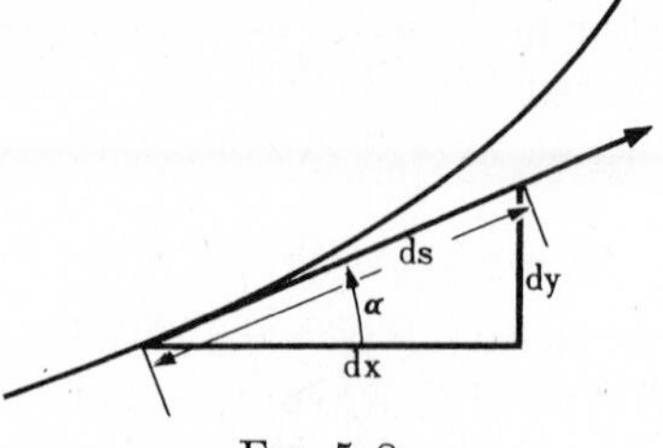

FIG. 5–8.

and

$$\int_C P\,dx + Q\,dy = \int_0^L \left(P\frac{dx}{ds} + Q\frac{dy}{ds}\right) ds = \int_C (P \cos \alpha + Q \sin \alpha)\, ds. \quad (5\text{–}17)$$

The basic properties of the line integral are analogous to those of ordinary definite integrals; the properties to be listed here can in fact be demonstrated by reducing the line integral to a definite integral by use of a parameter:

$$\int_{A_1\,C_1}^{A_2} (P\,dx + Q\,dy) + \int_{A_2\,C_2}^{A_3} (P\,dx + Q\,dy) = \int_{A_1\,C_3}^{A_3} (P\,dx + Q\,dy), \quad (5\text{–}18)$$

where C_3 is the combined path: from A_1 to A_2 via C_1 and from A_2 to A_3 via C_2;

$$\int_{A\,C}^{B} P\,dx + Q\,dy = -\int_{B\,C'}^{A} P\,dx + Q\,dy, \quad (5\text{–}19)$$

where C' denotes C traced in the reverse direction;

$$\int_C (P_1\,dx + Q_1\,dy) + \int_C (P_2\,dx + Q_2\,dy) = \int_C (P_1 + P_2)\,dx + (Q_1 + Q_2)\,dy; \quad (5\text{–}20)$$

$$K\int_C (P\,dx + Q\,dy) = \int_C (KP)\,dx + (KQ)\,dy, \quad K = \text{const}, \quad (5\text{–}21)$$

$$\int_C ds = L = \text{length of } C; \quad (5\text{–}22)$$

if $|f(x, y)| \leqq M$ on C, then

$$\left|\int_C f(x, y)\,ds\right| \leqq M \cdot L; \quad (5\text{–}23)$$

if C is a simple closed curve, as in Fig. 5–6(a) then

$$\oint_C x\,dy = -\oint_C y\,dx = \text{area enclosed by } C. \quad (5\text{–}24)$$

These theorems all hold under the assumptions that the paths are piecewise smooth and that the functions being integrated are continuous. All except the formula (5–24) follow at once from appropriate parametric representations; (5–24) will be proved in Section 5–5.

Problems

1. Evaluate the following integrals along the straight-line paths joining the end points:

(a) $\int_{(0,0)}^{(2,2)} y^2\,dx$, (b) $\int_{(2,1)}^{(1,2)} y\,dx$, (c) $\int_{(1,1)}^{(2,1)} x\,dy$.

2. Evaluate the following line integrals:

(a) $\int_{(0,-1)\,C}^{(0,1)} y^2\,dx + x^2\,dy$, where C is the semi-circle: $x = \sqrt{1-y^2}$;

(b) $\int_{(0,0)\,C}^{(2,4)} y\,dx + x\,dy$, where C is the parabola: $y = x^2$;

(c) $\int_{(1,0)\,C}^{(0,1)} \dfrac{y\,dx - x\,dy}{x^2+y^2}$, where C is the curve: $x = \cos^3 t$, $y = \sin^3 t$, $0 \leqq t \leqq \dfrac{\pi}{2}$.

(Hint: set $u = \tan^3 t$ in the integral for t.)

3. Evaluate the following line integrals:

(a) $\oint_C y^2\,dx + xy\,dy$, where C is the square with vertices $(1, 1)$, $(-1, 1)$, $(-1, -1)$, $(1, -1)$;

(b) $\oint_C y\,dx - x\,dy$, where C is the circle: $x^2 + y^2 = 1$ (cf. (5–24));

(c) $\oint_C x^2y^2\,dx - xy^3\,dy$, where C is the triangle with vertices $(0, 0)$, $(1, 0)$, $(1, 1)$.

4. Evaluate the following line integrals:

(a) $\oint (x^2 - y^2)\,ds$, where C is the circle: $x^2 + y^2 = 4$;

(b) $\int_{(0,0)\,C}^{(1,1)} x\,ds$, where C is the line $y = x$;

(c) $\int_{(0,0)\,C}^{(1,1)} ds$, where C is the parabola: $y = x^2$.

5. Let a path (5–1) be given and let a *change of parameter* be made by an equation $t = g(\tau)$, $\alpha \leqq \tau \leqq \beta$, where $g'(\tau)$ is continuous and positive in the interval and $g(\alpha) = h$, $g(\beta) = k$. As in (5–4) the line integral $\int f(x, y)\,dx$ on the path $x = \phi(g(\tau))$, $y = \psi(g(\tau))$ is given by

$$\int_\alpha^\beta f[\phi(g(\tau)),\ \psi(g(\tau))]\phi'(\tau)\,d\tau.$$

Show that this equals the integral in (5–4), so that such a change of parameter does not affect the value of the line integral.

6. *Numerical evaluation of line integrals.* If the parametric equations of the path C are explicitly known, Eqs. (5–4) and (5–5) reduce the evaluation of line integrals to a problem in ordinary definite integrals, to which the methods of Section 4–2 apply. One can in any case evaluate the integral directly from its definition as limit of a sum. Thus

$$\int_C P\,dx + Q\,dy \sim \sum_{i=1}^{n} (P(x_i^*, y_i^*)\,\Delta_i x + Q(x_i^*, y_i^*)\,\Delta_i y).$$

The points (x_i^*, y_i^*) can be chosen as the subdivision points (x_{i-1}, y_{i-1}) or as (x_i, y_i); one can also use a trapezoidal rule to obtain the sum:

$$\sum_{i=1}^{n} \{\tfrac{1}{2}[P(x_{i-1}, y_{i-1}) + P(x_i, y_i)]\,\Delta_i x + \tfrac{1}{2}[Q(x_{i-1}, y_{i-1}) + Q(x_i, y_i)]\,\Delta_i y\}. \quad \text{(a)}$$

As pointed out in Section 4–2, this is equivalent to making particular choices of the points (x_i^*, y_i^*). For example,

$$\int_{(0,0)\,C}^{(2,2)} y^2\,dx + x^2\,dy \sim [\tfrac{1}{2}(0+1)\cdot 1 + \tfrac{1}{2}(0+1)\cdot 1] + [\tfrac{1}{2}(1+4)\cdot 1 + \tfrac{1}{2}(1+4)\cdot 1] = 6,$$

where C is the straight line segment from (0, 0) to (2, 2) and the subdivision points are (0, 0), (1, 1), (2, 2).

Let the functions P and Q be given by the following table at the points $A \ldots S$.

	A	B	C	D	E	F	G	H	I	J	K	L	M	N	O	S
x	1	2	3	4	1	2	3	4	1	2	3	4	1	2	3	4
y	1	1	1	1	2	2	2	2	3	3	3	3	4	4	4	4
P	0	3	8	5	3	0	5	2	8	5	0	1	2	7	3	4
Q	1	2	3	4	2	4	6	8	3	6	9	2	4	8	2	6

Evaluate the integral $\int P\,dx + Q\,dy$ approximately by the trapezoidal rule, Eq. (a) above, on the following broken line paths: (a) $ABFG$ (b) $AFGKH$ (c) $ABCDH$–$LSONMIEA$ (d) $AFJNMIJFA$ (e) $ABFEAEFBA$.

7. Show that, if $f(x, y) > 0$ on C, the integral $\int_C f(x, y)\,ds$ can be interpreted as the area of the cylindrical surface $0 \leqq z \leqq f(x, y)$, (x, y) on C.

Answers

1. (a) $\frac{8}{3}$, (b) $-\frac{3}{2}$, (c) 0. 2. (a) $\frac{4}{3}$, (b) 8, (c) $-\pi/2$.
3. (a) 0, (b) -2π, (c) $-\frac{1}{4}$. 4. (a) 0, (b) $\sqrt{2}/2$,
(c) $\frac{1}{2}\sqrt{5} + \frac{1}{4}\log(2 + \sqrt{5})$. 6. (a) 7, (b) 5, (c) 8, (d) 5.5, (e) 0.

5–4 Line integrals as integrals of vectors. The functions $P(x,y)$ and $Q(x, y)$ appearing above can be regarded as the components of a vector $\mathbf{u}$:

$$\mathbf{u} = P(x, y)\mathbf{i} + Q(x, y)\mathbf{j}, \quad u_x = P(x, y), \quad u_y = Q(x, y). \quad (5\text{–}25)$$

The line integral

$$\int_C P\,dx + Q\,dy$$

has then a simple vector interpretation as follows:

$$\int_C P\,dx + Q\,dy = \int_C u_T\,ds, \qquad (5\text{–}26)$$

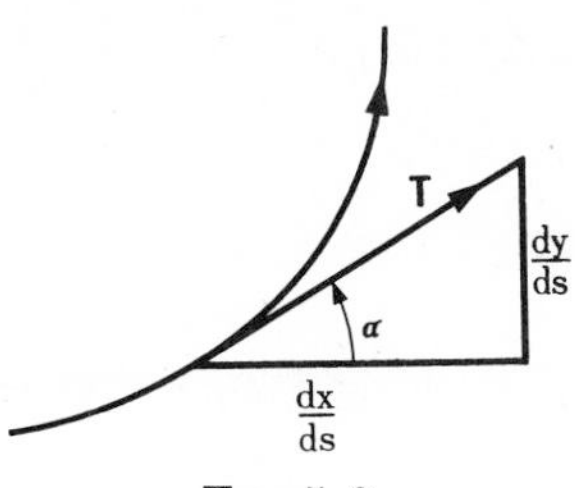

FIG. 5–9.

where u_T denotes the tangential component of **u**, i.e., the component of **u** in the direction of the unit tangent vector **T** in the direction of increasing s. For as Fig. 5–9 shows (cf. Section 2–12), **T** has components dx/ds, dy/ds:

$$\mathbf{T} = \frac{dx}{ds}\mathbf{i} + \frac{dy}{ds}\mathbf{j} = \cos\alpha\mathbf{i} + \sin\alpha\mathbf{j}. \qquad (5\text{–}27)$$

Hence

$$u_T = \mathbf{u}\cdot\mathbf{T} = P\cos\alpha + Q\sin\alpha, \qquad (5\text{–}28)$$

so that

$$\int_C u_T\,ds = \int_C (P\cos\alpha + Q\sin\alpha)\,ds = \int_C P\,dx + Q\,dy \qquad (5\text{–}29)$$

by (5–15) above.

In the case when $\mathbf{u} = P\mathbf{i} + Q\mathbf{j}$ is a *force* field, the integral $\int u_T\,ds = \int P\,dx + Q\,dy$ represents the *work* done by this force in moving a particle from one end of C to the other. For the work done is defined in mechanics as "force times distance" or, more precisely, as follows:

$$\text{work} = \int_C (\text{tangential force component})\,ds, \qquad (5\text{–}30)$$

and u_T is precisely this tangential force component.

The line integral $\int P\,dx + Q\,dy$ can be defined directly as a vector integral as follows:

$$\int_C P\,dx + Q\,dy = \int_C \mathbf{u}\cdot d\mathbf{r}, \quad d\mathbf{r} = dx\mathbf{i} + dy\mathbf{j}, \qquad (5\text{–}31)$$

where

$$\int_C \mathbf{u}\cdot d\mathbf{r} = \lim_{\substack{n\to\infty \\ \max \Delta_i s\to 0}} \sum_{i=1}^{n} \mathbf{u}(x_i^*, y_i^*)\cdot\Delta_i\mathbf{r} \qquad (5\text{–}32)$$

and $\Delta_i\mathbf{r} = \Delta_i x\mathbf{i} + \Delta_i y\mathbf{j}$. Equation (5–17) can then be written in the vector form:

$$\int_C \mathbf{u}\cdot d\mathbf{r} = \int_0^L \left(\mathbf{u}\cdot\frac{d\mathbf{r}}{ds}\right) ds = \int_C (\mathbf{u}\cdot\mathbf{T})\,ds. \qquad (5\text{–}33)$$

If C is represented in terms of a parameter t, then

$$\int_C \mathbf{u} \cdot d\mathbf{r} = \int_C P\,dx + Q\,dy = \int_h^k \left(P\frac{dx}{dt} + Q\frac{dy}{dt}\right) dt = \int_h^k \left(\mathbf{u} \cdot \frac{d\mathbf{r}}{dt}\right) dt. \tag{5–34}$$

If $\mathbf{r}$ is the position vector of a particle of mass m moving on C and $\mathbf{u}$ is the force applied, then by Newton's Second Law

$$\mathbf{u} = m\frac{d^2\mathbf{r}}{dt^2} = m\frac{d\mathbf{v}}{dt}. \tag{5–35}$$

Hence

$$\int_C \mathbf{u} \cdot d\mathbf{r} = \int_h^k \left(\mathbf{u} \cdot \frac{d\mathbf{r}}{dt}\right) dt = \int_h^k \left(m\frac{d\mathbf{v}}{dt} \cdot \mathbf{v}\right) dt = \int_h^k \frac{d}{dt}\left(\tfrac{1}{2}m\mathbf{v} \cdot \mathbf{v}\right) dt$$
$$= \int_h^k \frac{d}{dt}\left(\tfrac{1}{2}mv^2\right) dt.$$

One thus concludes that

$$\int_C u_T\,ds = \int_C \mathbf{u} \cdot d\mathbf{r} = \tfrac{1}{2}mv^2\Big|_{t=h}^{t=k}; \tag{5–36}$$

that is, the *work done equals the gain in kinetic energy.* This is a basic law of mechanics.

The line integral $\int_C P\,dx + Q\,dy$ can be interpreted as a vector integral in another way, namely as

$$\int_C \mathbf{v} \cdot \mathbf{n}\,ds = \int_C v_n\,ds,$$

where $\mathbf{v}$ is the vector $Q\mathbf{i} - P\mathbf{j}$ and $\mathbf{n}$ is the unit normal vector 90° behind $\mathbf{T}$, so that v_n is the normal component of $\mathbf{v}$. One has then (cf. Fig. 5–10)

$$\mathbf{n} = \mathbf{T} \times \mathbf{k} = \left(\frac{dx}{ds}\mathbf{i} + \frac{dy}{ds}\mathbf{j}\right) \times \mathbf{k} = \frac{dy}{ds}\mathbf{i} - \frac{dx}{ds}\mathbf{j}, \tag{5–37}$$

since the vector product $\mathbf{a} \times \mathbf{k}$ of a vector $\mathbf{a}$ in the xy plane with $\mathbf{k}$ is a vector $\mathbf{b}$ having the same magnitude as $\mathbf{a}$ and 90° behind $\mathbf{a}$. Accordingly,

$$v_n = \mathbf{v} \cdot \mathbf{n} = (Q\mathbf{i} - P\mathbf{j}) \cdot \left(\frac{dy}{ds}\mathbf{i} - \frac{dx}{ds}\mathbf{j}\right) = Q\frac{dy}{ds} + P\frac{dx}{ds}.$$

Thus one has

$$\int_C v_n\,ds = \int_C \left(Q\frac{dy}{ds} + P\frac{dx}{ds}\right) ds = \int_C P\,dx + Q\,dy, \tag{5–38}$$
$$\mathbf{v} = Q\mathbf{i} - P\mathbf{j}, \quad \mathbf{n} = \mathbf{T} \times \mathbf{k},$$

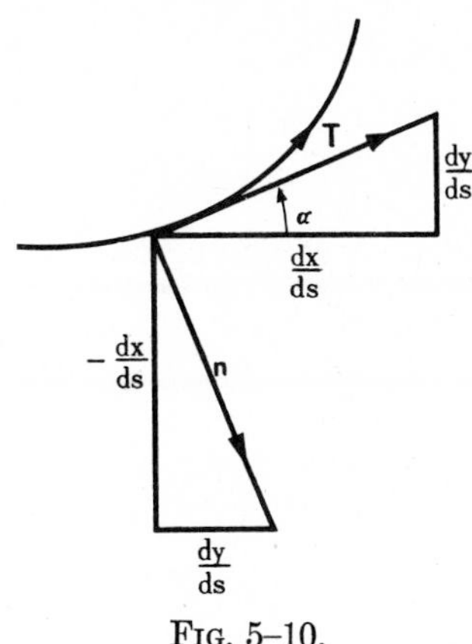

FIG. 5–10.

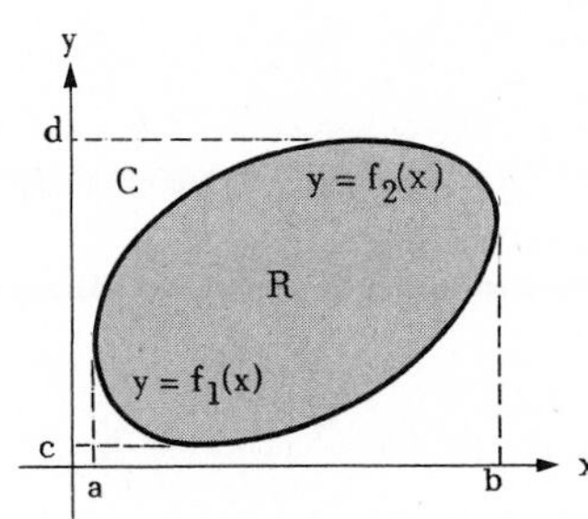

FIG. 5–11. Special region for Green's theorem.

as asserted. It should be noted that for $\mathbf{u} = P\mathbf{i} + Q\mathbf{j}$ one has then

$$\int_C u_n\, ds = \int_C \left(P\frac{dy}{ds} - Q\frac{dx}{ds}\right) ds = \int_C -Q\, dx + P\, dy, \qquad (5\text{–}39)$$
$$\mathbf{u} = P\mathbf{i} + Q\mathbf{j}, \quad \mathbf{n} = \mathbf{T} \times \mathbf{k}.$$

5–5 Green's theorem. The following theorem and its generalizations are fundamental in the theory of line integrals.

GREEN'S THEOREM. *Let D be a domain of the xy plane and let C be a piecewise smooth simple closed curve in D whose interior is also in D. Let $P(x, y)$ and $Q(x, y)$ be functions defined and continuous and having continuous first partial derivatives in D. Then*

$$\oint_C P\, dx + Q\, dy = \iint_R \left(\frac{\partial Q}{\partial x} - \frac{\partial P}{\partial y}\right) dx\, dy, \qquad (5\text{–}40)$$

where R is the closed region bounded by C.

The theorem will be proved first for the case in which R is representable in both of the forms:

$$a \leqq x \leqq b, \quad f_1(x) \leqq y \leqq f_2(x), \qquad (5\text{–}41)$$
$$c \leqq y \leqq d, \quad g_1(y) \leqq x \leqq g_2(y), \qquad (5\text{–}42)$$

as in Fig. 5–11.

The double integral

$$\iint_R \frac{\partial P}{\partial y}\, dx\, dy$$

can by (5–41) be written as an iterated integral:

$$\iint_R \frac{\partial P}{\partial y}\, dx\, dy = \int_a^b \int_{f_1(x)}^{f_2(x)} \frac{\partial P}{\partial y}\, dy\, dx.$$

One can now integrate out:

$$\iint_R \frac{\partial P}{\partial y}\,dx\,dy = \int_a^b \{P[x, f_2(x)] - P[x, f_1(x)]\}\,dx$$

$$= -\int_b^a P[x, f_2(x)]\,dx - \int_a^b P[x, f_1(x)]\,dx$$

$$= -\oint_C P(x, y)\,dx.$$

In the same way $\iint_R \frac{\partial Q}{\partial x}\,dx\,dy$ can be written as an iterated integral, with the aid of (5–42), and one concludes that

$$\iint_R \frac{\partial Q}{\partial x}\,dx\,dy = \oint_C Q\,dy.$$

Adding the two double integrals, one finds:

$$\iint_R \left(\frac{\partial Q}{\partial x} - \frac{\partial P}{\partial y}\right) dx\,dy = \oint_C P\,dx + Q\,dy.$$

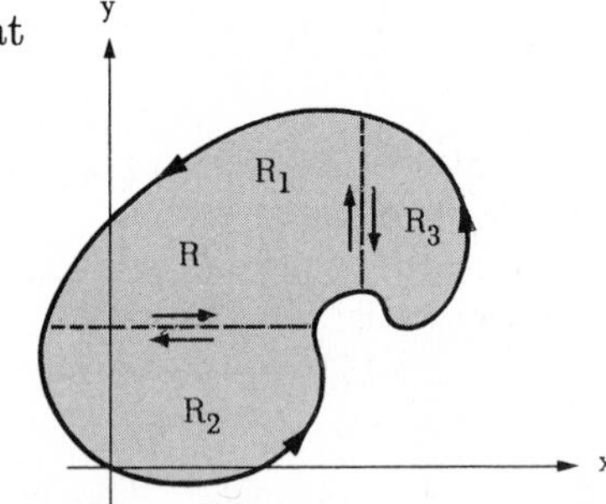

Fig. 5–12. Decomposition of region into special regions.

The theorem is thus proved for the special type of region R.

Suppose next that R is not itself of this form, but can be decomposed into a finite number of such regions: $R_1, R_2, \ldots, R_n$ by suitable lines or arcs, as in Fig. 5–12. Let $C_1, C_2, \ldots, C_n$ denote the corresponding boundaries. Then Eq. (5–40) can be applied to each region separately. Adding, one obtains the equation:

$$\oint_{C_1} (P\,dx + Q\,dy) + \oint_{C_2} (\ ) + \cdots \oint_{C_n} (\ ) = \iint_{R_1} \left(\frac{\partial Q}{\partial x} - \frac{\partial P}{\partial y}\right) dx\,dy + \cdots + \iint_{R_n} (\ )\,dx\,dy.$$

But the sum of the integrals on the left is just

$$\oint_C P\,dx + Q\,dy.$$

For the integrals along the added arcs are taken once in each direction and hence cancel each other; the remaining integrals add up to precisely the integral around C in the positive direction. The integrals on the right add up to

$$\iint_R \left(\frac{\partial Q}{\partial x} - \frac{\partial P}{\partial y}\right) dx\, dy$$

and hence

$$\oint_C P\, dx + Q\, dy = \iint_R \left(\frac{\partial Q}{\partial x} - \frac{\partial P}{\partial y}\right) dx\, dy.$$

The proof thus far covers all regions R of interest in practical problems. To prove the theorem for the most general region R, it is necessary to approximate this region by those of the special type just considered and then to use a limiting process. For details, see O. D. Kellogg, *Foundations of Potential Theory* (Berlin: Springer, 1929).

EXAMPLE 1. Let C be the circle: $x^2 + y^2 = 1$. Then

$$\oint_C 4xy^3\, dx + 6x^2y^2\, dy = \iint_R (12xy^2 - 12xy^2)\, dx\, dy = 0.$$

EXAMPLE 2. Let C be the ellipse: $x^2 + 4y^2 = 4$. Then

$$\oint_C (2x - y)\, dx + (x + 3y)\, dy = \iint_R (1 + 1)\, dx\, dy = 2A,$$

where A is the area of R. Since the ellipse has semi-axes $a = 2$, $b = 1$, the area is $\pi ab = 2\pi$ and the value of the line integral is 4π.

EXAMPLE 3. Let C be the ellipse: $4x^2 + y^2 = 4$. Then Green's theorem is not applicable to the integral:

$$\oint_C \frac{y}{x^2 + y^2}\, dx - \frac{x}{x^2 + y^2}\, dy,$$

since P and Q fail to be continuous at the origin.

EXAMPLE 4. Let C be the square with vertices $(1, 1)$, $(1, -1)$, $(-1, -1)$, $(-1, 1)$. Then

$$\oint_C (x^2 + 2y^2)\, dx = -\iint_R 4y\, dx\, dy = -4A\bar{y},$$

where $(\bar{x}, \bar{y})$ is the centroid of the square. Hence the value of the line integral is 0.

Vector interpretation of Green's theorem. If $\mathbf{u} = P(x, y)\mathbf{i} + Q(x, y)\mathbf{j}$, then as above

$$\oint_C P\, dx + Q\, dy = \oint_C u_T\, ds,$$

where $u_T = \mathbf{u} \cdot \mathbf{T}$ is the tangential component of $\mathbf{u}$. The integrand in the right-hand member of Green's theorem is the z component of curl $\mathbf{u}$; that is,

$$\operatorname{curl}_z \mathbf{u} = \frac{\partial Q}{\partial x} - \frac{\partial P}{\partial y}.$$

Hence Green's theorem states:

$$\oint_C u_T \, ds = \iint_R \operatorname{curl}_z \mathbf{u} \, dx \, dy. \tag{5-43}$$

This result is a special case of Stokes' theorem, to be considered below.

The integral $\oint_C P\,dx + Q\,dy$ can also be interpreted as above as the integral

$$\oint_C \mathbf{v} \cdot \mathbf{n} \, ds = \oint_C v_n \, ds,$$

where $\mathbf{v}$ is the vector $Q\mathbf{i} - P\mathbf{j}$. The right-hand member of Green's theorem is then the double integral of div $\mathbf{v}$. Thus one has, for an arbitrary vector field,

$$\oint_C v_n \, ds = \iint_R \operatorname{div} \mathbf{v} \, dx \, dy, \tag{5-44}$$

where $\mathbf{n}$ is the outer normal on C, as in Fig. 5–6(a). This result is the two-dimensional form of Gauss's theorem, to be considered below.

Application to area. Equation (5–24) above asserts that

$$\oint_C x \, dy = -\oint_C y \, dx = \text{area of } R.$$

This follows at once from Green's theorem, since

$$\oint_C x \, dy = -\oint_C y \, dx = \iint_R 1 \cdot dx \, dy.$$

If these two equal line integrals are averaged, another expression for area is obtained:

$$\tfrac{1}{2} \oint_C (-y \, dx + x \, dy) = \text{area of } R. \tag{5-45}$$

This can also be checked by Green's theorem.

Problems

1. If $\mathbf{v} = (x^2 + y^2)\mathbf{i} + (2xy)\mathbf{j}$, evaluate $\int_C v_T \, ds$ for the following paths:

(a) from (0, 0) to (1, 1) on the line $y = x$,
(b) from (0, 0) to (1, 1) on the line $y = x^2$;
(c) from (0, 0) to (1, 1) on the broken line with corner at (1, 0).

2. Evaluate $\int_C v_n\, ds$ for the vector $\mathbf{v}$ given in Prob. 1 on the paths (a), (b), (c) of Prob. 1, $\mathbf{n}$ being chosen as the normal 90° behind $\mathbf{T}$.

3. The gravitational force near a point on the earth's surface is represented approximately by the vector $-mg\mathbf{j}$, where the y axis points upward. Show that the work done by this force on a body moving in a vertical plane from height h_1 to height h_2 along any path is equal to $mg(h_1 - h_2)$.

4. Show that the earth's gravitational potential $U = -kMm/r$ is equal to the negative of the work done by the gravitational force $\mathbf{F} = -(kMm/r^2)(\mathbf{r}/r)$ in bringing the particle to its present position from infinite distance along the ray through the earth's center.

5. Evaluate by Green's theorem:

(a) $\oint_C ay\, dx + bx\, dy$ on any path;

(b) $\oint e^x \sin y\, dx + e^x \cos y\, dy$ around the rectangle with vertices $(0, 0)$, $(1, 0)$, $(1, \frac{1}{2}\pi)$, $(0, \frac{1}{2}\pi)$;

(c) $\oint (2x^3 - y^3)\, dx + (x^3 + y^3)\, dy$ around the circle: $x^2 + y^2 = 1$;

(d) $\oint_C u_T\, ds$, where $\mathbf{u} = \text{grad}\,(x^2y)$ and C is the circle: $x^2 + y^2 = 1$;

(e) $\oint_C v_n\, ds$, where $\mathbf{v} = (x^2 + y^2)\mathbf{i} - 2xy\mathbf{j}$, and C is the circle: $x^2 + y^2 = 1$, $\mathbf{n}$ being the outer normal;

(f) $\oint_C f(x)\, dx + g(y)\, dy$ on any path.

6. If $\mathbf{r} = x\mathbf{i} + y\mathbf{j}$ is the position vector of an arbitrary point (x, y), show that

$$\frac{1}{2}\oint_C r_n\, ds = \text{area enclosed by } C,$$

$\mathbf{n}$ being the outer normal to C.

7. Check the answers to Probs. 2 (a), 3 (a), (b), (c), 4 (a) following Section 5–3 by Green's theorem.

Answers

1. (a) $\frac{4}{3}$, (b) $\frac{4}{3}$, (c) $\frac{4}{3}$. 2. (a) 0, (b) $\frac{1}{3}$, (c) $\frac{4}{3}$.

5. (a) $(b - a)$ times area enclosed by C, (b) 0, (c) $\frac{3}{2}\pi$, (d) 0, (e) 0, (f) 0.

5–6 Independence of path. Simply connected domains. Let the functions $P(x, y)$ and $Q(x, y)$ be defined and continuous in a domain D. Then the line integral $\int P\, dx + Q\, dy$ is said to be *independent of path in D* if, for every pair of end points A and B in D, the value of the line integral

$$\int_A^B P\,dx + Q\,dy$$

is the same for all paths C from A to B. The value of the integral will then in general depend on the choice of A and B, but not on the choice of the path joining them. Thus, as in Fig. 5–13, the integrals on C_1, C_2, C_3 have the same value.

THEOREM I. *If the integral* $\int P\,dx + Q\,dy$ *is independent of path in* D, *then there is a function* $F(x, y)$ *defined in* D *such that*

$$\frac{\partial F}{\partial x} = P(x, y), \quad \frac{\partial F}{\partial y} = Q(x, y) \tag{5–46}$$

holds throughout D. *Conversely, if a function* $F(x, y)$ *can be found such that* (5–46) *holds in* D, *then* $\int P\,dx + Q\,dy$ *is independent of path in* D.

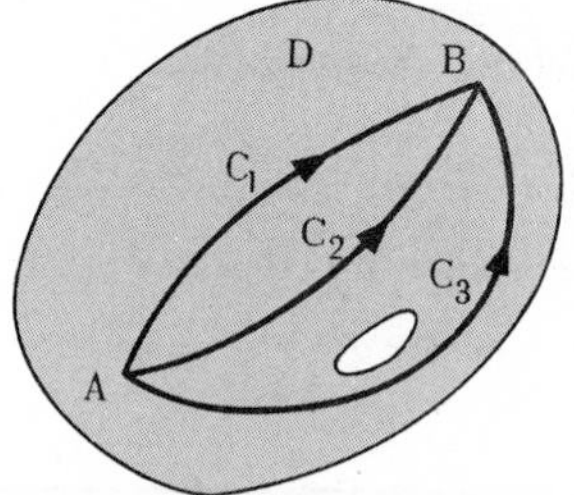

FIG. 5–13. Independence of path.

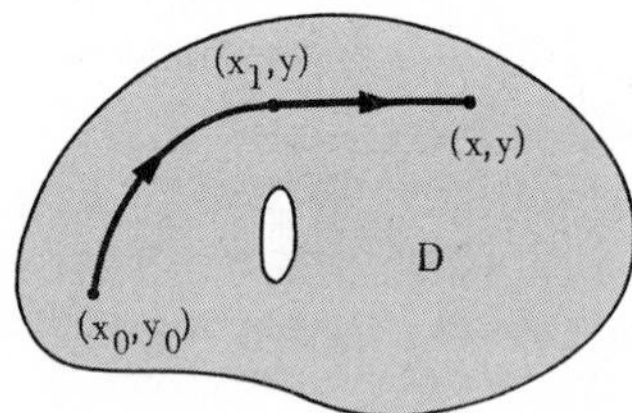

FIG. 5–14. Construction of function F such that $dF = P\,dx + Q\,dy$.

Proof: Suppose first that the integral is independent of path in D. Then choose a point (x_0, y_0) in D and let $F(x, y)$ be defined as follows:

$$F(x, y) = \int_{(x_0, y_0)}^{(x, y)} P\,dx + Q\,dy, \tag{5–47}$$

where the integral is taken on an arbitrary path in D joining (x_0, y_0) to (x, y). Since the integral is independent of path, the integral in (5–47) does indeed depend only on the end point (x, y) and defines a function $F(x, y)$. It remains to show that $\partial F/\partial x = P(x, y)$, $\partial F/\partial y = Q(x, y)$ in D.

For a particular (x, y) in D, choose (x_1, y) so that $x_1 \neq x$ and the line segment from (x_1, y) to (x, y) is in D, as shown in Fig. 5–14. Then, because of independence of path, one has

$$F(x, y) = \int_{(x_0, y_0)}^{(x_1, y)} (P\,dx + Q\,dy) + \int_{(x_1, y)}^{(x, y)} (P\,dx + Q\,dy).$$

Here x_1 and y are thought of as fixed, while (x, y) may vary along the line segment. Thus y has been restricted to a constant value and $F(x, y)$ is being considered as a function of x near a particular choice of x. The first integral on the right is then independent of x, while the second can be integrated along the line segment. Hence, for this fixed y,

$$F(x, y) = \text{const} + \int_{x_1}^{x} P(x, y)\, dx, \quad (y = \text{const})$$

or, with the dummy variable x replaced by a t,

$$F(x, y) = \text{const} + \int_{x_1}^{x} P(t, y)\, dt.$$

The fundamental theorem of calculus (Section 4–3) now gives

$$\frac{\partial F}{\partial x} = P(x, y).$$

Similarly, $\partial F/\partial y = Q(x, y)$ and the first part of the theorem is proved.

Conversely, let (5–46) hold in D for some F. Then, in terms of a parameter t,

$$\begin{aligned}\int_{(x_1, y_1)\atop C}^{(x_2, y_2)} P\, dx + Q\, dy &= \int_{t_1}^{t_2} \left(\frac{\partial F}{\partial x}\frac{dx}{dt} + \frac{\partial F}{\partial y}\frac{dy}{dt}\right) dt \\ &= \int_{t_1}^{t_2} \frac{dF}{dt}\, dt = F\Big|_{t=t_2} - F\Big|_{t=t_1} \\ &= F(x_2, y_2) - F(x_1, y_1).\end{aligned}$$

In other words, in simplified notation,

$$\int_{A\atop C}^{B} P\, dx + Q\, dy = \int_{A}^{B} dF = F(B) - F(A). \tag{5–48}$$

The value of the integral is simply the difference of the values of F at the two ends and is therefore independent of the path C.

Remark 1. Formula (5–48) is the analogue of the usual formula

$$\int_a^b f(x)\, dx = F(x)\Big|_a^b = F(b) - F(a),\ F'(x) = f(x)$$

for evaluating definite integrals and should be used whenever the function F can be found. Thus

$$\int_{(1,2)}^{(5,6)} y\, dx + x\, dy = \int_{(1,2)}^{(5,6)} d(xy) = xy\Big|_{(1,2)}^{(5,6)} = 30 - 2 = 28.$$

Remark 2. Suppose $\partial F/\partial x = P \equiv 0$ in D, $\partial F/\partial y = Q \equiv 0$ in D. Then (5–48) shows that

$$F(B) - F(A) = \int_{A\atop C}^{B} 0\, dx + 0\, dy = 0,$$

i.e., that $F(B) = F(A)$ for every two points of D. Hence $F \equiv \text{const}$ in D (cf. Section 2–22).

Remark 3. If P and Q are given such that $\int P\,dx + Q\,dy$ is independent of path, then the function F such that $dF = P\,dx + Q\,dy$ is determined only up to an additive constant:

$$F = \int_{(x_0,\,y_0)}^{(x,\,y)} P\,dx + Q\,dy + \text{const.} \tag{5–49}$$

For if G is a second function such that $\partial G/\partial x = P$, $\partial G/\partial y = Q$, then $(\partial/\partial x)(G - F) = P - P \equiv 0$, $(\partial/\partial y)(G - F) = Q - Q \equiv 0$; hence by Remark 2, $G - F$ is a constant or $G = F + \text{const.}$

It often happens that $P = \partial F/\partial x$ and $Q = \partial F/\partial y$ are known (perhaps in tabular form), while F is not; in this case (5–49) permits evaluation of F at any point in D, once its value at (x_0, y_0) has been chosen. For (5–49) can be written:

$$F = \int_{(x_0,y_0)}^{(x,\,y)} P\,dx + Q\,dy + F(x_0, y_0), \tag{5–50}$$

as the replacement of the upper limit (x, y) by (x_0, y_0) shows. In (5–50) any convenient path of integration can be used, e.g., one formed of lines parallel to the axes. This is very important in thermodynamics, in which the partial derivatives of such functions as internal energy, entropy, and free energy are the measured quantities, rather than the functions themselves (cf. Section 5–15).

Remark 4. It is to be emphasized that the function $F(x, y)$ in (5–46) must be *defined* and continuous in a domain containing the path. A function such as arc tan (y/x) is "many-valued" and hence useless for the theorem unless it is unambiguously defined and has the required continuity; these requirements cannot be satisfied when C is the circular path: $x = \cos t$, $y = \sin t$, $0 \leqq t \leqq 2\pi$. This point is considered further in connection with Example 2 below.

THEOREM II. *If the integral $\int P\,dx + Q\,dy$ is independent of path in D, then*

$$\oint_C P\,dx + Q\,dy = 0 \tag{5–51}$$

on every simple closed path in D. Conversely, if (5–51) *holds for every simple closed path in D, then $\int P\,dx + Q\,dy$ is independent of path in D.*

Proof: Suppose first that the integral is independent of path. Let C be a simple closed path in D and divide C into arcs AB and BA by points A, B as in Fig. 5–15. Then

$$\oint_C P\,dx + Q\,dy = \int_{\widehat{AB}} (P\,dx + Q\,dy) + \int_{\widehat{BA}} (P\,dx + Q\,dy).$$

Because of independence of path one can write this equation in the form

$$\oint_C P\,dx + Q\,dy = \int_A^B (P\,dx + Q\,dy) + \int_B^A (P\,dx + Q\,dy)$$

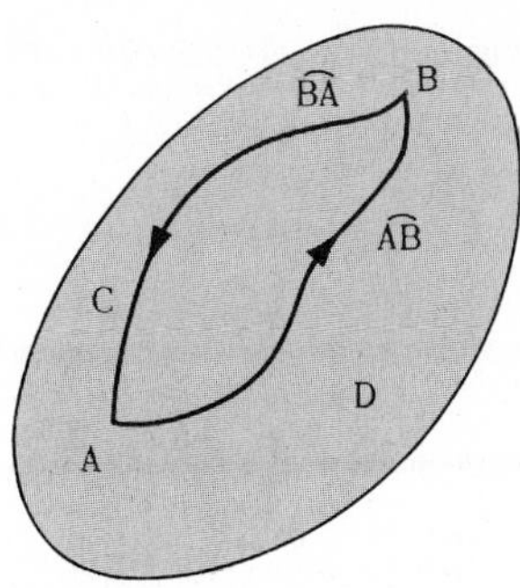

FIG. 5–15.

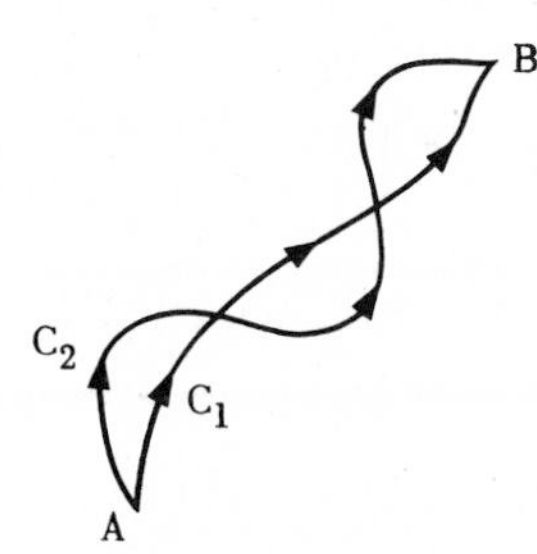

FIG. 5–16.

$$= \int_A^B (P\,dx + Q\,dy) - \int_A^B (P\,dx + Q\,dy) = 0.$$

Conversely, suppose $\int P\,dx + Q\,dy = 0$ on every simple closed path in D. Let A and B be two points of D and let C_1 and C_2 be two paths from A to B as in Fig. 5–16. We must show that

$$\int_{A\atop C_1}^B P\,dx + Q\,dy = \int_{A\atop C_2}^B P\,dx + Q\,dy$$

or, if C_2' denotes C_2 traced in the reverse direction,

$$\int_{A\atop C_1}^B (P\,dx + Q\,dy) + \int_{B\atop C_2'}^A P\,dx + Q\,dy = 0.$$

If the paths C_1 and C_2' do not cross, they together form a simple closed path C and

$$\int_{A\atop C_1}^B (P\,dx + Q\,dy) + \int_{B\atop C_2'}^A (P\,dx + Q\,dy) = \int_C P\,dx + Q\,dy = 0,$$

regardless of the direction of integration.

If the paths cross a finite number of times, as in Fig. 5–16, a repetition of the same argument gives the proof; this reasoning covers, in particular, the case when C_1 and C_2 are broken lines. If the paths cross infinitely often, it is necessary to show that the integrals can be arbitrarily closely approximated by integrals on broken line paths; a limiting process then gives the same result. For a complete discussion, the reader is referred to O. D. Kellogg, *Foundations of Potential Theory* (Berlin: Springer, 1929).

Remark. The first part of the preceding proof shows that, if $\int P\,dx + Q\,dy$ is independent of path, then

$$\int_C P\,dx + Q\,dy = 0$$

on every *closed path* C in D, regardless of whether or not C crosses itself.

THEOREM III. *If $P(x, y)$ and $Q(x, y)$ have continuous partial derivatives in D and $\int P\,dx + Q\,dy$ is independent of path in D, then*

$$\frac{\partial P}{\partial y} = \frac{\partial Q}{\partial x} \tag{5–52}$$

in D.

Proof. By Theorem I, one has

$$P = \frac{\partial F}{\partial x}, \quad Q = \frac{\partial F}{\partial y}$$

and, since P and Q have continuous derivatives in D,

$$\frac{\partial P}{\partial y} = \frac{\partial^2 F}{\partial y\,\partial x} = \frac{\partial^2 F}{\partial x\,\partial y} = \frac{\partial Q}{\partial x}.$$

Simply connected domains. The converse of Theorem III does not hold without a further restriction, namely, that the domain D be *simply connected.* In plain terms, a domain is simply connected if it has no "holes." Thus the domains of Figs. 5–13 and 5–14 are not simply connected, but rather are, as one says, *multiply connected.* More precisely, D is simply connected if, for every simple closed curve C in D, the region R formed of C plus interior lies wholly in D (cf. Fig. 5–17).

Examples of simply connected domains are the following: the interior of a circle, the interior of a square, a sector, a quadrant, the whole xy plane. The annular region between two circles is not simply connected, nor is the interior of a circle minus the center. Thus the holes may reduce to single points.

One can distinguish between types of multiply connected domains as follows: a domain with just one hole, as in Fig. 5–18, is doubly connected; one with two holes is triply connected; one with $n - 1$ holes is $n$tuply connected. One can even have infinitely many holes, in which case one speaks of a domain which is infinitely multiply connected.

After this preparation, we can state the converse of Theorem III:

THEOREM IV. *Let $P(x, y)$ and $Q(x, y)$ have continuous derivatives in D and let D be simply connected. If*

$$\frac{\partial P}{\partial y} = \frac{\partial Q}{\partial x} \tag{5–53}$$

in D, then $\int P\,dx + Q\,dy$ is independent of path in D.

This theorem and its consequences are fundamental in physical applications.

Proof. Suppose (5–53) to be satisfied. Then choose any simple closed curve C in D. By Green's theorem,

$$\oint_C P\,dx + Q\,dy = \iint_R \left(\frac{\partial Q}{\partial x} - \frac{\partial P}{\partial y}\right) dx\,dy = 0,$$

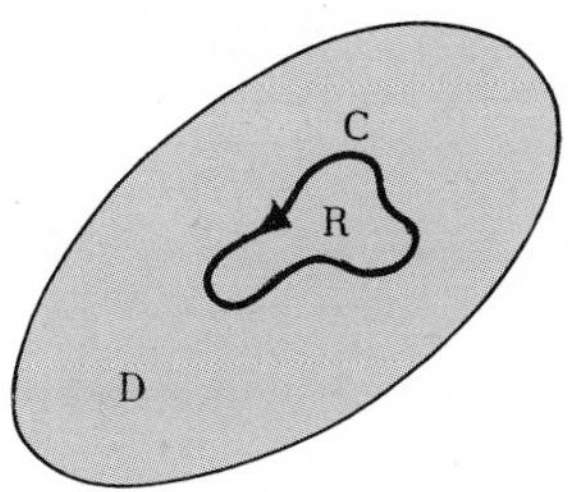

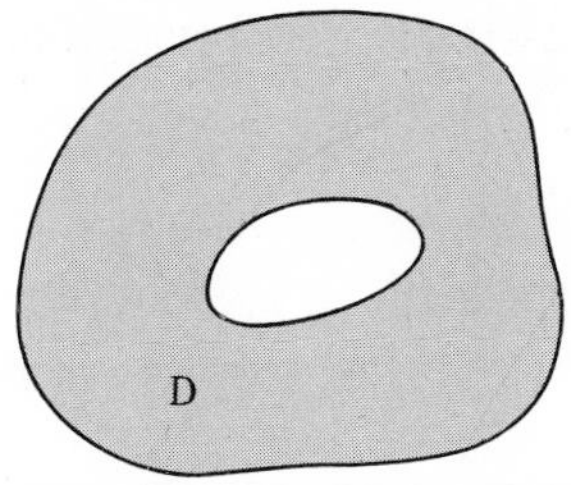

FIG. 5–17. Simply connected domain. FIG. 5–18. Doubly connected domain.

where R is the region formed of C plus interior. It should be remarked that Green's theorem is applicable because D is simply connected, so that R is in D and, in particular, $\partial Q/\partial x$ and $\partial P/\partial y$ are continuous in R. Since we have shown that

$$\oint_C P\,dx + Q\,dy = 0$$

for every simple closed path C in D, it follows from Theorem II that $\int P\,dx + Q\,dy$ is independent of path in D.

Vector interpretation. Theorem I asserts that $\int P\,dx + Q\,dy$ is independent of path precisely when $P = \partial F/\partial x$, $Q = \partial F/\partial y$, i.e.,

$$P\,\mathbf{i} + Q\,\mathbf{j} = \text{grad } F.$$

Thus $\int u_T\,ds$ is independent of path precisely when $\mathbf{u}$ is the gradient vector of some scalar F: $\mathbf{u} = \nabla F$. Equation (5–48) then states that

$$\int_A^B \nabla F \cdot d\mathbf{r} = F(B) - F(A). \tag{5–54}$$

Theorem III asserts that, if $\int P\,dx + Q\,dy$ is independent of path, then $\partial P/\partial y = \partial Q/\partial x$. If $\mathbf{u} = P\mathbf{i} + Q\mathbf{j}$, this is the statement that

$$\text{curl } \mathbf{u} = \mathbf{0},$$

for

$$\text{curl } \mathbf{u} = \nabla \times \mathbf{u} = \begin{vmatrix} \mathbf{i} & \mathbf{j} & \mathbf{k} \\ \dfrac{\partial}{\partial x} & \dfrac{\partial}{\partial y} & \dfrac{\partial}{\partial z} \\ P & Q & 0 \end{vmatrix} = \mathbf{k}\left(\frac{\partial Q}{\partial x} - \frac{\partial P}{\partial y}\right).$$

Since we know by Theorem I that $\mathbf{u}$ is a gradient, Theorem III is equivalent to the statement

$$\text{curl grad } F = \mathbf{0}.$$

This fundamental identity was pointed out in Chapter 3 [Eq. (3–31)].

Theorem IV then provides the converse: *if* curl $\mathbf{u} = \mathbf{0}$ *in* D, *then* $\mathbf{u} =$ grad F *for some* F, *provided* D *is simply connected.* The analogous statement for three dimensions will be proved in Section 5–13 below.

EXAMPLE 1. The integral

$$\int 2xy^3\,dx + 3x^2y^2\,dy$$

is independent of path in the whole plane, since

$$\frac{\partial P}{\partial y} - \frac{\partial Q}{\partial x} = 6xy^2 - 6xy^2 = 0.$$

To find a function F whose gradient is $P\mathbf{i} + Q\mathbf{j}$, set

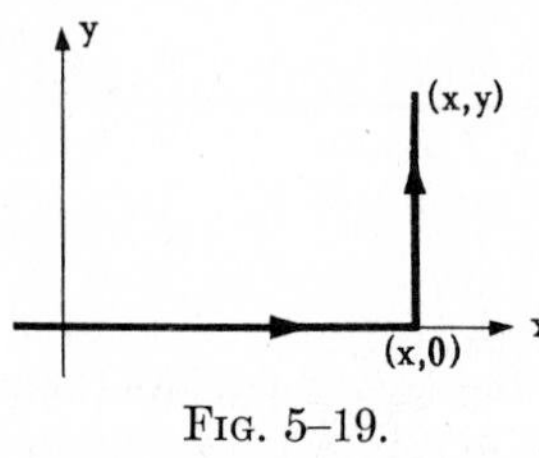

FIG. 5–19.

$$F = \int_{(0,0)}^{(x,y)} 2xy^3\,dx + 3x^2y^2\,dy$$
$$= \int_{(0,0)}^{(x,0)} 2xy^3\,dx + 3x^2y^2\,dy + \int_{(x,0)}^{(x,y)} 2xy^3\,dx + 3x^2y^2\,dy,$$

using a broken line path, as in Fig. 5–19. Along the first part $dy = 0$ and $y = 0$, so that the integral reduces to 0. Along the second part $dx = 0$ and x is constant, so that

$$F = \int_0^y 3x^2y^2\,dy = x^2y^3.$$

The general solution is then $x^2y^3 + C$. To evaluate the integral on any path from (1, 2) to (3, −2), one then writes

$$\int_{(1,2)}^{(3,-2)} 2xy^3\,dx + 3x^2y^2\,dy = \int_{(1,2)}^{(3,-2)} d(x^2y^3) = x^2y^3\Big|_{(1,2)}^{(3,-2)} = -80.$$

EXAMPLE 2. The integral

$$\int \frac{-y\,dx + x\,dy}{x^2 + y^2}$$

is independent of path in any simply connected domain D not containing the origin, for

$$\frac{\partial P}{\partial y} = \frac{\partial Q}{\partial x} = \frac{y^2 - x^2}{(x^2 + y^2)^2}$$

except at the origin. To find the function F whose differential is $P\,dx + Q\,dy$, one can proceed as in Example 1, using as path a broken line from (1, 0) to (x, y). However, $P\,dx + Q\,dy$ is here a familiar differential (cf. Example 3, Section 2–8), namely that of the polar coordinate angle θ:

$$d\theta = d\left(\arctan\frac{y}{x}\right) = \frac{-y\,dx + x\,dy}{x^2 + y^2}.$$

The arc tangent is rather to be avoided if possible, because one cannot in general restrict to principal values and also because of the awkwardness of arc tan ∞ for the y axis. One would therefore write

$$\int_A^B \frac{-y\,dx + x\,dy}{x^2 + y^2} = \int_A^B d\theta = \theta_B - \theta_A.$$

However, in order for θ to be a well-defined continuous function in a domain including the path (which in particular requires that θ be *single-valued*), so that Theorem I can be applied, it is necessary to restrict to a simply connected domain not containing the origin. Two such domains are shown in Figs. 5–20(a) and (b). In Fig. 5–20(c) the path is the circle $x^2 + y^2 = 1$ and the domain is doubly connected. Since $x^2 + y^2 = 1$ on the path

$$\oint \frac{-y\,dx + x\,dy}{x^2 + y^2} = \oint -y\,dx + x\,dy = \iint_R 2\,dx\,dy = 2\pi;$$

here Green's theorem, which is not applicable to the given integral, could be applied after the line integral had been simplified by using the relation: $x^2 + y^2 = 1$ on the path; one can also regard the line integral as a sum of integrals from $\theta = 0$ to $\theta = \pi$ and from $\theta = \pi$ to $\theta = 2\pi$, giving $\pi + \pi = 2\pi$. A similar procedure can be followed in general and one concludes that, for any path C not through $(0, 0)$,

$$\int_{A\atop C}^B \frac{-y\,dx + x\,dy}{x^2 + y^2} = \text{total increase in } \theta \text{ from } A \text{ to } B,$$

as θ varies *continuously* on the path C. The integral is not independent of path but depends on the number of times C goes around the origin.

Example 3. The integral

$$\int \frac{x\,dx + y\,dy}{x^2 + y^2}$$

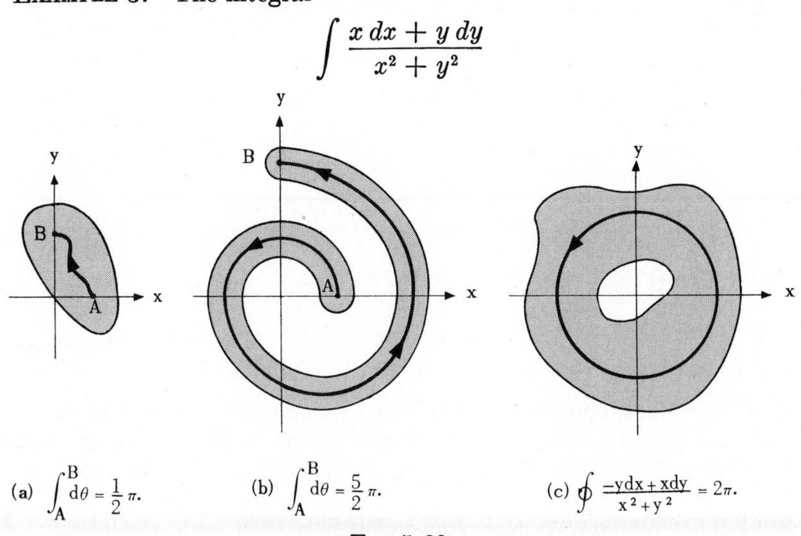

(a) $\int_A^B d\theta = \frac{1}{2}\pi$. (b) $\int_A^B d\theta = \frac{5}{2}\pi$. (c) $\oint \frac{-y\,dx + x\,dy}{x^2 + y^2} = 2\pi$.

Fig. 5–20.

is independent of path in the *doubly connected* domain consisting of the xy plane minus the origin. First of all,

$$\frac{\partial P}{\partial y} = \frac{\partial Q}{\partial x} = \frac{-2xy}{(x^2 + y^2)^2},$$

but, as Example 2 shows, this would guarantee independence of path only in a *simply connected* domain. However, for this example one has additional information:

$$\frac{x\,dx + y\,dy}{x^2 + y^2} = d \log \sqrt{x^2 + y^2},$$

and the function $\log \sqrt{x^2 + y^2} = \log r$ is well-defined (i.e., *single-valued*), with continuous derivatives except at the origin. Hence

$$\int_A^B \frac{x\,dx + y\,dy}{x^2 + y^2} = \int_A^B d \log r = \log r \Big|_A^B = \log \frac{r_B}{r_A}$$

for any path from A to B not through the origin — in particular, then, for the paths of Fig. 5–20(a) and (b). On the circle of Fig. 5–20(c) the integral is 0, in accordance with Theorem II.

5–7 Extension of results to multiply connected domains. If the closed curve C of Green's theorem encloses a point at which $\partial P/\partial y$ or $\partial Q/\partial x$ fails to exist, then the theorem cannot be applied. The following theorem shows that even in this case the reduction of a properly chosen line integral to a double integral is possible.

THEOREM. *Let $P(x, y)$ and $Q(x, y)$ be continuous and have continuous derivatives in a domain D of the plane. Let R be a closed region in D whose boundary consists of n distinct simple closed curves $C_1, C_2, \ldots, C_n$, where C_1 includes $C_2, \ldots, C_n$ in its interior. Then*

$$\oint_{C_1} (P\,dx + Q\,dy) + \oint_{C_2} (P\,dx + Q\,dy) + \cdots + \oint_{C_n} (P\,dx + Q\,dy)$$
$$= \iint_R \left(\frac{\partial Q}{\partial x} - \frac{\partial P}{\partial y}\right) dx\,dy. \qquad (5\text{–}55)$$

In particular, if $\partial Q/\partial x = \partial P/\partial y$ in D, then

$$\oint_{C_1} (P\,dx+Q\,dy)+\oint_{C_2} (P\,dx+Q\,dy)+\cdots+\oint_{C_n} (P\,dx+Q\,dy)=0. \qquad (5\text{–}55')$$

To prove the theorem, one first introduces auxiliary arcs from C_1 to C_2, C_1 to $C_3, \ldots$, two to each, as illustrated in Fig. 5–21. These decompose the region R into smaller regions, each of which is simply connected; there are four sub-regions in Fig. 5–21. If one integrates in a positive direction around the boundary of each sub-region and then adds the results, one finds that the integrals along the auxiliary arcs cancel out, leaving just the

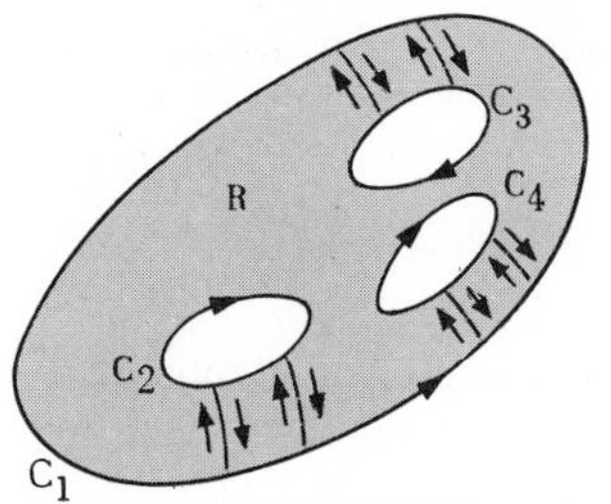

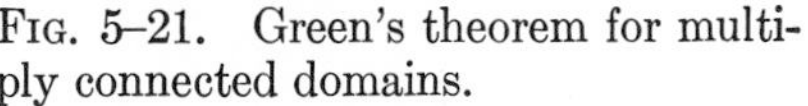

Fig. 5–21. Green's theorem for multiply connected domains.

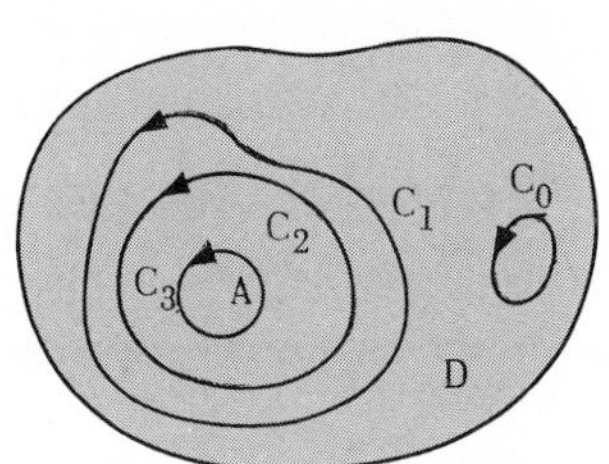

Fig. 5–22.

integral around C_1 in the positive direction plus the integrals around C_2, $C_3, \dots,$ in the negative direction. On the other hand the line integral around the boundary of each sub-region can be expressed as a double integral

$$\iint \left(\frac{\partial Q}{\partial x} - \frac{\partial P}{\partial y}\right) dx\, dy$$

over the sub-region by Green's theorem. Hence the sum of the line integrals is equal to the double integral over R. This gives (5–55); Eq. (5–55′) then follows as a special case.

If one denotes by B_R the directed boundary of R, i.e., the curves C_1, C_2, $\dots, C_n$ with the given directions, then Eqs. (5–55) and (5–55′) can be written in the concise form:

$$\int\limits_{B_R} P\, dx + Q\, dy = \iint\limits_{R} \left(\frac{\partial Q}{\partial x} - \frac{\partial P}{\partial y}\right) dx\, dy; \tag{5–56}$$

$$\int\limits_{B_R} P\, dx + Q\, dy = 0 \quad \left(\frac{\partial Q}{\partial x} = \frac{\partial P}{\partial y} \text{ in } D\right). \tag{5–56′}$$

It is to be noted that the correct direction of integration always keeps the region to the left, i.e., the normal pointing away from the region is 90° behind the tangent.

As an application of the theorem just proved, let us consider the case of a doubly connected region D with a hole which, for the sake of simplicity, we take to be a point A, as in Fig. 5–22. Let $\partial P/\partial y = \partial Q/\partial x$ in D. What are the possible values of an integral

$$\oint\limits_{C} P\, dx + Q\, dy$$

about a simple closed path in D? We assert that there are only two possible values: namely, the value 0 when C does not enclose the hole A (the curve C_0 of Fig. 5–22) and a value k *which is the same for all curves* C *enclosing the hole* A. That the value is 0 for a curve such as C_0 follows from the fact that C_0 lies in a simply connected part of D, so that Theorem IV of the preceding section is applicable. It remains to show that for paths C_1 and C_2 which both enclose A one has

$$\oint_{C_1} P\,dx + Q\,dy = \oint_{C_2} P\,dx + Q\,dy. \tag{5–57}$$

To prove this, let us assume first that C_1 and C_2 do not meet and that, for example, C_2 lies within C_1. Then C_1 and C_2 form the boundary of a region R to which (5–55′) is applicable:

$$\oint_{C_1} P\,dx + Q\,dy + \oint_{C_2} P\,dx + Q\,dy = 0$$

or

$$\oint_{C_1} P\,dx + Q\,dy = \oint_{C_2} P\,dx + Q\,dy,$$

as was to be shown. If C_1 and C_2 do intersect each other, a sufficiently small circle C_3 about A will meet neither of them. One has then

$$\oint_{C_1} P\,dx + Q\,dy = \oint_{C_3} P\,dx + Q\,dy, \quad \oint_{C_2} P\,dx + Q\,dy = \oint_{C_3} P\,dx + Q\,dy$$

as before. Thus again

$$\oint_{C_1} P\,dx + Q\,dy = \oint_{C_2} P\,dx + Q\,dy$$

and (5–57) is completely proved.

EXAMPLE 1. The integral

$$\oint_C \frac{-y\,dx + x\,dy}{x^2 + y^2}$$

of Example 2 of the preceding section is of the type considered, with the hole A at the origin. Hence the integral is 0 when C does not enclose the origin, has a certain value k when C does enclose the origin. To find k, we choose C to be the circle $x^2 + y^2 = 1$ and find that

$$k = \oint_C -y\,dx + x\,dy = \iint_R 2\,dx\,dy = 2\pi.$$

This is in agreement with the previous results.

EXAMPLE 2. Example 3 of the preceding section,

$$\oint_C \frac{x\,dx + y\,dy}{x^2 + y^2},$$

illustrates a case in which the constant value k is 0. Thus the integral is 0 on all simple closed paths not through the origin.

EXAMPLE 3. To evaluate

$$\oint \frac{y^3\,dx - xy^2\,dy}{(x^2+y^2)^2}$$

around the ellipse $x^2 + 3y^2 = 1$, one verifies that $\partial P/\partial y = \partial Q/\partial x$ except at (0, 0), where both functions are discontinuous. Then, since the ellipse is awkward for integration, one replaces it by the circle: $x^2 + y^2 = 1$, which also encloses the point of discontinuity. With the parametrization: $x = \cos t$, $y = \sin t$, the integral on the circle is

$$\int_0^{2\pi} (-\sin^3 t \sin t - \cos t \sin^2 t \cos t)\,dt = -\int_0^{2\pi} \sin^2 t\,dt = -\pi.$$

Thus the integral on the given ellipse equals $-\pi$.

The results (5–55) or (5–56) can be put in vector form. Thus, as in (5–43) and (5–44) (Section 5–5), one has

$$\int_{B_R} u_T\,ds = \iint_R \operatorname{curl}_z \mathbf{u}\,dx\,dy, \tag{5–58}$$

$$\int_{B_R} v_n\,ds = \iint_R \operatorname{div} \mathbf{v}\,dx\,dy, \tag{5–59}$$

where $\mathbf{T}$ is the unit tangent vector in the direction of integration and $\mathbf{n}$ is 90° behind $\mathbf{T}$, so that $\mathbf{n}$ is an *outer normal*, i.e., points away from R.

Remark. It can happen that the path C passes *through* a discontinuity of P or Q. If the line integral is reduced to a definite integral $\int_a^b f(t)\,dt$ in the parameter t, one has then the problem discussed in Section 4–4. If $f(t)$ is discontinuous only at a finite number of points and is bounded, then the integral continues to exist; if $f(t)$ is unbounded, the integral is improper and may or may not exist.

Problems

1. Determine by inspection a function $F(x, y)$ whose differential has the given value, and integrate the corresponding line integral:

(a) $dF = 2xy\,dx + x^2\,dy$, $\int_{(0,0)\,C}^{(1,1)} 2xy\,dx + x^2\,dy$, where C is the curve: $y = x^{\frac{3}{2}}$;

(b) $dF = ye^{xy}\,dx + xe^{xy}\,dy$, $\int_{(0,0)\,C}^{(\pi,0)} ye^{xy}\,dx + xe^{xy}\,dy$, where C is the curve. $y = \sin^3 x$;

(c) $dF = \dfrac{x\,dx + y\,dy}{(x^2+y^2)^{\frac{3}{2}}}$, $\int_{(1,0)\,C}^{(e^{2\pi},0)} \dfrac{x\,dx + y\,dy}{(x^2+y^2)^{\frac{3}{2}}}$, where C is the curve: $x = e^t \cos t$, $y = e^t \sin t$.

2. Test for independence of path and evaluate the following integrals:

(a) $\displaystyle\int_{(1,-2)}^{(3,4)} \frac{y\,dx - x\,dy}{x^2}$ on the line: $y = 3x - 5$;

(b) $\displaystyle\int_{(0,2)}^{(1,3)} \frac{3x^2}{y}\,dx - \frac{x^3}{y^2}\,dy$ on the parabola: $y = 2 + x^2$.

3. Evaluate the following integrals:

(a) $\displaystyle\oint [\sin(xy) + xy\cos(xy)]\,dx + x^2\cos(xy)\,dy$ on the circle: $x^2 + y^2 = 1$;

(b) $\displaystyle\oint \frac{y\,dx - (x-1)\,dy}{(x-1)^2 + y^2}$ on the circle: $x^2 + y^2 = 4$;

(c) $\displaystyle\oint y^3\,dx - x^3\,dy$ on the square: $|x| + |y| = 1$;

(d) $\displaystyle\oint xy^6\,dx + (3x^2y^5 + 6x)\,dy$ on the ellipse: $x^2 + 4y^2 = 4$.

4. Determine all values of the integral

$$\int_{(1,0)}^{(2,2)} \frac{-y\,dx + x\,dy}{x^2 + y^2}$$

on a path not passing through the origin.

5. Show that the following functions are independent of path in the xy plane and evaluate them:

(a) $\displaystyle\int_{(1,1)}^{(x,y)} 2xy\,dx + (x^2 - y^2)\,dy$; (b) $\displaystyle\int_{(0,0)}^{(x,y)} \sin y\,dx + x\cos y\,dy$.

6. Evaluate

$$\oint \frac{x^2y\,dx - x^3\,dy}{(x^2 + y^2)^2}$$

around the square with vertices $(\pm 1, \pm 1)$ (note that $\partial P/\partial y = \partial Q/\partial x$).

7. Let D be a domain which has a finite number of "holes" at points A_1, $A_2, \ldots, A_k$, so that D is $(k+1)$-tuply connected; cf. Fig. 5–23. Let P and Q be continuous and have continuous derivatives in D, with $\partial P/\partial y = \partial Q/\partial x$ in D. Let C_1 denote a circle about A_1 in D, enclosing none of the other A's. Let C_2 be chosen similarly for A_2, etc. Let

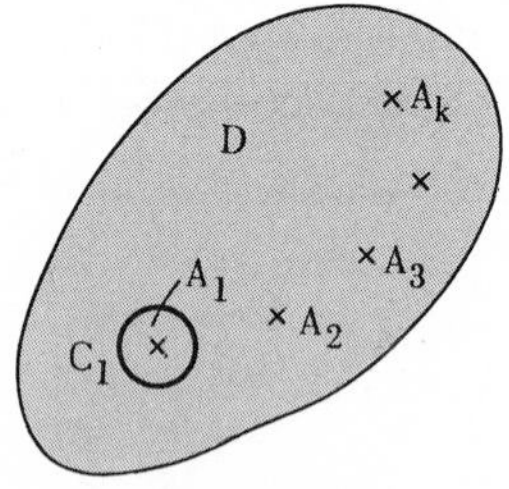

FIG. 5–23.

$$\oint_{C_1} P\,dx + Q\,dy = \alpha_1, \quad \oint_{C_2} P\,dx + Q\,dy = \alpha_2, \ldots, \quad \oint_{C_k} P\,dx + Q\,dy = \alpha_k.$$

(a) Show that, if C is an arbitrary simple closed path in D enclosing A_1, A_2, $\ldots$, A_k, then

$$\oint_C P\,dx + Q\,dy = \alpha_1 + \alpha_2 + \cdots + \alpha_k.$$

(b) Determine all possible values of the integral

$$\int_{(x_1, y_1)}^{(x_2, y_2)} P\,dx + Q\,dy$$

between two fixed points of D, if it is known that this integral has the value K for one particular path.

8. Let P and Q be continuous and have continuous derivatives, with $\partial P/\partial y = \partial Q/\partial x$, except at the points $(4, 0)$, $(0, 0)$, $(-4, 0)$. Let C_1 denote the circle: $(x-2)^2 + y^2 = 9$; let C_2 denote the circle: $(x+2)^2 + y^2 = 9$; let C_3 denote the circle: $x^2 + y^2 = 25$. Given that

$$\oint_{C_1} P\,dx + Q\,dy = 11, \quad \oint_{C_2} P\,dx + Q\,dy = 9, \quad \oint_{C_3} P\,dx + Q\,dy = 13,$$

find

$$\oint_{C_4} P\,dx + Q\,dy,$$

where C_4 is the circle: $x^2 + y^2 = 1$. [Hint: use the result of Prob. 7(a).]

9. Let $F(x, y) = x^2 - y^2$. Evaluate

(a) $\displaystyle\int_{(0,0)}^{(2,8)} \nabla F \cdot d\mathbf{r}$ on the curve: $y = x^3$;

(b) $\displaystyle\oint \frac{\partial F}{\partial n}\,ds$ on the circle $x^2 + y^2 = 1$, if $\mathbf{n}$ is the outer normal and $\dfrac{\partial F}{\partial n} = \nabla F \cdot \mathbf{n}$ is the directional derivative of F in the direction of $\mathbf{n}$ (Section 2–10).

10. Let $F(x, y)$ and $G(x, y)$ be continuous and have continuous derivatives in a domain D. Let R be a closed region in D with directed boundary B_R consisting of closed curves $C_1, \ldots, C_n$ as in Fig. 5–21. Let $\mathbf{n}$ be the unit outer normal of R and let $\partial F/\partial n$, $\partial G/\partial n$ denote the directional derivatives of F and G in the direction of $\mathbf{n}$: $\partial F/\partial n = \nabla F \cdot \mathbf{n}$, $\partial G/\partial n = \nabla G \cdot \mathbf{n}$.

(a) Show that $\displaystyle\int_{B_R} \frac{\partial F}{\partial n}\,ds = \iint_R \nabla^2 F\,dx\,dy.$

(b) Show that $\displaystyle\int_{B_R} \nabla F \cdot d\mathbf{r} = 0$

(c) Show that, if F is harmonic in D, then $\displaystyle\int_{B_R} \frac{\partial F}{\partial n}\,ds = 0.$

(d) Show that

$$\int_{B_R} F\frac{\partial G}{\partial n}\,ds = \iint_R F\nabla^2 G\,dx\,dy + \iint_R (\nabla F \cdot \nabla G)\,dx\,dy.$$

[Hint: use the identity: $\operatorname{div}(f\mathbf{u}) = f \operatorname{div}\mathbf{u} + \operatorname{grad} f \cdot \mathbf{u}$ of Section 3–6.]

11. Under the assumptions of Prob. 10 prove the identities:

(a) $\displaystyle\int_{B_R} \left(F\frac{\partial G}{\partial n} - G\frac{\partial F}{\partial n}\right) ds = \iint_R (F\,\nabla^2 G - G\,\nabla^2 F)\,dx\,dy.$

[Hint: use Prob. 10(d), applied to F and G and then with F and G interchanged];

(b) if F and G are harmonic in R, then

$$\int\limits_{B_R}\left(F\frac{\partial G}{\partial n} - G\frac{\partial F}{\partial n}\right) ds = 0.$$

[*Remark.* 10(d) and 11(a) are known as *identities of Green.* Generalizations to space and further applications are considered in Prob. 3 following Section 5–11.]

Answers

1. (a) $F = x^2y$, integral $= 1$; (b) $F = e^{xy}$, integral $= 0$; (c) $F = -1/\sqrt{x^2+y^2}$, integral $= 1 - e^{-2\pi}$.

2. (a) $-\dfrac{10}{3}$, (b) $\dfrac{1}{3}$. 3. (a) 0, (b) -2π, (c) -2, (d) 12π.

4. $\dfrac{\pi}{4} + 2n\pi$ $(n = 0, \pm 1, \pm 2, \ldots)$. 5. (a) $x^2y - \dfrac{1}{3}(y^3 + 2)$, (b) $x \sin y$.

6. $-\pi$. 7. $K + n_1\alpha_1 + n_2\alpha_2 + \cdots + n_k\alpha_k$, where the numbers $n_1, \ldots, n_k$ are positive or negative integers or 0. 8. 7. 9. (a) -60, (b) 0.

PART II. THREE-DIMENSIONAL THEORY AND APPLICATIONS

5–8 Line integrals in space. The basic definitions (Sections 5–2 and 5–3) of the line integrals

$$\int F(x, y)\, dx, \quad \int F(x, y)\, dy, \quad \int F(x, y)\, ds$$

can be generalized almost without change to give corresponding line integrals along a curve C in xyz space. Let C be given parametrically by equations:

$$x = \phi(t), \quad y = \psi(t), \quad z = \omega(t), \quad h \leqq t \leqq k, \tag{5–60}$$

where $\phi(t)$, $\psi(t)$, $\omega(t)$ are continuous and have continuous derivatives over the interval, so that C is "smooth." The interval $h \leqq t \leqq k$ is now subdivided by the successive points $t_0 = h, t_1, \ldots, t_i, \ldots, t_n = k$. The value t_i^* is chosen between t_{i-1} and t_i and one sets $x_i^* = \phi(t_i^*)$, $y_i^* = \psi(t_i^*)$, $z_i^* = \omega(t_i^*)$. One denotes by $\Delta_i t$ the difference $t_i - t_{i-1}$ and by $\Delta_i s$ the length of the corresponding piece of C. One defines

$$\Delta_i x = \phi(t_i) - \phi(t_{i-1}), \quad \Delta_i y = \psi(t_i) - \psi(t_{i-1}), \quad \Delta_i z = \omega(t_i) - \omega(t_{i-1}).$$

These quantities are indicated in Fig. 5–24. One can now make the definition

$$\int\limits_C X(x, y, z)\, dx = \lim_{\substack{n\to\infty \\ \max \Delta_i t \to 0}} \sum_{i=1}^{n} X(x_i^*, y_i^*, z_i^*)\, \Delta_i x. \tag{5–61}$$

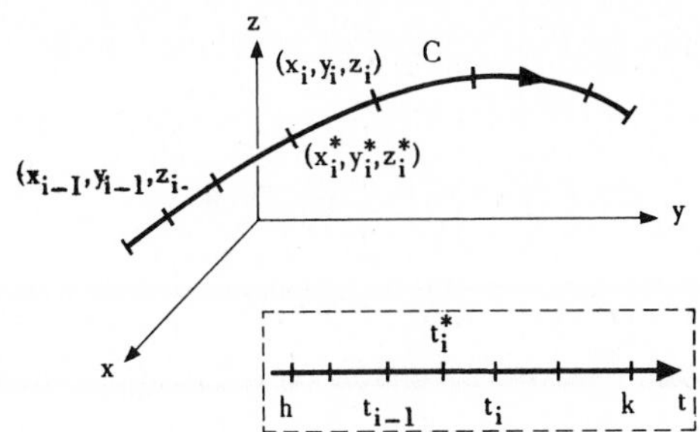

FIG. 5–24. Line integral in space.

The integrals

$$\int_C Y(x, y, z)\, dy, \quad \int_C Z(x, y, z)\, dz$$

are defined in similar fashion. Finally,

$$\int_C F(x, y, z)\, ds = \lim_{\substack{n\to\infty \\ \max \Delta_i s \to 0}} \sum_{i=1}^{n} F(x_i^*, y_i^*, z_i^*)\, \Delta_i s. \tag{5–62}$$

These definitions can again be shown to be meaningful if the functions X, Y, Z are continuous on C. The integrals can be reduced to ordinary definite integrals as in the plane:

$$\int_C X(x, y, z)\, dx = \int_h^k X[\phi(t), \psi(t), \omega(t)]\phi'(t)\, dt. \tag{5–63}$$

One can also extend the definitions and results to arbitrary "piecewise smooth" curves C, i.e., curves made up of a finite number of smooth pieces. The x, y, and z integrals will usually be considered together:

$$\int_C X\, dx + Y\, dy + Z\, dz,$$

and this combination is related to an s integral by the equation:

$$\int_C X\, dx + Y\, dy + Z\, dz = \int_C (X \cos\alpha + Y\cos\beta + Z\cos\gamma)\, ds, \tag{5–64}$$

where

$$\mathbf{T} = \cos\alpha\mathbf{i} + \cos\beta\mathbf{j} + \cos\gamma\mathbf{k}$$

is the unit tangent vector in the direction of increasing s. Equation (5–64) has a vector interpretation:

$$\int_C \mathbf{u}\cdot d\mathbf{r} = \int_C u_T\, ds, \tag{5–65}$$

where the vector line integral on the left is the limit of a sum

$$\sum \mathbf{u}(x_i^*, y_i^*, z_i^*) \cdot \Delta_i \mathbf{r}$$

and $u_T = \mathbf{u} \cdot \mathbf{T}$ is the tangential component of the vector

$$\mathbf{u} = X\mathbf{i} + Y\mathbf{j} + Z\mathbf{k}.$$

The integrals in (5–64) and (5–65) all measure the *work* done by the *force* $\mathbf{u}$ on the given path.

The interpretation $\int v_n \, ds$, in terms of a normal component, does not generalize to line integrals in space, but does to surface integrals, to be defined below.

5–9 Surfaces in space. Orientability. Let a surface S be given in space. The surface may be represented in the form:

$$z = f(x, y), \quad (x, y) \text{ in region } R_{xy}, \tag{5–66}$$

in the form

$$F(x, y, z) = 0, \tag{5–67}$$

or even in a parametric form (Section 4–9):

$$x = f(u, v), \quad y = g(u, v), \quad z = h(u, v), \quad (u, v) \text{ in } R_{uv}. \tag{5–68}$$

It will be assumed that S is "smooth," i.e., that in (5–66), for example, $f(x, y)$ has continuous derivatives in a domain containing the region R_{xy}. It will be assumed that R_{xy} is bounded and closed. It is then possible to evaluate the area of S or of a part of S.

If the surface is given in form (5–66), the area of S is given by the formula (Section 4–9):

$$\iint\limits_{R_{xy}} \sec \gamma \, dx \, dy = \iint\limits_{R_{xy}} \sqrt{1 + \left(\frac{\partial z}{\partial x}\right)^2 + \left(\frac{\partial z}{\partial y}\right)^2} \, dx \, dy, \tag{5–69}$$

where γ is the angle between the *upper* normal to S and the z direction, as in Fig. 5–25. Thus $0 \leq \gamma < \frac{\pi}{2}$ and $\sec \gamma > 0$. We can regard formula (5–69) as stating that the surface *area element* $d\sigma$ for a surface $z = f(x, y)$ is

$$d\sigma = \sec \gamma \, dx \, dy = \sqrt{1 + \left(\frac{\partial z}{\partial x}\right)^2 + \left(\frac{\partial z}{\partial y}\right)^2} \, dx \, dy. \tag{5–70}$$

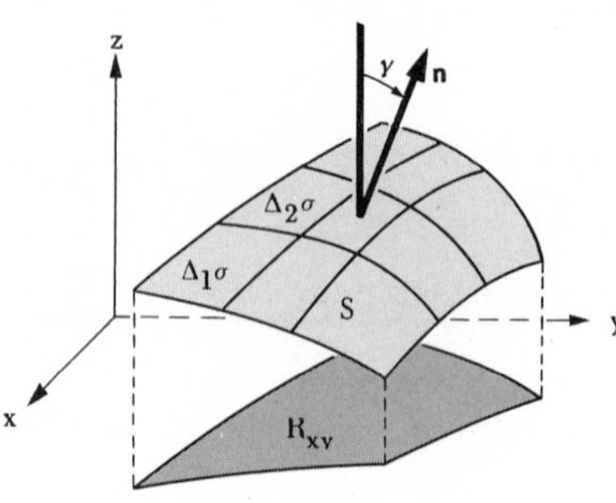

Fig. 5–25. Surface integral.

If the surface is given in form (5–68), the area is given by

$$\iint\limits_{R_{uv}} \sqrt{EG - F^2} \, du \, dv, \tag{5–71}$$

$$E = x_u^2 + y_u^2 + z_u^2, \qquad F = x_u x_v + y_u y_v + z_u z_v, \quad G = x_v^2 + y_v^2 + z_v^2,$$

where the subscripts denote derivatives (Section 4–9). It is assumed here that the correspondence between parameter points (u, v) and surface points (x, y, z) is one-to-one and that $EG - F^2 \neq 0$. Corresponding to (5–71) one has the *surface area element*

$$d\sigma = \sqrt{EG - F^2}\, du\, dv. \tag{5–72}$$

The preceding formulas can be extended to surfaces S which are *piecewise smooth*, i.e., pieced together of a finite number of smooth parts. One can compute the total area by adding the areas of the parts. For surfaces in implicit form: $F(x, y, z) = 0$, one can in general decompose the surface into smooth pieces in each of which a representation $z = z(x, y)$ or $x = x(y, z)$ or $y = y(x, z)$ is available, so that the area can be computed by several applications of the formula (5–69) (cf. Prob. 8 following Section 4–9).

In order to define a surface integral analogous to the line integral $\int X\, dx + Y\, dy + Z\, dz$, it is necessary to assume that the surface S has a "direction assigned." One does not ordinarily consider the possibility of assigning a direction to a surface, but something equivalent to this is not unfamiliar, namely the choice of a positive direction for angles, as in the xy plane. For a general smooth surface S in space, a simple way of achieving this is by choosing a unit normal vector $\mathbf{n}$ at each point of S in such a manner that $\mathbf{n}$ *varies continuously* on S. If this field of normal vectors has been chosen, we term the surface *oriented*. One can then define a positive direction for angles at each point of the surface, as suggested in Fig. 5–26(a).

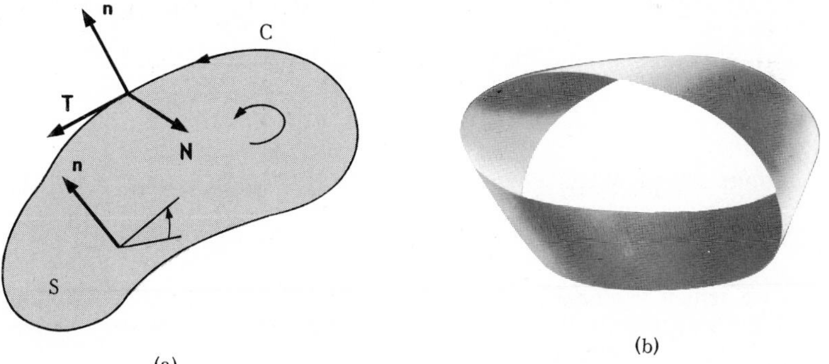

FIG. 5–26. (a) Oriented smooth surface. (b) Möbius strip, a nonorientable surface.

When this has been done, one can also assign positive directions to simple closed curves forming the boundary of S. For each boundary curve C one can choose a tangent $\mathbf{T}$ in the direction chosen and an *inner* normal $\mathbf{N}$, in a plane tangent to S, as in Fig. 5–26(a). The vectors $\mathbf{T}, \mathbf{N}, \mathbf{n}$ must then form a positive triple at each point of C.

If the surface S is only piecewise smooth, as is, for example, the surface of a cylinder, one cannot choose a continuously varying normal vector for all of S. In this case, we say that S has been oriented if an orientation has been chosen in each smooth piece of S, in such a manner that along each curve which is a common boundary of two pieces the positive direction relative to one piece is the opposite of the positive direction relative to the other piece. This is illustrated for a cylindrical surface in Fig. 5–27. If a piecewise smooth surface S can be so oriented, we term the surface S orientable.

It should be remarked that not every surface is orientable. Fig. 5–26(b) suggests a "Möbius strip," which is nonorientable. For this surface S one easily convinces one's self that the normal $\mathbf{n}$ cannot be chosen to vary continuously on S and that the surface has the peculiar property of having only "one side." One-sidedness and nonorientability go together for surfaces in space.

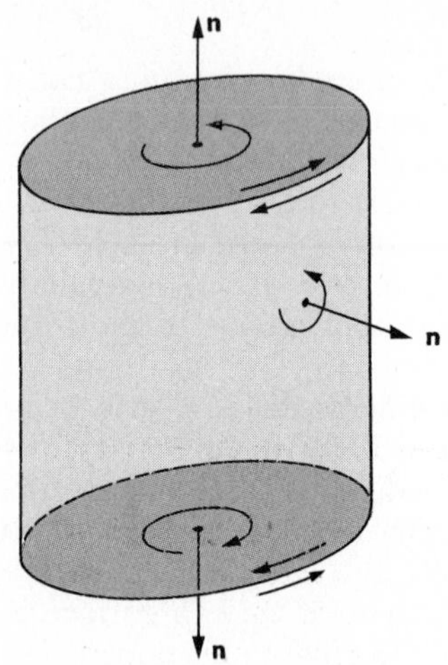

FIG. 5–27. Oriented piecewise smooth surface (cylinder).

For surfaces $z = f(x, y)$ there is no difficulty in this regard, since one can always choose $\mathbf{n}$ as the upper normal or, alternatively, as the lower normal. For surfaces in parametric form there is also no difficulty. It follows from the discussion of Section 4–9 that the vector

$$\mathbf{P}_1 \times \mathbf{P}_2 = (x_u\mathbf{i} + y_u\mathbf{j} + z_u\mathbf{k}) \times (x_v\mathbf{i} + y_v\mathbf{j} + z_v\mathbf{k})$$

is a normal vector having magnitude $\sqrt{EG - F^2}$. One can hence use

$$\mathbf{n} = \frac{\mathbf{P}_1 \times \mathbf{P}_2}{|\mathbf{P}_1 \times \mathbf{P}_2|} \tag{5–73}$$

throughout or the negative of this vector throughout, provided $EG - F^2 \neq 0$. For a surface defined by an implicit equation: $F(x, y, z) = 0$, one can choose $\mathbf{n}$ as $\nabla F/|\nabla F|$ (or its negative), provided $\nabla F \neq \mathbf{0}$.

5–10 Surface integrals. Let S be a smooth surface as above and let $H(x, y, z)$ be a function defined and continuous on S. Then the surface integral of H over S is defined as follows:

$$\iint\limits_S H(x, y, z)\, d\sigma = \lim_{n\to\infty} \sum_{i=1}^{n} H(x_i^*, y_i^*, z_i^*)\, \Delta_i\sigma. \tag{5–74}$$

The surface S is here assumed cut into n pieces as in Fig. 5–25, $\Delta_i\sigma$ denotes the area of the ith piece, and it is assumed that the ith piece shrinks to a point as $n \to \infty$ in an appropriate manner. The subdivision and the limit process are best described in terms of a representation (5–66) or (5–68), in which case they are precisely the same as the procedures used for the double integral in Section 4–6.

If the representation $z = f(x, y)$ is used, the integral is reducible to a double integral in the xy plane:

$$\iint\limits_S H\, d\sigma = \iint\limits_{R_{xy}} H[x, y, f(x, y)] \sec \gamma\, dx\, dy, \tag{5–75}$$

where γ is the angle between the upper normal and the z axis. This follows from the area formula (5–69) above or, more concisely, from the formula $d\sigma = \sec \gamma\, dx\, dy$.

If the parametric representation (5–68) is used, the integral has the expression:

$$\iint\limits_S H\, d\sigma = \iint\limits_{R_{uv}} H[f(u, v), g(u, v), h(u, v)]\sqrt{EG - F^2}\, du\, dv. \tag{5–76}$$

This follows similarly from the expression $d\sigma = \sqrt{EG - F^2}\, du\, dv$ for the area element.

The existence of the surface integral (5–74) and the expressions (5–75) and (5–76) can be established when H is continuous. As pointed out in Section 4–9, the notion of surface area is an awkward one. Once that has been settled and the formula for the area element justified, the proof of (5–75) and (5–76) parallels that for line integrals.

The integral $\iint H\, d\sigma$ is analogous to the line integral $\int F\, ds$. In order to obtain a surface integral analogous to the line integral $\int X\, dx + Y\, dy + Z\, dz$, we consider an oriented smooth surface S, with continuous unit normal vector $\mathbf{n}$:

$$\mathbf{n} = \cos \alpha\mathbf{i} + \cos \beta\mathbf{j} + \cos \gamma\mathbf{k}.$$

Let $L(x, y, z)$, $M(x, y, z)$, $N(x, y, z)$ be functions defined and continuous on S. Then by definition

$$\begin{aligned} \iint\limits_S L\, dy\, dz &= \iint\limits_S L \cos \alpha\, d\sigma, \\ \iint\limits_S M\, dz\, dx &= \iint\limits_S M \cos \beta\, d\sigma, \\ \iint\limits_S N\, dx\, dy &= \iint\limits_S N \cos \gamma\, d\sigma. \end{aligned} \tag{5–77}$$

If these are added, we obtain the general surface integral

$$\iint\limits_S L\, dy\, dz + M\, dz\, dx + N\, dx\, dy = \iint\limits_S (L \cos \alpha + M \cos \beta + N \cos \gamma)\, d\sigma. \tag{5–78}$$

This can at once be written in terms of vectors:

$$\begin{gathered} \iint\limits_S L\, dy\, dz + M\, dz\, dx + N\, dx\, dy = \iint\limits_S (\mathbf{v} \cdot \mathbf{n})\, d\sigma, \\ \mathbf{v} = L\mathbf{i} + M\mathbf{j} + N\mathbf{k}; \end{gathered} \tag{5–79}$$

thus the surface integral $\int\int L\,dy\,dz + M\,dz\,dx + N\,ax\,dy$ is the integral over the surface of the *normal* component of the vector $\mathbf{v} = L\mathbf{i} + M\mathbf{j} + N\mathbf{k}$.

For evaluation of the surface integral, one has the following formulas:

I. *If S is given in the form*

$$z = f(x, y), \quad (x, y) \text{ in } R_{xy},$$

with normal vector $\mathbf{n}$, *then*

$$\iint_S L\,dy\,dz + M\,dz\,dx + N\,dx\,dy = \pm \iint_{R_{xy}} \left(-L\frac{\partial z}{\partial x} - M\frac{\partial z}{\partial y} + N\right) dx\,dy, \tag{5-80}$$

with the + *sign when* $\mathbf{n}$ *is the upper normal and the* − *sign when* $\mathbf{n}$ *is the lower normal.*

II. *If S is given in parametric form*

$$x = f(u, v), \quad y = g(u, v), \quad z = h(u, v), \quad (u, v) \text{ in } R_{uv},$$

with normal vector $\mathbf{n}$, *then*

$$\iint_S L\,dy\,dz + M\,dz\,dx + N\,dx\,dy$$
$$= \pm \iint_{R_{uv}} \left[L\frac{\partial(y, z)}{\partial(u, v)} + M\frac{\partial(z, x)}{\partial(u, v)} + N\frac{\partial(x, y)}{\partial(u, v)}\right] du\,dv, \tag{5-81}$$

with the + *or* − *sign according as*

$$\mathbf{n} = \pm \frac{\mathbf{P}_1 \times \mathbf{P}_2}{|\mathbf{P}_1 \times \mathbf{P}_2|},$$
$$\mathbf{P}_1 = x_u\mathbf{i} + y_u\mathbf{j} + z_u\mathbf{k}, \quad \mathbf{P}_2 = x_v\mathbf{i} + y_v\mathbf{j} + z_v\mathbf{k}. \tag{5-82}$$

The proof of (5–80) is given here, the proof of (5–81) being left as an exercise [Prob. 7(c) below].

For a surface $z = f(x, y)$, the methods of Section 2–12 show that the tangent plane at (x_1, y_1, z_1) is

$$z - z_1 = \frac{\partial z}{\partial x}(x - x_1) + \frac{\partial z}{\partial y}(y - y_1)$$

and hence that the unit normal vector is

$$\mathbf{n} = \pm \frac{-\dfrac{\partial z}{\partial x}\mathbf{i} - \dfrac{\partial z}{\partial y}\mathbf{j} + \mathbf{k}}{\sqrt{1 + \left(\dfrac{\partial z}{\partial x}\right)^2 + \left(\dfrac{\partial z}{\partial y}\right)^2}},$$

with the + or − sign according as $\mathbf{n}$ is upper or lower. The denominator here is $\sec \gamma'$, where γ' is the angle between the *upper* normal and $\mathbf{k}$. Thus

$$\mathbf{n} = \pm \frac{-\dfrac{\partial z}{\partial x}\mathbf{i} - \dfrac{\partial z}{\partial y}\mathbf{j} + \mathbf{k}}{\sec \gamma'}.$$

By (5–79)

$$\begin{aligned}\iint_S L\,dy\,dz + M\,dz\,dx + N\,dx\,dy &= \iint_S [(L\mathbf{i} + M\mathbf{j} + N\mathbf{k}) \cdot \mathbf{n}]\,d\sigma \\ &= \pm \iint_S \frac{-L\dfrac{\partial z}{\partial x} - M\dfrac{\partial z}{\partial y} + N}{\sec \gamma'}\,d\sigma \\ &= \pm \iint_{R_{xy}} \left(-L\frac{\partial z}{\partial x} - M\frac{\partial z}{\partial y} + N\right) dx\,dy,\end{aligned}$$

since $d\sigma = \sec \gamma'\,dx\,dy$. Accordingly, (5–80) is proved.

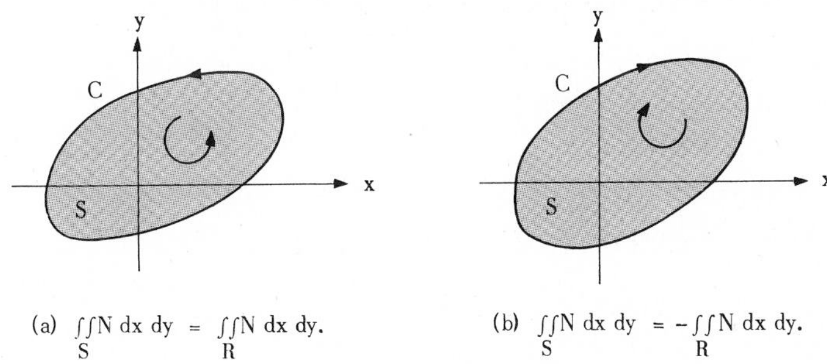

Fig. 5–28. Double integral over R vs. surface integral over S, $S = R$ with orientation.

When $L = 0$ and $M = 0$, (5–80) reduces to the formula

$$\iint_S N(x, y, z)\,dx\,dy = \pm \iint_{R_{xy}} N[x, y, f(x, y)]\,dx\,dy.$$

Thus the surface integral $\iint N\,dx\,dy$ over S is the same as plus or minus the double integral of N over the projection of S on the xy plane, with plus or minus sign according as the chosen normal on S is upper or lower. In particular, S can be chosen as a region R in the xy plane and $N = N(x, y)$. Thus the surface integral $\iint_S N(x, y)\,dx\,dy$ and the double integral $\iint_R N(x, y)\,dx\,dy$ are not the same; they are equal when S has the orientation determined by $\mathbf{k}$, are otherwise negatives of each other, as illustrated in Fig. 5–28. This situation indicates how one can enlarge the concept of double integral to permit a choice of direction of integration. If this is done, Green's theorem can be stated in the form

$$\int_C P\,dx + Q\,dy = \iint_S \left(\frac{\partial Q}{\partial x} - \frac{\partial P}{\partial y}\right) dx\,dy,$$

without a direction of integration indicated on C but with the understanding that the "directions" on C and S match; i.e., that when C is positively directed the normal associated with S is $\mathbf{k}$, but when C is negatively directed the normal associated with S is $-\mathbf{k}$. These two cases are shown in Fig. 5–28.

The surface integral $\iint L\,dy\,dz + \cdots$ can also be defined in a manner paralleling the vector line integral $\int \mathbf{u} \cdot d\mathbf{r}$ of (5–65) above. One introduces a *differential area vector* $d\boldsymbol{\sigma}$ by the equation

$$d\boldsymbol{\sigma} = \mathbf{n}\,d\sigma,$$

so that

$$d\boldsymbol{\sigma} = \cos\alpha\,d\sigma\mathbf{i} + \cos\beta\,d\sigma\mathbf{j} + \cos\gamma\,d\sigma\mathbf{k}$$

and

$$\iint_S L\,dy\,dz + M\,dz\,dx + N\,dx\,dy = \iint_S \mathbf{v}\cdot\mathbf{n}\,d\sigma = \iint_S \mathbf{v}\cdot d\boldsymbol{\sigma}.$$

One can also define this as the limit of a sum, as for the line integral. For the surface $z = f(x, y)$ one finds

$$d\boldsymbol{\sigma} = \pm\left(-\frac{\partial z}{\partial x}\mathbf{i} - \frac{\partial z}{\partial y}\mathbf{j} + \mathbf{k}\right) dx\,dy.$$

This can also be written:

$$d\boldsymbol{\sigma} = \mathbf{n}\,|\sec\gamma|\,dx\,dy = \frac{\mathbf{n}}{|\mathbf{n}\cdot\mathbf{k}|}\,dx\,dy.$$

From this and the corresponding expressions for surfaces $x = g(y, z)$, $y = h(x, z)$, one obtains the formulas

$$\iint_S (\mathbf{v}\cdot\mathbf{n})\,d\sigma = \iint_{R_{xy}} \frac{\mathbf{v}\cdot\mathbf{n}}{|\mathbf{n}\cdot\mathbf{k}|}\,dx\,dy = \iint_{R_{yz}} \frac{\mathbf{v}\cdot\mathbf{n}}{|\mathbf{n}\cdot\mathbf{i}|}\,dy\,dz = \iint_{R_{zx}} \frac{\mathbf{v}\cdot\mathbf{n}}{|\mathbf{n}\cdot\mathbf{j}|}\,dz\,dx. \tag{5–83}$$

For the surfaces in parametric form we have

$$d\boldsymbol{\sigma} = \pm\left(\frac{\partial(y, z)}{\partial(u, v)}\mathbf{i} + \frac{\partial(z, x)}{\partial(u, v)}\mathbf{j} + \frac{\partial(x, y)}{\partial(u, v)}\mathbf{k}\right) du\,dv$$

or, more concisely, by (5–82)

$$d\boldsymbol{\sigma} = \pm(\mathbf{P}_1 \times \mathbf{P}_2)\,du\,dv.$$

The formal properties of line and surface integrals are analogous to those for line integrals in the plane (Section 5–3) and need no special discussion here. Furthermore, the definitions and properties of line and surface integrals carry over without change to piecewise smooth curves and surfaces, provided the surfaces are orientable.

Problems

1. Evaluate the line integrals:

(a) $\displaystyle\int_{C\,(1,0,0)}^{(1,0,2\pi)} z\,dx + x\,dy + y\,dz$, where C is the curve: $x = \cos t,\ y = \sin t,$ $z = t,\ 0 \leqq t \leqq 2\pi$;

(b) $\displaystyle\int_{(1,0,1)}^{(2,3,2)} x^2\,dx - xz\,dy + y^2\,dz$ on the straight line joining the two points;

(c) $\displaystyle\int_{(1,1,0)}^{(0,0,\sqrt{2})} x^2yz\,ds$ on the curve: $x = \cos t,\ y = \cos t,\ z = \sqrt{2}\sin t,\ 0 \leqq t \leqq \dfrac{\pi}{2}$;

(d) $\displaystyle\int_C u_T\,ds$, where $\mathbf{u} = 2xy^2z\mathbf{i} + 2x^2yz\mathbf{j} + x^2y^2\mathbf{k}$ and C is the circle: $x^2 + y^2 = 1,$ $z = 2$.

2. If $\mathbf{u} = \operatorname{grad} F$ in a domain D, then show that

(a) $\displaystyle\int_{(x_1,y_1,z_1)}^{(x_2,y_2,z_2)} u_T\,ds = F(x_2, y_2, z_2) - F(x_1, y_1, z_1)$, where the integral is along any path in D joining the two points;

(b) $\displaystyle\int_C u_T\,ds = 0$ on any closed path in D.

3. Let a wire be given as a curve C in space. Let its density (mass per unit length) be $\delta = \delta(x, y, z)$, where (x, y, z) is a variable point in C. Justify the following formulas:

(a) length of wire $\displaystyle = \int_C ds = L$;

(b) mass of wire $\displaystyle = \int_C \delta\,ds = M$;

(c) center of mass of the wire is $(\bar{x}, \bar{y}, \bar{z})$, where

$$M\bar{x} = \int_C x\,\delta\,ds,\quad M\bar{y} = \int_C y\,\delta\,ds,\quad M\bar{z} = \int_C z\,\delta\,ds;$$

(d) moment of inertia of the wire about the z axis is

$$I_z = \int_C (x^2 + y^2)\,\delta\,ds.$$

4. Formulate and justify the formulas analogous to those of Prob. 3 for the surface area, mass, center of mass, and moment of inertia of a thin curved sheet of metal forming a surface S in space.

5. Evaluate the following surface integrals:

(a) $\iint\limits_S x\,dy\,dz + y\,dz\,dx + z\,dx\,dy$, where S is the triangle with vertices $(1, 0, 0)$, $(0, 1, 0)$, $(0, 0, 1)$ and the normal points away from $(0, 0, 0)$;

(b) $\iint\limits_S dy\,dz + dz\,dx + dx\,dy$, where S is the hemisphere: $z = \sqrt{1 - x^2 - y^2}$, $x^2 + y^2 \leqq 1$, and the normal is the upper normal;

(c) $\iint\limits_S (x \cos\alpha + y \cos\beta + z \cos\gamma)\,d\sigma$ for the surface of part (b);

(d) $\iint\limits_S x^2 z\,d\sigma$, where S is the cylindrical surface: $x^2 + y^2 = 1$, $0 \leqq z \leqq 1$.

6. Evaluate the surface integrals of Prob. 5, using the parametric representations:

(a) $x = u + v$, $y = u - v$, $z = 1 - 2u$;
(b) $x = \sin u \cos v$, $y = \sin u \sin v$, $z = \cos u$;
(c) same as (b);
(d) $x = \cos u$, $y = \sin u$, $z = v$.

7. (a) Let a surface $S: z = f(x, y)$ be defined by an implicit equation: $F(x, y, z) = 0$. Show that the surface integral $\iint H\,d\sigma$ over S becomes

$$\iint\limits_{R_{xy}} \sqrt{\left(\frac{\partial F}{\partial x}\right)^2 + \left(\frac{\partial F}{\partial y}\right)^2 + \left(\frac{\partial F}{\partial z}\right)^2}\,\frac{H}{\left|\dfrac{\partial F}{\partial z}\right|}\,dx\,dy,$$

provided $\dfrac{\partial F}{\partial z} \neq 0$.

(b) Prove that, for the surface of part (a) with $\mathbf{n} = \nabla F/|\nabla F|$,

$$\iint\limits_S (\mathbf{v}\cdot\mathbf{n})\,d\sigma = \iint\limits_{R_{xy}} (\mathbf{v}\cdot\nabla F)\,\frac{1}{\left|\dfrac{\partial F}{\partial z}\right|}\,dx\,dy.$$

(c) Prove (5–81).
(d) Prove that (5–81) reduces to (5–80) when $x = u$, $y = v$, $z = f(u, v)$.

8. Let S be an oriented surface in space which is planar; i.e., S lies in a plane. With S one can associate the vector $\mathbf{S}$ which has the direction of the normal chosen on S and has a length equal to the area of S.

(a) Show that if S_1, S_2, S_3, S_4 are the faces of a tetrahedron, oriented so that the normal is the exterior normal, then

$$\mathbf{S}_1 + \mathbf{S}_2 + \mathbf{S}_3 + \mathbf{S}_4 = 0.$$

(Hint: introduce vectors **a**, **b**, **c** along three concurrent edges of the tetrahedron and express $\mathbf{S}_1$, $\mathbf{S}_2$, $\mathbf{S}_3$, $\mathbf{S}_4$ in terms of **a**, **b**, and **c**.)

(b) Show that the result of (a) extends to an arbitrary convex polyhedron with faces $\mathbf{S}_1, \ldots, \mathbf{S}_n$, i.e., that

$$\mathbf{S}_1 + \mathbf{S}_2 + \cdots + \mathbf{S}_n = \mathbf{0},$$

when the orientation is that of the exterior normal.

(c) Using the result of (b), indicate a reasoning to justify the relation

$$\iint_S \mathbf{v} \cdot d\boldsymbol{\sigma} = 0$$

for any convex closed surface S (such as the surface of a sphere or ellipsoid), provided **v** is a constant vector.

(d) Apply the result of (b) to a triangular prism whose edges represent the vectors **a**, **b**, **a** + **b**, **c** to prove the *distributive law* (Equation (0–74))

$$\mathbf{c} \times (\mathbf{a} + \mathbf{b}) = \mathbf{c} \times \mathbf{a} + \mathbf{c} \times \mathbf{b}$$

for the vector product. This is the method used by Gibbs (cf. the book of Gibbs listed at the end of Chapter 3).

Answers

1. (a) 3π, (b) $-\frac{5}{3}$, (c) $\frac{1}{2}$, (d) 0. 5. (a) $\frac{1}{2}$, (b) π, (c) 2π, (d) $\frac{\pi}{2}$.

5–11 The divergence theorem. It was pointed out in Section 5–5 above that Green's theorem can be written in the form

$$\int_C v_n \, ds = \iint_R \operatorname{div} \mathbf{v} \, dx \, dy.$$

The following generalization thus appears natural:

$$\iint_S v_n \, d\sigma = \iiint_R \operatorname{div} \mathbf{v} \, dx \, dy \, dz,$$

where S is a surface forming the complete boundary of a bounded closed region R in space and **n** is the outer normal of S, i.e., the one pointing away from R. The theorem, known as the *divergence theorem* or *Gauss's theorem*, plays a role like that of Green's theorem for line integrals, in showing that certain surface integrals are 0 and that others are "independent of path." The theorem has a basic physical significance, which will be considered below. If one considers a continuous flow of matter in space, the left-hand side can be interpreted as the flux across the boundary S, i.e., the total mass leaving R per unit time; the right-hand side measures the rate at which the density is decreasing throughout R, i.e., the total loss of mass per unit time. Under the assumptions made, the only way mass can be lost is by crossing the boundary S. For this reason the flux equals the total divergence.

DIVERGENCE THEOREM (GAUSS'S THEOREM). *Let* $\mathbf{v} = L\mathbf{i} + M\mathbf{j} + N\mathbf{k}$ *be a vector field in a domain D of space; let L, M, N be continuous and have continuous derivatives in D. Let S be a piecewise smooth surface in D which forms the complete boundary of a bounded closed region R in D. Let* $\mathbf{n}$ *be the outer normal of S with respect to R. Then*

$$\iint_S v_n \, d\sigma = \iiint_R \operatorname{div} \mathbf{v} \, dx \, dy \, dz; \tag{5–84}$$

that is,

$$\iint_S L \, dy \, dz + M \, dz \, dx + N \, dx \, dy = \iiint_R \left(\frac{\partial L}{\partial x} + \frac{\partial M}{\partial y} + \frac{\partial N}{\partial z}\right) dx \, dy \, dz. \tag{5–85}$$

Proof. It will be proved that

$$\iint_S N \, dx \, dy = \iiint_R \frac{\partial N}{\partial z} \, dx \, dy \, dz, \tag{5–86}$$

the proofs of the two equations:

$$\iint_S L \, dy \, dz = \iiint_R \frac{\partial L}{\partial x} \, dx \, dy \, dz, \tag{5–87}$$

$$\iint_S M \, dz \, dx = \iiint_R \frac{\partial M}{\partial y} \, dx \, dy \, dz \tag{5–88}$$

being exactly the same.

It should first be remarked that the normal $\mathbf{n}$, defined as the outer normal of S with respect to R, necessarily varies continuously on each smooth part of S, so that the surface integral in (5–85) is well-defined. The orientation defined by $\mathbf{n}$ on each part of S in fact determines an orientation of all of S, so that S must be orientable.

We assume now that R is representable in the form:

$$f_1(x, y) \leqq z \leqq f_2(x, y), \quad (x, y) \text{ in } R_{xy}, \tag{5–89}$$

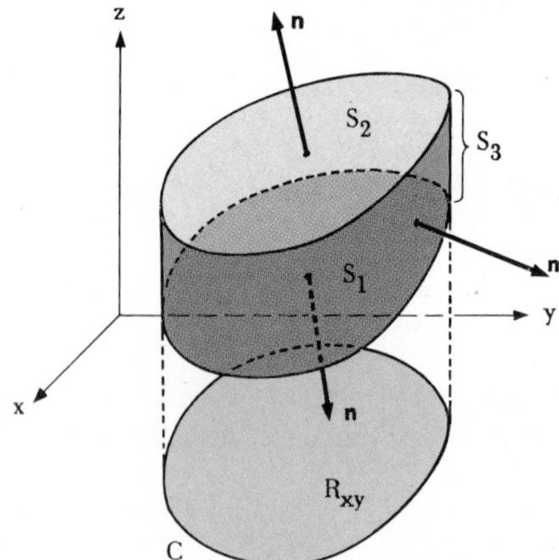

FIG. 5–29.

where R_{xy} is a bounded closed region in the xy plane (cf. Fig. 5–29) bounded by a simple closed curve C. The surface S is then composed of three parts:

$$S_1\colon z = f_1(x, y), \quad (x, y) \text{ in } R_{xy},$$
$$S_2\colon z = f_2(x, y), \quad (x, y) \text{ in } R_{xy},$$
$$S_3\colon f_1(x, y) \leqq z \leqq f_2(x, y), \quad (x, y) \text{ on } C.$$

S_2 forms the "top" of S, S_1 the "bottom," S_3 the "sides" (the portion S_3 may degenerate into a curve, as for a sphere). Now by definition (5–77) above

$$\iint\limits_S N\,dx\,dy = \iint\limits_S N \cos\gamma\,d\sigma,$$

where $\gamma = \measuredangle(\mathbf{n}, \mathbf{k})$. Along S_3, $\gamma = \pi/2$, so that $\cos\gamma = 0$; along S_2, $\gamma = \gamma'$ where γ' is the angle between the *upper* normal and $\mathbf{k}$; along S_1, $\gamma = \pi - \gamma'$. Since $d\sigma = \sec\gamma'\,dx\,dy$ on S_1 and S_2, one has

$$\begin{aligned}\iint\limits_S N\,dx\,dy &= \iint\limits_{S_1} N\,dx\,dy + \iint\limits_{S_2} N\,dx\,dy + \iint\limits_{S_3} N\,dx\,dy \\ &= -\iint\limits_{R_{xy}} N\cos\gamma'\sec\gamma'\,dx\,dy + \iint\limits_{R_{xy}} N\cos\gamma'\sec\gamma'\,dx\,dy,\end{aligned}$$

where $z = f_1(x, y)$ in the first integral and $z = f_2(x, y)$ in the second:

$$\iint\limits_S N\,dx\,dy = \iint\limits_{R_{xy}} \{N[x, y, f_2(x, y)] - N[x, y, f_1(x, y)]\}\,dx\,dy. \tag{5–90}$$

Now the triple integral on the right of (5–86) can be evaluated as follows:

$$\begin{aligned}\iiint\limits_R \frac{\partial N}{\partial z}\,dx\,dy\,dz &= \iint\limits_{R_{xy}} \left[\int_{f_1(x,y)}^{f_2(x,y)} \frac{\partial N}{\partial z}\,dz\right] dx\,dy \\ &= \iint\limits_{R_{xy}} \{N[x, y, f_2(x, y)] - N[x, y, f_1(x, y)]\}\,dx\,dy.\end{aligned}$$

This is the same as (5–90); hence

$$\iint\limits_S N\,dx\,dy = \iiint\limits_R \frac{\partial N}{\partial z}\,dx\,dy\,dz.$$

This is the result desired for the particular R assumed. For any region R which can be cut up into a finite number of pieces of this type by means of piecewise smooth auxiliary surfaces, the theorem then follows by adding the result for each part separately (cf. the proof of Green's theorem in Section 5–5). The surface integrals over the auxiliary surfaces cancel in pairs and the sum of the surface integrals is precisely that over the complete boundary S of R; the volume integrals over the pieces add up to that over R. The result can finally be extended to the most general R envisaged in the theorem by a limit process [see O. D. Kellogg, *Foundations of Potential Theory* (Berlin: Springer, 1929)].

Once (5–86) has been established, (5–87) and (5–88) then follow by merely relabeling the variables. Adding these three equations, one obtains Eq. (5–85).

As a first application of the divergence theorem, a new interpretation of the divergence of a vector field $\mathbf{v}$ will be given. Let S_r be a sphere with center (x_1, y_1, z_1) and radius r. Let R_r denote S_r plus interior. We recall that the law of the mean for integrals asserts that

$$\iiint\limits_R f(x, y, z)\,dx\,dy\,dz = f(x^*, y^*, z^*) \cdot V,$$

where (x^*, y^*, z^*) is a point of R and V is the volume of R, provided f is continuous in R (cf. the formulation for double integrals in Section 4–6). If this is applied to the integral

$$\iiint\limits_{R_r} \operatorname{div} \mathbf{v}\,dx\,dy\,dz,$$

one concludes that

$$\iiint\limits_{R_r} \operatorname{div} \mathbf{v}\,dx\,dy\,dz = \operatorname{div} \mathbf{v}(x^*, y^*, z^*) \cdot V_r,$$

for some (x^*, y^*, z^*) in R_r, V_r denoting the volume $(\frac{4}{3}\pi r^3)$ of R_r. If the divergence theorem is applied, one concludes that

$$\operatorname{div} \mathbf{v}(x^*, y^*, z^*) = \frac{1}{V_r} \iint\limits_{S_r} v_n\,d\sigma.$$

If now we let r approach 0, the point (x^*, y^*, z^*) must approach the center (x_1, y_1, z_1). Thus

$$\operatorname{div} \mathbf{v}(x_1, y_1, z_1) = \lim_{r \to 0} \frac{1}{V_r} \iint\limits_{S_r} v_n\,d\sigma. \tag{5–91}$$

The surface integral $\iint v_n\,d\sigma$ can be interpreted as *the flux* of the vector field $\mathbf{v}$ across S_r. Hence the *divergence of a vector field at a point equals the limiting value of the flux across a sphere about the point divided by the volume of the sphere, as the radius of the sphere approaches* 0. The spherical surface and solid used here can be replaced by a more general one, provided the "shrinking to zero" takes place in a suitable manner. Thus, in concise form, divergence equals flux per unit volume.

This result is of fundamental significance, for it assigns a meaning to the divergence which is independent of any coordinate system. It also provides considerable insight into the meaning of positive, negative, and zero divergence in a physical problem. One can use (5–91) as the *definition* of divergence and this is often done; one must then verify, with the aid of the divergence theorem in the form (5–85), that one obtains the usual formula

$$\frac{\partial v_x}{\partial x} + \frac{\partial v_y}{\partial y} + \frac{\partial v_z}{\partial z}$$

in each coordinate system.

One can apply (5–91) to the case when $\mathbf{v} = \mathbf{u}$, where $\mathbf{u}$ is the velocity vector of a fluid motion. The expression $u_n\, d\sigma$ can be interpreted as the volume filled up per unit time by the fluid crossing the surface element $d\sigma$ in the direction of $\mathbf{n}$ (see Fig. 5–30). The flux integral $\iint u_n\, d\sigma$ over the surface S_r then measures the total rate of filling volume (outgoing minus incoming volume per unit time). By (5–91), div $\mathbf{u}$ then measures the limiting value of this rate, per unit volume, at a point (x_1, y_1, z_1). If the fluid is *incompressible*, the outgoing volume precisely equals the incoming volume, so that the flux across S_r is always 0. By (5–91), this implies that div $\mathbf{u} = 0$; conversely, by (5–84), if div $\mathbf{u} = 0$ in D, then the flow is incompressible. Another interpretation of div $\mathbf{u}$ is given in Prob. 6 at the end of this chapter.

One can also let $\mathbf{v} = \rho\mathbf{u}$, where ρ is the density and $\mathbf{u}$ is the velocity of the fluid motion; the flux integral

$$\iint_S v_n\, d\sigma = \iint_S \rho u_n\, d\sigma$$

then measures the rate at which *mass* is leaving the solid R via the surface S. For $u_n\, d\sigma$ measures the volume of fluid crossing a surface element per unit of time and $\rho u_n\, d\sigma$ measures its mass. The divergence div $\mathbf{v}$ measures this rate of loss of mass per unit volume, by (5–91). But this is precisely the rate at which the density is *decreasing* at the point (x_1, y_1, z_1). Hence

$$\frac{\partial \rho}{\partial t} = -\text{div}\, \mathbf{v} = -\text{div}\, (\rho \mathbf{u})$$

or

$$\frac{\partial \rho}{\partial t} + \text{div}\, (\rho \mathbf{u}) = 0. \qquad (5\text{–}92)$$

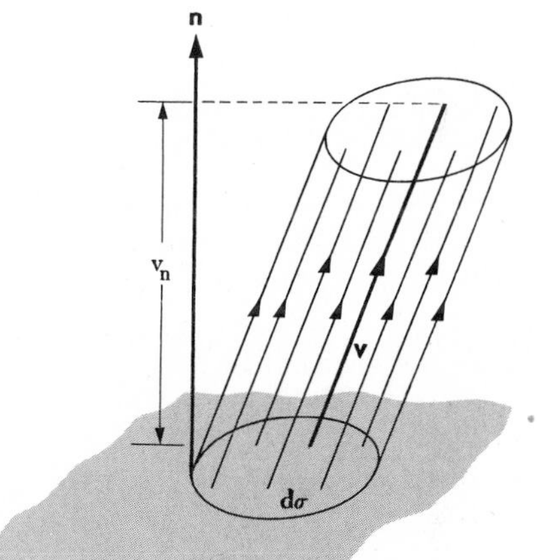

FIG. 5–30. Flux $= \iint v_n\, d\sigma$.

This is the *continuity equation* of hydrodynamics. It expresses the *conservation of mass*. Another derivation is given in Prob. 8 at the end of this chapter.

Problems

1. Evaluate by the divergence theorem:

(a) $\displaystyle\iint_S x\, dy\, dz + y\, dz\, dx + z\, dx\, dy$, where S is the sphere $x^2 + y^2 + z^2 = 1$ and $\mathbf{n}$ is the outer normal;

(b) $\displaystyle\iint_S v_n\, d\sigma$, where $\mathbf{v} = x^2\mathbf{i} + y^2\mathbf{j} + z^2\mathbf{k}$, $\mathbf{n}$ is the outer normal, and S is the surface of the cube: $0 \leqq x \leqq 1$, $0 \leqq y \leqq 1$, $0 \leqq z \leqq 1$.

2. Let S be the boundary surface of a region R in space and let $\mathbf{n}$ be its outer normal. Prove the formulas:

(a) $$V = \iint\limits_S x\,dy\,dz = \iint\limits_S y\,dz\,dx = \iint\limits_S z\,dx\,dy$$
$$= \tfrac{1}{3}\iint\limits_S x\,dy\,dz + y\,dz\,dx + z\,dx\,dy,$$

where V is the volume of R;

(b) $$\iint\limits_S x^2\,dy\,dz + 2xy\,dz\,dx + 2xz\,dx\,dy = 6V\bar{x},$$

where $(\bar{x}, \bar{y}, \bar{z})$ is the centroid of R;

(c) $$\iint\limits_S \operatorname{curl} \mathbf{v} \cdot \mathbf{n}\,d\sigma = 0,$$ where $\mathbf{v}$ is an arbitrary vector field.

3. Let S be the boundary surface of a region R, with outer normal $\mathbf{n}$, as in the divergence theorem above. Let $f(x, y, z)$ and $g(x, y, z)$ be functions defined and continuous, with continuous first and second derivatives, in a domain D containing R. Prove the following relations:

(a) $$\iint\limits_S f\frac{\partial g}{\partial n}\,d\sigma = \iiint\limits_R f\,\nabla^2 g\,dx\,dy\,dz + \iiint\limits_R (\nabla f \cdot \nabla g)\,dx\,dy\,dz;$$

[Hint: use the identity $\nabla \cdot (f\mathbf{u}) = \nabla f \cdot \mathbf{u} + f(\nabla \cdot \mathbf{u})$.]

(b) if g is harmonic in D, then

$$\iint\limits_S \frac{\partial g}{\partial n}\,d\sigma = 0;$$

[Hint: put $f = 1$ in (a).]

(c) if f is harmonic in D, then

$$\iint\limits_S f\frac{\partial f}{\partial n}\,d\sigma = \iiint\limits_R |\nabla f|^2\,dx\,dy\,dz;$$

(d) if f is harmonic in D and $f \equiv 0$ on S, then $f \equiv 0$ in R [cf. the last paragraph of Section 4–6);

(e) if f and g are harmonic in D and $f \equiv g$ on S, then $f \equiv g$ in R; [Hint: use (d).]

(f) if f is harmonic in D and $\partial f/\partial n = 0$ on S, then f is constant in R;

(g) if f and g are harmonic in D and $\partial f/\partial n = \partial g/\partial n$ on S, then $f = g +$ constant in R;

(h) if f and g are harmonic in R, and

$$\frac{\partial f}{\partial n} = -f + h, \quad \frac{\partial g}{\partial n} = -g + h \text{ on } S, \quad h = h(x, y, z),$$

then

$$f \equiv g \text{ in } R;$$

(i) if f and g both satisfy the same *Poisson equation* in R,

$$\nabla^2 f = -4\pi h, \quad \nabla^2 g = -4\pi h, \quad h = h(x, y, z),$$

and $f = g$ on S, then

$$f \equiv g \text{ in } R;$$

(j) $$\iint\limits_S \left(f\frac{\partial g}{\partial n} - g\frac{\partial f}{\partial n}\right) d\sigma = \iiint\limits_R (f\nabla^2 g - g\,\nabla^2 f)\, dx\, dy\, dz;$$

[Hint: use (a).]

(k) if f and g are harmonic in R, then

$$\iint\limits_S \left(f\frac{\partial g}{\partial n} - g\frac{\partial f}{\partial n}\right) d\sigma = 0;$$

(l) if f and g satisfy the equations:

$$\nabla^2 f = hf, \quad \nabla^2 g = hg, \quad h = h(x, y, z),$$

in R, then

$$\iint\limits_S \left(f\frac{\partial g}{\partial n} - g\frac{\partial f}{\partial n}\right) d\sigma = 0.$$

Remark. Parts (a) and (j) are known as *Green's first and second identities*, respectively.

Answers

1. (a) 4π, (b) 3.

5–12 Stokes' theorem. It was seen in Section 5–5 above that Green's theorem can be written in the form

$$\oint\limits_C u_T\, ds = \iint\limits_R \text{curl}_z\, \mathbf{u}\, dx\, dy.$$

This suggests that, for any simple closed plane curve C in space (Fig. 5–31),

$$\int\limits_C u_T\, ds = \iint\limits_S \text{curl}_n\, \mathbf{u}\, d\sigma, \qquad (5\text{–}93)$$

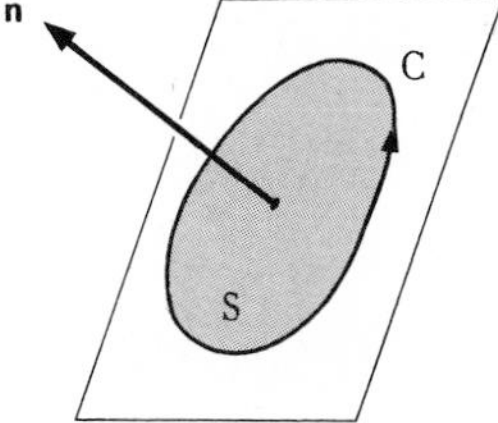

FIG. 5–31.

where $\mathbf{n}$ is normal to the plane in which C lies, S is the planar surface bounded by C, and the direction of C is positive in terms of the orientation of S determined by $\mathbf{n}$. It was pointed out in Section 5–9 above that choice of a continuously varying normal vector on a surface determines a positive sense for each simple closed curve C, when regarded as bounding a portion of the surface.

Equation (5–93) can be generalized further, namely by replacing S by an arbitrary smooth orientable surface whose boundary is a simple closed curve C, not necessarily a plane curve. Again choice of a continuous normal determines the direction on C, as in Fig. 5–32. It should be stressed that this relation between $\mathbf{n}$ and the direction of C, as defined in

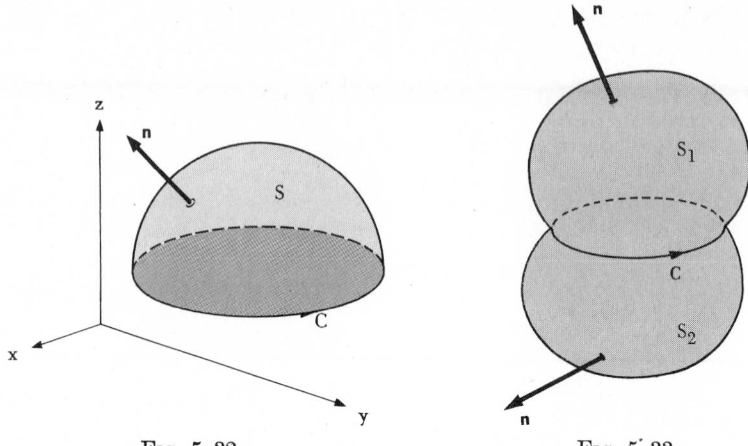

FIG. 5–32. FIG. 5–33.

Section 5–9, depends on the notion of *positive triple* of vectors, hence on a particular orientation of space (Section 0–6). If the orientation were reversed, the given normal $\mathbf{n}$ would correspond to the opposite direction on C.

Equation (5–93) is Stokes' theorem, which will be shown to be correct under very general assumptions. An additional reason for expecting such a relation as this is the fact that if two surfaces S_1 and S_2 have the common boundary C and no other points in common, as in Fig. 5–33, then

$$\iint_{S_1} \operatorname{curl}_n \mathbf{u}\, d\sigma = \iint_{S_2} \operatorname{curl}_n \mathbf{u}\, d\sigma, \tag{5–94}$$

provided S_1 and S_2 are oriented so as to determine the same direction on C. For

$$\iint_{S_1} \operatorname{curl}_n \mathbf{u}\, d\sigma - \iint_{S_2} \operatorname{curl}_n \mathbf{u}\, d\sigma = \iint_{S} \operatorname{curl}_n \mathbf{u}\, d\sigma,$$

where S is the oriented surface formed of S_1 and of S_2 with its normal reversed. S then bounds a certain solid region R and by the divergence theorem

$$\iint_{S} \operatorname{curl}_n \mathbf{u}\, d\sigma = \pm \iiint_{R} \operatorname{div} \operatorname{curl} \mathbf{u}\, dx\, dy\, dz = 0.$$

(The + or − depends on whether $\mathbf{n}$ is outer or inner normal.) Thus the surface integral $\iint \operatorname{curl}_n \mathbf{u}\, dA$ has the same value for all surfaces with boundary C and it is natural to expect the surface integrals to be expressible in terms of a line integral of $\mathbf{u}$ on C.

THEOREM OF STOKES. *Let S be a piecewise smooth oriented surface in space, whose boundary C is a piecewise smooth simple closed curve, directed in accordance with the given orientation in S. Let* $\mathbf{u} = L\mathbf{i} + M\mathbf{j} + N\mathbf{k}$

be a vector field, with continuous and differentiable components, in a domain D of space including S. Then

$$\int_C u_T\, ds = \iint_S (\text{curl}\, \mathbf{u} \cdot \mathbf{n})\, d\sigma, \tag{5–95}$$

where **n** *is the chosen unit normal vector on S; that is,*

$$\int_C L\, dx + M\, dy + N\, dz = \iint_S \left(\frac{\partial N}{\partial y} - \frac{\partial M}{\partial z}\right) dy\, dz + \left(\frac{\partial L}{\partial z} - \frac{\partial N}{\partial x}\right) dz\, dx + \left(\frac{\partial M}{\partial x} - \frac{\partial L}{\partial y}\right) dx\, dy. \tag{5–96}$$

Proof. Just as with the divergence theorem, it is sufficient to prove three separate equations:

$$\int_C L\, dx = \iint_S \frac{\partial L}{\partial z}\, dz\, dx - \frac{\partial L}{\partial y}\, dx\, dy, \quad \int M\, dy = \cdots, \cdots.$$

One can further restrict attention to the case when S is representable in the form

$$z = f(x, y), \quad (x, y) \text{ in } R_{xy},$$

as shown in Fig. 5–34; as in the proof of the divergence theorem, the general case is handled by decomposing into a finite number of such pieces, when possible, and by a limit process.

If S has the form $z = f(x, y)$, the curve C has as projection on the xy plane a curve C_{xy} as shown in Fig. 5–34; as (x, y, z) goes around C once in the given direction, its projection $(x, y, 0)$ goes once around C_{xy} in a corresponding direction. If the normal **n** chosen on S is the upper normal, the direction on C_{xy} is the positive direction and, by Green's theorem,

$$\int_C L(x, y, z)\, dx = \int_{C_{xy}} L[x, y, f(x, y)]\, dx = -\iint_{R_{xy}} \left(\frac{\partial L}{\partial y} + \frac{\partial L}{\partial z}\frac{\partial f}{\partial y}\right) dx\, dy.$$

Under the same assumption about the normal,

$$\iint_S \frac{\partial L}{\partial z}\, dz\, dx - \frac{\partial L}{\partial y}\, dx\, dy = \iint_{R_{xy}} \left(-\frac{\partial L}{\partial z}\frac{\partial f}{\partial y} - \frac{\partial L}{\partial y}\right) dx\, dy,$$

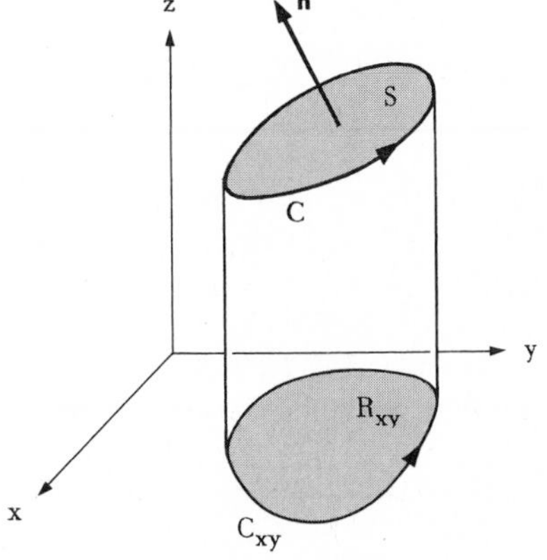

Fig. 5–34. Proof of Stokes' theorem.

by (5–80) above. It at once follows that

$$\int_C L\,dx = \iint_S \frac{\partial L}{\partial z}\,dz\,dx - \frac{\partial L}{\partial y}\,dx\,dy. \tag{5-97}$$

If the direction of $\mathbf{n}$ is reversed, both sides change sign, so that (5–97) holds in general. By the reasoning described above, one then finds that (5–97) holds for a general orientable S. In the same way, equations analogous to (5–97) are established for M and N for a general S. Upon adding the equations for L, M and N, one obtains Stokes' theorem in full generality.

Just as the divergence theorem gives a new interpretation for the divergence of a vector, so does the Stokes theorem give a new interpretation for the curl of a vector. To obtain this, we take S_r to be a circular disk in space of radius r and center (x_1, y_1, z_1) bounded by the circle C_r (see Fig. 5–35). By the Stokes theorem and the law of the mean for integrals

$$\int_{C_r} u_T\,ds = \iint_S \operatorname{curl}_n \mathbf{u}\,d\sigma = \operatorname{curl}_n \mathbf{u}(x^*, y^*, z^*) \cdot A_r,$$

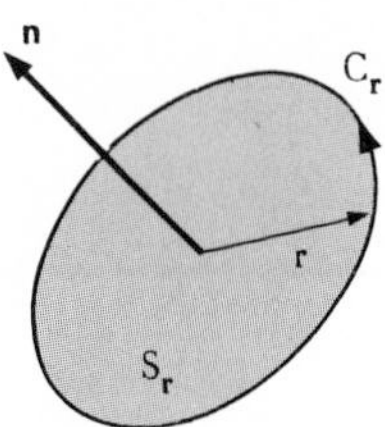

FIG. 5–35.

where A_r is the area (πr^2) of S_r and (x^*, y^*, z^*) is a suitably chosen point of S_r. One can now write

$$\operatorname{curl}_n \mathbf{u}(x^*, y^*, z^*) = \frac{1}{A_r}\int_{C_r} u_T\,ds.$$

In the case of a fluid motion with velocity $\mathbf{u}$, the integral $\int_{C_r} u_T\,ds$ is termed the *circulation* around C_r; it measures the extent to which the corresponding fluid motion is a rotation around the circle C_r in the given direction. If now r is allowed to approach 0, we find

$$\operatorname{curl}_n \mathbf{u}(x_1, y_1, z_1) = \lim_{r\to 0} \frac{1}{A_r}\int_{C_r} u_T\,ds; \tag{5-98}$$

that is, the component of curl $\mathbf{u}$ at (x_1, y_1, z_1) in the direction of $\mathbf{n}$ is the limiting ratio of circulation to area for a circle about (x_1, y_1, z_1) with $\mathbf{n}$ as normal. Briefly, curl equals circulation per unit area. In the limit process, the circular disk can be replaced by a more general surface with normal $\mathbf{n}$, provided the shrinking to zero is properly carried out. If $\mathbf{n}$ is taken as $\mathbf{i}$, $\mathbf{j}$, and $\mathbf{k}$ successively, one obtains the three components of curl $\mathbf{u}$ along the axes.

Since (5–98) has a significance independent of the coordinate system chosen, this equation proves that the curl of a vector field has a meaning independent of the particular (right-handed) coordinate system chosen in space. One could in fact use (5–98) to define the curl. [If the orientation of space is reversed, the direction of the curl will also be reversed (cf. also Section 3–8).]

5–13 Integrals independent of path. Irrotational fields and solenoidal fields. Since the generalization of Green's theorem to space takes two different forms, the divergence theorem and Stokes' theorem, one can generalize the discussions of Sections 5–6 and 5–7 above in two directions: to surface integrals and to line integrals. In the case of line integrals, the results for two dimensions carry over with minor modifications. The surface integrals require a somewhat different treatment.

Line integrals independent of path in space are defined just as in the plane. The following theorems then hold.

THEOREM I. *Let* $\mathbf{u} = X\mathbf{i} + Y\mathbf{j} + Z\mathbf{k}$ *be a vector field with continuous components in a domain D of space. The line integral*

$$\int u_T\,ds = \int X\,dx + Y\,dy + Z\,dz$$

is independent of path in D if and only if there is a function $F(x, y, z)$, *defined in D such that*

$$\frac{\partial F}{\partial x} = X, \quad \frac{\partial F}{\partial y} = Y, \quad \frac{\partial F}{\partial z} = Z$$

throughout D. In other words, the line integral is independent of path if and only if $\mathbf{u}$ *is a gradient vector:*

$$\mathbf{u} = \operatorname{grad} F.$$

The proof for two dimensions in Section 5–6 can be repeated without essential change. When the integral is independent of path, $X\,dx + Y\,dy + Z\,dz = dF$ for some F and

$$\int_A^B X\,dx + Y\,dy + Z\,dz = \int_A^B dF = F(B) - F(A)$$

as in the plane.

THEOREM II. *Let X, Y, Z be continuous in a domain D of space. The line integral*

$$\int X\,dx + Y\,dy + Z\,dz$$

is independent of path in D if and only if

$$\oint_C X\,dx + Y\,dy + Z\,dz = 0$$

on every simple closed curve C in D.

This is proved as in the plane.

THEOREM III. *Let* $\mathbf{u} = X\mathbf{i} + Y\mathbf{j} + Z\mathbf{k}$ *be a vector field in a domain* D *of space; let* X, Y, *and* Z *have continuous partial derivatives in* D. *If*

$$\int u_T\, ds = \int X\, dx + Y\, dy + Z\, dz$$

is independent of path in D, *then* $\operatorname{curl} \mathbf{u} \equiv \mathbf{0}$ *in* D; *that is,*

$$\frac{\partial Z}{\partial y} \equiv \frac{\partial Y}{\partial z}, \quad \frac{\partial X}{\partial z} \equiv \frac{\partial Z}{\partial x}, \quad \frac{\partial Y}{\partial x} \equiv \frac{\partial X}{\partial y}. \tag{5–99}$$

Conversely, if D *is simply connected and* (5–99) *holds, then* $\int X\, dx + Y\, dy + Z\, dz$ *is independent of the path in* D; *that is, if* D *is simply connected and* $\operatorname{curl} \mathbf{u} \equiv \mathbf{0}$ *in* D, *then*

$$\mathbf{u} = \operatorname{grad} F$$

for some F.

A domain D of space is called simply connected if every simple closed curve in D forms the boundary of a smooth orientable surface in D. Thus the interior of a sphere is simply connected, while the interior of a torus is not. The domain between two concentric spheres is simply connected, as is the interior of a sphere with a finite number of points removed.

In the first part of the theorem we assume that $\int u_T\, ds$ is independent of path. Hence, by Theorem I, $\mathbf{u} = \operatorname{grad} F$. Accordingly,

$$\operatorname{curl} \mathbf{u} = \operatorname{curl} \operatorname{grad} F \equiv \mathbf{0}$$

by the identity of Section 3–6.

In the second part of the theorem, we assume that D is simply connected and $\operatorname{curl} \mathbf{u} \equiv \mathbf{0}$. To show independence of path, it is sufficient, by Theorem II, to show that $\int_C u_T\, ds = 0$ on each simple closed curve C in D. By assumption, C forms the boundary of a piecewise smooth oriented surface S in D. Stokes' theorem is applicable and one finds

$$\int_C u_T\, ds = \iint_S \operatorname{curl}_n \mathbf{u}\, d\sigma = 0$$

for proper direction on C and normal $\mathbf{n}$ on S.

We remark that the Stokes theorem can be extended to an arbitrary oriented surface S whose boundary is formed of distinct simple closed curves $C_1, \ldots, C_n$. If B_S denotes this boundary, with proper directions, one has

$$\int_{B_S} u_T\, ds = \iint_S \operatorname{curl}_n \mathbf{u}\, d\sigma.$$

The proof is like that of Section 5–7. In particular, if $\operatorname{curl} \mathbf{u} \equiv \mathbf{0}$ in D,

$$\int_{BS} u_T \, ds = 0.$$

This can be applied to evaluate line integrals in "multiply connected" domains, as in Section 5-7.

A vector field $\mathbf{u}$ (whose components have continuous derivatives) such that

$$\operatorname{curl} \mathbf{u} \equiv 0$$

in a domain D, is called *irrotational* in D. By virtue of the above theorems, irrotationality in a simply connected domain is equivalent to each of the properties:

$$\int_C u_T \, ds = 0 \text{ for every simple closed curve in } D;$$

$$\int u_T \, ds \text{ is independent of the path in } D;$$

$$\mathbf{u} = \operatorname{grad} F \text{ in } D.$$

A theory similar to the preceding holds for surface integrals. Rather than give a full discussion here, we confine our attention to the counterpart of the last statement in Theorem III; for more details we refer the reader to the books of Brand, Phillips, and Kellogg listed at the end of the chapter.

THEOREM IV. *Let* $\mathbf{u} = L\mathbf{i} + M\mathbf{j} + N\mathbf{k}$ *be a vector field whose components have continuous partial derivatives in a spherical domain D. If* $\operatorname{div} \mathbf{u} \equiv 0$ *in D*:

$$\frac{\partial L}{\partial x} + \frac{\partial M}{\partial y} + \frac{\partial N}{\partial z} \equiv 0 \text{ in } D,$$

then a vector field $\mathbf{v} = X\mathbf{i} + Y\mathbf{j} + Z\mathbf{k}$ *in D can be found such that*

$$\operatorname{curl} \mathbf{v} \equiv \mathbf{u} \text{ in } D;$$

that is,

$$\frac{\partial Z}{\partial y} - \frac{\partial Y}{\partial z} = L, \quad \frac{\partial X}{\partial z} - \frac{\partial Z}{\partial x} = M, \quad \frac{\partial Y}{\partial x} - \frac{\partial X}{\partial y} = N \tag{5-100}$$

in D.

Remark. The theorem provides a converse to the theorem of Section 3-6:

$$\operatorname{div} \operatorname{curl} \mathbf{u} \equiv 0,$$

for it asserts that, if $\operatorname{div} \mathbf{u} \equiv 0$, then $\mathbf{u} \equiv \operatorname{curl} \mathbf{v}$ for some $\mathbf{v}$. However, while $\operatorname{div} \mathbf{u}$ may be 0 in an arbitrary domain D_1, the theorem provides $\mathbf{v}$ whose curl is $\mathbf{u}$ only in each spherical domain D contained in D_1, i.e., no one $\mathbf{v}$ serves for all of D_1. Actually the proof to follow gives one $\mathbf{v}$ for all of D_1 when D_1 is the interior of a cube or of an ellipsoid or of any "convex" surface. The existence of $\mathbf{v}$ for all of D_1 for more general cases can be established [see, for example, H. Lamb, *Hydrodynamics*, 6th ed., pages 203 ff. (Cambridge: Cambridge University Press, 1932)].

Vector fields $\mathbf{u}$ satisfying the condition div $\mathbf{u} \equiv 0$ are termed *solenoidal.* The theorem amounts to the assertion that solenoidal fields are (in suitable domains) fields of the form curl $\mathbf{v}$, provided the components of $\mathbf{u}$ have continuous partial derivatives. The field $\mathbf{v}$ is not unique (Prob. 5 below).

Proof of the theorem. Let the spherical domain D have center at P_0; for simplicity we assume P_0 to be the origin (0, 0, 0). If $P_1(x_1, y_1, z_1)$ is an arbitrary point of D, we set

$$\begin{aligned} X(x_1, y_1, z_1) &= \int_0^1 [zM(x, y, z) - yN(x, y, z)]\, dt, \\ Y(x_1, y_1, z_1) &= \int_0^1 [xN(x, y, z) - zL(x, y, z)]\, dt, \\ Z(x_1, y_1, z_1) &= \int_0^1 [yL(x, y, z) - xM(x, y, z)]\, dt, \end{aligned} \tag{5–101}$$

where, on the right-hand side, x, y, and z are the following functions of t:

$$x = x_1 t, \quad y = y_1 t, \quad z = z_1 t. \tag{5–102}$$

As t varies from 0 to 1, the point (x, y, z) varies from P_0 to P_1 on the line segment P_0P_1; hence (x, y, z) remains in D. We have now by Leibnitz's rule and the chain rule (Sections 4–12 and 2–7):

$$\begin{aligned} \frac{\partial Z}{\partial y_1} &= \int_0^1 \left[y \frac{\partial L}{\partial y} \frac{\partial y}{\partial y_1} + \frac{\partial y}{\partial y_1} L - x \frac{\partial M}{\partial y} \frac{\partial y}{\partial y_1} \right] dt \\ &= \int_0^1 \left[yt \frac{\partial L}{\partial y} + tL - xt \frac{\partial M}{\partial y} \right] dt \end{aligned}$$

and similarly

$$\frac{\partial Y}{\partial z_1} = \int_0^1 \left[xt \frac{\partial N}{\partial z} - zt \frac{\partial L}{\partial z} - tL \right] dt.$$

Accordingly,

$$\frac{\partial Z}{\partial y_1} - \frac{\partial Y}{\partial z_1} = \int_0^1 \left[2tL - xt \left(\frac{\partial M}{\partial y} + \frac{\partial N}{\partial z} \right) + yt \frac{\partial L}{\partial y} + zt \frac{\partial L}{\partial z} \right] dt.$$

Since $\operatorname{div}(L\mathbf{i} + M\mathbf{j} + N\mathbf{k}) = 0$, this can be written as follows:

$$\frac{\partial Z}{\partial y_1} - \frac{\partial Y}{\partial z_1} = \int_0^1 \left[2tL + xt \frac{\partial L}{\partial x} + yt \frac{\partial L}{\partial y} + zt \frac{\partial L}{\partial z} \right] dt.$$

Now

$$\begin{aligned} t \frac{\partial L}{\partial t} &= t \left(\frac{\partial L}{\partial x} \frac{\partial x}{\partial t} + \frac{\partial L}{\partial y} \frac{\partial y}{\partial t} + \frac{\partial L}{\partial z} \frac{\partial z}{\partial t} \right) \\ &= t \left(x_1 \frac{\partial L}{\partial x} + y_1 \frac{\partial L}{\partial y} + z_1 \frac{\partial L}{\partial z} \right) \\ &= x \frac{\partial L}{\partial x} + y \frac{\partial L}{\partial y} + z \frac{\partial L}{\partial z} \cdot \end{aligned}$$

Accordingly,

$$\frac{\partial Z}{\partial y_1} - \frac{\partial Y}{\partial z_1} = \int_0^1 \left(t^2 \frac{\partial L}{\partial t} + 2tL\right) dt = \int_0^1 \frac{\partial}{\partial t}(t^2 L)\, dt$$
$$= t^2 L \Big|_{t=0}^{t=1} = L(x_1, y_1, z_1).$$

This gives the first of (5–100). The other two equations are proved in the same way.

The solution $\mathbf{v}$ can be expressed in the compact form:

$$\mathbf{v}(x, y, z) = \int_0^1 t\mathbf{u}(xt, yt, zt) \times (x\mathbf{i} + y\mathbf{j} + z\mathbf{k})\, dt. \tag{5–103}$$

If the vector field $\mathbf{u}$ is *homogeneous of degree n*, that is,

$$\mathbf{u}(xt, yt, zt) = t^n \mathbf{u}(x, y, z)$$

(Prob. 9 following Section 2–8), the formula can be simplified further:

$$\mathbf{v} = \int_0^1 t^{n+1}\mathbf{u}(x, y, z) \times (x\mathbf{i} + y\mathbf{j} + z\mathbf{k})\, dt$$
$$= \frac{1}{n+2}(\mathbf{u} \times \mathbf{r}), \quad \mathbf{r} = x\mathbf{i} + y\mathbf{j} + z\mathbf{k}.$$

For an interesting discussion of this topic, the reader is referred to pages 487–489 of Vol. 58 (1951) of the *American Mathematical Monthly*.

Problems

1. Evaluate by Stokes' theorem:

(a) $\int_C u_T\, ds$, where C is the circle: $x^2 + y^2 = 1$, $z = 2$, directed so that y increases for positive x, and $\mathbf{u}$ is the vector $-3y\mathbf{i} + 3x\mathbf{j} + \mathbf{k}$;

(b) $\int_C 2xy^2z\, dx + 2x^2yz\, dy + (x^2y^2 - 2z)\, dz$ around the curve: $x = \cos t$, $y = \sin t$, $z = \sin t$, $0 \leqq t \leqq 2\pi$, directed with increasing t.

2. By showing that the integrand is an exact differential, evaluate

(a) $\int_{(1,1,2)}^{(3,5,0)} yz\, dx + xz\, dy + xy\, dz$ on any path;

(b) $\int_{(1,0,0)}^{(1,0,2\pi)} \sin yz\, dx + xz \cos yz\, dy + xy \cos yz\, dz$ on the helix: $x = \cos t$, $y = \sin t$, $z = t$.

3. Let C be a simple closed *plane* curve in space. Let $\mathbf{n} = a\mathbf{i} + b\mathbf{j} + c\mathbf{k}$ be a unit vector normal to the plane of C and let the direction on C match that of $\mathbf{n}$.

Prove that

$$\frac{1}{2}\int\limits_C (bz - cy)\,dx + (cx - az)\,dy + (ay - bx)\,dz$$

equals the plane area enclosed by C. What does the integral reduce to when C is in the xy plane?

4. Let $\mathbf{u} = \dfrac{-y}{x^2 + y^2}\mathbf{i} + \dfrac{x}{x^2 + y^2}\mathbf{j} + z\mathbf{k}$ and let D be the interior of the torus obtained by rotating the circle: $(x - 2)^2 + z^2 = 1$, $y = 0$ about the z axis. Show that curl $\mathbf{u} = \mathbf{0}$ in D but $\int\limits_C u_T\,ds$ is not zero when C is the circle: $x^2 + y^2 = 4$, $z = 0$.

Determine the possible values of the integral $\int_{(2,0,0)}^{(0,2,0)} u_T\,ds$ on a path in D.

5. (a) Show that, if $\mathbf{v}$ is one solution of the equation curl $\mathbf{v} = \mathbf{u}$ for given $\mathbf{u}$ in a simply connected domain D, then all solutions are given by $\mathbf{v}$ + grad f, where f is an arbitrary differentiable scalar in D.

(b) Find all vectors $\mathbf{v}$ such that curl $\mathbf{v} = \mathbf{u}$ if

$$\mathbf{u} = (2xyz^2 + xy^3)\mathbf{i} + (x^2y^2 - y^2z^2)\mathbf{j} - (y^3z + 2x^2yz)\mathbf{k}.$$

6. Show that, if f and g are scalars having continuous second partial derivatives in a domain D, then

$$\mathbf{u} = \nabla f \times \nabla g$$

is solenoidal in D. (It can be shown that every solenoidal vector has such a representation, at least in a suitably restricted domain.)

7. Show that, if $\iint\limits_S u_n\,d\sigma = 0$ for every oriented spherical surface S in a domain D and the components of $\mathbf{u}$ have continuous derivatives in D, then $\mathbf{u}$ is solenoidal in D. Does the converse hold?

Answers

1. (a) 6π, (b) 0. 2. (a) -2, (b) 0. 4. $\frac{\pi}{2} \pm 2n\pi$.

5. (b) $\mathbf{v}$ + grad f, where $\mathbf{v} = x^2y^2z\mathbf{i} - xy^3z\mathbf{j} + xy^2z^2\mathbf{k}$.

***5–14 Change of variables in a multiple integral.** The formula for change of variables in a double integral:

$$\iint\limits_{R_{xy}} F(x, y)\,dx\,dy = \iint\limits_{R_{uv}} F[f(u, v), g(u, v)]\left|\frac{\partial(x, y)}{\partial(u, v)}\right| du\,dv, \qquad (5\text{–}104)$$

is given in Section 4–8 above. In this section we shall give a proof of this formula under appropriate assumptions. We shall also indicate how widely the formula is applicable and shall explain the more general formula:

$$\delta \iint\limits_{R_{xy}} F(x, y)\, dx\, dy = \iint\limits_{R_{uv}} F[f(u, v), g(u, v)] \frac{\partial(x, y)}{\partial(u, v)}\, du\, dv, \quad (5\text{–}105)$$

where δ is the "degree" of the mapping of the boundary of R_{uv} into the boundary of R_{xy}.

THEOREM I. *The formula*

$$\iint\limits_{R_{xy}} F(x, y)\, dx\, dy = \pm \iint\limits_{R_{uv}} F[f(u, v), g(u, v)] \frac{\partial(x, y)}{\partial(u, v)}\, du\, dv \quad (5\text{–}106)$$

is valid under the following assumptions:
(a) *R_{xy} and R_{uv} are bounded closed regions in the xy and uv planes bounded by piecewise smooth simple closed curves C_{xy} and C_{uv} respectively.*
(b) *The function $F(x, y)$ is defined and has continuous first derivatives in a circular domain D_{xy} containing R_{xy}.*
(c) *The functions $x = f(u, v)$, $y = g(u, v)$ are defined and have continuous second derivatives in a domain D_{uv} including R_{uv}; when (u, v) is in D_{uv}, the point (x, y) is in D_{xy}.*
(d) *When (u, v) is on C_{uv}, the corresponding point (x, y), $x = f(u, v)$, $y = g(u, v)$, is on C_{xy}; as (u, v) traces C_{uv} once in the positive direction, (x, y) traces C_{xy} once in the positive direction [corresponding to the + sign in (5–106)] or negative direction [corresponding to the − sign in (5–106)].*

The situation is illustrated in Fig. 5–36. It is to be stressed that the mapping from the uv plane to the xy plane is not assumed to be one-to-one except on the boundary, that when (u, v) is in R_{uv} the corresponding point (x, y) need not be in R_{xy}, and that no assumptions are made about the sign of the Jacobian $\partial(x, y)/\partial(u, v)$.

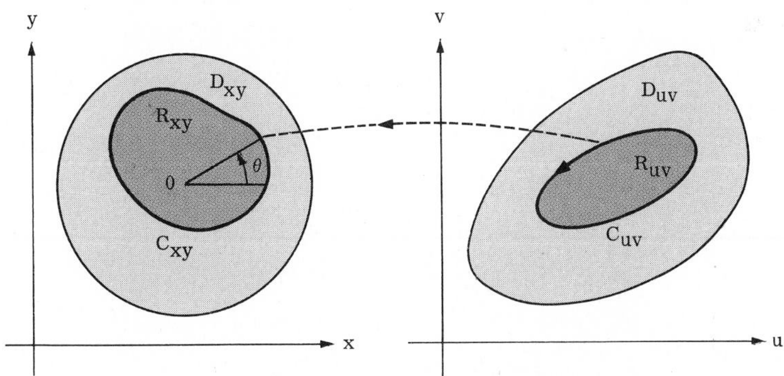

FIG. 5–36. Transformation of double integrals.

To prove the theorem we first remark that one can represent $F(x, y)$ as follows:

$$F(x, y) = \frac{\partial Q}{\partial x}$$

in D_{xy}, where Q has continuous first derivatives. One need only take

$$Q = \int_{x_0}^{x} F(x, y)\, dx,$$

where (x_0, y_0) is the center of D_{xy}.

Green's theorem now gives

$$\iint\limits_{R_{xy}} F(x, y)\, dx\, dy = \iint\limits_{R_{xy}} \frac{\partial Q}{\partial x}\, dx\, dy = \oint\limits_{C_{xy}} Q\, dy.$$

The line integral $\int Q\, dy$ can at once be written as a line integral in the uv plane:

$$\oint\limits_{C_{xy}} Q(x, y)\, dy = \pm \oint\limits_{C_{uv}} Q[f(u, v), g(u, v)]\left(\frac{\partial g}{\partial u}\, du + \frac{\partial g}{\partial v}\, dv\right)\cdot$$

The differentials here should all be thought of as expressed in terms of a parameter t: $dy = (dy/dt)\, dt$, $du = (du/dt)\, dt$, $dv = (dv/dt)\, dt$. As t goes from h to k, C_{uv} is to be traced just once in the positive direction, so that C_{xy} is traced just once in the positive or negative direction. The $\pm$ sign corresponds to these two cases.

The line integral on the right is of form

$$\oint\limits_{C_{uv}} P_1\, du + Q_1\, dv, \quad P_1 = Q\frac{\partial g}{\partial u}, \quad Q_1 = Q\frac{\partial g}{\partial v},$$

where P_1 and Q_1 are defined in D_{uv} and have continuous derivatives there. Hence Green's theorem is applicable:

$$\begin{aligned}
\oint\limits_{C_{uv}} P_1\, du + Q_1\, dv &= \iint\limits_{R_{uv}} \left(\frac{\partial Q_1}{\partial u} - \frac{\partial P_1}{\partial v}\right) du\, dv \\
&= \iint\limits_{R_{uv}} \left(\frac{\partial Q}{\partial u}\frac{\partial g}{\partial v} + Q\frac{\partial^2 g}{\partial u\, \partial v} - \frac{\partial Q}{\partial v}\frac{\partial g}{\partial u} - Q\frac{\partial^2 g}{\partial v\, \partial u}\right) du\, dv \\
&= \iint\limits_{R_{uv}} \left[\left(\frac{\partial Q}{\partial x}\frac{\partial f}{\partial u} + \frac{\partial Q}{\partial y}\frac{\partial g}{\partial u}\right)\frac{\partial g}{\partial v} - \left(\frac{\partial Q}{\partial x}\frac{\partial f}{\partial v} + \frac{\partial Q}{\partial y}\frac{\partial g}{\partial v}\right)\frac{\partial g}{\partial u}\right] du\, dv \\
&= \iint\limits_{R_{uv}} \frac{\partial Q}{\partial x}\left(\frac{\partial f}{\partial u}\frac{\partial g}{\partial v} - \frac{\partial f}{\partial v}\frac{\partial g}{\partial u}\right) du\, dv \\
&= \iint\limits_{R_{uv}} F[f(u, v), g(u, v)]\,\frac{\partial(x, y)}{\partial(u, v)}\, du\, dv,
\end{aligned}$$

since $\partial Q/\partial x = F(x, y)$. One thus concludes

$$\iint\limits_{R_{xy}} F(x, y)\, dx\, dy = \pm \iint\limits_{R_{uv}} F[f(u, v), g(u, v)]\,\frac{\partial(x, y)}{\partial(u, v)}\, du\, dv,$$

as was to be shown.

The restrictions made in the statement of the theorem can be relaxed considerably. First of all, it is not necessary to assume that, as (u, v) makes a circuit of C_{uv}, the point (x, y) moves steadily around C_{xy}; the point (x, y) can move ahead, then move backwards, then ahead again, etc., as long as one complete circuit is achieved. One can make clear what is allowed and, in fact, get even greater generality as follows: Let O be a point inside C_{xy} and let θ be a polar coordinate angle measured relative to O, as in Fig. 5–36. As the point (u, v) goes around C_{uv} in the positive direction, the angle θ for the corresponding point (x, y) can be chosen to vary *continuously*. When (u, v) has made a circuit of C_{uv}, θ will have increased by a certain multiple δ of 2π. This integer δ is known as the *degree* of the mapping of C_{uv} into C_{xy}. With the theorem as stated above $\delta = \pm 1$; for a general δ, the theorem must be restated:

THEOREM II. *The formula*

$$\delta \iint\limits_{R_{xy}} F(x, y)\, dx\, dy = \iint\limits_{R_{uv}} F[f(u, v), g(u, v)] \frac{\partial(x, y)}{\partial(u, v)}\, dx\, dv \qquad (5\text{–}107)$$

is valid under the assumptions (a), (b), (c), *of Theorem I and the assumption:*
(d′) *When (u, v) is on C_{uv}, the corresponding point (x, y) is on C_{xy}; the degree of the mapping of C_{uv} into C_{xy} is δ.*

The extension of the previous proof to this case causes no great difficulty. One need only verify that

$$\delta \oint\limits_{C_{xy}} Q(x, y)\, dy = \oint\limits_{C_{uv}} Q \left(\frac{\partial g}{\partial u}\, du + \frac{\partial g}{\partial v}\, dv \right);$$

this relation follows from the fact that, as (u, v) makes one circuit of C_{uv} in the positive direction, the point (x, y) effectively traces the curve C_{xy} δ times. If δ is negative, this means (x, y) traces C_{xy} $|\delta|$ times in the negative direction. It can happen that $\delta = 0$, in which case both sides of (5–107) are 0.

It is of interest to note that a careful treatment of the transformation from rectangular to polar coordinates requires a formula as general as (5–107) (see Prob. 4 below).

One can also consider the case of a multiply connected region R_{xy}, bounded by curves $C_{xy}^{(1)}, C_{xy}^{(2)}, \ldots, C_{xy}^{(k)}$. If the boundary of R_{uv} is a similar set of curves $C_{uv}^{(1)}, C_{uv}^{(2)}, \ldots, C_{uv}^{(k)}$, then the formula (5–107) continues to hold, provided each $C_{uv}^{(i)}$ is mapped on the corresponding $C_{xy}^{(i)}$ with the same degree δ. In particular, the points on these boundaries can be in one-to-one correspondence, with δ always equal to 1; one then obtains the formula (5–106) again, with the $+$ sign.

The requirement that $F(x, y)$ be defined in a *circular* domain D_{xy} was needed to show that a Q could be found such that $\partial Q/\partial x = F$. A more general domain would still permit this, but it is actually unnecessary to make any restriction whatsoever on the nature of D_{xy}.

The proof above clearly used the differentiability conditions on F, f, and g in an unavoidable manner. An entirely different method of proof can be devised which requires merely that F be continuous and that $f(u, v)$ and $g(u, v)$ have continuous first derivatives. Thus (5–107) remains valid, if assumptions (b) and (c) are replaced by the following:

(b′) *The function $F(x, y)$ is defined and continuous in a domain D_{xy} containing R_{xy}.*
(c′) *The functions $x = f(u, v)$, $y = g(u, v)$ are defined and have continuous derivatives in a domain D_{uv} including R_{uv}; when (u, v) is in D_{uv}, (x, y) is in D_{xy}.*

For proofs the reader is referred to an article, "The Transformation of Double Integrals," by R. G. Helsel and T. Radó, in *Transactions of the American Mathematical Society*, Vol. 54 (1943), pages 83–102. This article is quite advanced; the author regrets that he cannot cite a more elementary treatment.

When $F(x, y)$ is chosen identically equal to 1 in (5–106), the left-hand side gives the area A of R_{xy}. Thus

$$A = \pm \iint\limits_{R_{uv}} \frac{\partial(x, y)}{\partial(u, v)}\, du\, dv.$$

If the Jacobian $J = \partial(x, y)/\partial(u, v)$ is always positive or zero, the integral on the right is positive, hence only the + sign can hold. We therefore conclude, on the basis of (a), (b), (c), (d):

If the Jacobian J is always positive, then, as (u, v) traces C_{uv} once in the positive direction, the point (x, y) traces C_{xy} once in the positive direction.

A similar result holds when J is negative, the positive direction on C_{xy} being replaced by the negative direction. Combining the two cases, one obtains the theorem:

THEOREM III. *The formula*

$$\iint\limits_{R_{xy}} F(x, y)\, dx\, dy = \iint\limits_{R_{uv}} F[f(u, v), g(u, v)] \left| \frac{\partial(x, y)}{\partial(u, v)} \right| du\, dv \qquad (5\text{–}108)$$

holds under the assumptions (a), (b), (c), (d), *and*
(e) *The Jacobian $\partial(x, y)/\partial(u, v)$ does not change sign in R_{uv}.*

It can be shown that, if $J \neq 0$, the assumptions made for Theorem III imply that the mapping from the uv plane to the xy plane is one-to-one. Theorem III is thus essentially the standard form of the transformation theorem, as given in the books of Courant, Goursat, and Franklin listed at the end of the chapter.

As stated in Section 4–8, a formula analogous to (5–106) holds in three (or more) dimensions. The method of proof used here generalizes in a natural manner, the line integrals being replaced by surface integrals and

Green's theorem by the divergence theorem. One can also define a *degree* δ for the correspondence of boundaries, and a formula analogous to (5–107) then holds without any assumption about one-to-one correspondence on boundaries:

$$\delta \iiint\limits_{R_{xyz}} F(x, y, z)\, dx\, dy\, dz = \iiint\limits_{R_{uvw}} F[x(u, v, w), \ldots] \frac{\partial(x, y, z)}{\partial(u, v, w)}\, du\, dv\, dw. \tag{5–109}$$

The degree δ measures the effective number of times the bounding surface S_{xyz} is traced by (x, y, z) as the point (u, v, w) traces the bounding surface S_{uvw} of R_{uvw}. The surfaces S_{xyz} and S_{uvw} are both considered oriented, the normal being the outer normal for both; in measuring δ, one counts *negatively* the parts of S_{uvw} which are mapped into S_{xyz} with *reversal* of orientation. Thus for the mapping

$$x = -u, \quad y = -v, \quad z = -w$$

of the sphere $u^2 + v^2 + w^2 = 1$ onto the sphere $x^2 + y^2 + z^2 = 1$ one has $\delta = -1$. The formula (5–109) can also be extended to the case in which R_{uvw} and R_{xyz} are each bounded by several surfaces, as in two dimensions. In the case when R_{xyz} is simply connected and is bounded by a single surface S_{xyz} (which has then the structure of a sphere), the degree δ can be computed, by analogy with the planar case, by reference to a "solid angle" (see Prob. 6 below).

Problems

1. Transform the integrals, using the substitution given:

(a) $\displaystyle\int_0^1 \int_0^y (x^2 + y^2)\, dx\, dy, \quad u = y, \quad v = x;$

(b) $\displaystyle\iint\limits_{R_{xy}} (x - y)\, dx\, dy$, where R_{xy} is the region $x^2 + y^2 \leqq 1$, and $x = u + (1 - u^2 - v^2)$, $y = v + (1 - u^2 - v^2)$; (Hint: use as R_{uv} the region $u^2 + v^2 \leqq 1$.)

(c) $\displaystyle\iint\limits_{R_{xy}} xy\, dx\, dy$, where R_{xy} is the region $x^2 + y^2 \leqq 1$ and $x = u^2 - v^2$, $y = 2uv$. [Hint: choose R_{uv} as in (b).]

2. Let $x = f(u, v)$, $y = g(u, v)$ be given as in Theorem II above and let C_{xy} enclose the origin O. Show that the degree δ can be evaluated by the formula

$$2\pi\delta = \oint\limits_{C_{uv}} \frac{-y\, dx + x\, dy}{x^2 + y^2},$$

where x, y, dx, dy are expressed in terms of u, v:

$$x = f(u, v), \quad dx = \frac{\partial x}{\partial u}\, du + \frac{\partial x}{\partial v}\, dv \ldots.$$

(Hint: the line integral measures the change in $\theta = \arctan y/x$, as shown in Section 5–6 above.)

3. Apply the formula of Prob. 2 to evaluate the degree for the following mappings of the circle $u^2 + v^2 = 1$ into the circle $x^2 + y^2 = 1$:

(a) $x = \dfrac{3u + 4v}{5}, \quad y = \dfrac{4u - 3v}{5}$;
(b) $x = u^2 - v^2, \quad y = 2uv$;
(c) $x = u^3 - 3uv^2, \quad y = 3u^2v - v^3$.

4. *Polar coordinates.* Prove the validity of the transformation formula

$$\iint\limits_R F(x, y)\, dx\, dy = \int_0^{2\pi} \int_0^1 F(r\cos\theta, r\sin\theta) r\, dr\, d\theta,$$

where R is the circular region: $x^2 + y^2 \leqq 1$ and $x = r\cos\theta$, $y = r\sin\theta$. [Hint: consider first the semicircular region R_1: $x^2 + y^2 \leqq 1$, $y \geqq 0$ and the corresponding *rectangle*: $0 \leqq \theta \leqq \pi$, $0 \leqq r \leqq 1$ in the $r\theta$ plane. Show that the conditions of Theorem II are met, with $\delta = 1$. Note that the correspondence between the boundary of the rectangle and that of the semicircle is not one-to-one. Obtain a similar result for the semicircular region R_2: $x^2 + y^2 \leqq 1$, $y \leqq 0$ and add the results for R_1 and R_2.]

5. *The solid angle.* Let S be a plane surface, oriented in accordance with a unit normal, $\mathbf{n}$. The solid angle Ω of S with respect to a point O not in S is defined as

$$\Omega(O, S) = \pm\text{area of projection of } S \text{ on } S_1,$$

where S_1 is the sphere of radius 1 about O and the $+$ or $-$ sign is chosen according to whether $\mathbf{n}$ points away from or toward the side of S on which O lies. This is suggested in Fig. 5–37.

(a) Show that, if O lies in the plane of S but not in S, then $\Omega(O, S) = 0$.

(b) Show that, if S is a complete (i.e., infinite) plane, then $\Omega(O, S) = \pm 2\pi$.

(c) For a general oriented surface S, the surface can be thought of as made up of small elements, each of which is approximately planar and has a normal $\mathbf{n}$. Justify the following definition of *element of solid angle* for such a surface element:

$$d\Omega = \frac{\mathbf{r}\cdot\mathbf{n}}{r^3}\, d\sigma,$$

where $\mathbf{r}$ is the vector from O to the element.

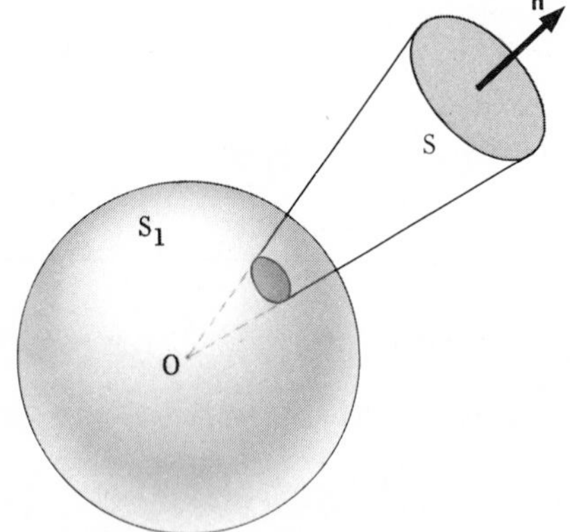

FIG. 5–37. Solid angle.

(d) On the basis of the formula of (c), one obtains as solid angle for a general oriented surface S the integral

$$\Omega(O, S) = \iint\limits_S \frac{\mathbf{r}\cdot\mathbf{n}}{r^3}\, d\sigma.$$

Show that for surfaces in parametric form, if O is the origin,

$$\Omega(O, S) = \iint\limits_{R_{uv}} \begin{vmatrix} x & y & z \\ \dfrac{\partial x}{\partial u} & \dfrac{\partial y}{\partial u} & \dfrac{\partial z}{\partial u} \\ \dfrac{\partial x}{\partial v} & \dfrac{\partial y}{\partial v} & \dfrac{\partial z}{\partial v} \end{vmatrix} \frac{1}{(x^2 + y^2 + z^2)^{\frac{3}{2}}}\, du\, dv.$$

This formula permits one to define solid angle for complicated surfaces which intersect themselves.

(e) Show that, if the normal of S_1 is the outer one, then $\Omega(O, S_1) = 4\pi$.

(f) Show that, if S forms the boundary of a bounded, closed, simply connected region R, then $\Omega(O, S) = \pm 4\pi$ when O is inside S and $\Omega(O, S) = 0$ when O is outside S.

(g) If S is a fixed circular disk and O is variable, show that $-2\pi \leqq \Omega(O, S) \leqq 2\pi$ and that $\Omega(O, S)$ jumps by 4π as O crosses S.

6. *Degree of mapping of one surface into another.* Let S_{uvw} and S_{xyz} be surfaces forming the boundaries of regions R_{uvw} and R_{xyz} respectively; it is assumed that R_{uvw} and R_{xyz} are bounded and closed and that R_{xyz} is simply connected. Let S_{uvw} and S_{xyz} be oriented by the outer normal. Let s, t be parameters for S_{uvw}:

$$u = u(s, t), \quad v = v(s, t), \quad w = w(s, t), \tag{a}$$

the normal having the direction of

$$(u_s\mathbf{i} + v_s\mathbf{j} + w_s\mathbf{k}) \times (u_t\mathbf{i} + v_t\mathbf{j} + w_t\mathbf{k}).$$

Let

$$x = x(u, v, w), \quad y = y(u, v, w), \quad z = z(u, v, w) \tag{b}$$

be functions defined and having continuous derivatives in a domain containing S_{uvw}, and let these equations define a mapping of S_{uvw} into S_{xyz}. The degree δ of this mapping is defined as $1/4\pi$ times the solid angle $\Omega(O, S)$ of the image S of S_{uvw} with respect to a point O interior to S_{xyz}. If O is the origin, the degree is hence given by the integral (Kronecker integral)

$$\delta = \frac{1}{4\pi} \iint\limits_{R_{st}} \begin{vmatrix} x & y & z \\ \dfrac{\partial x}{\partial s} & \dfrac{\partial y}{\partial s} & \dfrac{\partial z}{\partial s} \\ \dfrac{\partial x}{\partial t} & \dfrac{\partial y}{\partial t} & \dfrac{\partial z}{\partial t} \end{vmatrix} \frac{1}{(x^2 + y^2 + z^2)^{\frac{3}{2}}}\, ds\, dt,$$

where x, y, z are expressed in terms of s, t by (a) and (b). It can be shown that δ, as thus defined, is independent of the choice of the interior point O, that δ is a positive or negative integer or zero, and that δ does measure the effective number of times that S_{xyz} is covered.

Let S_{uvw} be the sphere: $u = \sin s \cos t$, $v = \sin s \sin t$, $w = \cos s$, $0 \leqq s \leqq \pi$, $0 \leqq t \leqq 2\pi$. Let S_{xyz} be the sphere: $x^2 + y^2 + z^2 = 1$. Evaluate the degree for the following mappings of S_{uvw} into S_{xyz}:

(a) $x = v, \quad y = -w, \quad z = u$;

(b) $x = u^2 - v^2, \quad y = 2uv, \quad z = w\sqrt{2 - w^2}$.

Answers

1. (a) $\displaystyle\int_0^1 \int_0^u (u^2 + v^2)\, dv\, du$, (b) $\displaystyle\iint\limits_{R_{uv}} (u - v)(1 - 2u - 2v)\, du\, dv$,

(c) $4\iint\limits_{R_{uv}} uv(u^4 - v^4)\, du\, dv.$

3. (a) -1, (b) 2, (c) 3. 6. (a) -1, (b) 2.

***5–15 Physical applications.** The following is a brief discussion of some of the important applications of the divergence and curl and of line and surface integrals.

(a) *Dynamics.* If **F** is a force field, then, as shown above, the work done by **F** on an arbitrary path C is

$$W = \int_C F_T\, ds. \tag{5–110}$$

In general, this will be dependent on the path. If, however, **F** is the gradient of a scalar, the work done will be independent of the path. If this holds, the scalar will be denoted by $-U$, so that

$$\mathbf{F} = -\text{grad } U, \quad U = U(x, y, z), \tag{5–111}$$

and **F** is said to be derived from the potential U; U is also termed the potential energy of the force field. U is determined uniquely up to an additive constant:

$$U = -\int_{(x_1, y_1, z_1)}^{(x, y, z)} F_T\, ds + \text{const} \tag{5–112}$$

as in Section 5–6 above. In practice the constant is often chosen so that U approaches 0 as $x^2 + y^2 + z^2$ becomes infinite.

The work done on a particle moving from A to B is then expressed in terms of U as follows:

$$W = \int_A^B F_T\, ds = U(A) - U(B); \tag{5–113}$$

that is, the *work done equals the loss in potential energy.*

The theorem: *work done equals gain in kinetic energy,* proved in Section 5–4 above, holds for a general force field. Combining the two results, one concludes:

$$\text{gain in kinetic energy} = \text{loss in potential energy}$$

or

$$(\text{gain in kinetic energy}) + (\text{gain in potential energy}) = 0,$$

since the gain in potential energy equals the negative of the loss. If one now defines the *total energy* of the particle to be E, where

$$E = (\text{kinetic energy}) + (\text{potential energy}) = \frac{mv^2}{2} + U, \tag{5–114}$$

then one concludes that, for an arbitrary motion of the particle under the given force,

$$E = \text{constant}; \tag{5–115}$$

that is, the total energy is conserved. This is the law of *conservation of energy* for a particle. This was established here under the assumption that $\mathbf{F}$ was a gradient vector; it can be shown that such a conservation law can hold only when $\mathbf{F}$ is a gradient vector. For this reason force fields which are gradients are termed *conservative* force fields.

(b) *Fluid dynamics.* If $\mathbf{u}$ is the velocity vector and ρ is the density for a fluid motion, then, as in Section 5–11, the *equation of continuity*

$$\frac{\partial \rho}{\partial t} + \text{div}\,(\rho \mathbf{u}) = 0 \tag{5–116}$$

holds. This can also be written, by virtue of an identity of Section 3–6, as follows:

$$\frac{\partial \rho}{\partial t} + \text{grad}\,\rho \cdot \mathbf{u} + \rho\,\text{div}\,\mathbf{u} = 0.$$

The first two terms are the Stokes total derivative of ρ:

$$\frac{D\rho}{Dt} = \frac{\partial \rho}{\partial t} + \frac{\partial \rho}{\partial x}\frac{dx}{dt} + \frac{\partial \rho}{\partial y}\frac{dy}{dt} + \frac{\partial \rho}{\partial z}\frac{dz}{dt} = \frac{\partial \rho}{\partial t} + \text{grad}\,\rho \cdot \mathbf{u},$$

and describe the rate of change of ρ as one stays with a particular particle of fluid in the motion. For an *incompressible* fluid, $D\rho/Dt = 0$, so that the equation of continuity becomes

$$\text{div}\,\mathbf{u} = 0 \tag{5–117}$$

and $\mathbf{u}$ is solenoidal.

Another interpretation of div $\mathbf{u}$ and a new proof of the continuity equation are given in the problems below.

The integral $\int_C u_T\,ds$ about a closed curve C has already been introduced as the *circulation* of the velocity field. If this is zero for every closed path C, then by Theorem III of Section 5–13

$$\text{curl}\,\mathbf{u} = \mathbf{0}. \tag{5–118}$$

If (5–118) holds, the flow is called *irrotational.* This implies, by Theorem III of Section 5–13, that the circulation is zero on every closed path, provided attention is restricted to a simply connected domain D. If this last assumption is made, then $\mathbf{u} = \text{grad}\,\phi$ for some scalar ϕ, termed the *velocity potential.*

If the flow is both irrotational and incompressible, then ϕ must satisfy the equation

$$\text{div grad}\,\phi = 0;$$

that is,

$$\frac{\partial^2 \phi}{\partial x^2} + \frac{\partial^2 \phi}{\partial y^2} + \frac{\partial^2 \phi}{\partial z^2} = 0, \tag{5–119}$$

and ϕ is *harmonic* in D.

(c) *Electromagnetism.* An electromagnetic field is described, in accordance with Maxwell's theory, by two vector fields $\mathbf{E}$ and $\mathbf{H}$, where $\mathbf{E}$ is the electric force and $\mathbf{H}$ is the magnetic field strength. Both $\mathbf{E}$ and $\mathbf{H}$ in general vary with time t. In the absence of conductors, $\mathbf{E}$ and $\mathbf{H}$ satisfy the *Maxwell equations:*

$$\begin{gathered}\operatorname{div} \mathbf{E} = 4\pi\rho, \quad \operatorname{div} \mathbf{H} = 0, \\ \operatorname{curl} \mathbf{E} = -\frac{1}{c}\frac{\partial \mathbf{H}}{\partial t}, \quad \operatorname{curl} \mathbf{H} = \frac{1}{c}\frac{\partial \mathbf{E}}{\partial t},\end{gathered} \tag{5–120}$$

where ρ is the charge density and c is a universal constant.

In the electrostatic case, $\mathbf{H} \equiv \mathbf{0}$, so that $\mathbf{E}$ does not depend on time and

$$\operatorname{curl} \mathbf{E} = \mathbf{0}. \tag{5–121}$$

Hence (in a simply connected domain), $\mathbf{E}$ is the gradient of a potential:

$$\mathbf{E} = -\operatorname{grad} \psi;$$

ψ is termed the *electrostatic potential.* The function ψ must then satisfy the *Poisson equation:*

$$\operatorname{div} \operatorname{grad} \psi = -4\pi\rho. \tag{5–122}$$

In any domain free of charge, ψ is therefore a *harmonic* function.

The function ψ can be computed by Coulomb's law for given charge distributions. Thus for a point charge e at the origin,

$$\psi = \frac{e}{r} + \text{const}, \quad r = \sqrt{x^2 + y^2 + z^2}, \tag{5–123}$$

and ψ is obtained by simple addition for a sum of point charges. If the charge is distributed along a wire C and ρ_s is the density (charge per unit length), then

$$\psi(x_1, y_1, z_1) = \int\limits_C \frac{\rho_s\, ds}{r_1} + \text{const}, \tag{5–124}$$

where $r_1 = \sqrt{(x - x_1)^2 + (y - y_1)^2 + (z - z_1)^2}$. If the charge is spread out over a surface S, then ψ is given by the surface integral

$$\psi(x_1, y_1, z_1) = \iint\limits_S \frac{\rho_a}{r_1}\, d\sigma + \text{const}, \tag{5–125}$$

where ρ_a is the charge density (charge per unit area).

(d) *Heat conduction.* Let $T(x, y, z)$ be the temperature at the point (x, y, z) of a body. If heat is being conducted in the body, the flow of heat can be represented by a vector $\mathbf{u}$ such that the flux integral

$$\iint\limits_S u_n\, d\sigma$$

for each oriented surface S represents the number of calories crossing S in the direction of the given normal per unit of time. The simplest law of thermal conduction postulates that

$$\mathbf{u} = -k \operatorname{grad} T, \tag{5–126}$$

with $k > 0$; k is usually treated as a constant. Equation (5–126) implies that heat flows in the direction of decreasing temperature and the rate of flow is proportional to the temperature gradient: $|\operatorname{grad} T|$.

If S is a closed surface, forming the boundary of a region R in the body, then

$$\iint_S u_n \, dA = \iiint_R \operatorname{div} \mathbf{u} \, dx \, dy \, dz, \tag{5–127}$$

by the divergence theorem. Hence the total amount of heat *entering* R is

$$-\iint_S u_n \, dA = \iiint_R k \operatorname{div} \operatorname{grad} T \, dx \, dy \, dz. \tag{5–128}$$

On the other hand the rate at which heat is being absorbed per unit mass can also be measured by $c \dfrac{\partial T}{\partial t}$, where c is the specific heat; the rate at which R is receiving heat is then

$$\iiint_R c\rho \frac{\partial T}{\partial t} \, dx \, dy \, dz, \tag{5–129}$$

where ρ is the density. Equating the two expressions, one finds

$$\iiint_R \left(c\rho \frac{\partial T}{\partial t} - k \operatorname{div} \operatorname{grad} T\right) dx \, dy \, dz = 0. \tag{5–130}$$

Since this must hold for an *arbitrary* solid region R, the function integrated (if continuous) must be zero everywhere. Hence

$$c\rho \frac{\partial T}{\partial t} - k \operatorname{div} \operatorname{grad} T = 0. \tag{5–131}$$

This is the fundamental equation for heat conduction. If the body is in temperature equilibrium, $\partial T/\partial t = 0$ and one concludes that

$$\operatorname{div} \operatorname{grad} T = 0; \tag{5–132}$$

that is, T is *harmonic*.

(e) *Thermodynamics*. Let a certain volume V of a gas be given, enclosed in a container, subject to a pressure p. It is known from experiment that, for each kind of gas, there is an "equation of state"

$$f(p, V, T) = 0, \tag{5–133}$$

connecting pressure, volume, and temperature T. For an "ideal gas" (low density and high temperature), Eq. (5–133) takes the special form:

$$pV = RT, \tag{5–134}$$

where R is constant (the same for all gases, if one mole of gas is used).

With each gas is also associated a scalar U, the total internal energy; this is analogous to the kinetic energy plus potential energy considered above. For each gas U is given as a definite function of the "state," hence of p and V:

$$U = U(p, V). \tag{5–135}$$

The particular equation (5–135) depends on the gas considered. For an ideal gas (5–135) takes the form:

$$U = c_V \frac{pV}{R} = c_V T, \tag{5–136}$$

where c_V is constant, the *specific heat* at constant volume.

A particular *process* gone through by a gas is a succession of changes in the state, so that p and V, and hence T and U, become functions of time t. The state at time t can be represented by a point (p, V) on a pV diagram (Fig. 5–38) and the process by a curve C, with t as parameter. During such a process it is possible to measure the *amount of heat*, Q, received by the gas. The first law of thermodynamics is equivalent to the statement that

$$\frac{dQ}{dt} = \frac{dU}{dt} + p\frac{dV}{dt}, \tag{5–137}$$

where $Q(t)$ is the amount of heat received up to time t.

Hence the amount of heat introduced in a particular process is given by an integral:

$$Q(h) - Q(0) = \int_0^h \left(\frac{dU}{dt} + p\frac{dV}{dt}\right) dt. \tag{5–138}$$

Now by (5–135) dU is expressible in terms of dp and dV:

$$dU = \left(\frac{\partial U}{\partial p}\right)_V dp + \left(\frac{\partial U}{\partial V}\right)_p dV \tag{5–139}$$

Hence (5–138) can be written as a line integral:

$$\begin{aligned} Q(h) - Q(0) &= \int_0^h \left\{\left(\frac{\partial U}{\partial p}\right)_V \frac{dp}{dt} + \left[\left(\frac{\partial U}{\partial V}\right)_p + p\right]\frac{dV}{dt}\right\} dt \\ &= \int_C \left(\frac{\partial U}{\partial p}\right)_V dp + \left[\left(\frac{\partial U}{\partial V}\right)_p + p\right] dV, \end{aligned} \tag{5–140}$$

or, with (5–139) understood, simply thus:

$$Q(h) - Q(0) = \int_C dU + p\,dV. \tag{5–141}$$

For (5–140) or (5–141) to be independent of the path C, one must have

$$\frac{\partial}{\partial V}\left(\frac{\partial U}{\partial p}\right) = \frac{\partial}{\partial p}\left(\frac{\partial U}{\partial V} + p\right);$$

that is,

$$\frac{\partial^2 U}{\partial V\,\partial p} = \frac{\partial^2 U}{\partial p\,\partial V} + 1.$$

Since this is impossible (when U has continuous second derivatives), the heat introduced is dependent on the path. For a simple closed path C_1, as in Fig. 5–38, the heat introduced is

$$\oint_{C_1} dU + p\,dV.$$

Since U is a given function of p and V, $\oint dU = 0$; thus the heat introduced reduces to

$$\int_{C_1} p\,dV.$$

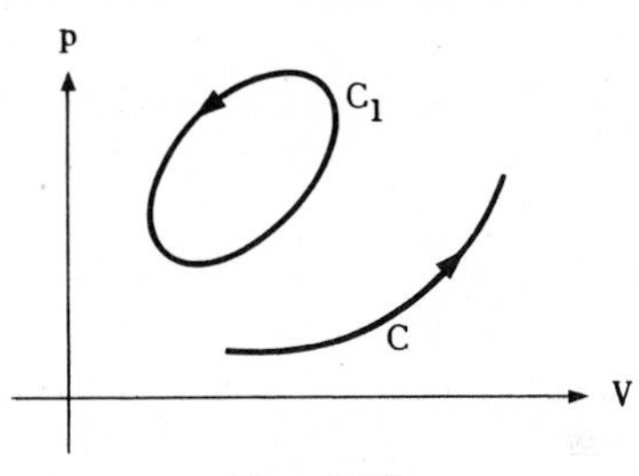

Fig. 5–38.

This integral is precisely the area integral $\int y\,dx$ considered in Section 5–5 above, with p replacing y and V replacing x. Hence for such a counterclockwise cycle the heat introduced is *negative;* there is a heat *loss*, equal to the area enclosed (a unit of area corresponding to a unit of *energy*). The integral $\int p\,dV$ can also be interpreted as the mechanical *work* done by the gas on the surrounding medium or as the negative of the work done on the gas by the surrounding medium. For the process of the curve C_1 considered above, the heat loss equals the work done on the gas; the total energy remains unchanged, in agreement with the law of conservation of energy expressed by the thermodynamic law (5–137).

While the integral $\int dU + p\,dV$ is not independent of path, it is an experimental law that the integral

$$\int \frac{1}{T}\,dU + \frac{p}{T}\,dV$$

is independent of path. One can accordingly introduce a scalar S whose differential is the expression being integrated:

$$\begin{aligned} dS &= \frac{1}{T}\,dU + \frac{p}{T}\,dV \\ &= \frac{1}{T}\frac{\partial U}{\partial p}\,dp + \frac{1}{T}\left(\frac{\partial U}{\partial V} + p\right)dV; \end{aligned} \qquad (5\text{–}142)$$

S is termed the *entropy*. In the first equation here, one can consider U and V as independent variables; in the second, p and V can be considered independent. Thus the first equation gives

$$\left(\frac{\partial S}{\partial U}\right)_V = \frac{1}{T}, \quad \left(\frac{\partial S}{\partial V}\right)_U = \frac{p}{T}$$

and hence

$$\frac{\partial^2 S}{\partial V\,\partial U} = \frac{\partial}{\partial V}\left(\frac{1}{T}\right) = \frac{\partial}{\partial U}\left(\frac{p}{T}\right) = \frac{\partial^2 S}{\partial U\,\partial V}.$$

Accordingly, one finds

$$T\frac{\partial p}{\partial U} - p\frac{\partial T}{\partial U} + \frac{\partial T}{\partial V} = 0 \quad (U, V \text{ indep.}). \tag{5–143}$$

A similar relation is obtainable from the second equation (5–142), and others are obtainable by varying the choice of independent variables. All these equations are simply different forms of the condition $\partial P/\partial y = \partial Q/\partial x$ for independence of path (see Prob. 14 following Section 2–11).

The second law of thermodynamics states first the existence of the entropy S and second the fact that, for any closed system

$$\frac{dS}{dt} \geqq 0;$$

that is, the entropy can never decrease.

Problems

1. (a) A particle of mass m moves on a straight line, the x axis, subject to a force $-k^2x$. Find the potential energy and determine the law of conservation of energy for this motion. Does the law hold if a resistance $-c\dfrac{dx}{dt}$ is added?

(b) A particle of mass m moves in the xy plane subject to a force $\mathbf{F} = -a^2x\mathbf{i} - b^2y\mathbf{j}$. Find the potential energy and determine the law of conservation of energy for this motion.

2. Let D be a simply connected domain in the xy plane and let $\mathbf{w} = u\mathbf{i} - v\mathbf{j}$ be the velocity vector of an irrotational incompressible flow in D. (This is the same as an irrotational incompressible flow in a three-dimensional domain whose projection is D and for which the z component of velocity is 0, while the x and y components of velocity are independent of z.) Show that the following properties hold:

(a) u and v satisfy the Cauchy-Riemann equations:

$$\frac{\partial u}{\partial x} = \frac{\partial v}{\partial y}, \quad \frac{\partial u}{\partial y} = -\frac{\partial v}{\partial x} \text{ in } D;$$

(b) u and v are harmonic in D;
(c) $\int u\,dx - v\,dy$ and $\int v\,dx + u\,dy$ are independent of the path in D;
(d) there is a vector $\mathbf{F} = \phi\mathbf{i} - \psi\mathbf{j}$ in D such that

$$\frac{\partial \phi}{\partial x} = u = \frac{\partial \psi}{\partial y}, \quad \frac{\partial \phi}{\partial y} = -v = -\frac{\partial \psi}{\partial x};$$

(e) $\operatorname{div}\mathbf{F} = 0$ and $\operatorname{curl}\mathbf{F} = \mathbf{0}$ in D;
(f) ϕ and ψ are harmonic in D;
(g) $\operatorname{grad}\phi = \mathbf{w}$, ψ is constant on each stream line.

The function ϕ is the *velocity potential*, ψ is the *stream function*.

3. Let a wire occupying the line segment from $(0, -c)$ to $(0, c)$ in the xy plane

have a constant charge density equal to ρ. Show that the electrostatic potential due to this wire at a point (x_1, y_1) of the xy plane is given by

$$\psi = \rho \log \frac{\sqrt{x_1^2 + (c - y_1)^2} + c - y_1}{\sqrt{x_1^2 + (c + y_1)^2} - c - y_1} + k,$$

where k is an arbitrary constant. Show that, if k is chosen so that $\psi(1, 0) = 0$, then, as c becomes infinite, ψ approaches the limiting value $-2\rho \log |x_1|$. This is the potential of an infinite wire with uniform charge.

4. Find the temperature distribution in a solid whose boundaries are two parallel planes, d units apart, kept at temperatures T_1, T_2 respectively. (Hint: take the boundaries to be the planes $x = 0$, $x = d$ and note that, by symmetry, T must be independent of y and z.)

5. Show that, on the basis of the laws of thermodynamics, the line integral

$$\int S\,dT + p\,dV$$

is independent of the path in the TV plane. The integrand is minus the differential of the *free energy* F.

6. Consider a fluid motion in space. A particle occupying position (x_0, y_0, z_0) at time $t = 0$ occupies position (x, y, z) at time t. Thus x, y, z become functions of x_0, y_0, z_0, t:

$$\begin{aligned} x &= \phi(x_0, y_0, z_0, t), \\ y &= \psi(x_0, y_0, z_0, t), \\ z &= \chi(x_0, y_0, z_0, t). \end{aligned} \qquad (*)$$

Let the ∇ symbol be used as follows:

$$\nabla = \frac{\partial}{\partial x_0}\mathbf{i} + \frac{\partial}{\partial y_0}\mathbf{j} + \frac{\partial}{\partial z_0}\mathbf{k}$$

and let J denote the Jacobian

$$\frac{\partial(x, y, z)}{\partial(x_0, y_0, z_0)}.$$

Let $\mathbf{v}$ denote the velocity vector:

$$\mathbf{v} = \frac{\partial x}{\partial t}\mathbf{i} + \frac{\partial y}{\partial t}\mathbf{j} + \frac{\partial z}{\partial t}\mathbf{k} = \frac{\partial \phi}{\partial t}\mathbf{i} + \frac{\partial \psi}{\partial t}\mathbf{j} + \frac{\partial \chi}{\partial t}\mathbf{k}.$$

(a) Show that $J = \nabla x \cdot \nabla y \times \nabla z$.
(b) Show that

$$\frac{\partial x_0}{\partial x} = \mathbf{i} \cdot \frac{\nabla y \times \nabla z}{J}, \quad \frac{\partial y_0}{\partial x} = \mathbf{j} \cdot \frac{\nabla y \times \nabla z}{J}, \quad \frac{\partial z_0}{\partial x} = \mathbf{k} \cdot \frac{\nabla y \times \nabla z}{J},$$

and obtain similar expressions for

$$\frac{\partial x_0}{\partial y}, \quad \frac{\partial y_0}{\partial y}, \quad \frac{\partial z_0}{\partial y}, \quad \frac{\partial x_0}{\partial z}, \quad \frac{\partial y_0}{\partial z}, \quad \frac{\partial z_0}{\partial z}.$$

(Hint: see Prob. 9 following Section 2–11).
(c) Show that

$$\frac{\partial J}{\partial t} = \nabla v_x \cdot \nabla y \times \nabla z + \nabla v_y \cdot \nabla z \times \nabla x + \nabla v_z \cdot \nabla x \times \nabla y.$$

(d) Show that

$$\operatorname{div} \mathbf{v} = \frac{1}{J}\frac{\partial J}{\partial t}.$$

[Hint: by the chain rule

$$\frac{\partial v_x}{\partial x} = \frac{\partial v_x}{\partial x_0}\frac{\partial x_0}{\partial x} + \frac{\partial v_x}{\partial y_0}\frac{\partial y_0}{\partial x} + \frac{\partial v_z}{\partial z_0}\frac{\partial z_0}{\partial x}.$$

Use the result of (b) to show that

$$\frac{\partial v_x}{\partial x} = \frac{\nabla v_x \cdot \nabla y \times \nabla z}{J}$$

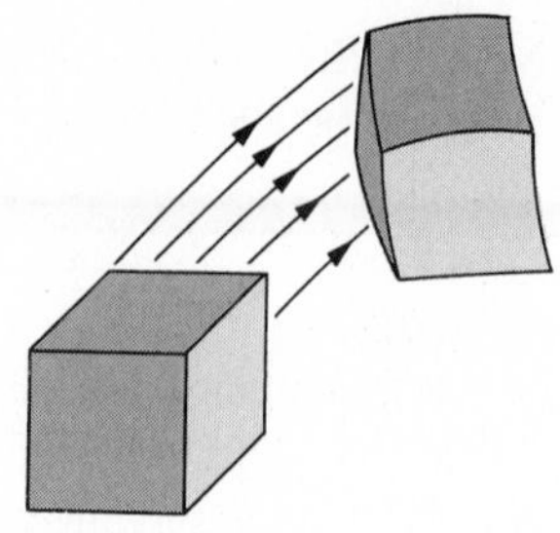

FIG. 5–39.

and obtain similar expressions for $\partial v_y/\partial y$, $\partial v_z/\partial z$. Add the results and use the result of (c).]

Remark. The Jacobian J can be interpreted as the ratio of the volume occupied by a small piece of the fluid at time t to the volume occupied by this piece when $t = 0$, as in Fig. 5–39. Hence by (d) the divergence of the velocity vector can be interpreted as measuring the percent of change in this ratio per unit time or simply as the *rate of change of volume per unit volume* of the moving piece of fluid.

7. Consider a piece of the fluid of Prob. 6 (not necessarily a "small" piece) occupying a region $R = R(t)$ at time t and a region $R_0 = R(0)$ when $t = 0$. Let $F(x, y, z, t)$ be a function differentiable throughout the part of space concerned.

(a) Show that

$$\iiint\limits_{R(t)} F(x, y, z, t)\, dx\, dy\, dz = \iiint\limits_{R_0} F[\phi(x_0, y_0, z_0, t), \ldots]J\, dx_0\, dy_0\, dz_0.$$

[Hint: use Eq. (5–109), noting that the degree δ must be 1 here.]

(b) Show that

$$\frac{d}{dt}\iiint\limits_{R(t)} F(x, y, z, t)\, dx\, dy\, dz = \iiint\limits_{R(t)} \left[\frac{\partial F}{\partial t} + \operatorname{div}(F\mathbf{v})\right] dx\, dy\, dz.$$

[Hint: use (a) and apply Leibnitz's rule of Section 4–11 to differentiate the right-hand side. Use the result of part (d) of Prob. 6 to simplify the result. Then return to the original variables by (a) again.]

8. Let $\rho = \rho(x, y, z, t)$ be the density of the fluid motion of Probs. 6 and 7. The integral

$$\iiint\limits_{R(t)} \rho\, dx\, dy\, dz$$

represents the mass of the fluid filling $R(t)$. The conservation of mass implies that this integral is constant:

$$\frac{d}{dt}\iiint\limits_{R(t)} \rho\, dx\, dy\, dz = 0.$$

Use this result and that of Prob. 7(b) to establish the *continuity equation*

$$\frac{\partial \rho}{\partial t} + \operatorname{div}(\rho\mathbf{v}) = 0.$$

[Hint: cf. the derivation of the heat equation (5–131) above.]

Answers

1. (a) Potential energy is $\frac{1}{2}k^2x^2$; $\frac{1}{2}mv^2 + \frac{1}{2}k^2x^2 = \text{const.}$ (b) Potential energy is $\frac{1}{2}(a^2x^2 + b^2y^2)$; $\frac{1}{2}(mv^2 + a^2x^2 + b^2y^2) = \text{const.}$

4. $T = T_1 + \dfrac{T_2 - T_1}{d}x.$

Suggested References

BRAND, LOUIS, *Vector and Tensor Analysis*. New York: John Wiley and Sons, Inc., 1947.

BUCK, R. C., *Advanced Calculus*, 2nd ed. New York: McGraw-Hill, 1965.

COURANT, RICHARD J., *Differential and Integral Calculus*, transl. by E. J. McShane, Vol. 2. New York: Interscience, 1947.

FRANKLIN, PHILIP, *A Treatise on Advanced Calculus*. New York: John Wiley and Sons, Inc., 1940.

GIBBS, J. W., *Vector Analysis*. New Haven: Yale University Press, 1913.

GOURSAT, ÉDOUARD, *A Course in Mathematical Analysis*, transl. by E. R. Hedrick, Vol. 1. New York: Ginn and Co., 1904.

KELLOGG, O. D., *Foundation of Potential Theory*. Berlin: Springer, 1929.

LAMB, H., *Hydrodynamics*, 6th ed. Cambridge: Cambridge University Press, 1932.

CHAPTER 6

Infinite Series

6–1 Introduction. An infinite series is an indicated sum of the form:

$$a_1 + a_2 + \cdots + a_n + \cdots,$$

going on to infinitely many terms. Such series are familiar even in the simplest operations with numbers. Thus one writes:

$$\frac{1}{3} = .33333\ldots;$$

this is the same as saying

$$\frac{1}{3} = \frac{3}{10} + \frac{3}{100} + \frac{3}{1000} + \frac{3}{10{,}000} + \frac{3}{100{,}000} + \cdots.$$

Of course, we do not interpret this as an ordinary addition problem, which would take forever to carry out. Instead we say, for example,

$$\frac{1}{3} = .33333,$$

to *very good accuracy*. We "round off" after a certain number of decimal places and use the resulting *rational number*

$$.33333 = \frac{33{,}333}{100{,}000}$$

as a sufficiently good *approximation* to the number $\frac{1}{3}$.

The procedure just used applies to the general series $a_1 + a_2 + \cdots + a_n + \cdots$. To evaluate it, we round off after k terms and replace the series by the finite sum

$$a_1 + a_2 + \cdots + a_k.$$

However, the rounding off procedure must be justified: we must be sure that taking more than k terms would not significantly affect the result. For the series

$$1 + 1 + \cdots + 1 + \cdots$$

such a justification is impossible. For one term gives 1 as sum, two terms give 2, three terms give 3, and so on; rounding off is of no help here. This series is an example of a *divergent series*.

On the other hand, for the series

$$1 + \frac{1}{4} + \frac{1}{9} + \frac{1}{16} + \cdots + \frac{1}{n^2} + \cdots$$

it seems *safe* to round off. Thus one has as sums of the first k terms:

$$1, \quad 1+\frac{1}{4}=\frac{5}{4}, \quad 1+\frac{1}{4}+\frac{1}{9}=\frac{49}{36}, \quad 1+\frac{1}{4}+\frac{1}{9}+\frac{1}{16}=\frac{205}{144}, \ldots .$$

The sums do not appear to change much and one would hazard the guess that the sum of 50 terms would not differ from that of 4 terms by more than, say, $\frac{1}{4}$. While it will be seen below that such appearances can be misleading, in this case our instinct happens to be right. This series is an example of a *convergent series*.

It is the purpose of the present chapter to systematize the procedure indicated and to formulate tests which enable one to decide when rounding off is meaningful (convergent case) or meaningless (divergent case). As the example of the decimal expansion of $\frac{1}{3}$ shows, the notion of infinite series lies right at the heart of the concept of the real number system. Accordingly, a complete theory of series would require a profound analysis of the real number system. It is not our purpose to carry this out here, so that some of the rules will have to be justified in an intuitive manner. For a complete treatment the reader is referred to the following two books: G. H. Hardy, *Pure Mathematics*, 9th ed. (Cambridge: Cambridge University Press, 1947); K. Knopp, *Theory and Application of Infinite Series* (Glasgow: Blackie and Son, 1928).

Because of the large number of theorems appearing in this chapter, the theorems will be numbered serially from 1 to 58 throughout the chapter.

6–2 Infinite sequences. If to each positive integer n there is assigned a number s_n, then the numbers s_n are said to form an *infinite sequence*. The numbers are thought of as arranged in order, according to subscript:

$$s_1, s_2, s_3, \ldots, s_n, s_{n+1}, \ldots .$$

Examples of such sequences are the following:

$$\frac{1}{2}, \frac{1}{4}, \ldots, \frac{1}{2^n}, \ldots; \tag{6–1}$$

$$2, \left(\frac{3}{2}\right)^2, \left(\frac{4}{3}\right)^3, \ldots, \left(\frac{n+1}{n}\right)^n, \ldots; \tag{6–2}$$

$$1, \quad 1+\frac{1}{2}, \quad 1+\frac{1}{2}+\frac{1}{3}, \ldots, \quad 1+\frac{1}{2}+\frac{1}{3}+\cdots+\frac{1}{n}, \ldots . \tag{6–3}$$

These are formed by the rules:

$$s_n=\frac{1}{2^n}, \quad s_n=\left(\frac{n+1}{n}\right)^n, \quad s_n=1+\frac{1}{2}+\frac{1}{3}+\cdots+\frac{1}{n}.$$

At times it is convenient to number the members of the sequence starting with 0, with 2, or with some other integer.

A sequence s_n is said to *converge* to the number s or to have the *limit* s:

$$\lim_{n\to\infty} s_n = s \tag{6–4}$$

if, to each number $\epsilon > 0$, a value N can be found such that

$$|s_n - s| < \epsilon \quad \text{for } n > N. \tag{6–5}$$

This is illustrated in Fig. 6–1(a). If s_n does not converge, it is said to *diverge*.

One can regard a sequence s_n as a function $s(n)$ of the *integer* variable n. The limit definition (6–4), (6–5) is then formally the same as that for a function $f(x)$ of the real variable x [Eq. (0-71), Section 0–6].

The sequences (6–1) and (6–2) can be shown to converge:

$$\lim_{n\to\infty} \frac{1}{2^n} = 0, \quad \lim_{n\to\infty} \left(\frac{n+1}{n}\right)^n = e,$$

while (6–3) diverges.

A sequence s_n is said to be *monotone increasing* if $s_1 \leqq s_2 \leqq \cdots \leqq s_n \leqq s_{n+1} \cdots$; this is illustrated in Figs. 6–1(b) and (d). Since the = sign is permitted, it is perhaps clearer to call these sequences *monotone nondecreasing*. Similarly, a sequence s_n is called *monotone decreasing* (or *monotone nonincreasing*) if $s_n \geqq s_{n+1}$ for all n; this is shown in Fig. 6–1(c).

A sequence s_n is said to be *bounded* if there are two numbers A and B such that $A \leqq s_n \leqq B$ for all n. All the sequences suggested in Fig. 6–1 are bounded except those of (d) and (f).

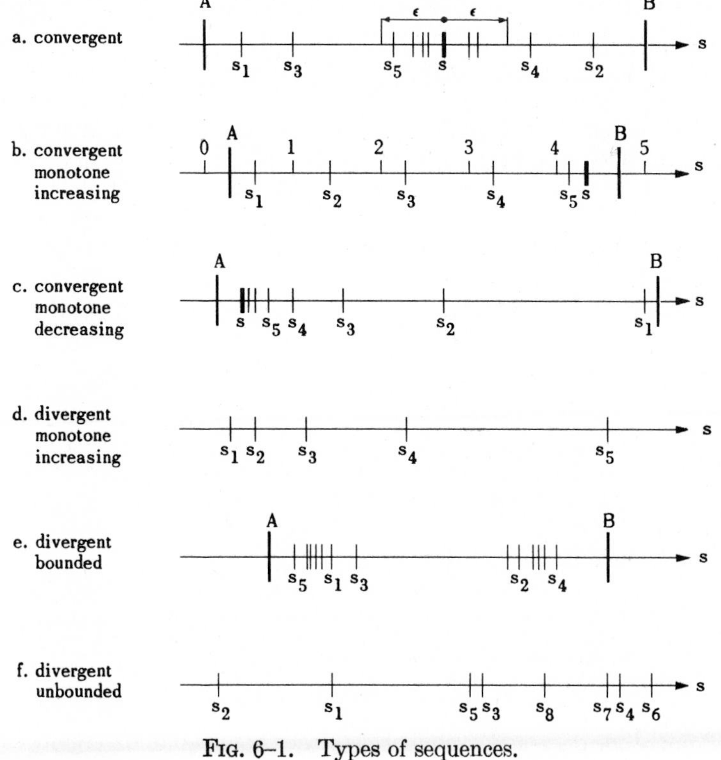

FIG. 6–1. Types of sequences.

THEOREM 1. *Every bounded monotone sequence converges.*

Thus let s_n be monotone increasing as in Fig. 6–1(b). The figure suggests that, since the numbers s_n move to the right as n increases but are not allowed to pass B, they must pile up on a number s to the left of B or at B. One can determine the decimal expansion of s as follows: Let k_1 be the largest integer such that $s_n \geqq k_1$ for n sufficiently large; in Fig. 6–1(b), $k_1 = 4$. Let k_2 be the largest integer between 0 and 9 (inclusive) such that

$$s_n \geqq k_1 + \frac{k_2}{10}$$

for n sufficiently large; in Fig. 6–1(b), $k_2 = 3$. Let k_3 be the largest integer between 0 and 9 such that

$$s_n \geqq k_1 + \frac{k_2}{10} + \frac{k_3}{100}$$

for n sufficiently large. Proceeding thus, we obtain the complete decimal expansion of the limit s:

$$s = k_1 + \frac{k_2}{10} + \frac{k_3}{100} + \frac{k_4}{1000} + \cdots.$$

If $s = 4\frac{1}{3}$, one would find

$$s = 4 + \frac{3}{10} + \frac{3}{100} + \frac{3}{1000} + \cdots.$$

A careful perusal of the discussion just given shows that the number s has actually been precisely defined as a real number and that the difference $s - s_n$ can be made as small as desired by making n sufficiently large. Thus Theorem 1 is proved.

If a sequence s_n is monotone increasing, there are then only two possibilities: the sequence is bounded and has a limit, or the sequence is unbounded. In the latter case, for n sufficiently large the sequence becomes larger than any given number K:

$$s_n > K, \quad \text{for } n > N; \tag{6–6}$$

this we also describe by writing:

$$\lim_{n\to\infty} s_n = \infty, \tag{6–7}$$

or "s_n diverges to ∞." This is illustrated by the sequence: $s_n = n$; see Fig. 6–1(d). It can happen that $\lim s_n = \infty$ without s_n being a monotone sequence; e.g., $s_n = n + (-1)^n$.

Similarly, one defines:

$$\lim_{n\to\infty} s_n = -\infty, \tag{6–8}$$

if, for every number K, an N can be found such that

$$s_n < K \quad \text{for } n > N. \tag{6–9}$$

Every monotone decreasing sequence is either bounded or divergent to $-\infty$.

THEOREM 2. *Every unbounded sequence diverges.*

Proof. Suppose on the contrary that the unbounded sequence s_n converges to s; then

$$s - \epsilon < s_n < s + \epsilon$$

for $n > N$. Thus all but a finite number of s_n lie in the interval between $s - \epsilon$ and $s + \epsilon$ and the sequence is necessarily bounded, contrary to assumption.

6–3 Upper and lower limits. It remains to consider sequences s_n which are bounded but not monotone. These can converge, as in Fig. 6–1(a), but can equally well diverge, as the example:

$$1,\ -1,\ 1,\ -1, \ldots,\ (-1)^{n+1}, \ldots$$

illustrates. This example also suggests the way in which a general bounded sequence can diverge, namely, by *oscillating* between various limiting values. In this example, the sequence oscillates between 1 and -1. The sequence

$$1,\ 0,\ -1,\ 0,\ 1,\ 0,\ -1,\ 0,\ 1, \ldots$$

oscillates with three limit values $1, 0, -1$; the oscillation suggests trigonometric functions. In fact, the general term of this sequence can be written thus:

$$s_n = \sin\left(\tfrac{1}{2}n\pi\right).$$

In general the bounded sequence can wander back and forth in its interval coming arbitrarily close to many (even all) values of the interval; it diverges not for lack of limiting values, but because it has too many, namely, more than one.

Of greatest interest for such bounded sequences are the largest and least limiting values, which are termed the *upper* and *lower limits* of the sequence s_n. The upper limit is defined as follows:

$$\overline{\lim_{n\to\infty}}\, s_n = k$$

if, for every positive ϵ,

$$|s_n - k| < \epsilon$$

for infinitely many values of n and if no number larger than k has this property. Similarly one defines the lower limit:

$$\underline{\lim_{n\to\infty}}\, s_n = h$$

if, for every positive ϵ,

$$|s_n - h| < \epsilon$$

for infinitely many values of n and if no number less than h has this property. It is a theorem that *every bounded sequence has an upper limit and a lower limit;* this is proved in much the same way as Theorem 1.

EXAMPLES:

$$\overline{\lim_{n\to\infty}}\,(-1)^n = 1, \quad \underline{\lim_{n\to\infty}}\,(-1)^n = -1,$$

$$\overline{\lim_{n\to\infty}}\,\sin\left(\tfrac{1}{3}n\pi\right) = \tfrac{1}{2}\sqrt{3}, \quad \underline{\lim_{n\to\infty}}\,\sin\left(\tfrac{1}{3}n\pi\right) = -\tfrac{1}{2}\sqrt{3},$$

$$\overline{\lim_{n\to\infty}}\,(-1)^n(1+1/n) = 1, \quad \underline{\lim_{n\to\infty}}\,(-1)^n(1+1/n) = -1,$$

$$\overline{\lim_{n\to\infty}}\,\sin n = 1, \quad \underline{\lim_{n\to\infty}}\,\sin n = -1.$$

These should be verified by graphing the sequences as in Fig. 6–1. The proofs of the relations are not difficult, with the exception of the last two, which are quite subtle.

The definitions of upper and lower limits can be extended to unbounded sequences. One must then consider ∞ and $-\infty$ as possible "limiting values"; thus ∞ is a limiting value if, for every number K, $s_n > K$ for infinitely many n; $-\infty$ is a limiting value if, for every number K, $s_n < K$ for infinitely many n. The upper limit is then defined as the "largest limiting value" and the lower limit as the "smallest limiting value."

EXAMPLES:

$$\overline{\lim_{n\to\infty}}\,(-1)^n n = \infty, \quad \underline{\lim_{n\to\infty}}\,(-1)^n n = -\infty,$$

$$\overline{\lim_{n\to\infty}}\,n^2\sin^2\left(\tfrac{1}{2}n\pi\right) = \infty, \quad \underline{\lim_{n\to\infty}}\,n^2\sin^2\left(\tfrac{1}{2}n\pi\right) = 0.$$

Every sequence will then possess upper and lower limits. It should be noted that, when $\lim s_n = -\infty$, the upper and lower limits are both $-\infty$. In general, the lower limit cannot exceed the upper limit.

THEOREM 3. *If the sequence s_n converges to s, then*

$$\overline{\lim_{n\to\infty}}\,s_n = \underline{\lim_{n\to\infty}}\,s_n = s;$$

conversely, if upper and lower limits are equal and finite, the sequence converges.

The first part of the theorem follows from the definition of convergence; for the limit s satisfies all conditions for both upper and lower limits. If the upper and lower limits are both finite and equal to s, the sequence is necessarily bounded, for otherwise the upper limit would be $+\infty$ or the lower limit would be $-\infty$. If s_n does not converge to s, then for some ϵ there are infinitely many members of the sequence such that $|s_n - s| > \epsilon$. One can then proceed as in the proof of Theorem 1 to produce a limiting value k other than s. This contradicts the fact that s is both the largest and least limiting value.

6–4 Further properties of sequences. We first point out connections between sequences and continuity of functions.

THEOREM 4. *Let $y = f(x)$ be defined for $a \leqq x \leqq b$. If $f(x)$ is continuous at x_0, then*

$$\lim_{n\to\infty} f(x_n) = f(x_0)$$

for every sequence x_n in the interval converging to x_0. Similarly, if $f(x, y)$ is defined in a domain D and is continuous at (x_0, y_0), then

$$\lim_{n\to\infty} f(x_n, y_n) = f(x_0, y_0)$$

for all sequences x_n, y_n such that (x_n, y_n) is in D and x_n converges to x_0, y_n converges to y_0.

We prove the statement for $f(x)$, the proof for a function of two (or more) variables being similar. Given $\epsilon > 0$, a δ can be found such that $|f(x) - f(x_0)| < \epsilon$ for $|x - x_0| < \delta$; this is simply the definition of continuity. One can then choose N so that $|x_n - x_0| < \delta$ for $n > N$, since x_n converges to x_0. Accordingly, $|f(x_n) - f(x_0)| < \epsilon$ for $n > N$; that is, $f(x_n)$ converges to $f(x_0)$.

We remark that there is a converse to Theorem 4: if $f(x)$ has the property that $f(x_n)$ converges to $f(x_0)$ for every sequence x_n which converges to x_0, then $f(x)$ must be continuous at x_0; cf. E. W. Hobson, *Functions of a Real Variable*, 3rd ed., Vol. 1, page 282 (Cambridge University Press, 1927).

THEOREM 5. *If*

$$\lim_{n\to\infty} x_n = x, \quad \lim_{n\to\infty} y_n = y,$$

then

$$\lim_{n\to\infty} (x_n \pm y_n) = x \pm y, \quad \lim_{n\to\infty} (x_n \cdot y_n) = x \cdot y, \quad \lim_{n\to\infty} \frac{x_n}{y_n} = \frac{x}{y},$$

provided, in the last case, there is no division by 0.

Proof. The function $f(x, y) = x + y$ is continuous for all x and y. Hence by Theorem 4

$$\lim_{n\to\infty} f(x_n, y_n) = \lim_{n\to\infty} (x_n + y_n) = x + y.$$

The other rules follow in the same way, with $f(x, y)$ chosen successively as $x - y$, $x \cdot y$, x/y.

THEOREM 6 (*Cauchy criterion*). *If the sequence s_n has the property that for every $\epsilon > 0$ an N can be found such that*

$$|s_m - s_n| < \epsilon \quad \textit{for } n > N \textit{ and } m > N, \tag{6–10}$$

then s_n converges. Conversely, if s_n converges, then for every $\epsilon > 0$ an N can be found such that (6–10) *holds.*

Proof. If condition (6–10) holds, then the sequence s_n is necessarily bounded. Indeed, if we choose a fixed ϵ and N according to (6–10), then for a fixed $m > N$,

$$s_m - \epsilon < s_n < s_m + \epsilon$$

for all n greater than N. Hence all but a finite number of the s_n lie in this interval, so that the sequence must be bounded. If the sequence fails to converge, then by Theorem 3 the upper limit $\bar{s}$ and lower limit $\underline{s}$ must be unequal. Let $\bar{s} - \underline{s} = a$. Then

$$|s_m - \bar{s}| < \frac{a}{3}, \quad |s_n - \underline{s}| < \frac{a}{3}$$

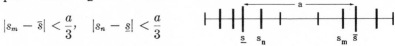

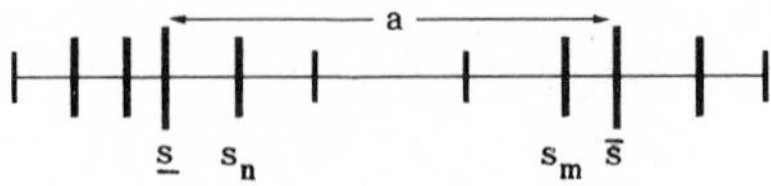

FIG. 6–2.

for infinitely many m and n. Hence $|s_m - s_n| > a/3$ for infinitely many m and n, as Fig. 6–2 shows. If ϵ is chosen as $a/3$, condition (6–10) cannot be satisfied for any N. This is a contradiction; hence $\bar{s} = \underline{s}$ and the sequence converges.

Conversely, if s_n converges to s, then, for a given $\epsilon > 0$, N can be chosen so that $|s_n - s| < \frac{1}{2}\epsilon$ for $n > N$. Hence, for $m > N$ and $n > N$

$$|s_m - s_n| = |s_m - s + s - s_n| \leqq |s_m - s| + |s - s_n| < \tfrac{1}{2}\epsilon + \tfrac{1}{2}\epsilon = \epsilon.$$

Accordingly, condition (6–10) is satisfied.

Problems

1. Show that the following sequences converge and find their limits:

(a) $\dfrac{(n^2 + 1)}{(n^3 + 1)}$, (b) $\dfrac{\log n}{n}$, (c) $\dfrac{n}{2^n}$, (d) $n \log\left(1 + \dfrac{1}{n}\right)$,
(e) $s_n = 1$ for $n = 1, 2, 3, \ldots$.

2. Determine the upper and lower limits of the sequences:

(a) $\cos n\pi$, (b) $\sin \frac{1}{5}n\pi$, (c) $n \sin \frac{1}{2}n\pi$.

3. Construct sequences having the properties stated:

(a) $\overline{\lim\limits_{n\to\infty}}\, s_n = 2$, $\underline{\lim\limits_{n\to\infty}}\, s_n = 0$;

(b) $\overline{\lim\limits_{n\to\infty}}\, s_n = 0$, $\underline{\lim\limits_{n\to\infty}}\, s_n = -\infty$;

(c) $\overline{\lim\limits_{n\to\infty}}\, s_n = +\infty$, $\underline{\lim\limits_{n\to\infty}}\, s_n = +\infty$.

4. Show that the Cauchy criterion is satisfied for the sequence $s_n = 1/n$.

5. Evaluate e to 2 decimal places from its definition as limit of the sequence $(1 + 1/n)^n$.

6. Evaluate $\overline{\lim\limits_{n\to\infty}}\, x^n$ and $\underline{\lim\limits_{n\to\infty}}\, x^n$.

7. The number π can be defined as the limit, as $n \to \infty$, of the area of a regular polygon of 2^n sides inscribed in a circle of radius 1. Show that the sequence is monotone and use it to evaluate π approximately.

Answers

1. (a) 0, (b) 0, (c) 0, (d) 1, (e) 1. 2. (a) 1, −1, (b) 0.951, −0.951, (c) ∞, $-\infty$.

6. Upper limit is ∞ for $|x| > 1$, 1 for $x = \pm 1$ and 0 for $|x| < 1$; lower limit is $-\infty$ for $x < -1$, -1 for $x = -1$, 0 for $-1 < x < 1$, 1 for $x = 1$, ∞ for $x > 1$.

6–5 Infinite series. An infinite series is an indicated sum:

$$a_1 + a_2 + \cdots + a_n + \cdots \tag{6–11}$$

of the members of a sequence a_n. The series can be abbreviated by the Σ sign:

$$a_1 + a_2 + \cdots + a_n + \cdots = \sum_{n=1}^{\infty} a_n, \tag{6–12}$$

and the Σ notation is to be preferred, except for very simple series.

Associated with each infinite series Σa_n is the *sequence of partial sums* S_n:

$$S_n = a_1 + a_2 + \cdots + a_n = \sum_{j=1}^{n} a_j. \tag{6–13}$$

(Note that the index j in the sum here is a *dummy* index, so that the result does not depend on j. Thus $\sum_{j=1}^{n} a_j = \sum_{m=1}^{n} a_m$. This is like the case of the integration variable in a definite integral.) Accordingly,

$$S_1 = a_1, \quad S_2 = a_1 + a_2, \ldots, \quad S_n = a_1 + \cdots + a_n.$$

Definition. The infinite series $\sum_{n=1}^{\infty} a_n$ is *convergent* if the sequence of partial sums is convergent; the series is *divergent* if the sequence of partial sums is divergent. If the series is convergent and the sequence of partial sums S_n converges to S, then S is called the *sum* of the series and one writes

$$\sum_{n=1}^{\infty} a_n = S. \tag{6–14}$$

Thus by definition

$$\sum_{n=1}^{\infty} a_n = \lim_{n \to \infty} \sum_{j=1}^{n} a_j$$

when the limit exists.

If $\lim S_n = +\infty$ or $\lim S_n = -\infty$, the series Σa_n is said to be *properly divergent*. From Theorem 1 of Section 6–2 one concludes:

THEOREM 7. *If $a_n \geqq 0$ for $n = 1, 2, \ldots$, then the series $\sum_{n=1}^{\infty} a_n$ is either convergent or properly divergent.*

For, since $a_n \geqq 0$,

$$S_1 = a_1 \leqq S_2 = a_1 + a_2 \leqq S_3 = a_1 + a_2 + a_3 \ldots;$$

i.e., the S_n form a monotone increasing sequence. If this sequence is bounded, it converges by Theorem 1; if it is unbounded, then necessarily $\lim S_n = \infty$, so that the series is properly divergent.

EXAMPLE 1. The series

$$\frac{1}{2}+\frac{3}{4}+\frac{7}{8}+\cdots+\frac{2^n-1}{2^n}+\cdots$$

is properly divergent. For each term is at least equal to $\frac{1}{2}$ and accordingly

$$S_n \geqq \frac{1}{2}+\frac{1}{2}+\cdots+\frac{1}{2}=n\cdot\frac{1}{2}.$$

The sequence S_n is monotone but unbounded, and $\lim S_n = \infty$.

It should be noted that the *terms* of this series form a convergent sequence; for

$$\lim_{n\to\infty}\left(\frac{2^n-1}{2^n}\right)=\lim_{n\to\infty}\left(1-\frac{1}{2^n}\right)=1.$$

This does not imply convergence of the series. In general, one must carefully distinguish the notions of series, sequence of terms of a series, and sequence of partial sums of a series. The series is simply another way of describing the sequence of partial sums; the terms of the series describe the steps taken from each partial sum to the next. This is suggested in Fig. 6–3.

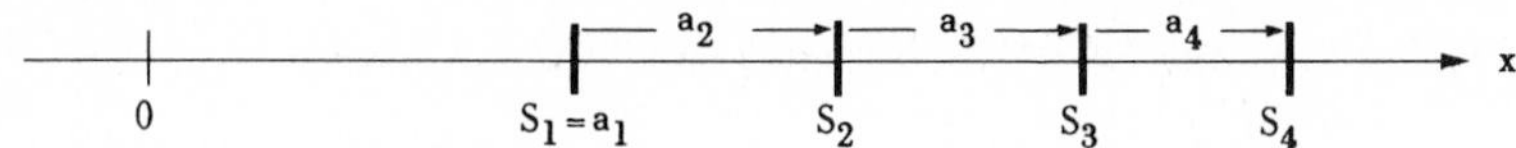

FIG. 6–3. Series Σa_n versus sequence S_n of partial sums.

EXAMPLE 2. The series

$$\frac{1}{2}+\frac{1}{4}+\frac{1}{8}+\cdots+\frac{1}{2^n}+\cdots$$

is convergent. Here the *partial sums* form the sequence

$$\frac{1}{2},\ \frac{3}{4},\ \frac{7}{8},\ldots,\ \frac{2^n-1}{2^n},\ldots$$

which converges, as above, to 1. Hence

$$\sum_{n=1}^{\infty}\frac{1}{2^n}=1.$$

From Theorem 5 of Section 6–4 one deduces a rule for addition and subtraction of convergent series; also a rule for multiplication by a constant:

THEOREM 8. *If* $\sum_{n=1}^{\infty} a_n$ *and* $\sum_{n=1}^{\infty} b_n$ *are convergent with sums* A *and* B *respectively and* k *is a constant, then*

$$\sum_{n=1}^{\infty}(a_n+b_n)=A+B,\quad \sum_{n=1}^{\infty}(a_n-b_n)=A-B,$$
$$\sum_{n=1}^{\infty}(ka_n)=k\sum_{n=1}^{\infty}a_n=kA. \tag{6–15}$$

For if one introduces the partial sums:

$$A_n = a_1 + \cdots + a_n, \quad B_n = b_1 + \cdots + b_n, \quad S_n = (a_1 + b_1) + \cdots + (a_n + b_n),$$

then $S_n = A_n + B_n$. Since A_n converges to A and B_n to B, S_n converges to $A + B$. A similar reasoning holds for the series $\Sigma(a_n - b_n)$, Σka_n.

The Cauchy criterion (Theorem 6 above) can also be interpreted in terms of series:

THEOREM 9. *The series* $\sum_{n=1}^{\infty} a_n$ *is convergent if and only if to each* $\epsilon > 0$ *an* N *can be found such that*

$$|a_{n+1} + a_{n+2} + \cdots + a_m| < \epsilon \quad \textit{for} \quad m > n > N. \tag{6–16}$$

This is a simple rewriting of condition (6–10). For, when $m > n$,

$$S_m - S_n = (a_1 + a_2 + \cdots + a_m) - (a_1 + \cdots + a_n) = a_{n+1} + \cdots + a_m.$$

6–6 Tests for convergence and divergence. A topic of prime importance is the formulation of rules or "tests" which permit one to determine whether a particular series converges or diverges. The problem is similar to and, in fact, closely related to that of improper integrals. This connection will be made clear below.

A few preliminary remarks will be helpful. The convergence or divergence of a series is unaffected if one modifies a *finite* number of terms of the series; thus the first ten terms can be replaced by zeros without affecting the convergence, though it will of course affect the sum if the series converges. The terms of the series can all be multiplied by the *same* nonzero constant k without affecting convergence or divergence, for the partial sums S_n converge or diverge according as kS_n converges or diverges. In particular, all terms can be multiplied by -1 without affecting convergence; however, other changes in sign can change the character of the series completely.

The operations of *grouping terms* and of *rearrangement* of the terms of a series require considerable care. They will be discussed briefly below.

THEOREM 10 (*The n-th term test*). *If it is not true that*

$$\lim_{n\to\infty} a_n = 0,$$

then $\sum_{n=1}^{\infty} a_n$ *diverges.*

Proof. If the series were to converge to S, then

$$\lim_{n\to\infty} a_n = \lim_{n\to\infty} (S_{n+1} - S_n) = \lim_{n\to\infty} S_{n+1} - \lim_{n\to\infty} S_n = S - S = 0.$$

Hence if a_n does not converge to 0, the series cannot converge.

It is to be emphasized that this test can be used only to prove divergence. If $\lim_{n\to\infty} a_n = 0$, the series Σa_n may converge or diverge. It should also be noted that, to show that a_n does not converge to 0, one need not show that a_n converges to a number different from 0; for the same conclusion is reached if one can show that the sequence a_n diverges.

EXAMPLES. 1. $\sum_{n=1}^{\infty} (-1)^n$. Here $a_n = \pm 1$, hence a_n diverges and the series diverges.

2. $\sum_{n=1}^{\infty} n$. $\lim_{n\to\infty} a_n = \lim_{n\to\infty} n = \infty$. The series diverges.

3. $\sum_{n=1}^{\infty} \frac{3n-1}{4n+5}$. $\lim_{n\to\infty} a_n = \frac{3}{4}$. The series diverges.

4. $\sum_{n=1}^{\infty} \frac{1}{n}$. $\lim_{n\to\infty} a_n = 0$. The test proves nothing. The series is the *harmonic* series and will be shown to diverge.

5. $\sum_{n=1}^{\infty} \frac{1}{2^n}$. $\lim_{n\to\infty} a_n = 0$. The test proves nothing. The series converges to 1, as shown in Section 6–5.

If a series Σa_n is such that $\Sigma\, |a_n|$ is convergent, then the series Σa_n is called *absolutely convergent.*

THEOREM 11 (*Theorem on absolute convergence*). *If* $\sum_{n=1}^{\infty} |a_n|$ *converges, then* $\sum_{n=1}^{\infty} a_n$ *converges; that is, every absolutely convergent series is convergent.*

Proof. The theorem follows from the Cauchy criterion (Theorem 9, Section 6–5). For

$$|a_{n+1} + \cdots + a_m| \leqq |a_{n+1}| + |a_{n+2}| + \cdots + |a_m|.$$

If $\sum_{n=1}^{\infty} |a_n|$ converges, then the sum on the right is less than ϵ when $n > N$, for suitable choice of N; hence the sum on the left is less than ϵ for $n > N$, so that the series Σa_n converges.

One can interpret this theorem as stating that introduction of minus signs before various terms of a positive term series tends to *help* convergence; if the original series diverges, introduction of enough minus signs may make the series converge; if the original series converges, introduction of the minus signs will make the series converge even more rapidly.

EXAMPLES. 6. $\sum_{n=0}^{\infty} \frac{(-1)^n}{2^n} = 1 - \frac{1}{2} + \frac{1}{4} \cdots$. Since the series $\sum_{n=1}^{\infty} \frac{1}{2^n}$ converges, this series converges.

7. $\sum_{n=1}^{\infty} (-1)^n$. The series of absolute values is $1 + 1 + \cdots + 1 + \cdots$. This series diverges, so that Theorem 11 gives no help. However the nth term fails to converge to 0, so that the series diverges.

8. $\sum_{n=1}^{\infty} \frac{(-1)^n}{-n} = 1 - \frac{1}{2} + \frac{1}{3} - \frac{1}{4} + \cdots$. The series of absolute values is the harmonic series of Example 4. While the harmonic series diverges, it will be seen that this series *converges*.

A series Σa_n which converges but is not absolutely convergent is called *conditionally convergent*. Such a series converges because the minus signs have been properly introduced. An example is the series of Example 8.

THEOREM 12 (*Comparison test for convergence*). *If* $|a_n| \leqq b_n$ *for* $n = 1, 2, \ldots$ *and* $\sum_{n=1}^{\infty} b_n$ *converges, then* $\sum_{n=1}^{\infty} a_n$ *is absolutely convergent.*

Proof. By Theorem 7, the series $\Sigma|a_n|$ is either convergent or properly divergent. If it were properly divergent, then

$$\lim_{n\to\infty} \sum_{j=1}^{n} |a_j| = \infty;$$

since $b_n \geqq |a_n|$, one would then have

$$\lim_{n\to\infty} \sum_{j=1}^{n} b_j = \infty,$$

so that Σb_n would diverge, contrary to assumption. Hence $\Sigma|a_n|$ converges and Σa_n is absolutely convergent.

EXAMPLE 9. $\sum_{n=1}^{\infty} \frac{1}{n2^n}$. Since

$$\left|\frac{1}{n2^n}\right| = \frac{1}{n2^n} \leqq \frac{1}{2^n}$$

and the series $\Sigma 1/2^n$ converges, the given series converges.

THEOREM 13 (*Comparison test for divergence*). *If* $a_n \geqq b_n \geqq 0$ *for* $n = 1, 2, \ldots$ *and* $\sum_{n=1}^{\infty} b_n$ *diverges, then* $\sum_{n=1}^{\infty} a_n$ *diverges.*

Proof. If Σa_n were to converge, then $\Sigma b_n = \Sigma|b_n|$ would converge by Theorem 12. Hence Σa_n diverges.

It should be emphasized that the test of Theorem 13 applies only to positive term series.

EXAMPLE 10. $\sum_{n=1}^{\infty} \frac{n-1}{n^2}$. It is apparent that the terms are close to those of the divergent harmonic series $\sum_{n=1}^{\infty} \frac{1}{n}$. However, the inequality

$$\frac{n-1}{n^2} > \frac{1}{n}$$

is false, since it is equivalent to the statement: $n^2 - n > n^2$. On the other hand,

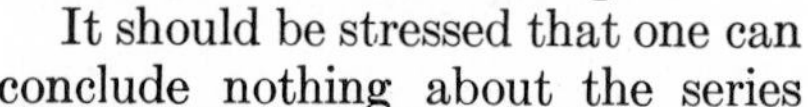

$$\frac{n-1}{n^2} > \frac{1}{2n}$$

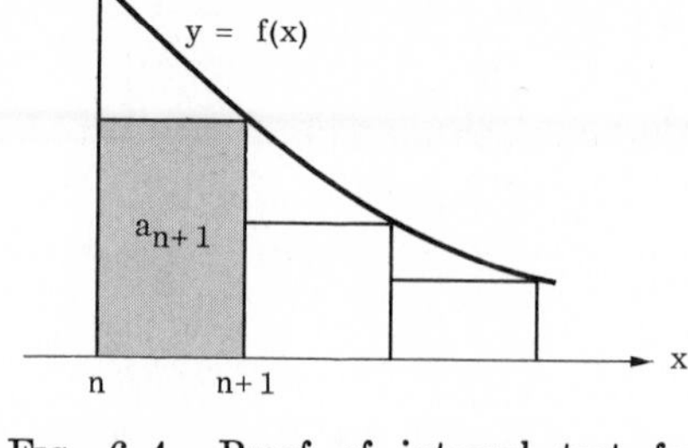

FIG. 6–4. Proof of integral test for convergence.

for $n > 2$, since this is equivalent to the inequality: $2n^2 - 2n > n^2$ or to the inequality $n > 2$. The constant factor $\frac{1}{2}$ and the failure of the condition for $n < 3$ have no effect on convergence or divergence. Hence Theorem 13 applies and one concludes that the series diverges.

It should be stressed that one can conclude nothing about the series Σa_n from the inequality: $|a_n| \leqq |b_n|$, where Σb_n *diverges;* nor can one conclude anything from the inequality: $a_n \geqq b_n > 0$, when Σb_n *converges.*

THEOREM 14 (*Integral test*). *Let $y = f(x)$ satisfy the following conditions:*

(a) *$f(x)$ is defined and continuous for $c \leqq x < \infty$;*
(b) *$f(x)$ decreases as x increases and $\lim_{x\to\infty} f(x) = 0$;*
(c) $f(n) = a_n$.

Then the series $\sum_{n=1}^{\infty} a_n$ converges or diverges according as the improper integral $\int_c^\infty f(x)\,dx$ converges or diverges.

Proof. Let us suppose the improper integral converges. Assumptions (b) and (c) imply that $a_n > 0$ for n sufficiently large. Hence, by Theorem 7 of Section 6–5, the series Σa_n must either converge or properly diverge. Let the integer m be chosen so that $m > c$. Then, since $f(x)$ is decreasing,

$$\int_n^{n+1} f(x)\,dx \geqq f(n+1) = a_{n+1} \quad \text{for } n \geqq m,$$

as illustrated in Fig. 6–4. Hence $0 < a_{m+1} + \cdots + a_{m+p} \leqq \int_m^{m+p} f(x)\,dx \leqq \int_m^\infty f(x)\,dx$. It is thus impossible for the series Σa_n to diverge properly and the series must be convergent.

The proof in the case of divergence is left as an exercise (Prob. 10 below).

THEOREM 15. *The harmonic series of order p:*

$$\sum_{n=1}^{\infty} \frac{1}{n^p} = 1 + \frac{1}{2^p} + \frac{1}{3^p} + \cdots$$

converges for $p > 1$ and diverges for $p \leqq 1$.

Proof. The nth term fails to converge to 0 when $p \leqq 0$, so that the series surely diverges when $p \leqq 0$. For $p > 0$, the integral test can be applied, with $f(x) = 1/x^p$. Now for $p \neq 1$

$$\int_1^\infty \frac{1}{x^p}\,dx = \lim_{b\to\infty}\int_1^b \frac{1}{x^p}\,dx = \lim_{b\to\infty}\left[\frac{1}{p-1}\left(1-\frac{1}{b^{p-1}}\right)\right].$$

This limit exists and has the value $1/(p-1)$ for $p > 1$. For $p < 1$ the integral diverges and for $p = 1$,

$$\int_1^\infty \frac{1}{x}\,dx = \lim_{b\to\infty}\log b = \infty,$$

so that again there is divergence. The theorem is now established.

When $p = 1$, one obtains the divergent series

$$1 + \frac{1}{2} + \cdots + \frac{1}{n} + \cdots,$$

commonly called the *harmonic series;* the connection with harmony (i.e., music) will be pointed out in connection with the Fourier series in the next chapter. For general (even complex) values of p, the series of Theorem 15 defines a function $\zeta(p)$, the *zeta-function of Riemann.*

THEOREM 16. *The geometric series*

$$a + ar + ar^2 + \cdots + ar^n + \cdots = \sum_{n=0}^{\infty} ar^n \quad (a \neq 0)$$

converges for $-1 < r < 1$:

$$\sum_{n=0}^{\infty} ar^n = \frac{a}{1-r}, \qquad -1 < r < 1, \tag{6–17}$$

and diverges for $|r| \geqq 1$.

Proof. As pointed out in Section 0–3 [Eq. (0–29)], the nth partial sum of the geometric series is

$$S_n = a\,\frac{1-r^n}{1-r}$$

for $r \neq 1$. For $-1 < r < 1$, r^n converges to 0; here S_n converges to $a/(1-r)$. For $|r| \geqq 1$, the nth term of the series does not converge to 0, so that the series diverges.

THEOREM 17 (*Ratio test*). *If* $a_n \neq 0$ *for* $n = 1, 2, \ldots$ *and*

$$\lim_{n\to\infty}\left|\frac{a_{n+1}}{a_n}\right| = L,$$

then

if $L < 1$, $\sum_{n=1}^{\infty} a_n$ *is absolutely convergent,*

if $L = 1$, *the test fails,*

if $L > 1$, $\sum_{n=1}^{\infty} a_n$ *is divergent.*

Proof. Let us assume first that $L < 1$. Let $r = L + (1 - L)/2 = (1 + L)/2$, so that $L < r < 1$. Let $b_n = |a_{n+1}/a_n|$. By assumption, b_n converges to L, so that $0 < b_n < r$ for $n > N$ and appropriate choice of N. Now one can write

$$\begin{aligned}|a_{N+1}| + |a_{N+2}| + \cdots + |a_{N+k}| + \cdots \\ &= |a_{N+1}|\left(1 + \left|\frac{a_{N+2}}{a_{N+1}}\right| + \left|\frac{a_{N+2}}{a_{N+1}}\right|\left|\frac{a_{N+3}}{a_{N+2}}\right| + \cdots\right) \\ &= |a_{N+1}|(1 + b_{N+1} + b_{N+1}b_{N+2} + b_{N+1}b_{N+2}b_{N+3} + \cdots).\end{aligned}$$

Since $0 < b_n < r$ for $n > N$, each term of the series in parentheses is less than the corresponding term of the geometric series

$$|a_{N+1}|(1 + r + r^2 + r^3 + \cdots).$$

This series converges, by Theorem 16, since $r < 1$. Hence by Theorem 12 the series $\sum_{n=N+1}^{\infty} |a_n|$ converges, so that $\sum_{n=1}^{\infty} |a_n|$ converges and Σa_n is absolutely convergent.

If $L > 1$, then $|a_{n+1}/a_n| \geqq 1$ for n sufficiently large, so that $|a_n| \leqq |a_{n+1}| \leqq |a_{n+2}| \ldots$. Since the terms are increasing in absolute value (and none is 0), it is impossible for a_n to converge to 0. Accordingly, the series diverges.

For $L = 1$ the series can converge or diverge. This is illustrated by the harmonic series of order p. For

$$\lim_{n\to\infty} \frac{n^p}{(n + 1)^p} = \lim_{n\to\infty} \left(\frac{n}{n + 1}\right)^p = 1.$$

The limit of the ratio is 1; however, the series converges when $p > 1$ and diverges when $p \leqq 1$ by Theorem 15.

THEOREM 18 (*Alternating series test*). *An alternating series*

$$a_1 - a_2 + a_3 - a_4 + \cdots = \sum_{n=1}^{\infty} (-1)^{n+1} a_n, \quad a_n > 0$$

converges if the following two conditions are satisfied:
(a) *its terms are decreasing in absolute value:*

$$a_{n+1} \leqq a_n \quad \textit{for} \quad n = 1, 2, \ldots,$$

(b) $\lim_{n\to\infty} a_n = 0$.

Proof. Let $S_n = a_1 - a_2 + a_3 - a_4 + \cdots + a_n$. Then $S_1 = a_1$, $S_2 = a_1 - a_2 < S_1$, $S_3 = S_2 + a_3 > S_2$, $S_3 = S_1 - (a_2 - a_3) < S_1$, so that $S_2 < S_3 < S_1$. Reasoning in this way, we conclude that

$$S_1 > S_3 > S_5 > S_7 > \cdots > S_6 > S_4 > S_2,$$

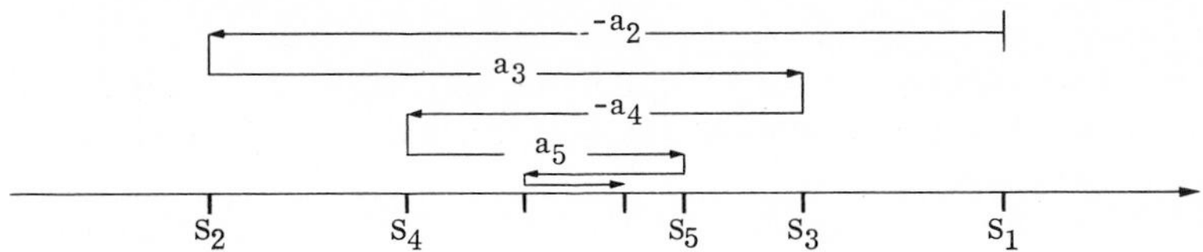

FIG. 6–5. Alternating series.

as shown in Fig. 6–5. Thus the odd partial sums form a bounded monotone decreasing sequence and the even partial sums form a bounded monotone increasing sequence. By Theorem 1 of Section 6–2 both sequences converge:

$$\lim_{n\to\infty} S_{2n+1} = S^*, \quad \lim_{n\to\infty} S_{2n} = S^{**}.$$

Now

$$\lim_{n\to\infty} a_{2n+1} = \lim_{n\to\infty} (S_{2n+1} - S_{2n}) = S^* - S^{**}.$$

By assumption (b), this limit is 0, so that $S^* = S^{**}$. It follows that the series converges to S^*.

THEOREM 19 (*Root test*). *Let a series* $\sum_{n=1}^{\infty} a_n$ *be given and let*

$$\lim_{n\to\infty} \sqrt[n]{|a_n|} = R.$$

Then

if $R < 1$, *the series is absolutely convergent;*
if $R > 1$, *the series diverges;*
if $R = 1$, *the test fails.*

Proof. If $R < 1$, then, as in the proof for the ratio test, one can choose r and N such that $r < 1$ and

$$\sqrt[n]{|a_n|} < r \quad \text{for} \quad n > N.$$

Hence

$$|a_n| < r^n \quad \text{for} \quad n > N$$

and the series $\Sigma |a_n|$ converges by comparison with the geometric series Σr^n. If $R > 1$, $\sqrt[n]{|a_n|} > 1$ for n sufficiently large, so that $|a_n| > 1$ and the nth term cannot converge to zero. The failure for $R = 1$ is again shown by the harmonic series: thus, for $a_n = 1/n^p$,

$$\lim_{n\to\infty} \sqrt[n]{\frac{1}{n^p}} = \lim_{n\to\infty} \frac{1}{n^{p/n}} = \lim_{n\to\infty} \frac{1}{e^{(p/n)\log n}} = \frac{1}{e^0} = 1,$$

since $(1/n) \log n$ converges to 0. These series converge for $p > 1$ and diverge for $p \leqq 1$; thus the root test can yield no information when $R = 1$.

6–7 Examples of applications of tests for convergence and divergence. The results of the preceding section provide the following tests:

(a) *n-th term test for divergence* (Theorem 10);
(b) *comparison test for convergence* (Theorem 12);
(c) *comparison test for divergence* (Theorem 13);
(d) *integral test* (Theorem 14);
(e) *ratio test* (Theorem 17);
(f) *alternating series test* (Theorem 18);
(g) *root test* (Theorem 19).

In addition, the theorem on absolute convergence relates the convergence of a series to that of the series of absolute values of the terms. It is simplest to remember this as a basic principle and then to regard the tests (b), (c), (d), (e), and (g) as tests for *positive* term series.

The general theorems on sequences and series in Sections 6–2 to 6–5 can also be of aid in testing convergence. In particular, one can in certain cases use the definition of convergence directly; that is, one can show that S_n converges to a definite number S. This was done above for the geometric series (Theorem 16). On occasion, the Cauchy criterion (Theorem 9) can also be used. Theorem 8 suggests a method of proving convergence by representing the given series as the "sum" of two convergent series; this idea can also be used to prove divergence, as shown in Example 12 below. On the basis of these remarks we enlarge the above list to include the tests:

(h) *partial sum test;*
(i) *Cauchy criterion* (Theorem 9);
(j) *addition of series* (Theorem 8).

In general, the problem of deciding convergence or divergence of a given series can require great ingenuity and resourcefulness. A vast number of particular series and classes of series have been studied; it is therefore important to be able to use the literature on the subject. The book of Knopp cited at the end of this chapter is suggested as a starting point for this. Some additional information is given in Section 6–8 below and in Chapters 7 and 9, but this can only be regarded as a sampling of a huge field.

It is recommended that the nth term test (Theorem 10) be used first; if this yields no information, other tests can then be tried.

EXAMPLE 1. $\sum\limits_{n=1}^{\infty} \frac{(-1)^{n+1}}{n} = 1 - \frac{1}{2} + \frac{1}{3} \cdots$. The series of absolute values is the harmonic series $\Sigma \frac{1}{n}$, which diverges by Theorem 15. The series is hence not absolutely convergent, so that tests (b), (d), (e) are of no use. Since the terms are of variable sign, (c) is of no use. The nth term does approach 0 as n becomes infinite, so that (a) yields no information. The alternating series test (f) is applicable, since $1 > \frac{1}{2} > \frac{1}{3} \ldots$ and the nth term approaches 0. Hence the series *converges*. Since it is not absolutely convergent, it is *conditionally convergent*.

It is of interest to remark that the sum of this series is $\log 2 = .69315\ldots$. There exist methods of applying (h), i.e., of showing that the nth partial sum does approach $\log 2$ as limit; however, these are too involved to describe here. The Cauchy criterion could also be applied. For one can show, without difficulty, that

$$\left|\frac{1}{n} - \frac{1}{n+1} + \frac{1}{n+2} - \cdots \pm \frac{1}{n+p}\right| < \frac{1}{n} \quad (p > 0).$$

As a matter of fact, the Cauchy criterion could also be used to prove Theorem 18 on alternating series; it will always apply when Theorem 18 applies (cf. Prob. 11 below).

The operations of addition and subtraction, at least in their simplest form, do not help here.

EXAMPLE 2. $\sum_{n=1}^{\infty} (-1)^{n+1} \frac{n+1}{n}$. This at first suggests the alternating series test. However, since $(n+1)/n$ converges to 1, the nth term does not converge to 0. The series diverges.

EXAMPLE 3. $\sum_{n=1}^{\infty} \frac{n+1}{3n^2+5n+2}$. The nth term converges to zero, so that (a) is of no help. For large values of n, the general term is approximately $n/3n^2 = 1/3n$, since the terms of lower degree in the numerator and denominator become small compared to those of highest degree. This suggests comparing with $\Sigma(1/3n)$ for divergence. The inequality

$$\frac{n+1}{3n^2+5n+2} > \frac{1}{3n}$$

is not correct. However, the inequality

$$\frac{n+1}{3n^2+5n+2} > \frac{1}{4n}$$

is correct for $n > 2$. Since $\Sigma(1/n)$ diverges, $\Sigma(1/4n)$ diverges and the given series diverges.

EXAMPLE 4. $\sum_{n=2}^{\infty} \frac{1}{n \log n}$. The terms are smaller than those of the harmonic series, so that one might hope for convergence. However,

$$\int_2^{\infty} \frac{dx}{x \log x} = \lim_{b\to\infty} \log \log x \Big|_2^b = \infty;$$

since all the conditions of the integral test are satisfied, the series diverges.

EXAMPLE 5. $\sum_{n=1}^{\infty} \frac{\log n}{n}$. Here there is no doubt concerning divergence, for

$$\frac{\log n}{n} > \frac{1}{n} \quad (n \geqq 3).$$

EXAMPLE 6. $\sum_{n=1}^{\infty} \frac{\log n}{n^2}$. This is like Example 5, but the higher power of n makes a considerable difference. The function $\log n$ grows very slowly as n increases; in fact,

$$\lim_{n\to\infty} \frac{\log n}{n^p} = 0$$

for $p > 0$, as a consideration of indeterminate forms (Prob. 19 at end of the Introduction) shows. Hence the inequality

$$\frac{\log n}{n^2} = \frac{\log n}{n^{\frac{1}{2}}} \cdot \frac{1}{n^{\frac{3}{2}}} < \frac{1}{n^{\frac{3}{2}}}$$

can be justified for n sufficiently large. Since a finite number of terms of the series can be disregarded in testing convergence, one concludes from Theorem 15 that the series converges.

EXAMPLE 7. $\sum_{n=1}^{\infty} \frac{2^n}{n!}$. Here the ratio test applies:

$$\lim_{n\to\infty} \frac{2^{n+1}}{(n+1)!} \cdot \frac{n!}{2^n} = \lim_{n\to\infty} \frac{2}{n+1} = 0,$$

since

$$\frac{n!}{(n+1)!} = \frac{1 \cdot 2 \cdot 3 \cdots n}{1 \cdot 2 \cdot 3 \cdots n(n+1)} = \frac{1}{n+1}.$$

Hence $L = 0$ and the series converges.

EXAMPLE 8. $\sum_{n=1}^{\infty} \frac{n^n}{n!}$. Again the ratio test applies:

$$\lim_{n\to\infty} \frac{(n+1)^{n+1}}{(n+1)!} \frac{n!}{n^n} = \lim_{n\to\infty} \frac{n+1}{n+1} \cdot \left(\frac{n+1}{n}\right)^n = \lim_{n\to\infty} \left(1 + \frac{1}{n}\right)^n.$$

The limit on the right is e [cf. Eq. (0–142)]; since $e > 1$, the series diverges. One can in fact conclude from this result that

$$\lim_{n\to\infty} \frac{n^n}{n!} = \infty;$$

i.e., the nth term approaches ∞ as n becomes infinite.

EXAMPLE 9. $\sum_{n=1}^{\infty} \frac{2^n}{1 \cdot 3 \cdot 5 \cdots (2n+1)}$. The ratio test applies:

$$\lim_{n\to\infty} \frac{2^{n+1}}{1 \cdot 3 \cdot 5 \cdots (2n+1)(2n+3)} \cdot \frac{1 \cdot 3 \cdot 5 \cdots (2n+1)}{2^n}$$

$$= \lim_{n\to\infty} \frac{2}{2n+3} = 0.$$

The series converges.

EXAMPLE 10. $\sum_{n=1}^{\infty} \log \frac{n}{n+1}$. Here the simplest procedure is to consider the partial sum:

$$S_n = \log\frac{1}{2} + \log\frac{2}{3} + \cdots + \log\frac{n}{n+1}$$
$$= \log\left(\frac{1}{2}\cdot\frac{2}{3}\cdot\frac{3}{4}\cdots\frac{n}{n+1}\right) = \log\frac{1}{n+1}.$$

Thus $\lim S_n = -\infty$ and the series diverges.

EXAMPLE 11. $\sum_{n=1}^{\infty} \frac{n^2+2^n}{2^n n^2}$. Here the principle of addition applies; for the general term is

$$\frac{n^2+2^n}{2^n n^2} = \frac{1}{2^n} + \frac{1}{n^2}.$$

Since $\Sigma(1/2^n)$ and $\Sigma(1/n^2)$ converge, the given series converges.

EXAMPLE 12. $\sum_{n=1}^{\infty}\left(\frac{1}{n} - \frac{1}{2^n}\right)$. Here the addition principle applies in reverse. If this series converges, then the sum

$$\sum_{n=1}^{\infty}\left(\frac{1}{n} - \frac{1}{2^n}\right) + \sum_{n=1}^{\infty}\frac{1}{2^n} = \sum_{n=1}^{\infty}\frac{1}{n}$$

would also have to converge. Hence the series diverges. In general, the "sum" of a convergent series and a divergent series must be divergent. However, the "sum" of two divergent series can be convergent; this is illustrated by the divergent series:

$$\sum_{n=1}^{\infty}\left(\frac{1}{n^2} + \frac{1}{n}\right), \quad \sum_{n=1}^{\infty}\left(\frac{1}{n^2} - \frac{1}{n}\right)$$

which, when added, become the convergent series

$$\sum_{n=1}^{\infty}\frac{2}{n^2}.$$

EXAMPLE 13. $\sum_{n=2}^{\infty}\frac{1}{(\log n)^n}$. Here the root test is easily applicable:

$$\sqrt[n]{|a_n|} = \frac{1}{\log n},$$

so that the root approaches 0 as limit and the series converges.

EXAMPLE 14. $\sum_{n=2}^{\infty}\left(\frac{n}{1+n^2}\right)^n$. Again the root test proves convergence:

$$\lim_{n\to\infty}\sqrt[n]{|a_n|} = \lim_{n\to\infty}\frac{n}{1+n^2} = 0.$$

Problems

1. Prove divergence by the nth term test:

(a) $\sum_{n=1}^{\infty} \sin\left(\frac{n^2\pi}{2}\right)$ (b) $\sum_{n=1}^{\infty} \frac{2^n}{n^3}$

2. Prove convergence by the comparison test:

(a) $\sum_{n=2}^{\infty} \frac{1}{n^3 - 1}$ (b) $\sum_{n=1}^{\infty} \frac{\sin n}{n^2}$

3. Prove divergence by the comparison test:

(a) $\sum_{n=1}^{\infty} \frac{n+5}{n^2 - 3n - 5}$ (b) $\sum_{n=2}^{\infty} \frac{1}{\sqrt{n}\log n}$

4. Prove convergence by the integral test:

(a) $\sum_{n=1}^{\infty} \frac{1}{n^2 + 1}$ (b) $\sum_{n=2}^{\infty} \frac{1}{n \log^2 n}$

5. Prove divergence by the integral test:

(a) $\sum_{n=1}^{\infty} \frac{n}{n^2 + 1}$ (b) $\sum_{n=10}^{\infty} \frac{1}{n \log n \log \log n}$

6. Determine convergence or divergence by the ratio test:

(a) $\sum_{n=1}^{\infty} \frac{(-1)^n}{n!}$ (b) $\sum_{n=1}^{\infty} \frac{2^n + 1}{3^n + n}$

7. Prove convergence by the alternating series test:

(a) $\sum_{n=2}^{\infty} \frac{(-1)^n}{\log n}$ (b) $\sum_{n=2}^{\infty} \frac{(-1)^n \log n}{n}$

8. Prove convergence by the root test:

(a) $\sum_{n=1}^{\infty} \frac{1}{n^n}$ (b) $\sum_{n=1}^{\infty} \left(\frac{n}{n+1}\right)^{n^2}$

9. Prove convergence by showing that the nth partial sum converges:

(a) $\sum_{n=1}^{\infty} \frac{1}{(n+2)(n+1)} = \sum_{n=1}^{\infty} \left(\frac{n+1}{n+2} - \frac{n}{n+1}\right)$,

(b) $\sum_{n=1}^{\infty} \frac{1-n}{2^{n+1}} = \sum_{n=1}^{\infty} \left(\frac{n+1}{2^{n+1}} - \frac{n}{2^n}\right)$.

10. Prove the validity of the integral test (Theorem 14) for divergence.

11. Prove the validity of the alternating series test (Theorem 18) by applying the Cauchy criterion (Theorem 9).

12. Determine convergence or divergence:

(a) $\sum_{n=1}^{\infty} \frac{n+4}{2n^3 - 1}$ (b) $\sum_{n=1}^{\infty} \frac{3n - 5}{n2^n}$

(c) $\sum_{n=1}^{\infty} \frac{e^n}{n+1}$

(d) $\sum_{n=1}^{\infty} \frac{n^2}{n!+1}$

(e) $\sum_{n=1}^{\infty} \frac{n!}{3 \cdot 5 \cdots (2n+3)}$

(f) $\sum_{n=1}^{\infty} \frac{(-1)^n \log n}{2n+3}$

(g) $\sum_{n=2}^{\infty} \frac{1+\log^2 n}{n \log^2 n}$

(h) $\sum_{n=1}^{\infty} \frac{\cos n\pi}{n+2}$

(i) $\sum_{n=1}^{\infty} \frac{\log n}{n+\log n}$

(j) $\sum_{n=1}^{\infty} \left(\frac{n+1}{2n}\right)^n$

Answers

12. (a) conv, (b) conv, (c) div, (d) conv, (e) conv, (f) conv, (g) div, (h) conv, (i) div, (j) conv.

***6–8 Extended ratio test and root test.** It may happen that the test ratio $|a_{n+1}/a_n|$ has no limit as n becomes infinite. The behavior of this ratio may still give information as to the convergence or divergence of the series.

THEOREM 20. *If $a_n \neq 0$ for $n = 1, 2, \ldots$ and there exists a number r such that $0 < r < 1$ and an integer N such that*

$$\left|\frac{a_{n+1}}{a_n}\right| \leqq r \quad \textit{for} \quad n > N,$$

then the series $\sum_{n=1}^{\infty} a_n$ is absolutely convergent. If, on the other hand,

$$\left|\frac{a_{n+1}}{a_n}\right| \geqq 1 \quad \textit{for} \quad n > N$$

for some integer N, then the series diverges.

The proof of Theorem 20 is the same as that of Theorem 17; for in that proof the existence of the limit L is not actually used. In the case of convergence, one uses only the existence of the number r such that

$$\left|\frac{a_{n+1}}{a_n}\right| \leqq r$$

for $n > N$; in the case of divergence, one uses only the fact that the terms are increasing in absolute value and cannot approach 0.

EXAMPLE. The series

$$1 + \frac{1}{3} + \frac{1}{4} + \frac{1}{12} + \frac{1}{16} + \frac{1}{48} + \cdots + \frac{5-(-1)^n}{3 \cdot 2^n} + \cdots,$$

in which the odd terms and even terms each form a geometric series with ratio $\frac{1}{4}$. The ratios of successive terms are

$$\tfrac{1}{3} \div 1 = \tfrac{1}{3}, \quad \tfrac{1}{4} \div \tfrac{1}{3} = \tfrac{3}{4}, \quad \tfrac{1}{12} \div \tfrac{1}{4} = \tfrac{1}{3}, \quad \tfrac{1}{16} \div \tfrac{1}{12} = \tfrac{3}{4}, \ldots;$$

in general the ratio of an even term to the preceding one is $\frac{1}{3}$, the ratio of an odd term to the preceding one is $\frac{3}{4}$. The ratios approach no definite limit, but are always less than or equal to $r = \frac{3}{4}$. Accordingly, the series converges.

It is of interest to note that part of Theorem 20 can be stated in terms of upper limits (Section 6–3):

THEOREM 20(a). *If $a_n \neq 0$ for $n = 1, 2, \ldots$ and*

$$\overline{\lim_{n\to\infty}} \left| \frac{a_{n+1}}{a_n} \right| < 1,$$

then the series Σa_n converges absolutely.

For if the upper limit is $k < 1$, then the ratio must remain below $k + \epsilon$, for n sufficiently large, for each given $\epsilon > 0$. One can choose ϵ so that $k + \epsilon = r < 1$, and Theorem 20 applies.

It should be remarked that no conclusions can be drawn from either of the relations:

$$\overline{\lim_{n\to\infty}} \left| \frac{a_{n+1}}{a_n} \right| = 1, \quad \underline{\lim_{n\to\infty}} \left| \frac{a_{n+1}}{a_n} \right| = 1.$$

The condition

$$\underline{\lim_{n\to\infty}} \left| \frac{a_{n+1}}{a_n} \right| > 1$$

implies divergence, but this is not as precise as Theorem 20.

The root test (Theorem 19) can be extended in the same way as the ratio test:

THEOREM 21. *If there exists a number r such that $0 < r < 1$ and an integer N such that*

$$\sqrt[n]{|a_n|} \leqq r \quad \text{for} \quad n > N,$$

then the series $\sum_{n=1}^{\infty} a_n$ converges absolutely. If, on the other hand,

$$\sqrt[n]{|a_n|} \geqq 1$$

for infinitely many values of n, then the series diverges.

The proof in the case of convergence is the same as that of Theorem 19. If $\sqrt[n]{|a_n|} \geqq 1$ for infinitely many values of n, then $|a_n| \geqq 1$ for infinitely many values of n, so that the nth term cannot converge to 0 and the series diverges.

The theorem has a counterpart in terms of upper limits:

THEOREM 21(a). *If*

$$\overline{\lim_{n\to\infty}} \sqrt[n]{|a_n|} < 1.$$

then the series $\sum_{n=1}^{\infty} a_n$ *is absolutely convergent. If*

$$\overline{\lim_{n\to\infty}} \sqrt[n]{|a_n|} > 1,$$

then the series $\sum_{n=1}^{\infty} a_n$ *diverges.*

No conclusion can be drawn from the relation

$$\overline{\lim_{n\to\infty}} \sqrt[n]{|a_n|} = 1,$$

as the harmonic series show.

Again the previous proof can be repeated.

***6–9 Computation with series — estimate of error.** Up to this point we have considered only the question of convergence or divergence of infinite series. If a series is known to be convergent, one still has the problem of evaluating the sum. In principle, the sum can always be found to desired accuracy by adding up enough terms of the series, i.e., by computing the partial sum S_n for n sufficiently large. However, "sufficiently large" can mean very different things for different series. In order to make the procedure precise, one must have information, for each particular series, as to how large n must be to ensure the accuracy desired. This amounts to saying that one must determine, for each series, a function $N(\epsilon)$ such that

$$|S_n - S| < \epsilon \quad \text{for} \quad n \geqq N;$$

in other words, the first N terms are sufficient to give the desired sum S, with an error less than ϵ.

One can phrase this in another manner. One can write

$$S = S_n + R_n,$$

where R_n is the "remainder." One then seeks $N(\epsilon)$ such that

$$|R_n| < \epsilon \quad \text{for} \quad n \geqq N(\epsilon).$$

It will be seen that, for certain convergent series, one can find a useful explicit *upper estimate* for R_n; more precisely, one can find a sequence T_n converging to 0 for which

$$|R_n| \leqq T_n \quad (n \geqq n_1).$$

If the sequence T_n is *monotone decreasing,* then we can choose $N(\epsilon)$ as the smallest integer n for which $T_n < \epsilon$. For, if $N(\epsilon)$ is so chosen and $n \geqq N(\epsilon)$, then

$$|R_n| \leqq T_n \leqq T_{N(\epsilon)} < \epsilon.$$

It will now be shown how such a monotone decreasing sequence T_n can be found when the series converges by one of the following tests: comparison test, integral test, ratio test, root test, alternating series test.

THEOREM 22. *If* $|a_n| \leqq b_n$ *for* $n \geqq n_1$ *and* $\sum_{n=1}^{\infty} b_n$ *converges, then*

$$|R_n| \leqq \sum_{m=n+1}^{\infty} b_m = T_n$$

for $n \geqq n_1$*; the sequence* T_n *is monotone decreasing and converges to* 0.

Proof. By definition

$$R_n = a_{n+1} + \cdots + a_{n+p} + \cdots = \lim_{p\to\infty} \sum_{m=n+1}^{n+p} a_m.$$

Since, for $n \geqq n_1$,

$$|a_{n+1} + \cdots + a_{n+p}| \leqq |a_{n+1}| + \cdots + |a_{n+p}|$$
$$\leqq b_{n+1} + \cdots + b_{n+p} \leqq \sum_{m=n+1}^{\infty} b_m,$$

one concludes that

$$|R_n| = \lim_{p\to\infty} \left| \sum_{m=n+1}^{n+p} a_m \right| \leqq \sum_{m=n+1}^{\infty} b_m = T_n.$$

Since T_n is the remainder, after n terms, of a convergent series of positive terms, it is necessarily a monotone sequence converging to 0.

The theorem can be stated as follows: if a series converges by the comparison test, then the remainder is in absolute value at most equal to that of the comparison series.

EXAMPLE 1. The series $\sum_{n=1}^{\infty} \frac{1}{n2^n}$ can be compared with the geometric series $\Sigma(1/2^n)$. Hence the remainder after 5 terms is at most

$$T_5 = \frac{1}{2^6} + \frac{1}{2^7} + \cdots = \frac{1}{2^6}\left(1 + \frac{1}{2} + \cdots\right) = \frac{1}{32}.$$

In general, $T_n = 2^{-n-1} + 2^{-n-2} + \cdots = 2^{-n}$. The condition: $T_n < \epsilon$ leads to the inequality: $2^n > 1/\epsilon$, so that

$$n > -(\log \epsilon / \log 2);$$

we can thus choose $N(\epsilon)$ as the smallest integer n satisfying this inequality.

THEOREM 23. *If the series* $\sum_{n=1}^{\infty} a_n$ *converges by the integral test of Theorem 14, with the function* $f(x)$ *decreasing for* $x \geqq c$*, then*

$$|R_n| < \int_n^{\infty} f(x)\, dx = T_n$$

for $n \geqq c$*; the sequence* T_n *is monotone decreasing and converges to* 0.

Proof. This theorem follows from Theorem 22. For one can interpret the improper integral as a series:

$$\int_n^\infty f(x)\,dx = \sum_{m=n+1}^{\infty} b_m, \quad b_m = \int_{m-1}^{m} f(x)\,dx;$$

one has then

$$\int_n^\infty f(x)\,dx = \lim_{p\to\infty} \int_n^{n+p} f(x)\,dx = \lim_{p\to\infty} \sum_{m=n+1}^{n+p} b_m.$$

As in the proof of Theorem 14,

$$|a_n| < \int_{n-1}^{n} f(x)\,dx = b_n,$$

so that

$$|R_n| < \sum_{m=n+1}^{\infty} b_m = \int_n^\infty f(x)\,dx.$$

EXAMPLE 2. For the harmonic series, with order $p > 1$, one finds

$$0 < R_n = \sum_{m=n+1}^{\infty} \frac{1}{m^p} < \int_n^\infty \frac{1}{x^p}\,dx = \frac{1}{(p-1)n^{p-1}}.$$

This result can now be used, by Theorem 22, for any series whose convergence is established by comparison with a harmonic series of order p.

If, for example, $p = 6$, then $T_n = 0.2\,n^{-5}$; it follows that we can choose $N(\epsilon)$ as the smallest integer n such that $n^5 > 0.2\epsilon^{-1}$. Accordingly, to evaluate the harmonic series of order 6 with an error less than 0.0001, 5 terms are sufficient:

$$\sum_{n=1}^{\infty} \frac{1}{n^6} \sim 1 + \frac{1}{2^6} + \frac{1}{3^6} + \frac{1}{4^6} + \frac{1}{5^6} = 1.0173.$$

THEOREM 24. *If*

$$\left|\frac{a_{n+1}}{a_n}\right| \leqq r < 1$$

for $n > n_1$, so that the series Σa_n converges by the ratio test, then

$$|R_n| \leqq \frac{|a_{n+1}|}{1-r} = T_n, \quad n \geqq n_1; \tag{6–18}$$

the sequence T_n is monotone decreasing and converges to 0. If

$$\lim_{n\to\infty} \left|\frac{a_{n+1}}{a_n}\right| = L < 1,$$

then r will be at least equal to L. If

$$1 > \left|\frac{a_{n+2}}{a_{n+1}}\right| \geqq \left|\frac{a_{n+3}}{a_{n+2}}\right|$$

for $n \geqq n_1$, then

$$|R_n| \leqq \frac{|a_{n+1}^2|}{|a_{n+1}| - |a_{n+2}|} = T_n^*, \quad n \geqq n_1; \tag{6–19}$$

the sequence T_n^* is monotone decreasing and converges to 0.

Proof. The first statement follows from Theorem 22 and the fact that the ratio test is a comparison of the given series with a geometric series. Thus one has, as in the proof of Theorem 17,

$$|R_n| \leqq |a_{n+1}| + |a_{n+2}| + \cdots \leqq |a_{n+1}|(1 + r + \cdots) = \frac{|a_{n+1}|}{1 - r} = T_n.$$

The second statement emphasizes the fact that, if the test ratio converges to L, then the ratio cannot remain less than or equal to a number r less than L. Thus $r \geqq L$. One can use $r = L$ only when

$$\left|\frac{a_{n+1}}{a_n}\right| \leqq L$$

for $n > N$, so that the limit L is approached from *below*.

The third statement concerns the case when the test ratios are steadily *decreasing* and hence approaching a limit L. One cannot use L in such a case; however, under the assumptions made, one can use $r = |a_{n+2}/a_{n+1}|$ for $n \geqq n_1$, so that

$$R_n \leqq |a_{n+1}| \frac{1}{1 - \dfrac{|a_{n+2}|}{|a_{n+1}|}} = \frac{|a_{n+1}|^2}{|a_{n+1}| - |a_{n+2}|} = T_n^*.$$

This gives formula (6–19).

Example 3. $\sum_{n=1}^{\infty} \frac{n+1}{n \cdot 2^n}$. The test ratio is found to be

$$\frac{n^2 + 2n}{n^2 + 2n + 1} \cdot \frac{1}{2}.$$

This converges to $\frac{1}{2}$, but is always less than $\frac{1}{2}$. Hence, we can use $r = \frac{1}{2}$ and, for example,

$$R_5 = \frac{7}{6 \cdot 2^6} + \frac{8}{7 \cdot 2^7} + \cdots < \frac{7}{6 \cdot 2^6}\left(1 + \frac{1}{2} + \cdots\right) = \frac{7}{192} = 0.037.$$

Example 4. $\sum_{n=1}^{\infty} \frac{n}{(n+1)2^n}$. Here the test ratio is

$$\frac{n^2 + 2n + 1}{n^2 + 2n} \cdot \frac{1}{2}.$$

Again the limit is $\frac{1}{2}$, but the ratio is always greater than $\frac{1}{2}$. The successive ratios are decreasing, since

$$\frac{n^2 + 2n + 1}{n^2 + 2n} \cdot \frac{1}{2} > \frac{(n+1)^2 + 2(n+1) + 1}{(n+1)^2 + 2(n+1)} \cdot \frac{1}{2},$$

as algebraic manipulation shows. Hence inequality (6–19) applies (for $n \geqq 1$) and one finds, for example,

$$R_5 = \frac{6}{7 \cdot 2^6} + \cdots < \frac{\left(\frac{6}{7 \cdot 2^6}\right)^2}{\frac{6}{7 \cdot 2^6} - \frac{7}{8 \cdot 2^7}} = \frac{9}{329} = 0.027.$$

THEOREM 25. *If*

$$\sqrt[n]{|a_n|} \leqq r < 1$$

for $n > n_1$, so that the series Σa_n converges by the root test, then

$$|R_n| \leqq \frac{r^{n+1}}{1 - r} = T_n, \quad n \geqq n_1. \tag{6–20}$$

If

$$\lim_{n\to\infty} \sqrt[n]{|a_n|} = R < 1,$$

then r will be at least equal to R. If

$$1 > |a_{n+1}|^{1/(n+1)} \geqq |a_{n+2}|^{1/(n+2)}$$

for $n \geqq n_1$, then

$$|R_n| \leqq \frac{|a_{n+1}|}{1 - |a_{n+1}|^{1/(n+1)}} = T_n^*, \quad n \geqq n_1. \tag{6–21}$$

The sequences T_n and T_n^* are monotone decreasing and converge to 0.

The proofs parallel those for Theorem 24, since again the test is based on comparison with a geometric series.

EXAMPLE 5. $\sum_{n=2}^{\infty} \frac{1}{(\log n)^n}$. Here

$$\sqrt[n]{|a_n|} = \frac{1}{\log n}.$$

This is decreasing and less than 1 for $n = 3, 4, \ldots$. Thus, for example,

$$R_5 = \frac{1}{(\log 6)^6} + \frac{1}{(\log 7)^7} + \cdots \leqq \frac{\frac{1}{(\log 6)^6}}{1 - \frac{1}{\log 6}} = 0.06\ldots.$$

THEOREM 26. *If the series*

$$a_1 - a_2 + a_3 - a_4 + \cdots = \sum_{n=1}^{\infty} (-1)^{n+1} a_n, \quad a_n > 0 \tag{6–22}$$

converges by the alternating series test, then

$$0 < |R_n| < a_{n+1} = T_n.$$

Hence $N(\epsilon)$ can be chosen as the smallest integer such that $a_{n+1} < \epsilon$.

The theorem can be stated in words as follows: when a series converges by the alternating series test, the error made in stopping at n terms is in absolute value less than the first term neglected.

Proof of Theorem 26. As pointed out in the proof of the alternating series test (Theorem 18), one has

$$S_2 < S_4 < \cdots < S_{2n} < \cdots < S_{2n-1} \cdots < S_3 < S_1.$$

It follows that the sum S lies between each two successive partial sums (one odd, one even):

$$S_{2n} < S < S_{2n+1}, \quad S_{2n} < S < S_{2n-1}.$$

Hence

$$0 < R_{2n} = S - S_{2n} < S_{2n+1} - S_{2n} = a_{2n+1},$$
$$0 > R_{2n-1} = S - S_{2n-1} > S_{2n} - S_{2n-1} = -a_{2n};$$

that is,

$$0 < R_{2n} < a_{2n+1}, \quad -a_{2n} < R_{2n-1} < 0.$$

This shows that, for every n,

$$0 < |R_n| < a_{n+1},$$

but actually tells more: R_n is positive, if n is even; R_n is negative, if n is odd.

EXAMPLE 6. $\sum_{n=1}^{\infty} \frac{(-1)^{n+1}}{n}$. Here the partial sums are $S_1 = 1, S_2 = 1 - \frac{1}{2} = \frac{1}{2}$, $S_3 = \frac{5}{6}$, $S_4 = \frac{7}{12}, \ldots$. The above theorem asserts that $|R_1| < \frac{1}{2}$, $|R_2| < \frac{1}{3}$, $|R_3| < \frac{1}{4}, \ldots$ or more precisely that $-\frac{1}{2} < R_1 < 0$, $0 < R_2 < \frac{1}{3}$, $-\frac{1}{4} < R_3 < 0$. Thus in particular, if 3 terms are used, the sum is between $\frac{5}{6}$ and $\frac{5}{6} - \frac{1}{4} = \frac{7}{12}$. In order to compute the sum with an error of less than 0.01, one would need 100 terms, for the 101st term: 1/101 is the first one less than 0.01.

Problems

1. Determine how many terms are sufficient to compute the sum with given allowed error ϵ and find the sum to this accuracy:

(a) $\sum_{n=1}^{\infty} \frac{1}{n^2}, \quad \epsilon = 1$

(b) $\sum_{n=1}^{\infty} \frac{(-1)^{n+1}}{n^2}, \quad \epsilon = 0.1$

(c) $\sum_{n=1}^{\infty} \frac{n}{n^3 + 5}, \quad \epsilon = 0.2$

(d) $\sum_{n=1}^{\infty} \frac{1}{n^2 + 1}, \quad \epsilon = 0.5$

(e) $\sum_{n=1}^{\infty} \frac{1}{n^n}, \quad \epsilon = 0.01$

(f) $\sum_{n=1}^{\infty} \frac{1}{n!}, \quad \epsilon = 0.01$

(g) $\sum_{n=1}^{\infty} \frac{(-1)^{n+1}}{(2n-1)!}, \quad \epsilon = 0.001$

(h) $\sum_{n=2}^{\infty} \frac{(-1)^n}{n \log n}, \quad \epsilon = 0.5$

(i) $\sum_{n=2}^{\infty} \frac{1}{n^3 \log n}, \quad \epsilon = 0.5$

(j) $\sum_{n=1}^{\infty} \frac{2^n}{3^n + 1}, \quad \epsilon = 0.1$

2. Let Σa_n be the geometric series $1 + r + r^2 + \cdots$.

(a) Determine how many terms are needed to compute the sum with error less than 0.01 when $r = \frac{1}{2}$, $r = 0.9$, $r = 0.99$.

(b) Show that, for any positive ϵ, $|R_n| < \epsilon$ when

$$n > \frac{\log \epsilon(1-r)}{\log |r|}, \quad -1 < r < 1,$$

and that no smaller value of n will suffice.

(c) Show that, as r approaches 1, the number of terms needed to compute the sum with error less than a fixed ϵ becomes infinite.

3. Show that, for $p > 0$, the sum of the series $1 - 1/2^p + 1/3^p - \cdots$ is positive.

Answers

1. (a) 1 term, 1, (b) 3 terms, 0.86, (c) 5 terms, 0.51, (d) 2 terms, 0.70, (e) 3 terms, 1.287, (f) 4 terms, 1.709, (g) 3 terms, 0.8417, (h) 1 term, 0.72, (i) 1 term, 0.18, (j) 8 terms, 1.70.

2. (a) 8 terms, 66 terms, 918 terms.

6–10 Operations on series.

It has been pointed out above (Theorem 8) that convergent series can be added and subtracted term by term:

$$\sum_{n=1}^{\infty} a_n + \sum_{n=1}^{\infty} b_n = \sum_{n=1}^{\infty} (a_n + b_n), \quad \sum_{n=1}^{\infty} a_n - \sum_{n=1}^{\infty} b_n = \sum_{n=1}^{\infty} (a_n - b_n),$$

also that a convergent series can be multiplied by a constant:

$$k \sum_{n=1}^{\infty} a_n = \sum_{n=1}^{\infty} (ka_n).$$

Three other operations will be considered in the present section: multiplication, grouping, and rearrangement.

We consider first the operation of *grouping:* that is, of inserting parentheses in a series Σa_n. For example, the series Σa_n might be replaced by the series $(a_1 + a_2) + \cdots + (a_{2n-1} + a_{2n}) + \cdots$; that is, by Σb_n, where $b_n = a_{2n-1} + a_{2n}$.

THEOREM 27. *If the series $\sum_{n=1}^{\infty} a_n$ is convergent, then insertion of parentheses yields a new convergent series having the same sum as $\sum_{n=1}^{\infty} a_n$. If $\sum_{n=1}^{\infty} a_n$ is properly divergent, insertion of parentheses yields a properly divergent series.*

Proof. The effect of insertion of parentheses is to cause one to *skip* certain partial sums. Thus the partial sums of the series

$$(a_1 + a_2) + (a_3 + a_4) + (a_5 + a_6) + \cdots + (a_{2n-1} + a_{2n}) + \cdots$$

are the partial sums $S_2, S_4, S_6, \ldots, S_{2n}, \ldots$ of the series Σa_n. If S_n converges to S, then the new sequence obtained by skipping must also converge to S; if $\lim S_n = +\infty$, then the sequence obtained by skipping must also diverge to $+\infty$. Thus the theorem follows.

Since a positive term series is either convergent or properly divergent, one can insert parentheses freely without affecting convergence or sum in such a series. This can be used to prove divergence of the harmonic series, for one has

$$\sum_{n=1}^{\infty}\frac{1}{n} = 1 + \frac{1}{2} + \left(\frac{1}{3} + \frac{1}{4}\right) + \left(\frac{1}{5} + \frac{1}{6} + \frac{1}{7} + \frac{1}{8}\right)$$
$$+ \left(\frac{1}{9} + \cdots + \frac{1}{16}\right) + \cdots + \left(\frac{1}{2^{n-1}+1} + \cdots + \frac{1}{2^n}\right) + \cdots;$$

each parenthesis contributes at least $\frac{1}{2}$, so that the general term of the grouped series does not converge to zero.

For a divergent series of variable signs, insertion of parentheses can on occasion produce a convergent series. Thus the series $\Sigma(-1)^{n+1}$ becomes

$$(1-1) + (1-1) + \cdots + (1-1) + \cdots = 0,$$

with the parentheses inserted as shown. Therefore, in testing a series of variable signs for convergence, parentheses should not be inserted.

A series $\sum_{m=1}^{\infty} b_m$ is said to be a *rearrangement* of a series $\sum_{m=1}^{\infty} a_n$ if there exists a one-to-one correspondence between the indices n and m such that $a_n = b_m$ for corresponding indices. Thus the series

$$1 + \tfrac{1}{3} + \tfrac{1}{2} + \tfrac{1}{5} + \tfrac{1}{4} + \tfrac{1}{7} + \tfrac{1}{6} + \cdots,$$
$$1 + \tfrac{1}{2} + \tfrac{1}{3} + \tfrac{1}{4} + \tfrac{1}{5} + \tfrac{1}{6} + \tfrac{1}{7} + \cdots,$$

are rearrangements of each other. Also

$$1 + \tfrac{1}{2} + \tfrac{1}{4} + \tfrac{1}{3} + \tfrac{1}{6} + \tfrac{1}{8} + \tfrac{1}{5} + \tfrac{1}{10} + \tfrac{1}{12} + \tfrac{1}{7} + \cdots$$

is a rearrangement of the harmonic series.

THEOREM 28. *If $\sum_{n=1}^{\infty} a_n$ is absolutely convergent and $\sum_{m=1}^{\infty} b_m$ is a rearrangement of $\sum_{n=1}^{\infty} a_n$, then $\sum_{m=1}^{\infty} b_m$ is absolutely convergent and has the same sum as $\sum_{n=1}^{\infty} a_n$.*

Proof. One has clearly

$$\sum_{m=1}^{N} |b_m| \leqq \sum_{n=1}^{\infty} |a_n|,$$

since every b_m equals an a_n for appropriate n and no two m's correspond to the same n. Hence the series $\sum_{m=1}^{\infty} |b_m|$ converges, so that Σb_m is absolutely convergent. Let S_n and S be the nth partial sum and sum for Σa_n, let S'_m and S' be the corresponding quantities for Σb_m. For given positive ϵ, let N be chosen so large that

$$|S_n - S| < \tfrac{1}{2}\epsilon \quad \text{and} \quad |a_{n+1}| + \cdots + |a_{n+p}| < \tfrac{1}{2}\epsilon$$

for $n > N$ and $p \geqq 1$; such an N can be found since Σa_n converges to S and since the Cauchy criterion holds for $\Sigma|a_n|$. For m sufficiently large, S'_m will be a sum of terms including all of $a_1 \ldots a_n$ and perhaps more:

$$S'_m = S_n + a_{k_1} + \cdots a_{k_s},$$

where $k_1, \ldots, k_s$ are all larger than n. Let $k_0 = n + p_0$ be the largest of these indices. Then

$$|S'_m - S_n| \leqq |a_{k_1}| + \cdots + |a_{k_s}| \leqq |a_{n+1}| + \cdots + |a_{n+p_0}| < \tfrac{1}{2}\epsilon.$$

Now

$$|S'_m - S| = |S'_m - S_n + S_n - S| \leqq |S'_m - S_n| + |S_n - S| \leqq \tfrac{1}{2}\epsilon + \tfrac{1}{2}\epsilon = \epsilon.$$

It follows that S'_m converges to S, so that $S' = S$ as asserted.

The preceding theorem is of importance in connection with the *multiplication* of series. If one multiplies two series as in algebra:

$$\begin{aligned}(a_1 + a_2 + a_3 + \cdots + a_n + \cdots)&(b_1 + b_2 + \cdots + b_n + \cdots)\\ &= a_1b_1 + a_1b_2 + \cdots + a_2b_1 + a_2b_2 + \cdots,\end{aligned}$$

one obtains a collection of terms of form a_nb_m without any special order. It will be seen that, when Σa_n and Σb_m are absolutely convergent, the products a_nb_m can be arranged in order to form an absolutely convergent infinite series. It follows from Theorem 28, that the same result is obtained no matter how the products are arranged and that the same sum is always obtained. Of the many possible arrangements, the following one, known as the *Cauchy product*, is the one most commonly used:

$$a_1b_1 + a_1b_2 + a_2b_1 + a_1b_3 + a_2b_2 + a_3b_1 + \cdots.$$

This is illustrated in Fig. 6–6. If the series Σa_n and Σb_m are indexed beginning with $n = 0$ and $m = 0$, the Cauchy product of the series appears as

$$a_0b_0 + a_0b_1 + a_1b_0 + a_0b_2 + a_1b_1 + a_2b_0 + \cdots.$$

This series is suggested by the multiplication of two power series (to be considered below):

$$\begin{aligned}(a_0 + a_1x + a_2x^2 + \cdots a_nx^n + \cdots)&(b_0 + b_1x + b_2x^2 + \cdots b_mx^m + \cdots)\\ &= a_0b_0 + x(a_0b_1 + a_1b_0) + x^2(a_0b_2 + a_1b_1 + a_2b_0) + \cdots,\end{aligned}$$

the terms of same degree in x being collected together.

THEOREM 29. *If the series* $\sum_{n=1}^{\infty} a_n$ *and* $\sum_{m=1}^{\infty} b_m$ *are absolutely convergent, then the products* a_nb_m *can be arranged to form an absolutely convergent series* $\sum_{i=1}^{\infty} c_i$ *and*

$$\sum_{n=1}^{\infty} a_n \cdot \sum_{m=1}^{\infty} b_m = \sum_{i=1}^{\infty} c_i. \tag{6–23}$$

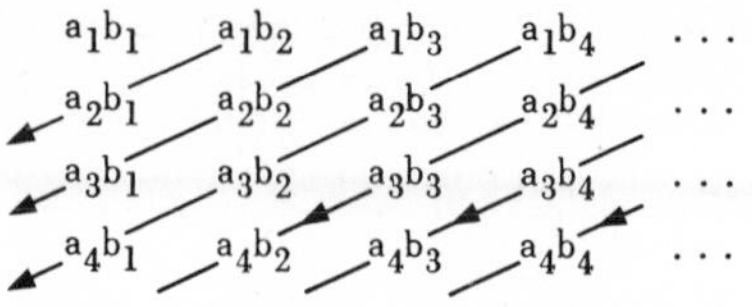

$a_1b_1 \quad a_1b_2 \quad a_1b_3 \quad a_1b_4 \quad \cdots$
$a_2b_1 \quad a_2b_2 \quad a_2b_3 \quad a_2b_4 \quad \cdots$
$a_3b_1 \quad a_3b_2 \quad a_3b_3 \quad a_3b_4 \quad \cdots$
$a_4b_1 \quad a_4b_2 \quad a_4b_3 \quad a_4b_4 \quad \cdots$

FIG. 6–6. Cauchy product of series.

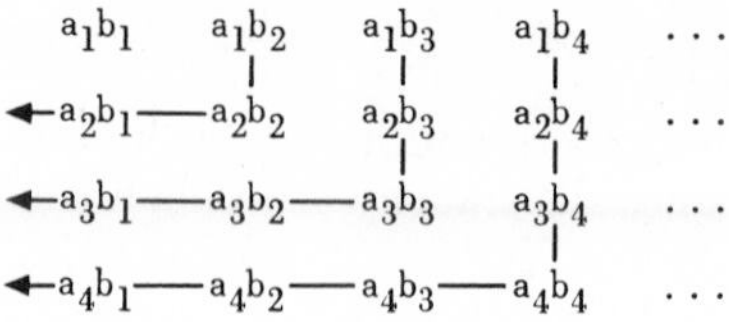

$a_1b_1 \quad a_1b_2 \quad a_1b_3 \quad a_1b_4 \quad \cdots$
$a_2b_1 — a_2b_2 \quad a_2b_3 \quad a_2b_4 \quad \cdots$
$a_3b_1 — a_3b_2 — a_3b_3 \quad a_3b_4 \quad \cdots$
$a_4b_1 — a_4b_2 — a_4b_3 — a_4b_4 \quad \cdots$

FIG. 6–7.

Proof. We choose the series Σc_i as the series

$$a_1b_1 + a_1b_2 + a_2b_2 + a_2b_1 + a_1b_3 + a_2b_3 + a_3b_3 + a_3b_2 + a_3b_1 + \cdots,$$

as suggested in Fig. 6–7. From this definition, one has in particular

$$c_1 = a_1b_1, \quad c_1 + c_2 + c_3 + c_4 = (a_1 + a_2)(b_1 + b_2), \ldots$$

and in general

$$c_1 + (c_2 + c_3 + c_4) + \cdots + (\cdots + c_{n^2}) = (a_1 + \cdots + a_n)(b_1 + \cdots + b_n).$$

Similarly,

$$|c_1| + (|c_2| + |c_3| + |c_4|) + \cdots + (\cdots + |c_{n^2}|) = (|a_1| + \cdots + |a_n|)(|b_1| + \cdots |b_n|).$$

Since the series $\Sigma|a_n|$ and $\Sigma|b_n|$ converge, the equation just written shows that the left-hand side approaches a limit as $n \to \infty$; dropping parentheses, on the basis of Theorem 27, we conclude that $\Sigma|c_n|$ converges, so that Σc_n is absolutely convergent. The preceding equation then shows that

$$\sum_{i=1}^{\infty} c_i = \lim_{n\to\infty} [c_1 + (\cdots) + \cdots + (\cdots + c_{n^2})] = \sum_{n=1}^{\infty} a_n \cdot \sum_{m=1}^{\infty} b_m,$$

the insertion of parentheses being justified by Theorem 27. The theorem is thus proved.

On the basis of Theorem 28, one can now use any arrangement desired for the c's, in particular that of the Cauchy product. One can also *group* the terms as suggested by the power series multiplication above:

$$\left(\sum_{n=0}^{\infty} a_n\right)\left(\sum_{m=0}^{\infty} b_m\right) = a_0b_0 + (a_0b_1 + a_1b_0) + (a_0b_2 + a_1b_1 + a_2b_0) + \cdots + (a_0b_n + a_1b_{n-1} + \cdots + a_{n-1}b_1 + a_nb_0) + \cdots.$$

The insertion of parentheses is justified by Theorem 27.

If the series Σa_n and Σb_n converge, but are not both absolutely convergent, one can nevertheless form the Cauchy product of the series as Σc_n, where $c_n = a_0b_n + a_1b_{n-1} + \cdots + a_{n-1}b_1 + a_nb_0$. It can be shown that, if the series Σc_n converges, then always

$$\sum_{n=0}^{\infty} a_n \cdot \sum_{n=0}^{\infty} b_n = \sum_{n=0}^{\infty} c_n. \tag{6–24}$$

Furthermore, if one of the two series Σa_n, Σb_n converges absolutely, then Σc_n must converge. For proofs, one is referred to K. Knopp, *Infinite Series* page 321 (London: Blackie, 1928).

Problems

1. Assuming the relations (proved in the next chapter):

$$\sum_{n=1}^{\infty}\frac{1}{n^2}=\frac{\pi^2}{6},\quad \sum_{n=1}^{\infty}\frac{1}{n^4}=\frac{\pi^4}{90},\quad \sum_{n=1}^{\infty}\frac{1}{n^6}=\frac{\pi^6}{945},$$

evaluate the series:

(a) $\displaystyle\sum_{n=1}^{\infty}\frac{6}{n^2}$ (b) $\displaystyle\sum_{n=1}^{\infty}\frac{n^2+1}{n^4}$ (c) $\displaystyle\sum_{n=1}^{\infty}\frac{2n^2-3}{n^4}$ (d) $\displaystyle\sum_{n=1}^{\infty}\frac{9+3n^2+5n^4}{n^6}$ (e) $\displaystyle\sum_{n=3}^{\infty}\frac{n^4-1}{n^6}$ (f) $\displaystyle\sum_{n=2}^{\infty}\frac{n^2+1}{(n^2-1)^2}$

2. Verify the following relations:

(a) $\displaystyle\sum_{n=1}^{\infty}\frac{1}{n^3}=\sum_{n=2}^{\infty}\frac{1}{(n-1)^3};$

(b) $\displaystyle\sum_{n=1}^{\infty}[f(n+1)-f(n)]=\lim_{n\to\infty}f(n)-f(1)$, if the limit exists;

(c) $\displaystyle\sum_{n=2}^{\infty}[f(n+1)-f(n-1)]=\lim_{n\to\infty}[f(n)+f(n+1)]-f(1)-f(2)$, if the limit exists.

3. Use $f(n) = 1/n$ and $f(n) = 1/n^2$ in 2(b) and 2(c) to prove

(a) $\displaystyle\sum_{n=1}^{\infty}\frac{2n+1}{n^2(n+1)^2}=1$ (b) $\displaystyle\sum_{n=1}^{\infty}\frac{1}{n(n+1)}=1$ (c) $\displaystyle\sum_{n=2}^{\infty}\frac{1}{n^2-1}=\frac{3}{4}$ (d) $\displaystyle\sum_{n=2}^{\infty}\frac{4n}{(n^2-1)^2}=\frac{5}{4}$

4. From the relation

$$\frac{1}{1-r}=1+r+\cdots+r^n+\cdots=\sum_{n=0}^{\infty}r^n,\quad -1<r<1,$$

prove the following:

(a) $\displaystyle\frac{1}{(1-r)^2}=1+2r+3r^2+\cdots+(n+1)r^n+\cdots,\quad -1<r<1;$

(b) $\displaystyle\frac{1}{(1-r)^3}=1+3r+\cdots+\frac{(n+2)(n+1)r^n}{2}+\cdots.\quad -1<r<1.$

5. Prove the general binomial formula for negative exponents:

$$(1-r)^{-k}=1+kr+\frac{k(k+1)}{1\cdot 2}r^2+\cdots+\frac{k(k+1)\cdots(k+n-1)}{1\cdot 2\cdots n}r^n+\cdots,$$
$$-1<r<1,\ k=1,2,\cdots$$

6. Assuming (as will be proved below) that $\sin x$ and $\cos x$ are representable for all x by the absolutely convergent series:

$$\sin x = x - \frac{x^3}{3!} + \frac{x^5}{5!} - \cdots + (-1)^{n+1}\frac{x^{2n-1}}{(2n-1)!} + \cdots,$$

$$\cos x = 1 - \frac{x^2}{2!} + \frac{x^4}{4!} - \cdots + (-1)^{n}\frac{x^{2n}}{(2n)!} + \cdots,$$

verify by means of series the identity:

$$2\sin x\cos x = \sin 2x.$$

Answers

1. (a) π^2, (b) $\frac{\pi^2}{6} + \frac{\pi^4}{90}$, (c) $\frac{\pi^2}{3} - \frac{\pi^4}{30}$, (d) $\frac{5\pi^2}{6} + \frac{\pi^4}{30} + \frac{\pi^6}{105}$, (e) $\frac{\pi^2}{6} - \frac{\pi^6}{945} - \frac{15}{64}$, (f) $\frac{\pi^2}{6} - \frac{5}{8}$.

6–11 Sequences and series of functions. If for each positive integer n a function $f_n(x)$ is given, then the functions $f_n(x)$ are said to form a *sequence of functions*. It will usually be assumed that the functions are all defined over the same interval (perhaps infinite) on the x axis.

For each fixed x of the interval, the sequence $f_n(x)$ may converge or diverge. The first question of interest for a particular sequence is the determination of the values of x for which it converges. Thus the sequence $f_n(x) = x^n/2^n$ converges for $-2 < x \leqq 2$ and diverges otherwise.

Similar remarks hold for infinite series whose terms are functions:

$$u_1(x) + u_2(x) + \cdots + u_n(x) + \cdots = \sum_{n=1}^{\infty} u_n(x).$$

The nth partial sum of such a series is itself a function of x:

$$S_n = S_n(x) = u_1(x) + \cdots + u_n(x).$$

Convergence of the series is by definition equivalent to convergence of the sequence of partial sums; if $S_n(x)$ converges to $S(x)$ for certain values of x, then

$$\sum_{n=1}^{\infty} u_n(x) = S(x)$$

for this range of x.

Examples of this have been considered above: the series $\sum_{n=1}^{\infty} 1/n^p$, in which the terms are functions of p, and the geometric series $\sum_{n=0}^{\infty} r^n$, in which the terms are functions of r. The first converges for $p > 1$, the second for $|r| < 1$. Other examples are the following:

$$\sum_{n=0}^{\infty}\frac{x^n}{n!}, \quad \sum_{n=1}^{\infty}\frac{x^n}{n^2}, \quad \sum_{n=1}^{\infty}\frac{\cos nx}{n^2},$$

which can be shown to converge for all x, for $|x| \leqq 1$, and for all x, respectively.

The tests for convergence of a series of functions are the same as those for a series of constants. For each particular choice of x (or whatever the independent variable may be), the series $\Sigma u_n(x)$ is nothing but a series of constants. Thus, when x ranges over an interval, the problem is really that of discussing the convergence of many (in fact, infinitely many) series. One could, of course, test a variety of x values separately; this is rarely advisable, although a particular application might make it unavoidable. It will be seen that for many series it is possible to determine by a single analysis all the values of x for which the series converges.

For power series precise results are stated below. In many cases the ratio test alone gives the answer; e.g., for the series

$$\sum_{n=0}^{\infty} \frac{x^n}{n!},$$

which represents e^x, one has

$$\lim_{n\to\infty} \left|\frac{a_{n+1}}{a_n}\right| = \lim_{n\to\infty} \left(\frac{|x^{n+1}|}{(n+1)!} \cdot \frac{n!}{|x^n|}\right) = \lim_{n\to\infty} \frac{|x|}{n+1}.$$

Since this limit is 0 for each x, the series converges for all x.

6–12 Uniform convergence. While it may thus be possible to prove that a series $\Sigma u_n(x)$ converges for all x of a certain range, another question remains to be answered: is the convergence equally rapid for all x of the range? This question turns out to be important both for practical applications and for basic theory. The practical application is as follows. Suppose it is known that a series $\Sigma u_n(x)$ converges for $a < x < b$ to a function $f(x)$. If the function $f(x)$ is a complicated one, while the functions $u_n(x)$ are simple, it would be a considerable simplification if we could replace the function $f(x)$ by a partial sum $S_n(x) = u_1(x) + \cdots + u_n(x)$; e.g., if the $u_n(x)$ are of the form $c_n x^n$, we would be replacing $f(x)$ by a *polynomial*. This amounts to "rounding off" the series of functions, just as we rounded off the decimal expansion of $\frac{1}{3}$ in Section 6–1. The procedure can be justified, if the error made is below a prescribed value ϵ; that is, if $|f(x) - S_n(x)| < \epsilon$ for $a < x < b$. However, it can happen, as will be seen below, that no matter how large n is chosen the error $|f(x) - S_n(x)|$ exceeds ϵ for some x of the interval $a < x < b$. This does not contradict the fact that the series converges for each x; it is simply due to the fact that the number of terms needed to reduce the error below ϵ varies from one x to another in such a manner that no one n will serve for the whole interval. When this difficulty arises, rounding off over the whole interval: $a < x < b$ cannot be justified. On the other hand, it may happen that, however ϵ is chosen, an N can be found such that, for each $n \geqq N$, $|S_n(x) - f(x)| < \epsilon$ for all x of the interval: $a < x < b$; if this is so, the series is termed *uniformly* convergent in the interval: $a < x < b$. A uniformly convergent series can be rounded off to any accuracy desired:

Definition. The series $\sum_{n=1}^{\infty} u_n(x)$ *is uniformly convergent to* $S(x)$ *for a set* E *of values of* x *if for each* $\epsilon > 0$ *an integer* N *can be found such that*

$$|S_n(x) - S(x)| < \epsilon$$

for $n \geqq N$ *and all* x *in* E.

The essential idea is that N can be found as soon as ϵ (or the number of decimal places desired) is known, without considering which x of the set E is going to be used. The set E must, of course, be a set of values of x for which the series converges, but it need not consist of all such values. In many cases it can be shown that only by restricting attention to a part of the range for which convergence is known can it be ensured that the convergence is uniform. In the worst case, one could restrict attention to a finite number of values of x, for which the convergence *must* be uniform.

The definition of uniform convergence is given above in terms of *series;* it can equally well be given in terms of *sequences:* the sequence $f_n(x)$ is *uniformly convergent* to $f(x)$ for a set E of values of x, if for each $\epsilon > 0$ an N can be found such that

$$|f_n(x) - f(x)| < \epsilon$$

for $n \geqq N$ and all x in E. Thus the uniform convergence of the series $\Sigma u_n(x)$ is equivalent to the uniform convergence of its sequence of partial sums $S_n(x)$. [Actually, sequences and series are simply two different ways of describing the same limit process; every series can be interpreted in terms of its partial sums and every sequence S_n can be interpreted as the partial sums of a series, namely, the series: $S_1+(S_2-S_1)+\cdots+(S_n-S_{n-1})+\cdots$.]

The most common case of uniform convergence is that in which the functions $u_n(x)$ are continuous in a closed interval $a \leqq x \leqq b$ and the series $\Sigma u_n(x)$ converges in this interval to a continuous function $f(x)$. [It is shown below, in Theorem 31, Section 6–14, that uniform convergence of the series guarantees the continuity of $f(x)$.] In such a case the absolute error $|R_n(x)|$ in replacement of $f(x)$ by the partial sum $S_n(x)$:

$$|R_n(x)| = |f(x) - S_n(x)| \tag{6–25}$$

is continuous for $a \leqq x \leqq b$ and has a definite maximum $\bar{R}_n$ in this interval:

$$\bar{R}_n = \max_{a \leqq x \leqq b} |f(x) - S_n(x)|; \tag{6–26}$$

$\bar{R}_n$ is simply the worst absolute error occurring in the interval. Uniform convergence of the series for $a \leqq x \leqq b$ is then equivalent to the statement:

$$\lim_{n\to\infty} \bar{R}_n = 0; \tag{6–27}$$

that is, *the series is uniformly convergent for* $a \leqq x \leqq b$ *if and only if the maximum error approaches* 0 *as* $n \to \infty$ (see Fig. 6–8). The new condition is simply a rewording of the definition above for the case considered, for convergence of $\bar{R}_n$ to 0 is the same as the condition: $\bar{R}_n < \epsilon$ for $n \geqq N(\epsilon)$ and hence as the condition: $|f(x) - S_n(x)| < \epsilon$ for $n \geqq N(\epsilon)$.

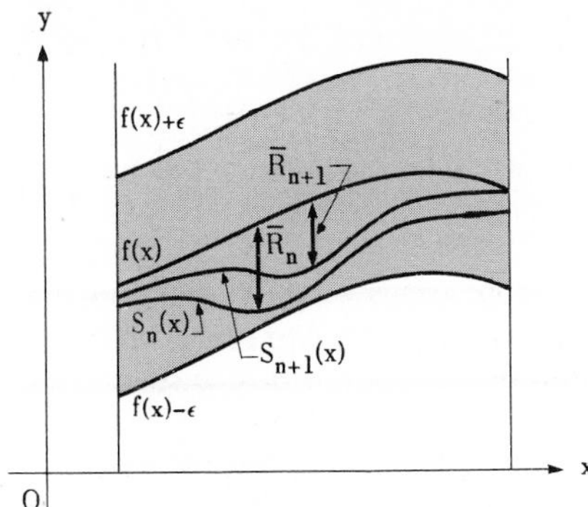

FIG. 6–8. Uniform convergence.

A similar formulation can be given for the general case (when, for example, E need not be a closed interval). One simply defines $\bar{R}_n$ as the *least upper bound* of $|f(x) - S_n(x)|$ on the set E i.e., as the smallest number K such that $|f(x) - S_n(x)| \leqq K$ for all x in E; the existence of such a number is established in the same way as for the upper limit of a sequence (Section 6–3). This reduces to the maximum of $|f(x) - S_n(x)|$ on the set E, if this function has a maximum; however, the function, even when continuous, need not have a maximum (e.g., when E is an open interval; cf. Section 2–15). Just as for upper limits, the least upper bound may be $+\infty$. Uniform convergence of $\Sigma u_n(x)$ to $f(x)$ in the set E is then equivalent to the statement: $\bar{R}_n \to 0$ as $n \to \infty$, i.e., to the statement:

$$\lim_{n\to\infty} \underset{x \text{ in } E}{\text{l.u.b.}} |f(x) - S_n(x)| = 0, \tag{6–27'}$$

where "l.u.b." stands for "least upper bound."

EXAMPLE 1. The geometric series $\sum_{n=0}^{\infty} x^n$. The series converges for $-1<x<1$. The nth partial sum is

$$S_n(x) = 1 + x + \cdots + x^{n-1} = \frac{1 - x^n}{1 - x}$$

and the sum is

$$S(x) = \frac{1}{1 - x};$$

these are plotted in Fig. 6–9. The remainder $R_n(x) = S(x) - S_n(x)$ satisfies the equation:

$$|R_n(x)| = \frac{|x^n|}{|1 - x|}.$$

We consider first a closed interval: $-\frac{1}{2} \leqq x \leqq \frac{1}{2}$. The convergence is uniform in this interval. We find

$$\bar{R}_n = \max_{-\frac{1}{2} \leqq x \leqq \frac{1}{2}} |R_n(x)| = \frac{(\frac{1}{2})^n}{1 - \frac{1}{2}} = \frac{1}{2^{n-1}};$$

for when $x = \frac{1}{2}$ the numerator $|x^n|$ has its largest value and the denominator $|1 - x|$ has its smallest value in the interval. Accordingly,

$$\lim_{n\to\infty} \bar{R}_n = \lim_{n\to\infty} \frac{1}{2^{n-1}} = 0;$$

the maximum absolute error tends to 0 as $n \to \infty$ and the convergence is uniform.

A similar reasoning applies to each closed interval: $-a \leqq x \leqq a$ $(0<a<1)$ within the interval of convergence. The worst absolute error occurs for $x = a$ and its value is $\bar{R}_n = a^n/(1-a)$; this tends to 0 as $n \to \infty$, so that the convergence is uniform in the interval.

One might expect from this that the convergence would be uniform over the entire *open* interval: $-1 < x < 1$. This is not the case. Indeed, for each n the absolute error $|R_n(x)|$ is unbounded for $-1 < x < 1$, since

$$\lim_{x \to 1-} |R_n(x)| = \lim_{x \to 1-} \left| \frac{x^n}{1-x} \right| = \infty .$$

The least upper bound $\bar{R}_n$ is always $+\infty$! We can see the difficulty in detail by asking how many terms are needed to compute the sum with an absolute error less than $\epsilon = 0.01$, for example. For $x = 0$, one term is sufficient; for $x = 0.5$, one must have

$$\frac{(0.5)^n}{0.5} < 0.01 \quad \text{or} \quad 2^{n-1} > 100;$$

hence n must be at least 8. For $x = 0.9$, one must have

$$\frac{(0.9)^n}{0.1} < 0.01 \quad \text{or} \quad (0.9)^n < 0.001;$$

this requires $n > 65$, as a computation with logarithms shows. As x increases towards 1, more and more terms are needed; one can easily show that, as $x \to 1$, the number of terms required $\to \infty$ (see Prob. 2 following Section 6–9).

It appears from this discussion that the difficulty is due to the fact that the sum $S(x) = 1/(1-x)$ becomes infinite as x approaches 1. However, the convergence is not uniform in the interval $-1 < x \leqq 0$. For the absolute error $|R_n(x)|$ lies between 0 and $\frac{1}{2}$ in this interval; this is suggested graphically in Fig. 6–9 and can be proved strictly by calculus. Since $|R_n(x)| \to \frac{1}{2}$ as $x \to -1$, the least upper bound $\bar{R}_n$ is always $\frac{1}{2}$; this is not the maximum absolute error, since $x = -1$ is excluded from the interval under consideration. Accordingly, $\bar{R}_n$ cannot converge to 0 as $n \to \infty$ and the conver-

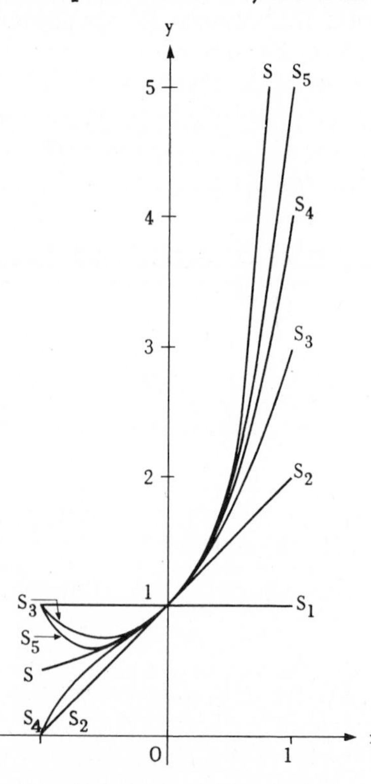

FIG. 6–9. Sequence of partial sums of the geometric series.

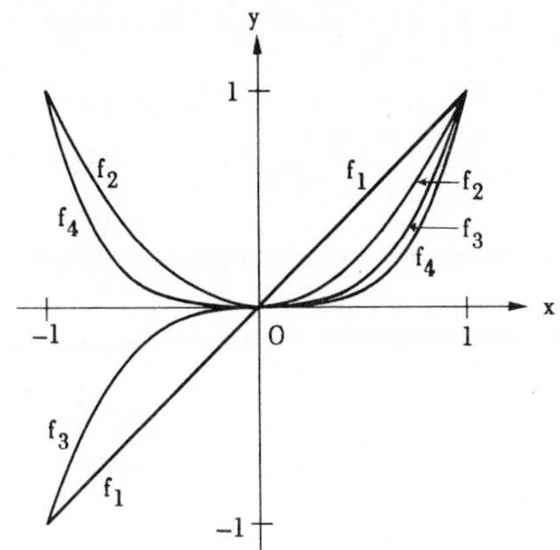

FIG. 6–10. The sequence $f_n(x) = x^n$.

gence is nonuniform for $-1 < x \leqq 0$. Again we can verify that the number of terms needed to compute $S(x)$ with an error less than $\epsilon = 0.01$ approaches ∞ as $x \to -1$.

EXAMPLE 2. The *sequence* $f_n(x) = x^n$ converges to 0 for $-1 < x < 1$, to 1 for $x = 1$; it diverges for all other values of x. The convergence for $-1 < x < 1$ is not uniform. For $0 \leqq x < 1$, the error is precisely x^n; for this to be less than ϵ one must have

$$x^n < \epsilon \quad \text{or} \quad n > \frac{\log \epsilon}{\log x}.$$

As x approaches 1, the value of n required approaches $+\infty$. For $x = 1$, the error is always 0. Successive members of the sequence $f_n(x)$ are plotted in Fig. 6–10. Since the worst errors occur near $x = \pm 1$, one can obtain uniform convergence by restricting x as in the preceding example: $-0.5 \leqq x \leqq 0.5$.

6–13 Weierstrass *M*-test for uniform convergence.

The following test is adequate for determining the uniform convergence of a large number of familiar series.

THEOREM 30 (*Weierstrass M-test*). *Let* $\sum_{n=1}^{\infty} u_n(x)$ *be a series of functions all defined for a set E of values of x. If there is a convergent series of constants* $\sum_{n=1}^{\infty} M_n$, *such that*

$$|u_n(x)| \leqq M_n \text{ for all } x \text{ in } E,$$

then the series $\sum_{n=1}^{\infty} u_n(x)$ *converges absolutely for each x in E and is uniformly convergent in E.*

Proof. The Weierstrass M-test is first of all a *comparison* test. For each fixed x, each term of the series $\Sigma|u_n(x)|$ is less than or equal to the nth term M_n of the convergent series ΣM_n. Hence, by the comparison test (Section 6–6, Theorem 12), the series $\Sigma u_n(x)$ is absolutely convergent.

But the comparison series ΣM_n is the *same* for all x of the range considered. It is from this fact that the uniform convergence follows. For, if $R_n = S - S_n(x)$ is the remainder after n terms of the series $\Sigma u_n(x)$, then

$$|R_n(x)| = |u_{n+1}(x) + u_{n+2}(x) + \cdots| \leqq |u_{n+1}(x)| + |u_{n+2}(x)| + \cdots \leqq M_{n+1} + M_{n+2} + \cdots,$$

as in Theorem 22 of Section 6–9. In other words, if T_n denotes the remainder after n terms of the convergent series ΣM_n:

$$T_n = M_{n+1} + M_{n+2} + \cdots,$$

then

$$|R_n(x)| \leqq T_n.$$

Since ΣM_n is a series of constants, for each given $\epsilon > 0$, N can be found such that $T_n < \epsilon$ for $n \geqq N$; for this same N one has

$$|R_n(x)| \leqq T_n < \epsilon \quad \text{for } n \geqq N.$$

Since N does not depend on x, but only on ϵ, the uniform convergence has been established. At the same time it has been shown that T_n serves as an upper estimate for the error committed in using only n terms of the series $\Sigma u_n(x)$, regardless of the x. The error will, of course, vary from one x to another, but the upper estimate is the same for all x.

EXAMPLE 1. $\displaystyle\sum_{n=1}^{\infty} \frac{x^n}{n^2}$. Here the ratio test gives

$$\lim_{n\to\infty} \left|\frac{a_{n+1}}{a_n}\right| = \lim_{n\to\infty} \left|\frac{x^{n+1}}{(n+1)^2} \cdot \frac{n^2}{x^n}\right| = \lim_{n\to\infty} |x| \cdot \frac{n^2}{(n+1)^2} = |x|.$$

Hence the series converges for $|x| < 1$ and diverges for $|x| > 1$. For $x = \pm 1$, the series converges by comparison with the harmonic series of order 2:

$$\left|\frac{(\pm 1)^n}{n^2}\right| \leqq \frac{1}{n^2}.$$

Hence the series converges for $-1 \leqq x \leqq 1$. The convergence is uniform for this range since the comparison

$$\left|\frac{x^n}{n^2}\right| \leqq M_n = \frac{1}{n^2}$$

holds for all x of the range and the series ΣM_n converges.

EXAMPLE 2. $\displaystyle\sum_{n=1}^{\infty} \frac{\cos nx}{2^n}$. This series converges uniformly for all x, since

$$\left|\frac{\cos nx}{2^n}\right| \leqq \frac{1}{2^n} = M_n$$

for all x and the series $\Sigma(1/2^n)$ is convergent.

EXAMPLE 3. $\displaystyle\sum_{n=1}^{\infty} \frac{x^n}{n}$. The ratio test shows, as in Example 1, that the series converges for $-1 < x < 1$ and diverges for $|x| > 1$. For $x = 1$, the series is the divergent harmonic series; for $x = -1$, the series is a convergent alternating series. One might try to prove uniform convergence for $0 \leqq x < 1$ by using the inequality

$$\left|\frac{x^n}{n}\right| \leqq x^n, \quad 0 \leqq x < 1$$

and the fact that the series $\sum_{n=1}^{\infty} x^n$ converges. However, this reasoning is incorrect, since the comparison series Σx^n *depends on* x, and is not a series of constants as required in Theorem 30. As it happens, the given series is *not* uniformly convergent for $0 \leqq x < 1$.

Problems

1. Determine the values of x for which each of the following series converges:

(a) $\sum_{n=1}^{\infty} \frac{x^n}{2n^2 - n}$

(b) $\sum_{n=1}^{\infty} \frac{nx^n}{2^n}$

(c) $\sum_{n=1}^{\infty} \frac{1}{nx^{2n}}$

(d) $\sum_{n=0}^{\infty} \frac{1}{2^{nx}}$

(e) $\sum_{n=1}^{\infty} \frac{x^n}{(1-x)^n}$

(f) $\sum_{n=1}^{\infty} \frac{2^n \sin^n x}{n^2}$

(g) $\sum_{n=1}^{\infty} \frac{(x-1)^n}{n^2}$

(h) $\sum_{n=1}^{\infty} \frac{1}{x^n \log (n+1)}$

(i) $\sum_{n=1}^{\infty} \frac{(x-2)^{3n}}{n!}$

(j) $\sum_{n=2}^{\infty} \frac{x^n}{(\log n)^n}$

2. Prove that each of the following series is uniformly convergent over the range of values of x given:

(a) $\sum_{n=1}^{\infty} \frac{x^n}{n^3}, \quad -1 \leqq x \leqq 1$

(b) $\sum_{n=1}^{\infty} \frac{(\tanh x)^n}{n!}$, all x

(c) $\sum_{n=1}^{\infty} \frac{\sin nx}{n^2+1}$, all x

(d) $\sum_{n=1}^{\infty} \frac{e^{nx}}{2^n}, \quad x \leqq \log \frac{3}{2}$

(e) $\sum_{n=0}^{\infty} \frac{x^n}{n!}, \quad -1 \leqq x \leqq 1$

(f) $\sum_{n=1}^{\infty} nx^n, \quad -\frac{1}{2} \leqq x \leqq \frac{1}{2}$

(g) $\sum_{n=1}^{\infty} nx^n, \quad -0.9 \leqq x \leqq 0.9$

(h) $\sum_{n=1}^{\infty} nx^n, \quad -a \leqq x \leqq a, \quad a < 1$

3. Prove: if $\sum_{n=1}^{\infty} u_n(x)$ is uniformly convergent for $a \leqq x \leqq b$, then the series is uniformly convergent in each smaller interval contained in the interval $a \leqq x \leqq b$. More generally, if a series is uniformly convergent for a given set E of values of x, then it is uniformly convergent for any set E_1 which is part of E.

4. Prove: if $\sum_{n=1}^{\infty} v_n(x)$ is uniformly convergent for a set E of values of x and $|u_n(x)| \leqq v_n(x)$ for x in E, then $\sum_{n=1}^{\infty} u_n(x)$ is uniformly convergent for x in E.

5. Prove: if $0 < u_n(x) < 1/n$ and $u_{n+1}(x) \leqq u_n(x)$ for $a \leqq x \leqq b$, then the series $\sum_{n=1}^{\infty} (-1)^n u_n(x)$ is uniformly convergent for $a \leqq x \leqq b$.

6. Prove: if the series $\sum_{n=1}^{\infty} M_n$ of constants M_n is convergent and $|f_{n+1}(x) - f_n(x)| \leqq M_n$ for x in E, then the *sequence* $f_n(x)$ is uniformly convergent for x in E.

7. Prove that the following sequences are uniformly convergent for the range of x given (cf. Prob. 6):

(a) $\dfrac{n+x}{n}$, $0 \leqq x \leqq 1$

(b) $\dfrac{x^n}{n!}$, $-1 \leqq x \leqq 1$

(c) $\dfrac{\log(1+nx)}{n}$, $1 \leqq x \leqq 2$

(d) $\dfrac{n}{e^{nx^2}}$, $\frac{1}{2} \leqq x \leqq 1$

Answers

1. (a) $|x| \leqq 1$, (b) $|x| < 2$, (c) $x > 1$ and $x < -1$, (d) $x > 0$, (e) $x < \frac{1}{2}$, (f) $|x - n\pi| \leqq \pi/6$ $(n = 0, \pm 1, \pm 2, \ldots)$, (g) $0 \leqq x \leqq 2$, (h) $x > 1$ and $x \leqq -1$, (i) all x, (j) all x.

6–14 Properties of uniformly convergent series and sequences. Let $\Sigma u_n(x)$ be a series of functions, each of which is defined for $a \leqq x \leqq b$. Let it be assumed further that this series converges to a sum $f(x)$ for $a \leqq x \leqq b$, so that one has

$$f(x) = \sum_{n=1}^{\infty} u_n(x), \quad a \leqq x \leqq b.$$

One can then ask questions such as the following: If each function $u_n(x)$ is continuous, is the sum $f(x)$ continuous? If each $u_n(x)$ has a derivative, does $f(x)$ have a derivative? The following theorems answer such questions.

THEOREM 31. *The sum of a uniformly convergent series of continuous functions is continuous; i.e., if each $u_n(x)$ is continuous for $a \leqq x \leqq b$, then so is $f(x) = \sum_{n=1}^{\infty} u_n(x)$, provided the series converges uniformly for $a \leqq x \leqq b$.*

Proof. Let x_0 be given, $a \leqq x_0 \leqq b$, and let $\epsilon > 0$ be given. We then seek a δ such that

$$|f(x) - f(x_0)| < \epsilon \quad \text{when} \quad |x - x_0| < \delta$$

and x is in the given interval. We choose N so large that

$$|S_n(x) - f(x)| < \tfrac{1}{3}\epsilon, \quad a \leqq x \leqq b, \quad n \geqq N, \tag{6–28}$$

where $S_n(x) = u_1(x) + \cdots + u_n(x)$; this is possible since the series is uniformly convergent. The function $S_N(x)$, as sum of a *finite* number of continuous functions, is itself continuous. One can hence choose a δ such that

$$|S_N(x) - S_N(x_0)| < \tfrac{1}{3}\epsilon \quad \text{for} \quad |x - x_0| < \delta. \tag{6–29}$$

By (6–28), one has

$$|S_N(x) - f(x)| < \tfrac{1}{3}\epsilon, \quad |S_N(x_0) - f(x_0)| < \tfrac{1}{3}\epsilon. \tag{6–30}$$

Hence

$$\begin{aligned} |f(x) - f(x_0)| &= |f(x) - S_N(x) + S_N(x) - S_N(x_0) + S_N(x_0) - f(x_0)| \\ &\leqq |f(x) - S_N(x)| + |S_N(x) - S_N(x_0)| + |S_N(x_0) - f(x_0)| \\ &< \tfrac{1}{3}\epsilon + \tfrac{1}{3}\epsilon + \tfrac{1}{3}\epsilon = \epsilon, \text{ for } |x - x_0| < \delta, \end{aligned}$$

by (6–29) and (6–30). Thus continuity is proved.

Remark 1. The property of convergence alone, for a series of continuous functions, does not guarantee continuity of the sum. This is seen by the example:

$$f(x) = x + \sum_{n=2}^{\infty} (x^n - x^{n-1}), \quad 0 \leqq x \leqq 1.$$

Here the nth partial sums form the sequence $S_n(x) = x^n$ plotted in Fig. 6–10. As pointed out in Section 6–12, this sequence does not converge uniformly. The sum of the series is 0 for $0 \leqq x < 1$ and 1 for $x = 1$; there is a jump discontinuity at $x = 1$.

Remark 2. Theorem 31 can be interpreted in terms of sequences as follows: if $S_n(x)$ is a sequence of functions all continuous for $a \leqq x \leqq b$ and this sequence converges uniformly to $f(x)$ for $a \leqq x \leqq b$, then $f(x)$ is continuous for $a \leqq x \leqq b$; furthermore, if $a \leqq x_0 \leqq b$, then

$$\lim_{x \to x_0} [\lim_{n \to \infty} S_n(x)] = \lim_{n \to \infty} [\lim_{x \to x_0} S_n(x)]. \tag{6–31}$$

The left-hand side is precisely $\lim_{x \to x_0} f(x)$ and the right-hand side is $\lim_{n \to \infty} S_n(x_0)$ $= f(x_0)$, by the continuity of $S_n(x)$; Eq. (6–31) is simply the assertion that

$$\lim_{x \to x_0} f(x) = f(x_0),$$

i.e., that $f(x)$ is continuous at x_0. Accordingly, as Eq. (6–31) shows, *uniform convergence permits one to interchange two limit processes.*

THEOREM 32. *A uniformly convergent series of continuous functions can be integrated term by term; i.e., if each $u_n(x)$ is continuous for $a \leqq x \leqq b$, then*

$$\int_a^b f(x)\,dx = \int_a^b u_1(x)\,dx + \int_a^b u_2(x)\,dx + \cdots + \int_a^b u_n(x)\,dx + \cdots. \tag{6–32}$$

Proof. As in the preceding proof, let $S_n(x)$ be the nth partial sum of the series $\Sigma u_n(x)$. Then

$$\int_a^b S_n(x)\,dx = \int_a^b u_1(x)\,dx + \cdots + \int_a^b u_n(x)\,dx.$$

To prove (6–32), one must show that the sequence $\int_a^b S_n(x)\,dx$ converges to $\int_a^b f(x)\,dx$, i.e., that, for each $\epsilon > 0$, an N can be found such that

$$\left|\int_a^b f(x)\,dx - \int_a^b S_n(x)\,dx\right| < \epsilon, \quad n \geqq N.$$

To establish this, we choose N so large that

$$|f(x) - S_n(x)| < \frac{\epsilon}{b-a}, \quad n \geqq N, \quad a \leqq x \leqq b;$$

this is possible because of the uniform convergence. Hence

$$\left|\int_a^b f(x)\,dx - \int_a^b S_n(x)\,dx\right| = \left|\int_a^b [f(x) - S_n(x)]\,dx\right|$$
$$\leqq \frac{\epsilon}{b-a}\cdot(b-a) = \epsilon, \quad n \geqq N,$$

by the theorem on estimate of error in Section 4–2. Theorem 32 is thus established.

Remark 3. This theorem can be formulated in terms of sequences and interchange of limit processes as in Remark 2 above.

Example. In Section 6–12 the series $\sum_{n=0}^{\infty} x^n$ was shown to converge uniformly to $1/(1-x)$ in each interval $-a \leqq x \leqq a$, where $a < 1$. On integrating the equation

$$\frac{1}{1-x} = 1 + x + \cdots + x^n + \cdots$$

from 0 to x_1, one finds

$$\int_0^{x_1} \frac{1}{1-x}\,dx = \log\frac{1}{1-x_1} = x_1 + \frac{1}{2}x_1^2 + \cdots + \frac{x_1^{n+1}}{n+1} + \cdots.$$

This holds for every x_1 between -1 and 1 (the ends excluded), so that one can write

$$\log\frac{1}{1-x} = x + \frac{1}{2}x^2 + \cdots + \frac{x^{n+1}}{n+1} + \cdots, \quad -1 < x < 1. \tag{6–33}$$

The same procedure could be phrased in terms of *indefinite* integrals. Thus

$$\int \frac{1}{1-x}\,dx = \int (1 + x + x^2 + \cdots + x^n + \cdots)\,dx + c, \quad -1 < x < 1;$$
$$\log\frac{1}{1-x} = x + \frac{1}{2}x^2 + \frac{1}{3}x^3 + \cdots + \frac{x^{n+1}}{n+1} + \cdots + c.$$

Of course c is no longer arbitrary; by choosing $x = 0$, we recognize that $c = 0$, so that (6–33) is again obtained.

THEOREM 33. *A convergent series can be differentiated term by term, provided that the functions of the series have continuous derivatives and that the series of derivatives is uniformly convergent; i.e., if* $u_n'(x) = du_n/dx$ *is continuous for* $a \leqq x \leqq b$, *if the series* $\sum_{n=1}^{\infty} u_n(x)$ *converges for* $a \leqq x \leqq b$ *to* $f(x)$, *and if the series* $\sum_{n=1}^{\infty} u_n'(x)$ *converges uniformly for* $a \leqq x \leqq b$, *then*

$$f'(x) = \sum_{n=1}^{\infty} u_n'(x), \quad a \leqq x \leqq b. \tag{6–34}$$

The derivatives at a and b are understood as right- and left-sided derivatives respectively.

Proof of Theorem 33. Let $g(x)$ be the sum of the series of derivatives:

$$g(x) = \sum_{n=1}^{\infty} u_n'(x), \quad a \leqq x \leqq b.$$

Then $g(x)$ is continuous by Theorem 31, and by Theorem 32

$$\int_a^{x_1} g(x)\,dx = \sum_{n=1}^{\infty} \int_a^{x_1} u_n'(x)\,dx, \quad a \leqq x_1 \leqq b.$$

Hence

$$\begin{aligned}\int_a^{x_1} g(x)\,dx &= \sum_{n=1}^{\infty} [u_n(x_1) - u_n(a)]\\ &= \sum_{n=1}^{\infty} u_n(x_1) - \sum_{n=1}^{\infty} u_n(a).\end{aligned}$$

Accordingly,

$$\int_a^{x_1} g(x)\,dx = f(x_1) - f(a).$$

If both sides are differentiated with respect to x_1, the fundamental theorem of calculus (Section 4–3) gives

$$g(x_1) = f'(x_1), \quad a \leqq x_1 \leqq b.$$

In other words,

$$f'(x) = g(x) = \sum_{n=1}^{\infty} u_n'(x), \quad a \leqq x \leqq b.$$

THEOREM 34. *If* $\sum_{n=1}^{\infty} u_n(x)$ *and* $\sum_{n=1}^{\infty} v_n(x)$ *are uniformly convergent for* $a \leqq x \leqq b$ *and* $h(x)$ *is continuous for* $a \leqq x \leqq b$, *then the series*

$$\sum_{n=1}^{\infty} [u_n(x) + v_n(x)], \quad \sum_{n=1}^{\infty} [u_n(x) - v_n(x)], \quad \sum_{n=1}^{\infty} [h(x)u_n(x)]$$

are uniformly convergent for $a \leqq x \leqq b$.

Proof. Let $f(x)$ and $g(x)$ be the sums of $\Sigma u_n(x)$ and $\Sigma v_n(x)$ respectively; let $S_n(x)$ and $Q_n(x)$ be the corresponding partial sums. The nth partial sum of $\Sigma(u_n + v_n)$ is $S_n + Q_n$. Let N be chosen, for given ϵ, so that

$$|S_n(x) - f(x)| < \tfrac{1}{2}\epsilon, \quad |Q_n(x) - g(x)| < \tfrac{1}{2}\epsilon,$$

for $n \geqq N$ and $a \leqq x \leqq b$. Then

$$\begin{aligned} &|\{S_n(x) + Q_n(x)\} - \{f(x) + g(x)\}| \\ &\qquad \leqq |S_n(x) - f(x)| + |Q_n(x) - g(x)| < \tfrac{1}{2}\epsilon + \tfrac{1}{2}\epsilon = \epsilon. \end{aligned}$$

Thus $\Sigma(u_n + v_n)$ converges uniformly to $f(x) + g(x)$. A similar proof applies to the difference.

Since $h(x)$ is continuous for $a \leqq x \leqq b$, it is necessarily bounded: $|h(x)| \leqq M$ for $a \leqq x \leqq b$. Hence

$$|h(x)S_n(x) - h(x)f(x)| = |h(x)|\,|S_n(x) - f(x)| < M \cdot \tfrac{1}{2}\epsilon < M\epsilon$$

for $n \geqq N$ as above. This shows that the series $\Sigma h(x)u_n$ converges uniformly to $h(x)f(x)$. It should be noted that actually only the *boundedness* of $h(x)$ was required.

6–15 Power series. By a *power series* in powers of x is meant a series of form

$$\sum_{n=0}^{\infty} c_n x^n = c_0 + c_1 x + \cdots + c_n x^n + \cdots, \tag{6–35}$$

where $c_0, c_1, \ldots, c_n, \ldots$ are constants. By a power series in powers of $(x - a)$ is meant a series:

$$\sum_{n=0}^{\infty} c_n (x - a)^n = c_0 + c_1(x - a) + \cdots + c_n(x - a)^n + \cdots. \tag{6–36}$$

By a power series in *negative* powers of x is meant a series:

$$\sum_{n=0}^{\infty} \frac{c_n}{x^n} = c_0 + \frac{c_1}{x} + \cdots + \frac{c_n}{x^n} + \cdots. \tag{6–37}$$

The term "power series" alone usually refers to (6–36), of which (6–35) is a special case ($a = 0$). The substitutions $t = x - a$ or $t = 1/x$ can be used to reduce series of form (6–36) or (6–37) to the form (6–35).

The power series (6–36) converges when $x = a$. It can happen that this is the only value of x for which the series converges. If there are other values of x for which the series converges, then it will be seen that these form an interval, the "convergence interval," having mid-point at $x = a$. This is pictured in Fig. 6–11. The interval can be infinite. These properties are summarized in the following basic theorem.

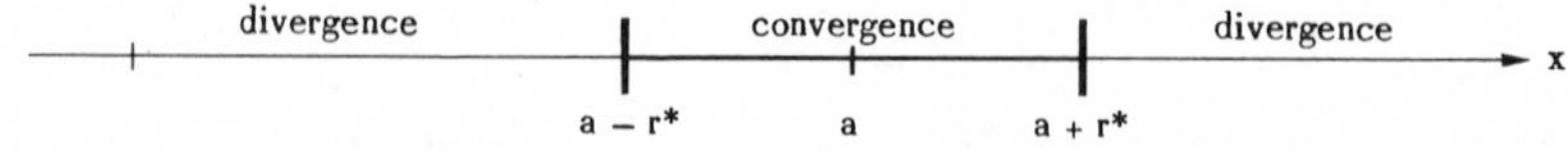

Fig. 6–11. Interval of convergence of a power series.

THEOREM 35. *Every power series*

$$c_0 + c_1(x-a) + \cdots + c_n(x-a)^n + \cdots$$

has a "radius of convergence" r^* such that the series converges absolutely when $|x-a| < r^*$ and diverges when $|x-a| > r^*$.
The number r^* can be 0 (in which case the series converges only for $x = a$), a positive number, or ∞ (in which case the series converges for all x).
If r^* is not zero and r_1 is such that $0 < r_1 < r^*$, then the series converges uniformly for $|x-a| \leqq r_1$.
The number r^* can be evaluated as follows:

$$r^* = \lim_{n\to\infty} \left|\frac{c_n}{c_{n+1}}\right|, \text{ if the limit exists,} \tag{6-38}$$

$$r^* = \lim_{n\to\infty} \frac{1}{\sqrt[n]{|c_n|}}, \text{ if the limit exists,} \tag{6-39}$$

and in any case by the formula:

$$r^* = \frac{1}{\overline{\lim\limits_{n\to\infty}} \sqrt[n]{|c_n|}}. \tag{6-40}$$

Proof. We begin by considering the case in which the limit (6-38) exists, as this at once suggests the general situation. The ratio test, applied to the given series, proceeds as follows:

$$\lim_{n\to\infty}\left|\frac{a_{n+1}}{a_n}\right| = \lim_{n\to\infty}\left|\frac{c_{n+1}(x-a)^{n+1}}{c_n(x-a)^n}\right| = \lim_{n\to\infty}\left|\frac{c_{n+1}}{c_n}\right| \cdot |x-a|$$

$$= \frac{|x-a|}{\lim\limits_{n\to\infty}\left|\frac{c_n}{c_{n+1}}\right|} = \frac{|x-a|}{r^*}.$$

The series converges absolutely when

$$\frac{|x-a|}{r^*} < 1, \quad \text{i.e.,} \quad |x-a| < r^*$$

and diverges when

$$\frac{|x-a|}{r^*} > 1, \quad \text{i.e.,} \quad |x-a| > r^*.$$

If $r^* = 0$, the test shows divergence except for $x = a$. If $r^* = \infty$, so that

$$\lim_{n\to\infty}\left|\frac{c_{n+1}}{c_n}\right| = 0,$$

the test shows that the series converges absolutely for all x.

The case in which the limit (6-39) exists is treated in the same way by the root test. This case is included in the final formula (6-40), which follows at once from Theorem 21(a) of Section 6–8, for

$$\overline{\lim_{n\to\infty}} \sqrt[n]{|a_n|} = \overline{\lim_{n\to\infty}} \left(\sqrt[n]{|c_n|} \cdot |x - a|\right) = \left(\overline{\lim_{n\to\infty}} \sqrt[n]{|c_n|}\right) \cdot |x - a|.$$

The series converges absolutely when the upper limit is less than 1, i.e., when

$$|x - a| < \frac{1}{\overline{\lim\limits_{n\to\infty}} \sqrt[n]{|c_n|}} = r^*,$$

and diverges when $|x - a| > r^*$.

Thus, however it may be determined, there is a number r^*, $0 \leqq r^* \leqq \infty$, with the properties described. It remains to prove that, if $0 < r_1 < r^*$, then the series converges uniformly for $|x - a| \leqq r_1$. This follows from the M-test, with $M_n = |c_n|\, r_1^n$, for

$$\sum_{n=0}^{\infty} M_n = \sum_{n=0}^{\infty} |c_n|\, |x_1 - a|^n, \quad x_1 = a + r_1,$$

and this converges, since $|x_1 - a| = r_1 < r^*$. For $|x - a| \leqq r_1$,

$$|c_n|\, |x - a|^n \leqq |c_n| r_1^n = M_n,$$

so that the convergence is uniform. The main idea of this proof is the one emphasized in Section 6–12: that the slowest convergence of a power series occurs towards the *ends* of the interval of convergence.

It should be emphasized that, when r^* is a finite positive number, the series may converge or diverge for each of the end values $x = a + r^*$, $x = a - r^*$. These must be investigated separately for each series.

EXAMPLE 1. $\sum\limits_{n=1}^{\infty} \dfrac{x^n}{n^2}$. Here $c_n = 1/n^2$ and (6–38) gives

$$r^* = \lim_{n\to\infty} \frac{(n+1)^2}{n^2} = 1;$$

similarly, (6–39) gives

$$r^* = \lim_{n\to\infty} \sqrt[n]{n^2} = \lim_{n\to\infty} n^{2/n} = \lim_{n\to\infty} e^{(2/n)\log n} = e^{\lim\limits_{n\to\infty} (2/n)\log n} = e^0 = 1,$$

by Theorem 4 of Section 6–4. Hence the series converges absolutely for $-1 < x < 1$ and diverges for $|x| > 1$. For $x = \pm 1$, the series converges absolutely, since

$$\left|\frac{(\pm 1)^n}{n^2}\right| = \frac{1}{n^2}.$$

In fact, the series converges absolutely and uniformly for $-1 \leqq x \leqq 1$, since

$$\left|\frac{x^n}{n^2}\right| \leqq \frac{1}{n^2} = M_n$$

in this range.

EXAMPLE 2. $\sum_{n=1}^{\infty} \frac{x^n}{n}$. The ratio formula and root formula apply as in Example 1 to give $r^* = 1$. Hence the series converges absolutely for $|x| < 1$, diverges for $|x| > 1$. For $x = 1$, the series is the harmonic series and diverges; for $x = -1$, the series converges conditionally by the alternating series test. Hence the series converges for $-1 \leqq x < 1$ and diverges otherwise.

It should be noted that Example 2 is obtained from Example 1 essentially by term-by-term differentiation:

$$\sum_{n=1}^{\infty} \frac{d}{dx}\left(\frac{x^n}{n^2}\right) = \sum_{n=1}^{\infty} \frac{nx^{n-1}}{n^2} = \sum_{n=1}^{\infty} \frac{x^{n-1}}{n};$$

an extra factor of x is needed. The effect of term-by-term differentiation is thus to multiply each term by n (while lowering the degree of x^n by 1). This cannot affect the radius of convergence, since

$$\lim_{n\to\infty} \sqrt[n]{n} = \lim_{n\to\infty} n^{1/n} = \lim_{n\to\infty} e^{(1/n)\log n} = 1.$$

However, it does slow down convergence at the ends of the interval, as the above example shows. If one differentiates the series of Example 2, one obtains the geometric series:

$$\sum_{n=1}^{\infty} \frac{nx^{n-1}}{n} = \sum_{n=1}^{\infty} x^{n-1} = \sum_{n=0}^{\infty} x^n = 1 + x + x^2 + \cdots,$$

which converges for $-1 < x < 1$ and *diverges* at the ends of the interval.

THEOREM 36. *A power series represents a continuous function within the interval of convergence; i.e., if r^* is the radius, then*

$$f(x) = \sum_{n=0}^{\infty} c_n(x-a)^n$$

is continuous for $a - r^* < x < a + r^*$.

Proof. By Theorem 35, the series converges uniformly for $a - r_1 \leqq x \leqq a + r_1$, $r_1 < r^*$. Each term $c_n(x-a)^n$ of the series is continuous for all x; hence, by Theorem 31, $f(x)$ is continuous for $a - r_1 \leqq x \leqq a + r_1$. This holds for every r_1 between 0 and r^*. Accordingly, $f(x)$ is continuous for $a - r^* < x < a + r^*$.

Remark. If the series converges at either end of the interval, then $f(x)$ is continuous at the end concerned. For a proof, the reader is referred to K. Knopp, *Infinite Series*, page 177 (London: Blackie, 1928).

THEOREM 37. *A power series can be integrated term by term within the interval of convergence; that is, if*

$$f(x) = \sum_{n=0}^{\infty} c_n(x-a)^n, \quad a - r^* < x < a + r^*,$$

then for $a - r^* < x_1 < x_2 < a + r^*$,

$$\int_{x_1}^{x} f(x)\,dx = \sum_{n=0}^{\infty} c_n \int_{x_1}^{x_2} (x-a)^n\,dx = \sum_{n=0}^{\infty} c_n \frac{(x_2-a)^{n+1} - (x_1-a)^{n+1}}{n+1}$$

or, in terms of indefinite integrals,

$$\int f(x)\,dx = \sum_{n=0}^{\infty} c_n \frac{(x-a)^{n+1}}{n+1} + C, \quad a - r^* < x < a + r^*.$$

Proof. By Theorem 35, the series converges uniformly in each interval $a - r_1 \leqq x \leqq a + r_1$. It is therefore uniformly convergent for each interval $x_1 \leqq x \leqq x_2$, for the latter interval will be included in the former for appropriate choice of r_1. One can now apply Theorem 32 to justify term-by-term integration. The indefinite integral can be regarded as a special case of this, for, if $x_1 = a$, then

$$\int_{a}^{x_2} f(x)\,dx = \sum_{n=0}^{\infty} c_n \frac{(x_2-a)^{n+1}}{n+1}.$$

This defines a function $F(x_2)$:

$$F(x_2) = \int_a^{x_2} f(x)\,dx$$

and

$$\frac{dF(x_2)}{dx_2} = f(x_2)$$

by the fundamental theorem of calculus; that is,

$$\frac{dF(x)}{dx} = f(x), \quad F(x) = \sum_{n=0}^{\infty} c_n \frac{(x-a)^{n+1}}{n+1}$$

or

$$\int f(x)\,dx = \sum_{n=0}^{\infty} c_n \frac{(x-a)^{n+1}}{n+1} + C$$

The number x_2 can be chosen anywhere in the interval $a - r^* < x < a + r^*$ and, accordingly, the last equation is valid for all x of this interval.

THEOREM 38. *A power series can be differentiated term by term within the interval of convergence; i.e., if*

$$f(x) = \sum_{n=0}^{\infty} c_n (x-a)^n, \qquad a - r^* < x < a + r^*,$$

then

$$f'(x) = \sum_{n=1}^{\infty} n c_n (x-a)^{n-1}, \quad a - r^* < x < a + r^*.$$

Proof. The general coefficient of the differentiated series is nc_n [or, more precisely, $(n+1)c_{n+1}$]. The extra factor of n has no effect on the upper limit of the nth root, as pointed out in connection with Example 2 above. Hence the *differentiated series has the same radius of convergence* r^*; this series is then uniformly convergent in each interval $a - r_1 \leqq x \leqq a + r_1$, $r_1 < r^*$. Hence by Theorem 33,

$$f'(x) = \sum_{n=1}^{\infty} nc_n(x-a)^{n-1}$$

for each x of such an interval. Every x such that $a - r^* < x < a + r^*$ lies in such an interval, so that the result is justified for all x within the interval of convergence.

Application. From the relation

$$\frac{1}{1-x} = 1 + x + \cdots + x^n + \cdots = \sum_{n=0}^{\infty} x^n, \quad -1 < x < 1,$$

one now obtains by successive differentiation the relations

$$\frac{1}{(1-x)^2} = 1 + 2x + \cdots + nx^{n-1} + \cdots = \sum_{n=0}^{\infty} (n+1)x^n, \quad -1 < x < 1,$$

$$\frac{2}{(1-x)^3} = 2 + 6x + \cdots + n(n-1)x^{n-2} + \cdots = \sum_{n=0}^{\infty} (n+2)(n+1)x^n,$$
$$-1 < x < 1.$$

The general case can be written thus:

$$\frac{1}{(1-x)^k} = (1-x)^{-k} = 1 + \frac{kx}{1} + \frac{k(k+1)}{1 \cdot 2} x^2 + \cdots + \frac{k(k+1)\cdots(k+n-1)}{1 \cdot 2 \cdots n} x^n + \cdots,$$
$$-1 < x < 1, \quad k = 1, 2, 3, \ldots. \tag{6–41}$$

This is termed the *binomial theorem for negative integral exponents.* Equation (6–41) is actually valid for every *real number* k, as will be seen.

6–16 Taylor and Maclaurin series.

Let $f(x)$ be the sum of a power series with convergence interval $a - r^* < x < a + r^*$ $(r^* > 0)$:

$$f(x) = \sum_{n=0}^{\infty} c_n(x-a)^n, \quad a - r^* < x < a + r^*. \tag{6–42}$$

This series is called the *Taylor series of $f(x)$ at $x = a$* if the coefficients c_n are given by the rule:

$$c_0 = f(a), \quad c_1 = \frac{f'(a)}{1!}, \quad c_2 = \frac{f''(a)}{2!}, \quad \ldots, \quad c_n = \frac{f^{(n)}(a)}{n!}, \quad \ldots,$$

so that

$$f(x) = f(a) + \frac{f'(a)}{1!}(x-a) + \cdots + \frac{f^{(n)}(a)}{n!}(x-a)^n + \cdots. \tag{6–43}$$

THEOREM 39. *Every power series with nonzero convergence radius is the Taylor series of its sum.*

Proof. Let $f(x)$ be given by (6–42). Then by repeated differentiation, on the basis of Theorem 38, one finds

$$\begin{aligned} f(x) &= c_0 + c_1(x-a) + \cdots + c_n(x-a)^n + \cdots, \\ f'(x) &= c_1 + 2c_2(x-a) + \cdots + n \cdot c_n(x-a)^{n-1} + \cdots, \\ f''(x) &= 2c_2 + 6c_3(x-a) + \cdots + n(n-1) \cdot c_n(x-a)^{n-2} + \cdots, \\ &\cdots\cdots\cdots, \\ f^{(n)}(x) &= n(n-1)(n-2)\cdots 2 \cdot 1 \cdot c_n \\ &\quad + (n+1)n(n-1)\cdots 2 \cdot c_{n+1}(x-a) + \cdots, \\ &\cdots\cdots\cdots \end{aligned}$$

Here all series converge for $a - r^* < x < a + r^*$. If one now sets $x = a$, one finds

$$f(a) = c_0, \quad f'(a) = c_1, \quad f''(a) = 2c_2, \ldots, \quad f^{(n)}(a) = n!\, c_n, \ldots.$$

This gives $c_0 = f(a)$ and

$$c_n = \frac{f^{(n)}(a)}{n!}, \quad n = 1, 2, \ldots,$$

as asserted.

In the case when $a = 0$, the expression (6–43) for the Taylor series of $f(x)$ becomes

$$f(x) = f(0) + \frac{f'(0)}{1!}x + \frac{f''(0)}{2!}x^2 + \cdots + \frac{f^{(n)}(0)x^n}{n!} + \cdots. \tag{6–44}$$

This is called the *Maclaurin series* of $f(x)$. For many purposes it is easier to employ. The substitution $t = x - a$ reduces the general Taylor series to the form of a Maclaurin series.

THEOREM 40. *If two power series*

$$\sum_{n=0}^{\infty} c_n(x-a)^n, \quad \sum_{n=0}^{\infty} C_n(x-a)^n$$

have nonzero convergence radii and have equal sums wherever both series converge, then the series are identical; that is,

$$c_n = C_n, \quad n = 0, 1, 2, \ldots.$$

Proof. Let the convergence radii be r^* and R^* respectively, where $0 < r^* \leqq R^*$. Then, by assumption,

$$\sum_{n=0}^{\infty} c_n(x-a)^n = \sum_{n=0}^{\infty} C_n(x-a)^n = f(x), \quad a - r^* < x < a + r^*.$$

By the preceding theorem, one must then have $c_0 = C_0 = f(a)$ and

$$c_n = C_n = \frac{f^{(n)}(a)}{n!}, \quad n = 1, 2, \ldots.$$

Thus the coefficients are equal for every value of n.

COROLLARY. *If a power series has a nonzero convergence radius and has a sum which is identically zero, then every coefficient of the series is zero.*

Problems

1. (a) Obtain the Maclaurin series

$$\log \frac{1}{1-x} = x + \frac{1}{2}x^2 + \cdots + \frac{x^n}{n} + \cdots, \quad -1 < x < 1$$

by integration of the series for $1/(1-x)$. Verify that (6–43) holds.

(b) Show that the series converges for $x = -1$ and hence prove (on the basis of the remark preceding Theorem 37) that

$$\log 2 = \sum_{n=1}^{\infty} \frac{(-1)^{n+1}}{n}.$$

2. Prove formula (6–41) by induction.

3. (a) Expand $1/x$ in a Taylor series about $x=1$. [Hint: write $1/[1-(1-x)] = 1/(1-r)$ and use the geometric series.]

(b) Expand $1/(x+2)$ in a Maclaurin series. [Hint: write $1/(x+2) = 1/\{2[1-(-\frac{1}{2}x)]\} = \frac{1}{2}\{1/(1-r)\}$.]

(c) Expand $1/(3x+5)$ in a Maclaurin series.

(d) Expand $1/(3x+5)$ in a Taylor series about $x = 1$. [Hint: write $1/(3x+5) = 1/[3(x-1)+8] = \frac{1}{8}/[1+\frac{3}{8}(x-1)]$.]

(e) Expand $1/(ax+b)$ in a Taylor series about $x = c$, where $ac + b \neq 0$, $a \neq 0$.

(f) Expand $1/(1-x^2)$ in a Maclaurin series.

(g) Expand $1/[(x-2)(x-3)]$ in a Maclaurin series. [Hint: write $1/[(x-2)(x-3)] = A/(x-2) + B/(x-3)$.]

(h) Expand $1/x^2$ in a Taylor series about $x = 1$. [Hint: write $1/x^2 = 1/[1-(1-x)]^2 = 1/(1-r)^2$ and use (6–41).]

(i) Expand $1/(3x+5)^2$ in a Taylor series about $x = 1$.

(j) Expand $1/(ax+b)^k$ in a Taylor series about $x = c$, where $ac + b \neq 0$, $a \neq 0$, and $k = 1, 2, \ldots$.

4. Let $f(x) = \sum_{n=1}^{\infty} \frac{x^n}{n^n}$.

(a) Show that $f(x)$ is defined for all x.

(b) Evaluate (approximately where necessary) $f(0), f(1), f'(0), f'(1), f''(0)$.

(c) Obtain Maclaurin series for $f'(x), f''(x)$.

5. Let $y = f(x)$ be a function (if there is one) such that $f(x)$ is defined for all x, $f(x)$ has a Maclaurin series valid for all x, $f(0) = 1$ and $dy/dx = y$ for all x. Show that necessarily

$$f(x) = 1 + x + x^2/2! + \cdots + x^n/n! + \cdots$$

and that this function satisfies all requirements. [It will be shown below that $f(x) = e^x$.]

Answers

3. (a) $\sum_{n=0}^{\infty} (-1)^n (x-1)^n, \quad 0 < x < 2;$ (b) $\sum_{n=0}^{\infty} (-1)^n \frac{x^n}{2^{n+1}}, \quad -2 < x < 2;$

(c) $\sum_{n=0}^{\infty} \frac{(-1)^n 3^n}{5^{n+1}} x^n, \quad -\frac{5}{3} < x < \frac{5}{3};$ (d) $\sum_{n=0}^{\infty} \frac{(-1)^n 3^n}{8^{n+1}} (x-1)^n, \quad -\frac{5}{3} < x < \frac{11}{3};$

(e) $\sum_{n=0}^{\infty} \frac{(-1)^n a^n}{(ac+b)^{n+1}} (x-c)^n$, $c - \left|\frac{ac+b}{a}\right| < x < c + \left|\frac{ac+b}{a}\right|$;

(f) $\sum_{n=0}^{\infty} x^{2n}$, $-1 < x < 1$; (g) $\sum_{n=0}^{\infty} \left(\frac{1}{2^{n+1}} - \frac{1}{3^{n+1}}\right) x^n$, $-2 < x < 2$;

(h) $\sum_{n=0}^{\infty} (-1)^n (n+1)(x-1)^n$, $0 < x < 2$; (i) $\sum_{n=0}^{\infty} \frac{(-1)^n 3^n (n+1)}{8^{n+2}} (x-1)^n$,

$-\frac{5}{3} < x < \frac{11}{3}$; (j) $\frac{1}{(ac+b)^k} + \sum_{n=1}^{\infty} (-1)^n \frac{k(k+1)\cdots(k+n-1)}{1\cdot 2 \cdots n} \frac{a^n (x-c)^n}{(ac+b)^{n+k}}$,

$c - \left|\frac{ac+b}{a}\right| < x < c + \left|\frac{ac+b}{a}\right|$.

4. (b) $f(0) = 0$, $f(1) = 1.29$, $f'(0) = 1$, $f'(1) = 1.63$, $f''(0) = \frac{1}{2}$;

(c) $f'(x) = \sum_{n=0}^{\infty} \frac{x^n}{(n+1)^n}$, $f''(x) = \sum_{n=0}^{\infty} \frac{n+1}{(n+2)^{n+1}} x^n$.

6–17 Taylor's formula with remainder. The preceding discussion has been concentrated on the power series, rather than on the functions which they represent. The opposite point of view is also of fundamental importance, the first question being the following: given a function $f(x)$, $a < x < b$, can $f(x)$ be represented by a power series in this interval? If $f(x)$ can be so represented, $f(x)$ is termed *analytic* in the given interval. More generally, $f(x)$ is termed analytic for $a < x < b$ if, for each x_0 of this interval, $f(x)$ can be represented by a power series in some interval $x_0 - \delta < x_0 < x_0 + \delta$. Most of the familiar functions: polynomials, rational functions, e^x, $\sin x$, $\cos x$, $\log x$, $\sqrt{x}$, and those constructed from them by operations of algebra and by substitutions are analytic in every interval in which the function considered is continuous. The exceptions are not too difficult to recognize. Thus $\sqrt{x^2} = |x|$ is continuous for all x but has a discontinuous derivative for $x = 0$. Hence the function cannot be analytic in an interval containing this value. The function $f(x) = e^{-1/x^2}$ is defined and continuous for all x other than 0. If one defines $f(0)$ to be 0, the function becomes continuous for all x and can in fact be shown to have derivatives of all orders for all x (Prob. 5 below). However, the function fails to be analytic in any interval containing $x = 0$. For example, one finds that, for this function, $f(0) = 0, f'(0) = 0, \ldots, f^{(n)}(0) = 0, \ldots$, so that the Maclaurin series would reduce identically to zero. The series converges but does not represent the function.

A satisfactory theory of analytic functions is most easily achieved with the aid of complex variables. The reader is referred to Chapter 9 for this topic. However, the following theorem does prove useful in establishing analyticity without recourse to complex numbers.

THEOREM 41. *(Taylor's formula with remainder). Let $f(x)$ be defined and continuous and have continuous derivatives up to the $(n+1)$-st order for $a - r_0 < x < a + r_0$. Then for each x of this interval except $x = a$*

$$f(x) = f(a) + \frac{f'(a)}{1}(x-a) + \cdots + \frac{f^{(n)}(a)}{n!}(x-a)^n + \frac{f^{(n+1)}(x_1)}{(n+1)!}(x-a)^{n+1}$$

for some x_1 such that $a < x_1 < x$ or (if $x < a$) $x < x_1 < a$.

It should be noted that, for $n = 0$, the theorem reduces to the law of the mean (Section 0–9): $f(x) = f(a) + f'(x_1)(x - a)$. For general n, it gives an expansion identical with the Taylor series up to the term in $(x - a)^n$, with the rest of the series replaced by a single term.

We prove the theorem for $n = 1$, leaving the general case as a problem (Prob. 3 below). Let x_2 be a fixed number in the given interval, $x_2 \neq a$, and let

$$F(x) = f(x_2) - f(x) - (x_2 - x)f'(x) - \left(\frac{x_2 - x}{x_2 - a}\right)^2 [f(x_2) - f(a) - (x_2 - a)f'(a)].$$

Then F is defined and continuous for x in the given interval and $F(a) = 0$, $F(x_2) = 0$. Hence by the law of the mean $F'(x_1) = 0$ for some x_1 between a and x_2. But a calculation shows that

$$F'(x) = \frac{2(x_2 - x)}{(x_2 - a)^2}\{f(x_2) - f(a) - (x_2 - a)f'(a) - \tfrac{1}{2}f''(x)\,(x_2 - a)^2\}.$$

The equation $F'(x_1) = 0$ thus becomes the equation

$$f(x_2) = f(a) + (x_2 - a)f'(a) + \tfrac{1}{2}f''(x_1)(x_2 - a)^2.$$

If x_2 is now replaced by a variable x, one has the desired result:

$$f(x) = f(a) + (x - a)f'(a) + \tfrac{1}{2}f''(x_1)(x - a)^2.$$

Now let $f(x)$ be a function having derivatives of all orders in the given interval so that one can form all terms of the (hypothetical) Taylor series of f about $x = a$. While this series may fail to converge, except for $x = a$, and, even if convergent, may fail to have $f(x)$ as sum, one can nevertheless write for each n:

$$f(x) = f(a) + \frac{f'(a)}{1!}(x-a) + \cdots + \frac{f^{(n)}(a)}{n!}(x-a)^n + R_n,$$

where R_n is the *remainder* term:

$$R_n = \frac{f^{(n+1)}(x_1)}{(n+1)!}(x-a)^{n+1}. \tag{6–45}$$

Since x_1 is not explicitly given, the remainder is not explicitly known. However, Eq. (6–45) can often be used to obtain an *upper estimate* for $|R_n|$. From this estimate, one may then be able to show that

$$\lim_{n\to\infty} R_n = 0$$

for all x of the chosen interval. If this has been demonstrated, then one concludes that

$$f(x) = f(a) + \frac{f'(a)}{1}(x-a) + \cdots + \frac{f^{(n)}(a)}{n!}(x-a)^n + \cdots,$$

i.e., that $f(x)$ is represented by a Taylor series over the given interval and is analytic; at the same time one has proved convergence of the series.

Example. Let $f(x) = e^x$. Then, for $a = 0$ and $x > 0$,

$$R_n = \frac{e^{x_1}x^{n+1}}{(n+1)!}, \quad 0 < x_1 < x.$$

Hence

$$0 < R_n < \frac{e^x x^{n+1}}{(n+1)!}.$$

This implies that R_n is less than the nth term of the series

$$\sum_{n=1}^{\infty} \frac{e^x x^{n+1}}{(n+1)!}$$

which converges, by the ratio test, for all x. Thus one has

$$\lim_{n\to\infty} \frac{e^x x^{n+1}}{(n+1)!} = 0$$

and accordingly $\lim R_n = 0$. A similar discussion applies for $x < 0$. Accordingly, e^x can be represented by a Taylor series:

$$e^x = 1 + \frac{x}{1!} + \frac{x^2}{2!} + \cdots + \frac{x^n}{n!} + \cdots = \sum_{n=0}^{\infty} \frac{x^n}{n!}, \quad \text{all } x. \tag{6–46}$$

In a similar manner one can prove that the following expansions are valid:

$$\sin x = \frac{x}{1!} - \frac{x^3}{3!} + \frac{x^5}{5!} + \cdots + \frac{(-1)^{n+1}x^{2n-1}}{(2n-1)!} + \cdots, \text{ all } x; \tag{6–47}$$

$$\cos x = 1 - \frac{x^2}{2!} + \frac{x^4}{4!} + \cdots + \frac{(-1)^n x^{2n}}{(2n)!} + \cdots, \text{ all } x; \tag{6–48}$$

$$(1+x)^m = 1 + \frac{m}{1!}x + \frac{m(m-1)}{2!}x^2 + \cdots + \frac{m(m-1)\cdots(m-n+1)}{n!}x^n + \cdots,$$

$$-1 < x < 1, \ m \text{ any real number.} \tag{6–49}$$

[For the derivation of (6–49), see *CLA*, Section 6–20.]

A variety of other expansions can be obtained from these by appropriate substitutions and combinations. The series in (6–49) becomes the geometric series when $m = -1$ and x is replaced by $-x$. This series can be used, as indicated in Prob. 3 following Section 6–16, to obtain expansions of other rational functions. Then by differentiation and integration further results can be obtained.

EXAMPLE 1. With $m = -1$ and x replaced by x^2, (6–49) becomes

$$\frac{1}{1+x^2} = 1 - x^2 + \cdots + (-1)^n x^{2n} + \cdots, \quad -1 < x < 1. \tag{6–50}$$

If this is integrated, one finds

$$\text{arc tan } x = x - \tfrac{1}{3}x^3 + \cdots + \frac{(-1)^n x^{2n+1}}{2n+1} + \cdots, \quad -1 < x < 1. \tag{6–51}$$

EXAMPLE 2. Since $\cosh x = \frac{1}{2}(e^x + e^{-x})$, one finds

$$\begin{aligned}\cosh x &= \frac{1}{2}\Bigg[\left(1 + \frac{x}{1!} + \frac{x^2}{2!} + \cdots + \frac{x^n}{n!} + \cdots\right) \\ &\quad + \left(1 - \frac{x}{1!} + \frac{x^2}{2!} + \cdots + \frac{(-1)^n x^n}{n!} + \cdots\right)\Bigg] \\ &= 1 + \frac{x^2}{2!} + \cdots + \frac{x^{2n}}{(2n)!} + \cdots, \quad -\infty < x < \infty.\end{aligned} \tag{6–52}$$

EXAMPLE 3. Since $\sin x \cos x = \frac{1}{2}\sin 2x$, one finds

$$\sin x \cos x = \frac{1}{2}\left\{\frac{2x}{1!} - \frac{2^3x^3}{3!} + \cdots + \frac{(-1)^{n-1}2^{2n-1}x^{2n-1}}{(2n-1)!} + \cdots\right\}, \quad -\infty < x < \infty. \tag{6–53}$$

Remark. Since the remainder formula does provide a method for estimating R_n, it enables one to estimate the error committed in computing the sum of a power series. Thus if $f(x)$ is known to be analytic in a given interval and it is known that in this interval

$$|f^{(n+1)}(x)| \leqq M_{n+1}$$

for a certain constant M_{n+1}, then

$$|R_n| \leqq \frac{M_{n+1}|x-a|^{n+1}}{(n+1)!}. \tag{6–54}$$

This formula can be added to those developed in Section 6–9.

As Theorem 41 shows, one does not require analyticity of the function $f(x)$ in order to be able to apply the remainder formula; $f(x)$ need only have continuous derivatives through the $(n+1)$-st order. Actually, the $(n+1)$-st derivative need only *exist* between a and x, continuity not being required. Thus, in principle, one can use the formula as a method for evaluating a nonanalytic $f(x)$ by a *finite* series, with remainder estimated by (6–54).

6–18 Further operations on power series. Four other operations by which new Taylor series can be obtained are described by the following theorems.

THEOREM 42. *Convergent power series can be multiplied; that is, if*

$$f(x) = \sum_{n=0}^{\infty} c_n(x-a)^n, \quad F(x) = \sum_{n=0}^{\infty} C_n(x-a)^n$$

are power series with convergence radii r_0^* and r_1^* respectively, $0 < r_0^* \leqq r_1^*$, then

$$f(x)F(x) = \sum_{n=0}^{\infty} k_n(x-a)^n, \quad a - r_0^* < x < a + r_0^*,$$

where

$$k_n = c_0C_n + c_1C_{n-1} + c_2C_{n-2} + \cdots + c_{n-1}C_1 + c_nC_0.$$

This is simply an application of the Cauchy product rule (Section 6–10) to the absolutely convergent series for $f(x)$ and $F(x)$.

THEOREM 43. *Convergent power series can be divided, provided there is no division by zero; that is, if $f(x)$ and $F(x)$ are given as in Theorem 42 and $F(a) = C_0 \neq 0$, then*

$$\frac{f(x)}{F(x)} = \sum_{n=0}^{\infty} p_n(x-a)^n, \quad a - r_2^* < x < a + r_2^*$$

for some positive number r_2^*, where the p_n satisfy the equations

$$c_n = p_0C_n + p_1C_{n-1} + \cdots + p_{n-1}C_1 + p_nC_0. \tag{6–55}$$

The rule (6–55) expresses the fact that the series $\Sigma p_n(x-a)^n$ multiplied by the series $\Sigma C_n(x-a)^n$ gives the series $\Sigma c_n(x-a)^n$. A proof of this theorem and a more precise determination of r_2^* require complex variables; this is taken up in Chapter 9 below. It should be noted that the equations (6–55) are implicit equations for the coefficients p_n:

$$c_0 = p_0C_0, \quad c_1 = p_0C_1 + p_1C_0, \ldots.$$

These can be solved in turn to obtain as many coefficients as are desired:

$$p_0 = \frac{c_0}{C_0}, \quad p_1 = \frac{c_1C_0 - c_0C_1}{C_0^2}, \ldots;$$

it will usually be difficult to obtain a formula for the general coefficient p_n.

THEOREM 44. *A Taylor series with constant term a can be substituted for the variable x in a Taylor series about $x = a$; that is, if*

$$f(x) = \sum_{n=0}^{\infty} c_n(x-a)^n, \quad g(x) = a + \sum_{n=1}^{\infty} d_n(x-b)^n,$$

have nonzero convergence radii r_0^* and r_1^* respectively and $|g(x) - a| < r_0^*$ for $|x - b| < r_2$, where $r_2 \leqq r_1^*$, then

$$f[g(x)] = \sum_{n=0}^{\infty} c_n \left\{ \sum_{m=1}^{\infty} d_m(x-b)^m \right\}^n = \sum_{n=0}^{\infty} q_n(x-b)^n, \quad |x-b| < r_2,$$

where the coefficients q_n are obtained by collecting terms of same degree.

An example will make clear the rule of formation of coefficients:

$$\begin{aligned} e^{\sin x} &= \sum_{n=0}^{\infty} \frac{(\sin x)^n}{n!} = \sum_{n=0}^{\infty} \frac{1}{n!}\left(x - \frac{x^3}{3!} + \frac{x^5}{5!} \cdots\right)^n \\ &= 1 + \left(x - \frac{x^3}{3!} + \cdots\right) + \frac{1}{2!}\left(x - \frac{x^3}{3!} + \cdots\right)^2 \\ &\quad + \frac{1}{3!}\left(x - \frac{x^3}{3!} + \cdots\right)^3 + \cdots \\ &= 1 + x + \frac{x^2}{2!} + x^3\left(-\frac{1}{3!} + \frac{1}{3!}\right) + \cdots. \end{aligned}$$

In this case $r_2 = \infty$.

The proof of the theorem is simplest in terms of complex variables and will not be considered here. The theorem is a special case of a theorem of Weierstrass on double series, for which the reader is referred to K. Knopp, *Infinite Series*, page 430 (London: Blackie, 1928).

THEOREM 45. *A power series can be inverted, provided the first degree term is not zero; that is, if*

$$y = f(x) = \sum_{n=0}^{\infty} c_n(x-a)^n, \quad |x-a| < r_0,$$

and $c_1 \neq 0$, then there is an inverse function

$$x = g(y) = a + \sum_{n=1}^{\infty} b_n(y-c_0)^n, \quad |y-c_0| < r_1, \quad r_1 > 0.$$

The coefficients b_n are determined from the identity

$$x - a \equiv \sum_{n=1}^{\infty} b_n \left[\sum_{m=1}^{\infty} c_m(x-a)^m\right]^n.$$

Again an example will clarify the method of determining coefficients. From the series (6–51):

$$\text{arc tan } x = x - \tfrac{1}{3}x^3 + \cdots,$$

one can seek to determine a series for $x = \tan y$:

$$x \equiv \sum_{n=1}^{\infty} b_n y^n = \sum_{n=1}^{\infty} b_n(x - \tfrac{1}{3}x^3 + \cdots)^n,$$

$$\begin{aligned} x &\equiv b_1(x - \tfrac{1}{3}x^3 + \cdots) + b_2(x - \tfrac{1}{3}x^3 + \cdots)^2 + b_3(x - \tfrac{1}{3}x^3 + \cdots)^3 + \cdots \\ &\equiv b_1 x + b_2 x^2 + x^3(-\tfrac{1}{3}b_1 + b_3) + \cdots. \end{aligned}$$

Hence

$$b_1 = 1, \quad b_2 = 0, \quad b_3 - \tfrac{1}{3}b_1 = 0, \ldots,$$

so that

$$x = \tan y = y + \tfrac{1}{3}y^3 + \cdots.$$

For a proof of the theorem, one is referred to page 184 of the book of Knopp referred to above.

This last theorem and the one on division suggest a principle which has a broad field of applications: in order to determine a function which is to satisfy a given condition, one can postulate that the function is expressible by a power series and then try to determine the coefficients of such a series in order to satisfy the given condition. If such a series can be found, one can then investigate the convergence of the series and determine whether it actually defines a function satisfying the given condition.

Problems

1. Obtain the following Taylor series expansions:

(a) $\sinh x = \sum_{n=1}^{\infty} \frac{x^{2n-1}}{(2n-1)!}$, all x;

(b) $\cos^2 x = 1 + \sum_{n=1}^{\infty} \frac{(-1)^n 2^{2n-1} x^{2n}}{(2n)!}$, all x;

(c) $\sin^2 x = \sum_{n=1}^{\infty} \frac{(-1)^{n+1} 2^{2n-1} x^{2n}}{(2n)!}$, all x;

(d) $\log x = \sum_{n=1}^{\infty} \frac{(-1)^{n+1}(x-1)^n}{n}$, $|x - 1| < 1$;

(e) $\sqrt{1-x} = 1 - \frac{x}{2} - \frac{1}{2^2 2!}x^2 - \frac{1 \cdot 3}{2^3 3!}x^3 - \cdots$, $|x| < 1$;

(f) $\frac{1}{\sqrt{1-x^2}} = 1 + \frac{x^2}{2} + \frac{1 \cdot 3}{2^2 2!}x^4 + \frac{1 \cdot 3 \cdot 5}{2^3 3!}x^6 + \cdots$, $|x| < 1$;

(g) $\text{arc} \sin x = x + \frac{x^3}{2 \cdot 3} + \frac{1 \cdot 3}{2^2 \cdot 2!}\frac{x^5}{5} + \frac{1 \cdot 3 \cdot 5}{2^3 \cdot 3!}\frac{x^7}{7} + \cdots$, $|x| < 1$.

2. Find the first three nonzero terms of the following Taylor series:

(a) $e^x \sin x$ about $x = 0$
(b) $\tan x$ about $x = 0$
(c) $\log^2(1 + x)$ about $x = 0$
(d) $\log(1 - x^2)$ about $x = 0$
(e) $x^3 + 3x + 1$ about $x = 2$
(f) $e^{\tan x}$ about $x = 0$
(g) $y = \sinh^{-1} x$ about $x = 0$
(h) $y = \tanh x$ about $x = 0$
(i) $y = \tanh^{-1} x$ about $x = 0$
(j) $y = \log \sec x$ about $x = 0$

3. Prove Taylor's remainder formula (Theorem 41) for general n. [Hint: replace the function $F(x)$ by the function $G(x) - [(x_2 - x)/(x_2 - a)]^n G(a)$, where

$$G(x) = f(x_2) - f(x) - (x_2 - x)f'(x) - \cdots - \frac{(x_2 - x)^{n-1}}{(n-1)!} f^{(n-1)}(x).]$$

4. Evaluate to three decimal places:

(a) $\int_0^1 e^{-x^2}\,dx$, (b) $\int_0^{0.5} \frac{dx}{\sqrt{1 + x^4}}$.

5. Let $f(x) = e^{-1/x^2}$ for $x \neq 0$ and let $f(0) = 0$.
(a) Prove that $f(x)$ is continuous for all x.
(b) Prove that $f'(x)$ is continuous for $x \neq 0$ and that

$$\lim_{x\to 0} f'(x) = f'(0) = 0,$$

so that $f'(x)$ is continuous for all x.
(c) Prove that $f^{(n)}(x)$ is continuous for all x and $f^{(n)}(0) = 0$.
(d) Graph the function $f(x)$.

6. Use the remainder formula to estimate the error in the following computations:

(a) $e = 1 + 1 + \dfrac{1}{2!} + \dfrac{1}{3!} + \dfrac{1}{4!} + \dfrac{1}{5!}$ [assume that $e < 3$ is known];

(b) $\sin 1 = 1 - \dfrac{1}{3!} + \dfrac{1}{5!}$;

(c) $\log \dfrac{3}{2} = \dfrac{1}{2} - \dfrac{1}{2\cdot 2^2} + \dfrac{1}{3\cdot 2^3}$.

7. Let $f(x)$ satisfy the conditions stated in Theorem 41. Let $f'(a) = f''(a) = \cdots = f^{(n)}(a) = 0$, but $f^{(n+1)}(a) \neq 0$. Show that $f(x)$ has a maximum, minimum, or horizontal inflection point at $x = a$ according as the function $f^{(n+1)}(a)(x - a)^{n+1}$ has a maximum, minimum or inflection point for $x = a$. (This gives another proof of the rule deduced in Section 2–19.)

Answers

2. (a) $x + x^2 + \frac{1}{3}x^3$; (b) $x + \frac{1}{3}x^3 + \frac{2}{15}x^5$; (c) $x^2 - x^3 + \frac{11}{12}x^4$; (d) $-x^2 - \frac{1}{2}x^4 - \frac{1}{3}x^6$; (e) $15 + 15(x-2) + 6(x-2)^2$; (f) $1 + x + \frac{1}{2}x^2$; (g) $x - \frac{1}{6}x^3 + \frac{3}{40}x^5$; (h) $x - \frac{1}{3}x^3 + \frac{2}{15}x^5$; (i) $x + \frac{1}{3}x^3 + \frac{1}{5}x^5$; (j) $\frac{1}{2}x^2 + \frac{1}{12}x^4 + \frac{1}{45}x^6$.

4. (a) 0.747, (b) 0.497.

***6–19 Sequences and series of complex numbers.** The basic properties of complex numbers are described in Section 0–2 of the Introduction. We recall the definition of *absolute value* of a complex number $z = x + iy$:

$$|z| = \sqrt{x^2 + y^2}$$

and that this represents the distance from z to the origin of the xy plane. Since addition or subtraction of complex numbers is like that of vectors, one can in particular interpret $|z_1 - z_2|$ as the *distance from z_1 to z_2*. This is illustrated in Fig. 6–12.

Sequences of complex numbers are defined as for real numbers. The following are examples:

$$z_n = i^n, \quad z_n = \frac{n(1-i)}{1+n^2}, \quad z_n = \left(1 + \frac{i}{n}\right)^n.$$

A sequence z_n is said to *converge to* z_0:

$$\lim_{n\to\infty} z_n = z_0$$

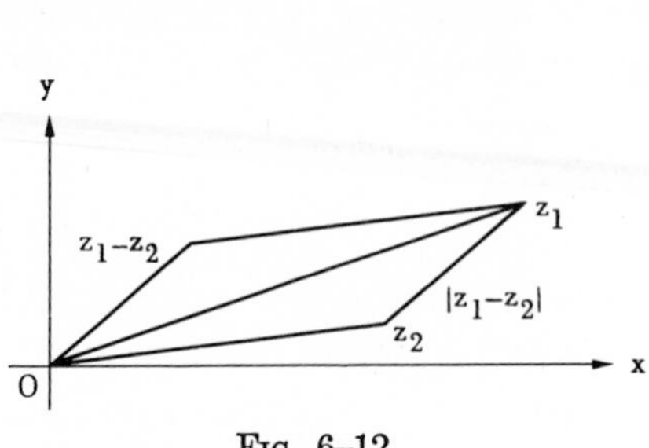

FIG. 6–12.

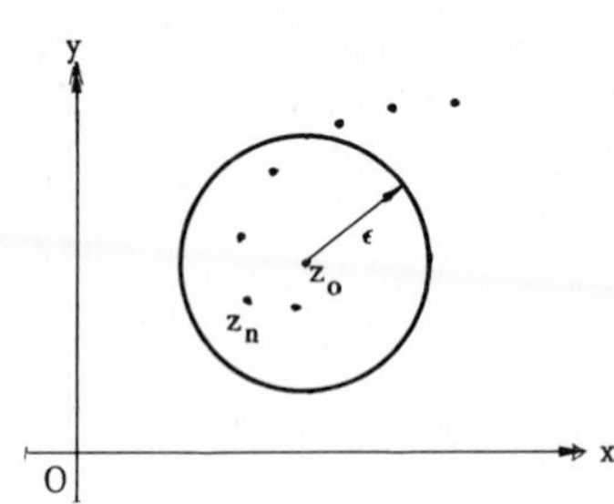

FIG. 6–13. Sequence converging to z_0.

if, given $\epsilon > 0$, an integer N can be found such that

$$|z_n - z_0| < \epsilon \quad \text{for} \quad n > N.$$

This resembles the definition for real numbers. The inequality states that z_n is within distance ϵ of z_0, or that z_n is within the circle of radius ϵ about z_0; this is illustrated in Fig. 6–13. If a sequence z_n does not converge, it is said to *diverge*.

THEOREM 46. *Let $z_n = x_n + iy_n$ $(n = 1, 2, \ldots)$ be a sequence of complex numbers. If this sequence converges to $z_0 = x_0 + iy_0$, then*

$$\lim_{n\to\infty} x_n = x_0, \quad \lim_{n\to\infty} y_n = y_0.$$

Conversely, if x_n converges to x_0 and y_n converges to y_0, then

$$\lim_{n\to\infty} z_n = \lim_{n\to\infty} (x_n + iy_n) = x_0 + iy_0.$$

This theorem shows that the convergence of sequences of complex numbers can be referred back to that of real numbers, simply by studying real and imaginary parts.

To prove the theorem, we remark that, if $|z_n - z_0| < \epsilon$, then (x_n, y_n) is within the circle of radius ϵ about (x_0, y_0), so that necessarily

$$|x_n - x_0| < \epsilon, \quad |y_n - y_0| < \epsilon.$$

Thus convergence of z_n to z_0 implies convergence of x_n to x_0 and y_n to y_0. Conversely, if x_n converges to x_0, y_n converges to y_0, then for given ϵ, N can be chosen so large that

$$|x_n - x_0| < \tfrac{1}{2}\epsilon, \quad |y_n - y_0| < \tfrac{1}{2}\epsilon \quad \text{for} \quad n > N.$$

These inequalities force (x_n, y_n) to lie within a *square* with center (x_0, y_0) and side ϵ; hence (x_n, y_n) must lie within the *circle* of radius ϵ about (x_0, y_0), so that $|z_n - z_0| < \epsilon$ for $n > N$. Accordingly, z_n must converge to z_0.

THEOREM 47 (*Cauchy criterion*). *A sequence z_n of complex numbers converges if and only if, for each $\epsilon > 0$ an N can be found such that*

$$|z_m - z_n| < \epsilon \quad \textit{for} \quad m > N, \quad n > N.$$

This is proved by referring the convergence of z_n back to that of the two real sequences x_n, y_n and then applying the Cauchy criterion (Theorem 6) to the real sequences.

An *infinite series* of complex numbers is defined as for real numbers:

$$\sum_{n=1}^{\infty} z_n = z_1 + z_2 + \cdots + z_n + \cdots.$$

The series is said to converge or diverge according as the nth partial sums:

$$S_n = z_1 + \cdots + z_n$$

form a convergent or divergent sequence. The *sum* of the series is then the limit

$$S = \lim_{n\to\infty} S_n = \lim_{n\to\infty} \sum_{m=1}^{n} z_m,$$

when the limit exists.

On the basis of Theorem 46, one can at once assert the following theorem.

THEOREM 48. *If* $z_n = x_n + iy_n$, *then the series*

$$\sum_{n=1}^{\infty} z_n = \sum_{n=1}^{\infty} (x_n + iy_n)$$

converges and has sum $S = A + Bi$ *if and only if*

$$\sum_{n=1}^{\infty} x_n = A, \quad \sum_{n=1}^{\infty} y_n = B.$$

Thus the convergence of series of complex numbers is also referred back to that for real numbers. This can also be done in a second way.

THEOREM 49. *If* $\sum_{n=1}^{\infty} |z_n|$ *converges, then* $\sum_{n=1}^{\infty} z_n$ *converges.*

In words: if a complex series is *absolutely convergent,* then it is convergent. The proof is the same as that of the corresponding theorem for real numbers (Theorem 11, Section 6–6).

On the basis of Theorems 48 and 49 one can now obtain tests for convergence and divergence of series of complex numbers. In particular, the following rules hold:

the n-th term test (Theorem 10);
the Cauchy criterion (Theorem 9);
the comparison test for convergence (Theorem 12);
the ratio test (Theorems 17 and 20);
the root test [Theorems 19, 21, 21(a)].

The rule for addition or subtraction of convergent series (Theorem 8) and the product rule (Theorem 29) also hold for complex series. The estimates of remainders (Theorems 22, 23, 24, 25) can also be used for complex series.

Example 1. The series

$$1 + \frac{i}{2!} + \cdots + \frac{i^n}{n!} + \cdots = \sum_{n=0}^{\infty} \frac{i^n}{n!}$$

is absolutely convergent, since the series of absolute values is the real series

$$\sum_{n=0}^{\infty} \frac{1}{n!}$$

which converges (e.g., by the ratio test).

Example 2. The series

$$\frac{i}{1} + \frac{i^2}{2} + \frac{i^3}{3} + \cdots + \frac{i^n}{n} + \cdots = \sum_{n=1}^{\infty} \frac{i^n}{n}$$

is not absolutely convergent, since $\Sigma 1/n$ diverges. However, the series of real parts is

$$0 - \tfrac{1}{2} + 0 + \tfrac{1}{4} + 0 - \tfrac{1}{6} + \cdots$$

and the series of imaginary parts is

$$1 + 0 - \tfrac{1}{3} + 0 + \tfrac{1}{5} + \cdots.$$

If the 0's are disregarded, these are convergent alternating series. Hence $\Sigma i^n/n$ converges.

The general theory of functions of a complex variable is developed in Chapter 9. Here we consider briefly the functions z^n, where n is a positive integer or 0, and the corresponding *power series:*

$$\sum_{n=0}^{\infty} c_n z^n = c_0 + c_1 z + c_2 z^2 + \cdots + c_n z^n + \cdots. \tag{6–56}$$

Since each term is defined for all z, the series may converge for some or all z. The basic theorem on power series (Theorem 35) can be restated and proved just as for real numbers and one concludes that each power series (6–56) has a *radius of convergence* r^* such that the series converges when $|z| < r^*$ and diverges when $|z| > r^*$. The use of the term "radius of convergence" now receives its justification; for the power series (6–56) converges within the circle with center O and radius r^*. This is illustrated in Fig. 6–14.

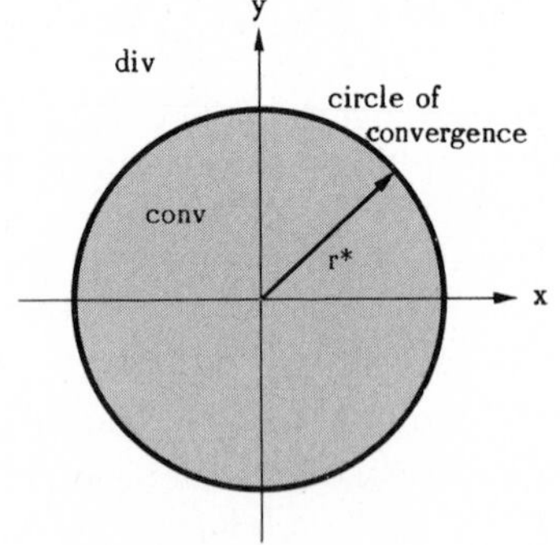

Fig. 6–14.

It is of special interest that the power series (6–46), (6–47), (6–48) for e^x, $\sin x$ and $\cos x$ continue to converge when x is replaced by an arbitrary complex number z. One can then use the equations

$$e^z = 1 + z + \cdots + \frac{z^n}{n!} + \cdots, \tag{6–57}$$

$$\sin z = z - \frac{z^3}{3!} + \cdots + (-1)^{n-1}\frac{z^{2n-1}}{(2n-1)!} + \cdots, \tag{6–58}$$

$$\cos z = 1 - \frac{z^2}{2!} + \cdots + (-1)^n \frac{z^{2n}}{(2n)!} + \cdots \tag{6–59}$$

to *define* these functions for complex z. From these series one derives the *Euler identity* (Prob. 4 below):

$$e^{iy} = \cos y + i \sin y, \tag{6–60}$$

or the more general relation:

$$e^{x+iy} = e^x(\cos y + i \sin y). \tag{6–61}$$

This last equation can also be used as a *definition* of e^z for complex z, the series expression being then a consequence of this definition. This is shown in Chapter 9.

Problems

1. Evaluate the limits:

(a) $\lim_{n\to\infty} \frac{i^n}{n}$, (b) $\lim_{n\to\infty} \frac{(1+i)n^3 - 2in + 3}{in^3 - 1}$.

2. Test for absolute convergence and for convergence:

(a) $\left(\frac{1+i}{2}\right) + \left(\frac{1+i}{2}\right)^2 + \cdots + \left(\frac{1+i}{2}\right)^n + \cdots$;

(b) $\sum_{n=1}^{\infty} ni^n$ (c) $\sum_{n=1}^{\infty} \frac{ni^n}{n^2+1}$ (d) $\sum_{n=1}^{\infty} \frac{1}{(n+i)^2}$.

3. Prove that the series (6–57), (6–58) and (6–59) converge for all z.

4. (a) Prove the Euler identity (6–60) from the series definition of e^z, $\cos z$, $\sin z$. (b) Prove (6–61) from the series definitions of e^z, $\cos z$, $\sin z$.

5. Use the series expressions (6–57), (6–58), (6–59) to prove the identities

(a) $\cos z = \frac{e^{iz} + e^{-iz}}{2}$

(b) $\sin z = \frac{e^{iz} - e^{-iz}}{2i}$

(c) $e^{z_1+z_2} = e^{z_1} \cdot e^{z_2}$

(d) $\sin(-z) = -\sin z$

(e) $\cos(-z) = \cos z$

(f) $\sin^2 z + \cos^2 z = 1$

(g) $\cos 2z = \cos^2 z - \sin^2 z$

(h) $\sin 2z = 2 \sin z \cos z$

6. Show that the following series converge for $|z| < 1$ and diverge for $|z| > 1$:

(a) $z + \frac{z^2}{2} + \cdots + \frac{z^n}{n} + \cdots$; (b) $1 + z + z^2 + \cdots + z^n + \cdots$.

Answers

1. (a) 0, (b) $1 - i$. 2. (a) absolutely convergent, (b) divergent, (c) convergent, not absolutely, (d) absolutely convergent.

***6–20 Sequences and series of functions of several variables.** The notions of sequence and series of functions extend at once to functions of several variables. Thus

$$\sum_{n=1}^{\infty}(xy)^n = xy + x^2y^2 + \cdots + x^ny^n + \cdots \tag{6–62}$$

is a series of functions of the two variables x and y. The notion of uniform convergence also extends at once, as well as the M-test and the properties described in Section 6–14.

One can in particular consider *power series* in several variables. For two variables x, y such a power series is a series

$$\sum_{n=0}^{\infty} f_n(x, y) = f_0(x, y) + f_1(x, y) + \cdots + f_n(x, y) + \cdots, \tag{6–63}$$

where

$$f_n(x, y) = c_{n,0}x^n + c_{n,1}x^{n-1}y + \cdots + c_{n,n-1}xy^{n-1} + c_{n,n}y^n, \tag{6–64}$$

the c's being constants. Thus f_n is a *homogeneous polynomial of degree n* in x and y. The series (6–62) is an example of this, with $f_{2n} = x^ny^n$ and $f_0 = 0$, $f_1 = f_3 = \cdots = 0$. This series also illustrates the fact that the values (x, y) for which a power series in x and y converges form a more complicated set than the convergence interval for series in x alone. For (6–62) the series converges when $|xy| < 1$; this region is pictured in Fig. 6–15. In general the convergence region can be quite complicated.

If a function $F(x, y)$ can be represented by such a power series in a neighborhood of the origin:

$$F(x, y) = c_{0,0} + (c_{1,0}x + c_{1,1}y) + (c_{2,0}x^2 + c_{2,1}xy + c_{2,2}y^2) + \cdots,$$

then a term-by-term differentiation (which can be justified) shows that

$$F(0, 0) = c_{0,0}, \quad \frac{\partial F}{\partial x}(0, 0) = c_{1,0},$$

$$\frac{\partial F}{\partial y}(0, 0) = c_{1,1}, \quad \frac{1}{2!}\frac{\partial^2 F}{\partial x^2} = c_{2,0},$$

$$\frac{2}{2!}\frac{\partial^2 F}{\partial x\,\partial y} = c_{2,1}, \quad \frac{1}{2!}\frac{\partial^2 F}{\partial y^2} = c_{2,2}.$$

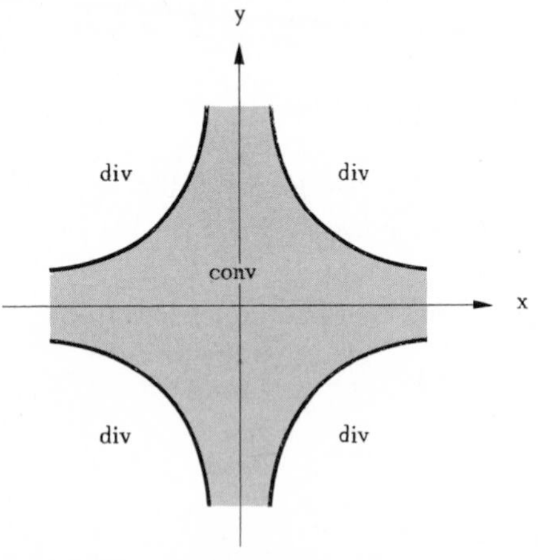

Fig. 6–15.

In general one finds

$$f_n(x, y) = \frac{1}{n!}\left(\frac{\partial^n F}{\partial x^n}x^n + n\frac{\partial^n F}{\partial x^{n-1}\,\partial y}x^{n-1}y + \frac{n(n-1)}{2!}\frac{\partial^n F}{\partial x^{n-2}\,\partial y^2}x^{n-2}y^2 + \cdots \right.$$
$$\left. + \frac{n(n-1)\cdots(n-k+1)}{k!}\frac{\partial^n F}{\partial x^{n-k}\,\partial y^k}x^{n-k}y^k + \cdots + \frac{\partial^n F}{\partial y^n}y^n\right), \tag{6–65}$$

all derivatives being evaluated at (0, 0). A series $\Sigma f_n(x, y)$, in which the f_n are given by (6–65), is known as a Taylor series in x and y, about (0, 0), and the function $F(x, y)$ which it represents is termed *analytic* in the corresponding region. The expansion about a general point (x_1, y_1) is obtained by a translation of origin:

$$\begin{aligned} F(x, y) = F(x_1, y_1) &+ \left[\frac{\partial F}{\partial x}(x - x_1) + \frac{\partial F}{\partial y}(y - y_1)\right] \\ &+ \frac{1}{2!}\left[\frac{\partial^2 F}{\partial x^2}(x - x_1)^2 + 2\frac{\partial^2 F}{\partial x\, \partial y}(x - x_1)(y - y_1) + \frac{\partial^2 F}{\partial y^2}(y - y_1)^2\right] \\ &+ \cdots + \frac{1}{n!}\left[\frac{\partial^n F}{\partial x^n}(x - x_1)^n + \cdots\right] + \cdots, \end{aligned} \tag{6–66}$$

all derivatives being evaluated at (x_1, y_1).

The general term of the series (6–66) can be interpreted in terms of an *n-th differential* d^nF of the function $F(x, y)$:

$$\begin{aligned} d^nF &= \frac{\partial^n F}{\partial x^n}(x - x_1)^n + \cdots \\ &= \sum_{r=0}^{n} \binom{n}{r} \frac{\partial^n F}{\partial x^r\, \partial y^{n-r}}(x_1, y_1)(x - x_1)^r (y - y_1)^{n-r}, \end{aligned}$$

where the $\binom{n}{r}$ are the binomial coefficients [Eq. (0–31)]. To indicate the dependence of d^nF on x_1, y_1, and the differences $x - x_1$, $y - y_1$, we write:

$$d^nF = d^nF(x_1, y_1; x - x_1, y - y_1).$$

When $n = 1$ and $x - x_1 = dx$, $y - y_1 = dy$, one finds:

$$d^1F(x_1, y_1; dx, dy) = \frac{\partial F}{\partial x}dx + \frac{\partial F}{\partial y}dy = dF;$$

this is the familiar expression for the first differential. The series (6–66) can now be written more concisely:

$$\begin{aligned} F(x, y) &= F(x_1, y_1) + dF(x_1, y_1; x - x_1, y - y_1) \\ &\quad + \frac{1}{2!}d^2F(x_1, y_1; x - x_1, y - y_1) + \cdots \\ &\quad + \frac{1}{n!}d^nF(x_1, y_1; x - x_1, y - y_1) + \cdots. \end{aligned} \tag{6–66$'$}$$

The theory of analytic functions of several variables is again best studied with the aid of complex numbers. As with one variable, the familiar functions are "in general" analytic. Thus

$$e^x \sin y = y + xy + \frac{3x^2y - y^3}{6} + \cdots,$$

the series converging for all x and y.

The subject of analytic functions of several variables has not been studied intensively until recent times, and most books on the subject are very advanced. In O. D. Kellogg's *Foundations of Potential Theory* (Berlin: Springer, 1929), a brief treatment is given on pages 135–140. An introduction to the topic is also given in the last chapter of the book *Introduction to Analytic Functions* by W. Kaplan (Reading, Mass.: Addison-Wesley, 1966).

***6–21 Taylor's formula for functions of several variables.** There is a Taylor formula with remainder for functions of several variables:

$$\begin{aligned} F(x, y) = F(x_1, y_1) &+ dF(x_1, y_1; x - x_1, y - y_1) \\ &+ \cdots + \frac{1}{n!} d^nF(x_1, y_1; x - x_1, y - y_1) \\ &+ \frac{1}{(n+1)!} d^{n+1}F(x^*, y^*; x - x_1, y - y_1); \qquad (6\text{–}67) \end{aligned}$$

$$x^* = x_1 + t^*(x - x_1), \quad y^* = y_1 + t^*(y - y_1), \quad 0 < t^* < 1.$$

The point (x^*, y^*) lies between (x_1, y_1) and (x, y) on the line segment joining these points, as in Fig. 6–16. For $n = 1$, the formula becomes

$$\begin{aligned} F(x, y) &= F(x_1, y_1) \\ &+ (x - x_1)F_x(x^*, y^*) \\ &+ (y - y_1)F_y(x^*, y^*). \qquad (6\text{–}68) \end{aligned}$$

This is known as the *law of the mean* for functions of two variables.

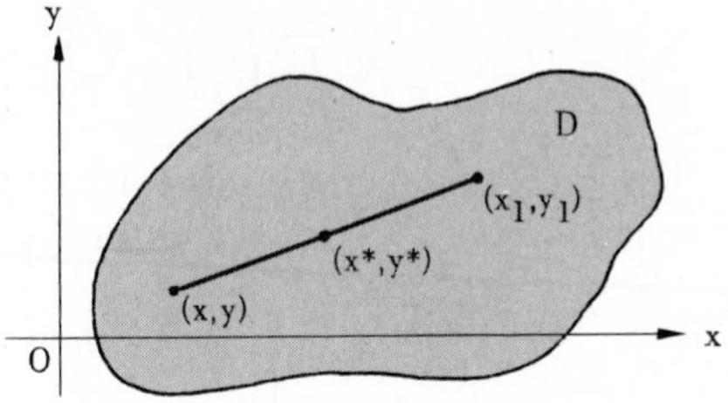

FIG. 6–16. Taylor's formula for $F(x,y)$.

To prove (6–68), one writes:

$$\phi(t) = F[x_1 + t(x - x_1), y_1 + t(y - y_1)], \quad 0 \leqq t \leqq 1.$$

Thus x and y are considered as fixed and ϕ depends only on t. By the law of the mean for ϕ,

$$\phi(1) = \phi(0) + \phi'(t^*), \quad 0 < t^* < 1.$$

But $\phi(1) = F(x, y)$, $\phi(0) = F(x_1, y_1)$ and

$$\begin{aligned} \phi'(t) = (x - x_1)F_x[x_1 + t(x - x_1), y_1 + t(y - y_1)] \\ + (y - y_1)F_y[x + t(x - x_1), y_1 + t(y - y_1)]. \end{aligned}$$

If t is replaced by t^*, one obtains (6–68). The general formula (6–67) is proved in the same way, on the basis of Taylor's formula for ϕ:

$$\phi(1) = \phi(0) + \phi'(0) + \cdots + \frac{\phi^{(n)}(0)}{n!} + \frac{\phi^{(n+1)}(t^*)}{(n+1)!},$$

where $0 < t^* < 1$. For one finds by induction that

$$\phi^{(n)}(t) = d^nF[x_1 + t(x - x_1), y_1 + t(y - y_1); x - x_1, y - y_1]; \quad (6\text{–}69)$$

the validity of this equation is ensured if $F(x, y)$ has continuous derivatives through the $(n + 1)$-st order in a domain D containing the line segment joining (x, y) to (x_1, y_1).

Taylor's series or Taylor's formula can be used to study the nature of a function near a particular point. As remarked above, the linear terms give dF, the best "linear approximation" to $F(x, y) - F(x_1, y_1)$. If $dF = 0$, the quadratic terms $d^2F/2!$ become of importance. In particular, if the quadratic expression

$$d^2F = A(x - x_1)^2 + 2B(x - x_1)(y - y_1) + C(y - y_1)^2$$

is positive, except for $x = x_1, y = y_1$, then $F(x, y)$ has a minimum at (x_1, y_1). Pursuing this further, we rediscover the criteria for maxima and minima developed in Section 2–15.

Problems

1. Expand in power series, stating the region of convergence:

(a) $e^{x^2-y^2}$, (b) $\sin(xy)$, (c) $\dfrac{1}{1 - x - y}$, (d) $\dfrac{1}{1 - x - y - z}$.

2. Prove Eq. (6–69) by induction.

3. Prove that, if a power series

$$c_{0,0} + (c_{1,0}x + c_{1,1}y) + (c_{2,0}x^2 + c_{2,1}xy + c_{2,2}y^2) + \cdots$$

converges at (x_0, y_0), then it converges at every point $(\lambda x_0, \lambda y_0)$, for $|\lambda| < 1$.

4. Evaluate $\int_0^1 \int_0^1 \sin(xy)\, dx\, dy$ with the aid of power series.

Answers

1. (a) $1 + (x^2 - y^2) + \dfrac{1}{2!}(x^4 - 2x^2y^2 + y^4) + \cdots + \dfrac{1}{n!}(x^2 - y^2)^n + \cdots$,

all (x, y); (b) $xy - \dfrac{1}{3!}x^3y^3 + \cdots + (-1)^{n-1}\dfrac{(xy)^{2n-1}}{(2n-1)!} + \cdots$, all (x, y);

(c) $1 + (x + y) + (x^2 + 2xy + y^2) + \cdots + (x + y)^n + \cdots$, $-1 < x + y < 1$;

(d) $1 + (x + y + z) + (x + y + z)^2 + \cdots + (x + y + z)^n + \cdots$,

$-1 < x + y + z < 1$.

4. 0.240 (3 significant figures).

***6–22 Improper integrals versus infinite series.** In Section 4–5 improper integrals are discussed and certain criteria for convergence and divergence are stated without proof. It will now be shown that there is a very close analogy between improper integrals and infinite series; an indication of the relationship is given in the integral test of Theorem 14. Once this analogy has been made clear, it becomes an easy matter to establish a variety of tests for convergence or divergence of improper integrals, including those of Section 4–5.

Let us consider first, as an example, the integral

$$\int_1^\infty \frac{dx}{x}.$$

This would be termed convergent if the limit

$$\lim_{b\to\infty}\int_1^b \frac{dx}{x}$$

exists. However, since the integrand $1/x$ is *positive*, the integral from 1 to b must *increase* as b increases. From this it follows that either the above limit is $+\infty$ (i.e., there is no limit) and the integral diverges, or else the limit is a finite number I. Now, in order to determine which of the two cases holds, it is clearly sufficient to let b approach ∞ through integral values n, that is to consider the limit of a sequence:

$$\lim_{n\to\infty}\int_1^n \frac{dx}{x} \quad (n = 1, 2, \ldots).$$

But one can write

$$\int_1^n \frac{dx}{x} = \int_1^2 \frac{dx}{x} + \int_2^3 \frac{dx}{x} + \cdots + \int_{n-1}^n \frac{dx}{x}.$$

In other words, the integral exists precisely when the series

$$\sum_{n=1}^\infty \int_n^{n+1} \frac{dx}{x} = \sum_{n=1}^\infty a_n$$

converges and the two have the same value. This is suggested in Fig. 6–17.

In this particular case

$$a_n = \int_n^{n+1} \frac{dx}{x} = \log\frac{n+1}{n}$$

and the series is

$$\sum_{n=1}^\infty \log\frac{n+1}{n}.$$

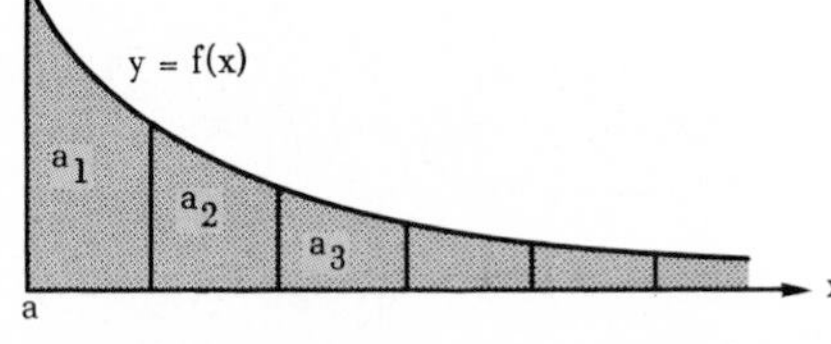

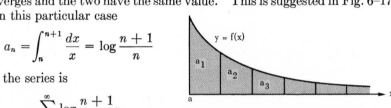

Fig. 6–17. Improper integral as a series.

One easily verifies that both series and integral diverge.

We now formulate the relationships in general terms.

Theorem 50. *Let $f(x)$ be continuous and let $f(x) \geqq 0$ for $a \leqq x < \infty$. Then the integral*

$$\int_a^\infty f(x)\,dx$$

converges and has value I if and only if the series

$$\sum_{n=1}^\infty a_n, \quad a_n = \int_{a+n-1}^{a+n} f(x)\,dx$$

converges and has sum I.

If $f(x)$ changes sign infinitely often, then convergence of the integral will certainly imply that of the series; however, the converse need not hold, as the example

$$\int_0^\infty \sin 2\pi x\, dx \tag{6–70}$$

shows (Prob. 2 below). The proper connection between series and integrals in this case is given by the following theorem.

THEOREM 51. *Let $f(x)$ be continuous for $a \leqq x < \infty$. Let $f(x) \geqq 0$ for $a = b_0 \leqq x \leqq b_1$, $f(x) \leqq 0$ for $b_1 \leqq x \leqq b_2$ and, in general, $(-1)^n f(x) \geqq 0$ for $b_n \leqq x \leqq b_{n+1}$, where b_n is a monotone sequence such that*

$$\lim_{n\to\infty} b_n = \infty .$$

Then the integral

$$\int_a^\infty f(x)\, dx$$

converges and has value I if and only if the alternating series

$$\sum_{n=1}^\infty a_n, \quad a_n = \int_{b_{n-1}}^{b_n} f(x)\, dx$$

converges and has sum I.

Proof. If the integral converges to I, then as above

$$\lim_{n\to\infty}\int_a^{b_n} f(x)\, dx = \lim_{n\to\infty}(a_1 + \cdots + a_n) = I,$$

so that the series converges and has sum I.

Conversely, let the series converge to I. Let $\epsilon > 0$ be given and choose N so large that

$$|a_1 + \cdots + a_n - I| < \tfrac{1}{2}\epsilon \quad \text{and} \quad |a_n| < \tfrac{1}{2}\epsilon \quad \text{for} \quad n \geqq N.$$

The latter condition can be satisfied by the nth term test. If $x_1 > b_N$, then $b_n \leqq x_1 \leqq b_{n+1}$ for some $n \geqq N$. Hence

$$\int_a^{x_1} f(x)\, dx = \int_a^{b_n} f(x)\, dx + \int_{b_n}^{x_1} f(x)\, dx = a_1 + \cdots + a_n + \int_{b_n}^{x_1} f(x)\, dx$$

and

$$\begin{aligned}\left|\int_a^{x_1} f(x)\, dx - I\right| &= \left|a_1 + \cdots + a_n - I + \int_{b_n}^{x_1} f(x)\, dx\right| \\ &\leqq |a_1 + \cdots + a_n - I| + \left|\int_{b_n}^{x_1} f(x)\, dx\right| \\ &\leqq |a_1 + \cdots + a_n - I| + \left|\int_{b_n}^{b_{n+1}} f(x)\, dx\right|;\end{aligned}$$

the last step is justified, since $f(x)$ does not change sign between b_n and b_{n+1}. Hence

$$\left|\int_a^{x_1} f(x)\,dx - I\right| \leqq |a_1 + \cdots + a_n - I| + |a_{n+1}| < \tfrac{1}{2}\epsilon + \tfrac{1}{2}\epsilon = \epsilon.$$

It follows that

$$\lim_{x_1\to\infty}\int_a^{x_1} f(x)\,dx = I.$$

Corollary. *Let $f(x)$ be continuous for $a \leqq x < \infty$; let $f(x)$ decrease as x increases and let $\lim_{x\to\infty} f(x) = 0$. Then the integrals*

$$\int_a^\infty f(x)\sin x\,dx,\quad \int_a^\infty f(x)\cos x\,dx$$

converge.

Proof. We consider the sine integral, the cosine integral being similar. Under the assumptions made, the alternating series of Theorem 51 is, except perhaps for the first term, of the form

$$\sum_{n=k}^{\infty} a_n,\quad a_n = \int_{n\pi}^{(n+1)\pi} f(x)\sin x\,dx.$$

Since $f(x)$ decreases as x increases, $|a_n|$ is decreasing; since $f(x)$ has limit 0 as $x\to\infty$, a_n converges to 0. The alternating series test (Theorem 18) then guarantees convergence.

Examples. The integrals

$$\int_1^\infty \frac{\sin x}{x}\,dx,\quad \int_0^\infty \frac{\cos x}{1+x}\,dx,\quad \int_1^\infty \frac{\sin x}{\sqrt{x}}\,dx$$

all exist, by virtue of the above corollary.

Theorem 52 (*Cauchy criterion*). *Let $f(x)$ be continuous for $a \leqq x < \infty$. The integral*

$$\int_a^\infty f(x)\,dx$$

exists if and only if, to each $\epsilon > 0$ a number B can be found such that

$$\left|\int_p^q f(x)\,dx\right| < \epsilon \quad \textit{for} \quad B < p < q.$$

Proof. If the integral converges to I, then for given $\epsilon > 0$ a B can be found such that

$$\left|\int_a^b f(x)\,dx - I\right| < \tfrac{1}{2}\epsilon \quad \text{for} \quad b > B.$$

Hence, for $B < p < q$,

$$\left|\int_a^p f\,dx - I\right| < \tfrac{1}{2}\epsilon, \quad \left|\int_a^q f\,dx - I\right| < \tfrac{1}{2}\epsilon$$

and

$$\left|\int_p^q f\,dx\right| = \left|\int_a^q f\,dx - \int_a^p f\,dx\right| = \left|\int_a^q f\,dx - I + I - \int_a^p f\,dx\right|$$
$$\leqq \left|\int_a^q f\,dx - I\right| + \left|\int_a^p f\,dx - I\right| < \tfrac{1}{2}\epsilon + \tfrac{1}{2}\epsilon = \epsilon.$$

Conversely, let the condition hold. Then the Cauchy criterion of Theorem 6 applies to the sequence

$$S_n = \int_a^n f(x)\,dx \quad (n > a);$$

hence this sequence converges to a number I. Let $\epsilon > 0$ be given and let a corresponding number B be chosen as in the theorem; let N be an integer larger than B and such that $|S_n - I| < \epsilon$ for $n \geqq N$. For $x_1 > N$,

$$\left|\int_a^{x_1} f(x)\,dx - I\right| = \left|\int_a^N f(x)\,dx - I + \int_N^{x_1} f(x)\,dx\right|$$
$$\leqq |S_N - I| + \left|\int_N^{x_1} f(x)\,dx\right|$$
$$< \epsilon + \epsilon = 2\epsilon.$$

It follows that

$$\lim_{x_1\to\infty} \int_a^{x_1} f(x)\,dx = I.$$

Remark. It follows from Theorem 52 that, for a convergent integral $\int_a^\infty f(x)\,dx$,

$$\lim_{b\to\infty} \int_b^{b+k} f(x)\,dx = 0$$

for every fixed k. However, it is not necessary that $f(x)$ itself approach 0. This is illustrated by the integral (Prob. 1 below):

$$\int_1^\infty \sin x^2\,dx.$$

THEOREM 53. *Let $f(x)$ be continuous for $a \leqq x < \infty$. If $\int_a^\infty |f(x)|\,dx$ converges, then*

$$\int_a^\infty f(x)\,dx$$

converges and

$$\left|\int_a^\infty f(x)\,dx\right| \leqq \int_a^\infty |f(x)|\,dx. \tag{6–71}$$

In words: an *absolutely convergent* improper integral is convergent.

Proof. Since $|f|$ is a continuous function of f, $|f(x)|$ is a continuous function of a continuous function and must hence be continuous. Now

$$\left|\int_p^q f(x)\,dx\right| \leqq \int_p^q |f(x)|\,dx.$$

Hence, if the Cauchy criterion holds for the integral of $|f|$, it must also hold for the integral of f itself; thus the convergence of $\int |f|\,dx$ implies that of $\int f\,dx$. Furthermore,

$$\left|\int_a^b f(x)\,dx\right| \leqq \int_a^b |f(x)|\,dx \leqq \int_a^\infty |f(x)|\,dx.$$

This is true for every b; hence, in the limit,

$$\left|\int_a^\infty f(x)\,dx\right| \leqq \int_a^\infty |f(x)|\,dx.$$

THEOREM 54 (*Comparison test*). *Let $f(x)$ and $g(x)$ be continuous for $a \leqq x < \infty$. If $0 \leqq |f(x)| \leqq g(x)$ and $\int_a^\infty g(x)\,dx$ converges, then*

$$\int_a^\infty f(x)\,dx$$

is absolutely convergent and

$$\left|\int_a^\infty f(x)\,dx\right| \leqq \int_a^\infty g(x)\,dx.$$

If $0 \leqq g(x) \leqq f(x)$ and

$$\int_a^\infty g(x)\,dx$$

diverges, then

$$\int_a^\infty f(x)\,dx$$

diverges.

The proof is just like that for series and need not be repeated.

The preceding discussion has been confined to integrals from a to ∞. A similar one holds for integrals from $-\infty$ to b and for integrals from a to b, which are improper at an end-point. Thus the convergence of integrals

$$\int_0^1 f(x)\,dx,$$

where f is unbounded near 0 and continuous for $0 < x \leqq 1$, is related to the convergence of series

$$\sum_{n=1}^{\infty} a_n, \quad a_n = \int_{b_{n+1}}^{b_n} f(x)\,dx,$$

where b_n is a monotone sequence converging to 0.

***6–23 Improper integrals depending on a parameter — uniform convergence.** The analogue of a series of functions

$$\sum_{n=1}^{\infty} f_n(x)$$

is an improper integral

$$\int_a^{\infty} f(t, x)\,dt;$$

thus the variable t replaces the index n. Both of these, when convergent, define a function $F(x)$. Because of the close relationship between series and improper integrals demonstrated in the preceding section, we can expect the discussion of functions defined by integrals to parallel that of Sections 6–11 to 6–14 for functions defined by series.

One defines the improper integral to be *uniformly convergent* to $F(x)$ for a given range of x, if, given $\epsilon > 0$, a number B can be found such that

$$\left|\int_a^b f(t, x)\,dt - F(x)\right| < \epsilon \quad \text{for} \quad b > B,$$

where B is independent of x.

THEOREM 55 (*M-test for integrals*). *Let $M(t)$ be continuous for $a \leqq t < \infty$; let $f(t, x)$ be continuous in t for $a \leqq t < \infty$ for each x of a set E. If*

$$|f(t, x)| \leqq M(t)$$

for x in E and

$$\int_a^{\infty} M(t)\,dt$$

converges, then

$$\int_a^{\infty} f(t, x)\,dt$$

is uniformly and absolutely convergent for x in E.

THEOREM 56. *If $f(t, x)$ is continuous in t and x for $a \leqq t < \infty$, $c \leqq x \leqq d$, and*

$$\int_a^{\infty} f(t, x)\,dt$$

is uniformly convergent for $c \leqq x \leqq d$, then the function $F(x)$ defined by this integral is continuous for $c \leqq x \leqq d$.

THEOREM 57. *If $f(t, x)$ is continuous in t and x for $a \leqq t < \infty$, $c \leqq x \leqq d$ and*

$$\int_a^\infty f(t, x)\, dt$$

is uniformly convergent to $F(x)$ for $c \leqq x \leqq d$, then

$$\int_c^d F(x)\, dx = \int_a^\infty \int_c^d f(t, x)\, dx\, dt.$$

THEOREM 58. *If $f(t, x)$ is continuous in t and x and has a derivative $\partial f/\partial x$ which is continuous in t and x for $a \leqq t < \infty$ and $c \leqq x \leqq d$, and the integrals*

$$\int_a^\infty f(t, x)\, dt, \quad \int_a^\infty \frac{\partial f}{\partial x}(t, x)\, dt$$

converge, the second one uniformly, for $c \leqq x \leqq d$, then

$$F(x) = \int_a^\infty f(t, x)\, dt$$

has a continuous derivative for $c \leqq x \leqq d$ and

$$F'(x) = \int_a^\infty \frac{\partial f}{\partial x}(t, x)\, dt.$$

These theorems are proved exactly as for series.

EXAMPLE. The integral

$$\int_0^\infty e^{-xt^2}\, dt \tag{6–72}$$

is uniformly convergent for $x \geqq 1$, since

$$0 \leqq e^{-xt^2} \leqq e^{-t^2} = M(t) \quad \text{for} \quad x \geqq 1$$

and the integral

$$\int_0^\infty e^{-t^2}\, dt$$

exists, as a comparison with

$$\int_0^\infty e^{-t}\, dt$$

reveals. Hence (6–72) defines a function $F(x)$. Theorem 56 shows that $F(x)$ is continuous for $x \geqq 1$. The integral

$$\int_0^\infty \frac{\partial}{\partial x}(e^{-xt^2})\, dt = -\int_0^\infty t^2 e^{-xt^2}\, dt$$

is also uniformly convergent, since

$$0 \leqq t^2 e^{-xt^2} \leqq t^2 e^{-t^2} < e^{-t}$$

for $x \geqq 1$ and t sufficiently large. Hence

$$\frac{d}{dx}\int_0^\infty e^{-xt^2}\,dt = -\int_0^\infty t^2 e^{-xt^2}\,dt, \quad x \geqq 1.$$

As in the preceding section, the theory extends to improper integrals over a finite interval without essential change.

***6–24 Principal value of improper integrals.** An integral from $-\infty$ to ∞ would normally be decomposed into integrals from $-\infty$ to 0 and from 0 to ∞, as in Section 4–5; if the last two integrals converge, the given integral converges; otherwise, it diverges. When this process leads to divergence, one may be able to salvage the integral by the following procedure, which has important applications to Laplace and Fourier transforms.

Let f be continuous for $-\infty < x < \infty$ (or, more generally, let f be such that f has an integral over each finite interval). Then the *Cauchy principal value* (briefly, *principal value*) of the integral of f from $-\infty$ to ∞ is the limit

$$\lim_{a\to\infty}\int_{-a}^{a} f(x)\,dx,$$

if this limit exists. When it does, one denotes the value by

$$(P)\int_{-\infty}^{\infty} f(x)\,dx.$$

One sees at once that if $\int_{-\infty}^{\infty} f(x)\,dx$ exists in the usual sense, then

$$\lim_{a\to\infty}\int_0^a f(x)\,dx \quad\text{and}\quad \lim_{a\to\infty}\int_{-a}^0 f(x)\,dx$$

both exist, and hence the principal value exists and equals the usual improper integral. However, the principal value may exist even though the usual value does not. For example,

$$\begin{aligned}(P)\int_{-\infty}^{\infty}\frac{x-1}{1+(x-1)^2}\,dx &= \lim_{a\to\infty}\int_{-a}^{a}\frac{x-1}{1+(x-1)^2}\,dx\\ &= \lim_{a\to\infty}\tfrac{1}{2}\ln\frac{1+(a-1)^2}{1+(a+1)^2} = 0.\end{aligned}$$

Here the integral from 0 to ∞ is $+\infty$, from $-\infty$ to 0 is $-\infty$. There is in effect a cancellation of the two infinities. This is related to a certain symmetry of the graph of the function.

The concept of principal value can be extended to the integral of a function f from a to b, where f is discontinuous only at c, with $a < c < b$.

Here one defines:

$$(P)\int_a^b f(x)\,dx = \lim_{\epsilon\to 0+}\left[\int_a^{c-\epsilon} f(x)\,dx + \int_{c+\epsilon}^b f(x)\,dx\right].$$

For example,

$$(P)\int_0^3 \frac{1}{(x-1)^3}\,dx = \lim_{\epsilon\to 0+}\left[\int_0^{1-\epsilon}\frac{1}{(x-1)^3}\,dx + \int_{1+\epsilon}^3 \frac{1}{(x-1)^3}\,dx\right]$$
$$= \lim_{\epsilon\to 0+}\left[\left(-\frac{1}{2\epsilon^2}+\frac{1}{2}\right)+\left(-\frac{1}{8}+\frac{1}{2\epsilon^2}\right)\right] = \frac{3}{8}.$$

Here the integrals from 0 to 1 and from 1 to 3 are respectively $+\infty$ and $-\infty$ and again there is a cancellation of the two infinities, made clear in the evaluation of the limit above. It is easily seen, as above, that whenever the integral of f from a to b exists, as usual, as an improper integral, the principal value exists and equals the previous value; the example just given shows that the principal value can exist even when the usual improper integral does not exist. Of course, the principal value can also fail to exist, as the following example shows:

$$\int_{-1}^1 \frac{1}{x^2}\,dx = \lim_{\epsilon\to 0+}\left[\int_{-1}^{-\epsilon}\frac{1}{x^2}\,dx + \int_\epsilon^1 \frac{1}{x^2}\,dx\right]$$
$$= \lim_{\epsilon\to 0+}\left[\left(\frac{1}{\epsilon}+1\right)+\left(-1+\frac{1}{\epsilon}\right)\right] = \infty.$$

Here there is no cancellation because the function is always positive.

If f is continuous for $-\infty < x < \infty$ except at c, then one can use a principal value both for c and for large x as follows:

$$(P)\int_{-\infty}^{\infty} f(x)\,dx = \lim_{a\to\infty}\lim_{\epsilon\to 0+}\left[\int_{-a}^{c-\epsilon} f(x)\,dx + \int_{c+\epsilon}^a f(x)\,dx\right].$$

This procedure can be adapted to the case of several discontinuities $c_1, \ldots, c_m$.

***6–25 Laplace transformation. Γ-function and B-function.** In view of the analogy between infinite series and improper integrals, it is natural to seek an improper integral corresponding to a power series. If we write the power series as

$$\sum_{n=0}^{\infty} f(n)x^n, \tag{6–73}$$

then a natural analogue is the improper integral

$$\int_0^{\infty} f(t)x^t\,dt. \tag{6–74}$$

Except for a minor change in notation, this is the *Laplace transform* of the function $f(t)$. More precisely, the Laplace transform of $f(t)$ is the integral

$$\int_0^\infty f(t)e^{-st}\,dt; \tag{6–75}$$

(6–75) is obtained from (6–74) by replacing x by e^{-s}. Just as (6–73) defines a function $G(x)$ within the interval of convergence, the integral (6–75) defines a function $F(s)$, for those values of s for which the integral converges:

$$F(s) = \int_0^\infty f(t)e^{-st}\,dt. \tag{6–76}$$

In fact, the integral (6–74) can be shown to have a "radius of convergence" r^* such that (6–74) converges for $0 \leqq x < r^*$; it is necessary to restrict to positive x, since x^t would be imaginary for x negative and $t = \frac{1}{2}, \frac{1}{4}$, etc. Accordingly, the integral (6–76) converges and defines a function $F(s)$ for $0 < e^{-s} < r^*$, that is, for $s > \log(1/r^*)$.

Integrals of form (6–76) have proved to be exceedingly useful in the theory of ordinary and partial differential equations. The integral can be regarded as an "operator" $\mathcal{L}$ transforming the function $f(t)$ into the function $F(s)$ and one writes

$$\mathcal{L}[f] = F.$$

This explains the word "transform" above and also the use of the term "operational method" in connection with the applications to differential equations. (See Section 7–19, Prob. 8 following Section 8–11, and Prob. 3 following Section 8–12.)

An important particular Laplace transform is the following:

$$\int_0^\infty t^k e^{-st}\,dt, \tag{6–77}$$

in which $f(t) = t^k$; the parameter k must be greater than -1, to avoid divergence of the integral at $t = 0$. With k so restricted, the integral converges for $s > 0$ (Prob. 1 below). When k is 0 or a positive integer, the integral is easily evaluated:

$$\int_0^\infty e^{-st}\,dt = \frac{1}{s}, \quad \int_0^\infty te^{-st}\,dt = \frac{1}{s^2}, \quad \cdots \quad (s > 0).$$

In general, an integration by parts (Prob. 11 below) shows that

$$\int_0^\infty t^k e^{-st}\,dt = \frac{k}{s}\int_0^\infty t^{k-1}e^{-st}\,dt, \quad s > 0. \tag{6–78}$$

Accordingly, we can apply induction to conclude that,

$$\mathcal{L}[t^k] = \int_0^\infty t^k e^{-st}\,dt = \frac{k}{s}\frac{k-1}{s}\cdots\frac{1}{s}\frac{1}{s} = \frac{k!}{s^{k+1}}, \quad s > 0. \tag{6–79}$$

For $s = 1$, (6–79) gives:

$$k! = \int_0^\infty t^k e^{-t}\,dt. \tag{6–80}$$

This suggests a method of generalizing the factorial; that is, we could use Eq. (6–80) to define $k!$ for k an arbitrary real number greater than -1. It is customary to denote this generalized factorial by $\Gamma(k+1)$; the Gamma function $\Gamma(k)$ is then defined by the equation

$$\Gamma(k) = \int_0^\infty t^{k-1}e^{-t}\,dt, \quad k > 0; \tag{6–81}$$

when k is a positive integer or 0,

$$\Gamma(k+1) = k! = \begin{cases} 1\cdot 2\cdots k, & k > 0, \\ 1, & k = 0. \end{cases} \tag{6–82}$$

The general integral (6–77) is expressible in terms of the Gamma function (Prob. 12 below):

$$\mathcal{L}[t^k] = \int_0^\infty t^k e^{-st}\,dt = \frac{\Gamma(k+1)}{s^{k+1}} \quad (s > 0). \tag{6–83}$$

Equation (6–78) then states that

$$\frac{\Gamma(k+1)}{s^{k+1}} = \frac{k}{s}\frac{\Gamma(k)}{s^k};$$

that is,

$$\Gamma(k+1) = k\Gamma(k). \tag{6–84}$$

This is the *functional equation of the Gamma function.* The functional equation can be used to define $\Gamma(k)$ for negative k; thus we write

$$\Gamma\left(\frac{1}{2}\right) = \left(-\frac{1}{2}\right)\Gamma\left(-\frac{1}{2}\right),$$

$$\Gamma\left(-\frac{1}{2}\right) = \left(-\frac{3}{2}\right)\Gamma\left(-\frac{3}{2}\right), \ldots$$

in order to define $\Gamma(-\frac{1}{2})$, $\Gamma(-\frac{3}{2})$, ... in terms of the known value of $\Gamma(\frac{1}{2})$. This procedure fails only for $k = -1, -2, \ldots$. In fact, we can show (Prob. 11 below) that

$$\lim_{k\to 0+} \Gamma(k) = +\infty,$$

and, if the Gamma function is extended to negative nonintegral k as above,

$$\lim_{k\to -n} |\Gamma(k)| = +\infty$$

for every negative integer $-n$. These properties are indicated in Fig. 6–18.

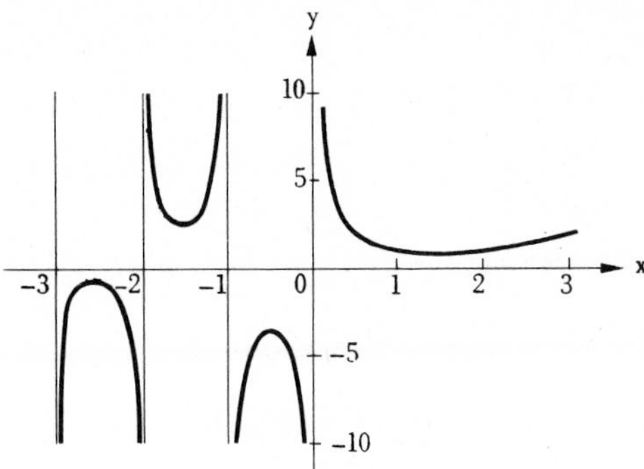

FIG. 6–18. $y = \Gamma(x)$.

The Gamma function has other important properties:

$$\frac{1}{\Gamma(k)} = ke^{\gamma k} \lim_{n\to\infty}\left[(1+k)\left(1+\frac{k}{2}\right)\cdots\left(1+\frac{k}{n}\right)e^{-k-\frac{k}{2}-\cdots-\frac{k}{n}}\right]; \tag{6–85}$$

here γ is the Euler-Mascheroni constant (Prob. 13 below):

$$\gamma = 1 + \sum_{n=2}^{\infty}\left(\frac{1}{n} + \log\frac{n-1}{n}\right) = 0.5772\ldots; \tag{6–86}$$

$$\Gamma(k) = k^{k-\frac{1}{2}}e^{-k}\sqrt{2\pi}\, e^{\frac{\theta(k)}{12k}},\ k > 0, \tag{6–87}$$

where $\theta(k)$ denotes a function of k such that $0 < \theta(k) < 1$. Proofs of (6–85) and (6–87) and other properties of the Gamma function are given in Chapter XII of the book of Whittaker and Watson listed at the end of this chapter. From (6–87) one can prove (Prob. 14 below) that

$$\lim_{k\to\infty}\frac{\Gamma(k+1)}{k^{k+\frac{1}{2}}\sqrt{2\pi}\, e^{-k}} = 1. \tag{6–88}$$

When k is an integer, this gives the *Stirling approximation to* $k!$:

$$k! \sim k^{k+\frac{1}{2}}\sqrt{2\pi}\, e^{-k}. \tag{6–89}$$

For $k = 10$, the left-hand side is 3.629×10^6, while the right-hand side is 3.600×10^6.

The Beta function $\mathrm{B}(p, q)$ is defined by the equation

$$\mathrm{B}(p, q) = \int_0^1 x^{p-1}(1-x)^{q-1}\, dx\ (p > 0, q > 0). \tag{6–90}$$

This can be expressed in terms of the Gamma function (Prob. 15 below):

$$\mathrm{B}(p, q) = \frac{\Gamma(p)\Gamma(q)}{\Gamma(p+q)}. \tag{6–91}$$

Problems

1. Determine convergence or divergence of the following improper integrals:

(a) $\displaystyle\int_1^\infty \frac{e^{\sin x}}{x}\,dx$, (b) $\displaystyle\int_2^\infty \frac{dx}{(\log x)^x}$, (c) $\displaystyle\int_1^\infty \frac{dx}{x^x}$, (d) $\displaystyle\int_0^\infty t^k e^{-st}\,dt$, $k > -1$,

(e) $\displaystyle\int_1^\infty \sin x^2\,dx$. (Hint: let $u = x^2$.)

2. Prove that, if $f(x) = \sin 2\pi x$, then the series

$$\sum_{n=1}^{\infty}\int_{n-1}^{n} f(x)\,dx$$

converges, but $\displaystyle\int_0^\infty f(x)\,dx$ diverges.

3. Prove the *ratio test for integrals:* if $f(x)$ is continuous for $a \leqq x < \infty$ and

$$\lim_{x\to\infty}\left|\frac{f(x+1)}{f(x)}\right| = k < 1,$$

then $\displaystyle\int_a^\infty f(x)\,dx$ is absolutely convergent.

4. Apply the ratio test of Prob. **3** to prove convergence of the integrals:

(a) $\displaystyle\int_1^\infty \frac{x^2}{e^x}\,dx$, (b) $\displaystyle\int_1^\infty \frac{1}{x^x}\,dx$.

5. Prove the *root test for integrals:* if $f(x)$ is continuous for $a \leqq x < \infty$ and

$$\lim_{x\to\infty}|f(x)|^{1/x} = k < 1,$$

then $\displaystyle\int_a^\infty f(x)\,dx$ is absolutely convergent.

6. Apply the root test of Prob. 5 to prove convergence of the integrals:

(a) $\displaystyle\int_a^\infty e^{-x^2}\,dx$, (b) $\displaystyle\int_2^\infty \frac{dx}{(\log x)^x}$.

7. Prove that the following integrals are uniformly convergent for $0 \leqq x \leqq 1$:

(a) $\displaystyle\int_1^\infty \frac{dt}{(x^2+t^2)^{\frac{5}{2}}}$, (b) $\displaystyle\int_1^\infty \frac{\sin t}{x^2+t^2}\,dt$.

8. (a) Prove that, for every choice of $x_1 > 0$, the integral

$$\int_0^\infty t^n e^{-xt^2}\,dt, \quad n > 0,$$

is uniformly convergent for $x \geqq x_1$.

(b) Use the known result that

$$\int_0^\infty e^{-x^2}\,dx = \tfrac{1}{2}\sqrt{\pi}$$

(Prob. 1 following Section 4–11) to prove that

$$\int_0^\infty e^{-xt^2}\,dt = \frac{1}{2}\sqrt{\frac{\pi}{x}}, \quad x > 0.$$

(c) Use the results of parts (a) and (b) and Theorem 58 to show that, for $n = 1, 2, \ldots$, and $x > 0$,

$$\int_0^\infty t^{2n}e^{-xt^2}\,dt = \tfrac{1}{2}\int_0^\infty t^{n-\frac{1}{2}}e^{-xt}\,dt = \tfrac{1}{2}\,\Gamma(n+\tfrac{1}{2})x^{-n-\frac{1}{2}} = \frac{\sqrt{\pi}}{2}\frac{1\cdot 3\cdot 5\cdots(2n-1)}{2^n x^{n+\frac{1}{2}}}.$$

9. (a) Prove that, if $n > 0$, the integrals

$$\int_0^\infty t^n e^{-t^2}\cos(tx)\,dt, \quad \int_0^\infty t^n e^{-t^2}\sin(tx)\,dt$$

are uniformly convergent for all x.

(b) Let

$$F(x) = \int_0^\infty e^{-t^2}\cos(tx)\,dt.$$

Use integration by parts (cf. Prob. 10) to show that $F'(x) = -\frac{1}{2}xF(x)$. From this deduce that $d\log F(x) = -\frac{1}{2}x\,dx$ and hence that $F(x) = ce^{-\frac{1}{4}x^2}$. Let $x = 0$ and use Prob. 8 to find c and thus prove that

$$F(x) = \tfrac{1}{2}\sqrt{\pi}e^{-\frac{1}{4}x^2}.$$

10. (a) Prove that, if $u(x)$, $u'(x)$, $v(x)$, $v'(x)$ are continuous for $a \leqq x < \infty$ and $\lim\limits_{x\to\infty}[u(x)v(x)]$ exists, then

$$\int_a^\infty u(x)v'(x)\,dx = \lim_{x\to\infty}[u(x)v(x)] - u(a)v(a) - \int_a^\infty u'(x)v(x)\,dx;$$

that is, if one of the two improper integrals converges, then the other converges and the equation holds.

(b) Use the result of part (a) to justify the derivation of Eq. (6–78).

(c) Prove Abel's formula:

$$\sum_{k=1}^{n} u_k(v_{k+1} - v_k) = u_n v_{n+1} - u_1 v_1 - \sum_{k=1}^{n-1} v_{k+1}(u_{k+1} - u_k)$$

and hence obtain the analogue of part (a) for infinite series:

$$\sum_{k=1}^{\infty} u_k(v_{k+1} - v_k) = \lim_{n\to\infty}(u_n v_{n+1}) - u_1 v_1 - \sum_{k=1}^{\infty} v_{k+1}(u_{k+1} - u_k);$$

that is, if $\lim\limits_{n\to\infty}(u_n v_{n+1})$ exists, then convergence of one of the two series implies convergence of the other and validity of the equation. This "integration by parts" for series provides valuable tests for convergence; cf. Chapter X of the book of Knopp listed below.

11. Prove: (a) if f is continuous for $-\infty < x < \infty$ and f is odd, then (P) $\int_{-\infty}^{\infty} f(x)\,dx = 0$; (b) if f is continuous for $-a \leqq x \leqq a$, except at $x = 0$, and f is odd, then (P) $\int_{-a}^{a} f(x)\,dx = 0$.

12. For each of the following integrals, obtain the principal value, if it exists; if it does, determine whether the usual value also exists. [Hint: note the result of Prob. 11.]

(a) $\displaystyle\int_{-\infty}^{\infty} \frac{x^3}{x^4+1}\,dx$

(b) $\displaystyle\int_{-\infty}^{\infty} \frac{x}{x^4+1}\,dx$

(c) $\displaystyle\int_{-\infty}^{\infty} \sin x\,dx$

(d) $\displaystyle\int_{-\infty}^{\infty} \frac{x^3}{x^2+1}\,dx$

(e) $\displaystyle\int_{-1}^{1} \frac{1}{x}\,dx$

(f) $\displaystyle\int_{-2}^{4} \frac{1}{(x+1)^{1/3}}\,dx$

(g) $\displaystyle\int_{-\infty}^{\infty} \frac{1}{x}\,dx$

(h) $\displaystyle\int_{-\infty}^{\infty} \frac{x}{(x-1)^2}\,dx$

13. (a) Prove from Eq. (6–81) that $\Gamma(k)$ is continuous and positive for $k > 0$. (b) Prove from Eq. (6–84) that

$$\lim_{k\to 0+} \Gamma(k) = +\infty;$$

(c) Prove that, if Eq. (6–84) is used to define $\Gamma(k)$ for k negative but nonintegral, then

$$\lim_{k\to -n} |\Gamma(k)| = +\infty \text{ for } n = 0, 1, 2, \cdots.$$

14. Prove (6–83).

15. (a) Prove by the integral test that the series (6–86) defining γ converges. (b) Use Theorem 23 to prove that $\gamma = 0.6$ to one significant figure.

16. Prove (6–88) from (6–87).

17. Prove the identity (6–91). [Hint: show that

$$\Gamma(p) = 2\int_0^{\infty} x^{2p-1}e^{-x^2}\,dx, \quad \Gamma(q) = 2\int_0^{\infty} y^{2q-1}e^{-y^2}\,dy$$

and hence, as in Prob. 1 following Section 4–11, that

$$\begin{aligned}\Gamma(p)\Gamma(q) &= 4\int_0^{\frac{1}{2}\pi}\int_0^{\infty} \sin^{2p-1}\theta\cos^{2q-1}\theta\, r^{2p+2q-1}e^{-r^2}\,dr\,d\theta \\ &= \left(2\int_0^{\infty} r^{2p+2q-1}e^{-r^2}\,dr\right)\left(2\int_0^{\frac{1}{2}\pi} \sin^{2p-1}\theta\cos^{2q-1}\theta\,d\theta\right).\end{aligned}$$

The first factor on the right is $\Gamma(p+q)$; the second reduces to $B(p, q)$ if one sets $x = \sin^2\theta$.]

18. Verify that the following functions $F(s)$ are Laplace transforms of the functions $f(t)$ given:

(a) $F(s) = \dfrac{1}{s-k}$, $s > k$; $f(t) = e^{kt}$;

(b) $F(s) = \dfrac{k}{s^2 + k^2}, \quad s > 0; \quad f(t) = \sin kt;$

(c) $F(s) = \dfrac{1}{s^k}, \quad s > 0, \quad k > 0; \quad f(t) = \dfrac{t^{k-1}}{\Gamma(k)};$

(d) $F(s) = \displaystyle\sum_{k=1}^{\infty} \frac{b_k}{s^k}, \quad s > s_1; \quad f(t) = \sum_{k=0}^{\infty} \frac{1}{k!} b_{k+1} t^k.$

Answers

1. (a) divergent, (b) convergent, (c) convergent, (d) convergent for $s > 0$, (e) convergent.

12. (a), (c), (d), (e), (g), (h) principal value 0, no usual value; (b) both values 0; (f) both values $(3/2)(5^{2/3} - 1)$.

***6–26 Series of vectors and matrices.** The concepts of sequences and series can be generalized to cover vectors and matrices. A sequence of vectors in V^n is a sequence $\mathbf{u}_1, \ldots, \mathbf{u}_m, \ldots$, each element of which is a vector of V^n. We write

$$\lim_{m\to\infty} \mathbf{u}_m = \mathbf{u}$$

if

$$\lim_{m\to\infty} |\mathbf{u}_m - \mathbf{u}| = 0;$$

that is, if, given $\epsilon > 0$, an integer N can be found such that $|\mathbf{u}_m - \mathbf{u}| < \epsilon$ for $m > N$. This definition is like that for complex numbers (Section 6–19 above); in fact, we can consider complex numbers as vectors in V^2. As for complex numbers, we have a theorem relating convergence of sequences of vectors to the convergence of the sequences of components:

Theorem 59. *Let $\mathbf{u}_m$ $(m = 1, 2, \ldots)$ be a sequence of vectors in V^n. Let $\mathbf{u}_m = (u_{m1}, \ldots, u_{mn})$, $\mathbf{u} = (u_1, \ldots, u_n)$. Then $\lim_{m\to\infty} \mathbf{u}_m = \mathbf{u}$ if and only if*

$$\lim_{m\to\infty} u_{m1} = u_1, \quad \ldots, \quad \lim u_{mn} = u_n. \tag{6–92}$$

The proof parallels that for Theorem 46 above. From this theorem one deduces the expected theorems on limit of a sum and limits of various kinds of products.

Example 1. The sequence

$$\left(\frac{m}{m+1}, 2^{-m}, \left(1 + \frac{1}{m}\right)^m\right), \quad m = 1, 2, \ldots$$

of vectors in V^3 has limit $(1, 0, e)$, since

$$\lim_{m\to\infty} \frac{m}{m+1} = 1, \quad \lim_{m\to\infty} 2^{-m} = 0, \quad \lim_{m\to\infty} \left(1 + \frac{1}{m}\right)^m = e.$$

In general, a vector function $\mathbf{f}(t)$ continuous at t_0 has the property:

$$\lim_{m\to\infty} \mathbf{f}(t_m) = \mathbf{f}(t_0)$$

for every sequence t_m converging to t_0. This follows at once from the definition of continuity.

An *infinite series of vectors in* V^n is a series

$$\sum_{m=1}^{\infty} \mathbf{u}_m,$$

whose terms are vectors of V^n. Such a series is said to converge and have sum $\mathbf{v}$ if

$$\lim_{m\to\infty} (\mathbf{u}_1 + \cdots + \mathbf{u}_m) = \mathbf{v}.$$

By Theorem 59, the convergence of the series can be reduced to that of n real series, obtained from the components.

EXAMPLE 2. The series

$$(1, 0) + (0, 1), + \left(-\frac{1}{2!}, 0\right) + \left(0, -\frac{1}{3!}\right) + \left(\frac{1}{4!}, 0\right) + \left(0, \frac{1}{5!}\right) + \cdots,$$

which corresponds to the series of complex numbers $1 + i + (i^2/2!) + \ldots$, converges to $(\cos 1, \sin 1)$, corresponding to the complex number e^i.

Sequences and series of matrices can be discussed in a similar manner, with the aid of a *matrix norm* $|A| = |(a_{ij})|$. If A is a p by q matrix (a_{ij}), we define:

$$|A| = |(a_{ij})| = \sum_{i=1}^{p} \sum_{j=1}^{q} |a_{ij}|.$$

For example, the matrix $\begin{bmatrix} 1 & 3 \\ 5 & -2 \end{bmatrix}$ has norm $1 + 3 + 5 + 2 = 11$.

A sequence of $p \times q$ matrices $A_1, A_2, \ldots, A_m, \ldots$ is said to converge to the $p \times q$ matrix A if the sequence $|A_m - A|$ of real numbers has limit 0. One then writes

$$\lim_{m\to\infty} A_m = A.$$

One then proves the analogue of Theorem 59:

THEOREM 60. *Let* $A_m = (a_{mij})$ *be a sequence of* $p \times q$ *matrices, let* $A = (a_{ij})$ *be a* $p \times q$ *matrix. Then the sequence converges to* A *if and only if*

$$\lim_{m\to\infty} a_{mij} = a_{ij}, \quad i = 1, \ldots, p, \quad j = 1, \ldots, q.$$

For example, a 2×2 matrix

$$A(t) = \begin{bmatrix} f_{11}(t) & f_{12}(t) \\ f_{21}(t) & f_{22}(t) \end{bmatrix},$$

whose elements are functions of t, all continuous at t_0, has the property that

$$\lim_{m \to \infty} A(t_m) = A(t_0)$$

for each sequence t_m converging to t_0.

An infinite series of $p \times q$ matrices is a series $\sum_{m=1}^{\infty} A_m$, whose terms are such matrices. The series is said to converge to A, the sum of the series, if

$$\lim_{m \to \infty} (A_1 + \cdots + A_m) = A.$$

By Theorem 60, the convergence of such a series of matrices can be referred back to the convergence of the series $\Sigma\, a_{mij}$ for the elements of the matrices.

A very important series of matrices is the series

$$I + A + \frac{1}{2!} A^2 + \cdots + \frac{1}{m!} A^m + \cdots \tag{6–93}$$

formed from a given *square* matrix A. It will be seen that this series converges for every A. Because of the resemblance of the series to that for e^x, the sum is denoted by e^A and is called the *exponential function of matrix* A.

Theorem 61. *The series* (6–93) *converges for every square matrix A and hence defines e^A for all A. Furthermore, for real s and t,*

$$e^{sA}e^{tA} = e^{(s+t)A}, \tag{6–94}$$

$$e^{A}e^{-A} = e^{0A} = I, \tag{6–95}$$

$$(e^A)^k = e^{kA}, \quad k = 0, \quad \pm 1, \quad \pm 2, \ldots \tag{6–96}$$

To prove the convergence, we let $A = (a_{ij})$ be $p \times p$ and choose a positive constant c such that $|a_{ij}| \leqq c$ for all i and j. Now each element of A^2 is of the form $a_{i1}a_{1j} + \cdots + a_{ip}a_{pj}$ and hence is in absolute value at most pc^2. Similarly, each element of A^3 is a sum of p terms, each in absolute value at most $pc^2 \cdot c$; therefore, each term of A^3 is in absolute value at most p^2c^3. By induction, we thus show that each element of A^m is in absolute value at most $p^{m-1}c^m$. Hence the series Σa_{mij} converges by comparison with the convergent series

$$1 + c + \frac{1}{2!} pc^2 + \cdots + \frac{1}{m!} p^{m-1}c^m + \cdots.$$

Thus by Theorem 60 the series (6–93) converges.

The rule (6–94) is proved in Section 8–16 below. The rules (6–95) and (6–96) follow from (6–94) (Problem 3 below). Rule (6–95) shows that e^A is always a *nonsingular* matrix.

Problems

1. Find the limit of the sequence, if it exists:

(a) $\left(\dfrac{\sin m}{m}, \dfrac{\cos m}{m}\right)$ (b) $((-1)^m, \ (-1)^{m+1})$

(c) $me^{-m}\mathbf{i} + \dfrac{\log m}{m}\mathbf{j} + \dfrac{e^m}{e^m + 1}\mathbf{k}$ (d) $\begin{bmatrix} m \sin(1/m) & 2^{-m} \\ e^{1/m} & \log(1/m) \end{bmatrix}$

(e) $\begin{bmatrix} 1 + \frac{1}{2} + \cdots + \frac{1}{2^m} & 1 + \frac{1}{3} + \cdots + \frac{1}{3^m} \\ 1 + \frac{1}{3} + \cdots + \frac{1}{3^m} & 1 + \frac{1}{4} + \cdots + \frac{1}{4^m} \end{bmatrix}$

(f) $A^m\begin{bmatrix} 1 \\ 2 \end{bmatrix}$, where $A = \begin{bmatrix} 0 & \frac{1}{2} \\ \frac{1}{2} & 0 \end{bmatrix}$ (g) $A^m\begin{bmatrix} 1 \\ 0 \end{bmatrix}$, where $A = \begin{bmatrix} 0 & -1 \\ 1 & 0 \end{bmatrix}$

(h) A^m, where A is as in part (f).

2. Test for convergence:

(a) $\displaystyle\sum_{m=1}^{\infty} \left(\frac{1}{2^m}, \frac{1}{3m}\right)$ (b) $\displaystyle\sum_{m=1}^{\infty} \left(\frac{m}{m^3 + 1}, \frac{m}{2^m}, \frac{1}{m! + m}\right)$

(c) $\displaystyle\sum_{m=1}^{\infty} A^m$, where $A = \begin{bmatrix} \frac{1}{3} & 1 \\ 0 & 0 \end{bmatrix}$ (d) $\displaystyle\sum_{m=1}^{\infty} \frac{1}{m} A^m$, with A as in part (c).

3. (a) Deduce (6–95) from (6–94). (b) Deduce (6–96) from (6–94).
(c) Prove that $e^{sA}e^{tA} = e^{tA}e^{sA}$ for all s, t.
(d) Prove: if $A = \operatorname{diag}(\lambda_1, \ldots, \lambda_p)$, then $e^A = \operatorname{diag}(e^{\lambda_1}, \ldots, e^{\lambda_p})$.

Answers

1. (a) $(0, 0)$, (b) none, (c) $\mathbf{k}$, (d) none, (e) $\begin{bmatrix} 2 & \frac{3}{2} \\ \frac{3}{2} & \frac{4}{3} \end{bmatrix}$, (f) $(0, 0)$, (g) none, (h) $\begin{bmatrix} 0 & 0 \\ 0 & 0 \end{bmatrix}$.

2. (a) diverges, (b), (c) and (d) converge.

Suggested References

COURANT, RICHARD J., *Differential and Integral Calculus*, transl. by E. J. McShane, 2 vols. New York: Interscience, 1947.

GREEN, J. A., *Sequences and Series*. Glencoe, Ill.: The Free Press, 1958.

HARDY, G. H., *Pure Mathematics*, 9th ed. Cambridge: Cambridge University Press, 1947.

KNOPP, K., *Theory and Application of Infinite Series*, transl. by Miss R. C. Young. Glasgow: Blackie and Son, 1928.

RUDIN, W., *Principles of Mathematical Analysis*, 2nd ed. New York: McGraw-Hill, 1964.

WHITTAKER, E. T., and WATSON, G. N., *Modern Analysis*, 4th ed. Cambridge: Cambridge University Press, 1940.

WIDDER, D. V., *The Laplace Transform*. Princeton: Princeton University Press, 1941.

CHAPTER 7

Fourier Series and Orthogonal Functions

7–1 Trigonometric series. A trigonometric series is a series of form

$$\tfrac{1}{2}a_0 + a_1 \cos x + b_1 \sin x + \cdots + a_n \cos nx + b_n \sin nx + \cdots, \tag{7–1}$$

where the coefficients a_n and b_n are constants. If these constants satisfy certain conditions, to be specified in Section 7–2, then the series is called a Fourier series. Almost all trigonometric series encountered in physical problems are Fourier series.

Each term in (7–1) has the property that it repeats itself in intervals of 2π:

$$\cos (x + 2\pi) = \cos x, \quad \sin (x + 2\pi) = \sin x, \ldots,$$
$$\cos [n(x + 2\pi)] = \cos (nx + 2n\pi) = \cos nx, \ldots .$$

It follows that if (7–1) converges for all x, then its sum $f(x)$ must also have this property:

$$f(x + 2\pi) = f(x). \tag{7–2}$$

We say: $f(x)$ has period 2π. In general a function $f(x)$ such that

$$f(x + p) = f(x) \quad (p \neq 0) \tag{7–3}$$

for all x is said to be *periodic* and have *period* p. It should be noted that $\cos 2x$ has, in addition to the period 2π, the period π and, in general, $\cos nx$ and $\sin nx$ have the periods $2\pi/n$. However, 2π is the smallest period shared by all terms of the series.

If $f(x)$ has period p, then the substitution:

$$x = p\frac{t}{2\pi} \tag{7–4}$$

converts $f(x)$ into a function of t having period 2π; for when t increases by 2π, x increases by p.

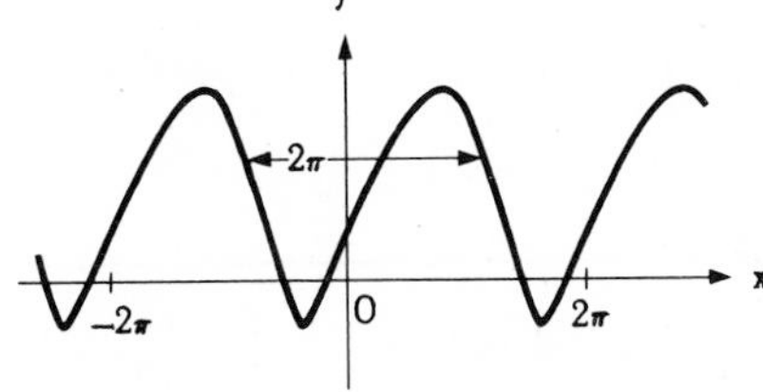

FIG. 7–1. Function with period 2π.

A function $f(x)$ having period 2π is illustrated in Fig. 7–1. Such periodic functions appear in a great variety of physical problems: the vibrations of a spring; the motion of the planets about the sun; the rotation of the earth about its axis; the motion of a pendulum; the tides and wave motion in general; vibrations of a violin string, of an air column (e.g., in a flute); and musical sounds in general. The modern theory of light is based on "wave mechanics," with periodic vibrations a characteristic feature; the spectrum of a molecule is simply a picture of the different vibrations taking place simultaneously within it. Electric circuits involve many periodically varying variables; e.g., the alternating current. The fact that a journey around the globe involves a total change in longitude of 360° is an expression of the fact that the rectangular coordinates of position on the globe are periodic functions of longitude, with period 360°; many other examples of such periodic functions of angular coordinates can be given.

Now it can be shown that *every* periodic function of x satisfying certain very general conditions can be represented in the form (7–1), that is, as a trigonometric series. This mathematical theorem is a reflection of a physical experience most vividly illustrated in the case of *sound*, for example, that of a violin string. The term $\frac{1}{2}a_0$ represents the neutral position, the terms $a_1 \cos x + b_1 \sin x$ the fundamental tone, the terms $a_2 \cos 2x + b_2 \sin 2x$ the first overtone (octave); the other terms represent higher overtones. The variable x must here be thought of as *time* and the function $f(x)$ as the displacement of an instrument, such as a phonograph needle, which is recording the sound, or of a point on the string. Thus the musical tone heard is a combination of simple harmonic vibrations: the terms $(a_n \cos nx + b_n \sin nx)$. Each such pair can be written in the form

$$A_n \sin (nx + \alpha),$$

where

$$A_n = \sqrt{a_n^2 + b_n^2}, \quad a_n = A_n \sin \alpha, \quad b_n = A_n \cos \alpha.$$

The "amplitude" A_{n+1} is a measure of the importance of the nth overtone in the whole sound. The differences in the tones of different musical instruments can be ascribed mainly to the differences in the weights A_n of the overtones.

7–2 Fourier series. Let us suppose now that a periodic function $f(x)$ is the sum of a trigonometric series (7–1), i.e., that

$$f(x) = \frac{a_0}{2} + \sum_{n=1}^{\infty} (a_n \cos nx + b_n \sin nx). \tag{7–5}$$

What is the relation between the coefficients a_n and b_n and the function $f(x)$? To answer this, we multiply $f(x)$ by $\cos mx$ and integrate from $-\pi$ to π:

$$\int_{-\pi}^{\pi} f(x) \cos mx \, dx$$

$$= \int_{-\pi}^{\pi} \left[\frac{a_0}{2} \cos mx + \sum_{n=1}^{\infty} (a_n \cos nx \cos mx + b_n \sin nx \cos mx)\right] dx.$$

If term-by-term integration of the series is allowed, then we find

$$\int_{-\pi}^{\pi} f(x) \cos mx \, dx = \frac{a_0}{2} \int_{-\pi}^{\pi} \cos mx \, dx$$

$$+ \sum_{n=1}^{\infty} \left\{ a_n \int_{-\pi}^{\pi} \cos nx \cos mx \, dx + b_n \int_{-\pi}^{\pi} \sin nx \cos mx \, dx \right\} \cdot \tag{7–6}$$

The integrals on the right are easily evaluated with the help of the identities for $\cos x \cos y$ and $\sin x \cos y$ of (0–81) in the Introduction. One finds

$$\int_{-\pi}^{\pi} \cos nx \cos mx \, dx = \begin{cases} 0, & n \neq m, \\ \pi, & n = m \neq 0 \end{cases}$$
$$\int_{-\pi}^{\pi} \sin nx \cos mx \, dx = 0. \tag{7–7}$$

If $m = 0$, then all terms on the right of (7–6) are 0 except the first one and one finds

$$\int_{-\pi}^{\pi} f(x) \, dx = \pi a_0. \tag{7–8'}$$

For any positive integer m only the term in a_m gives a result different from 0. Thus

$$\int_{-\pi}^{\pi} f(x) \cos mx \, dx = \pi a_m \quad (m = 1, 2, \ldots). \tag{7–8''}$$

Multiplying $f(x)$ by $\sin mx$ and proceeding in the same way, we find

$$\int_{-\pi}^{\pi} f(x) \sin mx \, dx = \pi b_m \quad (m = 1, 2, \ldots). \tag{7–8'''}$$

From the last three formulas one now concludes that

$$a_n = \frac{1}{\pi} \int_{-\pi}^{\pi} f(x) \cos nx \, dx \quad (n = 0, 1, 2, \ldots),$$
$$b_n = \frac{1}{\pi} \int_{-\pi}^{\pi} f(x) \sin nx \, dx \quad (n = 1, 2, \ldots). \tag{7–9}$$

This is the fundamental rule for coefficients in a Fourier series. Without concerning ourselves with the validity of the steps leading to (7–9), we *define* a Fourier series to be any trigonometric series

$$\tfrac{1}{2} a_0 + a_1 \cos x + b_1 \sin x + \cdots + a_n \cos nx + b_n \sin nx + \cdots \tag{7–10}$$

in which the coefficients a_n, b_n are computed from a function $f(x)$ by (7–9); the series is then called *the Fourier series of* $f(x)$. Concerning $f(x)$ we assume only that the integrals in (7–9) exist; for this it is sufficient that $f(x)$ be continuous except for a finite number of jumps between $-\pi$ and π.

No parentheses are used in the general definition (7–10). It is common practice to group the terms as in (7–5). However, the series will always be understood in the ungrouped form (7–10). We recall that *insertion* of parentheses in a *convergent* series is always permissible (Theorem 27, Section 6–10).

THEOREM 1. *Every uniformly convergent trigonometric series is a Fourier series. More precisely, if the series (7–10) converges uniformly for all x to $f(x)$, then $f(x)$ is continuous for all x, $f(x)$ has period 2π and the series (7–10) is the Fourier series of $f(x)$.*

Proof. Since the series converges uniformly for all x, its sum $f(x)$ is continuous for all x (Theorem 31, Section 6–14). The series remains uniformly convergent if all terms are multiplied by $\cos mx$ or by $\sin mx$ (Theorem 34, Section 6–14). Accordingly, the term-by-term integration of Eq. (7–6) is justified (Theorem 32, Section 6–14); (7–9) now follows as above, so that the series is the Fourier series of $f(x)$. The periodicity of $f(x)$ is a consequence of the periodicity of the terms of the series, as remarked in Section 7–1.

COROLLARY. *If two trigonometric series converge uniformly for all x and have the same sum for all x:*

$$\tfrac{1}{2}a_0 + \sum_{n=1}^{\infty} (a_n \cos nx + b_n \sin nx) \equiv \tfrac{1}{2}a_0' + \sum_{n=1}^{\infty} (a_n' \cos nx + b_n' \sin nx),$$

then the series are identical: $a_0 = a_0'$, $a_n = a_n'$, $b_n = b_n'$ for $n = 1, 2, \ldots$. In particular, if a trigonometric series converges uniformly to 0 for all x, then all coefficients are 0.

Proof. Let $f(x)$ denote the sum of both series. Then by Theorem 1

$$a_n = a_n' = \frac{1}{\pi}\int_{-\pi}^{\pi} f(x) \cos nx \, dx \quad (n = 0, 1, 2, \ldots)$$

and similarly $b_n = b_n'$ for all n. If $f(x) \equiv 0$, then all coefficients are 0.

7–3 Convergence of Fourier series. While the Fourier series of $f(x)$ is well-defined when $f(x)$ is merely "piecewise continuous," it is too much to expect that the series will converge to $f(x)$ under such general conditions. However, it turns out that very little more is required to ensure convergence to $f(x)$. In particular, if f is periodic with period 2π and has continuous first and second derivatives for all x, then the Fourier series of $f(x)$ will converge uniformly to $f(x)$ for all x. This result is in itself remarkable, when one considers the fact that expansion of f in a convergent power series requires continuous derivatives of all orders — plus the condition that the remainder R_n of Taylor's formula converges to 0. One can even go farther and guarantee uniform convergence of the Fourier series of $f(x)$ to $f(x)$ when $f(x)$ has "corners," i.e., points at which $f'(x)$ has a jump discontinuity, while f has continuous first and second derivatives between the corners; this is illustrated in Fig. 7-2. Indeed, one can enlarge the concept of corner to include jump discontinuities of $f(x)$, as illustrated in Fig. 7–3; one can hardly expect convergence of the series to $f(x)$ at the discontinuity points, where $f(x)$ may even be ambiguously defined. But the Fourier series makes up our minds for us, in a most reasonable way: it converges to the *average of the left- and right-hand limits*, i.e., to the number

$$\tfrac{1}{2}[\lim_{x \to x_1-} f(x) + \lim_{x \to x_1+} f(x)]$$

at the discontinuity x_1. One cannot expect the series to converge uniformly near the discontinuity (cf. Section 6–14), but it will converge uniformly in each closed interval containing no discontinuity.

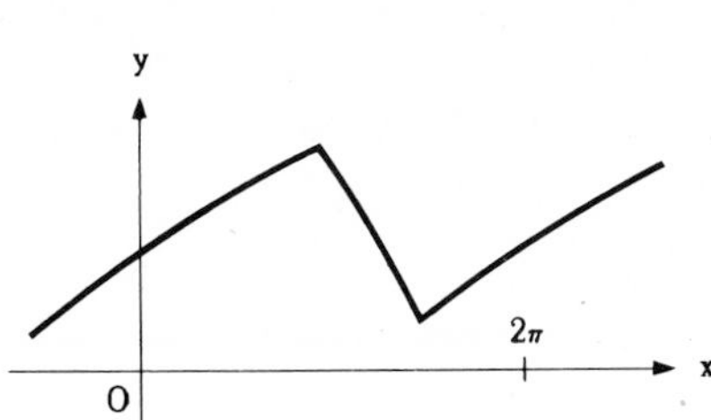

FIG. 7–2. Piecewise very smooth continuous function.

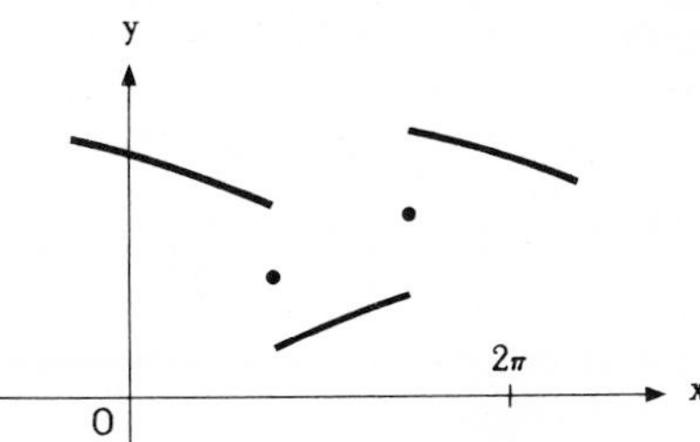

FIG. 7–3. Piecewise very smooth function with jump discontinuities.

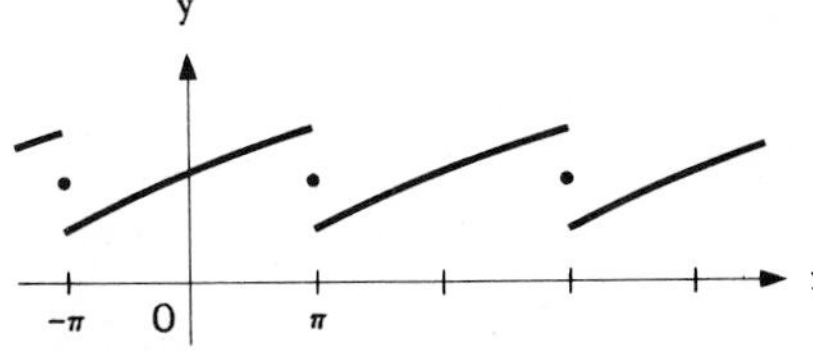

FIG. 7–4. Periodic extension of a function defined between $-\pi$ and π.

While we have up to this point considered only periodic functions $f(x)$ (with period 2π), it must be remarked that the basic coefficient formulas (7–9) use *only the values of* $f(x)$ *between* $-\pi$ *and* π. Thus if $f(x)$ is given only in this interval and is, for example, continuous, then the corresponding Fourier series can be formed and we continue to call the series the Fourier series of $f(x)$. If the series converges to $f(x)$ between $-\pi$ and π, then it will converge outside this interval to a function $F(x)$ which is the "periodic extension of $f(x)$"; this is illustrated in Fig. 7–4. It should be noted that, unless $f(\pi) = f(-\pi)$, the process of extension will introduce jump discontinuities at $x = \pi$ and $x = -\pi$. At these points the series will converge to the number midway between the two "values" of $F(x)$.

We term a function $f(x)$, defined for $a \leqq x \leqq b$, *piecewise continuous* in this interval, if the interval can be subdivided into a finite number of subintervals, inside each of which $f(x)$ is continuous and has finite limits at the left and right ends of the interval. Accordingly, inside the ith subinterval the function $f(x)$ coincides with a function $f_i(x)$ which is continuous in the *closed* subinterval; if, in addition, the functions $f_i(x)$ have continuous first derivatives, we term $f(x)$ *piecewise smooth;* if, in addition, the functions $f_i(x)$ have continuous second derivatives, we term $f(x)$ *piecewise very smooth.*

FUNDAMENTAL THEOREM. *Let* $f(x)$ *be piecewise very smooth in the interval* $-\pi \leqq x \leqq \pi$. *Then the Fourier series of* $f(x)$:

$$\frac{a_0}{2} + \sum_{n=1}^{\infty} (a_n \cos nx + b_n \sin nx),$$

$$a_n = \frac{1}{\pi} \int_{-\pi}^{\pi} f(x) \cos nx \, dx, \quad b_n = \frac{1}{\pi} \int_{-\pi}^{\pi} f(x) \sin nx \, dx,$$

converges to $f(x)$ wherever $f(x)$ is continuous inside the interval, to

$$\tfrac{1}{2}[\lim_{x\to x_1-} f(x) + \lim_{x\to x_1+} f(x)]$$

at each point of discontinuity x_1 inside the interval, and to

$$\tfrac{1}{2}[\lim_{x\to \pi-} f(x) + \lim_{x\to -\pi+} f(x)]$$

at $x = \pm\pi$. The convergence is uniform in each closed interval containing no discontinuity.

This theorem is adequate for most applications. The hypotheses can be weakened, "very smooth" being replaced by "smooth"; in fact, $f(x)$ need only be expressible as the difference of two functions, both of which are steadily increasing as x increases. For extensions to these cases, the reader is referred to the books of Jackson and Zygmund listed at the end of the chapter. The proof of the fundamental theorem will be given in Section 7–9 below.

7–4 Examples — minimizing of square error. We now proceed to consider several examples which will bring out more clearly the relation between $f(x)$ and its Fourier series.

EXAMPLE 1. Let $f(x)$ have the value -1 for $-\pi \leqq x < 0$ and the value $+1$ for $0 \leqq x \leqq \pi$. The periodic extension of $f(x)$ then gives the "square wave" of Fig. 7–5. One finds

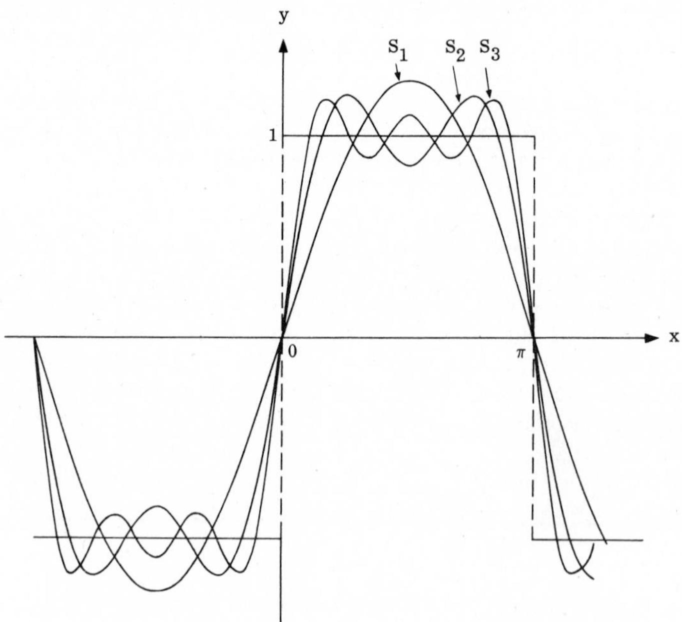

FIG. 7–5. Representation of square wave by Fourier series.

$$a_n = -\frac{1}{\pi}\int_{-\pi}^{0} \cos nx\,dx + \frac{1}{\pi}\int_{0}^{\pi} \cos nx\,dx = 0, \quad n = 0, 1, 2, \ldots,$$

$$b_n = -\frac{1}{\pi}\int_{-\pi}^{0} \sin nx\,dx + \frac{1}{\pi}\int_{0}^{\pi} \sin nx\,dx = \begin{cases} 0, & n = 2, 4, \ldots \\ \dfrac{4}{n\pi}, & n = 1, 3, 5, \ldots. \end{cases}$$

Hence for $0 < |x| < \pi$

$$f(x) = \frac{4}{\pi}\sin x + \frac{4}{3\pi}\sin 3x + \cdots = \frac{4}{\pi}\sum_{n=1}^{\infty}\frac{\sin(2n-1)x}{2n-1}.$$

The figure shows the first three partial sums:

$$S_1 = \frac{4}{\pi}\sin x, \quad S_2 = S_1 + \frac{4}{3\pi}\sin 3x, \quad S_3 = S_2 + \frac{4}{5\pi}\sin 5x$$

of this series. If the graphs are studied carefully, then it becomes clear that $f(x)$ is being approached as limit. For $x = 0$, each partial sum equals 0, so that the series does converge to the average value at the jump; there is a similar situation at $x = \pm\pi$. However, it should be noted that the approximation to $f(x)$ by each partial sum is poorest immediately to the left and right of the jump points.

Example 2. Let $f(x) = \frac{1}{2}\pi + x$ for $-\pi \leqq x \leqq 0$ and $f(x) = \frac{1}{2}\pi - x$ for $0 \leqq x \leqq \pi$. The periodic extension of $f(x)$ is the "triangular wave" of Fig. 7–6. In this example, the extended function is continuous for all x. One finds

$$\begin{aligned} a_n &= \frac{1}{\pi}\int_{-\pi}^{0}\left(\frac{\pi}{2} + x\right)\cos nx\,dx + \frac{1}{\pi}\int_{0}^{\pi}\left(\frac{\pi}{2} - x\right)\cos nx\,dx \\ &= \frac{1}{\pi}\left[\left(\frac{\pi}{2} + x\right)\frac{\sin nx}{n}\Bigg|_{-\pi}^{0} - \frac{1}{n}\int_{-\pi}^{0}\sin nx\,dx + \cdots\right] \\ &= \frac{1}{\pi}\left[\frac{1}{n^2}(1 - \cos n\pi) + \frac{1}{n^2}(1 - \cos n\pi)\right] = \frac{2}{n^2\pi}(1 - \cos n\pi) \end{aligned}$$

for $n = 1, 2, \ldots$. For $n = 0$ a separate computation is needed:

$$a_0 = \frac{1}{\pi}\int_{-\pi}^{0}\left(\frac{\pi}{2} + x\right)dx + \frac{1}{\pi}\int_{0}^{\pi}\left(\frac{\pi}{2} - x\right)dx = 0.$$

The computation of the b_n's is like that of the a_n's and one finds $b_n = 0$ for $n = 1, 2, \ldots$. Hence

$$f(x) = \frac{4}{\pi}\cos x + \frac{4}{9\pi}\cos 3x + \cdots = \frac{4}{\pi}\sum_{n=1}^{\infty}\frac{\cos(2n-1)x}{(2n-1)^2}.$$

The first two partial sums are plotted in Fig. 7–6. Since there are no jumps, one must expect convergence everywhere. It should, however, be noted that at the corners [where $f'(x)$ has a jump], the convergence is poorer than elsewhere.

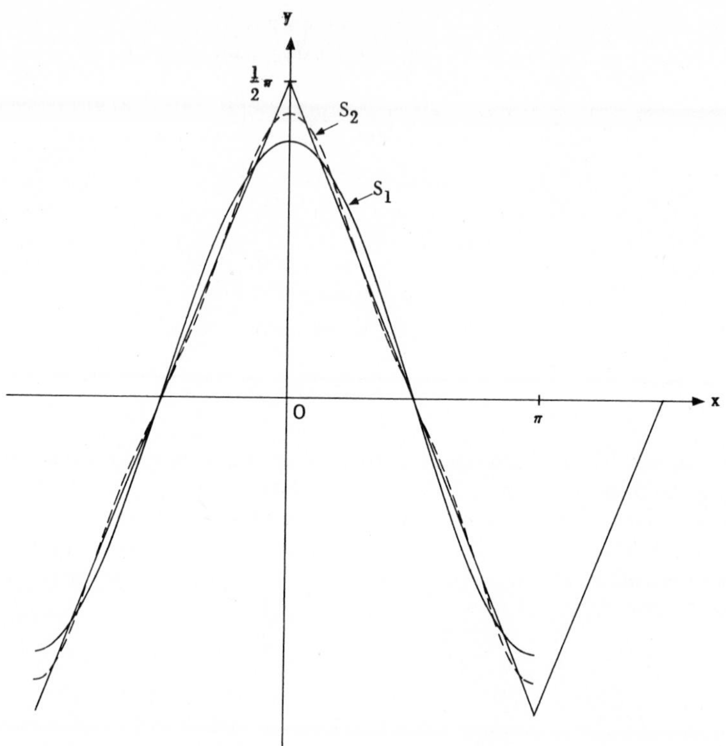

FIG. 7–6. Representation of triangular wave by Fourier series.

Thus far we have proceeded formally, evaluating coefficients and verifying graphically that the series converges to the function. We now proceed to examine the steps more carefully.

The constant term $a_0/2$ of the series is given by the formula

$$\frac{a_0}{2} = \frac{1}{2\pi} \int_{-\pi}^{\pi} f(x)\, dx.$$

The right-hand member is simply the *average* or *arithmetic mean* of $f(x)$ over the interval $-\pi \leqq x \leqq \pi$ (Section 4–2). One can also write

$$\int_{-\pi}^{\pi} \left[f(x) - \frac{a_0}{2} \right] dx = 0.$$

In words: the line $y = \frac{1}{2}a_0$ must be such that the *area between the line and the curve* $y = f(x)$ *lying above the line equals the area between the line and the curve* $y = f(x)$ *lying below the line*. Thus the line $y = \frac{1}{2}a_0$ is a sort of symmetry line for the graph of $y = f(x)$.

From either of these points of view, it is clear that in the two examples considered one must have $\frac{1}{2}a_0 = 0$; the average of $f(x)$ is 0 and there is as much area above the x axis as below.

Still another point of view yields the same formula for $\frac{1}{2}a_0$. We define the total *square error* of a function $g(x)$ relative to $f(x)$ as the integral

$$E = \int_{-\pi}^{\pi} [f(x) - g(x)]^2 \, dx. \tag{7–11}$$

This error is 0 when $g = f$ (or when $g = f$ except for a finite number of points), is otherwise positive. We now seek a constant function $y = g_0$ such that this error is as small as possible. In other words, we try to approximate $y = f(x)$ as accurately as possible, in terms of least square error, by a constant g_0. The error is now

$$E(g_0) = \int_{-\pi}^{\pi} [f(x) - g_0]^2 \, dx = \int_{-\pi}^{\pi} [f(x)]^2 \, dx - 2g_0 \int_{-\pi}^{\pi} f(x) \, dx + g_0^2 \cdot 2\pi$$
$$= A - 2Bg_0 + 2\pi g_0^2,$$

where A and B are constants. Thus $E(g_0)$ is a quadratic function of g_0, having a minimum when $dE/dg_0 = 0$:

$$-2B + 4\pi g_0 = 0.$$

Hence the error is minimized when

$$g_0 = \frac{B}{2\pi} = \frac{1}{2\pi} \int_{-\pi}^{\pi} f(x) \, dx = \frac{a_0}{2}.$$

Thus the constant function $y = \frac{1}{2}a_0$ is the best constant approximation, in the sense of least square error, to the function $f(x)$.

This last point of view holds for the coefficients of the general partial sum:

THEOREM 2. *Let $f(x)$ be piecewise continuous for $-\pi \leqq x \leqq \pi$. The coefficients of the partial sum*

$$\tfrac{1}{2}a_0 + a_1 \cos x + b_1 \sin x + \cdots + a_n \cos nx + b_n \sin nx$$

of the Fourier series of $f(x)$ are precisely those among all coefficients of the function

$$g_n(x) = p_0 + p_1 \cos x + q_1 \sin x + \cdots + p_n \cos nx + q_n \sin nx$$

which render the square error

$$\int_{-\pi}^{\pi} [f(x) - g_n(x)]^2 \, dx$$

a minimum. Furthermore, the minimum square error E_n satisfies the equation:

$$E_n = \int_{-\pi}^{\pi} [f(x)]^2 \, dx - \pi[\tfrac{1}{2}a_0^2 + \sum_{k=1}^{n} (a_k^2 + b_k^2)]. \tag{7–12}$$

The proof is left to Prob. 7 below.

COROLLARY. *If $f(x)$ is piecewise continuous for $-\pi \leqq x \leqq \pi$ and a_0, $a_1, \ldots, b_1, b_2, \ldots$ are the Fourier coefficients of $f(x)$, then*

$$\frac{1}{2}a_0^2 + \sum_{k=1}^{n}(a_k^2 + b_k^2) \leq \frac{1}{\pi}\int_{-\pi}^{\pi}[f(x)]^2\,dx, \tag{7-13}$$

so that the series $\sum_{n=1}^{\infty}(a_n^2 + b_n^2)$ *converges. Furthermore,*

$$\lim_{n\to\infty} a_n = 0, \quad \lim_{n\to\infty} b_n = 0. \tag{7-14}$$

Proof. Since the square error $\int (f - g)^2\,dx$ is always positive or 0, the minimum square error E_n is always positive or 0. Accordingly, (7–13) follows from (7–12). By Theorem 7 of Section 6–5, the series $\Sigma(a_n^2 + b_n^2)$ must converge or diverge properly; because of (7–13), the series cannot be properly divergent. Therefore the series converges; (7–14) then follows from the fact that the nth term of the series converges to 0.

It is shown in Section 7–12 below that E_n can be made as small as desired by choosing n sufficiently large; that is, the sequence E_n converges to 0. The relation (7–13) is *Bessel's inequality*.

Theorem 2 can be made the basis of a *graphical* estimation of Fourier coefficients. Thus for the function of Example 1, one first chooses the best fitting constant term $\frac{1}{2}a_0$; this is clearly 0. One then tries to add a function $p_1 \cos x + p_2 \sin x$ to make the square error as small as possible. It is apparent that the best approximation is achieved by taking a sine term alone. The function $\sin x$ itself fits fairly well, though the errors are large near $\pm\pi$ and 0. To reduce these, we *overshoot* at $\frac{1}{2}\pi$, taking, say, $1.3 \sin x$. New errors are introduced near $\frac{1}{2}\pi$, but the total square error will be less. If we subtract $1.3 \sin x$ from $f(x)$ graphically, we find the function of Fig. 7–7. To eliminate this error, it is clear that a function $p_6 \sin 3x$ is called for. Again we overshoot, and estimate $p_6 = 0.4$. We subtract $0.4 \sin 3x$ from the function graphed in Fig. 7–7 and so on. We thus obtain the expression

$$f(x) = 1.3 \sin x + 0.4 \sin 3x.$$

This agrees, up to the number of significant figures carried, with the first two terms of the Fourier series

$$\frac{4}{\pi}\sin x + \frac{4}{3\pi}\sin 3x + \cdots$$

obtained above.

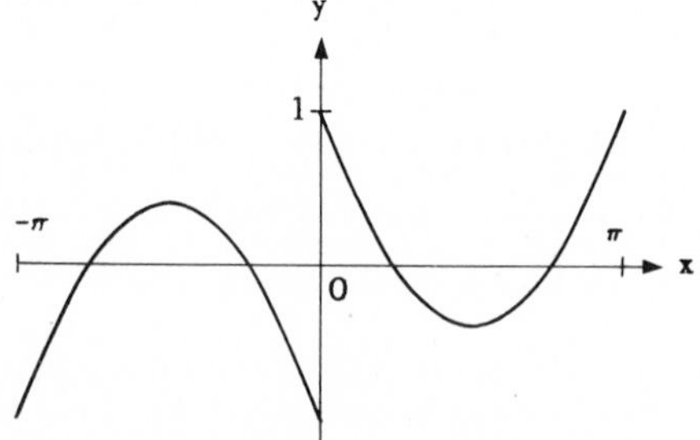

Fig. 7–7.

It is recommended that the problems below be solved first roughly by this graphical procedure. In this way considerable insight into the structure of the series can be gained. In constructing the series in this way it is best to think of the pair $a_n \cos nx + b_n \sin nx$ in the "amplitude-phase" form:

$$a_n \cos nx + b_n \sin nx = A_n \sin(nx + \alpha) \tag{7-15}$$

pointed out at the end of Section 7–1. The phase angle α effectively determines how much the curve $y = \sin nx$ has to be shifted to the left or right to match the given oscillation; the amplitude A_n merely adjusts the vertical scale.

It should be remarked that the process of decomposing a periodic phenomenon into its component simple harmonic parts is used in a great variety of common experiences. The rattling of an old-fashioned vehicle corresponds to a high frequency (large n) component with large amplitude; we automatically separate this from a low frequency vibration, or *swaying*. In a less precise sense, a large day-to-day fluctuation in weather conditions also corresponds to a high frequency component of large amplitude; the seasonal changes are of low frequency and are much less disturbing.

Problems

1. Find the Fourier series for each of the following functions:

(a) $f(x) = 0, \quad -\pi \leqq x < 0; \quad f(x) = 1, \quad 0 \leqq x \leqq \pi;$
(b) $f(x) = 0, \quad -\pi \leqq x < 0; \quad f(x) = x, \quad 0 \leqq x \leqq \pi;$
(c) $f(x) = -x, \quad -\pi \leqq x \leqq 0; \quad f(x) = x, \quad 0 \leqq x \leqq \pi;$
(d) $f(x) = x^2, \quad -\pi \leqq x \leqq \pi;$
(e) $F(x) = -\dfrac{1}{2} - \dfrac{x}{2\pi}, \quad -\pi \leqq x < 0; \quad F(x) = \dfrac{1}{2} - \dfrac{x}{2\pi}, \quad 0 < x \leqq \pi; \quad F(0) = 0;$
(f) $G(x) = \dfrac{\pi}{2} - \dfrac{x}{2} - \dfrac{x^2}{4\pi}, \quad -\pi \leqq x \leqq 0; \quad G(x) = \dfrac{\pi}{2} + \dfrac{x}{2} - \dfrac{x^2}{4\pi}, \quad 0 \leqq x \leqq \pi.$

(It is suggested that the first few partial sums be graphed and compared with the function in each case.)

2. It follows from the fundamental theorem of Section 7–3 that if $f(x)$ is defined between 0 and 2π and is piecewise very smooth in that interval, then $f(x)$ can be represented by a series of form (7–1) in that interval.

(a) Show that the coefficients a_n and b_n are given by the formulas:

$$a_n = \frac{1}{\pi}\int_0^{2\pi} f(x)\cos nx\,dx, \quad b_n = \frac{1}{\pi}\int_0^{2\pi} f(x)\sin nx\,dx.$$

(b) Extend this result to a function defined from $x = c$ to $x = c + 2\pi$, where c is any constant.

3. Using the results of Prob. 2(a) find Fourier series for the following functions:

(a) $f(x) = x, \quad 0 \leqq x \leqq 2\pi;$ (b) $f(x) = |\cos x|, \quad 0 \leqq x \leqq 2\pi.$

4. Determine which of the following functions are periodic and find the smallest period of those which are periodic:

(a) $\sin 5x$; (b) $\cos \dfrac{x}{3}$; (c) $\sin \pi x$; (d) $x \sin x$; (e) $\sin 3x + \sin 5x$;
(f) $\sin \dfrac{x}{3} + \sin \dfrac{x}{5}$; (g) $\sin x + \sin \pi x$.

5. The result of Prob. 1(a) implies that, for $0 < x < \pi$,

$$1 = \frac{1}{2} + \frac{2}{\pi}\left(\sin x + \frac{\sin 3x}{3} + \cdots\right).$$

Use $x = \frac{1}{2}\pi$ in this equation to show that

$$\frac{\pi}{4} = 1 - \frac{1}{3} + \frac{1}{5} - \frac{1}{7} + \cdots.$$

6. Use the method of Prob. 5 to obtain, with the aid of the series developed in Probs. 1 and 3, the relations

(a) $\dfrac{\pi}{\sqrt{8}} = 1 + \dfrac{1}{3} - \dfrac{1}{5} - \dfrac{1}{7} + \dfrac{1}{9} + \dfrac{1}{11} - \dfrac{1}{13} - \dfrac{1}{15} + \cdots;$

(b) $\dfrac{\pi^2}{8} = \dfrac{1}{1^2} + \dfrac{1}{3^2} + \dfrac{1}{5^2} + \cdots + \dfrac{1}{(2n-1)^2} + \cdots;$

(c) $\dfrac{\pi^2}{12} = 1 - \dfrac{1}{2^2} + \dfrac{1}{3^2} - \dfrac{1}{4^2} + \cdots;$

(d) $\dfrac{\pi^2}{6} = \dfrac{1}{1^2} + \dfrac{1}{2^2} + \dfrac{1}{3^2} + \dfrac{1}{4^2} + \cdots.$

7. Prove Theorem 2. [Hint: show that

$$\int (f-g)^2\,dx = \int f^2\,dx + \left(2\pi p_0^2 - 2p_0 \int f\,dx\right) + \left(\pi p_1^2 - 2p_1 \int f \cos x\,dx\right) + \cdots + \left(\pi q_n^2 - 2q_n \int f \sin nx\,dx\right);$$

accordingly, if $p_0, p_1, \ldots, q_n$ are chosen to give each term on the right its smallest value, the error will be minimized.]

8. (a) Prove the trigonometric identities:

$$\sin^3 x = \tfrac{3}{4}\sin x - \tfrac{1}{4}\sin 3x, \quad \cos^3 x = \tfrac{3}{4}\cos x + \tfrac{1}{4}\cos 3x.$$

(b) Obtain analogous expressions for $\sin^n x$ and $\cos^n x$. [Hint: use the identities: $\sin x = \frac{1}{2}(e^{ix} - e^{-ix})/i$, $\cos x = \frac{1}{2}(e^{ix} + e^{-ix})$.]

(c) Show that the identities of parts (a) and (b) can be interpreted as Fourier series expansions.

Answers

1. (a) $\dfrac{1}{2} + \dfrac{2}{\pi}\left(\sin x + \dfrac{\sin 3x}{3} + \dfrac{\sin 5x}{5} + \cdots\right);$

(b) $\dfrac{\pi}{4} - \dfrac{2}{\pi}\displaystyle\sum_{n=1}^{\infty} \dfrac{\cos(2n-1)x}{(2n-1)^2} - \sum_{n=1}^{\infty} (-1)^n \dfrac{\sin nx}{n};$

(c) $\dfrac{\pi}{2} - \dfrac{4}{\pi}\displaystyle\sum_{n=1}^{\infty} \dfrac{\cos(2n-1)x}{(2n-1)^2};$

(d) $\dfrac{\pi^2}{3} + 4\displaystyle\sum_{n=1}^{\infty} \dfrac{(-1)^n \cos nx}{n^2};$

(e) $\dfrac{1}{\pi}\displaystyle\sum_{n=1}^{\infty} \dfrac{\sin nx}{n};$

(f) $\dfrac{2\pi}{3} - \dfrac{1}{\pi}\displaystyle\sum_{n=1}^{\infty} \dfrac{\cos nx}{n^2}.$

3. (a) $\pi - 2\sum_{n=1}^{\infty} \frac{\sin nx}{n}$; (b) $\frac{2}{\pi} + \frac{4}{\pi}\sum_{n=1}^{\infty} \frac{(-1)^{n+1}}{4n^2 - 1} \cos 2nx$.

4. (a) $\frac{2\pi}{5}$, (b) 6π, (c) 2, (d) not periodic, (e) 2π, (f) 30π, (g) not periodic.

7–5 Generalizations; Fourier cosine series; Fourier sine series. The Fourier series up to this point have been considered only for functions of period 2π or, more restrictedly, for functions defined between $-\pi$ and π. We now proceed to enlarge the scope of the theory.

If $f(x)$ is a function of period 2π, one can use as basic interval any interval $c \leqq x \leqq c + 2\pi$, i.e., any interval of length 2π. For such an interval, the same reasoning as above leads to a Fourier series

$$\frac{a_0}{2} + \sum_{n=1}^{\infty} (a_n \cos nx + b_n \sin nx),$$

where

$$a_n = \frac{1}{\pi}\int_c^{c+2\pi} f(x) \cos nx\, dx, \qquad b_n = \frac{1}{\pi}\int_c^{c+2\pi} f(x) \sin nx\, dx. \tag{7–16}$$

If $f(x)$ is given for all x, with period 2π, this is merely another way of computing the coefficients a_n, b_n. If $f(x)$ is given only for $c \leqq x \leqq c + 2\pi$, the series can be used to represent f in this interval; it will then (if convergent) represent the periodic extension of f outside this interval.

The interval $-\pi \leqq x \leqq \pi$ has certain advantages for utilization of symmetry properties. Let $f(x)$ be defined in this interval and let

$$f(-x) = f(x), \quad -\pi \leqq x \leqq \pi. \tag{7–17}$$

Then f is called an *even* function of x (in the given interval). If, on the other hand,

$$f(-x) = -f(x), \quad -\pi \leqq x \leqq \pi, \tag{7–18}$$

then f is called an *odd* function of x. We note that the product of two even functions or of two odd functions is even, while the product of an odd function and an even function is odd. Furthermore,

$$\int_{-a}^{a} f(x)\, dx = \begin{cases} 0, & f \text{ odd} \\ 2\int_0^a f(x)\, dx, & f \text{ even}. \end{cases} \tag{7–19}$$

Let f now be even in the interval $-\pi \leqq x \leqq \pi$. Then $f(x) \cos nx$ is even (product of two even functions), while $f(x) \sin nx$ is odd (product of odd function and even function). Hence by (7–19)

$$a_n = \frac{2}{\pi}\int_0^{\pi} f(x) \cos nx\, dx \quad (n = 0, 1, 2, \ldots), \qquad b_n = 0 \quad (n = 1, 2, \ldots). \tag{7–20}$$

Similarly, if f is odd,

$$a_n = 0, \quad b_n = \frac{2}{\pi}\int_0^{\pi} f(x)\sin nx\, dx. \tag{7-21}$$

We have thus the expansions (for a function piecewise very smooth):

$$\begin{aligned} f(x) &= \frac{a_0}{2} + \sum_{n=1}^{\infty} a_n \cos nx \quad (f \text{ even}), \\ a_n &= \frac{2}{\pi}\int_0^{\pi} f(x)\cos nx\, dx \end{aligned} \tag{7-22}$$

and

$$\begin{aligned} f(x) &= \sum_{n=1}^{\infty} b_n \sin nx \quad (f \text{ odd}), \\ b_n &= \frac{2}{\pi}\int_0^{\pi} f(x)\sin nx\, dx. \end{aligned} \tag{7-23}$$

Now (7–22) uses only the values of $f(x)$ between $x = 0$ and $x = \pi$. Hence for any function $f(x)$ given only over this interval one can form the series (7–22). This is called the *Fourier cosine series* of $f(x)$. It follows from the fundamental theorem that the series will converge to $f(x)$ for $0 \leqq x \leqq \pi$ and outside this interval to the even periodic function which coincides with $f(x)$ for $0 \leqq x \leqq \pi$. This is illustrated in Fig. 7-8.

In the same way (7–23) defines the *Fourier sine series* of a function $f(x)$ defined only between 0 and π. The series represents an odd periodic function which coincides with $f(x)$ for $0 < x < \pi$, as illustrated in Fig. 7–9.

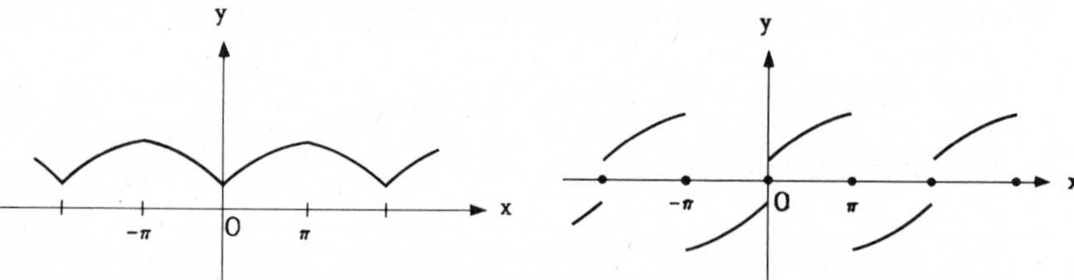

FIG. 7–8. Even periodic extension of function defined between 0 and π.

Fig. 7–9. Odd periodic extension of function defined between 0 and π.

EXAMPLE. Let $f(x) = \pi - x$. Then one can represent $f(x)$ by a Fourier series over the interval: $-\pi < x < \pi$. The formulas (7–9) give

$$a_0 = \frac{1}{\pi}\int_{-\pi}^{\pi} (\pi - x)\, dx = 2\pi, \quad a_1 = a_2 = \cdots = 0,$$

$$b_n = \frac{1}{\pi}\int_{-\pi}^{\pi} (\pi - x)\sin nx\, dx = \frac{2(-1)^n}{n}.$$

Hence one has

$$\pi - x = \pi + 2\sum_{n=1}^{\infty} \frac{(-1)^n \sin nx}{n}, \quad -\pi < x < \pi. \tag{a}$$

The same function, $\pi - x$, can be represented by a Fourier cosine series over the interval $0 \leqq x \leqq \pi$. The formulas (7–22) give

$$a_0 = \frac{2}{\pi}\int_0^{\pi} (\pi - x)\,dx = \pi,$$

$$a_n = \frac{2}{\pi}\int_0^{\pi} (\pi - x)\cos nx\,dx = \frac{2}{\pi n^2}(1 - (-1)^n), \quad n = 1, 2, \ldots.$$

Hence one has

$$\begin{aligned}\pi - x &= \frac{\pi}{2} + \frac{2}{\pi}\sum_{n=1}^{\infty}\frac{1-(-1)^n}{n^2}\cos nx \\ &= \frac{\pi}{2} + \frac{2}{\pi}\left(2\cos x + \frac{2\cos 3x}{3^2} + \frac{2\cos 5x}{5^2} + \cdots\right), \quad 0 \leqq x \leqq \pi.\end{aligned} \tag{b}$$

Finally, the same function, $\pi - x$, can be represented by a Fourier sine series over the interval $0 < x < \pi$. The formulas (7–23) give

$$\pi - x = 2\sum_{n=1}^{\infty}\frac{\sin nx}{n}, \quad 0 < x < \pi. \tag{c}$$

Figure 7–10 shows the graphs of the three functions represented by the series (a), (b), (c).

Change of period. If $f(x)$ has period p:

$$f(x + p) = f(x), \quad (p \neq 0)$$

then the substitution

$$x = \frac{p}{2\pi}t$$

transforms $f(x)$ into a function $g(t)$:

$$g(t) = f\left(\frac{p}{2\pi}t\right),$$

and $g(t)$ has period 2π, since

$$g(t + 2\pi) = f\left[\frac{p}{2\pi}(t + 2\pi)\right] = f\left(\frac{pt}{2\pi} + p\right) = f\left(\frac{pt}{2\pi}\right) = g(t).$$

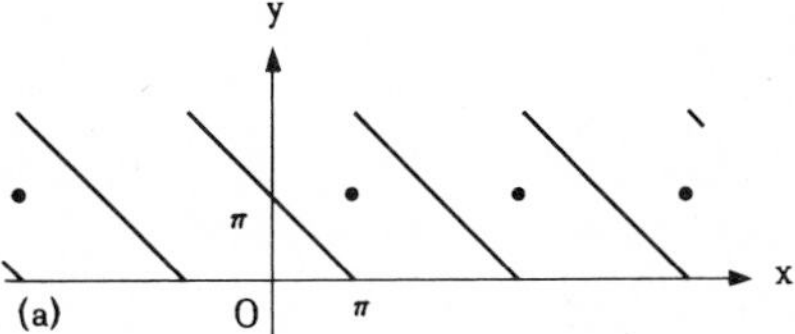

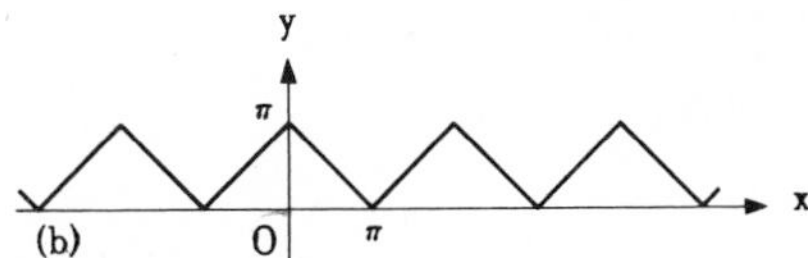

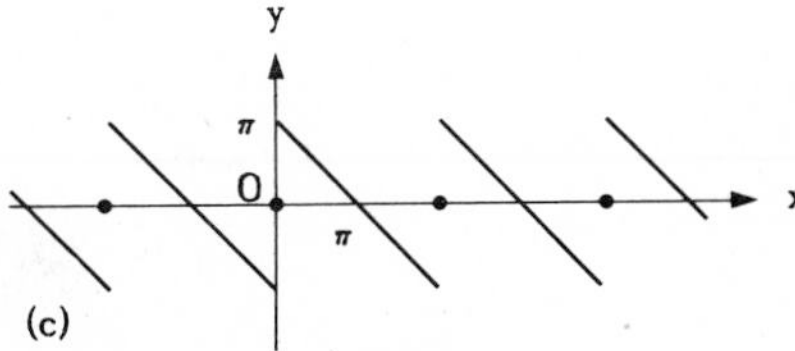

FIG. 7–10. (a) Fourier series. (b) Fourier cosine series. (c) Fourier sine series.

The change from x to t is simply a change of scale. Since g has period 2π, one has a Fourier series for g (assumed piecewise very smooth):

$$g(t) = \frac{a_0}{2} + \sum_{n=1}^{\infty}(a_n\cos nt + b_n\sin nt),$$

where, for example,

$$a_n = \frac{1}{\pi}\int_{-\pi}^{\pi} g(t)\cos nt\,dt, \quad b_n = \frac{1}{\pi}\int_{-\pi}^{\pi} g(t)\sin nt\,dt.$$

If now t is replaced by $(2\pi/p)x$, one finds a Fourier series for $f(x)$:

$$f(x) = \frac{a_0}{2} + \sum_{n=1}^{\infty}\left[a_n \cos\left(n\cdot\frac{2\pi}{p}x\right) + b_n \sin\left(n\cdot\frac{2\pi}{p}x\right)\right]. \tag{7–24}$$

The coefficients a_n, b_n can be expressed directly in terms of $f(x)$, thus:

$$\begin{aligned} a_n &= \frac{1}{L}\int_{-L}^{L} f(x)\cos\left(n\cdot\frac{2\pi}{p}x\right)dx, \\ b_n &= \frac{1}{L}\int_{-L}^{L} f(x)\sin\left(n\cdot\frac{2\pi}{p}x\right)dx, \end{aligned} \tag{7–25}$$

where $p = 2L$.

The Fourier cosine series can also be used in this case. One finds

$$f(x) = \frac{a_0}{2} + \sum_{n=1}^{\infty} a_n \cos\left(n\cdot\frac{2\pi}{p}x\right), \quad 0 \leqq x \leqq L, \tag{7–26}$$

where

$$a_n = \frac{2}{L}\int_0^L f(x)\cos\left(n\cdot\frac{2\pi}{p}x\right)dx. \tag{7–27}$$

Similarly, $f(x)$ has a Fourier sine series:

$$f(x) = \sum_{n=1}^{\infty} b_n \sin\left(n\cdot\frac{2\pi}{p}x\right), \quad 0 < x < L, \tag{7–28}$$

where

$$b_n = \frac{2}{L}\int_0^L f(x)\sin\left(n\cdot\frac{2\pi}{p}x\right)dx. \tag{7–29}$$

Example. Let $f(x) = 2x + 1$. Then $f(x)$ can be represented by a Fourier series over the interval $0 < x < 2$. Here $p = 2$ and $t = \pi x$, so that

$$f(x) = 2x + 1 = \frac{2t}{\pi} + 1 = g(t),$$

where $g(t)$ is defined for $0 \leqq t \leqq 2\pi$. Hence $g(t)$ can be represented by a Fourier series:

$$g(t) = \frac{a_0}{2} + \sum_{n=1}^{\infty}(a_n \cos nt + b_n \sin nt),$$

where

$$a_n = \frac{1}{\pi}\int_0^{2\pi}\left(\frac{2t}{\pi}+1\right)\cos nt\,dt, \quad b_n = \frac{1}{\pi}\int_0^{2\pi}\left(\frac{2t}{\pi}+1\right)\sin nt\,dt.$$

One finds

$$a_0 = 6, \quad a_1 = a_2 = \cdots = 0, \quad b_n = -\frac{4}{n\pi},$$

so that

$$g(t) = 3 - \frac{4}{\pi}\sum_{n=1}^{\infty} \frac{\sin nt}{n}$$

and hence

$$f(x) = g(\pi x) = 3 - \frac{4}{\pi}\sum_{n=1}^{\infty} \frac{\sin n\pi x}{n}.$$

One could use the formulas (7–25) directly, but the change to t simplifies the calculations.

Problems

1. Let $f(x) = 2x + 1$.
(a) Expand $f(x)$ in a Fourier series for $-\pi < x < \pi$.
(b) Expand $f(x)$ in a Fourier series for $0 < x < 2\pi$.
(c) Expand $f(x)$ in a Fourier cosine series for $0 \leqq x \leqq \pi$.
(d) Expand $f(x)$ in a Fourier sine series for $0 < x < \pi$.
(e) Expand $f(x)$ in a Fourier series for $0 < x < \pi$.
(f) Graph the functions represented by the series of parts (a), (b), (c), (d), (e).

2. Let $f(x) = x^2$.
(a) Expand $f(x)$ in a Fourier series for $\pi < x < 3\pi$.
(b) Expand $f(x)$ in a Fourier series for $1 < x < 2$.
(c) Graph the functions represented by the series of parts (a), (b).

3. Let $f(x) = \sin x$.
(a) Expand $f(x)$ in a Fourier series for $0 \leqq x \leqq 2\pi$.
(b) Expand $f(x)$ in a Fourier series for $0 \leqq x \leqq \pi$.
(c) Expand $f(x)$ in a Fourier cosine series for $0 \leqq x \leqq \pi$.

4. Let $f(x) = x$.
(a) Expand $f(x)$ in a Fourier cosine series for $0 \leqq x \leqq \pi$.
(b) Expand $f(x)$ in a Fourier sine series for $0 \leqq x < \pi$.
(c) Expand $f(x)$ in a Fourier cosine series for $0 \leqq x \leqq 1$.
(d) Expand $f(x)$ in a Fourier sine series for $0 \leqq x < 1$.

5. Let $f(x) = \dfrac{a_0}{2} + \displaystyle\sum_{n=1}^{\infty} (a_n \cos nx + b_n \sin nx)$,

where the series converges uniformly for all x. State what conclusions can be drawn concerning the coefficients a_n, b_n from each of the following properties of $f(x)$:

(a) $f(-x) = f(x)$
(b) $f(-x) = -f(x)$
(c) $f(\pi - x) = f(x)$
(d) $f\left(\frac{\pi}{2} - x\right) = f(x)$
(e) $f(-x) = f(x) = f\left(\frac{\pi}{2} - x\right)$
(f) $f(\pi - x) = -f(x)$
(g) $f(\pi + x) = f(x)$
(h) $f\left(\frac{\pi}{2} + x\right) = f(x)$

(i) $f\left(\frac{\pi}{3} + x\right) = f(x)$ (j) $f(x) = f(2x)$

[Hint: use the Corollary to Theorem 1.]

Answers

1. (a) $1 - 4\sum_{n=1}^{\infty} \frac{(-1)^n}{n} \sin nx$; (b) $1 + 2\pi - 4\sum_{n=1}^{\infty} \frac{\sin nx}{n}$;

(c) $1 + \pi - \frac{8}{\pi}\sum_{n=1}^{\infty} \frac{\cos (2n-1)x}{(2n-1)^2}$; (d) $\frac{2}{\pi}\sum_{n=1}^{\infty} \frac{1-(-1)^n(2\pi+1)}{n} \sin nx$;

(e) $\pi + 1 - 2\sum_{n=1}^{\infty} \frac{\sin 2nx}{n}$.

2. (a) $\frac{13\pi^2}{3} + 4\sum_{n=1}^{\infty} \frac{(-1)^n}{n^2} \cos nx - 8\pi \sum_{n=1}^{\infty} \frac{(-1)^n}{n} \sin nx$;

(b) $\frac{7}{3} + \frac{1}{\pi^2}\sum_{n=1}^{\infty} \frac{\cos 2n\pi x}{n^2} - \frac{3}{\pi}\sum_{n=1}^{\infty} \frac{\sin 2n\pi x}{n}$.

3. (a) $\sin x$; (b) and (c): $\frac{2}{\pi} - \frac{4}{\pi}\sum_{n=1}^{\infty} \frac{\cos 2nx}{4n^2-1}$.

4. (a) $\frac{\pi}{2} - \frac{4}{\pi}\sum_{n=1}^{\infty} \frac{\cos (2n-1)x}{(2n-1)^2}$; (b) $-2\sum_{n=1}^{\infty} (-1)^n \frac{\sin nx}{n}$;

(c) $\frac{1}{2} - \frac{4}{\pi^2}\sum_{n=1}^{\infty} \frac{\cos (2n-1)\pi x}{(2n-1)^2}$; (d) $-\frac{2}{\pi}\sum_{n=1}^{\infty} (-1)^n \frac{\sin n\pi x}{n}$.

5. (a) $b_n = 0$; (b) $a_n = 0$; (c) $a_{2n+1} = 0$, $b_{2n} = 0$;
(d) $a_{4n+2} = 0$, $b_{4n} = 0$, $a_{4n+1} = b_{4n+1}$, $a_{4n+3} = -b_{4n+3}$;
(e) $b_n = 0$; $a_n = 0$ except for $n = 0, 4, 8, \ldots, 4k, \ldots$;
(f) $a_{2n} = 0$, $b_{2n+1} = 0$; (g) $a_{2n+1} = 0$, $b_{2n+1} = 0$;
(h) $a_n = 0$ and $b_n = 0$ except for $n = 0, 4, 8, \ldots, 4k, \ldots$;
(i) $a_n = 0$ and $b_n = 0$ except for $n = 0, 6, 12, \ldots, 6k, \ldots$;
(j) $a_n = 0$ and $b_n = 0$ for $n = 1, 2, \ldots$.

7–6 Remarks on applications of Fourier series. The natural field of application of Fourier series is to periodic phenomena, as indicated in Section 7–1. The fact that a periodic function $f(t)$ can be resolved into its simple harmonic components $A_n \sin(nt + \alpha_n)$ is of fundamental physical significance. For all "linear" problems, this resolution permits one to reduce the problem to the simpler one of a single simple harmonic vibration and then to build up the general case by addition (superposition) of the simple ones.

The concrete application of Fourier series to such problems takes two main forms: a periodic function $f(t)$ may be given in graphical or tabulated form; an understanding of the physical mechanism leading to such a function requires a "harmonic analysis" of $f(t)$, or representation of $f(t)$ as a Fourier series. Second, the function $f(t)$ is known to be periodic and is known to satisfy some implicit relation, such as a differential equation; it is desired to determine $f(t)$ as a Fourier series on the basis of this information.

The first problem is thus one of *interpretation* of experimental data; the second is one of *prediction* of the result of an experiment, on the basis of a mathematical theory.

While the application of Fourier series to periodic phenomena is basic, there is a much wider field of application. As has been shown above, an "arbitrary" function $f(x)$, given for $a \leqq x \leqq b$, has a representation as a Fourier series over that interval. Thus, in any problem concerning a function over an interval, it may be advantageous to represent the function by the corresponding series. This permits a tremendous variety of applications. As before, the applications usually take the form of interpreting given data or of predicting functions satisfying given conditions.

In this book, particular applications will be considered in the chapters on ordinary and partial differential equations. In the following problems, several illustrations of the form of solutions to such differential equations are provided.

Problems

1. Show that the linear differential equation

$$\frac{d^2y}{dt^2} + 4\frac{dy}{dt} + y = p \cos \omega t + q \sin \omega t$$

is satisfied by

$$y = A \cos \omega t + B \sin \omega t,$$

where

$$A = \frac{p(1 - \omega^2) - 4\omega q}{(1 - \omega^2)^2 + 16\omega^2}, \quad B = \frac{4\omega p + q(1 - \omega^2)}{(1 - \omega^2)^2 + 16\omega^2},$$

and determine a solution, in series form, of the differential equation,

$$\frac{d^2y}{dt^2} + 4\frac{dy}{dt} + y = f(t) = \sum_{n=1}^{\infty} (a_n \cos nt + b_n \sin nt).$$

2. Show that, granting the correctness of the necessary term-by-term differentiations of series, the function

$$f(x, t) = \sum_{n=1}^{\infty} [A_n \cos nct + B_n \sin nct] \sin nx,$$

where the A_n and B_n are constants, satisfies the partial differential equation

$$\frac{\partial^2 f}{\partial t^2} = c^2 \frac{\partial^2 f}{\partial x^2}.$$

The differential equation is that of the vibrating string and the series represents the general solution when the ends are fixed, π units apart.

3. Show that, granting the correctness of the necessary term-by-term differentiations of series, the function

$$f(r, \theta) = A_0 + \sum_{n=1}^{\infty} (A_n r^n \cos n\theta + B_n r^n \sin n\theta)$$

satisfies the Laplace equation in polar coordinates:

$$\frac{\partial^2 f}{\partial r^2} + \frac{1}{r^2}\frac{\partial^2 f}{\partial \theta^2} + \frac{1}{r}\frac{\partial f}{\partial r} = 0.$$

As shown in Section 5–15, this equation describes equilibrium temperature distributions, electrostatic potentials, and velocity potentials. Every function f which is harmonic in a circular domain: $r < R$ can be represented by such a series, as will be shown in Section 9–11.

7–7 Uniqueness theorem. If two functions $f(x)$ and $f_1(x)$ have the same set of Fourier coefficients:

$$\begin{aligned} \int_{-\pi}^{\pi} f(x) \cos nx \, dx &= \int_{-\pi}^{\pi} f_1(x) \cos nx \, dx, \quad (n = 0, 1, 2, \ldots), \\ \int_{-\pi}^{\pi} f(x) \sin nx \, dx &= \int_{-\pi}^{\pi} f_1(x) \sin nx \, dx, \quad (n = 1, 2, \ldots), \end{aligned} \tag{7–30}$$

are the functions necessarily identical? In other words, is a function *uniquely determined* by its Fourier coefficients? The answer is in the affirmative:

THEOREM 3 (*Uniqueness theorem*). *Let $f(x)$ and $f_1(x)$ be piecewise continuous in the interval $-\pi \leqq x \leqq \pi$ and satisfy* (7–30), *so that the two functions have the same Fourier coefficients. Then $f(x) = f_1(x)$ except perhaps at points of discontinuity.*

Proof. Let $h(x) = f(x) - f_1(x)$. Then $h(x)$ is piecewise continuous and from (7–30) it at once follows that all Fourier coefficients of $h(x)$ are 0. We then show that $h(x) = 0$ except perhaps at discontinuity points.

Let us suppose $h(x_0) \neq 0$ at a point of continuity x_0, e.g., $h(x_0) = 2c > 0$. Then, by continuity, $h(x) > c$ for $|x - x_0| < \delta$ and δ sufficiently small (cf. Prob. 7 following Section 2–4).

We now achieve a contradiction by showing that there exists a "trigonometric polynomial"

$$P(x) = p_0 + p_1 \cos x + p_2 \sin x + \cdots + p_{2k-1} \cos kx + p_{2k} \sin kx$$

which represents a "pulse" at x_0 of arbitrarily large amplitude and arbitrarily small width. This is pictured in Fig. 7–11. If such a pulse can be constructed, then one has a contradiction. On the one hand

$$\int_{-\pi}^{\pi} h(x)P(x) \, dx = p_0 \int_{-\pi}^{\pi} h(x) \, dx + p_1 \int_{-\pi}^{\pi} h(x) \cos x \, dx + \cdots = 0.$$

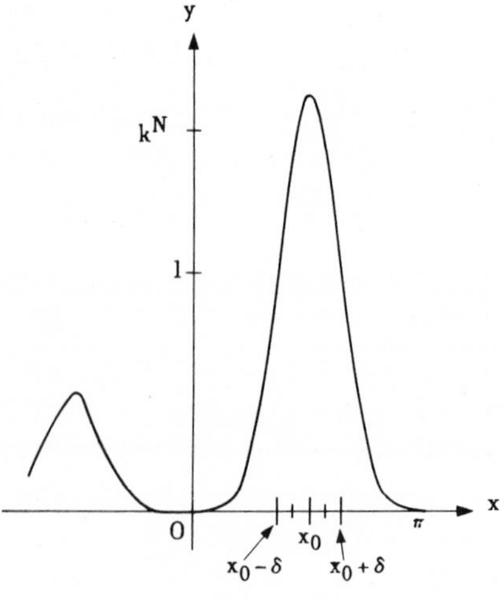

FIG. 7–11. Pulse function.

On the other hand, the major portion of the integral $\int h(x)P(x)\,dx$ is concentrated in the interval in which the pulse occurs; here $h(x)$ is positive and $P(x)$ is large and positive. Hence the integral is positive and cannot be 0.

To make this precise, we take

$$P(x) = [\psi(x)]^N, \quad \psi(x) = 1 + \cos(x - x_0) - \cos\delta$$

for an appropriate positive integer N. Since the functions $\sin^n x$ and $\cos^n x$ are expressible as trigonometric polynomials (Prob. 8 following Section 7–4), the function $P(x)$ is a trigonometric polynomial. Let

$$k = \psi\left(x_0 + \frac{\delta}{2}\right) = 1 + \cos\frac{\delta}{2} - \cos\delta.$$

Then $k > 1$ and $P \geqq k^N$ for $|x - x_0| \leqq \frac{1}{2}\delta$. Since ψ is positive (greater than 1) for $\frac{1}{2}\delta \leqq |x - x_0| < \delta$, P is positive in this range. On the other hand, $|\psi(x)| < 1$ for $-\pi \leqq x < x_0 - \delta$ and for $x_0 + \delta < x \leqq \pi$, so that $|P| < 1$ in this range.

Now the function $h(x)$, being piecewise continuous, is bounded by a constant M for $-\pi \leqq x \leqq \pi$: $|h(x)| \leqq M$. It follows from the properties of $P(x)$ just listed that $P(x)h(x) > -M$ for $-\pi \leqq x \leqq x_0 - \frac{1}{2}\delta$ and for $x_0 + \frac{1}{2}\delta \leqq x \leqq \pi$, while $P(x)h(x) \geqq ck^N$ for $x_0 - \frac{1}{2}\delta \leqq x \leqq x_0 + \frac{1}{2}\delta$. Accordingly, by rule (4–18) of Section 4–2,

$$\int_{-\pi}^{\pi} P(x)h(x)\,dx = \int_{-\pi}^{x_0 - \frac{1}{2}\delta} P(x)h(x)\,dx + \int_{x_0 + \frac{1}{2}\delta}^{\pi} P(x)h(x)\,dx$$
$$+ \int_{x_0 - \frac{1}{2}\delta}^{x_0 + \frac{1}{2}\delta} P(x)h(x)\,dx > -M(2\pi - \delta) + ck^N\delta.$$

Since $k^N \to +\infty$ as $N \to \infty$, the right-hand member of the inequality is surely positive when N is sufficiently large. Accordingly, the left-hand member is positive for appropriate choice of N. This contradicts the fact that the left-hand member is 0. Accordingly, $h(x) = f(x) - f_1(x) = 0$ wherever $f(x)$ and $f_1(x)$ are continuous.

Remarks. The uniqueness theorem can be looked at in another way, namely, as asserting that the system of functions

$$1, \quad \cos x, \quad \sin x, \ldots, \quad \cos nx, \quad \sin nx, \ldots$$

is "large enough": that is, that there are *enough* functions in this system to construct series for all the periodic functions envisaged. It should be noted that omission of any *one* function of the system would destroy this property. Thus, if $\cos x$ were omitted, one could still form a series

$$\tfrac{1}{2}a_0 + b_1 \sin x + a_2 \cos 2x + b_2 \sin 2x + \cdots$$

as before. But there are very smooth periodic functions whose "Fourier series" in this deficient form could never converge to the function, namely, all functions $A\cos x$ for constant A. For each such function would have all coefficients 0:

$$\int_{-\pi}^{\pi} A \cos x \, dx = 0, \quad \int_{-\pi}^{\pi} A \cos x \sin x \, dx = 0, \ldots .$$

The series reduces to 0 and cannot represent the function. The essence of the proof of the uniqueness theorem is the demonstration that there are *enough* functions in the system of sines and cosines to construct a pulse function $P(x)$.

THEOREM 4. *Let the function $f(x)$ be continuous for $-\pi \leqq x \leqq \pi$ and let the Fourier series of $f(x)$ converge uniformly in this interval. Then the series converges to $f(x)$ for $-\pi \leqq x \leqq \pi$.*

Proof. Let the sum of the Fourier series of $f(x)$ be denoted by $f_1(x)$:

$$f_1(x) = \tfrac{1}{2}a_0 + \sum_{n=1}^{\infty} (a_n \cos nx + b_n \sin nx).$$

Since the series converges uniformly, it follows from Theorem 1 that $f_1(x)$ is continuous and that a_n, b_n are the Fourier coefficients of $f_1(x)$. But the series was given as the Fourier series of $f(x)$. Hence $f(x)$ and $f_1(x)$ have the same Fourier coefficients and, by Theorem 3, $f(x) \equiv f_1(x)$; that is, $f(x)$ is the sum of its Fourier series for $-\pi \leqq x \leqq \pi$.

7-8 Proof of fundamental theorem for functions which are continuous, periodic, and piecewise very smooth.

THEOREM 5. *Let $f(x)$ be continuous and piecewise very smooth for all x and let $f(x)$ have period 2π. Then the Fourier series of $f(x)$ converges uniformly to $f(x)$ for all x.*

Proof. Let us first assume that $f(x)$ has continuous first and second derivatives for all x. One has (for $n \neq 0$)

$$a_n = \frac{1}{\pi}\int_{-\pi}^{\pi} f(x) \cos nx \, dx = \left.\frac{f(x) \sin nx}{n\pi}\right|_{-\pi}^{\pi} - \frac{1}{n\pi}\int_{-\pi}^{\pi} f'(x) \sin nx \, dx$$

by integration by parts. The first term on the right is zero. A second integration by parts gives

$$a_n = \left.\frac{f'(x) \cos nx}{n^2\pi}\right|_{-\pi}^{\pi} - \frac{1}{n^2\pi}\int_{-\pi}^{\pi} f''(x) \cos nx \, dx = -\frac{1}{n^2\pi}\int_{-\pi}^{\pi} f''(x) \cos nx \, dx,$$

the first term being zero because of the periodicity of $f'(x)$. The function $f''(x)$ is continuous in the interval $-\pi \leqq x \leqq \pi$ and hence $|f''(x)| \leqq M$ for an appropriate constant M. One concludes that

$$|a_n| = \left|\frac{1}{n^2\pi}\int_{-\pi}^{\pi} f''(x) \cos nx \, dx\right| \leqq \frac{2M}{n^2} \quad (n = 1, 2, \ldots).$$

In exactly the same way, we prove that $|b_n| \leqq 2M/n^2$ for all n. Hence each term of the Fourier series of $f(x)$ is in absolute value at most equal to the corresponding term of the convergent series

$$\tfrac{1}{2}|a_0| + \frac{2M}{1} + \frac{2M}{1} + \frac{2M}{2^2} + \frac{2M}{2^2} + \cdots.$$

Application of the Weierstrass M-test (Section 6–13) now establishes that the Fourier series converges uniformly for all x. By Theorem 4, the sum is $f(x)$.

We now consider the case of a function $f(x)$ which is periodic, continuous, and piecewise very smooth. The only step which requires re-examination is the proof that $|a_n| \leqq 2Mn^{-2}$ and that $|b_n| \leqq 2Mn^{-2}$. The integration by parts must now be carried out separately over each interval within which $f''(x)$ is continuous. If the results are added, one obtains, e.g., for b_n:

$$b_n = \left[\frac{-f(x)\cos nx}{n\pi}\Bigg|_{-\pi}^{x_1} + \frac{-f(x)\cos nx}{n\pi}\Bigg|_{x_1}^{x_2} + \cdots\right]$$
$$+ \left[\frac{f'(x)\sin nx}{n^2\pi}\Bigg|_{-\pi}^{x_1-} + \cdots\Bigg|_{x_1+}^{x_2} + \cdots\right] - \frac{1}{n^2\pi}\int_{-\pi}^{\pi} f''(x)\sin nx\,dx.$$

The integrals are technically improper, but exist as such. Since f is continuous and periodic, the terms in the first bracket add up to

$$\frac{-f(x)\cos nx}{n\pi}\Bigg|_{-\pi}^{\pi} = 0.$$

The functions $f'(x)$ and $f''(x)$ are bounded in each subinterval. Hence one constant M can be chosen so that $|f'(x)| \leqq M$ and $|f''(x)| \leqq M$ throughout. If there are k subintervals, we conclude that

$$|b_n| \leqq \frac{2kM}{n^2\pi} + \frac{2M}{n^2} = \frac{M_1}{n^2}.$$

A similar result holds for a_n. Hence *the Fourier series of a periodic, continuous, piecewise very smooth function converges uniformly to the function for all x.*

7–9 Proof of fundamental theorem. Before proceeding to the general case of a function with jump discontinuities, we consider an example illustrating the result just obtained. Let

$$G(x) = \frac{\pi}{2} - \frac{x}{2} - \frac{x^2}{4\pi}, \quad -\pi \leqq x \leqq 0,$$

$$G(x) = \frac{\pi}{2} + \frac{x}{2} - \frac{x^2}{4\pi}, \quad 0 \leqq x \leqq \pi,$$

and let G be repeated periodically outside this interval, as shown in Fig. 7–12. The resulting function $G(x)$ is continuous for all x and is piecewise smooth. Its Fourier series is found in Prob. 1(*f*) following Section 7–4 to be the series

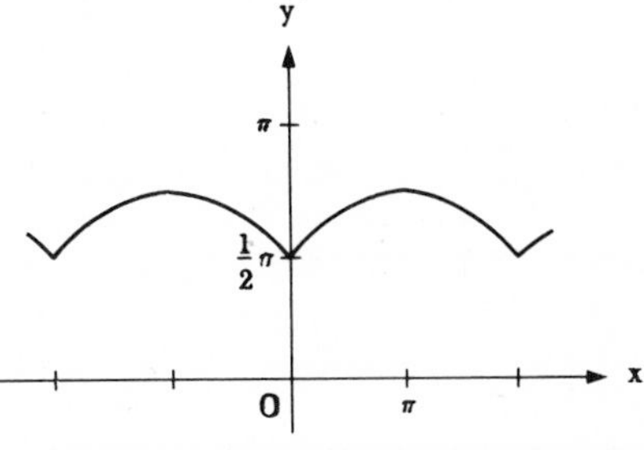

Fig. 7–12. Auxiliary function $G(x)$.

$$\frac{2\pi}{3} - \frac{1}{\pi}\sum_{n=1}^{\infty}\frac{\cos nx}{n^2}.$$

Hence $|a_n| \leqq Mn^{-2}$ as asserted, with $M = 1/\pi$; the b_n happen to be 0. By Theorem 5, this series converges uniformly to $G(x)$.

We now ask: is term-by-term differentiation of the series permissible? In other words, is

$$G'(x) = \frac{1}{\pi}\sum_{n=1}^{\infty}\frac{\sin nx}{n} \tag{7–31}$$

wherever $G'(x)$ is defined? By Theorem 33 of Section 6–14, this is correct if x lies within an interval within which the differentiated series converges uniformly. It is shown in Prob. 4 below that the series $\Sigma(\sin nx/n)$ converges uniformly for $a \leqq x \leqq \pi$, provided $a > 0$. Accordingly, (7–31) *is correct for* $-\pi \leqq x \leqq \pi$, *except for* $x = 0$. Now let $F(x)$ be the periodic function of period 2π such that $F(0) = 0$ and

$$F(x) = G'(x) = \begin{cases} -\frac{1}{2} - (x/2\pi), & -\pi \leqq x < 0, \\ \frac{1}{2} - (x/2\pi), & 0 < x \leqq \pi. \end{cases}$$

The function $F(x)$ is shown in Fig. 7–13. We have now proved that

$$F(x) = \frac{1}{\pi}\sum_{n=1}^{\infty}\frac{\sin nx}{n}$$

for all x, the convergence being uniform for $0 < a \leqq |x| \leqq \pi$. The series on the right was computed as the Fourier series of $F(x)$ in Prob. 1(e) following Section 7–4. Accordingly, we have shown that $F(x)$ is represented by its Fourier series for all x. The remarkable feature of this result is that $F(x)$ has a jump, from $-\frac{1}{2}$ to $\frac{1}{2}$ at $x = 0$. The series converges to the average value $F(0) = 0$. We have therefore verified a special case of the following theorem:

THEOREM 6. *Let $f(x)$ be defined and piecewise very smooth for $-\pi \leqq x \leqq \pi$ and let $f(x)$ be defined outside this interval in such a manner that $f(x)$ has period 2π. Then the Fourier series of $f(x)$ converges uniformly to $f(x)$ in each closed interval containing no discontinuity of $f(x)$. At each discontinuity x_0 the series converges to*

$$\tfrac{1}{2}[\lim_{x\to x_0+} f(x) + \lim_{x\to x_0-} f(x)].$$

Proof. For convenience we redefine $f(x)$ at each discontinuity x_0 as the average of left and right limit values. Let us suppose, for example, that the only discontinuity is at $x = 0$ (and the points $2k\pi$, $k = \pm 1, \pm 2, \ldots$) as in Fig. 7–13. Let

$$\lim_{x\to 0+} f(x) - \lim_{x\to 0-} f(x) = s,$$

so that s is precisely the "jump." We then proceed to eliminate the discontinuity by subtracting from $f(x)$ the function $sF(x)$, where $F(x)$ is the

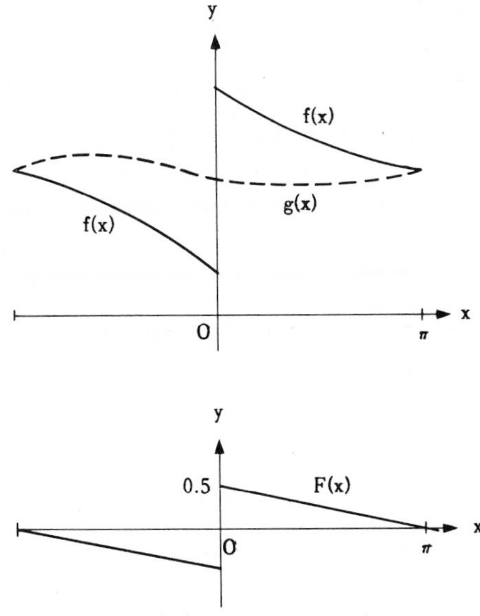

FIG. 7-13. Removal of jump discontinuity.

function defined above. Since $sF(x)$ has also the jump s at $x = 0$ (and $x = 2k\pi$), $g(x) = f(x) - sF(x)$ has jump 0 at $x = 0$ and is continuous for all x; for

$$\lim_{x\to 0-} g(x) = \lim_{x\to 0-} f(x) - s \lim_{x\to 0-} F(x)$$
$$= [f(0) - \tfrac{1}{2}s] + \tfrac{1}{2}s$$
$$= f(0) = g(0),$$

and a similar statement applies to the right-hand limit. Since $F(x)$ is piecewise linear, $g(x)$ is continuous and piecewise very smooth for all x, and has period 2π. Hence by Theorem 5, $g(x)$ is representable by a uniformly convergent Fourier series for all x:

$$g(x) = \tfrac{1}{2}A_0 + \sum_{n=1}^{\infty} (A_n \cos nx + B_n \sin nx)$$

Accordingly,

$$f(x) = g(x) + sF(x) = \frac{1}{2} A_0 + \sum_{n=1}^{\infty} (A_n \cos nx + B_n \sin nx) + \frac{s}{\pi} \sum_{n=1}^{\infty} \frac{\sin nx}{n}$$
$$= \frac{1}{2} A_0 + \sum_{n=1}^{\infty} \left[A_n \cos nx + \left(B_n + \frac{s}{n\pi}\right) \sin nx \right],$$

so that $f(x)$ is represented by a trigonometric series for all x. The series is the Fourier series of $f(x)$, for

$$b_n = \frac{1}{\pi} \int_{-\pi}^{\pi} f(x) \sin nx \, dx = \frac{1}{\pi} \int_{-\pi}^{\pi} [g(x) + sF(x)] \sin nx \, dx$$
$$= \frac{1}{\pi} \int_{-\pi}^{\pi} g(x) \sin nx \, dx + \frac{s}{\pi} \int_{-\pi}^{\pi} F(x) \sin nx \, dx = B_n + \frac{s}{n\pi}$$

and similarly $a_n = A_n$. Therefore the Fourier series of $f(x)$ converges to $f(x)$ for all x. At $x = 0$, the series converges to $f(0)$, which was defined to be the average of left and right limits at $x = 0$. Since the series for $g(x)$ is uniformly convergent for all x, while the series for $F(x)$ converges uniformly in each closed interval not containing $x = 0$ (or $x = 2k\pi$), the Fourier series of $f(x)$ converges uniformly in each such closed interval.

The theorem has now been proved for the case of just one jump discontinuity. If there are several jumps, at points $x_1, x_2, \ldots$, we simply remove them by subtracting from $f(x)$ the function

$$s_1F(x - x_1) + s_2F(x - x_2) + \cdots.$$

The resulting function $g(x)$ is again continuous and piecewise very smooth, so that the same conclusion holds. Thus the theorem is established in all generality.

Remarks. The proof just given uses the *principle of superposition:* the Fourier series of a linear combination of two functions is the same linear combination of the corresponding two series. This can be put to very good use to systematize the computation of Fourier series. Illustrations are given in Prob. 1 below.

The idea of subtracting off the series corresponding to a jump discontinuity has also a practical significance. If a function $f(x)$ is defined by its Fourier series, and is not otherwise explicitly known, one can, of course, use the series to tabulate the function. If $f(x)$ has a jump discontinuity, the convergence will be poor near the discontinuity; this will reveal itself in the presence of terms having coefficients approaching 0 like $1/n$. If the discontinuity x_1 and jump s_1 are known, as is often the case, one can subtract the corresponding function $s_1F(x - x_1)$ as above; the new series will converge much more rapidly.

The same idea can be applied to functions $f(x)$ which are continuous but for which $f'(x)$ has a jump discontinuity s_1 at x_1. One now subtracts from f the function $s_1G(x - x_1)$, for this continuous function has as derivative precisely the function $s_1F(x - x_1)$, with jump s_1 at x_1. By integrating $G(x) - 2\pi/3$ one obtains a periodic continuous function having a jump in its second derivative at $x = 0$. Continuing in this way, one provides a jump function for each derivative; each such function can be used to remove corresponding slowly converging terms from the Fourier series.

Problems

1. Let $f_1(x)$ and $f_2(x)$ be defined by the equations:

$$f_1(x) = 0, \quad -\pi \leqq x < 0; \quad f_1(x) = 1, \quad 0 \leqq x \leqq \pi;$$
$$f_2(x) = 0, \quad -\pi \leqq x < 0; \quad f_2(x) = x, \quad 0 \leqq x < \pi.$$

Then $f_1(x)$ and $f_2(x)$ can be represented by Fourier series:

$$f_1(x) = \frac{1}{2} + \frac{2}{\pi}\sum_{n=1}^{\infty} \frac{\sin (2n-1)x}{2n-1}, \quad 0 < |x| < \pi;$$

$$f_2(x) = \frac{\pi}{4} - \frac{2}{\pi}\sum_{n=1}^{\infty} \frac{\cos (2n-1)x}{(2n-1)^2} - \sum_{n=1}^{\infty} (-1)^n \frac{\sin nx}{n}, \quad -\pi < x < \pi.$$

Without further integration, find the Fourier series for the following functions:

(a) $f_3(x) = 1, \quad -\pi \leqq x < 0; \quad f_3(x) = 0, \quad 0 \leqq x \leqq \pi;$
(b) $f_4(x) = x, \quad -\pi \leqq x \leqq 0; \quad f_4(x) = 0, \quad 0 \leqq x \leqq \pi;$
(c) $f_5(x) = 1, \quad -\pi \leqq x < 0; \quad f_5(x) = x, \quad 0 \leqq x \leqq \pi;$
(d) $f_6(x) = 2, \quad -\pi \leqq x < 0; \quad f_6(x) = 0, \quad 0 \leqq x \leqq \pi;$
(e) $f_7(x) = 2, \quad -\pi \leqq x < 0; \quad f_7(x) = 3, \quad 0 \leqq x \leqq \pi;$
(f) $f_8(x) = 1, \quad -\pi \leqq x \leqq 0; \quad f_8(x) = 1 + 2x, \quad 0 \leqq x \leqq \pi;$
(g) $f_9(x) = a + bx, \quad -\pi \leqq x < 0; \quad f_9(x) = c + dx, \quad 0 \leqq x \leqq \pi.$

2. Let $f(x) = \dfrac{a_0}{2} + \sum_{n=1}^{\infty} (a_n \cos nx + b_n \sin nx), \; -\pi \leqq x \leqq \pi.$ By squaring this

series and integrating (assuming the operations are permitted) show that

$$\frac{1}{\pi}\int_{-\pi}^{\pi}[f(x)]^2\,dx = \frac{a_0^2}{2} + \sum_{n=1}^{\infty}(a_n^2 + b_n^2).$$

[This relation, known as *Parseval's equation*, can be justified for the most general function $f(x)$ considered above. See Sections 7–11 and 7–12.]

3. Use the result of Prob. 2 and the Fourier series found in previous problems to establish the formulas:

(a) $\dfrac{\pi^2}{8} = \dfrac{1}{1^2} + \dfrac{1}{3^2} + \cdots + \dfrac{1}{(2n-1)^2} + \cdots,$

(b) $\dfrac{\pi^2}{6} = \dfrac{1}{1^2} + \dfrac{1}{2^2} + \cdots + \dfrac{1}{n^2} + \cdots,$

(c) $\dfrac{\pi^4}{90} = \dfrac{1}{1^4} + \dfrac{1}{2^4} + \cdots + \dfrac{1}{n^4} + \cdots,$

(d) $\dfrac{\pi^6}{945} = \dfrac{1}{1^6} + \dfrac{1}{2^6} + \cdots + \dfrac{1}{n^6} + \cdots.$

4. Prove that the series

$$\sum_{n=1}^{\infty}\frac{\sin nx}{n} = \sin x + \frac{\sin 2x}{2}\cdots + \frac{\sin nx}{n} + \cdots$$

converges uniformly in each interval $-\pi \leqq x \leqq a$, $a \leqq x \leqq \pi$, provided $a > 0$. This can be established by the following procedure:

(a) Let $p_n(x) = \sin x + \cdots + \sin nx$. Prove the identity:

$$p_n(x) = \frac{\cos\frac{1}{2}x - \cos\left(n + \frac{1}{2}\right)x}{2\sin\frac{1}{2}x},$$

($x \neq 0$, $x \neq \pm 2\pi, \ldots$). [Hint: multiply $p_n(x)$ by $\sin\frac{1}{2}x$ and apply the identity (0–81) for $\sin x \sin y$ to each term of the result.]

(b) Show that, if $a > 0$ and $a \leqq |x| \leqq \pi$, then $|p_n(x)| \leqq 1/\sin\frac{1}{2}a$.

(c) Show that the nth partial sum $S_n(x)$ of the series

$$\sin x + \frac{\sin 2x}{2} + \cdots + \frac{\sin nx}{n} + \cdots$$

can be written as follows:

$$S_n(x) = \frac{p_1(x)}{1\cdot 2} + \frac{p_2(x)}{2\cdot 3} + \cdots + \frac{p_{n-1}(x)}{n(n-1)} + \frac{p_n(x)}{n}.$$

[Hint: write $\sin x = p_1$, $\sin 2x = p_2 - p_1$, etc.]

(d) Show that the series $\sum_{n=1}^{\infty}\dfrac{\sin nx}{n}$ is uniformly convergent for $|a| \leqq x \leqq \pi$, where $a > 0$. [Hint: by (c),

$$S_n(x) = S_n^*(x) + \frac{p_n(x)}{n}.$$

Hence uniform convergence of the sequences S_n^* and p_n/n implies uniform convergence of $S_n(x)$. The sequence S_n converges uniformly, since it is the nth partial sum of the series

$$\sum_{n=1}^{\infty} \frac{p_n(x)}{n(n+1)},$$

which converges uniformly, because of (b), by the M-test. The sequence p_n/n converges uniformly to 0 by (b).]

7–10 Orthogonal functions. If one reviews the theory of expansion of a function in a Fourier series, as presented in the preceding sections of this chapter, it is natural to ask why the trigonometric functions $\sin nx$ and $\cos nx$ play such a special part and whether these functions can be replaced by other functions. If one is interested only in periodic functions, there is indeed no natural alternative. However, if one is concerned with representation of a function over a given interval, it will be seen that a great variety of other series are available, in particular, series of Legendre polynomials, Bessel functions, Laguerre polynomials, Jacobi polynomials, Hermite polynomials, and general Sturm-Liouville series.

Let $f(x)$ be given in an interval $a \leqq x \leqq b$; this interval will be fixed throughout the following discussion. Let $\phi_1(x), \phi_2(x), \ldots, \phi_n(x), \ldots$ be functions all piecewise continuous in this interval; this sequence is to replace the system of sines and cosines. We then postulate a development:

$$f(x) = \sum_{n=1}^{\infty} c_n \phi_n(x), \tag{7–32}$$

just as in the case of Fourier series. Our next step, for Fourier series, was to multiply both sides by $\cos mx$ or $\sin mx$ and integrate from $-\pi$ to π; when we did this, all terms dropped out except the one in a_m or b_m respectively, because of the relations

$$\int_{-\pi}^{\pi} \cos mx \cos nx \, dx = 0, \quad m \neq n, \ldots .$$

By analogy, we multiply both sides of (7–32) by $\phi_m(x)$ and integrate term by term:

$$\int_a^b f(x)\phi_m(x)\, dx = \sum_{n=1}^{\infty} c_n \int_a^b \phi_m(x)\phi_n(x)\, dx. \tag{7–33}$$

In order to achieve a result analogous to that for Fourier series, we must postulate that

$$\int_a^b \phi_m(x)\phi_n(x)\, dx = 0, \quad m \neq n. \tag{7–34}$$

The series on the right of (7–33) then reduces to one term:

$$\int_a^b f(x)\phi_m(x)\, dx = c_m \int_a^b [\phi_m(x)]^2\, dx. \tag{7–35}$$

The integral on the right is a certain constant:

$$\int_a^b [\phi_m(x)]^2\, dx = B_m. \tag{7–36}$$

The constant B_m will be positive unless $\phi_m(x) \equiv 0$ (except at a finite number of points); to avoid this trivial case, we assume that no B_m is 0. Then

$$c_m = \frac{1}{B_m}\int_a^b f(x)\phi_m(x)\,dx. \tag{7–37}$$

Thus, under the simple conditions (7–34) and (7–36), we have a rule for formation of a series just like that for Fourier series and can hope that analogous convergence theorems can also be proved.

We now summarize the assumptions in formal definitions:

Definitions. Two functions $p(x)$, $q(x)$, which are piecewise continuous for $a \leqq x \leqq b$, are *orthogonal* in this interval if

$$\int_a^b p(x)q(x)\,dx = 0. \tag{7–38}$$

A system of functions $\{\phi_n(x)\}$ $(n = 1, 2, \ldots)$ is termed an *orthogonal system* in the interval $a \leqq x \leqq b$ if ϕ_n and ϕ_m are orthogonal for each pair of distinct indices m, n:

$$\int_a^b \phi_m(x)\phi_n(x)\,dx = 0 \quad (m \neq n), \tag{7–39}$$

and no $\phi_n(x)$ is identically 0 except at a finite number of points.

EXAMPLE. The trigonometric system in the interval $-\pi \leqq x \leqq \pi$:

$$1, \quad \cos x, \quad \sin x, \ldots, \quad \cos nx, \quad \sin nx, \ldots.$$

The function ϕ_1 is the constant 1, ϕ_2 is the function $\cos x$, ...

If $f(x)$ is piecewise continuous in the interval $a \leqq x \leqq b$ and $\{\phi_n(x)\}$ is an orthogonal system in this interval, then the series

$$\sum_{n=1}^{\infty} c_n\phi_n(x), \tag{7–40}$$

where

$$c_n = \frac{1}{B_n}\int_a^b f(x)\phi_n(x)\,dx, \quad B_n = \int_a^b [\phi_n(x)]^2\,dx, \tag{7–41}$$

is called the *Fourier series of f with respect to the system* $\{\phi_n(x)\}$. The numbers $c_1, c_2, \ldots$ are called the Fourier coefficients of $f(x)$ with respect to the system $\{\phi_n(x)\}$.

The preceding formulas can be simplified if one assumes that the constant B_n is always 1, i.e., that

$$\int_a^b [\phi_n(x)]^2 = 1 \quad (n = 1, 2, \ldots).$$

This can always be achieved by dividing the original $\phi_n(x)$ by appropriate constants. When the condition: $B_n = 1$ is satisfied for all n, the system of functions $\phi_n(x)$ is called *normalized*. A system which is both normalized and orthogonal is called *orthonormal*. This is illustrated by the functions:

$$\frac{1}{\sqrt{2\pi}}, \quad \frac{\cos x}{\sqrt{\pi}}, \quad \frac{\sin x}{\sqrt{\pi}}, \ldots, \quad \frac{\cos nx}{\sqrt{\pi}}, \quad \frac{\sin nx}{\sqrt{\pi}}, \ldots.$$

While the general theory is simpler for normalized systems, the advantages for applications are slight and we shall not use normalization in what follows.

The operations with orthogonal systems are strikingly similar to those with vectors. In fact, we can consider the piecewise continuous functions for $a \leqq x \leqq b$ as a sort of *vector space*, as in Section 1–15. The sum or difference of two such functions $f(x)$, $g(x)$ is again piecewise continuous, as is the product cf of $f(x)$ by a constant or *scalar* c. Equation (7–38) suggests a definition of *inner product* (or scalar product):

$$(f, g) = \int_a^b f(x)g(x)\,dx. \tag{7–42}$$

One can then define a *norm* (or absolute value):

$$\|f\| = \sqrt{(f, f)} = \left\{\int_a^b [f(x)]^2\,dx\right\}^{\frac{1}{2}}. \tag{7–43}$$

The zero function 0^* is a function which is 0 except at a finite number of points; in general, in this vector theory of functions, we consider two functions which differ only at a finite number of points to be the same function. It is now a straightforward exercise to verify that all the axioms for a Euclidean vector space, except the one concerning dimension, are satisfied. For convenience, we restate the axioms here [cf. (1–44) in Section 1–8]:

$$\begin{array}{ll}
\text{I. } f + g = g + f & \text{II. } (f + g) + h = f + (g + h) \\
\text{III. } c(f + g) = cf + cg & \text{IV. } (c_1 + c_2)f = c_1 f + c_2 f \\
\text{V. } (c_1 c_2)f = c_1(c_2 f) & \text{VI. } 1 \cdot f = f \\
\text{VII. } 0 \cdot f = 0^* & \text{VIII. } (f, g) = (g, f) \\
\text{IX. } (f + g, h) = (f, h) + (g, h) & \text{X. } (cf, g) = c(f, g) \\
\text{XI. } (f, f) \geqq 0 & \text{XII. } (f, f) = 0 \text{ if and only if } f = 0^*.
\end{array} \tag{7–44}$$

The proof is left as an exercise (Prob. 3 below).

We define k functions $f_1, f_2, \ldots, f_k$ to be *linearly independent* if the only scalars $c_1, \ldots, c_n$ for which

$$c_1 f_1 + \cdots + c_k f_k = 0^* \tag{7–45}$$

are the numbers $0, \ldots, 0$. In place of rule XIII of Section 1–9, we have now the theorem that, for every integer n, there exist n linearly independent functions. For example, the functions $\sin x$, $\sin 2x, \ldots, \sin nx$ are linearly independent for $-\pi \leqq x \leqq \pi$.

The theorem of Section 1–9 is proved as a consequence of the axioms (7–44) alone. Hence this theorem holds for functions:

THEOREM 7. *Let $f(x)$ and $g(x)$ be piecewise continuous for $a \leqq x \leqq b$. Then*

$$|(f, g)| \leqq \|f\| \cdot \|g\|. \tag{7–46}$$

The equality holds precisely when f and g are linearly dependent. Furthermore,

$$\|f + g\| \leqq \|f\| + \|g\|; \tag{7–47}$$

the equal sign holds precisely when $f = cg$ or $g = cf$, where c is a positive constant or 0.

The relation (7–46) is the *Schwarz inequality*. If we use the definitions (7–42), (7–43), it can be written as follows:

$$\left[\int_a^b f(x)g(x)\,dx\right]^2 \leqq \int_a^b [f(x)]^2\,dx \cdot \int_a^b [g(x)]^2\,dx. \tag{7–48}$$

The relation (7–47) is the *Minkowski inequality*. In explicit form, it reads:

$$\left\{\int_a^b [f(x) + g(x)]^2\,dx\right\}^{\frac{1}{2}} \leqq \left\{\int_a^b [f(x)]^2\,dx\right\}^{\frac{1}{2}} + \left\{\int_a^b [g(x)]^2\,dx\right\}^{\frac{1}{2}}. \tag{7–49}$$

***7–11. Fourier series of orthogonal functions. Completeness.** The definition of orthogonal functions can be restated in terms of inner products: $f(x)$ is orthogonal to $g(x)$ if $(f, g) = 0$. An orthogonal system is a system $\{\phi_n(x)\}$ $(n = 1, 2, \ldots)$ of functions all piecewise continuous for $a \leqq x \leqq b$ and such that

$$(\phi_m, \phi_n) = 0, \quad m \neq n; \quad (\phi_n, \phi_n) = \|\phi_n\|^2 = B_n > 0. \tag{7–50}$$

The system is *orthonormal* if (7–50) holds and $\|\phi_n\| = 1$ for $n = 1, 2, \ldots,$ so that the ϕ_n are "unit vectors."

The Fourier series of $f(x)$ with respect to the orthogonal system $\{\phi_n(x)\}$ is the series

$$\sum_{n=1}^{\infty} c_n\phi_n(x), \quad c_n = (f, \phi_n)/\|\phi_n\|^2. \tag{7–51}$$

If the system is orthonormal, the series becomes

$$\sum_{n=1}^{\infty} c_n\phi_n(x), \quad c_n = (f, \phi_n), \tag{7–52}$$

which is analogous to the expressions

$$\mathbf{v} = v_x\mathbf{i} + v_y\mathbf{j} + v_z\mathbf{k}, \qquad v_x = \mathbf{v}\cdot\mathbf{i}, \ldots,$$
$$\mathbf{v} = v_1\mathbf{e}_1 + \cdots + v_n\mathbf{e}_n, \quad v_j = \mathbf{v}\cdot\mathbf{e}_j$$

for a vector in terms of base vectors in 3-dimensional space and n-dimensional space. However, since (7–51) is an infinite series, complications arise through convergence questions.

THEOREM 8. *Let $\{\phi_n(x)\}$ be an orthogonal system of continuous functions for the interval $a \leqq x \leqq b$. If the series $\sum_{n=1}^{\infty} c_n\phi_n(x)$ converges uniformly to $f(x)$ for $a \leqq x \leqq b$, then*

$$c_n = (f, \phi_n)/\|\phi_n\|^2, \tag{7-53}$$

so that the series is the Fourier series of $f(x)$ with respect to $\{\phi_n(x)\}$. If the system is orthonormal, then $c_n = (f, \phi_n)$.

Proof. Just as in the proof of Theorem 1 in Section 7–2, we reason that $f(x)$ is continuous. Then

$$(f, \phi_m) = (c_1\phi_1 + \cdots + c_n\phi_n + \cdots, \phi_m) = (c_1\phi_1, \phi_m) + \cdots + (c_n\phi_n, \phi_m) + \cdots$$
$$= c_1(\phi_1, \phi_m) + \cdots + c_n(\phi_n, \phi_m) + \cdots = c_m(\phi_m, \phi_m) = c_m\|\phi_m\|^2,$$

so that (7–53) holds. The operations on the series are justified by the uniform convergence. If the system is orthonormal, then $\|\phi_n\| = 1$, so that $c_n = (f, \phi_n)$.

Corollary. *If, under the hypotheses of Theorem 8,*

$$\sum_{n=1}^{\infty} c_n\phi_n(x) \equiv \sum_{n=1}^{\infty} c_n'\phi_n(x), \quad a \leqq x \leqq b,$$

and both series converge uniformly over the interval, then $c_n = c_n'$ for $n = 1, 2, \ldots$.

Theorem 9. *Let $\{\phi_n(x)\}$ be an orthogonal system for the interval $a \leqq x \leqq b$ and let $f(x)$ be piecewise continuous for $a \leqq x \leqq b$. For each n, the coefficients $c_1, \ldots, c_n$ of the Fourier series of f with respect to $\{\phi_n(x)\}$ are those constants which give the square error $\|f - g\|^2$ its smallest value, when g ranges over all linear combinations $p_1\phi_1(x) + \cdots + p_n\phi_n(x)$. The minimum value of the error is*

$$E_n = \|f - (c_1\phi_1 + \cdots + c_n\phi_n)\|^2 = \|f\|^2 - c_1^2\|\phi_1\|^2 - \cdots - c_n^2\|\phi_n\|^2. \tag{7-54}$$

Corollary. *Under the hypotheses of Theorem 9, one has*

$$c_1^2\|\phi_1\|^2 + \cdots + c_n^2\|\phi_n\|^2 \leqq \|f\|^2, \tag{7-55}$$

so that the series $\Sigma c_k^2\|\phi_k\|^2$ converges. Furthermore,

$$\lim_{k\to\infty} c_k\|\phi_k\| = 0. \tag{7-56}$$

The proofs are left as exercises (Probs. 5, 6, 7 below); (7–55) is *Bessel's inequality.*

A crucial question is whether one can assert that E_n converges to 0, as $n \to \infty$. This is equivalent to asking whether $f(x)$ can be approximated, in the sense of least square error, as closely as desired by a linear combination of a finite number of the functions $\phi_n(x)$. If this is the case, the system $\{\phi_n(x)\}$ is termed complete:

Definition. An orthogonal system $\{\phi_n(x)\}$ for the interval $a \leqq x \leqq b$ is termed *complete* if, for every piecewise continuous function $f(x)$ in the interval $a \leqq x \leqq b$, the minimum square error $E_n = \|f - (c_1\phi_1 + \cdots + c_n\phi_n)\|^2$ converges to zero as n becomes infinite.

If the system is complete, then Theorem 9 shows that

$$\|f\|^2 = c_1^2\|\phi_1\|^2 + \cdots + c_n^2\|\phi_n\|^2 + \cdots. \tag{7–57}$$

This is *Parseval's equation.* Conversely, if Parseval's equation holds for every piecewise continuous f, then E_n must converge to 0 and the system $\{\phi_n(x)\}$ is complete. Therefore *the validity of Parseval's equation is equivalent to completeness.* If the system $\{\phi_n(x)\}$ is orthonormal, Parseval's equation becomes

$$\|f\|^2 = c_1^2 + \cdots + c_n^2 + \cdots.$$

This is analogous to the vector relations

$$|\mathbf{v}|^2 = v_x^2 + v_y^2 + v_z^2, \quad |\mathbf{v}|^2 = v_1^2 + \cdots + v_n^2.$$

If the orthogonal system $\{\phi_n(x)\}$ is complete, then the square error E_n converges to 0. This does not imply convergence of the Fourier series $\Sigma c_n\phi_n(x)$ to $f(x)$, although the successive partial sums do approach $f(x)$ in the sense of square error. We describe the situation by saying that the series *converges in the mean to* $f(x)$ and we write

$$\underset{n\to\infty}{\text{L.i.m.}}\, [c_1\phi_1(x) + \cdots + c_n\phi_n(x)] = f(x),$$

where "L.i.m." stands for "limit in the mean." In general, if functions $f(x)$ and $f_n(x)$ ($n = 1, 2, \ldots$) are all piecewise continuous for $a \leqq x \leqq b$, we write

$$\underset{n\to\infty}{\text{L.i.m.}}\, f_n(x) = f(x) \tag{7–58}$$

if the sequence $\|f_n(x) - f(x)\|$ converges to 0, that is, if

$$\lim_{n\to\infty} \int_a^b [f(x) - f_n(x)]^2\, dx = 0.$$

Remark. The following assertions can be proved: (a) if $f_n(x)$ converges uniformly to $f(x)$ for $a \leqq x \leqq b$, then $f_n(x)$ also converges in the mean to $f(x)$, provided all functions are piecewise continuous; (b) if $f_n(x)$ converges in the mean to $f(x)$, then $f_n(x)$ need not converge uniformly to $f(x)$; in fact, the sequence $f_n(x)$ need not converge for $a \leqq x \leqq b$; (c) if $f_n(x)$ converges to $f(x)$ for $a \leqq x \leqq b$, but not uniformly, then $f_n(x)$ need not converge in the mean to $f(x)$. For proofs, one is referred to Prob. 9 below.

Definition. An orthogonal system $\{\phi_n(x)\}$ for the interval $a \leqq x \leqq b$ has the *uniqueness property,* if every piecewise continuous function $f(x)$ for $a \leqq x \leqq b$ is uniquely determined by its Fourier coefficients with respect to $\{\phi_n(x)\}$; that is, if $f(x)$ and $g(x)$ are piecewise continuous for $a \leqq x \leqq b$ and $(f, \phi_n) = (g, \phi_n)$ for all n, then $f(x) - g(x) = 0^*$. This is equivalent to the statement that $h(x) = 0^*$ is the only piecewise continuous function orthogonal to all the functions $\phi_n(x)$; the system of orthogonal functions can therefore not be enlarged.

Theorem 10. *Let $\{\phi_n(x)\}$ be a complete system of orthogonal functions for the interval $a \leqq x \leqq b$. Then $\{\phi_n(x)\}$ has the uniqueness property.*

The proof is left as an exercise (Prob. 8 below). One might expect the converse to hold; i.e., that the uniqueness property implies completeness. However, examples can be given to show that this is not the case. Theorem 12 below gives more information on this point.

THEOREM 11. *Let* $\{\phi_n(x)\}$ *be an orthogonal system of continuous functions for the interval* $a \leqq x \leqq b$ *and let* $\{\phi_n(x)\}$ *have the uniqueness property. Let* $f(x)$ *be continuous for* $a \leqq x \leqq b$ *and let the Fourier series of* $f(x)$ *with respect to* $\{\phi_n(x)\}$ *converge uniformly for* $a \leqq x \leqq b$. *Then the Fourier series converges to* $f(x)$.

The proof is left as an exercise (Prob. 10 below).

*7–12 Sufficient conditions for completeness.

THEOREM 12. *Let* $\{\phi_n(x)\}$ *be an orthogonal system of continuous functions for the interval* $a \leqq x \leqq b$. *Let the following two properties hold: (a)* $\{\phi_n(x)\}$ *has the uniqueness property; (b) for some positive integer k, the Fourier series of* $g(x)$ *with respect to* $\{\phi_n(x)\}$ *is uniformly convergent for every* $g(x)$ *having continuous derivatives through the k-th order for* $a \leqq x \leqq b$ *and such that* $g(a) = g'(a) = \cdots = g^{(k)}(a) = 0$, $g(b) = g'(b) = \cdots = g^{(k)}(b) = 0$. *Then the system* $\{\phi_n(x)\}$ *is complete.*

Proof. Let $f(x)$ be piecewise continuous for $a \leqq x \leqq b$. We must then show that, given $\epsilon > 0$, a linear combination $c_1\phi_1(x) + \cdots + c_n\phi_n(x) = \psi(x)$ can be found such that $\|f - \psi\| < \epsilon$.

For simplicity, we assume the integer k of assumption (b) to be 2, as this case is typical of the general case.

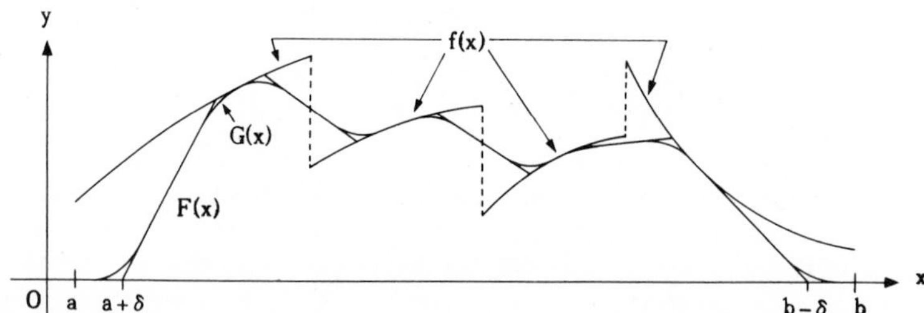

FIG. 7–14. Approximation of the piecewise continuous function $f(x)$ by a continuous function $F(x)$ and by a smooth function $G(x)$.

The construction of the function $\psi(x)$ proceeds in several stages. We first determine a continuous function $F(x)$ such that $\|F - f\| < \frac{1}{4}\epsilon$ and $F(x) \equiv 0$ when $a \leqq x \leqq a + \delta$, $b - \delta \leqq x \leqq b$ for a proper choice of $\delta > 0$. This is suggested graphically in Fig. 7–14. We denote by $f_1(x)$ the piecewise continuous function which coincides with $f(x)$ except for $a \leqq x \leqq a + 2\delta$ and for $b - 2\delta \leqq x \leqq b$, where $f_1(x)$ is identically zero. We now bridge the jumps in $f_1(x)$ by line segments. At each jump, the line joins $[x_0 - \delta, f_1(x_0 - \delta)]$ to $[x_0 + \delta, f_1(x_0 + \delta)]$. The function $F(x)$ coincides with $f_1(x)$ except between $x_0 - \delta$ and $x_0 + \delta$, where it follows the straight line. Accordingly, $F(x)$ is a continuous function and $\|F - f\|^2$ is a sum of a finite number K of integrals of form

$$\int_{x_0-\delta}^{x_0+\delta} [F(x) - f(x)]^2\, dx$$

plus two integrals of $[f(x)]^2$ from a to $a + \delta$ and from $b - \delta$ to b, where $F \equiv 0$. Since $f(x)$ is piecewise continuous, $|f(x)| \leqq M$ for some constant M. By its construction, $|F(x)| \leqq M$ also. Hence $|F(x) - f(x)| \leqq 2M$ for every x and

$$\int_{x_0-\delta}^{x_0+\delta} [F(x) - f(x)]^2\, dx \leqq 4M^2 \cdot 2\delta.$$

Adding the expressions for all K jumps and for the two ends, we find

$$\|F - f\|^2 \leqq K \cdot 4M^2 \cdot 2\delta + 2M^2 \cdot \delta.$$

As $\delta \to 0$, the expression on the right approaches 0. Accordingly, for δ sufficiently small, $\|F - f\| < \frac{1}{4}\epsilon$.

We next choose a function $G(x)$ having a continuous first derivative for $a \leqq x \leqq b$, such that $\|G - F\| < \frac{1}{4}\epsilon$ and $G(x) \equiv 0$ for $a \leqq x \leqq a + \frac{1}{2}\delta$ and for $b - \frac{1}{2}\delta \leqq x \leqq b$. We define $F(x)$ to be identically 0 outside the interval $a \leqq x \leqq b$ and then set

$$G(x_1) = \frac{1}{h}\int_{x_1-h}^{x_1+h} F(x)\, dx, \quad 0 < h < \tfrac{1}{2}\delta.$$

The constant h will be specified below. The function $G(x)$ is then defined for all x and $G(x) \equiv 0$ for $x \leqq a + \frac{1}{2}\delta$, and for $x \geqq b - \frac{1}{2}\delta$, as required. By the Fundamental Theorem of Calculus [Sections 4–3 and 4–12],

$$G'(x_1) = \frac{1}{h}[F(x_1 + h) - F(x_1 - h)].$$

Accordingly, $G(x)$ has a continuous derivative for all x. Now

$$G(x_1) - F(x_1) = \frac{1}{2h}\int_{x_1-h}^{x_1+h} F(x)\, dx - F(x_1) = \frac{1}{2h}\int_{x_1-h}^{x_1+h} [F(x) - F(x_1)]\, dx$$

and therefore

$$[G(x_1) - F(x_1)]^2 = \frac{1}{4h^2}\left\{\int_{x_1-h}^{x_1+h} [F(x) - F(x_1)]\, dx\right\}^2.$$

We apply the Schwarz inequality (7–48) to the integral on the right, with $f(x)$ replaced by $F(x) - F(x_1)$ and $g(x)$ by 1. Accordingly,

$$[G(x_1) - F(x_1)]^2 \leqq \frac{2h}{4h^2}\int_{x_1-h}^{x_1+h} [F(x) - F(x_1)]^2\, dx,$$

and

$$\|G - F\|^2 = \int_a^b [G(x_1) - F(x_1)]^2\, dx_1 \leqq \frac{1}{2h}\int_a^b \int_{x_1-h}^{x_1+h} [F(x) - F(x_1)]^2\, dx\, dx_1.$$

If we set $x_1 = u$, $x = u - v$ in the double integral on the right, it becomes (cf. Section 4–8)

$$\frac{1}{2h}\int_{-h}^{h}\int_{a}^{b}[F(u-v) - F(u)]^2\,du\,dv,$$

since the region of integration in the uv plane is the rectangle: $a \leqq u \leqq b$, $-h \leqq v \leqq h$. If we write $H(v)$ for the inner integral, then $H(v)$ is a continuous function of v (Section 4–6) and $H(0) = 0$. Now

$$\|G - F\|^2 \leqq \frac{1}{2h}\int_{-h}^{h} H(v)\,dv = H(v^*)\cdot\frac{2h}{2h} = H(v^*),$$

where $-h < v^* < h$, by the law of the mean. As $h \to 0$, $H(v^*)$ must approach 0; hence $\|G - F\| < \frac{1}{4}\epsilon$ for h sufficiently small.

Next we construct a function $g(x)$ such that $\|g - G\| < \frac{1}{4}\epsilon$, while g has continuous first and second derivatives and $g(x) \equiv 0$ for $a \leqq x \leqq a + \frac{1}{4}\delta$ and for $b - \frac{1}{4}\delta \leqq x \leqq b$. We need only repeat the averaging process of the preceding paragraph:

$$g(x_1) = \frac{1}{2p}\int_{x_1-p}^{x_1+p} G(x)\,dx, \quad 0 < p < \tfrac{1}{4}\delta.$$

The other steps can be repeated and the desired inequality is obtained by appropriate choice of p. Since

$$g'(x_1) = \frac{1}{2p}[G(x_1 + p) - G(x_1 - p)]$$

and G has a continuous derivative, $g(x)$ has continuous first and second derivatives for all x.

Finally, we construct a linear combination $\psi(x) = c_1\phi_1(x) + \cdots + c_n\phi_n(x)$ such that $\|g - \psi\| < \frac{1}{4}\epsilon$. Since $g(x)$ satisfies all the conditions of assumption (b) in the theorem, the Fourier series of $g(x)$ is uniformly convergent. By (a) and Theorem 11, the series converges to $g(x)$. By the remark preceding Theorem 10, the partial sums of this series also converge in the mean to $g(x)$. Accordingly, a partial sum $S_n(x) = \psi(x)$ can be chosen such that $\|g - \psi\| < \frac{1}{4}\epsilon$.

The function $\psi(x)$ is precisely the linear combination sought, for by (7–47)

$$\begin{aligned}\|f - \psi\| &= \|f - F + F - G + G - g + g - \psi\|\\ &\leqq \|f - F\| + \|F - G\| + \|G - g\| + \|g - \psi\|\\ &< \tfrac{1}{4}\epsilon + \tfrac{1}{4}\epsilon + \tfrac{1}{4}\epsilon + \tfrac{1}{4}\epsilon = \epsilon.\end{aligned}$$

Remark. The above proof shows that the assumption (a) can be omitted, if one replaces "uniformly convergent" in (b) by "convergent in the mean to $g(x)$."

THEOREM 13. *The trigonometric system:* 1, $\cos x$, $\sin x, \ldots, \cos nx$, $\sin nx, \ldots$ *is complete in the interval* $-\pi \leqq x \leqq \pi$.

Proof. The hypotheses (a) and (b) of Theorem 12 are satisfied by virtue of Theorems 3 and 5 above. Accordingly, the system is complete.

*7–13 Integration and differentiation of Fourier series.

THEOREM 14. *An orthogonal system $\{\phi_n(x)\}$, for the interval $a \leqq x \leqq b$, is complete if and only if, for every two functions $f(x)$ and $g(x)$ which are piecewise continuous for $a \leqq x \leqq b$, one has*

$$(f, g) = \int_a^b f(x)g(x)\,dx = \sum_{n=1}^{\infty} c_n c'_n \|\phi_n\|^2, \tag{7–59}$$

where c_n, c'_n are the Fourier coefficients of $f(x)$ and $g(x)$, respectively, with respect to $\{\phi_n(x)\}$.

Remark. Equation (7–59) is termed the *second form of Parseval's equation.* When the system is orthonormal, it reduces to the equation

$$(f, g) = c_1c'_1 + \cdots + c_nc'_n + \cdots, \quad c_n = (f, \phi_n),\; c'_n = (g, \phi_n),$$

which is analogous to the equations

$$\mathbf{u}\cdot\mathbf{v} = u_xv_x + u_yv_y + u_zv_z, \quad \mathbf{u}\cdot\mathbf{v} = u_1v_1 + \cdots + u_nv_n$$

for vectors.

Proof of Theorem 14. If (7–59) holds, then, by taking $g = f$, one concludes that Parseval's equation (7–57) holds; hence the system is complete. Conversely, let $\{\phi_n(x)\}$ be complete. Then the algebraic identity: $AB = \frac{1}{4}\{(A + B)^2 - (A - B)^2\}$ gives the equation

$$(f, g) = \frac{1}{4}\int_a^b (f + g)^2\,dx - \frac{1}{4}\int_a^b (f - g)^2\,dx.$$

Parseval's equation for $f + g$ and $f - g$ now gives

$$(f, g) = \frac{1}{4}\sum_{n=1}^{\infty} (c_n + c'_n)^2\|\phi_n\|^2 - \frac{1}{4}\sum_{n=1}^{\infty} (c_n - c'_n)^2\|\phi_n\|^2.$$

On adding the two series, we obtain (7–59).

THEOREM 15. *Let $\{\phi_n(x)\}$ be a complete orthonormal system for the interval $a \leqq x \leqq b$. Let $f(x)$ be piecewise continuous for $a \leqq x \leqq b$ and let $g(x)$ be piecewise continuous for $x_1 \leqq x \leqq x_2$, where $a \leqq x_1 < x_2 \leqq b$. Let $\Sigma c_n\phi_n(x)$ be the Fourier series of $f(x)$ with respect to $\{\phi_n(x)\}$. Then*

$$\int_{x_1}^{x_2} f(x)g(x)\,dx = \sum_{n=1}^{\infty} c_n \int_{x_1}^{x_2} g(x)\phi_n(x)\,dx. \tag{7–60}$$

Remark. The theorem asserts that the integral on the left can be computed by term-by-term integration of the series $\Sigma c_n g(x)\phi_n(x)$. This is remarkable, since there is no assumption of convergence — let alone uniform convergence — of the series before integration.

Proof of Theorem 15. We extend the definition of $g(x)$ to the whole of the interval $a \leqq x \leqq b$ by setting $g(x) \equiv 0$ for $a \leqq x \leqq x_1$ and for $x_2 \leqq x \leqq b$. Then Parseval's equation (7–59) gives

$$\int_a^b f(x)g(x)\,dx = \int_{x_1}^{x_2} f(x)g(x)\,dx = \sum_{n=1}^{\infty} c_n c_n' \|\phi_n\|^2,$$

$$\|\phi_n\|^2 c_n' = (g, \phi_n) = \int_a^b g(x)\phi_n(x)\,dx = \int_{x_1}^{x_2} g(x)\phi_n(x)\,dx,$$

so that Eq. (7–60) is valid. If we choose $g(x) \equiv 1$ for $x_1 \leqq x \leqq x_2$, we obtain the following result:

COROLLARY. *Under the hypotheses of Theorem* 15,

$$\int_{x_1}^{x_2} f(x)\,dx = \sum_{n=1}^{\infty} c_n \int_{x_1}^{x_2} \phi_n(x)\,dx; \tag{7–61}$$

that is, term-by-term integration is permissible for every Fourier series with respect to a complete orthogonal system $\{\phi_n(x)\}$.

While term-by-term integration causes no difficulties, term-by-term differentiation calls for great caution. For example, the series: $\sin x + \cdots + (\sin nx/n) + \cdots$ converges for all x, but the differentiated series: $\cos x + \cdots + \cos nx + \cdots$ diverges for all x. Differentiation *multiplies* the nth term by n, which interferes with convergence; integration *divides* the nth term by n, which aids convergence. The safest rule to follow is that of Theorem 33 of Section 6–14: term-by-term differentiation of a convergent Fourier series is allowed, *provided the differentiated series converges uniformly* in the interval considered.

Problems

1. Let $\phi_n(x) = \sin nx$ $(n = 1, 2, \ldots)$.

(a) Show that the functions $\{\phi_n(x)\}$ form an orthogonal system in the interval $0 \leqq x \leqq \pi$.

(b) Show that the functions $\phi_n(x)$ have the uniqueness property. [Hint: let $f(x)$ be a function which is orthogonal to all the ϕ_n. Let $F(x)$ be the odd function coinciding with $f(x)$ for $0 < x \leqq \pi$. Show that all Fourier coefficients of $F(x)$ are 0.]

(c) Show that the Fourier sine series of $f(x)$ is uniformly convergent for every $f(x)$ having continuous first and second derivatives for $0 \leqq x \leqq \pi$ and such that $f(0) = f(\pi) = 0$.

(d) Show that $\{\phi_n(x)\}$ is a complete system for the interval $0 \leqq x \leqq \pi$.

2. Carry out the steps (a), (b), (c), (d) of Prob. 1 for the functions $\phi_n(x) = \cos nx$ $(n = 0, 1, 2, \ldots)$. Show that the condition: $f(0) = f(\pi) = 0$ is not needed for (c).

3. Prove the validity of (7–44) for functions $f(x)$, $g(x)$ which are piecewise continuous for $a \leqq x \leqq b$.

4. Verify the correctness of the Schwarz inequality (7–48) and the Minkowski inequality (7–49) for $f(x) = x$ and $g(x) = e^x$ in the interval $0 \leqq x \leqq 1$.

5. Prove the Corollary to Theorem 8 [see the proof of the Corollary to Theorem 1, Section 7–2].

6. Prove Theorem 9 [cf. Prob. 7 following Section 7–4].

7. Prove the Corollary to Theorem 9 [cf. the proof of the Corollary to Theorem 2, Section 7–4].

8. Prove Theorem 10. [Hint: show by Parseval's equation (7–57) that, if $(h, \phi_n) = 0$ for all n, then $\|h\| = 0$.]

9. (a) Let the functions $f_n(x)$ be continuous for $a \leqq x \leqq b$ and let the sequence $f_n(x)$ converge uniformly to $f(x)$ for $a \leqq x \leqq b$. Prove that $\underset{n\to\infty}{\text{L.i.m.}}\, f_n(x) = f(x)$.

(b) Prove that the sequence $\cos^n x$ converges to 0 in the mean for $0 \leqq x \leqq \pi$ but does not converge for $x = \pi$. Prove that the sequence converges for $0 \leqq x \leqq \frac{1}{2}\pi$ but not uniformly.

(c) Let $f_n(x) = 0$ for $0 \leqq x \leqq 1/n$, $=n$ for $1/n \leqq x \leqq 2/n$, $=0$ for $2/n \leqq x \leqq 1$. Show that the sequence converges to 0 for $0 \leqq x \leqq 1$ but not uniformly and that the sequence does not converge in the mean to 0.

10. Prove Theorem 11 [cf. the proof of Theorem 4 in Section 7–7].

*7–14 Fourier-Legendre series.

Thus far we have considered only three examples of orthogonal systems: the trigonometric system, the Fourier cosine system, the Fourier sine system. In this section we present a fourth example, that of the Legendre polynomials; in the following section, other examples are considered.

The Legendre polynomials $P_n(x)$ $(n = 0, 1, 2, \ldots)$ can be defined by the formula of Rodrigues:

$$P_0(x) = 1, \quad P_n(x) = \frac{1}{2^n n!}\frac{d^n}{dx^n}(x^2-1)^n \quad (n = 1, 2, \ldots). \tag{7–62}$$

Thus

$$P_1(x) = \frac{1}{2}\frac{d}{dx}(x^2-1) = x,$$

$$P_2(x) = \frac{1}{8}\frac{d^2}{dx^2}(x^4-2x^2+1) = \frac{3}{2}x^2 - \frac{1}{2},$$

$$P_3(x) = \frac{5}{2}x^3 - \frac{3}{2}x, \quad P_4(x) = \frac{35}{8}x^4 - \frac{15}{4}x^2 + \frac{3}{8}, \quad \ldots .$$

THEOREM 16. *$P_n(x)$ is a polynomial of degree n. $P_n(x)$ is an odd function or an even function according as n is odd or even. The following identities hold for $n = 1, 2, \ldots$:*

$$\begin{aligned}&\text{(a)}\ P_n'(x) = xP_{n-1}'(x) + nP_{n-1}(x),\\ &\text{(b)}\ P_n(x) = xP_{n-1}(x) + \frac{x^2-1}{n}P_{n-1}'(x).\end{aligned} \tag{7–63}$$

Proof. The statements concerning the degree and the odd or even character can be verified from the definition. To prove the identity (a), we let $u = x^2 - 1$. Then

$$P_n'(x) = \frac{1}{2^n n!}\frac{d^{n+1}}{dx^{n+1}}u^n = \frac{1}{2^n n!}\frac{d^n}{dx^n}(2nxu^{n-1})$$

$$= \frac{1}{2^{n-1}(n-1)!}\left(x\frac{d^n}{dx^n}u^{n-1} + n\frac{d^{n-1}}{dx^{n-1}}u^{n-1}\right) = xP_{n-1}' + nP_{n-1},$$

by the Leibnitz rule for $(u \cdot v)^{(n)}$ [Eq. (0–94), Section 0–8]. To prove identity (b), we write $P_n(x)$ in two different ways:

$$P_n = \frac{1}{2^n n!}\frac{d^n}{dx^n}(u \cdot u^{n-1})$$

$$= \frac{1}{2^n n!}\left[u\frac{d^n}{dx^n}u^{n-1} + 2xn\frac{d^{n-1}}{dx^{n-1}}u^{n-1} + n(n-1)\frac{d^{n-2}}{dx^{n-2}}u^{n-1}\right];$$

$$P_n = \frac{1}{2^n n!}\frac{d^{n-1}}{dx^{n-1}}(2nxu^{n-1})$$

$$= \frac{1}{2^{n-1}(n-1)!}\left[x\frac{d^{n-1}}{dx^{n-1}}u^{n-1} + (n-1)\frac{d^{n-2}}{dx^{n-2}}u^{n-1}\right].$$

If the second equation is subtracted from twice the first, (b) is obtained.

It will be seen that the identities (a) and (b), plus the fact that $P_0(x) = 1$, are sufficient to establish all other properties needed, without reference to the definition (7–62).

THEOREM 17. *The Legendre polynomials satisfy the following identities and relations:*

$$\left.\begin{aligned}
&\text{(c)}\quad P_{n+1}'(x) - P_{n-1}'(x) = (2n+1)P_n(x) \quad (n \geqq 1)\\
&\text{(d)}\quad \frac{d}{dx}[(1-x^2)P_n'(x)] + n(n+1)P_n(x) = 0\\
&\text{(e)}\quad P_{n+1}(x) = \frac{(2n+1)xP_n(x) - nP_{n-1}(x)}{n+1} \quad (n \geqq 1)\\
&\text{(f)}\quad P_n(1) = 1, \quad P_n(-1) = (-1)^n\\
&\text{(g)}\quad \frac{1-x^2}{n^2}P_n'^2 + P_n^2 = \frac{1-x^2}{n^2}P_{n-1}'^2 + P_{n-1}^2 \quad (n \geqq 1)\\
&\text{(h)}\quad \frac{1-x^2}{n^2}P_n'^2 + P_n^2 \leqq 1 \quad (n \geqq 1, |x| \leqq 1)\\
&\text{(i)}\quad |P_n(x)| \leqq 1 \quad (|x| \leqq 1)\\
&\text{(j)}\quad \int_{-1}^{1} P_n(x)P_m(x)\,dx = 0 \quad (n \neq m)\\
&\text{(k)}\quad \int_{-1}^{1} [P_n(x)]^2\,dx = \frac{2}{2n+1}\\
&\text{(l)}\quad x^n \textit{ can be expressed as a linear combination of } P_0(x), \ldots, P_n(x).
\end{aligned}\right\} \quad (7\text{–}64)$$

The proofs are left to the exercises (Probs. 2 to 9 below). Identity (d) is the *differential equation* satisfied by each $P_n(x)$; its importance will be seen in Chapter 10. Identity (e) is known as the *recursion formula* for Legendre polynomials. It expresses each polynomial in terms of the two preceding members of the sequence; hence, knowing that $P_0 = 1$ and $P_1 = x$, one can successively determine $P_2, P_3, \ldots$ from (e) alone. By virtue of (j) and (k) we can state:

THEOREM 18. *The Legendre polynomials $P_n(x)$ $(n = 0, 1, 2, \ldots)$ form an orthogonal system for the interval $-1 \leqq x \leqq 1$ and*

$$\|P_n(x)\|^2 = \frac{2}{2n+1}. \tag{7–65}$$

Accordingly, we can form the Fourier series of an arbitrary function $f(x)$, piecewise continuous for $-1 \leqq x \leqq 1$, with respect to the system $\{P_n(x)\}$:

$$\sum_{n=0}^{\infty} c_n P_n(x), \quad c_n = \frac{(f, P_n)}{\|P_n\|^2} = \frac{2n+1}{2}\int_{-1}^{1} f(x)P_n(x)\,dx. \tag{7–66}$$

It will be seen that this "Fourier-Legendre" series behaves in essentially the same way as the trigonometric Fourier series.

THEOREM 19. *The system of Legendre polynomials in the interval $-1 \leqq x \leqq 1$ has the uniqueness property.*

The proof of Theorem 3 (Section 7–7) can be repeated with slight modifications. By (l) above, every polynomial in x is expressible as a finite linear combination of Legendre polynomials; hence it is sufficient to construct a polynomial pulse-function $P(x)$. One immediately verifies that the polynomial

$$P(x) = [1 - \tfrac{1}{2}(x - x_0)^2 + \tfrac{1}{2}\delta^2]^N$$

has the same properties as the function $P(x)$ used in Section 7–7, so that the proof can be completed in the same way.

THEOREM 20. *If $f(x)$ is very smooth for $-1 \leqq x \leqq 1$, then the Fourier-Legendre series of $f(x)$ converges uniformly for $-1 \leqq x \leqq 1$.*

Proof. By use of (c) and integration by parts we find

$$\begin{aligned} c_n &= \frac{2n+1}{2}\int_{-1}^{1} f(x)P_n(x)\,dx = \tfrac{1}{2}\int_{-1}^{1} f(x)(P'_{n+1} - P'_{n-1})\,dx \\ &= \tfrac{1}{2}[f(x)\{P_{n+1}(x) - P_{n-1}(x)\}]\Big|_{-1}^{1} - \tfrac{1}{2}\int_{-1}^{1} f'(x)(P_{n+1} - P_{n-1})\,dx. \end{aligned}$$

Because of (f), the first term is 0. Integrating by parts again, we find

$$\begin{aligned} c_n &= \tfrac{1}{2}\int_{-1}^{1} f''(x)\left[\frac{P_{n+2} - P_n}{2n+3} - \frac{P_n - P_{n-2}}{2n-1}\right]dx \\ &= \frac{1}{4n+6}(f'', P_{n+2}) - \frac{1}{4n+6}(f'', P_n) - \frac{1}{4n-2}(f'', P_n) \\ &\quad + \frac{1}{4n-2}(f'', P_{n-2}). \end{aligned}$$

Accordingly, since $4n - 2$ is the smallest denominator,

$$|c_n| \leqq \frac{1}{4n-2}\left[|(f'', P_{n+2})| + 2|(f'', P_n)| + |(f'', P_{n-2})|\right]$$
$$\leqq \frac{1}{4n-2}\left(\|f''\| \cdot \|P_{n+2}\| + 2\|f''\| \cdot \|P_n\| + \|f''\| \cdot \|P_{n-2}\|\right),$$

by the Schwarz inequality (7–46). By (7–65),

$$|c_n| \leqq \frac{\|f''\|}{4n-2}\left(\sqrt{\frac{2}{2n+5}} + 2\sqrt{\frac{2}{2n+1}} + \sqrt{\frac{2}{2n-3}}\right)$$
$$\leqq \frac{\|f''\|}{4n-2}\,\frac{4\sqrt{2}}{\sqrt{2n-3}} \leqq \frac{4\sqrt{2}\|f''\|}{(2n-3)^{\frac{3}{2}}} = M_n.$$

By (i) above, the Weierstrass M-test is applicable to the Fourier-Legendre series of $f(x)$:

$$|c_nP_n(x)| \leqq M_n = \text{const times } (2n-3)^{-\frac{3}{2}}.$$

The series ΣM_n converges by the integral test. Therefore the Fourier-Legendre series converges uniformly for $-1 \leqq x \leqq 1$.

Corollary. *If $f(x)$ is very smooth for $-1 \leqq x \leqq 1$, then the Fourier-Legendre series of $f(x)$ converges uniformly to $f(x)$ for $-1 \leqq x \leqq 1$.*

This is a consequence of Theorems 11, 19, and 20.

Remark. No condition of periodicity or other condition at $x = \pm 1$ is imposed on $f(x)$, as in the analogous theorem (Theorem 5) for trigonometric series. This is because of a symmetry of the Legendre polynomials, somewhat like that of the functions $\cos nx$ (Prob. 2 following Section 7–13).

Theorem 21. *The Legendre polynomials form a complete orthogonal system for the interval $-1 \leqq x \leqq 1$.*

This follows at once from the two preceding theorems, by virtue of the general Theorem 12 of Section 7–12.

Remark. Given a sequence $f_n(x)$ of functions continuous for $a \leqq x \leqq b$, no finite number of which are linearly dependent, one can construct linear combinations: $\phi_1 = f_1$, $\phi_2 = a_1f_1 + a_2f_2, \ldots$ such that the functions $\{\phi_n(x)\}$ form an orthogonal system in the interval $a \leqq x \leqq b$. This is carried out by the *Gram-Schmidt orthogonalization process*, described in Section 1–8. If we choose the sequence $f_n(x)$ to be $1, x, x^2, \ldots$ and the interval to be $-1 \leqq x \leqq 1$, the functions $\phi_n(x)$ turn out to be constants times the Legendre polynomials.

The case of a function $f(x)$ which is piecewise very smooth is covered as for Fourier series by studying a particular jump function $s(x)$. We do not carry through the details here but merely state the result: *just as does the Fourier series, the Fourier-Legendre series converges uniformly to the function in each closed interval containing no discontinuity and converges at each jump discontinuity to the half-way value*

$$\tfrac{1}{2}[\lim_{x\to x_1-} f(x) + \lim_{x\to x_1+} f(x)].$$

It is to be emphasized that there is no peculiar behavior at the points $x = \pm 1$; here the series converges to the function (provided the function remains continuous).

For a more complete treatment, the reader is referred to the book by Jackson listed at the end of the chapter.

***7–15 Fourier-Bessel series.** We consider first a special case of the Bessel functions.

The Bessel function of first kind of order 0 is defined by the equation

$$J_0(x) = 1 - \frac{x^2}{2^2} + \frac{x^4}{2^4(2!)^2} + \cdots + (-1)^n \frac{x^{2n}}{2^{2n}(n!)^2} + \cdots. \tag{7-67}$$

The ratio test shows at once that this series converges for all values of x. The series is somewhat suggestive of that for $\cos x$ and, as the graph of Fig. 7–15 shows, $J_0(x)$ does resemble a trigonometric function. *In particular, $J_0(x)$ has infinitely many roots.* The positive roots will be denoted, in increasing order, by $\lambda_1, \lambda_2, \ldots, \lambda_k, \ldots$. It can be shown that as k increases, these roots approach more and more closely the spacing at intervals of π of the roots of $\sin x$ or $\cos x$.

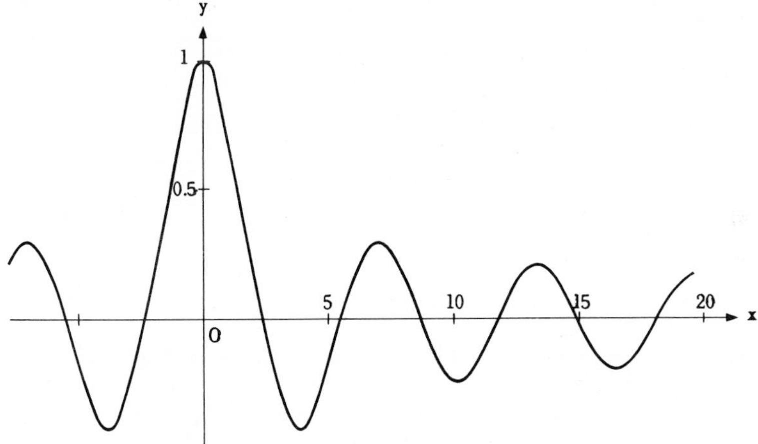

Fig. 7–15. The Bessel function $J_0(x)$.

The function $J_0(x)$ deviates from the pattern of the trigonometric functions in that

$$\lim_{x \to \infty} J_0(x) = 0.$$

The oscillation represented by $J_0(x)$ is "damped." The rate of damping is shown by the following "asymptotic formula":

$$J_0(x) = \sqrt{\frac{2}{\pi x}} \sin\left(x + \frac{\pi}{4}\right) + \frac{r(x)}{x^{\frac{3}{2}}}, \quad x > 0, \tag{7-68}$$

where $r(x)$ is bounded:

$$|r(x)| < K. \tag{7–69}$$

In other words, the function

$$\sqrt{\frac{\pi x}{2}}\, J_0(x)$$

differs from the function $\sin\,(x + \frac{1}{4}\pi)$ by a quantity $r(x)/x$ which approaches zero as x becomes infinite.

Because of the orthogonality of the trigonometric functions, it is not surprising that there is a corresponding orthogonality relation for the functions $\sqrt{x}\, J_0(\lambda_n x)$:

$$\int_0^1 [\sqrt{x}\, J_0(\lambda_n x) \cdot \sqrt{x}\, J_0(\lambda_m x)]\, dx = 0, \quad n \neq m. \tag{7–70}$$

The condition (7–70) is often described by stating that the functions $J_0(\lambda_n x)$ are *orthogonal with respect to the weight factor* $\rho(x) = x$:

$$\int_0^1 J_0(\lambda_n x) J_0(\lambda_m x) \rho(x)\, dx = 0, \quad n \neq m, \quad \rho(x) = x.$$

The functions $\sqrt{x}\, J_0(\lambda_n x)$ can now be made the basis of a Fourier-Bessel series. For simplicity, we write the function being expanded as $\sqrt{x}\, f(x)$, so that the series for $f(x)$ is

$$\sum_{n=1}^{\infty} c_n J_0(\lambda_n x).$$

Since it can be shown that

$$B_n = \int_0^1 x[J_0(\lambda_n x)]^2\, dx = \tfrac{1}{2}[J_0'(\lambda_n)]^2,$$

the series for $f(x)$ is given by

$$\sum_{n=1}^{\infty} c_n J_0(\lambda_n x), \quad c_n = \frac{2}{[J_0'(\lambda_n)]^2} \int_0^1 x f(x) J_0(\lambda_n x)\, dx. \tag{7–71}$$

This series is usually referred to as the Fourier-Bessel series of order 0 of $f(x)$.

Because of the close relationship between the functions $\sqrt{x}\, J_0(\lambda_n x)$ and trigonometric functions, it is natural to expect convergence theorems analogous to those for Fourier series. It can indeed be shown that the system $\{\sqrt{x}\, J_0(\lambda_n x)\}$ is *complete* for the interval $0 \leqq x \leqq 1$. Hence by Theorem 11 of Section 7–11 the Fourier-Bessel series of $f(x)$ converges to $f(x)$ whenever the series is uniformly convergent. The series can be shown to be uniformly convergent if $f(x)$ has, for example, a continuous second derivative and $f(1) = 0$. The peculiar requirement at $x = 1$ is due to the fact that, when $x = 1$, every function $J_0(\lambda_n x)$ is 0. If $f(x)$ is piecewise very smooth, an analysis of jump functions again shows that the series is uniformly convergent to $f(x)$ except near discontinuity points, where the series converges to the average of left- and right-hand limits.

For details on the theory of such expansions, the reader is referred to Chapter XVIII of the treatise of Watson listed at the end of the chapter.

The general *Bessel function of first kind and order m* is denoted by $J_m(x)$ and is defined by the series:

$$J_m(x) = \sum_{n=0}^{\infty} \frac{(-1)^n x^{m+2n}}{2^{m+2n} n!\,\Gamma(m+n+1)}. \tag{7–72}$$

The function $\Gamma(x)$ was defined for all x except $0, -1, -2, -3, \ldots$ in Section 6–24. The definition (7–72) therefore fails if m is a negative integer. However, it can be used for these values, if we interpret $1/\Gamma(k)$ to be 0 for $k = 0, -1, -2, \ldots$ If x^m is factored out of the series in (7–72), the resulting power series can be shown to converge for all x. However, if m is not an integer, the factor x^m gives difficulty for negative x and possibly for $x = 0$. Hence we shall concentrate on the functions $J_m(x)$ for $x \geqq 0$. These functions are continuous for $x > 0$ and, if $m \geqq 0$, for $x \geqq 0$.

We can formulate a theorem analogous to Theorem 17 for these Bessel functions:

THEOREM 18. *The Bessel functions of first kind satisfy the following identities and relations for $x > 0$:*

(a) $2J'_m(x) = J_{m-1}(x) - J_{m+1}(x)$

(b) $\dfrac{2m}{x} J_m(x) = J_{m-1}(x) + J_{m+1}(x)$

(c) $J_{m-1}(x) = J'_m(x) + \dfrac{m}{x} J_m(x)$, or $x^m J_{m-1}(x) = \dfrac{d}{dx}(x^m J_m(x))$

(d) $J_{m+1}(x) = \dfrac{m}{x} J_m(x) - J'_m(x)$, or $-x^{-m} J_{m+1}(x) = \dfrac{d}{dx}(x^{-m} J_m(x))$

(e) $x^2 J''_m(x) + xJ'_m(x) + (x^2 - m^2)J_m(x) = 0$ *(differential equation of the Bessel functions)*

(f) $\dfrac{d^2}{dx^2}[\sqrt{x}\, J_m(cx)] + \left(c^2 + \dfrac{1-4m^2}{4x^2}\right)\sqrt{x}\, J_m(cx) = 0$,
$c = const > 0$

(g) $(\alpha^2 - \beta^2)\displaystyle\int_0^1 xJ_m(\alpha x)J_m(\beta x)\,dx = \beta J_m(\alpha)J'_m(\beta) - \alpha J_m(\beta)J'_m(\alpha)$,
for $m \geqq 0,\ \alpha > 0,\ \beta > 0$

(h) $(\alpha^2 - \beta^2)\displaystyle\int_0^1 xJ_m(\alpha x)J_m(\beta x)\,dx = \alpha J_m(\beta)J_{m+1}(\alpha) - \beta J_m(\alpha)J_{m+1}(\beta)$,
for $m \geqq 0,\ \alpha > 0,\ \beta > 0$

(i) $2\alpha\displaystyle\int_0^1 xJ_m^2(\alpha x)\,dx = J_m(\alpha)J_{m+1}(\alpha) + \alpha[J_m(\alpha)J'_{m+1}(\alpha) - J_{m+1}(\alpha)J'_m(\alpha)]$,
for $m \geqq 0,\ \alpha > 0$

(j) *The equation* $J_m(x) = 0$ *has infinitely many positive roots* λ_{m1}, $\lambda_{m2}, \ldots,$ *where* $\lambda_{m1} < \lambda_{m2} < \cdots < \lambda_{mn} < \cdots$, $\lambda_{mn} \to \infty$ *as* $n \to \infty$, *and each root is simple (that is,* $J'_m(x) \neq 0$ *at each root).*

(k) *The equations* $J_m(x) = 0$ *and* $J_{m+1}(x) = 0$ *have no common positive roots.*

(l) *Between two successive positive roots of* $J_m(x) = 0$, *there is one and only one root of* $J_{m-1}(x) = 0$ *and one and only one root of* $J_{m+1}(x) = 0$.

(m) *For each fixed* $m \geqq 0$, *the functions* $\sqrt{x}\, J_m(\lambda_{mn}x)$, $n = 1, 2, 3, \ldots,$ *form an orthogonal system on the interval* $0 \leqq x \leqq 1$.

(n) $\displaystyle\int_0^1 xJ_m^2(\lambda_{mn}x)\, dx = \tfrac{1}{2}[J_{m+1}(\lambda_{mn})]^2.$

(o) *For* $m = 0, 1, 2, \ldots,$ $\quad J_m(x) = \dfrac{1}{\pi}\displaystyle\int_0^{\pi} \cos\,(x \sin\theta - m\theta)\, d\theta.$

(p) *For* $m = 0, 1, 2, \ldots,$ $\quad |J_m(x)| \leqq 1$ *for* $x \geqq 0$.

The proofs are left to Problem 10 below. It can be shown that the orthogonal system in (m) is complete, so that there is a corresponding Fourier-Bessel series of order m. This series is uniformly convergent for every function $\sqrt{x}\, f(x)$ such that $f(x)$ has a continuous second derivative in the interval and $f(1) = 0$. For proofs see Chapter XVIII of the treatise by Watson. On page 406 of the same book it is shown that assertion (p) can be extended to all $m \geqq 0$.

The Bessel functions of the *second kind* are the functions $Y_m(x)$ $(m = 0, 1, 2, \ldots)$ defined as follows:

$$\begin{aligned} Y_m(x) = {} & \frac{1}{\pi}\sum_{k=0}^{\infty} \frac{(-1)^k \left(\dfrac{x}{2}\right)^{m+2k}}{k!(m+k)!} \left\{2 \log \frac{x}{2} + 2\gamma - b_{m+k} - b_k\right\} \\ & - \frac{1}{\pi}\sum_{k=0}^{m-1} \left(\frac{x}{2}\right)^{-m+2k} \frac{(m-k-1)!}{k!}, \end{aligned} \qquad (7\text{–}73)$$

where $b_k = 1 + 2^{-1} + \cdots + k^{-1}$ and γ is defined in Equation (6–86). These functions satisfy the same differential equations (e) as the functions of first kind and have other analogous properties. The complex-valued functions

$$H_m^{(1)}(x) = J_m(x) + iY_m(x), \quad H_m^{(2)}(x) = J_m(x) - iY_m(x) \qquad (7\text{–}74)$$

are called Bessel functions of the *third kind* or *Hankel functions*. For details, see the book by Watson.

Tables and graphs of both Legendre polynomials and Bessel functions are available. The *Tables of Functions* of E. Jahnke and F. Emde (New York: Stechert, 1938) provide such data and also a convenient summary of properties of the functions.

We mention, without further discussion, several other important systems of orthogonal functions:

The Jacobi polynomials. For each $\alpha > -1$, $\beta > -1$, the system of polynomials $\{P_n^{(\alpha,\beta)}(x)\}$ $(n = 0, 1, 2, \ldots)$ is defined by the equations

$$P_n^{(\alpha,\beta)}(x) = \frac{(-1)^n}{2^n n!}(1-x)^{-\alpha}(1+x)^{-\beta}\frac{d^n}{dx^n}[(1-x)^{\alpha+n}(1+x)^{\beta+n}]. \tag{7–75}$$

The functions $\phi_n(x) = (1-x)^{\frac{1}{2}\alpha}(1+x)^{\frac{1}{2}\beta}P_n^{(\alpha,\beta)}(x)$ form a complete orthogonal system in the interval $-1 \leqq x \leqq 1$. When $\alpha = \beta = 0$, the functions $\phi_n(x)$ reduce to the Legendre polynomials.

The Hermite polynomials. The system of polynomials $\{H_n(x)\}$ is defined by the equations:

$$H_n(x) = (-1)^n e^{\frac{1}{2}x^2}\frac{d^n}{dx^n}e^{-\frac{1}{2}x^2} \quad (n = 0, 1, 2, \ldots). \tag{7–76}$$

They are orthogonal over the *infinite interval* $-\infty < x < \infty$ with respect to the weight function $e^{-\frac{1}{2}x^2}$.

The Laguerre polynomials. For each $\alpha > -1$, the system of polynomials $\{L_n^{(\alpha)}(x)\}$ is defined by the equations:

$$L_n^{(\alpha)}(x) = (-1)^n x^{-\alpha}e^x\frac{d^n}{dx^n}(x^{\alpha+n}e^{-x}) \quad (n = 0, 1, \ldots). \tag{7–77}$$

They are orthogonal over the *infinite interval* $0 \leqq x < \infty$ with respect to the weight function $x^\alpha e^{-x}$.

For the significance of the infinite interval and further information on these functions we refer to the book of Jackson listed at the end of the chapter.

All of these functions arise in a natural way in physical problems. This is illustrated in Chapter 10, where it will be shown that each of a large class of physical problems automatically provides a complete system of orthogonal functions.

Problems

1. Graph $P_0(x), P_1(x), \ldots, P_4(x)$.

2. Prove the following parts of Theorem 17:
(i) part (c); (ii) part (d); (iii) part (e). Each of these can be deduced from (7–63) alone.

3. Prove part (f) of Theorem 17 by induction, with the aid of the recursion formula (e).

4. Prove part (g) of Theorem 17 by squaring (a) and (b) of (7–63) and eliminating the terms in $P_{n-1}P'_{n-1}$ from the equations obtained.

5. Prove part (h) of Theorem 17. [Hint: show by induction that (g) gives the chain of inequalities:

$$\frac{1-x^2}{n^2}P_n'^2 + P_n^2 \leqq \frac{1-x^2}{(n-1)^2}P_{n-1}'^2 + P_{n-1}^2 \leqq \cdots \leqq 1, \quad |x| \leqq 1.]$$

6. Prove part (i) of Theorem 17 as a consequence of part (h).

7. Prove the orthogonality condition (j). {Hint: for every very smooth function $f(x)$ for $-1 \leqq x \leqq 1$, let $f^*(x) = [(1 - x^2)f'(x)]'$. Prove by integration by parts that $(f^*, g) - (f, g^*) = 0$. Take $f = P_n(x)$, $g = P_m(x)$ and use the differential equation (d) to replace f^* by $-n(n+1)f$ and g^* by $-m(m+1)g$, and conclude that $(f, g)[m(m+1) - n(n+1)] = 0$.}

8. Prove part (k) of Theorem 17. {Hint: use the recursion formula (e) to express P_n in terms of P_{n-1} and P_{n-2}, and use the orthogonality condition to show that $(P_n, P_n) = [(2n-1)/n](xP_n, P_{n-1})$. Apply the recursion formula to xP_n and use orthogonality to show that $(P_n, P_n) = (P_{n-1}, P_{n-1})[(2n-1)/(2n+1)]$. Now prove by induction that $(P_n, P_n) = 2/(2n+1)$.}

9. Prove part (l) of Theorem 17 by induction, using the fact that $P_n(x)$ is a polynomial of nth degree.

10. (a) . . . (p). Prove the corresponding parts of Theorem 18 with the aid of the following hints: For (a) and (b), use (7–72). For (c) and (d) use (a) and (b). For (e) use (a) through (d). For (f) use (e). For (g) let $u(x) = \sqrt{x}\, J_m(\alpha x)$, $v(x) = \sqrt{x}\, J_m(\beta x)$. Take $c = \alpha$ and then $c = \beta$ in (f) to obtain equations $u''(x) + \cdots = 0$, $v''(x) + \cdots = 0$. Multiply the first equation by $-v$, the second by u and add. Then integrate the result from b to 1, noting that $uv'' - vu'' = (uv' - vu')'$. Replace u and v by their expressions in terms of J_m and finally let $b \to 0+$. For (h) use (g) and (d). For (i) divide both sides of (h) by $\alpha - \beta$ and then, with α fixed, let $\beta \to \alpha$. For (j) note that, if $J_m(\alpha) = 0$ and $J_m'(\alpha) = 0$, then by (g) $\int_0^1 xJ_m(\alpha x)J_m(\beta x)\,dx = 0$ for all $\beta \neq \alpha$. Let $\beta \to \alpha$ to conclude that $J_m(\alpha x) \equiv 0$ for $0 \leqq x \leqq 1$, and this is impossible. Hence each root is simple. Deduce from this that there are only finitely many roots in each finite interval, so that the roots can be numbered in increasing order as stated and, if the sequence is infinite, it must diverge to ∞. If the sequence were to stop, then for example, $J_m(x) > 0$ for $x > x_0$. Take $c = 1$ in (f) to conclude that, with $g(x) = \sqrt{x}\, J_m(x)$, $g(x) > 0$ and $g''(x) < 0$ for $x > x_1$, for some x_1. If $g(x) \to \infty$ as $x \to \infty$, then by (f), $g''(x) \to -\infty$ as $x \to \infty$; show that this is impossible, so that $g(x)$ is bounded and that $g(x)$ must have a limit k as $x \to \infty$. Show by (f) that either $k > 0$ or $k = 0$ is impossible. Thus the sequence of roots must be infinite. For (k), use (h). For (l) use (c) (d) and Rolle's theorem. For (m) use (g). For (n) use (i) and (d). For (o) use induction. For $m = 0$, replace $\cos(x \sin\theta)$ by $1 - (x^2 \sin^2\theta/2!) + \cdots$ and use the formula $\int_0^\pi \sin^{2m}\theta\, d\theta = \pi \cdot 1 \cdot 3 \cdots (2m-1)/[2 \cdot 4 \cdots 2m]$. Show, with the aid of integration by parts, that $g_m(x) = \pi^{-1}\int_0^\pi \cos(x\sin\theta - m\theta)\,d\theta$ satisfies $g_{m+1}(x) = (m/x)g_m(x) - g_m'(x)$, which by (d) is the same recursion formula as is satisfied by $J_m(x)$. For (p) use (o).

11. Prove that for $m = 1, 2, 3, \ldots, J_{-m}(x) = (-1)^m J_m(x)$.

12. Prove: (a) $J_{\frac{1}{2}}(x) = \sqrt{\dfrac{2}{\pi x}} \sin x$. (b) $J_{-\frac{1}{2}}(x) = \sqrt{\dfrac{2}{\pi x}} \cos x$.

(c) $J_{\frac{3}{2}}(x) = \sqrt{\dfrac{2}{\pi x}}\left(\dfrac{\sin x}{x} - \cos x\right)$.

***7–16 Orthogonal systems of functions of several variables.** The theory of Sections 7–10 to 7–12 can be generalized with minor changes to functions of several variables; one need only replace the interval by a bounded closed region R and the definite integral by a multiple integral over R. For example, in two dimensions, the inner product and norm are defined as follows:

$$(f, g) = \iint_R f(x, y)g(x, y)\,dx\,dy,$$
$$\|f\| = (f, f)^{\frac{1}{2}} = \left\{\iint_R [f(x, y)]^2\,dx\,dy\right\}^{\frac{1}{2}}. \tag{7–78}$$

Orthogonal systems are defined as before, and one can consider the corresponding Fourier series. Discussion of discontinuities becomes more involved, and it is sufficient for most purposes to restrict attention to continuous functions. The analogues of Theorems 7 to 11 can then be proved essentially without alteration. The generalization of Theorem 12 can be carried out, but this requires more care.

Corresponding to the trigonometric system, one has the following system in two dimensions:

$$1,\quad \sin x,\quad \cos x,\quad \sin y,\quad \cos y,\quad \sin x\cos y,\ \ldots,\quad \sin px\sin qy,$$
$$\cos px\sin qy,\quad \sin px\cos qy,\quad \cos px\cos qy,\ \ldots. \tag{7–79}$$

This system can be arranged in a definite order to form a system $\{\phi_n(x, y)\}$, which is orthogonal and complete for the rectangle R: $-\pi \leqq x \leqq \pi$, $-\pi \leqq y \leqq \pi$ (see Prob. 1 below). One can also write the Fourier series as a "double Fourier series":

$$\sum_{q=0}^{\infty}\sum_{p=0}^{\infty}\{a_{pq}\sin px\sin qy + b_{pq}\cos px\sin qy$$
$$+ c_{pq}\sin px\cos qy + d_{pq}\cos px\cos qy\}. \tag{7–80}$$

This amounts to a special sort of rearrangement and grouping of the series; if the series is absolutely convergent, then the reasoning of Section 6–10 shows that the double series has the same sum as the single series in any order. If $f(x, y)$ has continuous first and second derivatives for all (x, y) and is periodic in both variables:

$$f(x, y) = f(x + 2\pi, y) = f(x, y + 2\pi),$$

then an argument similar to that of Section 7–8 shows that the series is absolutely and uniformly convergent to $f(x, y)$.

***7–17 Complex form of Fourier series.** From the identity

$$e^{ix} = \cos x + i\sin x \quad (i = \sqrt{-1})$$

of Section 6–19, one derives the relations

$$\cos x = \frac{e^{ix} + e^{-ix}}{2}, \quad \sin x = \frac{e^{ix} - e^{-ix}}{2i}. \tag{7–81}$$

A Fourier series

$$\frac{a_0}{2} + \sum_{n=1}^{\infty} (a_n \cos nx + b_n \sin nx)$$

can hence be written in the form

$$\frac{a_0}{2} + \sum_{n=1}^{\infty} (c_n e^{inx} + d_n e^{-inx}) = \sum_{n=-\infty}^{\infty} c_n e^{inx}, \tag{7–82}$$

$$c_0 = \frac{a_0}{2}, \quad c_n = \frac{a_n - ib_n}{2}, \quad d_n = \frac{a_n + ib_n}{2} = c_{-n} \quad (n = 1, 2, \ldots).$$

The summation from $-\infty$ to ∞ is understood to mean an addition of two series:

$$\sum_{n=-\infty}^{\infty} c_n e^{inx} = \sum_{n=0}^{\infty} c_n e^{inx} + \sum_{n=1}^{\infty} c_{-n} e^{-inx}.$$

If both series converge, the result is clearly the same as the single series on the left of (7–82).

The form (7–82) has various advantages. The coefficients c_n can be defined directly in terms of $f(x)$:

$$c_n = \frac{1}{2\pi} \int_{-\pi}^{\pi} f(x) e^{-inx}\, dx \quad (n = 0, \pm 1, \pm 2, \ldots), \tag{7–83}$$

for the integral on the right is interpreted as

$$\frac{1}{2\pi} \int_{-\pi}^{\pi} f(x)(\cos nx - i \sin nx)\, dx = \frac{1}{2\pi} \int_{-\pi}^{\pi} f(x) \cos nx\, dx - \frac{i}{2\pi} \int_{-\pi}^{\pi} f(x) \sin nx\, dx.$$

See Section 9–2 for the theory of such integrals of complex functions. When $n = 0$, this gives $\frac{1}{2}a_0$; when $n > 0$, the integral equals $\frac{1}{2}(a_n - ib_n)$; and when $n < 0$ it equals $\frac{1}{2}(a_{-n} + ib_{-n})$. One has thus the concise statement:

$$f(x) = \sum_{n=-\infty}^{\infty} c_n e^{inx}, \quad c_n = \frac{1}{2\pi} \int_{-\pi}^{\pi} f(x) e^{-inx}\, dx, \tag{7–84}$$

whenever the series converges to $f(x)$.

For formal work with Fourier series, and even for computation of coefficients, the series (7–84) provides a considerable simplification.

***7–18 Fourier integral.** By a suitable limiting process the equations (7–84) lead to the relations

$$f(x) = \frac{1}{\sqrt{2\pi}} \int_{-\infty}^{\infty} g(t) e^{ixt}\, dt, \quad g(t) = \frac{1}{\sqrt{2\pi}} \int_{-\infty}^{\infty} f(x) e^{-ixt}\, dx. \tag{7–85}$$

Thus, under appropriate hypotheses, a function $f(x)$ defined for $-\infty < x < \infty$ can be represented as a "continuous sum" of sines and cosines

($e^{ixt} = \cos xt + i \sin xt$). The integral representing $f(x)$ is termed the *Fourier integral* of $f(x)$. The Fourier coefficients of $f(x)$ in this integral representation are the numbers $g(t)$, which form a new function. Equations (7–85) show that the relation between f and g is nearly symmetrical.

Equations (7–85) can also be written in real form as follows:

$$\begin{aligned} f(x) &= \int_0^\infty \alpha(t) \cos xt \, dt + \int_0^\infty \beta(t) \sin xt \, dt, \\ \alpha(t) &= \frac{1}{\pi} \int_{-\infty}^\infty f(x) \cos xt \, dx, \quad \beta(t) = \frac{1}{\pi} \int_{-\infty}^\infty f(x) \sin xt \, dx. \end{aligned} \tag{7–86}$$

This is in direct analogy with the real form for Fourier series.

The validity of the formulas (7–85) or the equivalent formulas (7–86) can be established if $f(x)$ is piecewise smooth in each finite interval and the integral

$$\int_{-\infty}^\infty |f(x)| \, dx$$

converges. The Fourier integral must in general be taken as a *principal value* (Section 6–24). It converges to $f(x)$ wherever f is continuous and to the average of left and right limits at jump discontinuities. For a proof the reader is referred to Chapter 5 of the book by Kaplan listed at the end of the chapter.

The equations (7–85) are often rephrased in the language of transforms. If

$$F(t) = \int_{-\infty}^\infty f(x) e^{-ixt} \, dx, \quad -\infty < t < \infty, \tag{7–87}$$

then one calls the function F the *Fourier transform* of the function f and writes

$$F = \Phi[f]. \tag{7–88}$$

The equations (7–85) then state that

$$f(x) = \frac{1}{2\pi} \int_{-\infty}^\infty F(t) e^{ixt} \, dt = \frac{1}{2\pi} \int_{-\infty}^\infty F(-t) e^{-ixt} \, dt. \tag{7–89}$$

Thus, by interchanging the roles of x and t, we see that f is itself the Fourier transform of $F(-t)/(2\pi)$.

The Fourier transform has a number of properties which make it exceptionally useful in applications to differential equations. We state several such properties formally here and refer to the books on Fourier theory listed at the end of the chapter for details. First of all, Φ is a *linear operator:*

$$\Phi[c_1 f_1 + c_2 f_2] = c_1 \Phi[f_1] + c_2 \Phi[f_2]. \tag{7–90}$$

Next

$$\Phi[f^{(k)}(x)] = (it)^k \Phi[f], \quad k = 1, 2, \ldots \tag{7–91}$$

This property is crucial for differential equations. Here it is assumed, in particular, that f has continuous derivatives through order k. The result must be modified when discontinuities occur. (See (7–101) below for such a rule for Laplace transforms.) If c is a real constant, then

$$\Phi[f(x - c)] = e^{-ict}\Phi[f], \tag{7–92}$$

$$\Phi[e^{icx}f(x)] = F(t - c), \quad \text{where} \quad F = \Phi[f]. \tag{7–93}$$

From two functions $f(x)$ and $g(x)$ (under appropriate hypotheses) one forms their *convolution* $f * g = h$ as the function $h(x)$ such that

$$h(x) = \int_{-\infty}^{\infty} f(u)g(x - u)\, du. \tag{7–94}$$

Then

$$\Phi[f * g] = \Phi[f]\Phi[g]. \tag{7–95}$$

Examples of Fourier transforms. Let $f_1(x) = e^{-x}$ for $x > 0$ and $f_1(x) = 0$ for $x \leqq 0$. Then

$$\Phi[f_1] = \int_0^{\infty} e^{-x}e^{-ixt}\, dx = \frac{1}{1 + it}\cdot$$

Similarly, if $f_2(x) = e^x$ for $x \leqq 0$ and $f_2(x) = 0$ for $x > 0$, then

$$\Phi[f_2] = \frac{1}{1 - it}\cdot$$

Now $f_1(x) + f_2(x) = e^{-|x|} = f(x)$ and hence, by linearity,

$$\Phi[f] = \frac{1}{1 + it} + \frac{1}{1 - it} = \frac{2}{1 + t^2} = F(t).$$

Since f is the Fourier transform of $(2\pi)^{-1}F(-t) = [\pi(1 + t^2)]^{-1}$, we conclude (after interchanging x and t) that

$$\Phi[g] = e^{-|t|}, \quad \text{where} \quad g(x) = \frac{1}{\pi(1 + x^2)}$$

or, by linearity,

$$\Phi[g_1] = \pi e^{-|t|}, \quad \text{where} \quad g_1(x) = \frac{1}{1 + x^2}\cdot$$

***7–19 The Laplace transform as a case of the Fourier transform.** The Laplace transform of a function $f(x)$ is defined (Section 6–25) as $\mathcal{L}[f] = F$, where

$$F(s) = \int_0^{\infty} f(x)e^{-sx}\, dx. \tag{7–96}$$

Here f need only be defined for $x \geqq 0$. However, it will be convenient here to define $f(x)$ to be 0 for negative x. The Laplace transform F is a function of s, defined wherever the integral in (7–96) exists. We now allow s to be complex and write $s = \sigma + it$, where σ and t are real. Since

$f(x) = 0$ for $x < 0$, we can write

$$F(s) = \int_{-\infty}^{\infty} f(x)e^{-(\sigma+it)x}\,dx = \int_{-\infty}^{\infty} f(x)e^{-\sigma x}e^{-itx}\,dx.$$

Thus we see that, for each fixed σ, $F(s)$ is the *Fourier transform* of $f(x)e^{-\sigma x}$:

$$\mathfrak{L}[f] = \Phi[f(x)e^{-\sigma x}]. \tag{7–97}$$

It can be shown that, if f is piecewise continuous on each finite interval and $|f(x)| < ke^{bx}$, $x \geqq 0$, for some constants k and b, then for each $\sigma > b$, the Laplace transform $F(s) = F(\sigma + it)$ is defined for $-\infty < t < \infty$.

Example. Let $f(x) = 1$ for $x = 0$ (and $f(x) = 0$ for $x < 0$). Then

$$\mathfrak{L}[f] = \Phi[f(x)e^{-\sigma x}] = \int_0^{\infty} e^{-\sigma x}e^{-itx}\,dx = \frac{1}{\sigma + it} = \frac{1}{s}.$$

Here $|f(x)| < 2e^{0x}$ for $x \geqq 0$, so that we can take $b = 0$, and $\mathfrak{L}[f] = F$, where $F(s) = 1/s = 1/(\sigma + it)$ for $\sigma > 0$.

Since the Laplace transform is related to the Fourier transform by (7–97), we can apply the theory of the Fourier transform to Laplace transforms. In particular, if $\mathfrak{L}[f] = F(s) = F(\sigma + it)$, then Eq. (7–89) allows us to write, under appropriate conditions on f,

$$f(x)e^{-\sigma x} = \frac{1}{2\pi}\int_{-\infty}^{\infty} F(\sigma + it)e^{ixt}\,dt$$

(where, as for (7–89), the integral is a principal value). We can rewrite this result as follows:

$$f(x) = \frac{1}{2\pi}\int_{-\infty}^{\infty} F(\sigma + it)e^{(\sigma+it)x}\,dt = \frac{1}{2\pi}\int_{-\infty}^{\infty} F(s)e^{sx}\,dt. \tag{7–98}$$

This is the *inversion formula* for Laplace transforms. It permits us to recover a function $f(x)$ from its Laplace transform $F(s)$. The existence of such a formula, as of the analogous formula for Fourier transforms, shows that we are dealing with *one-to-one* mappings.

For $f(x)$ as in the Example, we found $F(s) = 1/s$ for $\sigma > 0$. Hence (with principal values understood)

$$\frac{1}{2\pi}\int_{-\infty}^{\infty} \frac{1}{s}e^{sx}\,dt = \frac{1}{2\pi}\int_{-\infty}^{\infty} \frac{e^{(\sigma+it)x}}{\sigma + it}\,dt = \begin{cases} 1, & x > 0 \\ 0, & x < 0 \end{cases}.$$

At $x = 0$, f has a jump discontinuity and the integral converges to the expected average value, namely, $\frac{1}{2}$.

We can list formally other properties of Laplace transforms, all deducible from the corresponding properties of Fourier transforms: The Laplace transform is a *linear operator*. Corresponding to the formula

$$\Phi[f^{(k)}] = (it)^k\Phi[f],$$

one has

$$\mathcal{L}[f^{(k)}(x)] = s^k\mathcal{L}[f]. \tag{7–100}$$

However, this is valid only if f (regarded as 0 for $x < 0$) and the derivatives occurring have no discontinuities on the infinite x-axis. If f has continuous derivatives through order n, when regarded as a function on the interval $x \geqq 0$, then the formula becomes

$$\mathcal{L}[f^{(k)}(x)] = s^k\mathcal{L}[f] - [f^{(k-1)}(0) + sf^{(k-2)}(0) + \cdots + s^{k-1}f(0)], \tag{7–101}$$

where all derivatives are regarded as *right-hand* derivatives.

The formulas (7–92) and (7–93) become

$$\mathcal{L}[f(x - c)] = e^{-cs}\mathcal{L}[f], \quad c > 0, \tag{7–102}$$

$$\mathcal{L}[e^{bx}f(x)] = F(s - b), \quad \text{where} \quad F(s) = \mathcal{L}[f]. \tag{7–103}$$

The formula (7–95) becomes

$$\mathcal{L}[f * g] = \mathcal{L}[f]\mathcal{L}[g], \tag{7–104}$$

where, since $f = 0$ and $g = 0$ for $x < 0$, the convolution (7–94) can be simplified to $f * g = h$, where

$$h(x) = \int_0^x f(u)g(x - u)\,du. \tag{7–105}$$

This automatically makes $h(x) = 0$ for $x < 0$.

For further information on Laplace transforms, see the books by Churchill, Kaplan and Widder listed at the end of the chapter.

Problems

1. (a) Prove that the functions (7–77) form an orthogonal system in the rectangle R: $-\pi \leqq x \leqq \pi$, $-\pi \leqq y \leqq \pi$.

(b) Expand $f(x, y) = x^2y^2$ in a double Fourier series in R.

2. Represent the following functions as Fourier integrals:

(a) $f(x) = 0, \quad x < 0; \quad f(x) = e^{-x}, \quad x \geqq 0;$

(b) $f(x) = 0, \quad x < 0; \quad f(x) = 1, \quad 0 \leqq x \leqq 1; \quad f(x) = 0, \quad x > 1.$

3. Show that, if f is even, then its Fourier integral reduces to

$$\int_0^\infty \alpha(t) \cos xt\,dt, \quad \alpha(t) = \frac{2}{\pi}\int_0^\infty f(x) \cos xt\,dx$$

and that, if f is odd, its Fourier integral reduces to

$$\int_0^\infty \beta(t) \sin xt\,dt, \quad \beta(t) = \frac{2}{\pi}\int_0^\infty f(x) \sin xt\,dx.$$

4. Find the Fourier transform of each function:

(a) $f(x) = 1, \quad -1 \leqq x \leqq 1; \qquad f(x) = 0$ otherwise;

(b) $f(x) = xe^{-x}, \quad x \geqq 0; \qquad f(x) = 0$ otherwise;

(c) $f(x) = x^2, \quad -1 \leqq x \leqq 1; \qquad f(x) = 0$ otherwise.

5. (a) Prove the rule (7–92). (b) Prove the rule (7–93).

6. It is shown in Section 6–25 that, for $k = 0, 1, 2, \ldots$, $\mathcal{L}[x^k] = k!/s^{k+1}$ for real positive s. (a) Show that the result is valid for complex $s = \sigma + it$, provided $\sigma > 0$. (b) Write out the corresponding inversion formula (7–98) for $f(x) = x^k$.

7. For each of the following choices of f and g evaluate $h = f * g$ and verify that (7–104) holds. (Throughout $f = 0$ and $g = 0$ for $x < 0$.)
(a) $f(x) = g(x) = 1$ for $x \geqq 0$.
(b) $f(x) = 1$ and $g(x) = x$ for $x \geqq 0$.
(c) $f(x) = 1$ and $g(x) = e^x$ for $x \geqq 0$.

Answers

1. (b) $$\frac{\pi^4}{9} + \frac{4\pi^2}{3}\sum_{n=1}^{\infty}(-1)^n\frac{\cos nx}{n^2} + \frac{4\pi^2}{3}\sum_{m=1}^{\infty}(-1)^m\frac{\cos my}{m^2} + 16\sum_{n=1}^{\infty}\sum_{m=1}^{\infty}(-1)^{m+n}\frac{\cos nx\cos my}{n^2m^2}.$$

2. (a) $$\frac{1}{\pi}\int_0^{\infty}\frac{\cos xt + t\sin xt}{1+t^2}\,dt,$$

(b) $$\frac{1}{\pi}\int_0^{\infty}\frac{\sin t\cos xt + (1-\cos t)\sin xt}{t}\,dt.$$

4. (a) $(2\sin t)/t$, (b) $(1+it)^{-2}$, (c) $(2t^{-1} - 4t^{-3})\sin t + 4t^{-2}\cos t$.

6. (b) for $\sigma > 0$, $(2\pi)^{-1}\int_{-\infty}^{\infty} k!e^{(\sigma+it)x}(\sigma+it)^{-k-1}\,dt = x^k$ for $x > 0$, $= 0$ for $x < 0$.

7. For $x \geqq 0$, (a) $h(x) = x$, (b) $h(x) = x^2/2$, (c) $h(x) = e^x - 1$.

***7–20 Generalized functions.** A number of physical problems lead one to attempt to extend the concept of "function" to include new mathematical objects, called *generalized functions* or *distributions*, differing markedly from traditional functions. We here give an intuitive introduction to the subject and give references for a general discussion.

A simple example is that of density, say for mass distributed along a line, the x-axis. If $\rho(x)$ is the density (in units of mass per unit length), then $\int_a^b \rho(x)\,dx$ should give the total mass in the interval $a \leqq x \leqq b$. Now suppose that all our mass is concentrated in a single particle of mass 1, located at the origin. It is tempting to assign a density $\rho(x)$ such that $\int_a^b \rho(x)\,dx = 0$ if the interval $a \leqq x \leqq b$ does not contain the origin, and such that $\int_a^b \rho(x)\,dx = 1$ if the interval does contain the origin. However, these conditions would make $\rho(x) = 0$ except "near the origin", where $\rho(x)$ must suddenly become infinite so rapidly that the integral of $\rho(x)$ over each small interval containing the origin is 1. This particular density "function" is the *Dirac delta function* $\delta(x)$. Thus we require:

$$\delta(x) = 0, \quad x \neq 0; \qquad \delta(0) = \infty; \qquad \int_a^b \delta(x)\,dx = 1 \quad \text{if} \quad a < 0 < b. \tag{7–106}$$

The apparently contradictory nature of these properties led to their rejection as meaningless by mathematicians, until recently. Then a number of mathematicians, notably L. Schwartz, developed systematic theories of generalized functions, which satisfy all the desired properties in a rigorous manner. The function $\delta(x)$ is such a generalized function.

We can arrive at the delta function in another way, which suggests a limit process for obtaining this and other generalized functions. Let $g(b)$ denote the total mass on the interval $-\infty \leqq x < b$, for a distribution of mass (with finite total mass) on the x-axis. Thus, for a continuous density $\rho(x)$, $g(b) = \int_{-\infty}^{b} \rho(x)\,dx$ and $g'(b) = \rho(b)$. Now for the case of a single particle of mass 1 at the origin, we have difficulty in defining $\rho(x)$ but can easily find $g(x)$ for each x. For $g(x)$ is simply the total mass to the left of x and hence, for the case of the single particle,

$$g(x) = 0 \quad \text{for} \quad x < 0, \quad g(x) = 1 \quad \text{for} \quad x \geqq 0. \tag{7–107}$$

Thus $g(x)$ has a jump discontinuity at $x = 0$. We call the function $g(x)$ in (7–107) the *Heaviside unit function.*

Now from this $g(x)$ we can again try to obtain the density $\rho(x)$ as the *derivative* of $g(x)$. It is clear that $g'(x) = 0$ for $x \neq 0$ and it is reasonable to write $g'(0) = \infty$, because of the jump. Thus we have

$$g'(x) = 0 = \delta(x) \quad \text{for} \quad x \neq 0, \quad g'(0) = \infty = \delta(0). \tag{7–108}$$

Accordingly, we interpret the "function" $\delta(x)$ as the *derivative of the Heaviside unit function.* Now the unit function really has no derivative at $x = 0$, so that we have not completely clarified the meaning of $\delta(x)$. To go further, we *approximate* $g(x)$ by a smooth function $g_\epsilon(x)$ having a derivative for all x: we choose an $\epsilon > 0$ and set

$$g_\epsilon(x) = 1 - \tfrac{1}{2}e^{-x/\epsilon} \quad \text{for} \quad x \geqq 0, \quad g_\epsilon(x) = \tfrac{1}{2}e^{x/\epsilon} \quad \text{for} \quad x < 0. \tag{7–109}$$

(See Fig. 7–16). The function $g_\epsilon(x)$ has been chosen so that, as $\epsilon \to 0+$, $g_\epsilon(x) \to g(x)$ (except at $x = 0$, where $g_\epsilon(x) = \frac{1}{2}$ for all ϵ). Furthermore, $g_\epsilon(x)$ has a derivative

$$g'_\epsilon(x) = \frac{1}{2\epsilon}\,e^{-x/\epsilon} \quad \text{for} \quad x \geqq 0, \quad g'_\epsilon(x) = \frac{1}{2\epsilon}\,e^{x/\epsilon} \quad \text{for} \quad x < 0, \tag{7–110}$$

and, since $g_\epsilon(x) \to g(x)$ as $\epsilon \to 0+$, it is reasonable to consider $\delta(x)$ as the limit of $g'_\epsilon(x)$ as $\delta \to 0+$. Thus $\delta(x)$ is considered to be the limit of the "pulse" graphed in Fig. 7–17 (a smooth approximation to $g'_\epsilon(x)$), as $\epsilon \to 0+$. However, if we do indeed pass to the limit, we again obtain the value 0 for $x \neq 0$, the "value" ∞ for $x = 0$. It is thus not clear what we have gained.

Operations on generalized functions. We *add* or *subtract* generalized functions of the type considered in the obvious way. For example,

$$(e^x + \delta(x)) + (\sin x + \delta(x) + \delta'(x - 1)) \\ = e^x + \sin x + 2\delta(x) + \delta'(x - 1).$$

We can also *multiply* such a function by a scalar constant, by multiplying each term by the scalar. Multiplication of arbitrary generalized functions is not defined; in particular, $\delta(x) \cdot \delta(x)$ is not defined. However, if one of the two factors is an ordinary function $f(x)$ satisfying certain continuity conditions, then one can multiply f by each term of the other factor. In particular, one defines $f(x)\,\delta(x)$ to equal $f(0)\,\delta(x)$, provided $f(x)$ is continuous at $x = 0$; one defines $f(x)\,\delta'(x)$ to equal $f(0)\,\delta'(x) - f'(0)\,\delta(x)$, provided f and f' are continuous at $x = 0$. These rules can be justified by the limit process described above; the latter rule can also be justified by the rule for differentiating products. In general, one is led to the definition:

$$f(x)\,\delta^{(k)}(x - c) = \delta^{(k)}(x - c)f(x) = \sum_{r=0}^{k} (-1)^r \binom{k}{r} f^{(r)}(c)\,\delta^{(k-r)}(x - c), \tag{7–113}$$

provided that $f, \ldots, f^{(k)}$ are continuous at $x = c$. For example,

$$f(x)\,\delta''(x - c) = f(c)\,\delta''(x - c) - 2f'(c)\,\delta'(x - c) + f''(c)\,\delta(x - c).$$

One can differentiate a generalized function of the type considered, by differentiating each term: the derivative of $a\delta^{(k)}(x - c)$ is $a\delta^{(k+1)}(x - c)$; the derivative of $f(x)$ is the ordinary derivative plus contributions from the jump discontinuities. If c is such a discontinuity, then we add a term

$$\left[\lim_{x \to c+} f(x) - \lim_{x \to c-} f(x)\right] \delta(x - c).$$

Thus if $f(x) = 0$ for $x < 0$, $f(x) = \cos x$ for $0 < x < \pi/2$, and $f(x) = -1$ for $x > \pi/2$, then $f'(x) = g(x) + \delta(x) - \delta(x - (\pi/2))$, where $g(x) = 0$ for $x < 0$, $g(x) = -\sin x$ for $0 < x < \pi/2$, $g(x) = 0$ for $x > \pi/2$. One can verify that the usual rules for derivatives of sums and products hold.

Generalized functions can also be integrated. We consider here only definite integrals from a to b (where a may be $-\infty$ and b may be ∞) and exclude the case where terms of form $A\ \delta^{(k)}(x - c)$ occur, with $c = a$ or b. Then we integrate each term separately, with

$$\int_a^b \delta(x - c)\,dx = \begin{cases} 1 & \text{if } a < c < b \\ 0 & \text{otherwise} \end{cases} \tag{7–114}$$

$$\int_a^b \delta^{(k)}(x - c)\,dx = 0, \quad k = 1, 2, \ldots$$

From these definitions, we can now integrate certain products of form

$g(x)\,\delta^{(k)}(x-c)$: if $a < c < b$, then

$$\int_a^b g(x)\,\delta^{(k)}(x-c)\,dx = (-1)^k g^{(k)}(c), \tag{7-115}$$

provided that $g^{(k)}(x)$ is continuous at c. This rule follows from the previous rules, especially the rule (7–113) for multiplication. The rule (7–115) is a crucial one and is itself often made the starting point for the theory of generalized functions. It shows that $\delta^{(k)}(x-c)$ acts as a "linear functional", assigning real numbers to sufficiently smooth ordinary functions $g(x)$.

Fourier and Laplace transforms of generalized functions. From the rule (7–115) we deduce that $\delta^{(k)}(x-c)$ has a Fourier transform:

$$\int_{-\infty}^{\infty} e^{-ixt}\,\delta^{(k)}(x-c)\,dx = (it)^k e^{-ict}. \tag{7-116}$$

In particular, $\Phi[\delta(x)] = 1$ (constant function). If we could write $\delta(x)$ in terms of its Fourier integral, we would have

$$\delta(x) = \frac{1}{2\pi}\int_{-\infty}^{\infty} e^{ixt}\,dt. \tag{7-117}$$

This equation states that the constant function 1 has a Fourier transform $2\pi\,\delta(t)$. One normally *defines* the Fourier transform of 1 to be $2\pi\,\delta(t)$, even though the reasoning leading to the definition is purely formal. The definition has been shown to lead to no inconsistencies and is a useful one. In a similar manner, from (7–116) one is led to assign to $x^k e^{-icx}$ the Fourier transform $2\pi i^k\,\delta^{(k)}(t+c)$.

The Fourier transform can be regarded as a one-to-one linear mapping of a certain vector space of functions onto a second vector space of functions. The process we are following can be regarded as an extension of the linear mapping to *larger* vector spaces. The extension is chosen to preserve linearity and one-to-one-ness.

There is a similar discussion of the Laplace transform, which we know to be essentially a special case of the Fourier transform. We find

$$\mathfrak{L}[\delta^{(k)}(x-c)] = s^k e^{-cs}, \quad c \geqq 0. \tag{7-118}$$

It is a remarkable fact that, for both Fourier and Laplace transforms, generalized functions permit one to circumvent the awkward rules for transforms of derivatives. One finds:

$$\Phi[f^{(k)}(x)] = (it)^k \Phi[f], \tag{7-119}$$

$$\mathfrak{L}[f^{(k)}(x)] = s^k \mathfrak{L}[f], \tag{7-120}$$

where f is as in the previous rules, except for a finite number of jump discontinuities, as allowed for generalized functions. In (7–119) and (7–120), $f^{(k)}(x)$ is treated as a *generalized* function, so that it will contain terms in delta functions arising from jumps in f or its derivatives.

Convolutions can also be defined for generalized functions. If the rule "transform of convolution of two functions equals product of two transforms" is to hold, then we must require:

$$g(x) * \delta^{(k)}(x) = \delta^{(k)}(x) * g(x) = g^{(k)}(x), \tag{7–121}$$

where $g^{(k)}(x)$ is a generalized function. For, in the case of Laplace transforms, (7–121) gives

$$\mathcal{L}[g(x) * \delta^{(k)}(x)] = s^k \mathcal{L}[g],$$

in agreement with (7–120) above. More generally, we are led to

$$g(x) * \delta^{(k)}(x - c) = \delta^{(k)}(x - c) * g(x) = g^{(k)}(x - c) \tag{7–122}$$

and then, by term-by-term operations (as for multiplication), we can obtain the convolution of two arbitrary generalized functions.

The fact that convolution becomes multiplication in "transform language" has also been used as a starting point for the theory of generalized functions. For details, see the book by Mikusiński listed below.

Problems

1. (a) Verify the graph of $g_\epsilon(x)$, shown in Fig. 7–16.
(b) Graph $g'_\epsilon(x)$, approximating $\delta(x)$ [cf. Fig. 7–17].
(c) Graph the derivative of the pulse of Fig. 7–17, to approximate $\delta'(x)$ [cf. Fig. 7–18].
(d) Graph the derivative of the function of Fig. 7–18 to approximate $\delta''(x)$.

2. Another smooth approximation of the Heaviside unit function is the function $h_\epsilon(x) = \frac{1}{2}(1 + \sin(x/\epsilon))$ for $-\pi/(2\epsilon) \leqq x \leqq \pi/(2\epsilon)$, $= 1$ for $x > \pi/(2\epsilon)$, $= 0$ for $x < -\pi/(2\epsilon)$.
(a) Graph $h_\epsilon(x)$.
(b) Graph $h'_\epsilon(x)$ and consider this as an approximation to $\delta(x)$ for small positive ϵ.
(c) Show that $\int_{-1}^{1} h'_\epsilon(x)\,dx$ has limit 1 as $\epsilon \to 0+$.

3. Let $f(x) = 0$ for $x < 0$, $f(x) = 3 - x$ for $x > 0$. Let $g(x) = \cos x$ for $-\pi \leqq x \leqq \pi$, $g(x) = 0$ otherwise. Evaluate the following:
(a) $5f'(x) + 2g'(x)$. (b) $f''(x)$. (c) $f(x)\delta(x - 1)$. (d) $g(x)\delta'(x)$.
(e) $f(x)g''(x)$. (f) $\int_{-1}^{2} f(x)\delta'(x - 1)\,dx$. (g) $\int_{-1}^{\pi/2} g(x)\delta''(x)\,dx$. (h) $\mathcal{L}[f''(x)]$.
(i) $\Phi[g''(x)]$. (j) $f(x) * \delta'(x)$. (k) $f(x) * f''(x)$.

Answers

3. (a) $p(x) + q(x) + 15\delta(x) - 2\delta(x + \pi) + 2\delta(x - \pi)$, where $p(x) = -1$ for $x > 0$, $q(x) = -\sin x$, $-\pi \leqq x \leqq \pi$, $p = q = 0$ otherwise. (b) $3\delta'(x) - \delta(x)$. (c) $2\delta(x - 1)$. (d) $\delta'(x)$. (e) $-f(x)g(x) + (3 - \pi)\delta'(x - \pi) + \delta(x - \pi)$. (f) 1. (g) -1. (h) $3s - 1$. (i) $2t^3 \sin \pi t/(1 - t^2)$ (value π^3 for $t = \pm 1$). (j) $p(x) + 3\delta(x)$, with $p(x)$ as in answer to (a). (k) $9\delta(x) + r(x)$, where $r(x) = x - 6$ for $x \geqq 0$, $r = 0$ otherwise.

Suggested References

Churchill, Ruel V., *Fourier Series and Boundary Value Problems*, 2nd ed. New York: McGraw-Hill, 1963.

Churchill, Ruel V., *Operational Mathematics.* New York: McGraw-Hill, 1958.

Franklin, Philip, *Fourier Methods.* New York: McGraw-Hill, 1949.

Franklin, Philip, *A Treatise on Advanced Calculus.* New York: John Wiley and Sons, Inc., 1940.

Jackson, Dunham, *Fourier Series and Orthogonal Polynomials* (Carus Mathematical Monographs, No. 6). Menasha, Wisconsin: Mathematical Association of America, 1941.

Jahnke, E., and Emde, F., *Tables of Functions.* Leipzig: B. G. Teubner, 1938

Kaplan, Wilfred, *Operational Methods for Linear Systems.* Reading: Mass., Addison-Wesley, 1962.

Lighthill, M. J., *An Introduction to Fourier Analysis and Generalized Functions.* Cambridge: Cambridge University Press, 1958.

Mikusiński, Jan, *Operational Calculus.* New York: Pergamon Press, 1959.

Rogosinski, Werner, *Fourier Series* (transl. by H. Cohn and F. Steinhardt). New York: Chelsea Publishing Co., 1950.

Szegö, Gabor, *Orthogonal Polynomials* (American Mathematical Society Colloquium Publications, Vol. 23). New York: American Mathematical Society, 1939

Titchmarsh, E. C., *Eigenfunction Expansions.* Oxford: Oxford University Press, 1946.

Titchmarsh, E. C., *Theory of Fourier Integrals.* Oxford: Oxford University Press, 1937.

Watson, G. N., *Theory of Bessel Functions*, 2nd ed. New York: Macmillan, 1944.

Whittaker, E. T., and Watson, G. N., *Modern Analysis*, 4th ed. Cambridge: Cambridge University Press, 1940.

Widder, D. V., *The Laplace Transform.* Princeton: Princeton University Press, 1941.

Wiener, Norbert, *The Fourier Integral.* Cambridge: Cambridge University Press, 1933.

Zygmund, A., *Trigonometric Series.* 2 Vols. London: Cambridge University Press, 1959.

CHAPTER 8

Ordinary Differential Equations

8–1 Differential equations. An ordinary differential equation of order n is an equation of form

$$F(x, y, y', \ldots, y^{(n)}) = 0 \tag{8–1}$$

expressing a relation between x, an unspecified function $y(x)$, and its derivatives $y', y'', \ldots$ through the nth order. For example,

$$y'' + 3y' + 2y - 6e^x = 0, \tag{8–2}$$

$$(y''')^2 - 2y'y''' + (y'')^3 = 0 \tag{8–3}$$

are ordinary differential equations of orders 2 and 3 respectively.

In order for the differential equation (8–1) to have significance, it is necessary that the function F be defined in some domain of the space of the variables on which it depends. In this chapter we shall consider only equations which can be solved for the highest derivatives and written in the form:

$$y^{(n)} = F(x, y, y', \ldots, y^{(n-1)}). \tag{8–4}$$

Equation (8–2) is at once reducible to this form. Equation (8–3) is a *quadratic* equation in y'''; by solving this quadratic equation, we obtain two different equations for y''', i.e., two equations of form (8–4). We say: (8–3) is a differential equation of *degree* 2, whereas (8–2) is of degree 1.

Systems of differential equations will also be considered (Section 8–12).

The term "ordinary" is used here to emphasize that no partial derivatives appear, there being just one independent variable. An equation such as

$$\frac{\partial^2 z}{\partial x^2} - \frac{\partial^2 z}{\partial y^2} = 0 \tag{8–5}$$

would be called a *partial differential equation*. Chapter 10 below is devoted to such equations.

The applications of ordinary differential equations to physical problems are numerous. The equations of dynamics are relations between coordinates, velocities, accelerations, and time and hence give differential equations of second order or systems of higher order. Electric circuits obey laws described by differential equations relating currents and their time derivatives. Servomechanisms, or control systems, are combinations of mechanical and electrical (and perhaps other) components and can be described by differential equations. Problems involving continuous media: fluid dynamics, elasticity, heat conduction, etc., lead to *partial* differential equations.

8–2 Solutions. By a *particular solution* of (8–1) is meant a function $y = f(x)$, $a < x < b$, having derivatives up to the nth order throughout the interval and such that (8–1) becomes an identity when y and its derivatives are replaced by $f(x)$ and its derivatives. Thus $y = e^x$ is a particular solution of (8–2) and $y = x$ is a particular solution of (8–3). For most of the differential equations to be considered here it will be found that all particular solutions can be included in one formula:

$$y = f(x, c_1, \ldots, c_n), \tag{8–6}$$

where $c_1, \ldots, c_n$ are "arbitrary" constants. Thus, for each special assignment of values to the c's, (8–6) gives a solution of (8–1) and all solutions can be so obtained. (The range of the c's and of x may have to be restricted in some cases to avoid imaginary expressions or other degeneracies.) For example, all solutions of (8–2) are given by the formula

$$y = c_1 e^{-x} + c_2 e^{-2x} + e^x; \tag{8–7}$$

the solution $y = e^x$ is obtained when $c_1 = 0$, $c_2 = 0$. When a formula such as (8–6) is obtained, providing *all* solutions, it is called the *general solution* of (8–1). For the equations considered here it will be found that the *number of arbitrary constants equals the order n*.

The presence of arbitrary constants should not be surprising, for they occur in the simplest differential equation:

$$y' = F(x). \tag{8–8}$$

All solutions of (8–8) are obtained by integration:

$$y = \int F(x)\, dx + C. \tag{8–9}$$

Here there is one arbitrary constant: $c_1 = C$. This can be generalized to equations of higher order, as the following example shows:

$$y'' = 20\, x^3. \tag{8–10}$$

Since y'' is the derivative of y', one concludes by integrating twice in succession, that

$$y' = 5x^4 + c_1, \quad y = x^5 + c_1 x + c_2. \tag{8–11}$$

8–3 The basic problems. Fundamental theorem. For a given differential equation or a given system of equations, the fundamental problem is that of finding all solutions. This is a formidable task and one which can be satisfactorily completed only for a very few simple differential equations. Moreover, it will become clear from the examples to be considered that the meaning of the phrase "finding all solutions" is not as obvious as it might seem at first. It will, in fact, be seen that as many particular solutions as are required for a specific practical problem can be obtained to any desired accuracy. The only real obstacle is one of *time*, for lengthy calculations may be unavoidable. Furthermore, in many practical cases it is not the explicit expression for a general solution which is required but a knowledge of certain qualitative properties of the family of solutions. It may be possible to determine these properties without explicitly solving the differential equations. This will be illustrated below.

Besides the general problem of finding all solutions, there are two special problems of great importance: the *initial value* problem and the *boundary value* problem.

For the equation (8–1) the initial value problem is as follows: Given a value x_0 of x and n constants:

$$y_0, y_0', \ldots, y_0^{(n-1)},$$

a solution: $y = f(x)$ of (8–1) is sought such that $y = f(x)$ is defined in an interval $|x - x_0| < \delta$ $(\delta > 0)$ and

$$f(x_0) = y_0, \quad f'(x_0) = y_0', \ldots, \quad f^{(n-1)}(x_0) = y_0^{(n-1)}. \tag{8–12}$$

If x is a variable representing time and $x_0 = 0$, then (8–12) imposes n conditions on the solution at time 0 or "initial conditions."

Fundamental Theorem. *Let an ordinary differential equation of order n be given in the form*

$$y^{(n)} = F(x, y, y', \ldots, y^{(n-1)}), \tag{8–13}$$

and let the function F be defined and have continuous first partial derivatives in a domain D of the space of its variables. Let $(x_0, y_0, y_0', \ldots, y_0^{(n-1)})$ be a point of D. Then there exists a function $y = f(x)$, $x_0 - \delta < x < x_0 + \delta$ $(\delta > 0)$, which is a particular solution of (8–13) *and satisfies the initial conditions* (8–12). *Furthermore, the solution is unique: that is, if $y = g(x)$ is a second solution of* (8–13) *satisfying* (8–12), *then $f(x) = g(x)$ wherever both functions are defined.*

For a proof of this theorem (under somewhat more general conditions), one is referred to Chapter 12 of the book by Kaplan listed at the end of the chapter.†

When the general solution is explicitly known, the solution of the initial value problem is relatively simple. For example, from the general solution (8–7) of (8–2) one obtains a particular solution such that $y = 1$ and $y' = 2$ for $x = 0$ by solving the equations

$$1 = c_1 + c_2 + 1, \quad 2 = -c_1 - 2c_2 + 1.$$

Accordingly, $c_1 = 1$, $c_2 = -1$ and $y = e^{-x} - e^{-2x} + e^x$ is the particular solution sought. In general, the Fundamental Theorem gives a way of testing a formula which is believed to give the general solution (*all* solutions): *if the formula defines solutions and provides one solution for each allowable set of initial conditions, then it is indeed the general solution.*

It should be noted that the initial conditions (8–12) impose n conditions on the function $f(x)$ for a chosen value x_0 of x. Since the general solution (when it can be found) has n arbitrary constants, one is always led to n

† This book will be cited as *ODE*.

simultaneous equations in n unknowns, which "in general" have one and only one solution. One can impose some of the n conditions at one value, x_0, of x and the others at a second value, x_1. Then one seeks a particular solution $y = f(x)$ defined for $x_0 \leqq x \leqq x_1$ and satisfying the given conditions at x_0 and x_1. This is precisely the *boundary value* problem for (8–13).

For example, let the problem be to find a solution $y = f(x)$ of (8–2) which satisfies the two conditions: $f(0) = 0, f(1) = 0$. One is immediately led to the two equations

$$0 = c_1 + c_2 + 1, \quad 0 = c_1e^{-1} + c_2e^{-2} + e;$$

these have the one solution: $c_1 = -1 - e - e^2$, $c_2 = e + e^2$, so that

$$y = (-1 - e - e^2)e^{-x} + (e + e^2)e^{-2x} + e^x$$

is the solution sought.

The conditions under which the general boundary value problem has a unique solution are quite involved and will not be considered here.

Problems

1. Find the general solution of each of the following differential equations:

(a) $\dfrac{dy}{dx} = e^{2x} - x$ (c) $\dfrac{d^3y}{dx^3} = x$ (e) $\dfrac{d^ny}{dx^n} = 1$

(b) $\dfrac{d^2y}{dx^2} = 0$ (d) $\dfrac{d^ny}{dx^n} = 0$ (f) $\dfrac{dy}{dx} = \dfrac{1}{x}$

2. Find a particular solution satisfying the given initial conditions for each of the following differential equations:

(a) $\dfrac{dy}{dx} = \sin x; \quad y = 1$ for $x = 0$;

(b) $\dfrac{d^2y}{dx^2} = e^x; \quad y = 1$ and $y' = 0$ for $x = 1$;

(c) $\dfrac{dy}{dx} = y; \quad y = 1$ for $x = 0$.

3. Find a particular solution of the differential equation satisfying the given boundary conditions:

(a) $\dfrac{d^2y}{dx^2} = 1; \quad y = 1$ for $x = 0$; $\quad y = 2$ for $x = 1$;

(b) $\dfrac{d^4y}{dx^4} = 0; \quad y = 1$ for $x = -1$ and $x = 1$; $\quad y' = 0$ for $x = -1$ and $x = 1$.

4. Verify that the following are particular solutions of the differential equations given:

(a) $y = \sin x$, for $y'' + y = 0$;
(b) $y = e^{2x}$, for $y'' - 4y = 0$;
(c) $y = c_1 \cos x + c_2 \sin x$ (c_1 and c_2 any constants), for $y'' + y = 0$;
(d) $y = c_1e^{2x} + c_2e^{-2x}$ for $y'' - 4y = 0$.

5. State the order and degree of each of the following differential equations:

(a) $\dfrac{dy}{dx} = x^2 - y^2$ (c) $\left(\dfrac{dy}{dx}\right)^2 + x\dfrac{dy}{dx} - y^2 = 0$

(b) $\dfrac{d^2y}{dx^2} - \left(\dfrac{dy}{dx}\right)^2 + xy = 0$ (d) $\left(\dfrac{d^2y}{dx^2}\right)^4 - 2\dfrac{d^2y}{dx^2} + x\dfrac{dy}{dx} = 0$

Answers

1. (a) $\frac{1}{2}(e^{2x} - x^2) + c$, (b) $c_1x + c_2$, (c) $\frac{1}{24}x^4 + c_1x^2 + c_2x + c_3$,
(d) $c_1x^{n-1} + c_2x^{n-2} + \cdots + c_{n-1}x + c_n$, (e) $\dfrac{x^n}{n!} + c_1x^{n-1} + \cdots + c_n$,
(f) $\log|x| + c$ $(x \neq 0)$.

2. (a) $2 - \cos x$, (b) $e^x - ex + 1$, (c) e^x. 3. (a) $\frac{1}{2}(x^2 + x) + 1$, (b) 1.

5. Order: (a) 1, (b) 2, (c) 1, (d) 2. Degree: (a) 1, (b) 1, (c) 2, (d) 4.

8–4 Equations of first order and first degree. The general equation of first order and first degree will be considered in the form

$$\frac{dy}{dx} = F(x, y). \tag{8–14}$$

After multiplication by dx, this becomes

$$dy = F(x, y)\, dx;$$

if we now multiply by a function $g(x, y)$, it can be written in the form

$$P(x, y)\, dx + Q(x, y)\, dy = 0. \tag{8–15}$$

This last multiplication may introduce discontinuities, at points (x, y) which are discontinuities of $g(x, y)$, or extraneous solutions: curves along which $g(x, y) \equiv 0$.

The basic method for obtaining the general solution of (8–14) is to convert it by the steps described into an equation (8–15) which has the form

$$du = 0, \quad u = u(x, y). \tag{8–16}$$

The solutions of (8–16) are then given in implicit form by the equation

$$u(x, y) = c, \quad c = \text{const}, \tag{8–17}$$

that is, by the level curves of $u(x, y)$. An equation (8–15) which can be interpreted as an equation (8–16), so that $P\,dx + Q\,dy = du$, is called an *exact* equation.

From the fact that the solutions of (8–14) are level curves of a function $u(x, y)$, one obtains a geometric picture of the solutions as a *family of curves* in the xy plane (Sections 2–3 and 2–17). This is illustrated in Fig. 8–1. The Fundamental Theorem then asserts that through each point of a domain in which $\partial F/\partial x$ and $\partial F/\partial y$ are continuous there passes precisely one such curve.

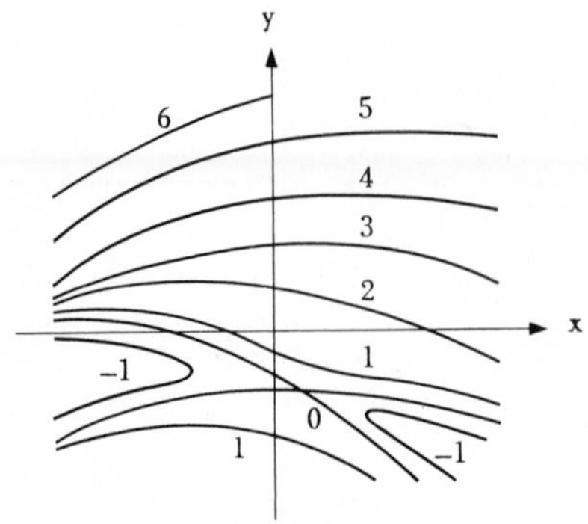

FIG. 8-1. Solutions of $y' = F(x, y)$ as level curves of $u = f(x, y)$.

Equations with variables separable. If the function $F(x, y)$ can be written as a product of a function of x by a function of y: $F(x, y) = X(x)Y(y)$, then (8-14) is said to have *variables separable*. For example,

$$\frac{dy}{dx} = -\frac{x}{y} \tag{8-18}$$

is of this type, with $X(x) = -x$, $Y(y) = 1/y$. After multiplication by ydx, (8-18) becomes

$$x\,dx + y\,dy = 0.$$

The equation is exact, since $x\,dx + y\,dy = d(\frac{1}{2}x^2 + \frac{1}{2}y^2)$. Accordingly, the solutions are the circles

$$\tfrac{1}{2}(x^2 + y^2) = \tfrac{1}{2}c, \quad \text{i.e.,} \quad x^2 + y^2 = c. \tag{8-19}$$

Equation (8-18) would give y' an infinite value along the x axis; if we allow for this as a limiting case, (8-19) provides solutions for all (x, y) except $(0, 0)$.

The general equation with variables separable can be written in the exact form:

$$P(x)\,dx + Q(y)\,dy = 0, \tag{8-20}$$

and the corresponding solutions are

$$\int P(x)\,dx + \int Q(y)\,dy = c. \tag{8-21}$$

EXAMPLE 1. $y' = y/x$.

Solution. $\dfrac{dy}{y} - \dfrac{dx}{x} = 0.$

$$\log y - \log x = c \quad (x > 0, y > 0), \quad \log\frac{y}{x} = c.$$

$$y/x = c', \quad y = c'x.$$

Here $c' = e^c$ is a new arbitrary constant and $c' > 0$. However, the restrictions: $x > 0, y > 0, c' > 0$ can be removed, since the solutions are valid for all (x, y) except when $x = 0$; this difficulty could have been avoided by writing

$$\int\frac{dy}{y} = \log|y| + \text{const}, \qquad \int\frac{dx}{x} = \log|x| + \text{const}.$$

The line $x = 0$ can be considered as a limiting solution for which $c' = \infty$.

EXAMPLE 2. $y' = x\sqrt{1 - y^5}$. The solutions can be written in the form

$$\int \frac{dy}{\sqrt{1-y^5}} = \frac{1}{2}x^2 + c \quad (y < 1).$$

The integral on the left does define a function of y, as is explained in Section 4–3. The line $y = 1$ is also a solution.

Homogeneous equations. If, in the differential equation (8–14), the function $F(x, y)$ can be expressed in terms of the one variable $v = y/x$: $F(x, y) = G(v)$, then the equation (8–14) is said to be a *homogeneous* first order equation. The following equations are of this type:

$$y' = \frac{y}{x}, \qquad y' = \frac{x^2 - y^2}{xy} = \frac{1 - v^2}{v}, \quad y' = \sin\left(\frac{y}{x}\right) = \sin v.$$

A homogeneous equation can be reduced to exact form as follows: since $y = xv$, $dy = x\,dv + v\,dx$ and

$$x\,dv + v\,dx = dy = F(x, y)\,dx = G(v)\,dx,$$

$$\frac{dv}{v - G(v)} + \frac{dx}{x} = 0 \quad \left(v = \frac{y}{x}\right).$$

In other words, the substitution $v = y/x$ and elimination of y leads to a separation of variables.

EXAMPLE 3. $y' = \dfrac{x^2 + y^2}{xy}$.

Solution.

$$y' = \frac{x}{y} + \frac{y}{x} = \frac{1}{v} + v,$$

$$x\,dv + v\,dx = dy = \left(\frac{1}{v} + v\right)dx,$$

$$v\,dv - \frac{dx}{x} = 0, \quad \frac{1}{2}v^2 - \log|x| = c,$$

$$y^2 = x^2 \log x^2 + cx^2 \quad (y \neq 0, x \neq 0).$$

8–5 The general exact equation. In the examples considered thus far, the exactness and form of the function u were immediately verifiable. The equation

$$(3x^2y + 2xy)\,dx + (x^3 + x^2 + 2y)\,dy = 0 \tag{8–22}$$

is also exact, but this is not so evident.

If an equation $P\,dx + Q\,dy = 0$ is exact, then

$$du = P(x, y)\,dx + Q(x, y)\,dy, \quad u = u(x, y).$$

Hence (Sections 2–6 and 2–11),

$$\frac{\partial u}{\partial x} = P(x, y), \quad \frac{\partial u}{\partial y} = Q(x, y),$$

$$\frac{\partial P}{\partial y} = \frac{\partial^2 u}{\partial y\,\partial x} = \frac{\partial^2 u}{\partial x\,\partial y} = \frac{\partial Q}{\partial x},$$

provided P and Q have continuous first derivatives. Therefore, if an equation $P\,dx + Q\,dy = 0$ is exact, then

$$\frac{\partial P}{\partial y} = \frac{\partial Q}{\partial x}. \tag{8–23}$$

Conversely, if (8–23) holds, then the equation $P\,dx + Q\,dy = 0$ is exact, as will be made clear. Accordingly, (8–23) is a perfect *test for exactness.*

EXAMPLE 4. The differential equation (8–22). Here

$$P = 3x^2y + 2xy, \quad Q = x^3 + x^2 + 2y;$$
$$\frac{\partial P}{\partial y} = 3x^2 + 2x, \qquad \frac{\partial Q}{\partial x} = 3x^2 + 2x.$$

The test is satisfied. To find the function u, we write

$$\frac{\partial u}{\partial x} = 3x^2y + 2xy,$$
$$u = x^3y + x^2y + C(y);$$

that is, we integrate with respect to x, but allow for an arbitrary "constant" which depends on y. The condition $\partial u/\partial y = Q$ will be satisfied if

$$x^3 + x^2 + C'(y) = x^3 + x^2 + 2y.$$

Accordingly, $C'(y) = 2y$ and $C(y) = y^2$; we add no constant to y^2, since this constant would be absorbed in the final constant. The function u is therefore the function $x^3y + x^2y + y^2$ and the equation (8–22) is equivalent to the equation

$$d(x^3y + x^2y + y^2) = 0.$$

The solutions are the curves

$$x^3y + x^2y + y^2 = c.$$

EXAMPLE 5. $x\,dy + y\,dx = 0$. By inspection this can be written: $d(xy) = 0$. Accordingly, the solutions are the curves $xy = c$.

EXAMPLE 6. $(xy \cos xy + \sin xy)\,dx + (x^2 \cos xy + e^y)\,dy = 0$. This can be solved by the method used for Example 4 or by inspection, but it will be instructive to use a different procedure. We first verify exactness:

$$\frac{\partial P}{\partial y} = 2x \cos xy - x^2y \sin xy = \frac{\partial Q}{\partial x}.$$

We then reason, as in Section 5–6, that the line integral $\int P\,dx + Q\,dy$ is independent of path and that a function u whose differential is $P\,dx + Q\,dy$ is given by

$$\int_{(x_0,\,y_0)}^{(x,\,y)} P(x, y)\,dx + Q(x, y)\,dy,$$

where (x_0, y_0) is a fixed reference point and the line integral is taken on any convenient path. If we choose (x_0, y_0) to be $(0, 0)$ and the path to be the broken line from $(0, 0)$ to $(x, 0)$ to (x, y), then we find

$$\begin{aligned} u &= \int_{(0,0)}^{(x,0)} (xy \cos xy + \sin xy)\, dx + (x^2 \cos xy + e^y)\, dy \\ &\quad + \int_{(x,0)}^{(x,y)} (xy \cos xy + \sin xy)\, dx + (x^2 \cos xy + e^y)\, dy \\ &= x \sin xy + e^y - 1, \end{aligned}$$

since $y = dy = 0$ for the first integral, while $dx = 0$ for the second. Accordingly, the solutions sought are the curves

$$x \sin xy + e^y - 1 = c.$$

Since, from its definition as a line integral, $u(x, y) = 0$ when $(x, y) = 0$, the solution through the origin (initial value problem) is

$$x \sin xy + e^y - 1 = 0.$$

The example just considered shows that the problem of exact equations is completely covered by the theory of line integrals (Chapter 5) and we can at once conclude:

If $P(x, y)$ and $Q(x, y)$ have continuous first partial derivatives in a simply connected domain D and $\partial P/\partial y = \partial Q/\partial x$ in D, then the differential equation $P\,dx + Q\,dy = 0$ is exact in D and all solutions are given by the equation $u(x, y) = c$, where

$$u = \int_{(x_0, y_0)}^{(x, y)} P\,dx + Q\,dy \tag{8–24}$$

and (x_0, y_0) is a point of D. The solution through (x_0, y_0) is defined by the equation

$$\int_{(x_0, y_0)}^{(x, y)} P\,dx + Q\,dy = 0. \tag{8–25}$$

If the domain D is not simply connected, the line integral can be used to obtain the solutions in each simply connected part of D; all solutions can then be obtained by combining those for the different portions of D.

When the variables are separable, (8–25) takes the form

$$\int_{x_0}^{x} P(x)\,dx + \int_{y_0}^{y} Q(y)\,dy = 0. \tag{8–26}$$

Integrating factors. The differential equation

$$y\,dx - x\,dy = 0$$

is not exact. However, it becomes exact when the equation is divided by y^2, for

$$\frac{y\,dx - x\,dy}{y^2} = d\left(\frac{x}{y}\right).$$

In this case the factor y^{-2} is termed an *integrating* factor of the given differential equation. In the regions $y \neq 0$, the given equation is equivalent to the equation $d(x/y) = 0$. The solutions are hence given by the lines $x/y = \text{const}$.

The question arises whether or not such an integrating factor can always be found. In a general sense, an integrating factor exists, in a suitably restricted domain; but this fact does not itself aid in finding integrating factors for particular equations.

The following differentials are worth noting, as clues for determination of integrating factors:

$$d(xy) = y\,dx + x\,dy; \tag{8–27a}$$

$$d\left(\frac{y}{x}\right) = \frac{x\,dy - y\,dx}{x^2}; \tag{8–27b}$$

$$d \arctan \frac{y}{x} = \frac{x\,dy - y\,dx}{x^2 + y^2}; \tag{8–27c}$$

$$\frac{1}{2}\,d \log (x^2 + y^2) = \frac{x\,dx + y\,dy}{x^2 + y^2}. \tag{8–27d}$$

Also it should be noted that, if $df = P\,dx + Q\,dy$, then $P\,dx + Q\,dy$ remains exact when multiplied by any function of the function f. Thus

$$2x\,dx + 2y\,dy = d(x^2 + y^2),$$

$$(x^2 + y^2)(2x\,dx + 2y\,dy) = f\,df = d\left(\frac{f^2}{2}\right) \quad (f = x^2 + y^2),$$

$$\frac{2x\,dx + 2y\,dy}{x^2 + y^2} = \frac{df}{f} = d \log f.$$

This explains (8–27d). Also (8–27c) is obtained from (8–27b) by multiplying by

$$\frac{x^2}{x^2 + y^2} = \frac{1}{1 + \left(\frac{y}{x}\right)^2} = \frac{1}{1 + f^2} \quad \left(f = \frac{y}{x}\right).$$

8–6 Linear equation of first order. A differential equation of first order is said to be linear if it can be written in the form

$$y' + p(x)y = q(x). \tag{8–28}$$

It will be convenient to choose a function $s(x)$ such that

$$p(x) = \frac{s'(x)}{s(x)} = \frac{d}{dx} \log s(x);$$

one need only choose

$$s(x) = e^{\int p(x)\,dx}. \tag{8–29}$$

The differential equation (8–28) now becomes

$$y' + \frac{s'(x)}{s(x)}\, y = q(x).$$

If both sides are multiplied by $s(x)$, it has the form

$$\frac{d}{dx}\,[s(x)y] = q(x)s(x),$$

so that the general solution is

$$s(x)y = \int q(x)s(x)\, dx + c, \quad s(x) = e^{\int p\, dx}. \tag{8–30}$$

We have actually shown that $s(x)$ is an integrating factor for the equation $dy + (py - q)\, dx = 0$. There is no need to carry an arbitrary constant in the integral $\int p\, dx$, since it can be absorbed in the final constant.

EXAMPLE. $y' + xy = x$. Here $p = x$ and $s = e^{\frac{1}{2}x^2}$. Multiplying by s, one finds

$$e^{\frac{1}{2}x^2}y' + xe^{\frac{1}{2}x^2}y = xe^{\frac{1}{2}x^2},$$

$$\frac{d}{dx}\,(e^{\frac{1}{2}x^2}y) = xe^{\frac{1}{2}x^2}.$$

Accordingly,

$$e^{\frac{1}{2}x^2}y = \int xe^{\frac{1}{2}x^2}\, dx + c = e^{\frac{1}{2}x^2} + c, \quad y = 1 + ce^{-\frac{1}{2}x^2}.$$

Problems

1. Find all solutions by separation of variables:

(a) $y' = e^{x+y}$ (c) $y' = (y-1)(y-2)$

(b) $y' = \sin x \cos y$ (d) $y' = y^{-2}$

2. Find all solutions of the following homogeneous equations:

(a) $y' = \dfrac{x-y}{x+y}$; (b) $xy' - y = xe^{y/x}$;

(c) $(3x^2y + y^3)\, dx + (x^3 + 3xy^2)\, dy = 0$.

3. Verify that the following equations are exact and find all solutions:

(a) $2xy\, dx + (x^2 + 1)\, dy = 0$;

(b) $(2x + y)\, dx + (x - 2y)\, dy = 0$;

(c) $[x \cos(x + y) + \sin(x + y)]\, dx + x \cos(x + y)\, dy = 0$.

4. Find integrating factors for each of the following differential equations and obtain the general solutions:

(a) $(x + 2y)\, dx + x\, dy = 0$;

(b) $(x + 3y)\, dx + x\, dy = 0$;

(c) $y\, dx + (y - x)\, dy = 0$;

(d) $2y^2\, dx + (2x + 3xy)\, dy = 0$;

(e) $(x^2 + y^2 + x)\, dx + y\, dy = 0$.

5. Find the general solution of each of the following linear differential equations:

(a) $\dfrac{dy}{dx} + \dfrac{1}{x+1}\,y = \sin x;$
(b) $(\sin^2 x - y)\,dx - \tan x\,dy = 0;$
(c) $(y^2 - 1)\,dx + (y^3 - y + 2x)\,dy = 0;$
(d) $\dfrac{dx}{dt} + x = e^{2t}.$

6. For each of the following differential equations, determine whether the equation has variables separable, is homogeneous, is linear, or is exact; then solve by all methods applicable:

(a) $y' = \dfrac{x+1}{y};$ (b) $y' + y = 2x + 1;$
(c) $(2xy - y + 2x)\,dx + (x^2 - x)\,dy = 0;$
(d) $y' = \dfrac{x^2 - 1}{y^2 + 1};$
(e) $\left(\dfrac{y}{xy+1} + x^2\right)dx + \dfrac{x\,dy}{xy+1} = 0;$
(f) $y \sin \log x\,dx - \tan y\,dy = 0;$
(g) $y' = \dfrac{x + \sqrt{x^2 - y^2}}{y};$
(h) $(2x \sin xy + x^2 y \cos xy)\,dx + x^3 \cos xy\,dy = 0;$
(i) $y' = y + e^y;$
(j) $(2x - y)\,dx + (x + 2y)\,dy = 0.$

7. Find the particular solution specified, for the equation given:

(a) $(2x + y + 1)\,dx + (x + 3y + 2)\,dy = 0,\quad y = 0$ when $x = 0;$
(b) $y' = \dfrac{x(y^2+1)}{(x-1)y^3},\quad y = 2$ when $x = 2;$
(c) $(3xy + 2)\,dx + x^2\,dy = 0,\quad y = 1$ when $x = 1;$
(d) $\dfrac{dx}{dt} = \dfrac{xt}{x^2 + t^2},\quad x = 1$ when $t = 0;$
(e) $e^{-x^2} y\,dx + \left(\displaystyle\int_0^x e^{-x^2}\,dx + y\right) dy = 0,\quad y = 1$ when $x = 1.$

8. Choose n so that x^n is an integrating factor of each of the following equations and obtain the general solution:

(a) $(x + y^3)\,dx + 6xy^2\,dy = 0,$ (b) $(x^2 + 2y)\,dx - x\,dy = 0.$

9. Show that $g(x, y)$ is an integrating factor of the differential equation

$$P\,dx + Q\,dy = 0$$

if and only if

$$g\left(\frac{\partial Q}{\partial x} - \frac{\partial P}{\partial y}\right) = -Q\frac{\partial g}{\partial x} + P\frac{\partial g}{\partial y}.$$

10. Given a family of curves in the xy plane, one curve through each point of a domain D, a second such family is called the family of *orthogonal trajectories* of the first family if the curves of the second family cut those of the first family at right

angles. Thus, if $y' = F(x, y)$ is a differential equation describing the first family, then

$$y' = -\frac{1}{F(x, y)}$$

is a differential equation for the second family. Find the orthogonal trajectories of the following families and graph the results:

(a) $x^2 + y^2 = c^2$
(b) $x^2 + y^2 + cx = 0$
(c) $y^2 = 4cx$
(d) $x^2 + y^2 + 2cy - 1 = 0$

11. *Method of substitution.* If new variables u, v are introduced by equations

$$x = g(u, v), \quad y = h(u, v),$$

so that

$$dx = \frac{\partial g}{\partial u} du + \frac{\partial g}{\partial v} dv, \quad dy = \frac{\partial h}{\partial u} du + \frac{\partial h}{\partial v} dv,$$

the differential equation

$$P(x, y)\, dx + Q(x, y)\, dy = 0$$

becomes an equation

$$R(u, v)\, du + S(u, v)\, dv = 0.$$

If the new equation is exact:

$$R\, du + S\, dv = df, \quad f = f(u, v),$$

then the original equation was exact:

$$P\, dx + Q\, dy = df, \quad f = f[u(x, y), v(x, y)].$$

If the new equation is converted to an exact form by an integrating factor μ, then the same reasoning shows that μ, when expressed in terms of x and y, must be an integrating factor for the original equation. If the solutions of the new equation are obtained in the form

$$\phi(u, v) = c,$$

then the solutions of the original equation are the curves

$$\phi[u(x, y), v(x, y)] = c.$$

Obtain all solutions of the following equations with the aid of the substitution indicated:

(a) $2xy\, dx + (x^2 + 2y)\, dy = 0, \quad u = x^2, \quad v = y;$
(b) $(x^2 + y^2)(x\, dx + y\, dy) + y\, dx - x\, dy = 0, \quad x = r\cos\theta, \quad y = r\sin\theta;$
(c) $(x + y - 1)\, dx + (2x + 2y + 1)\, dy = 0, \quad u = x + y, \quad v = y.$

12. Show that an appropriate translation of axes: $x = u + h$, $y = v + k$, converts the equation

$$(ax + by + c)\, dx + (px + qy + r)\, dy = 0$$

into a homogeneous equation, provided $aq - bp \neq 0$. Show that, when $aq - bp = 0$, the substitution: $u = ax + by$, $v = y$ leads to a separation of variables unless $a = 0$; discuss the exceptional case.

13. Find all solutions (cf. Prob. 12):

(a) $(x + 2y - 1)\, dx + (2x - y - 7)\, dy = 0,$
(b) $(x + y + 1)\, dx + (2x + 2y + 1)\, dy = 0.$

14. Show that the introduction of polar coordinates: $x = r\cos\theta$, $y = r\sin\theta$, leads to a separation of variables in a homogeneous equation. $y' = F(x, y)$.

Answers

1. (a) $e^x + e^{-y} = c$, (b) $(1 + \sin y) = c\cos y\, e^{-\cos x}$, (c) $y - 2 = ce^x(y - 1)$, $y = 1$, (d) $y^3 = 3x + c$.

2. (a) $x^2 - 2xy - y^2 = c$, (b) $e^{-\frac{y}{x}} + \log|x| = c$, (c) $x^3y + xy^3 = c$.

3. (a) $x^2y + y = c$, (b) $x^2 + xy - y^2 = c$, (c) $x\sin(x + y) = c$.

4. (a) $x^3 + 3x^2y = c$; (b) $x^4 + 4x^3y = c$; (c) $x + y\log|y| = cy$ and $y = 0$; (d) $y\log|x^2y^3| - 2 = cy$, $x = 0$ and $y = 0$; (e) $2x + \log(x^2 + y^2) = c$.

5. (a) $(x + 1)y = \sin x - (x + 1)\cos x + c$ and $x + 1 = 0$;
(b) $3y\sin x = \sin^3 x + c$;
(c) $x = \dfrac{y + 1}{2(y - 1)}[4y - y^2 - \log(y + 1)^4 + c]$ and $y = \pm 1$;
(d) $x = \frac{1}{3}e^{2t} + ce^{-t}$.

6. (a) $y^2 = (x + 1)^2 + c$; (b) $y = 2x - 1 + ce^{-x}$; (c) $x^2y - xy + x^2 = c$; (d) $y^3 + 3y = x^3 - 3x + c$; (e) $(xy + 1)^3 = ce^{-x^3}$;
(f) $\displaystyle\int \sin\log x\,dx - \int \frac{\tan y}{y}\,dy = c$; (g) $x + \sqrt{x^2 - y^2} = c$; (h) $x^2\sin xy = c$;
(i) $\displaystyle\int \frac{dy}{y + e^y} = x + c$; (j) $r = ce^{-\frac{\theta}{2}}$ (in polar coordinates).

7. (a) $2x^2 + 2xy + 3y^2 + 2x + 4y = 0$,
(b) $y^2 - \log(y^2 + 1) = 2x + \log(x - 1)^2 - \log 5$,
(c) $x^3y + x^2 = 2$, (d) $x = e^{\frac{\frac{1}{2}t^2}{x^2}}$,
(e) $\displaystyle 2y\int_0^x e^{-x^2}\,dx + y^2 = 2\int_0^1 e^{-x^2}\,dx + 1$.

8. (a) $x(x + 3y^3)^2 = c$, $c \neq 0$; (b) $x^2\log|x| - y = cx^2$, $x = 0$.

10. (a) $y = cx$, $x = 0$; (b) $x^2 + y^2 + cy = 0$; (c) $2x^2 + y^2 = c^2$; (d) $x^2 + y^2 - cx + 1 = 0$.

11. (a) $x^2y + y^2 = c$; (b) $x^2 + y^2 = 2\arctan\dfrac{y}{x} + c$;
(c) $x + 2y - 3\log|x + y + 2| = c$ and $x + y = -2$.

13. (a) $x^2 + 4xy - y^2 - 2x - 14y = c$;
(b) $x + 2y + \log|x + y| = c$ and $x + y = 0$.

8–7 Properties of the solutions of the linear equation. Linear differential equations of first and higher orders are of fundamental importance for applications. In fact, it can be stated that with a few exceptions the only mechanisms which are really well understood are those which obey linear equations.

The first order linear equation can be used to illustrate many basic properties of the linear equations. To emphasize the fact that it is usually the *time* which is the independent variable, the equation will be written in the form

$$a\frac{dx}{dt} + x = F(t), \tag{8–31}$$

Throughout the following, except where otherwise indicated, it will be assumed that a is a *positive constant.*

Case I. $F(t) \equiv 0$. The general solution is found from (8–30) or by separation of variables to be $x = ce^{-t/a}$ These curves are plotted for the case $a = 1$ in Fig. 8–2. They illustrate a phenomenon of very common occurrence, known as *exponential decay*. Examples are the fading out of a light bulb when the current is turned off, the cooling of a thermometer to the temperature of the surrounding medium, the decay of radium, and the reaction rates in various chemical reactions. In all cases the system is approaching a state of equilibrium, represented by the line $x = 0$. The rate of approach to the equilibrium state is $|dx/dt| = |x/a|$, and is proportional to the difference $|x|$ between the present state and the equilibrium state. For $t = 0$, $x = c$, so that $c = x_0$, the *initial value* of x. For $t = a$, $x = x_0e^{-1}$ or e^{-1} times the initial value; for $t = 2a$, $x = x_0e^{-2}$ or e^{-2} times the initial value. In general, the values of x at equally spaced times form a *geometric progression*, approaching 0 as t increases.

The number a measures the rate of approach to 0; the larger a is, the slower the rate. The number a has the dimensions of *time* and is often termed the *time constant* or *solution time*. It has been assumed here that a is positive. If a is negative, the solutions increase in absolute value as time increases, and the system is *unstable*. This case has applications in many practical problems: growth of population, growth of bacteria, growth of money at compound interest, etc.

Case II. $F(t) = K =$ constant. The solutions are found to be $x = K + ce^{-t/a}$. It is clear that the only effect is a translation of the picture of Fig. 8–2 along the x direction, as shown in Fig. 8–3. The equilibrium solution is now the line $x = K$.

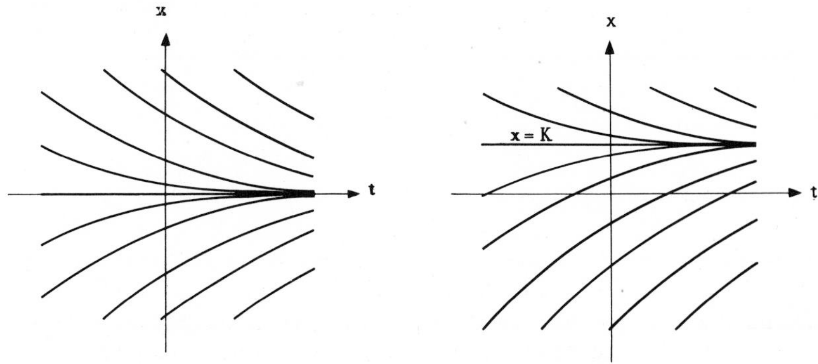

FIG. 8–2. Exponential decay to 0.

FIG. 8–3. Exponential approach to $x = K$.

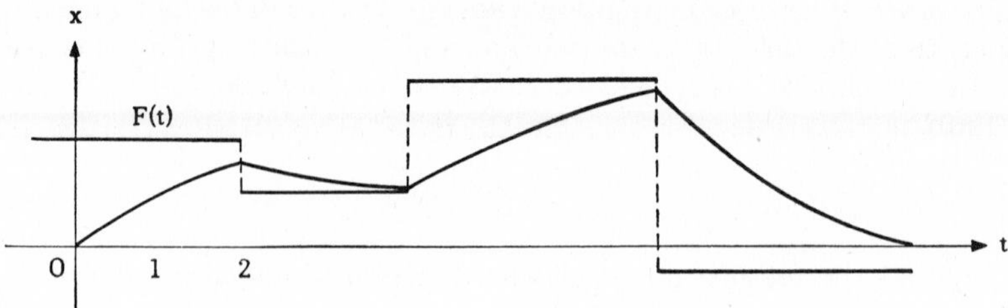

FIG. 8–4. Follow-up of step-function input.

Case III. $F(t)$ a discontinuous function, piecewise constant, as shown in Fig. 8–4. Such a function is known as a *step-function.*

Let us assume a fixed a and consider a solution $x = x(t)$ with $x(0) = 0$. From a physical point of view (not using the theory of relativity!), the mechanism described by x knows only the values of $F(t)$ in the present and past. It cannot anticipate that $F(t)$ will jump at $t = 2$. Accordingly, between $t = 0$ and $t = 2$, the solution is the same as that of Case II, with the constant $K = 2$. As t increases from 0 to 2, x increases, approaching the line $x = 2$ exponentially. When $t = 2$, F jumps to the new value 1; the mechanism proceeds to approach this value of x exponentially. The process repeats itself in the other intervals.

The situation can be described in the following way. The function $F(t)$ serves as an *input*, the solution $x(t)$ as an *output*. If there were no "mechanism," a would equal 0 and the output would equal the input. In any case, the output $x(t)$ tries to stay with the input $F(t)$. Thus, *if one ignores the term in dx/dt, the approximate solution $x = F(t)$ is obtained.* The accuracy of the approximation depends on the size of a, which governs the speed of "follow-up."

If $F(t)$ is now replaced by an arbitrary continuous, or piecewise continuous, function of t, it can be approximated as accurately as desired by a step-function and the same qualitative conclusions are reached.

If a is allowed to vary with time, the results are similar. The speed of follow-up is fluctuating, rather than steady. If a becomes negative, the solutions are unstable and are receding from, rather than following, $F(t)$.

Further information can be obtained from the general formula (8–30). For the case at hand, with $a = \text{const}$, the formula becomes:

$$x = e^{-\frac{t}{a}} \int e^{\frac{t}{a}} F(t)\,dt + ce^{-\frac{t}{a}}$$

or, in terms of an initial condition $x(0) = x_0$,

$$x = e^{-\frac{t}{a}} \int e^{\frac{t}{a}} F(t)\,dt + x_0 e^{-\frac{t}{a}}. \tag{8–32}$$

Varying the initial value of x affects the second term, but does not affect the first term. If a is positive, the second term approaches 0 exponentially, is a "transient." Thus for large values of t the solution is effectively independent of the initial conditions; the differential equation has, in this sense, only *one solution*. One can describe the situation by saying that the mechanism has a poor "memory." For any particular solution, it becomes more and more difficult as time goes on to determine what the initial value x_0 was.

If $F(t)$ is multiplied by a constant k, then the solution $x(t)$ is also, except for a transient, multiplied by the constant k. If $F(t)$ is a sum of two functions $F_1(t)$ and $F_2(t)$, then the solution $x(t)$ is (apart from transients) the sum of two solutions $x_1(t)$ and $x_2(t)$, corresponding to F_1 and F_2 respectively. In other words, *the output depends linearly on the input.* These conclusions follow immediately from (8–32).

If $F(t)$ is simple harmonic: $F(t) = A_i \sin \omega t$, then so is the solution $x(t)$, except for a transient:

$$x(t) = A_0 \sin (\omega t - \alpha) + ce^{-\frac{t}{a}}.$$

The frequency ω is the same for input and output, but the output amplitude A_0 is less than the input amplitude A_i and there is a lag α in phase. This is shown in Fig. 8–5. More specifically, one finds

$$A_0 = \frac{A_i}{\sqrt{1 + a^2\omega^2}}, \qquad \alpha = \arctan (a\omega), \quad 0 < \alpha < \frac{\pi}{2};$$

these relations are portrayed in Fig. 8–6, in which the "amplification factor" A_0/A_i and phase lag α are graphed against ω. Since $A_0 < A_i$, the amplification factor is less than 1, so that the input is diminished rather than amplified. The results stated here follow from (8–32) (see Prob. 2 below).

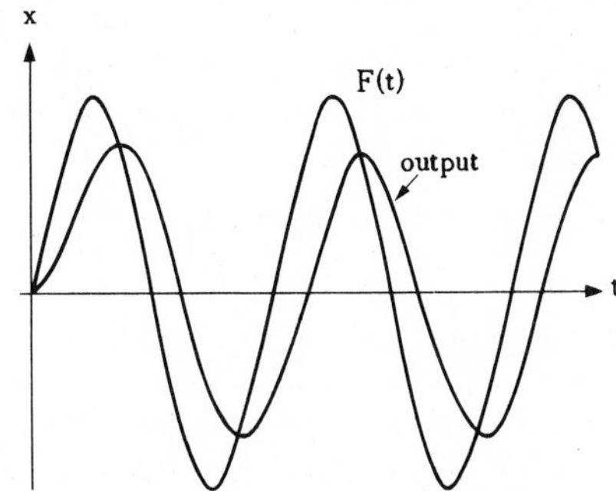

Fig. 8–5. Response to a sinusoidal input.

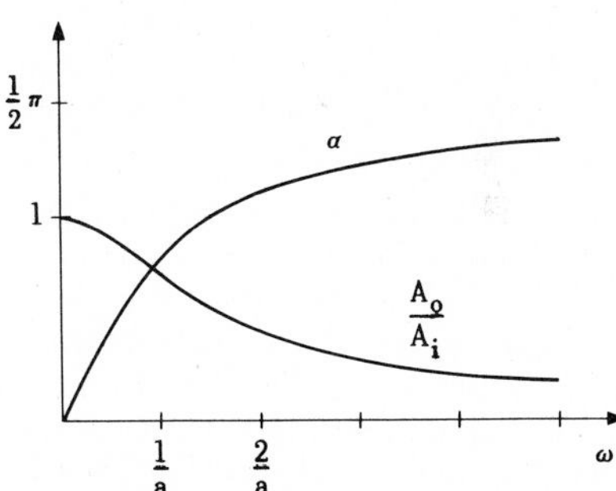

Fig. 8–6. Amplification and phase lag against input frequency.

From the linearity property, it then follows that, if F is represented by a Fourier series of period $p = 2\pi/\omega$:

$$F = \tfrac{1}{2}a_0 + \sum_{n=1}^{\infty} \{a_n \cos (n\omega t) + b_n \sin (n\omega t\},$$

then so is $x(t)$, except for a transient:

$$x(t) = \tfrac{1}{2}\bar{a}_0 + \sum_{n=1}^{\infty} \{\bar{a}_n \cos (n\omega t) + \bar{b}_n \sin (n\omega t)\} + ce^{-\frac{t}{a}}.$$

If the series for F is uniformly convergent, then this follows at once from (8–32) (see Prob. 3 below); it can be shown to hold more generally, for example, when F is continuous. Each term in the series for F represents a simple harmonic oscillation with frequency ω:

$$a_n \cos(n\omega t) + b_n \sin(n\omega t) = A_n^i \sin(n\omega t + \beta_n).$$

To this term corresponds the term of same frequency in $x(t)$:

$$\bar{a}_n \cos(n\omega t) + \bar{b}_n \sin(n\omega t) = A_n^0 \sin(n\omega t + \beta_n - \alpha_n).$$

The amplitudes and phase lags are related as before:

$$A_n^0 = \frac{A_n^i}{\sqrt{1 + n^2\omega^2 a^2}}, \quad \alpha_n = \arctan(an\omega), \quad 0 < \alpha_n < \tfrac{1}{2}\pi. \quad (8\text{–}33)$$

For the constant terms $\omega = 0$ and one finds $a_0 = \bar{a}_0$.

Because of the term in n^2 in the denominator, the terms in $F(t)$ of high frequency are diminished greatly in amplitude. The mechanism is sensitive mainly to low frequencies. This can be predicted on a qualitative basis (Prob. 4 below).

Problems

1. Evaluate and plot the solution such that $x = 0$ for $t = 0$ for the following differential equations:

(a) $\dfrac{dx}{dt} + x = 1$ (c) $\dfrac{dx}{dt} + x = \sin t$

(b) $10\dfrac{dx}{dt} + x = 1$ (d) $10\dfrac{dx}{dt} + x = \sin t$

Compare output with input in each case and discuss the amount of lag.

2. (a) Show that, if a is a positive constant, then the general solution of the differential equation

$$a\frac{dx}{dt} + x = A_i \sin \omega t$$

is given by

$$x = A_0 \sin(\omega t - \alpha) + ce^{-\frac{t}{a}},$$

where

$$A_0 = \frac{A_i}{\sqrt{1 + a^2\omega^2}}, \quad \tan \alpha = a\omega, \quad 0 \leqq \alpha < \frac{1}{2}\pi.$$

(b) Verify the graphs of Fig. 8–6.

3. Find the general solution of the differential equation

$$a\frac{dx}{dt} + x = \frac{1}{2}a_0 + \sum_{n=1}^{\infty}\{a_n \cos(n\omega t) + b_n \sin(n\omega t)\},$$

where a is a constant and the Fourier series on the right is uniformly convergent for all t [compare with the results above — especially (8–33)].

4. Let $F(t)$ be equal to 1 for $0 < t < b$, to -1 for $b < t < 2b$, to 1 for $2b < t < 3b$, to -1 for $3b < t < 4b$, etc., so that $F(t)$ is a "square wave" with period

$2b$ and amplitude 1. Discuss the qualitative features of the solutions of the differential equation

$$a\frac{dx}{dt} + x = F(t)$$

and their dependence on a and b. In particular, show that the ratio of output amplitude to input amplitude decreases and approaches 0 as b decreases.

Answers

1. (a) $1 - e^{-t}$; (b) $1 - e^{-0.1t}$; (c) $\frac{1}{2}(\sin t - \cos t + e^{-t})$; (d) $(\sin t - 10\cos t + 10e^{-t})/101$.

8–8 Graphical and numerical procedures for the first order equation. The differential equation $y' = F(x, y)$ prescribes the slope of the tangent to the solution $y = f(x)$ at the point (x, y). Accordingly, even though the solutions have not been found, we can draw the *tangents* to the solutions. If many short tangent lines are drawn, one obtains a *line element* diagram, as in Fig. 8–7, for which the differential equation is $y' = -x/y$; for example, at (1, 1) the slope is -1, at (3, 2) it is $-3/2$, etc. If enough line elements are drawn, the solutions themselves begin to appear as smooth curves.

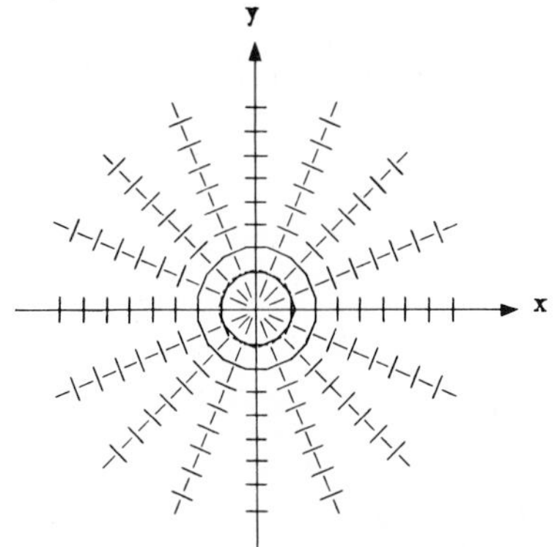

FIG. 8–7. Line elements for $y' = -x/y$.

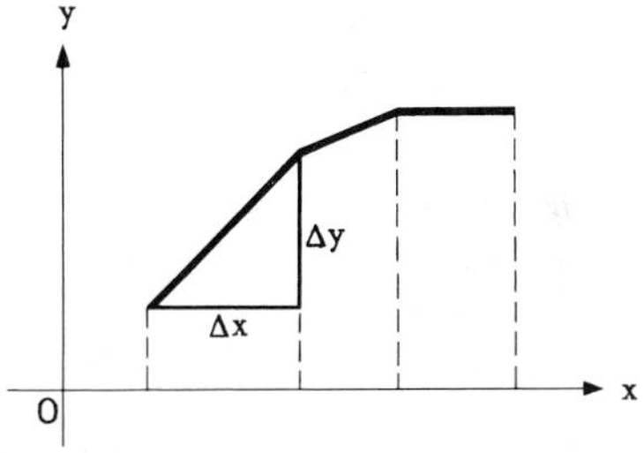

FIG. 8–8. Step-by-step integration.

The procedure of plotting the line elements can be shortened by plotting together all elements with the same slope; that is, we plot the loci $F(x, y) = m = \text{const}$ for different choices of m; these curves are not the solutions — they are called *isoclines*; the line elements through the points of the isocline $F(x, y) = m$ all have the slope m. The isoclines for Fig. 8–7, shown as broken lines, are straight lines through the origin, whereas the solutions are circles.

If one seeks just one solution through a particular point (x_0, y_0) (initial value problem), then one need not draw a complete field of line elements. One can simply draw a short segment through (x_0, y_0), with slope $F(x_0, y_0)$, and follow this to a nearby point (x_1, y_1). At this point one evaluates the slope $F(x_1, y_1)$ and draws a line segment with this slope from (x_1, y_1) to a nearby point (x_2, y_2). Repeating the process many times, one obtains a broken line, as shown in Fig. 8–8, which is an approximation to the solution sought. It is clear that, the shorter the segments used, the more accurate the solution will be. The Fundamental Theorem of Section 8–3 can in fact be proved by demonstrating that a unique solution through (x_0, y_0) is obtained by passage to the limit in the process described.

The numerical work involved in computing the particular broken-line solution can be arranged in tabular form in four columns, listing the values of x, y, $F(x, y)$, and Δy. The increment Δx is chosen at discretion, while Δy is computed by the formula

$$\Delta y = F(x, y)\,\Delta x.$$

(Thus it is really dy which is being computed, and the approximation $dy = \Delta y$ is being used.) The increment Δx can be varied from step to step, although it is simpler to keep it constant.

The example: $y' = x^2 - y^2$, with $x_0 = 1$, $y_0 = 1$, $\Delta x = 0.1$, is worked out in Table 8–1.

TABLE 8–1

x	y	$y' = x^2 - y^2$	Δy
1	1	0	0
1.1	1	0.21	0.021
1.2	1.021	0.40	0.040
1.3	1.061	...	....

The numerical procedure described here is known as *step-by-step integration* of the differential equation. It should be noted that, for the differential equation $y' = F(x)$, the procedure is equivalent to integrating $F(x)$ by a rectangular formula, as in Section 4–2. For the value of y for $x = x_0 + n\,\Delta x$ on the solution through (x_0, y_0) is given exactly by

$$y = \int_{x_0}^{x_0+n\Delta x} F(x)\,dx + y_0;$$

the above numerical procedure gives

$$\begin{aligned} y &= y_0 + F(x_0)\,\Delta x + F(x_0 + \Delta x)\,\Delta x + \cdots \\ &= y_0 + \Delta x\{F(x_0) + F(x_0 + \Delta x) + \cdots + F[x_0 + (n-1)\,\Delta x]\}. \end{aligned}$$

This is precisely the rectangular sum with F evaluated at left-hand end points.

The generalization of step-by-step integration to equations of higher order is indicated in Probs. 5 and 6 below. Other numerical procedures are described in the books of Henrici, Hildebrand and Ralston listed at the end of the chapter. Such methods are widely used in solving differential equations on *digital computers*, with whose aid astonishing speed and accuracy can be achieved.

Problems

1. For each of the following families of curves plot the family (i.e., plot a number of curves of the family) and find the slope of the curve of the family through an arbitrary point (x, y):

(a) $y = 2x + c$
(b) $y = x^2 + c$
(c) $x^2 - y^2 = c$
(d) $y = cx + 1$
(e) $x^2 + cy = 0$
(f) $y = ce^{-x}$

2. Plot a number of line elements for each of the following differential equations:

(a) $y' = \dfrac{y}{x}$; (b) $y' = x - y$; (c) $y' = x + y^2$.

Also try to sketch a few solution curves in each case. For (a) and (b) compare the results with the general solutions which are as follows:

(a) $y = cx$, $x \neq 0$; (b) $y = ce^{-x} + x - 1$.

3. Using step-by-step integration with $\Delta x = 0.1$, find the value of y at $x = 1.5$ on the solution of $y' = x - y^2$ such that $y = 1$ when $x = 1$. Plot the solution obtained as a broken line.

4. Using step-by-step integration with $\Delta x = 0.1$, find the value of y at $x = 0.5$ on the solution of $y' = \sqrt{1 - y^2}$ such that $y = 0$ for $x = 0$. Compare the result with the exact solution, which is $y = \sin x$.

5. The equations

$$\frac{dy}{dx} = f(x, y, z), \quad \frac{dz}{dx} = g(x, y, z)$$

would be termed a *system of two simultaneous ordinary differential equations*. By a particular solution of such a system is meant a pair of functions $y(x)$, $z(x)$ satisfying the equations identically. The Fundamental Theorem of Section 8–3 can be extended to this case and under appropriate assumptions guarantees existence of a unique solution $y(x)$, $z(x)$ satisfying given initial conditions $y(x_0) = y_0$, $z(x_0) = z_0$. An approximate solution can be obtained by step-by-step integration as follows. One selects Δx and then computes $y_1 = y_0 + \Delta y$, $z_1 = z_0 + \Delta z$ by the equations: $\Delta y = f(x_0, y_0, z_0)\,\Delta x$, $\Delta z = g(x_0, y_0, z_0)\,\Delta x$. The process can then be repeated with the new initial values $x_1 = x_0 + \Delta x$, y_1, z_1. Apply the process described, with $x_0 = 0$, $y_0 = 1$, $z_0 = 0$, $\Delta x = 0.5$, to the differential equations

$$\frac{dy}{dx} = yz - x, \quad \frac{dz}{dx} = x + y;$$

Compute the solution $y(x)$, $z(x)$ up to $x = 3$.

6. A second order equation: $y'' = F(x, y, y')$ is equivalent to a pair of equations:

$$\frac{dy}{dx} = z, \quad \frac{dz}{dx} = F(x, y, z),$$

where $z = y'$. Accordingly, step-by-step integration can be applied as in Prob. 5. Determine by step-by-step integration the solution of the equation:

$$y'' = yy' + x$$

such that $y = 1$ and $y' = 0$ for $x = 0$. Use $\Delta x = 0.5$ and compute the solution up to $x = 3$.

Answers

1. (a) $y' = 2$, (b) $y' = 2x$, (c) $y' = \frac{x}{y}$, (d) $y' = \frac{(y-1)}{x}$,
(e) $y' = \frac{2y}{x}$, (f) $y' = -y$.

3. 1.082. 4. 0.4850. 5. $y = 27.1$, $z = 10.8$ when $x = 3$.

6. $y = 5.00$ when $x = 3$.

8–9 Linear differential equations of arbitrary order. An ordinary linear differential equation of order n is a differential equation of form

$$a_0(x)y^{(n)} + a_1(x)y^{(n-1)} + \cdots + a_{n-1}(x)y' + a_n(x)y = Q(x). \quad (8\text{–}34)$$

It will be assumed here that the coefficients $a_0(x), a_1(x), \ldots, a_n(x)$ and the right-hand member $Q(x)$ are defined and continuous in an interval $a \leqq x \leqq b$ of the x axis and that $a_0(x) \neq 0$ in this interval.

The following are examples of linear equations:

$$y'' + y = \sin 2x, \quad \text{(a)}$$

$$x^2y''' - xy' + e^xy = \log x \ (x > 0), \quad \text{(b)}$$

$$\frac{d^5y}{dx^5} - x\frac{d^3y}{dx^3} + x^3\frac{dy}{dx} = 0, \quad \text{(c)}$$

$$y'' + y = 0, \quad \text{(d)}$$

$$\frac{dy}{dx} + x^2y = e^x. \quad \text{(e)}$$

If $Q(x) = 0$, the equation (8–34) is said to be *homogeneous* (with respect to y and its derivatives). Thus (c) and (d) are homogeneous; the other examples are *nonhomogeneous*. If $Q(x)$ is replaced by 0 in a general equation (8–34), a new homogeneous equation is obtained, termed the *homogeneous equation corresponding to the given differential equation.*

If the coefficients $a_0(x), a_1(x), \ldots, a_n(x)$ are all constants, hence independent of x, the equation (8–34) is said to have *constant coefficients,* even though $Q(x)$ depends on x. Thus (a) and (d) have constant coefficients; the other examples do not.

It will be convenient to write the differential equation (8–34) in "operational" form:

$$\left[a_0(x)\frac{d^n}{dx^n} + \cdots + a_{n-1}(x)\frac{d}{dx} + a_n(x)\right][y] = Q(x)$$

or, with the abbreviation:

$$L = a_0(x)\frac{d^n}{dx^n} + \cdots + a_{n-1}\frac{d}{dx} + a_n(x), \quad (8\text{–}35)$$

simply as follows:

$$L[y] = Q(x).$$

For example, the equation: $xy'' + 2xy' - 3y = 5$ would be abbreviated: $L[y] = 5$, $L = x(d^2/dx^2) + 2x(d/dx) - 3$.

From the basic rules of differentiation, we conclude that L is a *linear operator* (Section 3–6); that is,

$$L[c_1y_1(x) + c_2y_2(x)] = c_1L[y_1] + c_2L[y_2], \tag{8–36}$$

where $y_1(x)$ and $y_2(x)$ are functions having derivatives through the nth order for $a \leqq x \leqq b$ and c_1, c_2 are constants. From this we conclude: *if $y_1(x)$ and $y_2(x)$ are solutions of the homogeneous equation $L[y] = 0$, then so also is $c_1y_1(x) + c_2y_2(x)$.* Hence from known solutions $y_1(x), \ldots, y_n(x)$ of the homogeneous equation we can construct solutions $y = cy_1(x) + \cdots + c_ny_n(x)$ containing n arbitrary constants. This has the appearance of a general solution. However, it might be the case that $y_n(x)$, for example, is a *linear combination* of $y_1(x), \ldots, y_{n-1}(x)$:

$$y_n(x) = k_1y_1(x) + \cdots + k_{n-1}y_{n-1}(x), \quad a \leqq x \leqq b,$$

where $k_1, \ldots, k_{n-1}$ are constants. The hypothetical general solution would then really involve only $n - 1$ constants:

$$y = (c_1 + c_nk_1)y_1(x) + \cdots + (c_{n-1} + c_nk_{n-1})\, y_{n-1}(x)$$

and could not be expected to be the general solution. In order to rule this out, we assume that the functions $y_1(x), \ldots, y_n(x)$ are *linearly independent* (Sections 1–15, 7–10); that is, that no one of the functions is expressible as a linear combination of the others or, equivalently, that an identity

$$c_1y_1(x) + \cdots + c_ny_n(x) \equiv 0, \quad a \leqq x \leqq b,$$

can hold only if the constants $c_1, \ldots, c_n$ are all 0. If this is the case, then $y = c_1y_1(x) + \cdots + c_ny_n(x)$ is indeed the general solution:

THEOREM A. *There exist n linearly independent solutions of the homogeneous differential equation $L[y] = 0$ in the given interval $a \leqq x \leqq b$. If $y_1(x), \ldots, y_n(x)$ are linearly independent solutions of the equation $L[y] = 0$ for $a \leqq x \leqq b$, then $y = c_1y_1(x) + \cdots + c_ny_n(x)$ is the general solution.*

For a proof we refer to Chapter 12 of *ODE*.

The general solution of the nonhomogeneous equation $L[y] = Q(x)$ can be constructed from the general solution $c_1y_1(x) + \cdots + c_ny_n(x)$ of $L[y] = 0$ and one particular solution $y^*(x)$ of $L[y] = Q$, namely, as the expression

$$y = y^*(x) + c_1y_1(x) + \cdots + c_ny_n(x). \tag{8–37}$$

For, by linearity,

$$\begin{aligned} L[y] &= L[y^* + c_1y_1 + \cdots + c_ny_n] = L[y^*] + c_1L[y_1] + \cdots + c_nL[y_n] \\ &= L[y^*] + 0 = Q(x); \end{aligned}$$

furthermore, if $y(x)$ is any solution of $L[y] = Q$, then $L[y - y^*] = L[y] -$

$L[y^*] = Q - Q = 0$. Hence $y - y^*$ is a solution of the homogeneous equation and y has form (8–37):

THEOREM B. *There exists a solution of the nonhomogeneous equation* $L[y] = Q(x)$ *in the given interval* $a \leqq x \leqq b$. *If* $y^*(x)$ *is one such solution and* $c_1y_1(x) + \cdots + c_ny_n(x)$ *is the general solution of* $L[y] = 0$, *then* $y = y^*(x) + c_1y_1(x) + \cdots + c_ny_n(x)$ *is the general solution of* $L[y] = Q(x)$ *for* $a \leqq x \leqq b$.

For a proof of existence of a particular solution $y^*(x)$ we again refer to *ODE*. In general, existence of solutions is guaranteed by the Fundamental Theorem of Section 8–3; however, one must also show that each solution is defined throughout the given interval $a \leqq x \leqq b$.

8–10 Linear differential equations with constant coefficients. Homogeneous case. The discussion of the preceding section gives no clue as to how the functions $y_1(x), \ldots, y_n(x)$ and $y^*(x)$ are to be found. For the general linear equation this is quite difficult, although infinite power series prove to be very helpful; this is discussed in Section 8–14 below.

For the special case of *constant coefficients*, the problem has been completely solved. In this section we give the solution for the homogeneous case.

Let the given differential equation be

$$L[y] \equiv a_0 \frac{d^ny}{dx^n} + \cdots + a_{n-1} \frac{dy}{dx} + a_ny = 0, \tag{8–38}$$

where $a_0, \ldots, a_n$ are constants and $a_0 \neq 0$. Now by direct substitution

$$L[e^{rx}] = e^{rx}(a_0r^n + \cdots + a_{n-1}r + a_n) = f(r)e^{rx},$$
$$f(r) = a_0r^n + \cdots + a_{n-1}r + a_n.$$

Accordingly, e^{rx} will be a solution, provided r satisfies the algebraic equation of degree n: $f(r) = 0$. The equation $f(r) = 0$ is called the *characteristic equation* associated with (8–38). If the equation has n distinct real roots $r_1, \ldots, r_n$, then the functions

$$y_1 = e^{r_1x}, \quad y_2 = e^{r_2x}, \ldots, \quad y_n = e^{r_nx}$$

are all solutions and are linearly independent for all x (cf. Prob. 3 below). Hence in this case

$$y = c_1e^{r_1x} + \cdots + c_ne^{r_nx} \tag{8–39}$$

is the general solution of (8–38).

While it is shown in algebra that an equation of degree n has n roots (cf. Section 0–3), some of the n roots may be complex numbers and some roots may be equal. It will be seen that (8–39) can be modified appropriately to cover these cases.

For example, the equation $y'' + y = 0$ has the characteristic equation: $r^2 + 1 = 0$, with roots $\pm i$. Proceeding formally, we would obtain the "solutions" e^{ix} and e^{-ix}. From these imaginary solutions, we can form linear combinations which are real:

$$\cos x = \frac{e^{ix} + e^{-ix}}{2}, \quad \sin x = \frac{e^{ix} - e^{-ix}}{2i}, \tag{8–40}$$

as follows from the identity (Section 6–19)

$$e^{ix} = \cos x + i \sin x. \tag{8–41}$$

The functions $y_1(x) = \cos x$ and $y_2(x) = \sin x$ are indeed linearly independent solutions of the given equation: $y'' + y = 0$, and

$$y = c_1 \cos x + c_2 \sin x$$

is the general solution.

A similar analysis applies to complex roots in general. As is shown in algebra, such roots come in pairs $a \pm bi$ (complex conjugates). From the two functions

$$e^{(a \pm bi)x} = e^{ax}(\cos bx \pm i \sin bx)$$

one obtains the two real functions

$$e^{ax} \cos bx, \quad e^{ax} \sin bx$$

as solutions of the differential equation.

If the equation has a multiple real root r, of multiplicity k, then it can be shown (Prob. 9 below) that the functions $e^{rx}, xe^{rx}, \ldots, x^{k-1}e^{rx}$ are solutions. If $a \pm bi$ is a pair of multiple complex roots, then one obtains solutions $e^{ax} \cos bx$, $e^{ax} \sin bx$, $xe^{ax} \cos bx$, $xe^{ax} \sin bx, \ldots$, $x^{k-1}e^{ax} \cos bx$, $x^{k-1}e^{ax} \sin bx$.

We summarize the results. After finding the n roots of the characteristic equation, one assigns:

(I) to each simple real root r the function e^{rx};
(II) to each pair $a \pm bi$ of simple complex roots the functions $e^{ax} \cos bx$, $e^{ax} \sin bx$;
(III) to each real root r of multiplicity k the functions $e^{rx}, xe^{rx}, \ldots, x^{k-1}e^{rx}$;
(IV) to each pair $a \pm bi$ of complex roots of multiplicity k the functions $e^{ax} \cos bx$, $e^{ax} \sin bx$, $xe^{ax} \cos bx$, $xe^{ax} \sin bx, \ldots$, $x^{k-1}e^{ax} \cos bx$, $x^{k-1}e^{ax} \sin bx$.

If the n functions $y_1(x), \ldots, y_n(x)$ thus obtained are multiplied by arbitrary constants and added, then the general solution of (8–38) is obtained in the form

$$y = c_1y_1(x) + c_2y_2(x) + \cdots + c_ny_n(x).$$

Example 1. $y'' - y = 0$. The characteristic equation

$$r^2 - 1 = 0$$

has the distinct roots $1, -1$. Hence the general solution is

$$y = c_1e^x + c_2e^{-x}.$$

EXAMPLE 2. $y''' - y'' + y' - y = 0$. The characteristic equation

$$r^3 - r^2 + r - 1 = 0$$

has the distinct roots $1, i, -i$. Hence the general solution is

$$y = c_1e^x + c_2 \cos x + c_3 \sin x.$$

EXAMPLE 3. $y'' + y' + y = 0$. The characteristic equation is

$$r^2 + r + 1 = 0,$$

with roots $-\frac{1}{2} \pm \frac{\sqrt{3}}{2} i$. Hence by (II) the general solution is

$$y = e^{-\frac{1}{2}x}\left(c_1 \cos \frac{\sqrt{3}}{2} x + c_2 \sin \frac{\sqrt{3}}{2} x\right).$$

EXAMPLE 4. $\dfrac{d^6y}{dx^6} + 8\dfrac{d^4y}{dx^4} + 16\dfrac{d^2y}{dx^2} = 0$. The characteristic equation is

$$r^6 + 8r^4 + 16r^2 = 0$$

or

$$r^2(r^2 + 4)^2 = 0.$$

The roots are $0, 0, \pm 2i, \pm 2i$. By (III) and (IV) the general solution is

$$y = c_1 + c_2x + c_3 \cos 2x + c_4 \sin 2x + c_5x \cos 2x + c_6x \sin 2x.$$

It thus appears that the homogeneous linear equation with constant coefficients is completely solved; once one has solved a certain algebraic equation, one can write down all solutions of the differential equation. However, in a practical sense this is not sufficient, for the solution of an algebraic equation need be neither simple nor brief. For information on this we refer to the books by Henrici and Ralston listed at the end of the chapter.

Problems

1. Verify that the function $y = c_1x^{-2} + c_2x^{-1}$, $x > 0$, satisfies the differential equation

$$x^2y'' + 4xy' + 2y = 0$$

for every choice of c_1 and c_2, and that c_1 and c_2 can be chosen uniquely so that y satisfies the initial conditions $y = y_0$ and $y' = y_0'$ for $x = x_0$ ($x_0 > 0$). Hence $y = c_1x^{-2} + c_2x^{-1}$ is the general solution for $x > 0$.

2. Verify that the function

$$y = x^2 + e^{-x}(c_1 \cos 2x + c_2 \sin 2x)$$

satisfies the differential equation

$$y'' + 2y' + 5y = 5x^2 + 4x + 2$$

for all x, and that c_1 and c_2 can be chosen uniquely so that y satisfies the initial conditions $y = y_0$, $y' = y_0'$ for $x = x_0$. Hence the given expression is the general solution.

3. Show that the functions e^x, e^{2x}, e^{3x} are linearly independent for all x. [Hint: if an identity

$$c_1e^x + c_2e^{2x} + c_3e^{3x} \equiv 0$$

holds, then differentiate twice. This gives 3 equations for c_1, c_2, c_3 whose only solution is $c_1 = 0$, $c_2 = 0$, $c_3 = 0$.]

4. Show that the following functions are linearly independent for all x (cf. Prob. 3):

(a) x, x^2, x^3
(b) $\sin x$, $\cos x$, $\sin 2x$
(c) e^x, xe^x, $\sinh x$
(d) e^x, e^{-x}

5. Determine which of the following sets of functions are linearly independent for all x:

(a) $\sinh x$, e^x, e^{-x}
(b) $\cos 2x$, $\cos^2 x$, $\sin^2 x$
(c) $1 + x$, $1 + 2x$, x^2
(d) $x^2 - x + 1$, $x^2 - 1$, $3x^2 - x - 1$

6. Find the general solution:

(a) $y'' - 4y = 0$
(b) $y'' + 4y' = 0$
(c) $y''' - 3y'' + 3y' - y = 0$
(d) $\dfrac{d^4y}{dx^4} + y = 0$
(e) $\dfrac{d^5y}{dx^5} + 2\dfrac{d^3y}{dx^3} + \dfrac{dy}{dx} = 0$
(f) $\dfrac{dx}{dt} + 3x = 0$
(g) $\dfrac{d^2x}{dt^2} + \dfrac{dx}{dt} + 7x = 0$
(h) $\dfrac{d^5y}{dx^5} = 0$

7. Find the particular solution of the differential equation satisfying the given initial conditions for each of the following cases:

(a) $y'' + y = 0$, $y = 1$ and $y' = 0$ for $x = 0$;

(b) $\dfrac{d^2x}{dt^2} + \dfrac{dx}{dt} - 3x = 0$, $x = 0$ and $\dfrac{dx}{dt} = 1$ for $t = 0$.

8. Find a particular solution of the equation satisfying the given boundary conditions for each of the following cases:

(a) $y'' - y' - 6y = 0$, $y = 1$ for $x = 0$, $y = 0$ for $x = 1$;

(b) $y'' + y = 0$, $y = 1$ for $x = 0$, $y = 2$ for $x = \dfrac{\pi}{2}$;

(c) $y'' + y = 0$, $y = 0$ for $x = 0$, $y = 0$ for $x = \pi$.

9. (a) Prove that, if r_1 is a double root of the characteristic equation $f(r) = 0$, then xe^{r_1x} is a solution of the corresponding homogeneous linear differential equation. [Hint: if $f(r) = (r - r_1)^2(a_0r^{n-2} + \cdots)$, then $f(r_1) = 0$ and $f'(r_1) = 0$. Now e^{rx} can be considered as a function of r and x and from the rule

$$\frac{\partial^2 u}{\partial r\,\partial x} = \frac{\partial^2 u}{\partial x\,\partial r}$$

for such a function we conclude that

$$L[xe^{rx}] = L\left[\frac{\partial}{\partial r}e^{rx}\right] = \frac{\partial}{\partial r}L[e^{rx}] = \frac{\partial}{\partial r}[f(r)e^{rx}] = f'(r)e^{rx} + xf(r)e^{rx}.$$

Now set $r = r_1$.]

(b) Extend the result of part (a) to the case of a multiple root of multiplicity k.

Answers

5. (a), (b), (d) are linearly dependent, (c) is a linearly independent set.

6. (a) $c_1e^{2x} + c_2e^{-2x}$; (b) $c_1 + c_2e^{-4x}$; (c) $c_1e^x + c_2xe^x + c_3x^2e^x$; (d) $e^{\frac{1}{2}\sqrt{2}x}[c_1 \cos(\frac{1}{2}\sqrt{2}x) + c_2 \sin(\frac{1}{2}\sqrt{2}x)] + e^{-\frac{1}{2}\sqrt{2}x}[c_3 \cos(\frac{1}{2}\sqrt{2}x) + c_4 \sin(\frac{1}{2}\sqrt{2}x)]$; (e) $c_1 + c_2 \cos x + c_3 \sin x + c_4x \cos x + c_5x \sin x$; (f) c_1e^{-3t}; (g) $e^{-\frac{1}{2}t}[c_1 \cos(\frac{1}{2}3\sqrt{3}t) + c_2 \sin(\frac{1}{2}3\sqrt{3}t)]$; (h) $c_1 + c_2x + c_3x^2 + c_4x^3 + c_5x^4$.

7. (a) $\cos x$; (b) $\dfrac{2}{\sqrt{13}}\, e^{-\frac{1}{2}t} \sinh(\frac{1}{2}\sqrt{13}t)$.

8. (a) $\dfrac{1}{1 - e^5}(e^{3x} - e^{5-2x})$; (b) $\cos x + 2 \sin x$; (c) $c \sin x$.

8–11 Linear differential equations, nonhomogeneous case. In view of Theorem B of Section 8–9, the solution of a nonhomogeneous linear equation

$$L[y] = a_0(x)y^{(n)} + \cdots + a_{n-1}(x)y' + a_n(x)y = Q(x) \tag{8–42}$$

is reduced to two separate problems: (a) determination of the general solution of the corresponding homogeneous equation $L[y] = 0$; (b) determination of *one* particular solution $y^*(x)$ of the nonhomogeneous equation $L[y] = Q$.

The solution of (a) is a function

$$y_c(x) = c_1y_1(x) + \cdots + c_ny_n(x) \tag{8–43}$$

depending on x and n arbitrary constants; we term $y_c(x)$ the *complementary function*. When the coefficients are constant, this function can be explicitly found by the methods of Section 8–10.

The present section is devoted to the method of *variation of parameters*, by which a particular solution $y^*(x)$ can be found, whenever the complementary function is known. The method is therefore applicable whether the coefficients are constant or not, provided always we can determine the complementary function. Methods of determining the complementary function when the coefficients are variable are described in Section 8–14. Other methods of finding $y^*(x)$ when the coefficients are constant are described in Probs. 5 and 6 below.

Let the equation (8–42) be given, with known complementary function $y_c(x)$. Then a solution $y^*(x)$ of (8–42) will be sought, having the form

$$y^*(x) = v_1(x)y_1(x) + \cdots + v_n(x)y_n(x). \tag{8–44}$$

Thus the constants (or "parameters") $c_1, \ldots, c_n$ in (8–43) have been replaced by functions $v_1(x), \ldots, v_n(x)$. The v's must be chosen so that (8–44) satisfies (8–42); that is *one* condition and there are n functions

Hence $n-1$ additional conditions must be imposed.

We choose as these $n-1$ conditions the following equations:

$$\begin{gathered} y_1v_1' + \cdots + y_nv_n' = 0, \quad y_1'v_1' + \cdots + y_n'v_n' = 0, \ldots, \\ y_1^{(n-2)}v_1' + \cdots + y_n^{(n-2)}v_n' = 0. \end{gathered} \tag{8–45}$$

They are chosen so that *the successive derivatives up to order* $n-1$ *of the function* (8–44) *depend on* $v_1(x), \ldots, v_n(x)$ *and not on the derivatives of the* v's. Indeed,

$$y^{*\prime} = v_1y_1' + \cdots + v_ny_n' + y_1v_1' + \cdots + y_nv_n' = v_1y_1' + \cdots + v_ny_n',$$

because of the first of (8–45). In general,

$$y^{*(k)} = v_1y_1^{(k)} + \cdots + v_ny_n^{(k)} \quad (k = 1, \ldots, n-1), \tag{8–46}$$

$$y^{*(n)} = v_1y_1^{(n)} + \cdots + v_ny_n^{(n)} + y_1^{(n-1)}v_1' + \cdots + y_n^{(n-1)}v_n'. \tag{8–47}$$

From these relations we find that

$$\begin{aligned} L[y^*] &= v_1L[y_1] + \cdots + v_nL[y_n] + a_0(y_1^{(n-1)}v_1' + \cdots + y_n^{(n-1)}v_n') \\ &= a_0(y_1^{(n-1)}v_1' + \cdots + y_n^{(n-1)}v_n'), \end{aligned}$$

since $L[y_1] = 0, \ldots, L[y_n] = 0$. Accordingly, substitution of (8–44) in (8–42) gives the equation

$$a_0(y_1^{(n-1)}v_1' + \cdots + y_n^{(n-1)}v_n') = Q(x). \tag{8–48}$$

This is the nth condition on the v's.

In all, then, the functions $v_1(x), \ldots, v_n(x)$ are required to satisfy the n equations

$$\begin{aligned} &y_1v_1' + y_2v_2' + \cdots + y_nv_n' = 0, \\ &y_1'v_1' + y_2'v_2' + \cdots + y_n'v_n' = 0, \\ &\ldots, \\ &y_1^{(n-2)}v_1' + \cdots + y_n^{(n-2)}v_n' = 0, \\ &y_1^{(n-1)}v_1' + \cdots + y_n^{(n-1)}v_n' = \frac{Q(x)}{a_0(x)}. \end{aligned} \tag{8–49}$$

These are n simultaneous linear equations for the derivatives $v_1', \ldots, v_n'$. They can be solved by determinants (Section 0–3). Thus one finds

$$v_1' = \frac{D_1}{D}, \quad v_2' = \frac{D_2}{D}, \quad \ldots, \quad v_n' = \frac{D_n}{D}, \tag{8–50}$$

where

$$D = \begin{vmatrix} y_1 & y_2 & \cdot & y_n \\ y_1' & y_2' & \cdot & y_n' \\ \cdot & \cdot & \cdot & \cdot \\ y_1^{(n-1)} & y_2^{(n-1)} & \cdot & y_n^{(n-1)} \end{vmatrix}, \quad D_1 = \begin{vmatrix} 0 & y_2 & \cdot & y_n \\ 0 & y_2' & \cdot & y_n' \\ \cdot & \cdot & \cdot & \cdot \\ \dfrac{Q}{a_0} & y_2^{(n-1)} & \cdot & y_n^{(n-1)} \end{vmatrix}, \ldots. \tag{8–51}$$

When $v_1', \ldots, v_n'$ are known, $v_1, \ldots, v_n$ can be found by integration. Substitution in (8–44) then gives the particular solution sought.

It should be noted that the determinant D cannot equal 0, for it can be shown that, if $D = 0$ at one point of the given interval, then the functions $y_1(x), \ldots, y_n(x)$ are linearly dependent. A proof is given on pages 118–119 of *ODE*. The determinant D is termed the *Wronskian determinant* of the functions $y_1(x), \ldots, y_n(x)$.

Example. $y''' - y'' + y' - y = x$.

Solution. The complementary function is found to be

$$y_c(x) = c_1 e^x + c_2 \cos x + c_3 \sin x.$$

Hence the particular solution $y^*(x)$ will be sought in the form

$$y^*(x) = v_1(x)e^x + v_2(x) \cos x + v_3(x) \sin x.$$

The equations (8–49) become

$$\begin{aligned} e^x v_1' + \cos x\, v_2' + \sin x\, v_3' &= 0, \\ e^x v_1' - \sin x\, v_2' + \cos x\, v_3' &= 0, \\ e^x v_1' - \cos x\, v_2' - \sin x\, v_3' &= x. \end{aligned}$$

Solving the equations simultaneously, we find

$$v_1' = \tfrac{1}{2}xe^{-x}, \quad v_2' = \tfrac{1}{2}x(\sin x - \cos x), \quad v_3' = -\tfrac{1}{2}x(\sin x + \cos x).$$

Hence after integration (arbitrary constants being ignored), we find

$$v_1 = -\tfrac{1}{2}(x+1)e^{-x}, \quad v_2 = -\tfrac{1}{2}[x(\sin x + \cos x) - \sin x + \cos x],$$
$$v_3 = \tfrac{1}{2}[x(\cos x - \sin x) - \sin x - \cos x].$$

Accordingly,

$$y^* = v_1 e^x + v_2 \cos x + v_3 \sin x = -x - 1.$$

The general solution is

$$y = c_1 e^x + c_2 \cos x + c_3 \sin x - x - 1 = y_c(x) + y^*(x).$$

Problems

1. Find the general solution by the method of variation of parameters:

(a) $y'' - y = e^x$
(b) $y''' - 6y'' + 11y' - 6y = e^{4x}$
(c) $y'' + y = \cot x$
(d) $y'' + 4y = \sec 2x$
(e) $y'' - y = \log x$

2. Solve the first order linear equation: $y' + P(x)y = Q(x)$ by solving the corresponding homogeneous equation first and then obtaining the general solution by variation of parameters.

3. Verify that $y = c_1 x + c_2 x^2$ is the general solution of the equation

$$x^2 y'' - 2xy' + 2y = 0,$$

and find the general solution of the equation

$$x^2y'' - 2xy' + 2y = x^3.$$

4. Verify that $y = c_1e^x + c_2x^{-1}$ is the general solution of the homogeneous equation corresponding to

$$x(x+1)y'' + (2 - x^2)y' - (2 + x)y = (x+1)^2,$$

and find the general solution.

5. *Operational method.* We write $D = d/dx$, so that the operator L can be written: $L = a_0D^n + \cdots + a_{n-1}D + a_n$, i.e., as a polynomial in the operator D. The *sum* and *product* of two operators L_1, L_2 are defined by the equations: $(L_1 + L_2)[y] = L_1[y] + L_2[y]$, $(L_1L_2)[y] = L_1\{L_2[y]\}$. *If the coefficients are constant, then these operations can be carried out as for ordinary polynomials.* For example,

$$\begin{aligned}(2D+1)(D-2)[y] &= (2D+1)[y' - 2y] = 2y'' - 3y' - 2y\\ &= (2D^2 - 3D - 2)[y].\end{aligned}$$

The general rule can be established by induction. To find a solution $y^*(x)$ of the nonhomogeneous equation

$$2y'' - 3y' - 2y = Q(x),$$

we write the equation in operator form:

$$(2D+1)(D-2)[y] = Q(x).$$

We then set $(D - 2)[y] = u$, so that $(2D + 1)[u] = Q(x)$. Accordingly, by Section 8–6,

$$y = e^{2x}\int e^{-2x}u\,dx, \quad u = \tfrac{1}{2}e^{-\frac{1}{2}x}\int e^{\frac{1}{2}x}Q(x)\,dx.$$

Obtain the general solutions of the equations given:

(a) $D(D^2 - 9)[y] = 0$
(b) $(D^5 - D^3)[y] = 0$
(c) $(D^2 - 9)[y] = 8e^x$
(d) $(D-1)^3[y] = 1$
(e) Prob. 1(a) above
(f) Prob. 1(b) above

The use of operators has been developed into a highly efficient technique. For more information one is referred to Chapter 4 of *ODE*.

6. *Method of undetermined coefficients.* To obtain a solution $y^*(x)$ of the equation

$$(D^2 + 1)[y] = e^x \tag{a}$$

we can multiply both sides by $D - 1$:

$$(D-1)(D^2+1)[y] = (D-1)[e^x] = 0. \tag{b}$$

The right-hand member is 0 because e^x satisfies the homogeneous equation $(D-1)[y] = 0$. From (b) we obtain the characteristic equation: $(r-1)(r^2+1) = 0$. Accordingly, every solution of (b) has the form

$$y = c_1\cos x + c_2\sin x + c_3e^x.$$

If we substitute this expression in (a), the first two terms give 0, since they form the complementary function of (a). One obtains an equation for the "undetermined coefficient" c_3: $2c_3e^x = e^x$, $c_3 = \frac{1}{2}$. Accordingly, $y^* = \frac{1}{2}e^x$ is the particular solution sought. The method depends upon finding an operator which "annihilates" the right-hand member $Q(x)$. Such an operator can always be found if Q is a solution of a homogeneous equation with constant coefficients; the operator L such that $L[Q] = 0$ is the operator sought. If Q is of form $(p_0 + p_1 x + \cdots + p_k x^{k-1})e^{ax}$ $(A \cos bx + B \sin bx)$, the annihilator is $L = \{(D - a)^2 + b^2\}^k$. For a sum of such terms, one can multiply the corresponding operators. Use the method described to find particular solutions of the following equations:

(a) $y'' + y = \sin x$ (annihilator is $D^2 + 1$);
(b) $y'' + y = e^{2x}$ (annihilator is $D - 2$);
(c) $y'' + y' - y = x^2$ (annihilator is D^3);
(d) $y'' + 2y' + y = \sin 2x + e^x$; (e) $y'' + 4y = 25xe^x$;
(f) $y''' - 2y' - 4y = e^{-x} \sin x$; (g) $y'' + 2y' + y = 6xe^{-x}$.

7. Prove by variation of parameters or by the method of Prob. 6 that a particular solution of the equation

$$\frac{d^2x}{dt^2} + 2h\frac{dx}{dt} + \lambda^2 x = B \sin \omega t \quad (h > 0, \omega > 0)$$

is given by

$$x = \frac{B \sin (\omega t - \alpha)}{\sqrt{(\lambda^2 - \omega^2)^2 + 4\omega^2 h^2}}, \quad \tan \alpha = \frac{2\omega h}{\lambda^2 - \omega^2}, \quad 0 < \alpha < \pi.$$

8. *Application of Laplace transforms.* (See Sections 6–25 and 7–19.) For the equation $a_0 y^{(n)} + \cdots + a_n y = g(x)$ with constant coefficients, one can find the solution for $x \geqq 0$, with given initial values $y_0, y_0, \ldots, y_0^{(n-1)}$, with the aid of Laplace transforms. By (7–101),

$$\mathfrak{L}[y^{(k)}] = s^k \mathfrak{L}[y] - [y_0^{(k-1)} + s y_0^{(k-2)} + \cdots + s^{k-1} y_0]$$

for $k = 1, \ldots, n$. Thus if $y(x)$ is the solution of $y'' - y = g(x)$ with $y(0) = 1$ and $y'(0) = 3$, then $\mathfrak{L}[y'' - y] = \mathfrak{L}[y''] - \mathfrak{L}[y] = \mathfrak{L}[g(x)] = G(s)$, so that $s^2\mathfrak{L}[y] - (3 + s) - \mathfrak{L}[y] = G(s)$ and $\mathfrak{L}[y] = (3 + s)(s^2 - 1)^{-1} + G(s)(s^2 - 1)^{-1}$. Now, by partial fractions, $(3 + s)(s^2 - 1)^{-1} = 2(s - 1)^{-1} - (s + 1)^{-1} = \mathfrak{L}[2e^x - e^{-x}]$. For given $g(x)$ we may be able to find explicitly $\phi(x)$ such that $\mathfrak{L}[\phi] = G(s)(s^2 - 1)^{-1}$. Then $\mathfrak{L}[y] = \mathfrak{L}[2e^x - e^{-x} + \phi(x)]$, so that $y = 2e^x - e^{-x} + \phi(x)$. In any case, $(s^2 - 1)^{-1} = (\frac{1}{2})(s - 1)^{-1} - (\frac{1}{2})(s + 1)^{-1} = \mathfrak{L}[(\frac{1}{2})e^x - (\frac{1}{2})e^{-x}] = \mathfrak{L}[W(x)]$ and hence $G(s)(s^2 - 1)^{-1} = \mathfrak{L}[W(x)]\mathfrak{L}[g(x)] = \mathfrak{L}[W * g]$ by (7–104). Hence $\phi(x) = W * g = \int_0^x W(u)g(x - u)\,du$. The method can be applied generally to equations with constant coefficients, provided $G(s)$ exists. Partial fraction expansions lead one to terms of form $a(s - b)^{-k} = \mathfrak{L}[ax^{k-1}e^{bx}/(k - 1)!]$. Apply the method described to obtain the solution with given initial values, for $x \geq 0$, for the following equations.

(a) $y'' - 4y = e^{3x}$, $y = 1$, $y' = -1$ for $x = 0$;
(b) $y'' - 2y' - 3y = x$, $y = 0$, $y' = 0$ for $x = 0$;
(c) $y'' - 4y = g(x)$, $y = 0$, $y' = 0$ for $x = 0$;
(d) $y'' + y = g(x)$, $y = 0$, $y' = 0$ for $x = 0$.

Answers

1. (a) $c_1 e^x + c_2 e^{-x} + \frac{1}{2}xe^x$; (b) $c_1 e^x + c_2 e^{2x} + c_3 e^{3x} + \frac{1}{6}e^{4x}$;
(c) $c_1 \cos x + c_2 \sin x - \sin x - \sin x \log (\csc x + \cot x)$;
(d) $c_1 \cos 2x + c_2 \sin 2x + \frac{1}{4} \cos 2x \log |\cos 2x| + \frac{1}{2}x \sin 2x$;
(e) $c_1 e^x + c_2 e^{-x} + \frac{1}{2}e^x \int e^{-x} \log x\,dx - \frac{1}{2}e^{-x} \int e^x \log x\,dx$.

3. $c_1x + c_2x^2 + \frac{1}{2}x^3$. 4. $c_1e^x + \dfrac{c_2}{x} - \frac{1}{2}(x+2)$.

5. (a) $c_1 + c_2e^{3x} + c_3e^{-3x}$, (b) $c_1 + c_2x + c_3x^2 + c_4e^x + c_5e^{-x}$, (c) $-e^x + c_1e^{3x} + c_2e^{-3x}$, (d) $-1 + c_1e^x + c_2xe^x + c_3x^2e^x$.

6. (a) $-\frac{1}{2}x\cos x$, (b) $\frac{1}{5}e^{2x}$, (c) $-x^2 - 2x - 4$, (d) $\frac{1}{7}(4\cos 2x + 3\sin 2x) + \frac{1}{4}e^x$, (e) $e^x(5x-2)$, (f) $\frac{1}{20}xe^{-x}(3\cos x - \sin x)$, (g) x^3e^{-x}.

8. (a) $(\frac{1}{5})(e^{3x} + 4e^{-2x})$, (b) $(\frac{1}{36})(e^{3x} - 9e^{-x} - 12x + 8)$, (c) $(\frac{1}{4})\displaystyle\int_0^x (e^{2u} - e^{-2u})g(x-u)\,du$, (d) $\displaystyle\int_0^x \sin u\, g(x-u)\,du$.

8–12 Systems of linear equations with constant coefficients.

We consider systems such as the following:

$$\begin{aligned}\frac{dx}{dt} &= a_1x + b_1y + p(t),\\ \frac{dy}{dt} &= a_2x + b_2y + q(t);\end{aligned} \tag{8–52}$$

$$\begin{aligned}\frac{dx}{dt} &= a_1x + b_1y + c_1z + p(t),\\ \frac{dy}{dt} &= a_2x + b_2y + c_2z + q(t),\\ \frac{dz}{dt} &= a_3x + b_3y + c_3z + r(t).\end{aligned} \tag{8–53}$$

Here $a_1, \ldots, c_3$ are constants. The general case would concern n variables $x_1, \ldots, x_n$ which are functions of t:

$$\frac{dx_i}{dt} = a_{i1}x_1 + \cdots + a_{in}x_n + p_i(t) \quad (i = 1, \ldots, n). \tag{8–54}$$

There are more general systems which can be reduced to this form (Prob. 2 below).

Such systems are treated most simply with the aid of vectors and matrices. We replace (8–54) by the single vector equation

$$\frac{d\mathbf{x}}{dt} = A\mathbf{x} + \mathbf{p}(t), \tag{8–55}$$

where $\mathbf{x} = (x_1, \ldots, x_n)$, $A = (a_{ij})$, $\mathbf{p}(t) = (p_1(t), \ldots, p_n(t))$ and all vectors are column vectors. Each solution of (8–55) is a vector function $\mathbf{x} = \mathbf{f}(t) = (f_1(t), \ldots, f_n(t))$ satisfying the equation over some interval. An *existence theorem* asserts that, if $\mathbf{p}(t)$ is continuous for $a \leq t \leq b$, then there is a unique solution $\mathbf{x} = \mathbf{f}(t)$ of (8–55) over this interval with prescribed *initial value* $\mathbf{x}_0 = \mathbf{f}(t_0)$, where t_0 is a chosen number in the interval.

A proof is given in Chapter 12 of *ODE*. There is a similar assertion for other types of intervals, including infinite intervals.

We first consider the *homogeneous equation*

$$\frac{d\mathbf{x}}{dt} = A\mathbf{x}. \tag{8–56}$$

An example is the pair of equations

$$\frac{dx}{dt} = x - 2y, \quad \frac{dy}{dt} = -2x + y \quad \text{or} \quad \frac{d\mathbf{x}}{dt} = \begin{bmatrix} 1 & -2 \\ -2 & 1 \end{bmatrix} \mathbf{x}, \tag{8–57}$$

where $\mathbf{x}$ is the vector col (x, y). It will be seen that the general solution of (8–56) on the interval $-\infty < t < \infty$ has the form

$$\mathbf{x} = c_1\mathbf{x}_1(t) + \cdots + c_n\mathbf{x}_n(t), \tag{8–58}$$

where $\mathbf{x}_1(t), \ldots, \mathbf{x}_n(t)$ are solutions and are *linearly independent:* that is,

$$c_1\mathbf{x}_1(t) + \cdots + c_n\mathbf{x}_n(t) \equiv \mathbf{0} \quad \text{implies} \quad c_1 = 0, \ldots, c_n = 0. \tag{8–59}$$

Indeed if $\mathbf{x}_1(t), \ldots, \mathbf{x}_n(t)$ are such linearly independent solutions, then $d\mathbf{x}_1/dt = A\mathbf{x}_1, \ldots, d\mathbf{x}_n/dt = A\mathbf{x}_n$, so that, for every choice of the constants $c_1, \ldots, c_n$, the function $\mathbf{x}(t)$ defined by (8–58) satisfies

$$\begin{aligned} \frac{d\mathbf{x}}{dt} &= c_1\frac{d\mathbf{x}_1}{dt} + \cdots + c_n\frac{d\mathbf{x}_n}{dt} = c_1A\mathbf{x}_1 + \cdots + c_nA\mathbf{x}_n \\ &= A(c_1\mathbf{x}_1 + \cdots + c_n\mathbf{x}_n) = A\mathbf{x}, \end{aligned}$$

so that (8–56) holds. Furthermore, for each fixed t_0, the vectors $\mathbf{x}_1(t_0), \ldots, \mathbf{x}_n(t_0)$ are linearly independent. For if

$$c_1\mathbf{x}_1(t_0) + \cdots + c_n\mathbf{x}_n(t_0) = \mathbf{0},$$

then the solution $\mathbf{x}(t)$ defined by (8–58) would satisfy: $\mathbf{x}(t_0) = \mathbf{0}$. By the existence theorem, there is just one solution $\mathbf{x}(t)$ with $\mathbf{x}(t_0) = \mathbf{0}$—namely, $\mathbf{x}(t) \equiv \mathbf{0}$! Hence $\mathbf{x}(t) = c_1\mathbf{x}_1(t) + \cdots + c_n\mathbf{x}_n(t) \equiv \mathbf{0}$ and thus, by (8–59), $c_1 = 0, \ldots, c_n = 0$. Therefore $\mathbf{x}_1(t_0), \ldots, \mathbf{x}_n(t_0)$ are linearly independent vectors and form a basis for V^n. Thus for given $\mathbf{x}_0$ there are unique constants $c_1, \ldots, c_n$ such that $\mathbf{x}_0 = c_1\mathbf{x}_1(t_0) + \cdots + c_n\mathbf{x}_n(t_0)$; that is, (8–58) provides a unique solution satisfying the initial condition $\mathbf{x}(t_0) = \mathbf{x}_0$ and (8–58) is indeed the general solution.

We still have to show that n linearly independent solutions exist. To that end, we choose t_0 and apply the existence theorem to obtain a solution $\mathbf{x}_1(t)$ such that $\mathbf{x}_1(t_0) = \mathbf{e}_1 = (1, 0, \ldots, 0)$, then a solution $\mathbf{x}_2(t)$ such that $\mathbf{x}_2(t_0) = \mathbf{e}_2, \ldots,$ finally a solution $\mathbf{x}_n(t)$ such that $\mathbf{x}_n(t_0) = \mathbf{e}_n$. Then these n solutions are linearly independent. For if $c_1\mathbf{x}_1(t) + \cdots + c_n\mathbf{x}_n(t) \equiv \mathbf{0}$, then $c_1\mathbf{x}_1(t_0) + \cdots + c_n\mathbf{x}_n(t_0) = \mathbf{0}$ or $c_1\mathbf{e}_1 + \cdots + c_n\mathbf{e}_n = \mathbf{0}$, which implies $c_1 = 0, \ldots, c_n = 0$.

Thus it only remains to show how we can explicitly find n linearly independent solutions of (8–56). To find them, we seek solutions of form

$$\mathbf{x} = e^{\lambda t}\mathbf{u}, \tag{8–60}$$

where λ is a constant scalar and $\mathbf{u}$ is a constant nonzero vector. Substitution in (8–56) gives

$$\lambda e^{\lambda t}\mathbf{u} = Ae^{\lambda t}\mathbf{u} \quad \text{or} \quad A\mathbf{u} = \lambda\mathbf{u}.$$

Thus (8–60) is a solution precisely when $\mathbf{u}$ is an eigenvector of A, associated with the eigenvalue λ. By the theory of Section 1–5, λ must be a root of the *characteristic equation:*

$$\det(A - \lambda I) = 0, \tag{8–61}$$

and each real root λ of this equation is an eigenvalue. If there are n distinct real eigenvalues $\lambda_1, \ldots, \lambda_n$, then we obtain n corresponding eigenvectors $\mathbf{u}_1, \ldots, \mathbf{u}_n$ and n solutions

$$\mathbf{x}_1(t) = e^{\lambda_1 t}\mathbf{u}_1, \ldots, \mathbf{x}_n(t) = e^{\lambda_n t}\mathbf{u}_n. \tag{8–62}$$

Furthermore, the eigenvectors $\mathbf{u}_1, \ldots, \mathbf{u}_n$ are *linearly independent vectors* (Problem 12 following Section 1–10) and hence the solutions (8–62) are linearly independent solutions of (8–56) (Problem 6 below). Therefore the general solution of (8–58) is

$$\mathbf{x} = c_1 e^{\lambda_1 t}\mathbf{u}_1 + \cdots + c_n e^{\lambda_n t}\mathbf{u}_n. \tag{8–63}$$

EXAMPLE 1. The differential equations (8–57). Here $x_1 = x$, $x_2 = y$, $A = \begin{bmatrix} 1 & -2 \\ -2 & 1 \end{bmatrix}$, the characteristic equation is $\begin{vmatrix} 1-\lambda & -2 \\ -2 & 1-\lambda \end{vmatrix} = 0$ or $\lambda^2 - 2\lambda - 3 = 0$ and $\lambda_1 = 3$, $\lambda_2 = -1$ are eigenvalues with associated eigenvectors $\mathbf{u}_1 = (1, -1)$, $\mathbf{u}_2 = (1, 1)$. The general solution is

$$\mathbf{x} = c_1 e^{3t}\begin{bmatrix} 1 \\ -1 \end{bmatrix} + c_2 e^{-t}\begin{bmatrix} 1 \\ 1 \end{bmatrix}$$

or, in the original notations,

$$x = c_1 e^{3t} + c_2 e^{-t}, \quad y = -c_1 e^{3t} + c_2 e^{-t}.$$

If the eigenvalues $\lambda_1, \ldots, \lambda_n$ are distinct but some are complex, then one obtains n linearly independent solutions by taking real and imaginary parts of the corresponding expressions $e^{\lambda t}\mathbf{u}$ for complex roots. (See Section 6–3 of *ODE* for a proof.)

EXAMPLE 2. $\dfrac{d\mathbf{x}}{dt} = \begin{bmatrix} 0 & 1 & 0 \\ 0 & 0 & 1 \\ 1 & -1 & 1 \end{bmatrix}\mathbf{x}.$

Here we find the eigenvalues to be $1, \pm i$. Corresponding to 1 we obtain the eigenvector $\mathbf{u}_1 = (1, 1, 1)$ and the solution $\mathbf{x}_1(t) = e^t\mathbf{u}_1$. Corresponding to $+i$ we obtain the eigenvector $\mathbf{u} = (1, i, -1)$ and the solution

$$e^{it}\begin{bmatrix} 1 \\ i \\ -1 \end{bmatrix} = \begin{bmatrix} e^{it} \\ ie^{it} \\ -e^{it} \end{bmatrix} = \begin{bmatrix} \cos t + i\sin t \\ -\sin t + i\cos t \\ -\cos t - i\sin t \end{bmatrix} = \begin{bmatrix} \cos t \\ -\sin t \\ -\cos t \end{bmatrix} + i\begin{bmatrix} \sin t \\ \cos t \\ -\sin t \end{bmatrix}.$$

Hence we obtain the real solutions

$$\mathbf{x}_2(t) = \begin{bmatrix} \cos t \\ -\sin t \\ -\sin t \end{bmatrix}, \quad \mathbf{x}_3(t) = \begin{bmatrix} \sin t \\ \cos t \\ -\sin t \end{bmatrix}.$$

(The complex root $-i$ leads to the same two solutions.) The general real solution is

$$\mathbf{x} = c_1\mathbf{x}_1(t) + c_2\mathbf{x}_2(t) + c_3\mathbf{x}_3(t) = c_1 \begin{bmatrix} e^t \\ e^t \\ e^t \end{bmatrix} + c_2 \begin{bmatrix} \cos t \\ -\sin t \\ -\sin t \end{bmatrix} + c_3 \begin{bmatrix} \sin t \\ \cos t \\ -\sin t \end{bmatrix}.$$

This can be verified directly (Problem 4 below).

When repeated roots occur, the procedure has to be modified. If λ is an eigenvalue of multiplicity k, one replaces the expression $e^{\lambda t}\mathbf{u}$ by $e^{\lambda t}\mathbf{q}(t)$ where $\mathbf{q}(t)$ is a column vector function whose components are polynomials of degree at most $k - 1$. It can be shown that n linearly independent solutions can be found in this form (with complex roots treated as above). For details see Section 6–22 of *ODE*. The method of Laplace transforms also handles all cases without difficulty (see Prob. 3 below).

Nonhomogeneous equation. Variation of parameters. We return to the nonhomogeneous equation (8–55) and assume that we have found the general solution (8–58) of the *related* homogeneous equation $d\mathbf{x}/dt = A\mathbf{x}$. We observe that we can write this general solution as

$$\mathbf{x} = X(t)\mathbf{c}, \tag{8–64}$$

where $X(t)$ is the $n \times n$ matrix whose column vectors are $\mathbf{x}_1(t), \ldots, \mathbf{x}_n(t)$ and $\mathbf{c} = \operatorname{col}(c_1, \ldots, c_n)$. Also dX/dt is obtained (Prob. 5 below) by differentiating each entry of $X(t)$, hence by replacing each column by A times that column. Accordingly,

$$\frac{dX}{dt} = AX. \tag{8–65}$$

(We say that $X(t)$ satisfies a *matrix differential equation.*)

To find the general solution of (8–55), we now replace the arbitrary constants $c_1, \ldots, c_n$ by unknown functions $v_1(t), \ldots, v_n(t)$; that is, in (8–64) we replace $\mathbf{c}$ by $\mathbf{v}(t)$ to obtain

$$\mathbf{x} = X(t)\mathbf{v}(t). \tag{8–66}$$

The usual rules of calculus apply to matrix functions (Prob. 5 below). Hence if we substitute in (8–55) and use (8–65), we obtain successively

$$\frac{d\mathbf{x}}{dt} = X(t)\frac{d\mathbf{v}}{dt} + \frac{dX}{dt}\mathbf{v} = AX(t)\mathbf{v} + \mathbf{p}(t), \quad X(t)\frac{d\mathbf{v}}{dt} + AX\mathbf{v} = AX\mathbf{v} + \mathbf{p}(t),$$

$$X(t)\frac{d\mathbf{v}}{dt} = \mathbf{p}(t).$$

Since $\mathbf{x}_1(t), \ldots, \mathbf{x}_n(t)$ are linearly independent vectors for each fixed t,

$X(t)$ is nonsingular for all t. Therefore we can write

$$\frac{d\mathbf{v}}{dt} = X^{-1}(t)\mathbf{p}(t), \quad \mathbf{v}(t) = \int X^{-1}(t)\mathbf{p}(t)\,dt$$

and by (8–66)

$$\mathbf{x}(t) = X(t)\int X^{-1}(t)\mathbf{p}(t)\,dt. \tag{8–67}$$

Here we make a definite choice for each indefinite integral and thereby obtain a particular solution of (8–55). If $\mathbf{p}(t)$ is continuous for $a \leqq t \leqq b$, then we can write

$$\mathbf{x} = \mathbf{x}^*(t) = X(t)\int_{t_0}^{t} X^{-1}(u)\mathbf{p}(u)\,du, \quad a \leqq t \leqq b, \tag{8–68}$$

and have a particular solution of (8–55) reducing to $\mathbf{0}$ for $t = t_0$. This equation is known as the *variation of parameters formula*.

The general solution of (8–55) is now given by

$$\mathbf{x} = \mathbf{x}^*(t) + c_1\mathbf{x}_1(t) + \cdots + c_n\mathbf{x}_n(t). \tag{8–69}$$

For if $\mathbf{x}(t)$ is a solution, then

$$\frac{d\mathbf{x}}{dt} = A\mathbf{x} + \mathbf{p}(t), \quad \frac{d\mathbf{x}^*}{dt} = A\mathbf{x}^* + \mathbf{p}(t),$$

$$\frac{d}{dt}(\mathbf{x} - \mathbf{x}^*) = A(\mathbf{x} - \mathbf{x}^*).$$

Hence $\mathbf{x} - \mathbf{x}^*$ is a solution of the related homogeneous equation and $\mathbf{x} - \mathbf{x}^* = c_1\mathbf{x}_1(t) + \cdots + c_n\mathbf{x}_n(t)$ for appropriate c_i, so that (8–69) holds. Conversely, every function (8–69) satisfies (8–55), as we verify by direct substitution.

EXAMPLE 3. $\dfrac{dx}{dt} = x - 2y + \cos t,\ \dfrac{dy}{dt} = -2x + y - \sin t.$ The related homogeneous system is that of Example 1 and therefore we can take

$$X(t) = \begin{bmatrix} e^{3t} & e^{-t} \\ -e^{3t} & e^{-t} \end{bmatrix}, \quad \mathbf{p}(t) = \begin{bmatrix} \cos t \\ -\sin t \end{bmatrix}.$$

We also find $\int X^{-1}(t)\mathbf{p}(t)\,dt$ to be

$$\tfrac{1}{2}\int \begin{bmatrix} e^{-3t} & -e^{-3t} \\ e^{t} & e^{t} \end{bmatrix} \begin{bmatrix} \cos t \\ \sin t \end{bmatrix} dt = \tfrac{1}{2}\begin{bmatrix} \int e^{-3t}(\cos t + \sin t)\,dt \\ \int e^{t}\ (\cos t - \sin t)\,dt \end{bmatrix}$$

$$= \tfrac{1}{2}\begin{bmatrix} \dfrac{e^{-3t}}{10}(-4\cos t \quad -2\sin t) \\ e^{t}\cos t \end{bmatrix},$$

and therefore find

$$\mathbf{x}^*(t) = X(t)\int X^{-1}(t)\mathbf{p}(t)\,dt = \tfrac{1}{10}\begin{bmatrix} 3\cos t - \sin t \\ 7\cos t + \sin t \end{bmatrix}.$$

The general solution is

$$\mathbf{x} = \begin{bmatrix} x \\ y \end{bmatrix} = c_1\begin{bmatrix} e^{3t} \\ -e^{3t} \end{bmatrix} + c_2\begin{bmatrix} e^{-t} \\ e^{-t} \end{bmatrix} + \tfrac{1}{10}\begin{bmatrix} 3\cos t - \sin t \\ 7\cos t + \sin t \end{bmatrix},$$

or

$$\begin{aligned} x &= c_1e^{3t} + c_2e^{-t} + \tfrac{1}{10}(3\cos t - \sin t), \\ y &= -c_1e^{3t} + c_2e^{-t} + \tfrac{1}{10}(7\cos t + \sin t). \end{aligned}$$

Remark. The theory can be extended to simultaneous linear equations with variable coefficients—that is to the vector equation

$$\frac{d\mathbf{x}}{dt} = A(t)\mathbf{x} + \mathbf{p}(t). \tag{8–70}$$

If $A(t) = (a_{ij}(t))$ and $\mathbf{p}(t)$ are continuous on an interval of t, then the existence theorem continues to hold. Furthermore, the general solution of the related homogeneous equation has the form (8–58), where (8–59) holds. Finding this general solution is generally much more difficult than in the case of constant coefficients; one may be forced to use series solutions (Section 8–15 below). Once the general solution of the related homogeneous equation is known, the method of variation of parameters can be applied as above to obtain the general solution of the nonhomogeneous equation.

Problems

1. Find the general solutions:

(a) $\dfrac{dx}{dt} = x + 2y, \quad \dfrac{dy}{dt} = 12x - y;$

(b) $\dfrac{d\mathbf{x}}{dt} = \begin{bmatrix} 1 & -2 \\ 5 & -1 \end{bmatrix}\mathbf{x};$

(c) $\dfrac{d\mathbf{x}}{dt} = \begin{bmatrix} 1 & -2 \\ 5 & -1 \end{bmatrix}\mathbf{x} + \begin{bmatrix} t^2 + 2t \\ 2t - 4t^2 \end{bmatrix};$

(d) $\dfrac{d\mathbf{x}}{dt} = \begin{bmatrix} 1 & -1 & 0 \\ 0 & 1 & -1 \\ -1 & 0 & 1 \end{bmatrix}\mathbf{x};$

(e) $\dfrac{d\mathbf{x}}{dt} = \begin{bmatrix} 1 & 2 \\ -1 & 4 \end{bmatrix}\mathbf{x} + \begin{bmatrix} e^t \\ -2 \end{bmatrix};$

(f) $\dfrac{d\mathbf{x}}{dt} = \begin{bmatrix} 2 & -1 & 3 \\ -1 & 1 & -1 \\ 0 & 1 & -1 \end{bmatrix}\mathbf{x} + \begin{bmatrix} t \\ -1 \\ 0 \end{bmatrix}.$

2. (a) Find the general solution of the system:

$$\frac{d^2x}{dt^2} = x - 4y, \quad \frac{d^2y}{dt^2} = y - x$$

by solving the equivalent system:

$$\frac{dx}{dt} = z, \quad \frac{dy}{dt} = w, \quad \frac{dz}{dt} = x - y, \quad \frac{dw}{dt} = y - x.$$

(b) Find the general solution of the equation

$$\frac{d^2x}{dt^2} + 2\frac{dx}{dt} - 3x = 3t^2 - 4t - 2$$

by solving the equivalent system:

$$\frac{dx}{dt} = y, \quad \frac{dy}{dt} = 3x - 2y + 3t^2 - 4t - 2.$$

By generalizing the method indicated, essentially every system of simultaneous differential equations can be reduced to a system of first order equations:

$$\frac{dx_i}{dt} = f_i(x_1, \ldots x_n, t) \quad (i = 1, \ldots, n).$$

3. One may be able to obtain the solution of (8–55) for $t \geqq 0$ with initial value $\mathbf{x}_0$ for $t = 0$ by applying Laplace transforms as in Problem 8 following Section 8–11. For Example 1 in the text, let $(x(t), y(t))$ be the solution with initial value $(1, 2)$ and let $\mathcal{L}[x(t)] = X(s)$, $\mathcal{L}[y(t)] = Y(s)$. Then we obtain $sX(s) - 1 = X(s) - 2Y(s)$, $sY(s) - 2 = -2X(s) + Y(s)$. We solve these simultaneous equations for $X(s)$, $Y(s)$, obtaining $X(s) = (\frac{3}{2})(s+1)^{-1} - (\frac{1}{2})(s-3)^{-1}$, $Y(s) = (\frac{1}{2})(s-3)^{-1} + (\frac{3}{2})(s+1)^{-1}$, so that $x = (\frac{3}{2})e^{-t} - (\frac{1}{2})e^{3t}$, $y = (\frac{1}{2})e^{3t} + (\frac{3}{2})e^{-t}$. Apply the method to the following equations with initial conditions. [The rule $\mathcal{L}[t^k e^{at}] = k!(s-a)^{-k-1}$ will be found helpful. In parts (e) and (f) there are multiple eigenvalues.]

(a) Equations of Prob. 1(a), with $x(0) = 2$, $y(0) = -1$.
(b) Equation of Prob. 1(b), with $x_1(0) = 3$, $x_2(0) = 2$.
(c) Equation of Prob. 1(c) with $\mathbf{x}(0) = \mathbf{0}$.
(d) Equation of Prob. 1(d), with $\mathbf{x}(0) = \text{col}\,(1, 1, 1)$.

(e) $\dfrac{d\mathbf{x}}{dt} = \begin{bmatrix} 1 & 2 \\ -2 & 5 \end{bmatrix}\mathbf{x}, \quad \mathbf{x}(0) = \begin{bmatrix} c_1 \\ c_2 \end{bmatrix}.$

(f) $\dfrac{d\mathbf{x}}{dt} = \begin{bmatrix} 2 & 1 & 0 \\ 0 & -3 & 1 \\ -1 & -13 & 4 \end{bmatrix}\mathbf{x}, \quad \mathbf{x}(0) = \text{col}\,(c_1, c_2, c_3).$

4. In the text a general solution for Example 2 is obtained. Verify that this is indeed the general solution by showing that the vector functions $\mathbf{x}_1(t)$, $\mathbf{x}_2(t)$, $\mathbf{x}_3(t)$ are linearly independent for $-\infty < t < \infty$ and that each function is a solution of the differential equation for $-\infty < t < \infty$.

5. For a matrix function $A(t) = (a_{ij}(t))$, $a \leqq t \leqq b$, one defines the derivative dA/dt to be the matrix function (da_{ij}/dt), provided all $a_{ij}(t)$ have derivatives. One defines $\int A(t)\,dt$ to be the matrix $\left(\int a_{ij}(t)\,dt\right)$ and $\int_a^b A(t)\,dt$

to be $\left(\int_a^b a_{ij}(t)\,dt\right)$; here we assume all $a_{ij}(t)$ to be continuous over the interval considered. Verify the following rules, for appropriate hypotheses on the sizes of the matrices $X(t)$, $Y(t)$ and on the continuity or differentiability of their elements.

(a) $\dfrac{d}{dt}(X(t) + Y(t)) = \dfrac{dX}{dt} + \dfrac{dY}{dt}$;

(b) $\dfrac{d}{dt} X(t)Y(t) = X(t)\dfrac{dY}{dt} + Y(t)\dfrac{dX}{dt}$;

(c) If $X(u) = (x_{ij}(u))$ and $u = \phi(t)$, then $\dfrac{dX}{dt} = \phi'(t)\dfrac{dX}{du}$;

(d) $\displaystyle\int_a^b [X(t) + Y(t)]\,dt = \int_a^b X(t)\,dt + \int_a^b Y(t)\,dt.$

6. Prove: if $\mathbf{u}_1, \ldots, \mathbf{u}_n$ are linearly independent vectors of V^n, then the solutions (8–62) of (8–56) are linearly independent for $-\infty < t < \infty$.

Answers

1. (a) $x = c_1e^{5t} + c_2e^{-5t}$, $y = 2c_1e^{5t} - 3c_2e^{-5t}$;
(b) $c_1(2\cos 3t, \cos 3t + 3\sin 3t) + c_2(2\sin 3t, -3\cos 3t + \sin 3t)$;
(c) $c_1(2\cos 3t, \cos 3t + 3\sin 3t) + c_2(2\sin 3t, -3\cos 3t + \sin 3t) + (t^2, t^2)$;
(d) $c_1(1, 1, 1) + c_2e^{at}(2\cos bt, -\cos bt + 2b\sin bt, -\cos bt - 2b\sin bt) + c_3e^{at}(2\sin bt, -\sin bt - 2b\cos bt, 2b\cos bt - \sin bt)$, where $a = \frac{3}{2}$, $b = \sqrt{\frac{3}{4}}$.
(e) $c_1e^{3t}(1, 1) + c_2e^{2t}(2, 1) + (\frac{1}{6})(-9e^t - 4, -3e^t + 2)$;
(f) $c_1e^t(1, -2, -1) + c_2e^{-t}(1, 0, -1) + c_3e^{2t}(4, -3, -1) + (\frac{1}{4})(-4, 1 - 2t, 3 - 2t)$.

2. (a) $x = 2c_1\cos t + 2c_2\sin t + 2c_3e^{at} + 2c_4e^{-at}$, $y = c_1\cos t + c_2\sin t - c_3e^{at} - c_4e^{-at}$, where $a = \sqrt{3}$; (b) $c_1e^{-3t} + c_2e^t - t^2$.

3. (a) $x = e^{5t} + e^{-5t}$, $y = 2e^{5t} - 3e^{-5t}$; (b) $(\frac{1}{6})(18\cos 3t - 2\sin 3t, 12\cos 3t + 26\sin 3t)$; (c) (t^2, t^2); (d) $(1, 1, 1)$; (e) $e^{3t}[c_1(1 - 2t, -2t) + c_2(2t, 1 + 2t)]$; (f) $e^t[c_1(1 + t + (t^2/2), -t^2/2, -t - 2t^2) + c_2(t - 3(t^2/2), 1 - 4t + 3(t^2/2), -13t + 6t^2) + c_3(t^2/2, t - (t^2/2), 1 + 3t - 2t^2)]$.

8–13 Applications of linear differential equations. The applications to be considered here will be mainly those represented by the second order linear differential equation:

$$m\frac{d^2x}{dt^2} + 2c\frac{dx}{dt} + k^2x = F(t). \tag{8–71}$$

The notation suggests a problem in mechanics. A particle of mass m moves on a line, the x axis, subject to three forces: a "frictional" force $-2c(dx/dt)$, opposing the motion (if $c > 0$); a "restoring" force $-k^2x$; an "applied force" $F(t)$. This is illustrated in Fig. 8–9.

This problem has a parallel in electric circuit theory, namely, that represented by the differential equation:

$$L\frac{d^2i}{dt^2} + R\frac{di}{dt} + \frac{1}{C}i = F(t), \qquad (8\text{–}72)$$

where L is the inductance, R the resistance, C the capacitance, and $F(t)$ the time derivative of the applied emf.

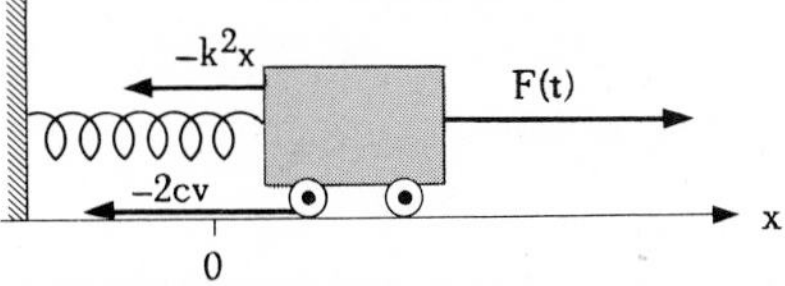

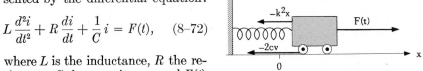

FIG. 8–9. Model for linear equation of second order.

It will be assumed that m, c, k are constants and that $m > 0$, $c \geqq 0$, $k > 0$. Hence the characteristic equation of (8–71).

$$mr^2 + 2cr + k^2 = 0, \qquad (8\text{–}73)$$

has roots which are either negative or complex. If the abbreviations

$$h = \frac{c}{m}, \quad \lambda = \frac{k}{\sqrt{m}} \qquad (8\text{–}74)$$

are used, the roots are given by

$$r = -h \pm \sqrt{h^2 - \lambda^2}. \qquad (8\text{–}75)$$

Let it first be assumed that $F(t) = 0$. Then the following cases arise, depending on the value of h:

Case I. $h = 0$. $r = \pm\lambda i$. *Simple harmonic motion.*
Case II. $0 < h < \lambda$. $r = -h \pm i\sqrt{\lambda^2 - h^2}$. *Damped vibrations.*
Case III. $h = \lambda$. $r = -h, -h$. *Critical damping.*
Case IV. $h > \lambda$. $r = -h \pm \sqrt{h^2 - \lambda^2}$. *Over-critical damping.*

In Case I there is no damping through friction. The mass vibrates freely according to the equation

$$x = A\sin(\lambda t + \alpha), \qquad (8\text{–}76)$$

where A and α are arbitrary constants. The frequency λ of these vibrations is termed the *natural frequency* of the system. In Case II there is damping through friction, as a result of which the mass oscillates, but with amplitude approaching zero:

$$x = Ae^{-ht}\sin(\beta t + \alpha), \quad \beta = \sqrt{\lambda^2 - h^2}. \qquad (8\text{–}77)$$

The frequency of these oscillations is less than the natural frequency. In Cases III and IV the oscillations have disappeared because of excessive friction, and in both cases, aside from a possible initial period of rise or fall, the motion is essentially that of exponential decay, as described in Section 8–7:

$$\begin{aligned} x &= c_1e^{-ht} + c_2te^{-ht} \quad \text{[Case III]};\\ x &= c_1e^{-a_1t} + c_2e^{-a_2t} \quad \text{[Case IV]}; \qquad (8\text{–}78)\\ a_1 &= h + \sqrt{h^2 - \lambda^2}, \quad a_2 = h - \sqrt{h^2 - \lambda^2}. \end{aligned}$$

The types of motions in all cases are pictured in Fig. 8–10.

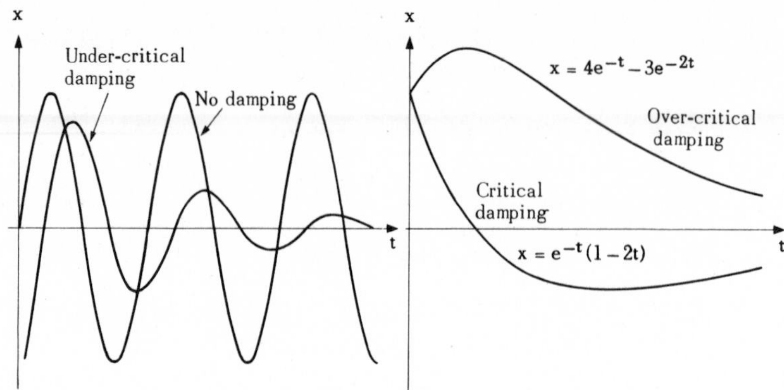

FIG. 8–10. Solutions of $m\dfrac{d^2x}{dt^2} + 2c\dfrac{dx}{dt} + k^2x = 0.$

One can discuss the qualitative properties of the solutions when a forcing term $F(t)$ is present with the aid of the input-output point of view of Section 8–7. The input is defined, as in the first order case, as the solution $x_i(t)$ obtained by neglecting the terms in dx/dt and d^2x/dt^2. Thus

$$\text{input} = x_i(t) = \frac{F(t)}{k^2}. \tag{8–79}$$

The actual solution $x(t)$, obtained by taking into account the derivative terms, will be termed the *output*. Just as in the first order case, the approximation

$$\text{input} = \text{output}$$

can be used here; the accuracy of this approximation depends on the smallness of m/k^2 and c/k^2.

Let us suppose, for example, that one has Case II for the homogeneous equation and that $F(t) = F_1$, a constant. Then the general solution of (8–71) is of the form

$$x = \frac{F_1}{k^2} + Ae^{-ht}\sin(\beta t + \alpha),$$

as in Fig. 8–11. Thus the output approaches the input, while oscillating less and less. If now F is a step-function, as in Fig. 8–12, the solution (output) attempts to follow the input, while oscillating. As the figure illustrates, after an initial transient period any two solutions effectively coincide; in this sense, the output is one definite function of x.

It will be noted that, if F jumps to a positive value when the input is increasing and to a negative value when the input is decreasing, then initially small oscillations steadily increase in size. This is typified by a person pushing a child in a swing; if one pushes in the direction of motion, the amplitude is increased. This effect of synchronization of forcing term and frequency of the response is known as *resonance*. If there were no damping present, a perfect synchronization would lead to an amplitude approaching infinite size. Now the *work* done by the applied force is on the one hand equal to the *energy* contributed to the system and on the other

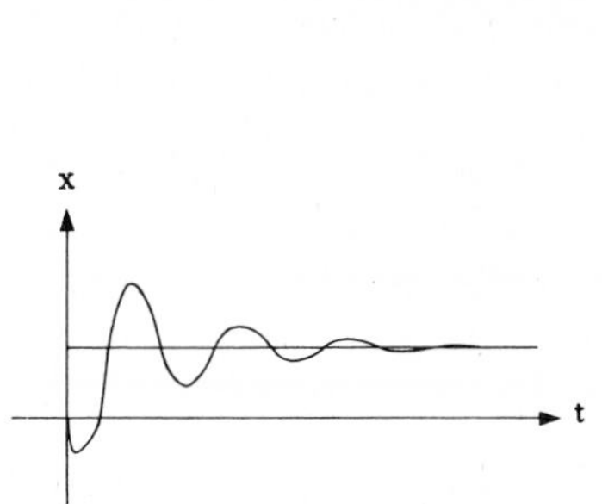

FIG. 8-11. Response of second order system to constant input.

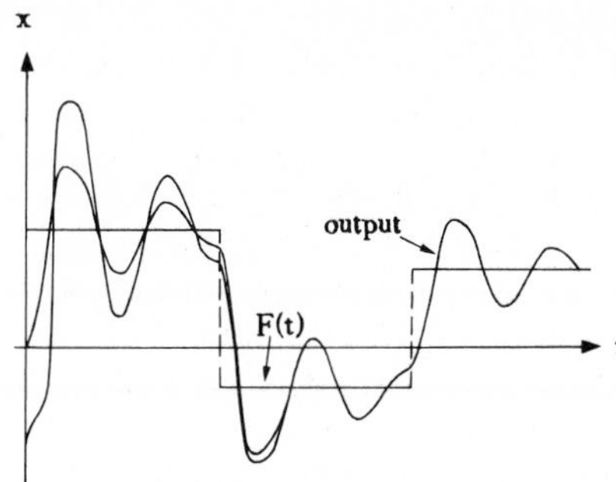

FIG. 8-12. Response of second order system to step-function input.

to the product of the force component in the direction of motion by the distance moved (Section 5-4). Hence resonance is simply a case of a force always employed to add energy to the system.

To obtain a mathematical description of resonance, one can take the case of a function $F(t) = F_1 \sin \omega t$. In this case the input is

$$x_i(t) = A_i \sin \omega t = \frac{F_1}{k^2} \sin \omega t, \tag{8-80}$$

where A_i is the input amplitude. The output is found (see Prob. 7 following Section 8-11) to be

$$x_0 = A_0 \sin (\omega t - \alpha) + (\text{transient}), \tag{8-81}$$

where

$$A_0 = \frac{\lambda^2 A_i}{\sqrt{(\lambda^2 - \omega^2)^2 + 4\omega^2 h^2}}, \quad \tan \alpha = \frac{2\omega h}{\lambda^2 - \omega^2}, \quad 0 \leqq \alpha < \pi. \tag{8-82}$$

The larger h is, the stronger the damping and the smaller the ratio of A_0 to A_i. For fixed h and λ, the ratio of A_0 to A_i depends only on the ratio of the *input frequency* ω to the natural frequency λ; the relationship is graphed in Fig. 8-13. The ratio A_0/A_i, the *amplification factor*, is large when h is small, with a maximum, for fixed h, at a frequency ω known as the *resonant frequency*. If $h = 0$, the above formulas give $A_0 = \infty$ when $\lambda = \omega$; this is the case of pure resonance. When h is large, A_0/A_i remains less than 1; amplitudes are diminished, especially for large frequencies.

The phenomenon of resonance or "sympathetic vibrations" is one familiar in daily experience. While it often results in unpleasant noise or in physical damage, it can also be used to great advantage in designing a device which should be "tuned" to respond to a given frequency.

The remarks about the second order equation here can easily be extended to equations of higher order or (what is the same) systems of linear equations. This is discussed in Chapter 10, where it is shown that such systems form a natural transition to partial differential equations. For all systems the point of view of input and output can be maintained. As long as the system is stable, that is, as long as all solutions approach 0 as t increases in the absence of forcing terms, there is essentially a unique output for given input. Furthermore, there is a superposition principle: the output varies linearly with the input. All realizations of such systems can be considered as types of "servo-mechanisms," which are designed to respond to a given input in a specified way. The differential analyzer is an exceptionally flexible type of servo-mechanism and one which can transform input into output in accordance with a wide class of differential equations.

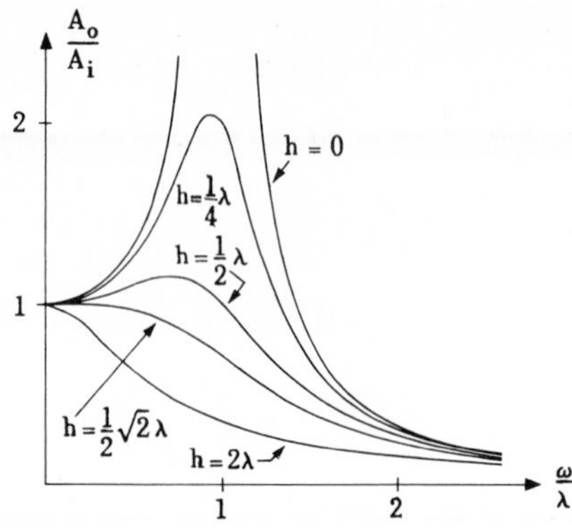

Fig. 8–13. Amplification versus input frequency for second order system.

For the case of a system described by a vector equation

$$\frac{d\mathbf{x}}{dt} = A\mathbf{x} + \mathbf{p}(t)$$

the stability is determined by the nature of the solutions of the related homogeneous equation

$$\frac{d\mathbf{x}}{dt} = A\mathbf{x}.$$

For a constant matrix A we saw in Section 8–12 that the solutions were vector functions whose components are linear combinations of terms of the form $e^{at}q(t) \cos \omega t$ or $e^{at}q(t) \sin \omega t$, where $a \pm \omega i$ are eigenvalues of A and $q(t)$ is a polynomial. Each such term approaches 0 as $t \to \infty$ precisely when $a < 0$. Hence *the system is stable when all eigenvalues have negative real parts.* If one eigenvalue has positive or zero real part, there are solutions not approaching 0 as $t \to \infty$ and the system is not stable.

The case when all eigenvalues are pure imaginary (real part 0) is of considerable interest, even though stability fails. In the following section we consider an important class of such equations.

8–14 A class of vibration problems. In many physical problems one is led to simultaneous differential equations of the form

$$m_i \frac{d^2 x_i}{dt^2} + \frac{\partial V}{\partial x_i} = 0, \quad i = 1, \ldots, n. \tag{8–83}$$

Here $m_1, \ldots, m_n$ are positive constants (typically, masses), $V(x_1, x_2, \ldots, x_n)$ is the potential energy of the system and V has a minimum at $(0, \ldots, 0)$. Hence $\partial V/\partial x_i = 0$ at $(0, \ldots, 0)$ for $i = 1, \ldots, n$ and if we expand V by Taylor's formula (Section 6–21), we obtain an expression of form

$$V = V(0, \ldots, 0) + \tfrac{1}{2} \sum_{i,j=1}^{n} a_{ij} x_i x_j + \cdots,$$

where $A = (a_{ij})$ is a *symmetric matrix*. We assume further that the quadratic form $\Sigma a_{ij}x_ix_j$ is *positive definite*, which ensures that V has a minimum at $(0, \ldots, 0)$; accordingly, all eigenvalues of A are positive (see Section 2–21). To study the behavior system of the system near $(0, \ldots, 0)$ we drop the higher degree terms and replace V by

$$V(0, \ldots, 0) + \tfrac{1}{2} \sum a_{ij}x_ix_j$$

in the differential equation. Since $\partial V/\partial x_i = \Sigma a_{ij}x_j$, the new equation can be written in vector form as follows:

$$N^2 \frac{d^2\mathbf{x}}{dt^2} + A\mathbf{x} = \mathbf{0}, \quad N = \operatorname{diag}(m_1^{\frac{1}{2}}, \ldots, m_n^{\frac{1}{2}}). \tag{8–84}$$

Here we make a change of variable, setting $\mathbf{y} = N\mathbf{x}$, so that

$$\frac{d^2\mathbf{y}}{dt^2} = N\frac{d^2\mathbf{x}}{dt^2} = -N^{-1}A\mathbf{x} = -N^{-1}AN^{-1}\mathbf{y} = -B\mathbf{y},$$

where $B = N^{-1}AN^{-1}$. Since A is symmetric, we verify that B is also symmetric (Prob. 7 below). Therefore B is similar to a diagonal matrix $D = \operatorname{diag}(\lambda_1, \ldots, \lambda_n)$, where $\lambda_1, \ldots, \lambda_n$ are the eigenvalues of B, which are also positive (Prob. 7 below); we write

$$B = C^{-1}DC, \quad \text{so that} \quad D = CBC^{-1}.$$

If we now let $\mathbf{y} = C^{-1}\mathbf{z}$, then we obtain $\mathbf{z} = C\mathbf{y}$ and

$$\frac{d^2\mathbf{z}}{dt^2} = C\frac{d^2\mathbf{y}}{dt^2} = -CB\mathbf{y} = -CBC^{-1}\mathbf{z} = -D\mathbf{z};$$

that is

$$\frac{d^2z_i}{dt^2} + \lambda_i z_i = 0, \quad i = 1, \ldots, n. \tag{8–85}$$

Since each λ_i is positive, we can write $\lambda_i = \omega_i^2$ and the solutions of (8–85) are given by

$$z_i = A_i \sin(\omega_i t + \alpha_i), \quad i = 1, \ldots, n, \tag{8–86}$$

where $A_1, \ldots, A_n, \alpha_1, \ldots, \alpha_n$ are arbitrary real constants. Now $\mathbf{x} = N^{-1}\mathbf{y} = N^{-1}C^{-1}\mathbf{z} = E\mathbf{z}$, for an appropriate matrix $E = (e_{ij})$. Hence

$$x_i = \sum_{j=1}^{n} e_{ij}z_j = \sum_{j=1}^{n} e_{ij}A_j \sin(\omega_j t + \alpha_j), \quad i = 1, \ldots, n. \tag{8–87}$$

This is the general solution of (8–84). If, for example, $A_2 = \cdots = A_n = 0$, then

$$x_i = e_{i1}A_1 \sin(\omega_1 t + \alpha_1), \quad i = 1, \ldots, n, \tag{8–88}$$

and all coordinates x_i oscillate synchronously with frequency ω_1. This is called a *normal mode of oscillation* of the system. The general solution (8–87) can be regarded as a superposition of normal modes.

For physical examples of such vibrations, see Sections 10–3 and 10–4.

EXAMPLE. $4\dfrac{d^2x_1}{dt^2} + \dfrac{\partial V}{\partial x_1} = 0, \quad 4\dfrac{d^2x_2}{dt^2} + \dfrac{\partial V}{\partial x_2} = 0$, with

$$V = 5x_1^2 - 6x_1x_2 + 5x_2^2.$$

Here $A = \begin{bmatrix} 10 & -6 \\ -6 & 10 \end{bmatrix}$ and A has eigenvalues 16 and 4, so that V is positive definite and has a minimum at (0, 0). To find the solutions, we can simply set $x_i = u_i \sin(\lambda t + \alpha_i)$ in the differential equations, since we know that there are solutions of this form—the normal modes. We obtain the equations

$$(10 - 4\lambda^2)u_1 - 6u_2 = 0, \quad -6u_1 + (10 - 4\lambda^2)u_2 = 0.$$

They are satisfied for $\lambda = \pm 1$, $\lambda = \pm 2$. For $\lambda = 1$ we obtain $u_1 = 1$, $u_2 = 1$; for $\lambda = 2$, we obtain $u_1 = 1$, $u_2 = -1$. Hence we obtain the normal modes $\mathbf{x} = \sin(t + \alpha_1)(1, 1)$ and $\mathbf{x} = \sin(2t + \alpha_2)(1, -1)$. The general solution is an arbitrary linear combination of these normal modes:

$$\begin{aligned} x_1 &= c_1 \sin(t + \alpha_1) + c_2 \sin(2t + \alpha_2), \\ x_2 &= c_1 \sin(t + \alpha_1) - c_2 \sin(2t + \alpha_2). \end{aligned}$$

(We remark that the choices $\lambda = -1$ and $\lambda = -2$ yield no additional solutions.)

Problems

1. Find a particular solution and compare input with output graphically for the following equations:

(a) $\dfrac{d^2x}{dt^2} + 4x = \sin t$

(b) $\dfrac{d^2x}{dt^2} + 4x = t$

(c) $\dfrac{d^2x}{dt^2} + 4x = \sin 2t$

(d) $\dfrac{d^2x}{dt^2} + 2\dfrac{dx}{dt} + 5x = \sin 2t$

(e) $\dfrac{d^2x}{dt^2} + 5\dfrac{dx}{dt} + 6x = \sin 2t$

(f) $\dfrac{d^2x}{dt^2} + 2\dfrac{dx}{dt} + 5x = t^3 - 3t^2 + 2t$

2. Let $F(t)$ be a "square wave": $F(t)$ has period $2t_1$ and $F(t) = 1$ for $0 < t < t_1$, $F(t) = -1$ for $t_1 < t < 2t_1$. Discuss the nature of the solutions of the equations:

(a) $\dfrac{d^2x}{dt^2} + \lambda^2 x = F(t)$

(b) $\dfrac{d^2x}{dt^2} + 2h\dfrac{dx}{dt} + \dfrac{\pi^2}{t_1^2}x = F(t)$

3. Find the equation of the locus of maxima of A_0/A_i as a function of ω/λ for fixed h according to (8–79); compare with Fig. 8–13.

4. Discuss the significance of the phase lag α given by (8–82) and analyze its dependence on λ, h, and ω.

5. Discuss the stability of the system described by the equation $d\mathbf{x}/dt = A\mathbf{x} + \mathbf{p}(t)$ for each of the following choices of A:

(a) $\begin{bmatrix} -1 & -3 \\ 1 & -2 \end{bmatrix}$, (b) $\begin{bmatrix} 0 & -1 & -2 \\ 1 & 0 & -1 \\ 0 & 2 & 3 \end{bmatrix}$, (c) $\begin{bmatrix} 0 & 1 & 0 \\ 0 & 0 & 1 \\ -1 & -1 & -2 \end{bmatrix}$.

6. Find the normal modes of oscillation:

(a) $\dfrac{d^2\mathbf{x}}{dt^2} + \begin{bmatrix} 4 & 0 \\ 0 & 9 \end{bmatrix} \mathbf{x} = \mathbf{0}$, (b) $\dfrac{d^2\mathbf{x}}{dt^2} + \begin{bmatrix} 2 & 1 \\ 1 & 2 \end{bmatrix} \mathbf{x} = \mathbf{0}$.

7. Show that the matrix $B = N^{-1}AN^{-1}$ of Section 8–14 is symmetric and that the associated quadratic form $\sum b_{ij}y_iy_j$ is obtained from $\sum a_{ij}x_ix_j$ by replacing each x_i by $y_i/m_i^{\frac{1}{2}}$, so that $\sum b_{ij}y_iy_j$ is also positive definite.

Answers

5. (a) stable, (b) unstable, (c) stable.

6. (a) $x_1 = A_1 \sin(2t + \alpha_1)$, $x_2 = 0$ and $x_1 = 0$, $x_2 = A_2 \sin(3t + \alpha_2)$.
(b) $x_1 = -x_2 = A_1 \sin(t + \alpha_1)$ and $x_1 = x_2 = A_2 \sin(\sqrt{3}\, t + \alpha_2)$.

8–15 Solution of differential equations by means of Taylor series. The Taylor series

$$f(a) + \frac{f'(a)(x-a)}{1!} + \frac{f''(a)(x-a)^2}{2!} + \cdots + \frac{f^{(n)}(a)(x-a)^n}{n!} + \cdots$$

(Section 6–16) of a function $f(x)$ can be formed when all its derivatives at $x = a$ are known. If, for example, f satisfies the differential equation

$$y' = F(x, y) \tag{8–89}$$

with given initial condition $f(a) = y_0$, then the equation itself gives $y' = f'(a)$. Thus

$$f'(a) = F[a, f(a)].$$

If F has continuous derivatives for the values of the variables concerned, then one can differentiate (8–89) to obtain a formula for the second derivative:

$$y'' = \frac{\partial F}{\partial x} + \frac{\partial F}{\partial y} y',$$

so that

$$f''(a) = \frac{\partial F}{\partial x}[a, f(a)] + \left\{\frac{\partial F}{\partial y}[a, f(a)]\right\} f'(a).$$

By proceeding in this way, one obtains a definite Taylor series for f. The fact that the series obtained does converge in an interval about $x = a$ and represents a solution $y = f(x)$ of the differential equation can be justified under suitable assumptions about the function $F(x, y)$. In most applications, F is a rational function of x and y and the method is applicable except when the denominator is zero at (a, y_0). More generally, the method is applicable when F is *analytic*, that is, can be expanded in a power series in powers of $x - a$ and $y - y_0$ in a neighborhood of (a, y_0) (Section 6–20). For proofs the reader is referred to Chapter 12 of *ODE*.

EXAMPLE. $y' = x^2 - y^2$; $y = 1$ for $x = 0$. Here

$$y'' = 2x - 2yy', \quad y''' = 2 - 2y'^2 - 2yy'', \quad y^{\text{iv}} = -6y'y'' - 2yy''', \ldots.$$

Hence at $x = 0$ one has, for the solution sought,

$$y = 1, \quad y' = -1, \quad y'' = 2, \quad y''' = -4, \quad y^{\text{iv}} = 20, \ldots.$$

Thus the solution is given by

$$y = 1 - x + x^2 - \tfrac{2}{3}x^3 + \tfrac{5}{6}x^4 + \cdots.$$

It is essentially hopeless to attempt to find the general term here; this is a characteristic defect of the method, when applied to a nonlinear differential equation. However, the general theorem referred to above gives assurance that the series does converge for x in some interval $-a < x < a$ and is a solution. It is, furthermore, possible to obtain estimates for the interval of convergence and for the remainder after n terms. Hence the series can be used for carefully controlled numerical work.

The method described applies equally well to higher order differential equations and to systems of equations. Thus for the equation

$$y'' = x + y^2,$$

with initial conditions: $y = 1$ and $y' = 2$ for $x = 1$, one has

$$y''' = 2yy' + 1, \quad y^{\text{iv}} = 2y'^2 + 2yy'', \ldots,$$

so that

$$y = 1 + 2(x - 1) + (x - 1)^2 + \tfrac{5}{6}(x - 1)^3 + \tfrac{1}{2}(x - 1)^4 + \cdots.$$

For linear differential equations with variable coefficients, another way of obtaining the series is available and will often make it easy to find an expression for the general term of the series. This is a method of "undetermined coefficients." For example, let the equation be

$$y'' + xy' + y = 0,$$

and let the solution be sought in the form of a series about $x = 0$:

$$y = c_0 + c_1x + c_2x^2 + \cdots + c_nx^n + \cdots.$$

Upon substituting this series in the equation and collecting terms according to powers of x, one obtains the equation:

$$(c_0 + 2c_2) + x(2c_1 + 6c_3) + x^2(3c_2 + 12c_4) + \cdots \\ + x^n[(n + 1)c_n + (n + 1)(n + 2)c_{n+2}] + \cdots = 0.$$

By the Corollary to Theorem 40 of Section 6–16, each coefficient must equal 0. Hence one obtains the equations:

$$c_0 + 2c_2 = 0, \quad 2c_1 + 6c_3 = 0, \quad 3c_2 + 12c_4 = 0, \ldots, \\ c_n + (n + 2)c_{n+2} = 0, \ldots.$$

Hence there is a *recursion formula* for the coefficients:

$$c_{n+2} = -\frac{c_n}{n+2}.$$

Thus

$$c_2 = -\frac{c_0}{2}, \quad c_4 = \frac{c_0}{2 \cdot 4}, \quad c_6 = -\frac{c_0}{2 \cdot 4 \cdot 6}, \ \ldots$$

$$c_3 = -\frac{c_1}{3}, \quad c_5 = \frac{c_1}{3 \cdot 5}, \quad c_7 = -\frac{c_1}{3 \cdot 5 \cdot 7}, \ \ldots.$$

It appears that c_0 and c_1 are arbitrary constants; they are, in fact, simply the initial values of y and y' at $x = 0$. The solutions found can now be written in the form:

$$y = c_0\left[1 - \frac{1}{2}x^2 + \frac{1}{2 \cdot 4}x^4 - \frac{1}{2 \cdot 4 \cdot 6}x^6 + \cdots + \frac{(-1)^n}{2 \cdot 4 \cdot 6 \cdots 2n}x^{2n} + \cdots\right]$$

$$+ c_1\left[x - \frac{1}{3}x^3 + \frac{1}{3 \cdot 5}x^5 + \cdots + \frac{(-1)^{n+1}}{3 \cdot 5 \cdot 7 \cdots (2n-1)}x^{2n-1} + \cdots\right].$$

Here an application of the ratio test shows that the series converge for all x. Furthermore, y satisfies the differential equation for all x. Since the functions

$$y_1(x) = \sum_{n=0}^{\infty} \frac{(-1)^n x^{2n}}{2^n n!}, \quad y_2(x) = \sum_{n=1}^{\infty} \frac{(-1)^{n+1} x^{2n-1}}{1 \cdot 3 \cdots (2n-1)}$$

are clearly linearly independent, the functions

$$y = c_0 y_1(x) + c_1 y_2(x)$$

actually give the general solution of the differential equation.

This method can also be used for nonlinear equations, but the algebraic processes usually become highly involved and the chances of obtaining the general term of the series or an expression for the general solution are very small.

For many applications it is important to have a series solution of a differential equation $y' = F(x, y)$, even when the function $F(x, y)$ is not analytic in a neighborhood of the initial point considered. The most common case is that in which F is a rational function whose denominator is 0 at the initial point. The initial point is then a *singular point* of the differential equation and no general statement can be made about solutions. However, it is possible in many cases to obtain the solutions through and near the singular point in the form of series of appropriate types. For example, the series

$$y = x^m(c_0 + c_1 x + c_2 x^2 + \cdots), \tag{8–90}$$

where m is not necessarily positive or an integer, can be used in certain cases. In other cases, the solution can be expressed as a series

$$x^m \sum_{n=0}^{\infty} c_n x^{pn}, \tag{8–91}$$

where m and p are quite general.

Similar remarks apply to equations of higher order and to systems of equations. Series of type (8–90) are of special importance for linear equations. For example, *Bessel's differential equation of order n:*

$$\frac{d^2y}{dx^2} + \frac{1}{x}\frac{dy}{dx} + \left(1 - \frac{n^2}{x^2}\right)y = 0, \tag{8–92}$$

has solutions of this form (with $m = n$). One solution is the function $J_n(x)$, the Bessel function of order n, considered in Section 7–15. Further examples are given in Probs. 7 to 9 below. For nonlinear equations, one can often express the solutions near the singular point by means of series of type (8–91) in terms of a parameter t.

For a full discussion of this subject, one is referred to the books of Ince, Picard, and Whittaker and Watson listed at the end of this chapter. It will be found that the theory of functions of a complex variable is essential for a complete analysis of the problem.

Problems

1. Obtain the first four nonzero terms of the series solution for the following equations with initial conditions:

(a) $y' = x^2y^2 + 1$, $y = 1$ for $x = 1$;
(b) $y' = \sin(xy) + x^2$, $y = 3$ for $x = 0$;
(c) $y'' = x^2 - y^2$, $y = 1$ and $y' = 0$ for $x = 0$;
(d) $y''' = xy + yy'$, $y = 0$, $y' = 1$, $y'' = 2$ for $x = 0$.

2. Obtain the general solution in series form:

(a) $y'' + 2xy' + 4y = 0$ (b) $y'' - x^2y = 0$

3. Obtain a solution such that $y = 1$ and $y' = 0$ for $x = 0$ for the equation: $y'' + y' + xy = 0$.

4. Obtain in series form, up to terms in x^3, a solution of the system

$$\frac{dy}{dx} = yz, \quad \frac{dz}{dx} = xz + y$$

such that $y = 1$ and $z = 0$ for $x = 0$.

5. Show that

$$J_0(x) = \sum_{n=0}^{\infty} \frac{(-1)^n x^{2n}}{4^n(n!)^2}$$

satisfies Bessel's equation of order 0 in the form

$$xy'' + y' + xy = 0$$

for all x.

6. Determine, if possible, a solution of Bessel's equation of order 1:

$$x^2y'' + xy' + (x^2 - 1)y = 0,$$

having the form

$$y = \sum_{n=0}^{\infty} c_n x^n.$$

7. Determine, if possible, a solution of Bessel's equation of order n having the form

$$y = x^m \sum_{k=0}^{\infty} c_k x^k.$$

8. (a) Determine a solution of the *hypergeometric equation:*

$$(x^2 - x)y'' + [(\alpha + \beta + 1)x - \gamma]y' + \alpha\beta y = 0 \quad (\alpha, \beta, \gamma \text{ constants}),$$

having the form given in Prob. 7.

(b) Determine a solution of the hypergeometric equation having the form

$$y = x^m \sum_{k=0}^{\infty} c_k x^{-k}.$$

9. (a) Find solutions of *Legendre's equation:* $(1 - x^2)y'' - 2xy' + \alpha(\alpha + 1)y = 0$ having the form

$$y = \sum_{k=0}^{\infty} c_k x^k.$$

(b) Show that the *Legendre polynomial* $P_n(x)$ (Section 7–14):

$$P_n(x) = \frac{1}{2^n n!} \frac{d^n}{dx^n} (x^2 - 1)^n \quad (n = 1, 2, \ldots), \quad P_n(x) = 1 \text{ for } n = 0,$$

is a solution when $\alpha = n$, and that every polynomial solution is a constant times a $P_n(x)$. [Hint: set

$$z = (x^2 - 1)^n;$$

show that

$$(x^2 - 1)z' = 2nxz.$$

Differentiate both sides $n + 1$ times with the aid of the Leibnitz rule (0–133).]

(c) By regarding the Legendre polynomial as a particular power series solution, obtain the formulas (for $n \geqq 1$):

$$P_n(x) = (-1)^{\frac{n}{2}} \frac{1 \cdot 3 \cdots (n-1)}{2 \cdot 4 \cdots (n)} \left[1 - \frac{(n+1)n}{2!} x^2 + \frac{(n+1)(n+3)n(n-2)}{4!} x^4 + \cdots + (-1)^{\frac{n}{2}} \frac{(n+1)(n+3)\cdots(2n-1)n(n-2)\cdots 2}{n!} x^n\right], \quad n \text{ even};$$

$$P_n(x) = (-1)^{\frac{n-1}{2}} \frac{1 \cdot 3 \cdots n}{2 \cdot 4 \cdots (n-1)} \left[x - \frac{(n+2)(n-1)}{3!} x^3 + \frac{(n+2)(n+4)(n-1)(n-3)}{5!} x^5 + \cdots + (-1)^{\frac{n-1}{2}} \frac{(n+2)(n+4)\cdots(2n-1)(n-1)(n-3)\cdots 2}{n!} x^n\right], \quad n \text{ odd}.$$

Answers

1. (a) $1 + 2(x-1) + 3(x-1)^2 + \frac{19}{3}(x-1)^3 + \cdots$; (b) $3 + \frac{3}{2}x^2 + \frac{1}{3}x^3 - \frac{3}{4}x^4 + \cdots$; (c) $1 - \frac{1}{2}x^2 + \frac{1}{6}x^4 - \frac{7}{360}x^6$; (d) $x + x^2 + \frac{1}{24}x^4 + \frac{1}{15}x^5$.

2. (a) $c_0\left[1 - 2x^2 + \frac{2^2}{1 \cdot 3}x^4 + \cdots + (-1)^n \frac{2^n x^{2n}}{1 \cdot 3 \cdots (2n-1)} + \cdots\right]$

$+ c_1\left[x - \frac{2}{2}x^3 + \frac{2^2}{2 \cdot 4}x^5 + \cdots + (-1)^{n-1}\frac{2^{n-1}x^{2n-1}}{2 \cdot 4 \cdots (2n-2)} + \cdots\right];$

(b) $c_0\left[1 + \frac{x^4}{4 \cdot 3} + \frac{x^8}{(8 \cdot 7)(4 \cdot 3)} + \cdots \right.$

$\left. + \frac{x^{4n}}{(4n)(4n-1)(4n-4)(4n-5) \cdots (4 \cdot 3)} + \cdots\right]$

$+ c_1\left[x + \frac{x^5}{5 \cdot 4} + \frac{x^9}{(9 \cdot 8)(5 \cdot 4)} + \cdots \right.$

$\left. + \frac{x^{4n+1}}{(4n+1)(4n)(4n-3)(4n-4) \cdots (5 \cdot 4)} + \cdots\right].$

3. $1 - \frac{x^3}{6} + \frac{x^4}{24} - \frac{x^5}{120} + \frac{x^6}{144} + \cdots + c_n x^n + \cdots$, where $c_n + (n+2)c_{n+2} + (n+3)(n+2)c_{n+3} = 0$.

4. $y = 1 + \frac{x^2}{2} + \cdots, \quad z = x + \frac{x^3}{2} + \cdots.$

6. $c\left[x - \frac{x^3}{2 \cdot 4} + \frac{x^5}{2 \cdot 4 \cdot 4 \cdot 6} + \cdots + (-1)^n \frac{x^{2n+1}}{2 \cdot 4^2 \cdot 6^2 \cdots (2n)^2(2n+2)} + \cdots\right].$
[This is $2cJ_1(x)$, where J_1 is the Bessel function of first kind of order 1.]

7. $cx^m \sum_{k=0}^{\infty} \frac{(-1)^k x^{2k}}{4^k k!(m+1)(m+2)\cdots(m+k)}$ where $m = \pm n$, but m not a negative integer. [For m a positive integer and $c = \frac{1}{2^m m!}$, this is $J_m(x)$, the Bessel function of first kind and order m.]

8. (a)

$$cx^m\left[1+\sum_{k=1}^{\infty}\frac{(\alpha+m)(\beta+m)(\alpha+m+1)(\beta+m+1)\cdots(\alpha+m+k-1)(\beta+m+k-1)}{(m+1)(m+\gamma)(m+2)(m+\gamma+1)\cdots(m+k)(m+\gamma+k-1)}x^k\right],$$

where $m = 0$ (unless γ is 0 or a negative integer) or $m = 1 - \gamma$ (unless γ is a positive integer); the series converges for $|x| < 1$. [For $m = 0$, the solution is $cF(\alpha, \beta, \gamma, x)$, where F is the *hypergeometric series*.]

(b)

$$cx^m\left[1+\sum_{k=1}^{\infty}\frac{(-m)(1-\gamma-m)(1-m)(2-\gamma-m)\cdots(k-1-m)(k-\gamma-m)x^{-k}}{(1-\alpha-m)(1-\beta-m)(2-\alpha-m)(2-\beta-m)\cdots(k-\alpha-m)(k-\beta-m)}\right],$$

where $m = -\alpha$ (unless $\alpha - \beta$ is a negative integer) or $m = -\beta$ (unless $\beta - \alpha$ is a negative integer); the series converges for $|x| > 1$.

9. (a) $c_1\left[1+\sum_{n=1}^{\infty}(-1)^n x^{2n}\frac{\alpha(\alpha-2)\cdots(\alpha-2n+2)(\alpha+1)\cdots(\alpha+2n-1)}{(2n)!}\right]$

$+c_2\left[x+\sum_{n=1}^{\infty}(-1)^n x^{2n+1}\frac{(\alpha-1)(\alpha-3)\cdots(\alpha-2n+1)(\alpha+2)(\alpha+4)\cdots(\alpha+2n)}{(2n+1)!}\right].$

The series converge for $|x| < 1$ or else reduce to polynomials.

***8–16 The exponential function of a matrix.** Let A be an $n \times n$ matrix. In Section 6–26 we defined e^A to be the $n \times n$ matrix which is the sum of the convergent series $\Sigma A^n/n!$. This series arises in a natural way if we solve the vector differential equation

$$\frac{d\mathbf{x}}{dt} = A\mathbf{x} \tag{8–93}$$

by power series. For if $\mathbf{x} = \mathbf{f}(t)$ is a solution of the differential equation then the corresponding Taylor's series about $t = 0$ is

$$\mathbf{f}(0) + \frac{t}{1!}\mathbf{f}'(0) + \cdots + \frac{t^k}{k!}\mathbf{f}^{(k)}(0) + \cdots.$$

Now here $\mathbf{f}'(t) = A\mathbf{f}(t)$, $\mathbf{f}''(t) = A\mathbf{f}'(t), \ldots$, so that, if $\mathbf{f}(0) = \mathbf{x}_0$, then

$$\mathbf{f}'(0) = A\mathbf{x}_0, \quad \mathbf{f}''(0) = A^2\mathbf{x}_0, \ldots, \quad \mathbf{f}^{(k)}(0) = A^k\mathbf{x}_0, \ldots$$

and the Taylor series becomes

$$\mathbf{x}_0 + tA\mathbf{x}_0 + \frac{t^2}{2!}A^2\mathbf{x}_0 + \cdots = \left(I + tA + \frac{t^2}{2!}A^2 + \cdots\right)\mathbf{x}_0 = e^{tA}\mathbf{x}_0,$$

and we have obtained a series for $\mathbf{f}(t)$ (that is, a power series for each component of the vector function $\mathbf{f}(t)$) converging for all t and hence satisfying the differential equation (8–93) for all t.

Now let us expand the same solution $\mathbf{f}(t)$ in a power series about t_1. At t_1 the solution has the value $\mathbf{x}_1 = \mathbf{f}(t_1) = e^{t_1A}\mathbf{x}_0$. The derivatives of $\mathbf{f}$ at t_1 can now be computed as before from the differential equation (8–93) and we obtain

$$\mathbf{x} = \mathbf{f}(t) = \mathbf{x}_1 + (t - t_1)A\mathbf{x}_1 + \cdots = e^{(t-t_1)A}\mathbf{x}_1.$$

In particular, $\mathbf{f}(t_1 + h) = e^{hA}\mathbf{x}_1$. But $\mathbf{f}(t_1 + h) = e^{(t_1+h)A}\mathbf{x}_0$. Hence

$$e^{hA}e^{t_1A}\mathbf{x}_0 = e^{(t_1+h)A}\mathbf{x}_0.$$

This holds for all $\mathbf{x}_0$, and we conclude that

$$e^{hA}e^{t_1A} = e^{(t_1+h)A} = e^{(h+t_1)A}. \tag{8–94}$$

This important identity was discussed in Section 6–26 (Theorem 61).

We remark that $e^{tA}\mathbf{e}_1 = e^{tA}(1, 0, \ldots, 0)$ is the solution $\mathbf{x}(t)$ of (8–93) such that $\mathbf{x}(0) = \mathbf{e}_1$. But $e^{tA}\mathbf{e}_1$ is simply the first column of e^{tA}. In this way we see that $X(t) = e^{tA}$ is the $n \times n$ matrix whose successive columns are the solutions of (8–93) with initial values $\mathbf{e}_1, \ldots, \mathbf{e}_n$ for $t = 0$. Thus $X(0) = I = e^{0A}$ and, as in Section 8–12, $X(t) = e^{tA}$ can be regarded as the solution of the matrix differential equation $dX/dt = AX$ with $X(0) = I$. Accordingly,

$$\frac{d}{dt}e^{tA} = Ae^{tA}. \tag{8–95}$$

Example. $\dfrac{d\mathbf{x}}{dt} = A\mathbf{x}$, with $A = \begin{bmatrix} 1 & -2 \\ -2 & 1 \end{bmatrix}$. This is Example 1 of

Section 8–12. Computing $A^2, A^3, \ldots$, we find that

$$e^{tA} = I + t\begin{bmatrix} 1 & -2 \\ -2 & 1 \end{bmatrix} + \frac{t^2}{2!}\begin{bmatrix} 5 & -4 \\ -4 & 5 \end{bmatrix} + \frac{t^3}{3!}\begin{bmatrix} 13 & -14 \\ -14 & 13 \end{bmatrix} + \cdots.$$

Now we already know the general solution of the given differential equation: $\mathbf{x} = c_1 \operatorname{col}(e^{3t}, -e^{3t}) + c_2 \operatorname{col}(e^{-t}, e^{-t})$, and hence easily obtain the solutions with initial values $\mathbf{e}_1 = (1, 0)$ and $\mathbf{e}_2 = (0, 1)$:

$$\mathbf{x} = \frac{1}{2}\begin{bmatrix} e^{3t} + e^{-t} \\ -e^{3t} + e^{-t} \end{bmatrix}, \quad \mathbf{x} = \frac{1}{2}\begin{bmatrix} -e^{3t} + e^{-t} \\ e^{3t} + e^{-t} \end{bmatrix}.$$

Therefore,

$$X(t) = e^{tA} = \begin{bmatrix} \frac{1}{2}e^{3t} + \frac{1}{2}e^{-t} & -\frac{1}{2}e^{3t} + \frac{1}{2}e^{-t} \\ -\frac{1}{2}e^{3t} + \frac{1}{2}e^{-t} & \frac{1}{2}e^{3t} + \frac{1}{2}e^{-t} \end{bmatrix}$$

If we expand the entries in power series in powers of t, we can verify term by term that the two expressions for e^{tA} are equal. The second expression is far simpler. For some purposes, the series expression is more useful and it is a valuable theoretical tool.

Problems

1. For each of the following matrices A, find e^{tA} by finding $X(t)$ such that $dX/dt = AX$ and $X(0) = I$. Then expand $X(t)$ in a power series in t and verify that the terms through t^2 agree with the corresponding terms of the series $I + tA + (t^2/2!)A^2 + \cdots$. (The answers to Prob. 1 following Section 8–12 will be found helpful).

(a) $\begin{bmatrix} 1 & 2 \\ 12 & -1 \end{bmatrix}$, (b) $\begin{bmatrix} 1 & -2 \\ 5 & -1 \end{bmatrix}$.

2. Prove, on the basis of the theory of differential equations, that $e^{tA} \neq O$, $-\infty < t < \infty$.

Answers

1. (a) $\frac{1}{5}\begin{bmatrix} 3e^{5t} + 2e^{-5t} & e^{5t} - e^{-5t} \\ 6e^{5t} - 6e^{-5t} & 2e^{5t} + 3e^{-5t} \end{bmatrix}$,

(b) $\frac{1}{3}\begin{bmatrix} 3\cos 3t + \sin 3t & -2\sin 3t \\ 5\sin 3t & 3\cos 3t - \sin 3t \end{bmatrix}$.

Suggested References

AGNEW, RALPH P., *Differential Equations.* New York: McGraw-Hill, 1942.

ANDRONOW, A., and CHAIKIN, C. E., *Theory of Oscillations.* Princeton: Princeton University Press, 1949.

CODDINGTON, EARL A. and LEVINSON, NORMAN, *Theory of Ordinary Differential Equations.* New York: McGraw-Hill, 1955.

Cohen, A., *Differential Equations*, 2nd ed. Boston: Heath, 1933.

Forsyth, A. R., *Theory of Differential Equations*, Vols. 1–6. Cambridge: Cambridge University Press, 1890–1906.

Goursat, Édouard, *A Course in Mathematical Analysis*, Vol. II, Part II, transl. by E. R. Hedrick and O. Dunkel. New York: Ginn and Co., 1917.

Henrici, Peter K., *Elements of Numerical Analysis*. New York: John Wiley and Sons, 1964.

Hildebrand, F. B., *Introduction to Numerical Analysis*. New York McGraw-Hill, 1956.

Ince, E. L., *Ordinary Differential Equations*. London: Longmans, Green, 1927.

Kamke, E., *Differentialgleichungen, Lösungsmethoden und Lösungen*, Vol. 1, 2nd ed. Leipzig: Akademische Verlagsgesellschaft, 1943.

Kamke, E., *Differentialgleichungen reeller Funktionen*. Leipzig: Akademische Verlagsgesellschaft, 1933.

Kaplan, Wilfred, *Ordinary Differential Equations*, Reading: Mass., Addison-Wesley, 1958.

McLachlan, N. W., *Ordinary Non-linear Differential Equations in Engineering and Physical Sciences*. Oxford: Oxford University Press, 1950.

Milne, W. E., *Numerical Calculus*. Princeton: Princeton University Press, 1949.

Morris, M., and Brown, O. E., *Differential Equations*, Rev. ed. New York: Prentice-Hall, 1942.

Picard, Emile, *Traité d'Analyse* (3 Vols.), 3d ed. Paris: Gauthier-Villars, 1922.

Rainville, Earl D., *Intermediate Differential Equations*. New York: John Wiley and Sons, Inc., 1943.

Ralston, Anthony, *First Course in Numerical Analysis*. New York: McGraw-Hill, 1965.

Scarborough, James B., *Numerical Mathematical Analysis*, 2nd ed. Baltimore: Johns Hopkins Press, 1950.

Whittaker, E. T., and Watson, G. N., *A Course of Modern Analysis*, 4th ed. Cambridge: Cambridge University Press, 1940.

Willers, F. A., *Practical Analysis*, transl. by R. T. Beyer. New York: Dover, 1948.

CHAPTER 9

Functions of a Complex Variable

In this chapter we give an introduction to the theory of analytic functions of a complex variable. The principal topics are series expansions, integrals and residues. For a more thorough treatment of these topics and a discussion of conformal mapping and its applications, the reader is referred to the author's book *Introduction to Analytic Functions* (Reading, Mass.: Addison-Wesley, 1966).

9–1 Complex functions. We assume familiarity with the complex number system (Section 0–2). Figure 9–1 (next page) reviews standard notations for the complex z-plane, where $z = x + iy$. We shall denote real and imaginary parts as follows:

$$\text{for } z = x + iy, \qquad x = \operatorname{Re} z, \qquad y = \operatorname{Im} z.$$

We shall make steady use of all these notations, as well as the analogous notations in a complex w-plane, where $w = u + iv$.

Complex-valued functions of z. If to each value of the complex number $z = x + iy$, with certain exceptions, there is assigned a value of the complex number $w = u + iv$, then w is given as a *complex-valued function* of z and we write $w = f(z)$. For example,

$$w = z^2, \qquad w = z^3 + 5z + 7, \qquad w = \frac{z+1}{z-2} \quad (z \neq 2)$$

are such functions. Important functions of this type are

polynomials: $w = a_0 z^n + \cdots + a_{n-1}z + a_n$,

rational functions: $w = \dfrac{a_0 z^n + \cdots + a_n}{b_0 z^m + \cdots + b_m}$,

exponential function: $\exp z = e^z = e^{x+iy} = e^x(\cos y + i \sin y)$,

trigonometric functions: $\sin z = \dfrac{e^{iz} - e^{-iz}}{2i}$, $\qquad \cos z = \dfrac{e^{iz} + e^{-iz}}{2}$,

hyperbolic functions: $\sinh z = \dfrac{e^z - e^{-z}}{2}$, $\qquad \cosh z = \dfrac{e^z + e^{-z}}{2}$,

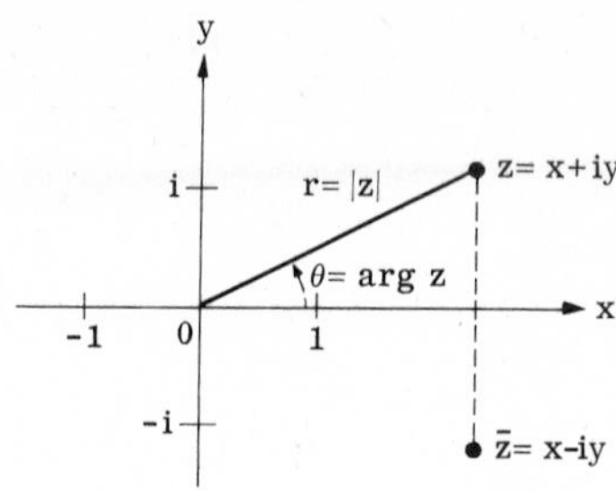

FIG. 9–1. The complex z-plane.

The definition of the exponential function is motivated by interpreting e^z for complex z as the sum of its power series $\Sigma z^n/n!$; see Section 6–19. From the definitions it follows that, for real y,

$$e^{iy} = \cos y + i \sin y \qquad \text{and} \qquad e^{-iy} = \cos y - i \sin y,$$

so that

$$\sin y = \frac{e^{iy} - e^{-iy}}{2i}, \qquad \cos y = \frac{e^{iy} + e^{-iy}}{2}.$$

These equations suggest the definitions given above for $\sin z$ and $\cos z$. The definitions of the hyperbolic functions are based on the usual definitions for real variables. By the reasoning described, it follows that, when z is real ($z = x + i0$), each of the five functions reduces to the familiar real function. For example, $e^{x+i0} = e^x$.

The other trigonometric and hyperbolic functions are defined in terms of the sine, cosine, sinh and cosh in the usual way.

9–2 Complex-valued functions of a real variable. It will be convenient to represent paths for line integrals in the complex plane by equations of form

$$z = F(t), \qquad a \leq t \leq b. \tag{9–1}$$

Here t is a real variable and F is a function whose values are complex, so that we are dealing with a complex-valued function of a real variable. Examples are the following functions:

$$z = e^{it}, \qquad 0 \leqq t \leqq 2\pi; \qquad z = t + it^2, \qquad 0 \leqq t \leqq 1.$$

In (9–1) we can write $z = x + iy$ and $F(t) = f(t) + ig(t)$, where f and g are real-valued. Then (9–1) is equivalent to the *pair* of equations

$$x = f(t), \qquad y = g(t), \qquad a \leqq t \leqq b. \tag{9–2}$$

We can also consider (9–1) as an alternative to the familiar vector function representation of a path.

The path $z = e^{it}$ given above is equivalent to the path $x = \cos t$, $y = \sin t$, and hence its graph is a circle.

The calculus can be developed for complex-valued functions of t in strict analogy with the development for real-valued functions. We write

$$\lim_{t \to t_0} F(t) = c = x_0 + iy_0 \tag{9–3}$$

if, given $\epsilon > 0$, we can choose $\delta > 0$ so that $|F(t) - c| < \epsilon$ when $0 < |t - t_0| < \delta$. It is assumed here that $F(t)$ is defined for t sufficiently close to t_0, but not necessarily for $t = t_0$; if $F(t)$ is defined only in an interval $t_0 < t < \beta$, the limit is interpreted as a limit *to the right* and is written

$$\lim_{t \to t_0+} F(t).$$

Limits to the left are defined similarly. Limits as $t \to \infty$ or $t \to -\infty$ are defined as for real functions. [However,

$$\lim_{t \to t_0} F(t) = \infty$$

is defined to mean

$$\lim_{t \to t_0} |F(t)| = \infty;$$

there is no concept of $+\infty$ or $-\infty$ for complex numbers; see Section 9–14.] If $F(t)$ is defined for $\alpha < t < \beta$ and t_0 lies in this interval, then $F(t)$ is said to be *continuous* at t_0 if

$$\lim_{t \to t_0} F(t) = F(t_0).$$

If $F(t)$ is also defined at $t = \alpha$ and

$$\lim_{t \to \alpha+} F(t) = F(\alpha),$$

then $F(t)$ is said to be continuous to the right at $t = \alpha$. Continuity to the left and continuity in an interval are defined as for real functions.

If $F(t) = f(t) + ig(t)$ then Eq. (9–3) is equivalent to the two equations

$$\lim_{t \to t_0} f(t) = x_0, \qquad \lim_{t \to t_0} g(t) = y_0. \tag{9–4}$$

For Eq. (9–3) signifies that $F(t)$ is as close to c as desired, for t sufficiently close to t_0; by a geometric argument we see that this is equivalent to the requirement that $f = \operatorname{Re} F$ be as close to a as desired and $g = \operatorname{Im} F$ as close to b as desired, for t sufficiently close to t_0. Similarly, continuity of $F(t)$ at t_0 is equivalent to continuity of $f(t)$ and $g(t)$ at t_0. Accordingly, such functions as $t^2 + it^3$ and e^{it} are continuous for all t. Furthermore, the rules for limits of sums, products and quotients, and the analogous continuity theorems must carry over to the complex case. For example, if $F_1(t) = f_1(t) + ig_1(t)$ and $F_2(t) = f_2(t) + ig_2(t)$ are continuous in an interval, then so also is

$$\begin{aligned} F_1(t) \cdot F_2(t) &= [f_1(t) + ig_1(t)] \cdot [f_2(t) + ig_2(t)] \\ &= f_1(t)f_2(t) - g_1(t)g_2(t) + i[f_1(t)g_2(t) + f_2(t)g_1(t)]. \end{aligned}$$

For $f_1(t)$, $g_1(t)$, $f_2(t)$, $g_2(t)$ must be continuous, so that the real and imaginary parts of $F_1(t) \cdot F_2(t)$ are continuous, and hence $F_1(t) \cdot F_2(t)$ is continuous.

For the discussion thus far, we are to some extent repeating the theory of vector functions as given in Section 0–7. However, we remark that complex multiplication and division have no analogue for vectors.

Derivatives and integrals. The derivative of $F(t)$ can be defined as for real functions:

$$F'(t_0) = \lim_{\Delta t \to 0} \frac{F(t_0 + \Delta t) - F(t_0)}{\Delta t}. \tag{9–5}$$

By taking real and imaginary parts and applying (9–4), we conclude that

$$F'(t_0) = f'(t_0) + ig'(t_0); \tag{9–6}$$

that is, $F(t)$ has a derivative at t_0 precisely when $f(t)$, $g(t)$ have derivatives at t_0, and the derivatives are related by Eq. (9–6). Derivatives to the left or right are defined by requiring that $\Delta t < 0$ or $\Delta t > 0$, respectively in Eq. (9–5); Eq. (9–6) also applies to these derivatives.

The rules for derivative of sum, product, quotient, constant, and constant times function all carry over, and the proofs for real functions can be repeated. We also note the rules

$$\frac{d}{dt}[F(t)]^n = n[F(t)]^{n-1}F'(t) \quad (n = 1, 2, \ldots), \tag{9–7}$$

$$\frac{d}{dt}e^{(a+bi)t} = (a + bi)e^{(a+bi)t} \quad (a, b \text{ real}). \tag{9–8}$$

Furthermore, if $F'(t) \equiv 0$ for $\alpha < t < \beta$, then $F(t)$ is identically constant for $\alpha < t < \beta$. The proofs are left as exercises (Problems 9 to 11 below).

Higher derivatives are obtained by repeated differentiation:

$$F''(t) = [F'(t)]' = D^2F, \; F'''(t) = [F''(t)]' = D^3F, \ldots$$

The first derivative of $F(t)$ can be thought of as the *velocity vector* of the point (x, y) as it moves on the path $x = f(t)$, $y = g(t)$, with t as *time*. The second derivative can be interpreted as *acceleration*.

The *definite integral* of $F(t)$ over an interval $\alpha \leqq t \leqq \beta$ is defined as a limit of a sum $\sum F(t_k^*) \Delta_k t$ as for real functions. However, again the limit theorem permits us to take real and imaginary parts:

$$\int_\alpha^\beta F(t)\,dt = \int_\alpha^\beta f(t)\,dt + i\int_\alpha^\beta g(t)\,dt. \tag{9–9}$$

If $F(t)$ is continuous over the interval, then $f(t)$ and $g(t)$ are continuous so that the integral exists. More generally, the integral exists if $f(t)$ and $g(t)$ are *piecewise continuous* for $\alpha \leqq t \leqq \beta$. When f and g are piecewise continuous, we term $F = f + ig$ piecewise continuous.

An *indefinite integral* of $F(t)$ is defined as a function $G(t)$ whose derivative is $F(t)$. As in ordinary calculus, we find that, if $G(t)$ is one indefinite integral, then $G(t) + c$ provides all indefinite integrals:

$$\int F(t)\,dt = G(t) + c,$$

c being an arbitrary complex constant. If an indefinite integral G of F is known, then it can be used to evaluate definite integrals of F as in calculus:

$$\int_\alpha^\beta F(t)\,dt = \int_\alpha^\beta G'(t)\,dt = G(\beta) - G(\alpha). \tag{9–10}$$

The proof is left as an exercise (Problem 12 below).

From Eq. (9–9), we can verify the familiar rules for the integral of a sum, the integral of constant times function, the combination of integrals from α to β and from β to γ, and integration by parts. We also have the basic inequality

$$\left|\int_\alpha^\beta F(t)\,dt\right| \leqq \int_\alpha^\beta |F(t)|\,dt \leqq M(\beta - \alpha); \tag{9–11}$$

this is valid if $\alpha < \beta$, if $F(t)$ is, for example, piecewise continuous for $\alpha \leqq t \leqq \beta$, and if $|F(t)| \leqq M$ on this interval. The inequality is most easily obtained from the definition of the integral as limit of a sum; for we have, by repeated application of the triangle inequality [see Eq. (0–10)],

$$\left|\sum_{k=1}^n F(t_k^*)\,\Delta_k t\right| \leqq \sum_{k=1}^n |F(t_k^*)|\,\Delta_k t \leqq M(\beta - \alpha),$$

and passage to the limit gives (9–11). Definite integrals and indefinite integrals are related by the rule

$$\frac{d}{dt}\int_\alpha^t F(u)\,du = F(t) \tag{9–12}$$

(see Problem 12 below).

EXAMPLE 1. $\displaystyle\int_1^2 (t + it^2)\,dt = \left(\frac{t^2}{2} + i\frac{t^3}{3}\right)\Big|_1^2 = \frac{3}{2} + \frac{7}{3}i.$

EXAMPLE 2. $\displaystyle\int_0^1 e^{(a+bi)t}\,dt = \frac{e^{(a+bi)t}}{a+bi}\Big|_0^1 = \frac{e^{a+bi} - 1}{a+bi} \quad (a + bi \neq 0).$

EXAMPLE 3.

$$\int p(t)e^{-at}\,dt = -e^{-at}\left[\frac{p(t)}{a} + \frac{p'(t)}{a^2} + \cdots + \frac{p^{(n)}(t)}{a^{n+1}}\right] + C,$$

where $p(t)$ is a polynomial of degree n and a is a complex constant, not 0. The equation is established by integration by parts [Problem 8(f)].

Problems

1. (a) Let $z_1 = r_1(\cos\theta_1 + i\sin\theta_1)$, $z_2 = r_2(\cos\theta_2 + i\sin\theta_2)$. Show that

$$z_1 \cdot z_2 = r_1r_2[\cos(\theta_1 + \theta_2) + i\sin(\theta_1 + \theta_2)]$$

(b) Show that $e^{i(\theta_1+\theta_2)} = e^{i\theta_1}\cdot e^{i\theta_2}$ [see part (a)].

2. Prove the following identities:

(a) $e^{z_1}\cdot e^{z_2} = e^{z_1+z_2}$ [see Problem 1(b)]
(b) $(e^z)^n = e^{nz}$ $(n = 0, \pm 1, \pm 2, \ldots)$
(c) $\sin^2 z + \cos^2 z = 1$
(d) $\sin(z_1 + z_2) = \sin z_1 \cos z_2 + \cos z_1 \sin z_2$
(e) $\operatorname{Re}(\sin z) = \sin x \cosh y$, $\operatorname{Im}(\sin z) = \cos x \sinh y$
(f) $\sin iz = i\sinh z$, $\cos iz = \cosh z$
(g) $\overline{e^z} = e^{\bar z}$, $\overline{\sin z} = \sin\bar z$, $\overline{\cos z} = \cos\bar z$

3. (a) Prove that $e^z \neq 0$ for all z.
(b) Prove that $\sin z$ and $\cos z$ are 0 only for appropriate real value of z.

4. Represent the following functions graphically:

(a) $w = (1 + t) + i(1 - t)$
(b) $w = t^4 + i(t^2 + 1)$
(c) $w = e^{3it}$
(d) $w = 2e^{(-1+2i)t}$
(e) $w = te^{(-1+2i)t}$
(f) $w = e^{-t} - ie^{it}$

5. Find the derivatives of the functions of Problem 4.

6. Graph $w = 3e^{2it}$ and indicate the first and second derivatives graphically for $t = 0$, $t = \pi/2$, $t = \pi$.

7. Integrate the functions of Problem 4 from 0 to 1.

8. Use integration by parts to evaluate each of the following:

(a) $\int (1 + it)^2 \sin t \, dt$ (b) $\int t^n e^{-at} \, dt \quad (n = 1, 2, \ldots)$

(c) $\int t^n \sin bt \, dt = \frac{1}{2i} \int t^n (e^{bit} - e^{-bit}) \, dt \quad (n = 1, 2, \ldots)$

(d) $\int t^n \cos at \, dt = \operatorname{Re} \int t^n e^{iat} \, dt \quad (a \text{ real}, n = 1, 2, \ldots)$

(e) $\int t^n \cos at \cos bt \cos ct \, dt \quad (n = 1, 2, \ldots)$

(f) $\int p(t) e^{-at} \, dt$, where $p(t)$ is a polynomial of degree n (Example 3 in text)

9. Prove Eq. (9–7) by induction (repeated application of rule for differentiation of a product).

10. Prove (9–8) with the aid of (9–6).

11. Prove that if $F'(t) \equiv 0$, $\alpha < t < \beta$, then $F(t) \equiv$ constant for $\alpha < t < \beta$.

12. (a) Prove (9–12) by taking real and imaginary parts.
(b) Prove (9–10) either directly or as a consequence of (9–12).

Answers

5. (a) $1 - i$ (b) $4t^3 + 2it$ (c) $3ie^{3it}$ (d) $(-2 + 4i)e^{(-1+2i)t}$ (e) $e^{(-1+2i)t}[1 + t(-1 + 2i)]$ (f) $-e^{-t} + e^{it}$

6. $w' = 6ie^{2it}$, $w'' = -12e^{2it}$

7. (a) $(3 + i)/2$ (b) $(3 + 20i)/15$ (c) $(e^{3i} - 1)/3i$ (d) $2(e^{-1+2i} - 1)/(-1 + 2i)$ (e) $[1 + (2i - 2)e^{-1+2i}]/(-3 - 4i)$ (f) $2 - e^{-1} - e^{i}$

8. (a) $(t^2 - 2it - 3) \cos t + (2i - 2t) \sin t + c$

(b) $-e^{-at}\left(\frac{t^n}{a} + \frac{nt^{n-1}}{a^2} + \cdots + \frac{n!}{a^{n+1}}\right) + c$

(c) $$\cos bt\left[-\frac{t^n}{b} + \frac{n(n-1)t^{n-2}}{b^3} - \frac{n(n-1)(n-2)(n-3)t^{n-4}}{b^5} + \cdots\right] + \sin bt\left[\frac{nt^{n-1}}{b^2} - \frac{n(n-1)(n-2)t^{n-3}}{b^4} + \cdots\right] + c$$

(d) $\operatorname{Re}\left\{-e^{ait}\left[\frac{t^n}{-ai} + \frac{nt^{n-1}}{(-ai)^2} + \cdots + \frac{n!}{(-ai)^{n+1}}\right]\right\} + c$

(e) $-\frac{1}{8}\sum_{k=1}^{8}\left[e^{-a_k t}\left(\frac{t^n}{a_k} + \frac{nt^{n-1}}{a_k^2} + \cdots + \frac{n!}{a_k^{n+1}}\right)\right] + C,$

where the a_k are the 8 numbers $(\pm a \pm b \pm c)i$.

9–3 Complex-valued functions of a complex variable. Limits and continuity. We return to the general complex-valued function of a complex variable. These functions will be our principal concern for the remainder of this chapter. We write:

$$w = f(z),$$

where $z = x + iy$, $w = u + iv$, to indicate such a function. An example is the function

$$w = z^2 \quad (\text{all } z).$$

Here we can also write:

$$u + iv = (x + iy)^2 = x^2 - y^2 + 2ixy,$$

so that (on taking real and imaginary parts)

$$u = x^2 - y^2, \qquad v = 2xy.$$

In a similar manner, *every* complex function $w = f(z)$ is equivalent to a pair of real functions:

$$u = u(x, y) = \text{Re}[f(z)], \qquad v = v(x, y) = \text{Im}[f(z)],$$

of the two real variables x, y. Also from such a pair of real functions, defined on the same set, we obtain a complex function of z. For example,

$$u = x^2 + xy + y^2, \qquad v = xy^3$$

is equivalent to the complex function

$$w = f(z) = x^2 + xy + y^2 + xy^3 i,$$

for which $f(1 + 2i) = 1 + 2 + 4 + 8i = 7 + 8i$.

The functions e^z, $\sin z$, $\cos z$, $\sinh z$, $\cosh z$ were defined in Section 9–1. For these the corresponding pairs of real functions are as follows:

$$\begin{aligned}
w &= e^z: & u &= e^x \cos y, & v &= e^x \sin y,\\
w &= \sin z: & u &= \sin x \cosh y, & v &= \cos x \sinh y,\\
w &= \cos z: & u &= \cos x \cosh y, & v &= -\sin x \sinh y,\\
w &= \sinh z: & u &= \sinh x \cos y, & v &= \cosh x \sin y,\\
w &= \cosh z: & u &= \cosh x \cos y, & v &= \sinh x \sin y.
\end{aligned} \tag{9–13}$$

The proofs are left as exercises (Problem 1 below). In (9–13) each function is defined for all z; that is, for all (x, y).

In general, we assume $w = f(z)$ to be defined in a *domain* (open region) D in the z-plane, as suggested in Fig. 9–2. If z_0 is a point of D, we can then find a circular *neighborhood* $|z - z_0| < k$ about z_0 in D. If $f(z)$ is defined in such a neighborhood, except perhaps at z_0, then we write

$$\lim_{z \to z_0} f(z) = w_0 \qquad (9\text{–}14)$$

if, for every $\epsilon > 0$, we can choose $\delta > 0$, so that

$$|f(z) - w_0| < \epsilon \qquad (9\text{–}15)$$
$$\text{for } 0 < |z - z_0| < \delta.$$

If $f(z_0)$ is defined and equals w_0, and (9–14) holds, then we call $f(z)$ *continuous at* z_0.

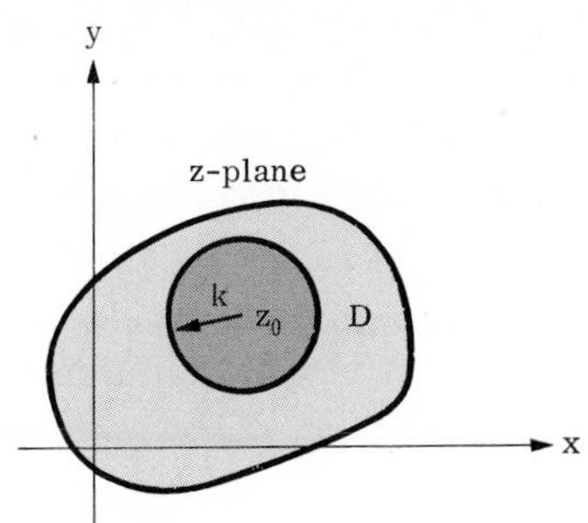

FIG. 9–2. Domain and neighborhood.

THEOREM 1. *The function* $w = f(z)$ *is continuous at* $z_0 = x_0 + iy_0$ *if and only if* $u(x, y) = \text{Re}\,[f(z)]$ *and* $v(x, y) = \text{Im}\,[f(z)]$ *are continuous at* (x_0, y_0).

Thus $w = z^2 = x^2 - y^2 + 2ixy$ is continuous for all z, since $u = x^2 - y^2$ and $v = 2xy$ are continuous for all (x, y). The proof of Theorem 1 is left as an exercise (Problem 5 below).

THEOREM 2. *The sum, product, and quotient of continuous functions of* z *are continuous, except for division by zero; a continuous function of a continuous function is continuous. Similarly, if the limits exist,*

$$\lim_{z \to z_0} [f(z) + g(z)] = \lim_{z \to z_0} f(z) + \lim_{z \to z_0} g(z), \quad \ldots \qquad (9\text{–}16)$$

These properties are proved as for real variables. (It is assumed in Theorem 2 that the functions are defined in appropriate domains.)

It follows from Theorem 2 that polynomials in z are continuous for all z, and each rational function is continuous except where the denominator is zero. From Theorem 1 it follows that

$$e^z = e^x \cos y + ie^x \sin y$$

is continuous for all z. Hence, by Theorem 2, so also are the functions

$$\sin z = \frac{e^{iz} - e^{-iz}}{2i}, \qquad \cos z = \frac{e^{iz} + e^{-iz}}{2}.$$

We write

$$\lim_{z \to z_0} f(z) = \infty \quad \text{if} \lim_{z \to z_0} |f(z)| = +\infty;$$

that is, if for each real number K there is a positive δ such that $|f(z)| > K$

for $0 < |z - z_0| < \delta$. Similarly, if $f(z)$ is defined for $|z| > R$, for some R, then $\lim_{z\to\infty} f(z) = c$ if for each $\epsilon > 0$ we can choose a number R_0 such that $|f(z) - c| < \epsilon$ for $|z| > R_0$. All these definitions emphasize that there is but *one* complex number ∞ and that "approaching ∞" is equivalent to receding from the origin.

9–4 Derivatives and differentials. Let $w = f(z)$ be given in D and let z_0 be a point of D. Then w is said to have a derivative $f'(z_0)$ if

$$\lim_{\Delta z\to 0} \frac{f(z_0 + \Delta z) - f(z_0)}{\Delta z} = f'(z_0).$$

In appearance this definition is the same as that for functions of a real variable, and it will be seen that the derivative does have the usual properties. However, it will also be shown that if $w = f(z)$ has a continuous derivative in a domain D, then $f(z)$ has a number of additional properties; in particular, the second derivative $f''(z)$, third derivative $f'''(z), \ldots,$ must also exist in D.

The reason for the remarkable consequences of possession of a derivative lies in the fact that the increment Δz is allowed to approach zero in any manner. If we restricted Δz so that $z_0 + \Delta z$ approached z_0 along a particular line, then we would obtain a "directional derivative." But here the limit obtained is required to be the *same for all directions*, so that the "directional derivative" has the same value in all directions. Moreover, $z_0 + \Delta z$ may approach z_0 in a quite arbitrary manner, for example along a spiral path. The limit of the ratio $\Delta w/\Delta z$ must be the same for all manners of approach.

We say that $f(z)$ has a *differential* $dw = c\,\Delta z$ at z_0 if $f(z_0 + \Delta z) - f(z_0) = c\,\Delta z + \epsilon\,\Delta z$, where ϵ depends on Δz and is continuous at $\Delta z = 0$, with value zero when $\Delta z = 0$.

THEOREM 3. *If $w = f(z)$ has a differential $dw = c\,\Delta z$ at z_0, then w has a derivative $f'(z_0) = c$. Conversely, if w has a derivative at z_0, then w has a differential at z_0: $dw = f'(z_0)\,\Delta z$.*

This is proved just as for real functions. We also write $\Delta z = dz$, as for real variables, so that

$$dw = f'(z)\,dz, \qquad \frac{dw}{dz} = f'(z). \tag{9–17}$$

From Theorem 3 we see that existence of the derivative $f'(z_0)$ implies continuity of f at z_0, for

$$f(z_0 + \Delta z) - f(z_0) = c\,\Delta z + \epsilon\,\Delta z \to 0$$

as $\Delta z \to 0$.

THEOREM 4. *If w_1 and w_2 are functions of z which have differentials in D, then*

$$d(w_1 + w_2) = dw_1 + dw_2,$$

$$d(w_1 w_2) = w_1\, dw_2 + w_2\, dw_1, \tag{9–18}$$

$$d\frac{w_1}{w_2} = \frac{w_2\, dw_1 - w_1\, dw_2}{w_2^2} \quad (w_2 \neq 0).$$

If w_2 is a differentiable function of w_1, and w_1 is a differentiable function of z, then wherever $w_2[w_1(z)]$ is defined

$$\frac{dw_2}{dz} = \frac{dw_2}{dw_1} \cdot \frac{dw_1}{dz}. \tag{9–19}$$

These rules are proved as in elementary calculus. We can now prove as usual the basic rule:

$$\frac{d}{dz} z^n = nz^{n-1} \quad (n = 1, 2, \ldots). \tag{9–20}$$

Furthermore, the derivative of a constant is zero.

Problems

1. For each of the following write the given function as two real functions of x and y and determine where the given function is continuous:

(a) $w = (1 + i)z^2$
(b) $w = \dfrac{z}{z + i}$
(c) $w = \tan z = \dfrac{\sin z}{\cos z}$
(d) $w = \dfrac{e^{-z}}{z + 1}$
(e) $w = e^z$
(f) $w = \sin z$
(g) $w = \cos z$
(h) $w = \sinh z$
(i) $w = \cosh z$
(j) $w = e^z \cos z$

2. Evaluate each of the following limits:

(a) $\lim\limits_{z \to \pi i} \dfrac{\sin z + z}{e^z + 2}$
(b) $\lim\limits_{z \to 0} \dfrac{z^2 - z}{2z}$
(c) $\lim\limits_{z \to 0} \dfrac{\cos z}{z}$
(d) $\lim\limits_{z \to \infty} \dfrac{z}{z^2 + 1}$

3. Differentiate each of the following complex functions:

(a) $w = z^3 + 5z + 1$
(b) $w = \dfrac{1}{z - 1}$
(c) $w = [1 + (z^2 + 1)^3]^7$
(d) $w = \dfrac{z^2}{(z + 1)^3}$

4. Prove the rule (9–20).

5. Prove Theorem 1.

Answers

1. (a) $u = x^2 - y^2 - 2xy$, $v = x^2 - y^2 + 2xy$, all z
(b) $u = (x^2 + y^2 + y)[x^2 + (y+1)^2]^{-1}$
$v = -x[x^2 + (y+1)^2]^{-1}$ $(z \neq -i)$
(c) $u = \tan x \operatorname{sech}^2 y\,[1 + \tan^2 x \tanh^2 y]^{-1}$
$v = \tanh y \sec^2 x\,[1 + \tan^2 x \tanh^2 y]^{-1}$
$z \neq (\pi/2) + n\pi$, $n = 0, \pm 1, \pm 2, \ldots$
(d) $u = e^{-x}[(1+x)\cos y - y \sin y][(1+x)^2 + y^2]^{-1}$
$v = -e^{-x}[(1+x)\sin y + y\cos y][(1+x)^2 + y^2]^{-1}$ $(z \neq -1)$
(e) . . . (i) See (9–13), continuous for all z
(j) $u = e^x(\cos x \cos y \cosh y + \sin x \sin y \sinh y)$
$v = e^x(\cos x \sin y \cosh y - \sin x \cos y \sinh y)$, all z

2. (a) $i(\pi + \sinh \pi)$ (b) $-\frac{1}{2}$ (c) ∞ (d) 0

3. (a) $3z^2 + 5$ (b) $-(z-1)^{-2}$ (c) $42[1 + (z^2+1)^3]^6(z^2+1)^2 z$
(d) $(z+1)^{-4}(2z - z^2)$

9–5 Integrals. The complex integral $\int f(z)\,dz$ is defined as a line integral, and its properties are closely related to those of the integral $\int P\,dx + Q\,dy$ (see Chapter 5).

Let C be a path from A to B in the complex plane: $x = x(t)$, $y = y(t)$, $a \leqq t \leqq b$. We assume C to have a direction, usually that of increasing t. We subdivide the interval $a \leqq t \leqq b$ into n parts by $t_0 = a, t_1, \ldots, t_n = b$. We let $z_j = x(t_j) + iy(t_j)$ and $\Delta_j z = z_j - z_{j-1}$, $\Delta_j t = t_j - t_{j-1}$. We choose an arbitrary value t_j^* in the interval $t_{j-1} \leqq t \leqq t_j$ and set $z_j^* = x(t_j^*) + iy(t_j^*)$. These quantities are all shown in Fig. 9–3. We then write

$$\int_C f(z)\,dz = \int_{A\,C}^{B} f(z)\,dz = \lim_{\substack{n\to\infty \\ \max \Delta_j t \to 0}} \sum_{j=1}^{n} f(z_j^*)\,\Delta_j z. \tag{9–21}$$

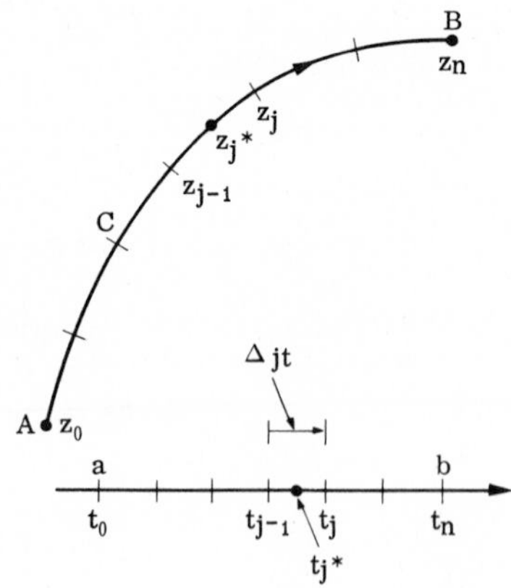

Fig. 9–3. Complex line integral.

If we take real and imaginary parts in (9–21), we find

$$\int_C f(z)\,dz = \lim \sum (u + iv)\,(\Delta x + i\,\Delta y)$$
$$= \lim \{\sum(u\,\Delta x - v\,\Delta y) + i\sum(v\,\Delta x + u\,\Delta y)\};$$

that is,

$$\int_C f(z)\,dz = \int_C (u + iv)\,(dx + i\,dy)$$
$$= \int_C (u\,dx - v\,dy) + i\int_C (v\,dx + u\,dy). \tag{9–22}$$

The complex line integral is thus simply a combination of two real line integrals. Hence we can apply all the theory of real line integrals. In the following, each path is assumed to be *piecewise smooth;* that is, $x(t)$ and $y(t)$ are to be continuous with piecewise continuous derivatives.

THEOREM 5. *If $f(z)$ is continuous in domain D, then the integral* (9–21) *exists and*

$$\int_C f(z)\,dz = \int_a^b \left(u\frac{dx}{dt} - v\frac{dy}{dt}\right)dt + i\int_a^b \left(v\frac{dx}{dt} + u\frac{dy}{dt}\right)dt. \tag{9–23}$$

We now write our path as $z = z(t)$ as in Section 9–2. If we introduce the derivative

$$\frac{dz}{dt} = \frac{dx}{dt} + i\frac{dy}{dt}$$

of z with respect to the real variable t, and also use the theory of integrals of such functions (Section 9–2), we can write (9–23) more concisely:

$$\int_C f(z)\,dz = \int_a^b f[z(t)]\,\frac{dz}{dt}\,dt. \tag{9–24}$$

EXAMPLE 1. Let C be the path $x = 2t$, $y = 3t$, $1 \leqq t \leqq 2$. Let $f(z) = z^2$. Then

$$\int_C z^2\,dz = \int_1^2 (2t + 3it)^2(2 + 3i)\,dt = (2 + 3i)^3\int_1^2 t^2\,dt$$
$$= -107\tfrac{1}{3} + 21i.$$

EXAMPLE 2. Let C be the circular path $x = \cos t$, $y = \sin t$, $0 \leqq t \leqq 2\pi$. This can be written more concisely thus: $z = e^{it}$, $0 \leqq t \leqq 2\pi$. Since $dz/dt = ie^{it}$,

$$\int_C \frac{1}{z}\,dz = \int_0^{2\pi} e^{-it}ie^{it}\,dt = i\int_0^{2\pi} dt = 2\pi i.$$

Further properties of complex integrals follow from those of real integrals:

THEOREM 6. *Let $f(z)$ and $g(z)$ be continuous in a domain D. Let C be a piecewise smooth path in D. Then*

$$\int_C [f(z) + g(z)]\,dz = \int_C f(z)\,dz + \int_C g(z)\,dz,$$

$$\int_C kf(z)\,dz = k\int_C f(z)\,dz \quad (k = \text{const}),$$

$$\int_C f(z)\,dz = \int_{C_1} f(z)\,dz + \int_{C_2} f(z)\,dz,$$

where C is composed of a path C_1 from z_0 to z_1 and a path C_2 from z_1 to z_2, and

$$\int_C f(z)\,dz = -\int_{C'} f(z)\,dz,$$

where C' is obtained from C by reversing direction on C.

Upper estimates for the absolute value of a complex integral are obtained by the following theorem.

THEOREM 7. *Let $f(z)$ be continuous on C, let $|f(z)| \leqq M$ on C, and let*

$$L = \int_C ds = \int_a^b \sqrt{(dx/dt)^2 + (dy/dt)^2}\,dt$$

be the length of C. Then

$$\left|\int_C f(z)\,dz\right| \leqq \int_C |f(z)|\,ds \leqq M \cdot L. \tag{9–25}$$

Proof. The line integral $\int |f(z)|\,ds$ is defined as a limit:

$$\int_C |f(z)|\,ds = \lim \sum |f(z_j^*)|\,\Delta_j s,$$

where $\Delta_j s$ is the length of the jth arc of C. Now

$$|f(z_j^*)\,\Delta_j z| = |f(z_j^*)| \cdot |\Delta_j z| \leqq |f(z_j^*)| \cdot \Delta_j s,$$

for $|\Delta_j z|$ represents the *chord* of the arc $\Delta_j s$. Hence

$$\left|\sum f(z_j^*)\,\Delta_j z\right| \leqq \sum |f(z_j^*)\,\Delta_j z| \leqq \sum |f(z_j^*)|\,\Delta_j s$$

by repeated application of the triangle inequality (0–10). Passing to the limit, we conclude that

$$\left|\int_C f(z)\,dz\right| \leqq \int_C |f(z)|\,ds. \tag{9–26}$$

Also, if $|f| \leqq M = \text{const}$,

$$\sum |f(z_j^*)|\, \Delta_j s \leqq \sum M\, \Delta_j s = M \cdot L.$$

Hence

$$\int_C |f(z)|\, ds \leqq M \cdot L. \tag{9–27}$$

Inequalities (9–25) follow from (9–26) and (9–27).

Problems

1. Evaluate the following integrals:

(a) $\displaystyle\int_0^{1+i} (x^2 - iy^2)\, dz$ on the straight line from 0 to $1 + i$.

(b) $\displaystyle\int_0^{\pi} z\, dz$ on the curve $y = \sin x$. (c) $\displaystyle\int_1^{1+i} \frac{dz}{z}$ on the line $x = 1$.

2. Write each of the following integrals in the form $\int u\, dx - v\, dy + i\int v\, dx + u\, dy$; then show that each of the two real integrals is independent of path in the xy-plane.

(a) $\int (z + 1)\, dz$ (b) $\int e^z\, dz$
(c) $\int z^4\, dz$ (d) $\int \sin z\, dz$

3. (a) Evaluate

$$\oint \frac{1}{z}\, dz$$

on the circle $|z| = R$.
(b) Show that

$$\oint \frac{1}{z}\, dz = 0$$

on every simple closed path not meeting or enclosing the origin.
(c) Show that

$$\oint \frac{1}{z^2}\, dz = 0$$

on every simple closed path not passing through the origin.

Answers

1. (a) $\frac{2}{3}$ (b) $\pi^2/2$ (c) $\frac{1}{2}\log 2 + i(\pi/4)$
3. (a) $2\pi i$

9–6 Analytic functions. Cauchy-Riemann equations. A function $w = f(z)$, defined in a domain D, is said to be an *analytic function* in D if w has a continuous derivative in D. Almost the entire theory of functions of a complex variable is confined to the study of such functions. Furthermore, almost all functions used in the applications of mathematics to physical problems are analytic functions or are derived from such.

It will be seen that possession of a continuous derivative implies possession of a continuous second derivative, third derivative, . . . , and, in fact, convergence of the Taylor series

$$f(z_0) + f'(z_0)\frac{(z - z_0)}{1!} + f''(z_0)\frac{(z - z_0)^2}{2!} + \cdots$$

in a neighborhood of each z_0 of D. Thus one could define an analytic function as one so representable by Taylor series, and this definition is often used. The two definitions are equivalent, for convergence of the Taylor series in a neighborhood of each z_0 implies continuity of the derivatives of all orders.

While it is possible to construct continuous functions of z which are not analytic (examples will be given below), it is impossible to construct a function $f(z)$ possessing a derivative, but not a continuous one, in D. In other words, if $f(z)$ has a derivative in D, the derivative is necessarily continuous, so that $f(z)$ is analytic. Therefore we could define an analytic function as one merely possessing a derivative in domain D, and this definition is also often used. For a proof that existence of the derivative implies its continuity, refer to Vol. I of the book by Knopp listed at the end of the chapter.

THEOREM 8. *If $w = u + iv = f(z)$ is analytic in D, then u and v have continuous first partial derivatives in D and satisfy the Cauchy-Riemann equations*

$$\frac{\partial u}{\partial x} = \frac{\partial v}{\partial y}, \qquad \frac{\partial u}{\partial y} = -\frac{\partial v}{\partial x} \tag{9–28}$$

in D. Furthermore,

$$\frac{dw}{dz} = \frac{\partial u}{\partial x} + i\frac{\partial v}{\partial x} = \frac{\partial v}{\partial y} + i\frac{\partial v}{\partial x} = \frac{\partial u}{\partial x} - i\frac{\partial u}{\partial y} = \frac{\partial v}{\partial y} - i\frac{\partial u}{\partial y}. \tag{9–29}$$

Conversely, if $u(x, y)$ and $v(x, y)$ have continuous first partial derivatives in D and satisfy the Cauchy-Riemann equations (9–28), *then $w = u + iv = f(z)$ is analytic in D.*

Proof. Let z_0 be a fixed point of D and let

$$\Delta w = \Delta u + i\,\Delta v = f(z_0 + \Delta z) - f(z_0), \qquad \Delta z = \Delta x + i\,\Delta y,$$

as in Fig. 9–4. We consider several equivalent formulations of the condition that $f'(z_0)$ exists. Throughout, ϵ, ϵ_1, ϵ_2, ϵ_3, ϵ_4 denote functions of $\Delta z = \Delta x + i\,\Delta y$, continuous and equal to zero at $\Delta z = 0$. By Theorem 3, existence of $f'(z_0)$ is equivalent to the statement

$$\Delta w = c \cdot \Delta z + \epsilon \cdot \Delta z, \qquad c = f'(z_0), \qquad c = a + ib; \tag{9–30}$$

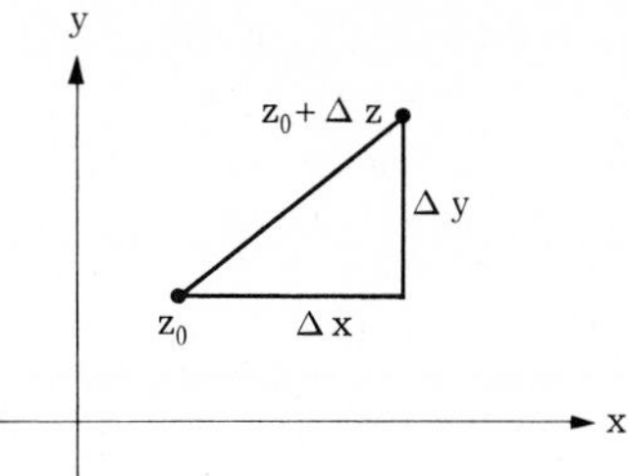

FIG. 9–4. Complex derivative.

this is equivalent to

$$\Delta w = c\,\Delta z + \epsilon\,\Delta x + i\epsilon\,\Delta y \tag{9–30$'$}$$

and also to

$$\Delta w = c\,\Delta z + \epsilon_1\,\Delta x + \epsilon_2\,\Delta y + i(\epsilon_3\,\Delta x + \epsilon_4\,\Delta y), \tag{9–30$''$}$$

where $\epsilon_1, \epsilon_2, \epsilon_3, \epsilon_4$ are real. For if (9–30′) holds, then (9–30″) holds with $\epsilon_1 = \operatorname{Re}(\epsilon)$, $\epsilon_2 = -\operatorname{Im}(\epsilon)$, $\epsilon_3 = \operatorname{Im}(\epsilon)$, $\epsilon_4 = \operatorname{Re}(\epsilon)$. Conversely, if (9–30″) holds, then (9–30′) holds with $\epsilon = 0$ for $\Delta z = 0$ and

$$\epsilon = (\epsilon_1 + i\epsilon_3)\frac{\Delta x}{\Delta z} + (\epsilon_2 + i\epsilon_4)\frac{\Delta y}{\Delta z} \quad (\Delta z \neq 0). \tag{9–31}$$

As Fig. 9–4 shows,

$$\left|\frac{\Delta x}{\Delta z}\right| \leqq 1, \qquad \left|\frac{\Delta y}{\Delta z}\right| \leqq 1,$$

so that we deduce from (9–31) that $\epsilon \to 0$ as $\Delta z \to 0$. Thus (9–30), (9–30′) and (9–30″) are all equivalent to existence of $f'(z_0) = c = a + ib$. By taking real and imaginary parts in (9–30″), we obtain one more equivalent condition:

$$\begin{aligned} \Delta u &= a\,\Delta x - b\,\Delta y + \epsilon_1\,\Delta x + \epsilon_2\,\Delta y, \\ \Delta v &= b\,\Delta x + a\,\Delta y + \epsilon_3\,\Delta x + \epsilon_4\,\Delta y; \end{aligned} \tag{9–30$'''$}$$

these equations state that u, v have differentials $du = a\,dx - b\,dy$, $dv = b\,dx + a\,dy$ at (x_0, y_0), and hence at this point

$$\frac{\partial u}{\partial x} = a = \frac{\partial v}{\partial y}, \qquad \frac{\partial u}{\partial y} = -b = -\frac{\partial v}{\partial x}.$$

Thus differentiability of $f'(z)$ at any z is equivalent to differentiability of u, v along with validity of the Cauchy-Riemann equations. Furthermore, $f'(z)$ and $\partial u/\partial x, \ldots$ are related by (9–29). By Theorem 1, these equations show that continuity of $f'(z)$ in D is equivalent to continuity of $\partial u/\partial x, \ldots$ Thus the theorem is proved.

The theorem provides a perfect test for analyticity: if $f(z)$ is analytic, then the Cauchy-Riemann equations hold; if the equations hold (and the derivatives concerned are continuous), then $f(z)$ is analytic.

EXAMPLE 1. $w = z^2 = x^2 - y^2 + i \cdot 2xy$. Here $u = x^2 - y^2, v = 2xy$. Thus

$$\frac{\partial u}{\partial x} = 2x = \frac{\partial v}{\partial y}, \qquad \frac{\partial u}{\partial y} = -2y = -\frac{\partial v}{\partial x}$$

and w is analytic for all z.

EXAMPLE 2. $w = \dfrac{x}{x^2 + y^2} - \dfrac{iy}{x^2 + y^2}\cdot$ Here

$$\frac{\partial u}{\partial x} = \frac{y^2 - x^2}{(x^2 + y^2)^2} = \frac{\partial v}{\partial y}, \qquad \frac{\partial u}{\partial y} = \frac{-2xy}{(x^2 + y^2)^2} = -\frac{\partial v}{\partial x}\cdot$$

Hence w is analytic except for $x^2 + y^2 = 0$, that is, for $z = 0$.

EXAMPLE 3. $w = x - iy = \bar{z}$. Here $u = x$, $v = -y$ and

$$\frac{\partial u}{\partial x} = 1, \qquad \frac{\partial v}{\partial y} = -1, \qquad \frac{\partial u}{\partial y} = 0 = \frac{\partial v}{\partial x}\cdot$$

Thus w is not analytic in any domain.

EXAMPLE 4. $w = x^2y^2 + 2x^2y^2i$. Here

$$\frac{\partial u}{\partial x} = 2xy^2, \qquad \frac{\partial v}{\partial y} = 4x^2y, \qquad \frac{\partial u}{\partial y} = 2x^2y, \qquad \frac{\partial v}{\partial x} = 4xy^2.$$

The Cauchy-Riemann equations give $2xy^2 = 4x^2y$, $2x^2y = -4xy^2$. These equations are satisfied only along the lines $x = 0$, $y = 0$. There is *no domain* in which the Cauchy-Riemann equations hold, hence no domain in which $f(z)$ is analytic. One does not consider functions analytic only at certain points unless these points form a domain.

The terms "analytic at a point" or "analytic along a curve" are used, apparently in contradiction to the remark just made. However, we say that $f(z)$ is *analytic at the point* z_0 only if there is a domain containing z_0 within which $f(z)$ is analytic. Similarly, $f(z)$ is *analytic along a curve* C only if $f(z)$ is analytic in a domain containing C.

THEOREM 9. *The sum, product, and quotient of analytic functions is analytic (provided in the last case the denominator is not equal to zero at any point of the domain under consideration). All polynomials are analytic for all z. Every rational function is analytic in each domain containing no root of the denominator. An analytic function of an analytic function is analytic.*

This follows from Theorem 4.

We readily verify (Problem 1 below) that the Cauchy-Riemann equations are satisfied for $u = \text{Re}\,(e^z)$, $v = \text{Im}\,(e^z)$. Hence e^z is analytic for all z. It then follows from Theorem 9 that $\sin z$, $\cos z$, $\sinh z$, and $\cosh z$ are analytic for all z, while $\tan z$, $\sec z$, and $\csc z$ are analytic except for certain points (Problem 6 below). Furthermore, the usual formulas for derivatives hold:

$$\frac{d}{dz} e^z = e^z, \qquad \frac{d}{dz} \sin z = \cos z, \qquad \ldots \tag{9–32}$$

(Problem 3).

Two basic theorems of more advanced theory are useful at this point. Proofs are given in Chapter IV of the book by Goursat listed at the end of the chapter.

THEOREM 10. *Given a function $f(x)$ of the real variable x, $a \leqq x \leqq b$, there is at most one analytic function $f(z)$ which reduces to $f(x)$ when z is real.*

THEOREM 11. *If $f(z)$, $g(z)$, ... are functions which are all analytic in a domain D which includes part of the real axis, and $f(z)$, $g(z)$, ... satisfy an algebraic identity when z is real, then these functions satisfy the same identity for all z in D.*

Theorem 10 implies that our definitions of e^z, $\sin z$, ... are the only ones which yield analytic functions and agree with the definitions for real variables.

Because of Theorem 11, we can be sure that all familiar identities of trigonometry, namely,

$$\sin^2 z + \cos^2 z = 1, \qquad \sin\left(\frac{\pi}{2} - z\right) = \cos z, \qquad \ldots \tag{9–33}$$

continue to hold for complex z. A general algebraic identity is formed by replacing the variables $w_1, \ldots, w_n$ in an algebraic equation by functions $f_1(z), \ldots, f_n(z)$. Thus, in the two examples given, one has

$$w_1^2 + w_2^2 - 1 = 0 \quad (w_1 = \sin z,\, w_2 = \cos z),$$

$$w_1 - w_2 = 0 \quad \left[w_1 = \sin\left(\frac{\pi}{2} - z\right),\, w_2 = \cos z\right].$$

To prove identities such as

$$e^{z_1} \cdot e^{z_2} = e^{z_1 + z_2}, \tag{9–34}$$

it may be necessary to apply Theorem 11 several times. (See Problems 4 and 5 below.)

It should be remarked that while e^z is written as a power of e, it is best not to think of it as such. Thus $e^{1/2}$ has only one value, not two, as would a usual complex root. To avoid confusion with the general power function, to be defined below, we often write $e^z = \exp z$ and refer to e^z as the *exponential function of z*.

To obtain the real and imaginary parts of $\sin z$, we use the identity

$$\sin (z_1 + z_2) = \sin z_1 \cos z_2 + \cos z_1 \sin z_2,$$

which holds, by the reasoning described above, for all complex z_1 and z_2. Hence $\sin (x + iy) = \sin x \cos iy + \cos x \sin iy$. Now from the definitions (Section 9–1),

$$\begin{aligned} \sinh y &= -i \sin iy, \\ \cosh y &= \cos iy. \end{aligned} \tag{9–35}$$

Hence

$$\sin z = \sin x \cosh y + i \cos x \sinh y. \tag{9–36}$$

Similarly, we prove, as in (9–13) above,

$$\begin{aligned} \cos z &= \cos x \cosh y - i \sin x \sinh y, \\ \sinh z &= \sinh x \cos y + i \cosh x \sin y, \\ \cosh z &= \cosh x \cos y + i \sinh x \sin y. \end{aligned} \tag{9–37}$$

Conformal mapping. A complex function $w = f(z)$ can be considered as a *mapping* from the xy-plane to the uv-plane as in Section 2–7. In the case of an analytic function $f(z)$, this mapping has a special property: it is a *conformal mapping*. By this we mean that two curves in the xy-plane, meeting at (x_0, y_0) at angle α, correspond to two curves meeting at the corresponding point (u_0, v_0) at the *same angle* α (in value and in sense—positive or negative). This means that a small triangle in the xy-plane corresponds to a *similar* small (curvilinear) triangle in the uv-plane. (The properties described fail at the exceptional points where $f'(z) = 0$.) Furthermore, every conformal mapping from the xy-plane to the uv-plane is given by an analytic function. For a discussion of conformal mapping and its applications, see Chapter 7 of the book by Kaplan listed at the end of the chapter.

Problems

1. Verify that the following are analytic functions of z:

(a) $2x^3 - 3x^2y - 6xy^2 + y^3 + i(x^3 + 6x^2y - 3xy^2 - 2y^3)$
(b) $w = e^z = e^x \cos y + ie^x \sin y$
(c) $w = \sin z = \sin x \cosh y + i \cos x \sinh y$

2. Test each of the following for analyticity:

(a) $x^3 + y^3 + i(3x^2y + 3xy^2)$
(b) $\sin x \cos y + i \cos x \sin y$
(c) $3x + 5y + i(3y - 5x)$

3. Prove the following properties directly from the definitions of the functions:

(a) $\dfrac{d}{dz} e^z = e^z$ (b) $\dfrac{d}{dz} \sin z = \cos z$, $\dfrac{d}{dz} \cos z = -\sin z$

(c) $\sin (z + \pi) = -\sin z$ (d) $\sin (-z) = -\sin z$, $\cos (-z) = \cos z$

4. Prove the identity $e^{z_1+z_2} = e^{z_1} \cdot e^{z_2}$ by application of Theorem 11. [*Hint:* Let $z_2 = b$, a fixed real number, and $z_1 = z$, a variable complex number. Then $e^{z+b} = e^z \cdot e^b$ is an identity connecting analytic functions which is known to be true for z real. Hence it is true for all complex z. Now proceed similarly with the identity $e^{z_1+z} = e^{z_1} \cdot e^z$.]

5. Prove the following identities by application of Theorem 11 (see Problem 4):

(a) $\cos (z_1 + z_2) = \cos z_1 \cos z_2 - \sin z_1 \sin z_2$
(b) $e^{iz} = \cos z + i \sin z$
(c) $(e^z)^n = e^{nz}$ $(n = 0, 1, 2, \ldots)$

6. Determine where the following functions are analytic (see Problem 3 following Section 9–2):

(a) $\tan z = \dfrac{\sin z}{\cos z}$ (b) $\cot z = \dfrac{\cos z}{\sin z}$

(c) $\tanh z = \dfrac{\sinh z}{\cosh z}$ (d) $\dfrac{\sin z}{z}$

(e) $\dfrac{e^z}{z \cos z}$ (f) $\dfrac{e^z}{\sin z + \cos z}$

Answers

2. (a) Analytic nowhere, (b) analytic nowhere, (c) analytic for all z.
6. The functions are analytic except at the following points: (a) $\frac{1}{2}\pi + n\pi$; (b) $n\pi$; (c) $\frac{1}{2}\pi i + n\pi i$; (d) 0; (e) 0, $\frac{1}{2}\pi + n\pi$; (f) $-\frac{1}{4}\pi + n\pi$, where $n = 0$, $\pm 1, \pm 2, \ldots$

9–7 The functions $\log z$, a^z, z^a, $\sin^{-1} z$, $\cos^{-1} z$. The function $w = \log z$ is defined as the inverse of the exponential function $z = e^w$. We write $z = re^{i\theta}$, in terms of polar coordinates r, θ, and $w = u + iv$, so that

$$re^{i\theta} = e^{u+iv} = e^u e^{iv},$$

$$e^u = r, \qquad v = \theta + 2k\pi \quad (k = 0, \pm 1, \ldots).$$

Accordingly,

$$w = \log z = \log r + i(\theta + 2k\pi) = \log |z| + i \arg z, \tag{9–38}$$

where $\log r$ is the real logarithm of r Thus $\log z$ is a multiple-valued function of z, with infinitely many values except for $z = 0$. We can select one value of θ for each z and obtain a single-valued function, $\log z = \log r + i\theta$; however, θ cannot be chosen to depend continuously on z for all $z \neq 0$, since θ will increase by 2π each time one encircles the origin in the positive direction.

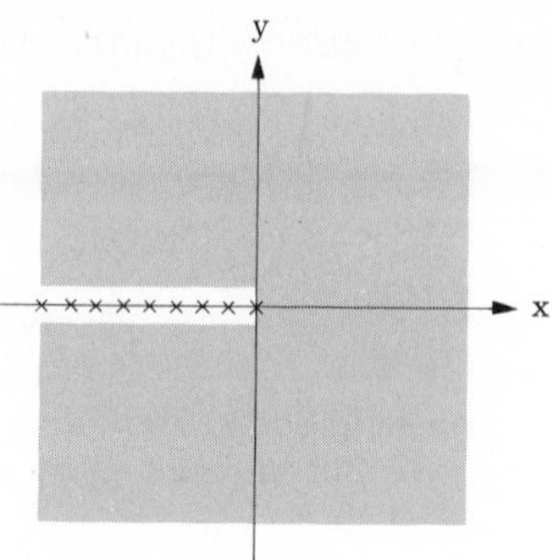

FIG. 9–5. Domain for log z.

If we concentrate on an appropriate portion of the z-plane, we can choose θ to vary continuously within the domain. For example, the inequalities

$$-\pi < \theta < \pi, \qquad r > 0$$

together describe a domain (Fig. 9–5) and also tell how to assign the values of θ within the domain. With θ so restricted, $\log r + i\theta$ then defines a *branch* of $\log z$ in the domain chosen; this particular branch is called the *principal value* of $\log z$ and is denoted by $\operatorname{Log} z$. The points on the negative real axis are excluded from the domain, but we usually assign the values $\operatorname{Log} z = \log |x| + i\pi$ on this line. Within the domain of Fig. 9–5, $\operatorname{Log} z$ is *an analytic function of* z (Problem 4 below). Other branches of $\log z$ are obtained by varying the choice of θ or of the domain. For example, in the domain of Fig. 9–5, we might choose θ so that $\pi < \theta < 3\pi$, or so that $-3\pi < \theta < -\pi$. The inequalities $0 < \theta < 2\pi$, $\pi/2 < \theta < 5\pi/2, \ldots$ also suggest other domains and choices of θ. We can verify that so long as θ varies continuously in the domain, $\log z = \log r + i\theta$ is analytic there. The most general domain possible here is an arbitrary simply-connected domain not containing the origin.

As a result of this discussion, it appears that $\log z$ is formed of many branches, each analytic in some domain not containing the origin. The branches fit together in a simple way; in general, we can get from one branch to another by moving around the origin a sufficient number of times, while varying the choice of $\log z$ continuously. We say that the branches form "analytic continuations" of each other.

We can further verify that for each branch of $\log z$, the rule

$$\frac{d}{dz} \log z = \frac{1}{z} \tag{9–39}$$

remains valid. The familiar identities are also satisfied (Problems 4 and 5 below).

The *general exponential function* a^z is defined, for $a \neq 0$, by the equation

$$a^z = e^{z \log a} = \exp(z \log a). \tag{9–40}$$

Thus for $z = 0$, $a^0 = 1$. Otherwise, $\log a = \log|a| + i \arg a$, and we obtain many values: $a^z = \exp[z(\log|a| + i(\alpha + 2n\pi))]$, $(n = 0, \pm 1, \pm 2, \ldots)$, where α denotes one choice of arg a. For example,

$$\begin{aligned}(1+i)^i &= \exp\left[i\left\{\log\sqrt{2} + i\left(\frac{\pi}{4} + 2n\pi\right)\right\}\right]\\ &= e^{-(\pi/4)-2n\pi}(\cos\log\sqrt{2} + i\sin\log\sqrt{2}).\end{aligned}$$

If z is a positive integer m, a^z reduces to a^m and has only one value. The same holds for $z = -m$, and we have

$$a^{-m} = \frac{1}{a^m}\cdot \tag{9–41}$$

If z is a fraction p/q (in lowest terms), we find that a^z has q distinct values, which are the qth roots of a^p. (See Eq. (0–14).)

If a fixed choice of $\log a$ is made in (9–40), then a^z is simply e^{cz}, $c = \log a$, and is hence an analytic function of z for all z. Each choice of $\log a$ determines such a function.

If a and z are interchanged in (9–40), we obtain the *general power function*,

$$z^a = e^{a \log z}. \tag{9–42}$$

If an analytic branch of $\log z$ is chosen as above, then this function becomes an analytic function of an analytic function and is hence analytic in the domain chosen. In particular, the *principal value* of z^a is defined as the analytic function $z^a = e^{a \operatorname{Log} z}$, in terms of the principal value of $\log z$.

For example, if $a = \frac{1}{2}$, we have

$$\begin{aligned}z^{1/2} = e^{(1/2)\log z} &= e^{(1/2)(\log r + i\theta)} = e^{(1/2)\log r}e^{(1/2)i\theta}\\ &= \sqrt{r}\left(\cos\frac{\theta}{2} + i\sin\frac{\theta}{2}\right),\end{aligned}$$

as in Eq. (0–14). If $\operatorname{Log} z$ is used, then $\sqrt{z} = f_1(z)$ becomes analytic in the domain of Fig. 9–5. A second analytic branch $f_2(z)$ in the same domain is obtained by requiring that $\pi < \theta < 3\pi$. These are the only two analytic branches which can be obtained in this domain. It should be remarked that these two branches are related by the equation $f_2(z) = -f_1(z)$. For f_2 is obtained from f_1 by increasing θ by 2π, which replaces $e^{(1/2)i\theta}$ by

$$e^{(1/2)i(\theta+2\pi)} = e^{\pi i}e^{(1/2)i\theta} = -e^{(1/2)i\theta}.$$

The functions $\sin^{-1} z$ and $\cos^{-1} z$ are defined as the inverses of $\sin z$ and $\cos z$. We then find

$$\begin{aligned}\sin^{-1} z &= \frac{1}{i}\log\,[iz \pm \sqrt{1-z^2}],\\ \cos^{-1} z &= \frac{1}{i}\log\,[z \pm i\sqrt{1-z^2}].\end{aligned} \tag{9–43}$$

The proofs are left to the exercises (Problem 2). It can be shown that analytic branches of both these functions can be defined in each simply-connected domain not containing the points ± 1. For each z other than ± 1, one has two choices of $\sqrt{1-z^2}$ and then an infinite sequence of choices of the logarithm, differing by multiples of $2\pi i$.

Problems

1. Obtain all values of each of the following:
(a) $\log 2$ (b) $\log i$ (c) $\log (1 - i)$ (d) i^i (e) $(1 + i)^{2/3}$ (f) $i^{\sqrt{2}}$ (g) $\sin^{-1} 1$ (h) $\cos^{-1} 2$

2. Prove the formulas (9–43). [*Hint:* If $w = \sin^{-1} z$, then $2iz = e^{iw} - e^{-iw}$; multiply by e^{iw} and solve the resulting equation as a quadratic for e^{iw}.]

3. (a) Evaluate $\sin^{-1} 0$, $\cos^{-1} 0$.
(b) Find all roots of $\sin z$ and $\cos z$ [compare part (a)].

4. Show that each branch of $\log z$ is analytic in each domain in which θ varies continuously and that

$$(d/dz) \log z = 1/z.$$

[*Hint:* Show from the equations $x = r\cos\theta$, $y = r\sin\theta$ that $\partial\theta/\partial x = -y/r^2$, $\partial\theta/\partial y = x/r^2$. Show that the Cauchy-Riemann equations hold for $u = \log r$, $v = \theta$.]

5. Prove the following identities in the sense that, for proper selection of values of the multiple-valued functions concerned, the equation is correct for each allowed choice of the variables:
(a) $\log (z_1 \cdot z_2) = \log z_1 + \log z_2 \quad (z_1 \neq 0,\ z_2 \neq 0)$
(b) $e^{\log z} = z \quad (z \neq 0)$
(c) $\log e^z = z$
(d) $\log z_1^{z_2} = z_2 \log z_1 \quad (z_1 \neq 0)$

6. For each of the following determine all analytic branches of the multiple-valued function in the domain given:
(a) $\log z,\quad x < 0$ (b) $\sqrt[3]{z},\quad x > 0$

7. Prove that for the analytic function z^a (principal value),

$$(d/dz)z^a = (az^a)/z = az^{a-1}.$$

8. Plot the functions $u = \operatorname{Re}(\sqrt{z})$ and $v = \operatorname{Im}(\sqrt{z})$ as functions of x and y and show the two branches described in the text.

Answers

1. (a) $0.693 + 2n\pi i$ (b) $i(\frac{1}{2}\pi + 2n\pi)$ (c) $0.347 + i(\frac{7}{4}\pi + 2n\pi)$

(d) $\exp(-\frac{1}{2}\pi - 2n\pi)$ (e) $\sqrt[3]{2}\exp\left(\frac{1}{6}\pi i + \frac{4n\pi}{3}i\right)$

(f) $\exp\left(\frac{\sqrt{2}}{2}\pi i + 2\sqrt{2}\,n\pi i\right)$ (g) $\frac{1}{2}\pi + 2n\pi$ (h) $2n\pi \pm 1.317i$

The range of n is 0, ±1, ±2, . . . , except in (e), where it is 0, 1, 2.

3. (a) and (b) $n\pi$ and $(\pi/2) + n\pi$ $(n = 0, \pm 1, \pm 2, \ldots)$
6. (a) $\log r + i\theta$, $\frac{1}{2}\pi + 2n\pi < \theta < \frac{3}{2}\pi + 2n\pi$ $(n = 0, \pm 1, \pm 2, \ldots)$
   (b) $\sqrt[3]{r}\exp(i\theta/3)$, $-(\pi/2) + 2n\pi < \theta < (\pi/2) + 2n\pi$ $(n = 0, 1, 2)$

9–8 Integrals of analytic functions. Cauchy integral theorem. All paths in the integrals concerned here, as elsewhere in the chapter, are assumed to be piecewise smooth.

The following theorem is fundamental for the theory of analytic functions:

THEOREM 12 (Cauchy integral theorem). *If $f(z)$ is analytic in a simply-connected domain D, then*

$$\oint_C f(z)\,dz = 0$$

on every simple closed path C in D (Fig. 9–6).

Proof. We have, by (9–22) above,

$$\oint_C f(z)\,dz = \oint_C u\,dx - v\,dy + i\oint_C v\,dx + u\,dy.$$

The two real integrals are equal to zero (see Section 5–6 above)

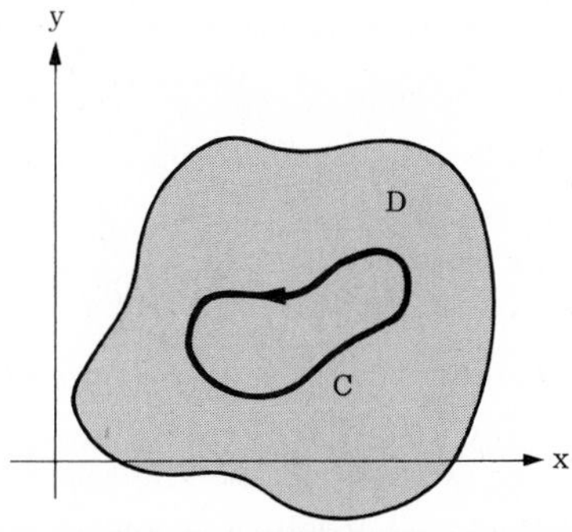

FIG. 9–6. Cauchy integral theorem.

provided u and v have continuous derivatives in D and

$$\frac{\partial u}{\partial y} = -\frac{\partial v}{\partial x}, \qquad \frac{\partial v}{\partial y} = \frac{\partial u}{\partial x}.$$

These are just the Cauchy-Riemann equations. Hence

$$\oint_C f(z)\,dz = 0 + i \cdot 0 = 0.$$

This theorem can be stated in an equivalent form:

THEOREM 12′. *If $f(z)$ is analytic in the simply-connected domain D, then $\int f(z)\,dz$ is independent of the path in D.*

For independence of path and equaling zero on closed paths are equivalent properties of line integrals. If C is a path from z_1 to z_2, we can now write

$$\int_C f(z)\,dz = \int_{z_1}^{z_2} f(z)\,dz,$$

the integral being the same for all paths C from z_1 to z_2.

THEOREM 13. *Let $f(z) = u + iv$ be defined in domain D and let u and v have continuous partial derivatives in D. If*

$$\oint_C f(z)\,dz = 0 \tag{9–44}$$

on every simple closed path C in D, then $f(z)$ is analytic in D.

Proof. The condition (9–44) implies that

$$\oint_C u\,dx - v\,dy = 0, \qquad \oint_C v\,dx + u\,dy = 0$$

on all simple closed paths C; that is, the two real line integrals are independent of path in D. Therefore, by Theorem III in Section 5–6,

$$\frac{\partial u}{\partial y} = -\frac{\partial v}{\partial x}, \qquad \frac{\partial v}{\partial y} = \frac{\partial u}{\partial x};$$

since the Cauchy-Riemann equations hold, f is analytic.

This theorem can be proved with the assumption that u and v have continuous derivatives in D replaced by the assumption that f is continuous in D; it is then known as *Morera's* theorem. For a proof, see Chapter 5 of Vol. I of the book by Knopp listed at the end of the chapter.

THEOREM 14. *If $f(z)$ is analytic in D, then*

$$\int_{z_1}^{z_2} f'(z)\, dz = f(z)\Big|_{z_1}^{z_2} = f(z_2) - f(z_1) \tag{9–45}$$

on every path in D from z_1 to z_2. In particular,

$$\oint f'(z)\, dz = 0$$

on every closed path in D.

Proof. By (9–29) above,

$$\begin{aligned}\int_{z_1}^{z_2} f'(z)\, dz &= \int_{z_1}^{z_2} \left(\frac{\partial u}{\partial x} + i\,\frac{\partial v}{\partial x}\right)(dx + i\, dy) \\ &= \int_{z_1}^{z_2} \frac{\partial u}{\partial x}\, dx + \frac{\partial u}{\partial y}\, dy + i\int_{z_1}^{z_2} \frac{\partial v}{\partial x}\, dx + \frac{\partial v}{\partial y}\, dy \\ &= \int_{z_1}^{z_2} du + i\, dv = (u + iv)\Big|_{z_1}^{z_2} = f(z_2) - f(z_1).\end{aligned}$$

This rule is the basis for evaluation of simple integrals, just as in elementary calculus. Thus we have

$$\int_i^{1+i} z^2\, dz = \frac{z^3}{3}\Big|_i^{1+i} = \frac{(1+i)^3 - i^3}{3} = -\tfrac{2}{3} + i,$$

$$\int_i^{-i} \frac{1}{z^2}\, dz = -\frac{1}{z}\Big|_i^{-i} = -i - i = -2i.$$

In the first of these any path can be used; in the second, any path not through the origin.

THEOREM 15. *If $f(z)$ is analytic in D and D is simply-connected, then*

$$F(z) = \int_{z_1}^{z} f(z)\, dz \quad (z_1 \textit{ fixed in } D) \tag{9–46}$$

is an indefinite integral of $f(z)$; that is, $F'(z) = f(z)$. Thus $F(z)$ is itself analytic.

Proof. Since $f(z)$ is analytic in D and D is simply-connected, $\int_{z_1}^{z} f(z)\, dz$ is independent of path and defines a function F which depends only on the upper limit z. We have, further, $F = U + iV$, where

$$U = \int_{z_1}^{z} u\, dx - v\, dy, \qquad V = \int_{z_1}^{z} v\, dx + u\, dy$$

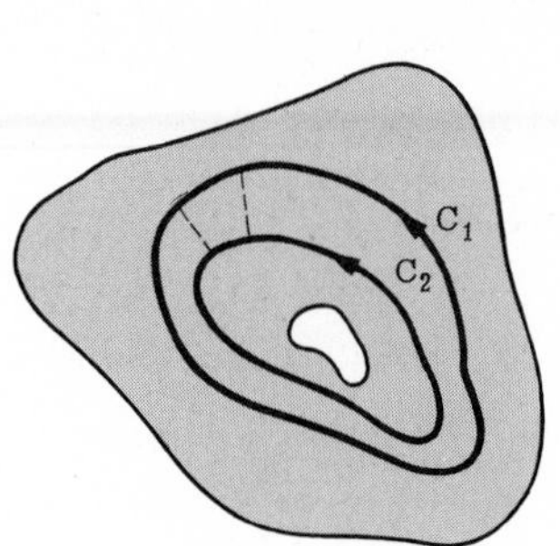

FIG. 9–7. Cauchy theorem for doubly-connected domain.

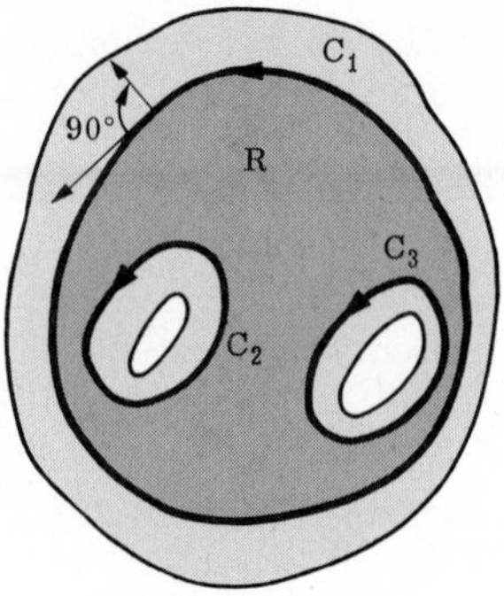

FIG. 9–8. Cauchy theorem for triply-connected domain.

and both integrals are independent of path. Hence $dU = u\,dx - v\,dy$, $dV = v\,dx + u\,dy$. Thus U and V satisfy the Cauchy-Riemann equations, so that $F = U + iV$ is analytic and

$$F'(z) = \frac{\partial U}{\partial x} + i\frac{\partial V}{\partial x} = u + iv = f(z).$$

Cauchy's theorem for multiply-connected domains. If $f(z)$ is analytic in a multiply-connected domain D, then we cannot conclude that

$$\oint f(z)\,dz = 0$$

on every simple closed path C in D. Thus, if D is the doubly-connected domain of Fig. 9–7 and C is the curve C_1 shown, then the integral around C need not be zero. However, by introducing cuts, we can reason that

$$\oint_{C_1} f(z)\,dz = \oint_{C_2} f(z)\,dz; \tag{9–47}$$

that is, the integral has the same value on all paths which go around the inner "hole" once in the positive direction. For a triply-connected domain, as in Fig. 9–8, we obtain the equation

$$\oint_{C_1} f(z)\,dz = \oint_{C_2} f(z)\,dz + \oint_{C_3} f(z)\,dz. \tag{9–48}$$

This can be written in the form

$$\oint_{C_1} f(z)\,dz + \oint_{C_2} f(z)\,dz + \oint_{C_3} f(z)\,dz = 0; \tag{9–49}$$

Eq. (9–49) states that the integral around the complete boundary of a certain region in D is equal to zero. More generally, we have the following theorem:

THEOREM 16 (Cauchy's theorem for multiply-connected domains). *Let $f(z)$ be analytic in a domain D and let $C_1, \ldots, C_n$ be n simple closed curves in D which together form the boundary B of a region R contained in D. Then*

$$\int_B f(z)\,dz = 0,$$

where the direction of integration on B is such that the outer normal is 90° behind the tangent vector in the direction of integration.

9–9 Cauchy's integral formula. Now let D be a simply-connected domain and let z_0 be a fixed point of D. If $f(z)$ is analytic in D, the function $f(z)/(z - z_0)$ will fail to be analytic at z_0. Hence

$$\oint \frac{f(z)}{z - z_0}\,dz$$

will in general not be zero on a path C enclosing z_0. However, as above, this integral will have the same value on all paths C about z_0. To determine this value, we reason that if C is a very small circle of radius R about z_0, then $f(z_0)$ has, by continuity, approximately the constant value $f(z_0)$ on the path. This suggests that

$$\oint_C \frac{f(z)}{z - z_0}\,dz = f(z_0) \cdot \oint_{|z-z_0|=R} \frac{dz}{z - z_0} = f(z_0) \cdot 2\pi i,$$

since we find

$$\oint_{|z-z_0|=R} \frac{dz}{z - z_0} = \int_0^{2\pi} \frac{Rie^{i\theta}}{Re^{i\theta}}\,d\theta = i\int_0^{2\pi} d\theta = 2\pi i,$$

with the aid of the substitution: $z - z_0 = Re^{i\theta}$. The correctness of the conclusion reached is the content of the following fundamental result:

THEOREM 17 (Cauchy integral formula). *Let $f(z)$ be analytic in a domain D. Let C be a simple closed curve in D, within which $f(z)$ is analytic and let z_0 be inside C. Then*

$$f(z_0) = \frac{1}{2\pi i} \oint_C \frac{f(z)}{z - z_0}\,dz. \tag{9–50}$$

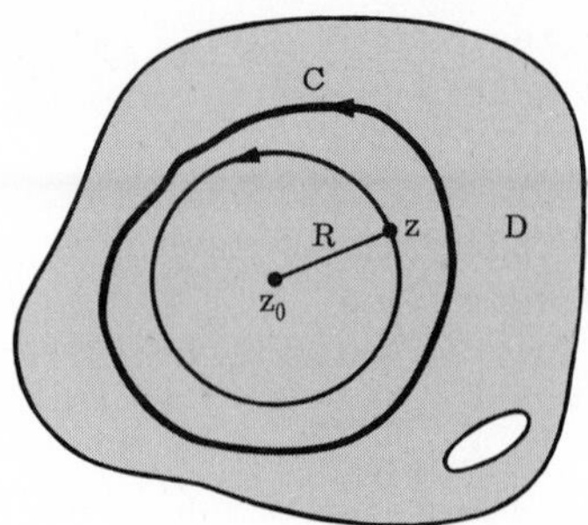

FIG. 9–9. Cauchy integral formula.

Proof. The domain D is not required to be simply-connected, but since f is analytic within C, the theorem concerns only a simply-connected part of D, as shown in Fig. 9–9. We reason as above to conclude that

$$\oint_C \frac{f(z)}{z - z_0}\,dz = \oint_{|z-z_0|=R} \frac{f(z)}{z - z_0}\,dz.$$

It remains to show that the integral on the right is indeed $f(z_0) \cdot 2\pi i$. Now, since $f(z_0) = \text{const}$,

$$\oint \frac{f(z_0)}{z - z_0}\,dz = f(z_0) \oint \frac{dz}{z - z_0} = f(z_0) \cdot 2\pi i,$$

where we integrate always on the circle $|z - z_0| = R$. Hence, on the same path,

$$\oint \frac{f(z)}{z - z_0}\,dz - f(z_0) \cdot 2\pi i = \oint \frac{f(z) - f(z_0)}{z - z_0}\,dz. \tag{9–51}$$

Now $|z - z_0| = R$ on the path, and since $f(z)$ is continuous at z_0, $|f(z) - f(z_0)| < \epsilon$ for $R < \delta$, for each preassigned $\epsilon > 0$. Hence, by Theorem 7,

$$\left|\oint \frac{f(z) - f(z_0)}{z - z_0}\,dz\right| < \frac{\epsilon}{R} \cdot 2\pi R = 2\pi\epsilon.$$

Thus the absolute value of the integral can be made as small as desired by choosing R sufficiently small. But the integral has the same value for all choices of R. This is possible only if the integral is zero for all R. Hence the left side of (9–51) is zero and (9–50) follows.

The integral formula (9–50) is remarkable in that it expresses the values of the function $f(z)$ at points z_0 inside the curve C in terms of the values along C alone. If C is taken as a circle $z = z_0 + Re^{i\theta}$, then (9-50) reduces to the following:

$$f(z_0) = \frac{1}{2\pi}\int_0^{2\pi} f(z_0 + Re^{i\theta})\, d\theta. \tag{9-52}$$

Thus the *value of an analytic function at the center of a circle equals the average (arithmetic mean) of the values on the circumference.*

Just as with the Cauchy integral theorem, the Cauchy integral formula can be extended to multiply-connected domains. Under the hypotheses of Theorem 16,

$$f(z_0) = \frac{1}{2\pi i}\int_B \frac{f(z)}{z - z_0}\, dz = \frac{1}{2\pi i}\left(\oint_{C_1} \frac{f(z)}{z - z_0}\, dz + \oint_{C_2} \frac{f(z)}{z - z_0}\, dz + \cdots\right), \tag{9-53}$$

where z_0 is any point inside the region R bounded by C_1 (the outer boundary), $C_2, \ldots, C_n$. The proof is left as an exercise (Problem 6 below).

Problems

1. Evaluate the following integrals:

(a) $\oint z^2 \sin z\, dz$ on the ellipse $x^2 + 2y^2 = 1$

(b) $\oint \dfrac{z^2}{z+1}\, dz$ on the circle $|z - 2| = 1$

(c) $\int_1^{2i} ze^z\, dz$ on the line segment joining the endpoints

(d) $\int_{1+i}^{1-i} \dfrac{1}{z^2}\, dz$ on the parabola $2y^2 = x + 1$

2. (a) Evaluate $\int_{-i}^{i} (dz/z)$ on the path $z = e^{it}$, $-\pi/2 \leqq t \leqq \pi/2$, with the aid of the relation $(\log z)' = 1/z$, for an appropriate branch of $\log z$.

(b) Evaluate $\int_i^{-i} (dz/z)$ on the path $z = e^{it}$, $\pi/2 \leqq t \leqq 3\pi/2$, as in part (a).

(c) Why does the relation $(\log z)' = 1/z$ not imply that the sum of the two integrals of parts (a) and (b) is zero?

3. A certain function $f(z)$ is known to be analytic except for $z = 1$, $z = 2$, $z = 3$, and it is known that

$$\oint_{C_k} f(z)\, dz = a_k \quad (k = 1, 2, 3),$$

where C_k is a circle of radius $\frac{1}{2}$ with center at $z = k$. Evaluate

$$\oint f(z)\,dz$$

on each of the following paths:

(a) $|z| = 4$ (b) $|z| = 2.5$ (c) $|z - 2.5| = 1$

4. A certain function $f(z)$ is analytic except for $z = 0$, and it is known that

$$\lim_{z\to\infty} zf(z) = 0.$$

Show that

$$\oint f(z)\,dz = 0$$

on every simple closed path not passing through the origin. [*Hint:* Show that the value of the integral on a path $|z| = R$ can be made as small as desired by making R sufficiently large.]

5. Evaluate each of the following with the aid of the Cauchy integral formula:

(a) $\displaystyle\oint \frac{z}{z-3}\,dz$ on $|z| = 5$ (b) $\displaystyle\oint \frac{e^z}{z^2 - 3z}\,dz$ on $|z| = 1$

(c) $\displaystyle\oint \frac{z+2}{z^2-1}\,dz$ on $|z| = 2$ (d) $\displaystyle\oint \frac{\sin z}{z^2+1}\,dz$ on $|z| = 2$

[*Hint for* (c) *and* (d): Expand the rational function in partial fractions.]

6. Prove (9–53) under the hypotheses stated.

7. Prove that if $f(z)$ is analytic in domain D and $f'(z) \equiv 0$, then $f(z) \equiv$ constant. [*Hint:* Apply Theorem 14.]

Answers

1. (a) 0 (b) 0 (c) $(2i - 1)e^{2i}$ (d) $-i$
2. (a) πi (b) πi
3. (a) $a_1 + a_2 + a_3$ (b) $a_1 + a_2$ (c) $a_2 + a_3$
5. (a) $6\pi i$ (b) $-2\pi i/3$ (c) $2\pi i$ (d) $2\pi i \sinh 1$

9–10 Power series as analytic functions. We now proceed to enlarge the class of specific analytic functions still further by showing that every power series

$$\sum_{n=0}^{\infty} c_n(z - z_0)^n = c_0 + c_1(z - z_0) + \cdots + c_n(z - z_0)^n + \cdots$$

converging for some values of z other than $z = z_0$ represents an analytic function.

For the theory of series of complex numbers, see Section 6–19. The following fundamental theorem for complex power series is proved just as for real series (Theorem 35 in Section 6–15).

THEOREM 18. *Every power series* $\sum_{n=0}^{\infty} c_n(z - z_0)^n$ *has a radius of convergence* r^* *such that the series converges absolutely when* $|z - z_0| < r^*$, *and diverges when* $|z - z_0| > r^*$. *The series converges uniformly for* $|z - z_0| \leqq r_1$, *provided* $r_1 < r^*$.

The number r^* can be zero, in which case the series converges only for $z = z_0$, a positive number, or ∞, in which case the series converges for all z.

The number r^* can be evaluated as follows:

$$r^* = \lim_{n\to\infty} \left|\frac{c_n}{c_{n+1}}\right|, \quad \text{if the limit exists,}$$

$$r^* = \lim_{n\to\infty} \frac{1}{\sqrt[n]{|c_n|}}, \quad \text{if the limit exists,} \tag{9–54}$$

and in any case by the formula

$$r^* = \frac{1}{\overline{\lim\limits_{n\to\infty}} \sqrt[n]{|c_n|}}. \tag{9–55}$$

As for real variables, no general statement can be made about convergence on the boundary of the domain of convergence. This boundary (when $r^* \neq 0$, $r^* \neq \infty$) is a circle $|z - z_0| = r^*$, termed the *circle of convergence* (Fig. 9–10). The series may converge at some points, all points, or no points of this circle.

EXAMPLE 1. $\sum_{n=1}^{\infty} (z^n/n^2)$. The first formula (9–54) gives

$$r^* = \lim_{n\to\infty} \frac{(n+1)^2}{n^2} = 1.$$

The series converges absolutely on the circle of convergence, for when $|z| = 1$, the series of absolute value is the convergent series $\sum(1/n^2)$.

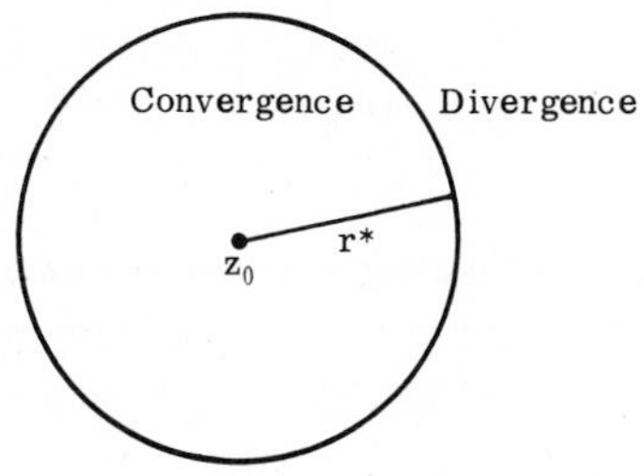

FIG. 9–10. Circle of convergence of a power series.

EXAMPLE 2. $\sum_{n=0}^{\infty} z^n$. This complex geometric series converges for $|z| < 1$, as (9–54) shows. We have further

$$\sum_{n=0}^{\infty} z^n = \frac{1}{1-z}, \quad (|z| < 1),$$

as for real variables. On the circle of convergence, the series diverges everywhere, since the nth term fails to converge to zero.

The following theorems are proved as for real variables.

THEOREM 19. *A power series with nonzero convergence radius represents a continuous function within the circle of convergence.*

THEOREM 20. *A power series can be integrated term by term within the circle of convergence; that is, if $r^* \neq 0$ and*

$$f(z) = \sum_{n=0}^{\infty} c_n (z - z_0)^n, \quad (|z - z_0| < r^*),$$

then, for every path C inside the circle of convergence

$$\int_{z_1 \atop C}^{z_2} f(z)\, dz = \sum_{n=0}^{\infty} c_n \int_{z_1}^{z_2} (z - z_0)^n\, dz = \sum_{n=0}^{\infty} c_n \left. \frac{(z - z_0)^{n+1}}{n+1} \right|_{z_1}^{z_2},$$

or in terms of indefinite integrals,

$$\int f(z)\, dz = \sum_{n=0}^{\infty} c_n \frac{(z - z_0)^{n+1}}{n+1} + \text{const}, \quad (|z - z_0| < r^*).$$

THEOREM 21. *A power series can be differentiated term by term; that is, if $r^* \neq 0$ and*

$$f(z) = \sum_{n=0}^{\infty} c_n (z - z_0)^n, \quad (|z - z_0| < r^*),$$

then

$$f'(z) = \sum_{n=1}^{\infty} n c_n (z - z_0)^{n-1}, \quad (|z - z_0| < r^*),$$

$$f''(z) = \sum_{n=2}^{\infty} n(n-1) c_n (z - z_0)^{n-2}, \quad (|z - z_0| < r^*),$$

$$\vdots$$

Hence every power series with nonzero convergence radius defines an analytic function $f(z)$ within the circle of convergence, and the power series is the Taylor series of $f(z)$:

$$c_n = \frac{f^{(n)}(z_0)}{n!}.$$

THEOREM 22. *If two power series* $\sum_{n=0}^{\infty} c_n(z - z_0)^n$, $\sum_{n=0}^{\infty} C_n(z - z_0)^n$ *have nonzero convergence radii and have equal sums wherever both series converge, then the series are identical; that is,*

$$c_n = C_n \quad (n = 0, 1, 2, \ldots).$$

9–11 Power series expansion of general analytic function. In Section 9–10 it was shown that every power series with nonzero convergence radius represents an analytic function. We now proceed to show that all analytic functions are obtainable in this way. If a function $f(z)$ is analytic in a domain D of general shape, we cannot expect to represent $f(z)$ by one power series; for the power series converges only in a circular domain. However, we can show that for each circular domain D_0 in D, there is a power series converging in D_0 whose sum is $f(z)$. Thus several (perhaps infinitely many) power series are needed to represent $f(z)$ throughout all of D.

THEOREM 23. *Let $f(z)$ be analytic in the domain D. Let z_0 be in D and let R be the radius of the largest circle with center at z_0 and having its interior in D. Then there is a power series*

$$\sum_{n=0}^{\infty} c_n(z - z_0)^n$$

which converges to $f(z)$ for $|z - z_0| < R$. Furthermore,

$$c_n = \frac{f^{(n)}(z_0)}{n!} = \frac{1}{2\pi i} \oint_C \frac{f(z)}{(z - z_0)^{n+1}}\, dz, \tag{9–56}$$

where C is a simple closed path in D enclosing z_0 and within which $f(z)$ is analytic.

Proof. For simplicity we take $z_0 = 0$. The general case can then be obtained by the substitution $z' = z - z_0$. Let the circle $|z| = R$ be the largest circle with center at z_0 and having its interior within D; the radius R is then positive or $+\infty$ (in which case D is the whole z-plane). Let z_1 be a point within this circle, so that $|z_1| < R$. Choose R_2 so that $|z_1| < R_2 < R$ (see Fig. 9–11). Then $f(z)$ is analytic in a domain including the circle C_2: $|z| = R_2$ plus interior. Hence by the Cauchy integral formula,

$$f(z_1) = \frac{1}{2\pi i} \oint_{C_2} \frac{f(z)}{z - z_1}\, dz.$$

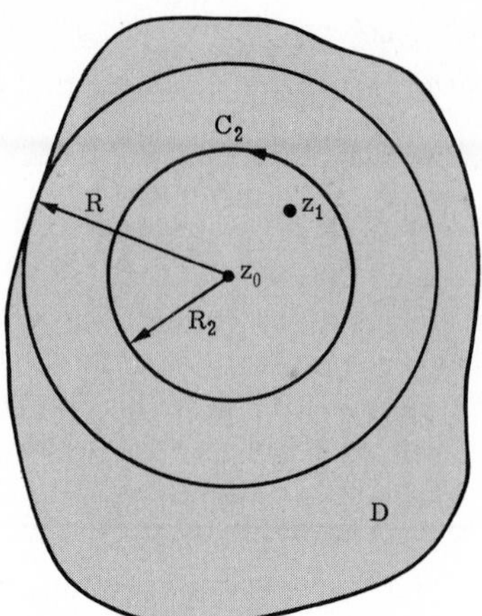

FIG. 9–11. Taylor series of an analytic function.

Now the factor $1/(z - z_1)$ can be expanded in a geometric series:

$$\frac{1}{z - z_1} = \frac{1}{z\left(1 - \frac{z_1}{z}\right)} = \frac{1}{z}\left(1 + \frac{z_1}{z} + \cdots + \frac{z_1^n}{z^n} + \cdots\right).$$

The series can be considered as a power series in powers of $1/z$, for fixed z_1. It converges for $|z_1/z| < 1$ and converges uniformly for $|z_1/z| \leqq |z_1|/R_2 < 1$.

If we multiply by $f(z)$, we find

$$\frac{f(z)}{z - z_1} = \frac{f(z)}{z} + z_1 \frac{f(z)}{z^2} + \cdots + z_1^n \frac{f(z)}{z^{n+1}} + \cdots;$$

since $f(z)$ is continuous for $|z| = R_2$, the series remains uniformly convergent on C_2. Hence we can integrate term by term on C_2:

$$\frac{1}{2\pi i} \oint_{C_2} \frac{f(z)}{z - z_1}\, dz = \frac{1}{2\pi i} \oint_{C_2} \frac{f(z)}{z}\, dz + \frac{z_1}{2\pi i} \oint_{C_2} \frac{f(z)}{z^2}\, dz + \cdots + \frac{z_1^n}{2\pi i} \oint_{C_2} \frac{f(z)}{z^{n+1}}\, dz + \cdots$$

The left-hand side is precisely $f(z_1)$, by the integral formula. Hence

$$f(z_1) = \sum_{n=0}^{\infty} c_n z_1^n, \qquad c_n = \frac{1}{2\pi i} \oint_{C_2} \frac{f(z)}{z^{n+1}}\, dz.$$

The path C_2 can be replaced by any path C as described in the theorem, since $f(z)/z^{n+1}$ is analytic in D except for $z = z_0 = 0$.

By Theorem 21, the series obtained is the Taylor series of f, so that

$$c_n = \frac{f^{(n)}(z_0)}{n!} \quad (z_0 = 0).$$

The theorem is now completely proved.

The consequences of this theorem are far-reaching. First of all, not only does it guarantee that every analytic function is representable by power series, but it ensures that the Taylor series converges to the function within each circular domain within the domain in which the function is given. Thus, *without further analysis*, we at once conclude that

$$e^z = 1 + z + \frac{z^2}{2!} + \cdots + \frac{z^n}{n!} + \cdots,$$

$$\sin z = z - \frac{z^3}{3!} + \frac{z^5}{5!} + \cdots + (-1)^n \frac{z^{2n+1}}{(2n+1)!} + \cdots,$$

$$\cos z = 1 - \frac{z^2}{2!} + \cdots + (-1)^n \frac{z^{2n}}{(2n)!} + \cdots$$

for all z. A variety of other familiar expansions can be obtained in the same way.

It should be recalled that a function $f(z)$ is defined to be analytic in a domain D if $f(z)$ has a continuous derivative $f'(z)$ in D (Section 9–6). By Theorem 23, $f(z)$ must have derivatives of all orders at every point of D. In particular, the derivative of an analytic function is itself analytic:

THEOREM 24. *If $f(z)$ is analytic in domain D, then $f'(z)$, $f''(z)$, . . . , $f^{(n)}(z)$, . . . exist and are analytic in D. Furthermore, for each n*

$$f^{(n)}(z_0) = \frac{n!}{2\pi i} \oint_C \frac{f(z)}{(z - z_0)^{n+1}}\, dz, \tag{9–57}$$

where C is any simple closed path in D enclosing z_0 and within which $f(z)$ is analytic.

Equation (9–57) is a restatement of (9–56).

Circle of convergence of the Taylor series. Theorem 23 guarantees convergence of the Taylor series of $f(z)$ about each z_0 in D in the largest circular domain $|z - z_0| < R$ in D, as shown in Fig. 9–11. However, this does not mean that R is the radius of convergence r^* of the series, for r^* can be larger than R, as suggested in Fig. 9–12. When this happens, the function $f(z)$ can be prolonged into a larger domain, while retaining

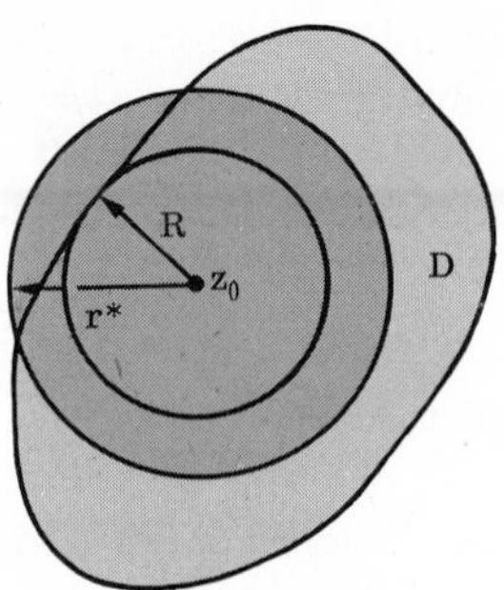

FIG. 9–12. Analytic continuation.

analyticity. For example, if $f(z) = \text{Log}\, z$ $(0 < \theta < \pi)$ is expanded in a Taylor series about the point $z = -1 + i$, the series has convergence radius $\sqrt{2}$, whereas $R = 1$ [Problem 4(c) below].

The process of prolonging the function suggested here is called *analytic continuation.*

Harmonic functions. Let $w = f(x)$ be analytic in D and let

$$u(x, y) = \text{Re}\,[f(z)], \qquad v(x, y) = \text{Im}\,[f(z)].$$

By Theorem 8 (Section 9–6), $f'(z) = u_x + iv_x = v_y - iu_y = \cdots$ and, by Theorem 24, $f'(z)$ is analytic in D, so that

$$f''(z) = u_{xx} + iv_{xx} = v_{yx} - iu_{yx} = \cdots.$$

Thus u_{xx}, u_{xy}, v_{xx}, ... are also continuous in D. In general, Theorem 24 shows that u and v have partial derivatives of all orders in D. Furthermore, by taking real and imaginary parts in the Taylor series of $f(z)$, as given in Theorem 23, we obtain convergent power series for $u(x, y)$ and $v(x, y)$. For example, if $f(z) = \sum c_n z^n$ for $|z| < R$, with $c_n = a_n + ib_n$, then

$$u(x, y) = a_0 + (a_1 x - b_1 y) + (a_2 x^2 - 2b_2 xy - a_2 y^2) + \cdots, x^2 + y^2 < R^2.$$

Also, by the Cauchy-Riemann equations, $u_x = v_y$, $u_y = -v_x$, so that

$$\frac{\partial^2 u}{\partial x^2} + \frac{\partial^2 u}{\partial y^2} = \frac{\partial^2 v}{\partial x\, \partial y} - \frac{\partial^2 v}{\partial y\, \partial x} = 0;$$

that is, $u(x, y)$ is *harmonic* in D. Similarly $v(x, y)$ is harmonic in D. Conversely, it can be shown that, in a simply-connected domain D, each harmonic function $\phi(x, y)$ can be interpreted as the real or imaginary part of an analytic function in D. (See Section 5–5 of the author's book listed at the end of the chapter.)

Problems

1. Determine the radius of convergence of each of the following series:

(a) $\sum_{n=0}^{\infty} \frac{z^n}{n^2+1}$ (b) $\sum_{n=0}^{\infty} \frac{(z-1)^n}{3^n}$ (c) $\sum_{n=0}^{\infty} n!z^n$ (d) $\sum_{n=0}^{\infty} \frac{z^n}{n!}$

2. Given the series $\sum_{n=1}^{\infty}(z^n/n)$, show that

(a) the series has radius of convergence 1;

(b) the series diverges for $z = 1$;

(c) the series converges for $z = i$ and for $z = -1$. It can be shown that the series converges for $|z| = 1$, except for $z = 1$.

3. By means of (9–57), evaluate each of the following:

(a) $\oint \frac{ze^z}{(z-1)^4}\,dz$ on $|z| = 2$ (b) $\oint \frac{\sin z}{z^4}\,dz$ on $|z| = 1$

(c) $\oint \frac{dz}{z^3(z+4)}$ on $|z| = 2$

4. Expand in a Taylor series about the point indicated, and determine the radius of convergence r^* and the radius R of the largest circle within which the series converges to the function:

(a) $\sin z$ about $z = 0$ (b) $1/(z-1)$ about $z = 2$

(c) $\operatorname{Log} z$ $(0 < \theta < \pi)$ about $z = -1 + i$

5. *Cauchy's inequalities.* Let $f(z)$ be analytic in a domain including the circle $C: |z - z_0| = R$ and interior, and let $|f(z)| \leqq M = \text{const}$ on C. Prove that

$$|f^{(n)}(z_0)| \leqq \frac{Mn!}{R^n} \quad (n = 0, 1, 2, \ldots).$$

[*Hint:* Apply (9–57).]

6. A function $f(z)$ which is analytic in the whole z-plane is termed an *entire* function or an *integral* function. Examples are polynomials, e^z, $\sin z$, $\cos z$. Prove *Liouville's theorem:* If $f(z)$ is an entire function and $|f(z)| \leqq M$ for all z, where M is constant, then $f(z)$ reduces to a constant. [*Hint:* Take $n = 1$ in the Cauchy inequalities of Problem 5 to show that $f'(z_0) = 0$ for every z_0.]

Answers

1. (a) 1 (b) 3 (c) 0 (d) ∞

3. (a) $4\pi ei/3$ (b) $-\pi i/3$ (c) $\pi i/32$

4. (a) $\sum_{n=0}^{\infty} \frac{(-1)^n z^{2n+1}}{(2n+1)!}$ $(r^* = R = \infty)$

(b) $\sum_{n=0}^{\infty} (-1)^n (z-2)^n$ $(r^* = R = 1)$

(c) $\frac{1}{2}\log 2 + \frac{3}{4}\pi i - \sum_{n=1}^{\infty} \left(\frac{1+i}{2}\right)^n \frac{(z+1-i)^n}{n}$ $(r^* = \sqrt{2},\ R = 1)$

9–12 Power series in positive and negative powers; Laurent expansion. We have shown that every power series $\sum a_n(z - z_0)^n$ with nonzero convergence radius represents an analytic function and that every analytic function can be built up out of such series. It thus appears unnecessary to seek other explicit expressions for analytic functions. However, the power series represent functions only in circular domains and are hence awkward for representing a function in a more complicated type of domain. It is therefore worthwhile to consider other types of representations. A series of form

$$\sum_{n=1}^{\infty} \frac{b_n}{(z - z_0)^n} = \frac{b_1}{z - z_0} + \cdots + \frac{b_n}{(z - z_0)^n} + \cdots \tag{9–58}$$

will also represent an analytic function in a domain in which the series converges. For the substitution $z_1 = 1/(z - z_0)$ reduces the series to an ordinary power series,

$$\sum_{n=1}^{\infty} b_n z_1^n.$$

If this series converges for $|z_1| < r_1^*$, then its sum is an analytic function $F(z_1)$; hence the series (9–58) converges for

$$|z - z_0| > \frac{1}{r_1^*} = r_0^* \tag{9–59}$$

to the analytic function $g(z) = F(1/(z - z_0))$. The value $z_1 = 0$ corresponds to $z = \infty$, in a limiting sense; accordingly we can also say that $g(z)$ is analytic at ∞ and $g(\infty) = 0$. This will be justified more fully in Section 9–14.

The domain of convergence of the series (9–58) is the region (9–59), which is the *exterior* of a circle. It can happen that $r_1^* = \infty$, in which case the series converges for all z except z_0; if $r_1^* = 0$, the series diverges for all z (except $z = \infty$, as above).

If we add to a series (9–58) a usual power series

$$\sum_{n=0}^{\infty} a_n(z - z_0)^n = a_0 + a_1(z - z_0) + \cdots,$$

converging for $|z - z_0| < r_2^*$, we obtain a sum

$$\sum_{n=1}^{\infty} \frac{b_n}{(z - z_0)^n} + \sum_{n=0}^{\infty} a_n(z - z_0)^n. \tag{9–60}$$

If $r_0^* < r_2^*$, the sum converges and represents an analytic function $f(z)$ in the *annular domain:* $r_0^* < |z - z_0| < r_2^*$; for each series has an analytic sum in this domain, so that the sum of the two series is analytic there.

We can write this sum in the more compact form (after some relabeling)

$$f(z) = \sum_{n=-\infty}^{\infty} a_n(z - z_0)^n, \tag{9–61}$$

though this should be interpreted as the sum of two series, as in (9–60).

In this way we build up a new class of analytic functions, each defined in a ring-shaped domain. Every function analytic in such a domain can be obtained in this way:

THEOREM 25 (Laurent's theorem). *Let $f(z)$ be analytic in the ring:*

$$R_1 < |z - z_0| < R_2.$$

Then

$$f(z) = \sum_{n=-\infty}^{\infty} a_n(z - z_0)^n = [a_0 + a_1(z - z_0) + \cdots] + \left[\frac{a_{-1}}{z - z_0} + \frac{a_{-2}}{(z - z_0)^2} + \cdots\right],$$

where

$$a_n = \frac{1}{2\pi i} \oint_C \frac{f(z)}{(z - z_0)^{n+1}}\, dz \tag{9–62}$$

and C is any simple closed curve separating $|z| = R_1$ from $|z| = R_2$. The series converges uniformly for $R_1 < k_1 \leqq |z - z_0| \leqq k_2 < R_2$.

Proof. For simplicity we take $z_0 = 0$. Let z_1 be any point of the ring and choose r_1, r_2 so that $R_1 < r_1 < |z_1| < r_2 < R_2$, as in Fig. 9–13. We then apply the Cauchy integral formula in general form [Eq. (9–53)] to the region bounded by C_1: $|z| = r_1$ and C_2: $|z| = r_2$. Hence

$$f(z_1) = \frac{1}{2\pi i} \oint_{C_2} \frac{f(z)}{z - z_1}\, dz - \frac{1}{2\pi i} \oint_{C_1} \frac{f(z)}{z - z_1}\, dz.$$

The first term can be replaced by a power series

$$\sum_{n=0}^{\infty} a_n z_1^n, \qquad a_n = \frac{1}{2\pi i} \oint_{C_2} \frac{f(z)}{z^{n+1}}\, dz$$

as in the proof of Theorem 23 (Section 9–11). For the second term, the series expansion

$$\frac{1}{z - z_1} = -\frac{1}{z_1}\left(\frac{1}{1 - z/z_1}\right) = -\frac{1}{z_1} - \frac{z}{z_1^2} - \frac{z^2}{z_1^3} - \cdots,$$

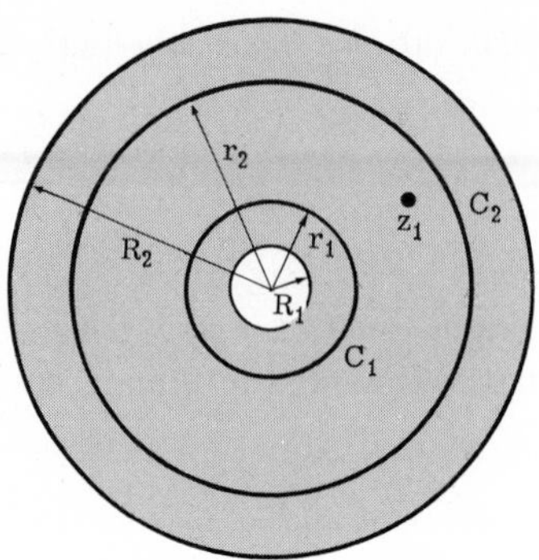

FIG. 9–13. Laurent's theorem.

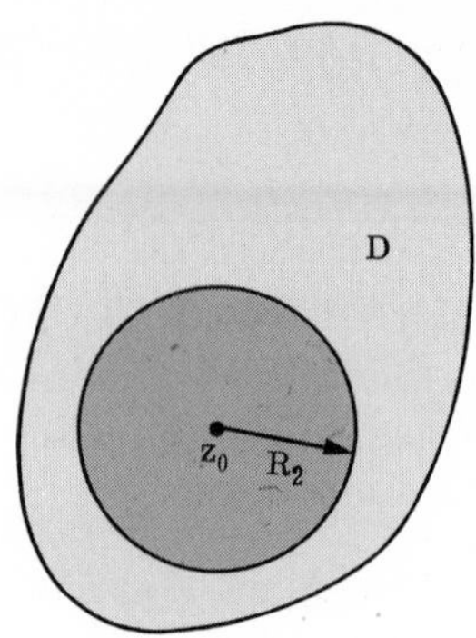

FIG. 9–14. Isolated singularity.

valid for $|z_1| > |z| = r_1$, leads similarly to the series

$$\sum_{n=1}^{\infty} \frac{b_n}{z_1^n} = \sum_{n=-\infty}^{-1} a_n z_1^n, \qquad a_n = \frac{1}{2\pi i} \oint_{C_1} \frac{f(z)}{z^{n+1}}\, dz.$$

Hence

$$f(z_1) = \sum_{n=-\infty}^{\infty} a_n z_1^n, \qquad a_n = \frac{1}{2\pi i} \oint_{C_1} \frac{f(z)}{z^{n+1}}\, dz;$$

the path C_2 or C_1 can be replaced by any path C separating $|z| = R_1$ from $|z| = R_2$, since the function integrated is analytic throughout the annulus. The uniform convergence follows as for ordinary power series (Theorem 18). The theorem is now established.

Laurent's theorem continues to hold when $R_1 = 0$ or $R_2 = \infty$ or both. In the case $R_1 = 0$, the Laurent expansion represents a function $f(z)$ analytic in a *deleted neighborhood* of z_0, that is, in the circular domain $|z - z_0| < R_2$ minus its center z_0. If $R_2 = \infty$, we can say similarly that the series represents $f(z)$ in a *deleted neighborhood* of $z = \infty$.

9–13 Isolated singularities of an analytic function. Zeros and poles. Let $f(z)$ be defined and analytic in domain D. We say that $f(z)$ has an *isolated singularity* at the point z_0 if $f(z)$ is analytic throughout a neighborhood of z_0 except at z_0 itself; that is, to use the term mentioned at the end of the preceding section, $f(z)$ is analytic in a deleted neighborhood of z_0, but not at z_0. The point z_0 is then a boundary point of D and would be called an *isolated boundary point* (see Fig. 9–14).

A deleted neighborhood $0 < |z - z_0| < R_2$ forms a special case of the annular domain for which Laurent's theorem is applicable. Hence in this deleted neighborhood $f(z)$ has a representation as a Laurent series:

$$f(z) = \sum_{n=-\infty}^{\infty} a_n(z - z_0)^n.$$

The form of this series leads to a classification of isolated singularities into three fundamental types:

Case I. No terms in negative powers of $z - z_0$ appear. In this case the series is a Taylor series and represents a function analytic in a neighborhood of z_0. Thus the singularity can be removed by setting $f(z_0) = a_0$. We call this a *removable singularity* of $f(z)$. It is illustrated by

$$\frac{\sin z}{z} = 1 - \frac{z^2}{3!} + \frac{z^4}{5!} - \cdots$$

at $z = 0$. In practice, we automatically remove the singularity by defining the function properly.

Case II. A finite number of negative powers of $z - z_0$ appear. Thus we have

$$f(z) = \frac{a_{-N}}{(z - z_0)^N} + \cdots + \frac{a_{-1}}{z - z_0} + a_0 + \cdots + a_n(z - z_0)^n + \cdots,$$
$$(N \geqq 1,\ a_{-N} \neq 0). \qquad (9\text{–}63)$$

Here $f(z)$ is said to have a *pole of order* N at z_0. We can write

$$f(z) = \frac{1}{(z - z_0)^N} g(z), \qquad g(z) = a_{-N} + a_{-N+1}(z - z_0) + \cdots, \qquad (9\text{–}64)$$

so that $g(z)$ is analytic for $|z - z_0| < R_2$ and $g(z_0) \neq 0$. Conversely, every function $f(z)$ representable in the form (9–64) has a pole of order N at z_0. Poles are illustrated by rational functions of z, such as

$$f(z) = \frac{z - 2}{(z^2 + 1)(z - 1)^3}, \qquad (9\text{–}65)$$

which has poles of order 1 at $\pm i$ and of order 3 at $z = 1$.

The rational function

$$\frac{a_{-N}}{(z - z_0)^N} + \cdots + \frac{a_{-1}}{z - z_0} = p(z) \qquad (9\text{–}66)$$

is called the *principal part* of $f(z)$ at the pole z_0. Thus $f(z) - p(z)$ is analytic at z_0.

EXAMPLE 1.

$$f(z) = \frac{e^z \cos z}{z^3} \quad \text{at } z = 0.$$

To obtain the Laurent series, we expand the numerator in a Taylor series:

$$e^z \cos z = \left(1 + z + \frac{z^2}{2!} + \cdots\right)\left(1 - \frac{z^2}{2!} + \cdots\right) = 1 + z - \frac{z^3}{3} + \cdots$$

Hence

$$\frac{e^z \cos z}{z^3} = \frac{1}{z^3} + \frac{1}{z^2} - \frac{1}{3} + \cdots$$

Here the first two terms form the principal part; the pole is of order 3.

EXAMPLE 2.

$$f(z) = \frac{z}{(z+1)^2(z^3+2)} \quad \text{at } z = -1.$$

We expand $z/(z^3 + 2)$ in a Taylor series about $z = -1$.

$$w = \frac{z}{z^3+2}, \qquad w' = \frac{2 - 2z^3}{(z^3+2)^2}, \qquad w'' = \frac{6(z^5 - 4z^2)}{(z^3+2)^3}, \ldots,$$

$$\frac{z}{z^3+2} = -1 + 4(z+1) - 15(z+1)^2 + \cdots,$$

$$\frac{z}{(z+1)^2(z^3+2)} = \frac{-1}{(z+1)^2} + \frac{4}{z+1} - 15 + \cdots$$

The first two terms form the principal part; the pole is of order 2.

Case III. Infinitely many negative powers of $z - z_0$ appear. In this case $f(z)$ is said to have an *essential singularity* at z_0. This is illustrated by the function

$$f(z) = e^{1/z} = 1 + \frac{1}{z} + \frac{1}{2!}\frac{1}{z^2} + \frac{1}{3!}\frac{1}{z^3} + \cdots,$$

which has an essential singularity at $z = 0$.

In Case I, $f(z)$ has a finite limit at z_0 and accordingly $|f(z)|$ is bounded near z_0; that is, there is a real constant M such that $|f(z)| < M$ for z sufficiently close to z_0.

In Case II, $\lim_{z\to z_0} f(z) = \infty$, and it is customary to assign the value ∞ (complex) to $f(z)$ at a pole. At an essential singularity, $f(z)$ has a very complicated discontinuity. In fact, for every complex number c, we can find a sequence z_n converging to z_0 such that $\lim_{n\to\infty} f(z_n) = c$ (see Problem 8 below). It follows from this that if $|f(z)|$ is bounded near z_0, then z_0 must be a removable singularity, and if $\lim f(z) = \infty$ at z_0, then z_0 must be a pole.

If $f(z)$ is analytic at a point z_0, and $f(z_0) = 0$, then z_0 is termed a *root* or *zero* of $f(z)$. Thus the zeros of $\sin z$ are the numbers $n\pi$ ($n = 0, \pm 1, \pm 2, \ldots$). The Taylor series about z_0 has the form

$$f(z) = a_N(z - z_0)^N + a_{N+1}(z - z_0)^{N+1} + \cdots,$$

where $N \geqq 1$ and $a_N \neq 0$, or else $f(z) \equiv 0$ in a neighborhood of z_0. It will be seen that the latter case can occur only if $f(z) \equiv 0$ throughout the domain in which it is given. If now $f(z)$ is not identically zero, then

$$f(z) = (z - z_0)^N \phi(z),$$

$$\phi(z) = a_N + a_{N+1}(z - z_0) + \cdots,$$

$$\phi(z_0) = \frac{f^{(N)}(z_0)}{N!} = a_N \neq 0.$$

We say that $f(z)$ has a zero of *order* N or *multiplicity* N at z_0. For example, $1 - \cos z$ has a zero of order 2 at $z = 0$, since

$$1 - \cos z = \frac{z^2}{2} - \frac{z^4}{24} + \cdots$$

If $f(z)$ has a zero of order N at z_0, then $F(z) = 1/f(z)$ has a pole of order N at z_0, and conversely. For if f has a zero of order N, then

$$f(z) = (z - z_0)^N \phi(z)$$

as above, with $\phi(z_0) \neq 0$. It follows from continuity that $\phi(z) \neq 0$ in a sufficiently small neighborhood of z_0. Hence $g(z) = 1/\phi(z)$ is analytic in the neighborhood and $g(z_0) \neq 0$. Now in this neighborhood, except for z_0,

$$F(z) = \frac{1}{f(z)} = \frac{1}{(z - z_0)^N \phi(z)} = \frac{g(z)}{(z - z_0)^N},$$

so that F has a pole at z_0. The converse is proved in the same way.

It remains to consider the case when $f \equiv 0$ in a neighborhood of z_0. This is covered by the following theorem.

THEOREM 26. *The zeros of an analytic function are isolated, unless the function is identically zero; that is, if $f(z)$ is analytic in domain D and $f(z)$ is not identically zero, then for each zero z_0 of $f(z)$ there is a deleted neighborhood of z_0 in which $f(z) \neq 0$.*

The proof is given in Section 6–2 of the book by Kaplan listed at the end of this chapter.

9–14 The complex number ∞. The complex number ∞ has been introduced several times in connection with limiting processes; for example, in the discussion of poles in the preceding section. In each case ∞ has appeared in a natural way as the limiting position of a point receding indefinitely from the origin. We can incorporate this number into the complex number system with special algebraic rules:

$$\frac{z}{\infty} = 0 \quad (z \neq \infty), \qquad z \pm \infty = \infty \quad (z \neq \infty), \qquad \frac{z}{0} = \infty \quad (z \neq 0),$$
$$z \cdot \infty = \infty \quad (z \neq 0), \qquad \frac{\infty}{z} = \infty \quad (z \neq \infty). \tag{9–67}$$

Expressions such as $\infty + \infty$, $\infty - \infty$, and ∞/∞ are not defined.

A function $f(z)$ is said to be analytic in a deleted neighborhood of ∞ if $f(z)$ is analytic for $|z| > R_1$ for some R_1. In this case the Laurent expansion with $R_2 = \infty$ and $z_0 = 0$ is available, and we have

$$f(z) = \sum_{n=-\infty}^{\infty} a_n z^n \quad (|z| > R_1).$$

If there are no *positive* powers of z here, $f(z)$ is said to have a *removable singularity* at ∞ and we make *f analytic at ∞* by defining $f(\infty) = a_0$:

$$f(z) = a_0 + \frac{a_{-1}}{z} + \cdots + \frac{a_{-n}}{z^n} + \cdots \quad (|z| > R_1). \tag{9–68}$$

This is clearly equivalent to the statement that if we set $z_1 = 1/z$, then $f(z)$ becomes a function of z_1 with removable singularity at $z_1 = 0$.

If a finite number of positive powers occurs, we have, with $N \geqq 1$,

$$\begin{aligned} f(z) &= a_N z^N + \cdots + a_1 z + a_0 + \frac{a_{-1}}{z} + \cdots \\ &= z^N \phi(z), \\ \phi(z) &= a_N + \frac{a_{N-1}}{z} + \cdots, \end{aligned} \tag{9–69}$$

where $\phi(z)$ is analytic at ∞ and $\phi(\infty) = a_N \neq 0$. In this case $f(z)$ is said to have a *pole of order N at ∞*. The same holds for $f(1/z_1)$ at $z_1 = 0$. Furthermore,

$$\lim_{z \to \infty} f(z) = \infty. \tag{9–70}$$

If infinitely many positive powers appear, $f(z)$ is said to have an *essential singularity* at $z = \infty$.

If $f(z)$ is analytic at ∞ as in (9–68) and $f(\infty) = a_0 = 0$, then $f(z)$ is said to have a *zero* at $z = \infty$. If f is not identically zero, then necessarily

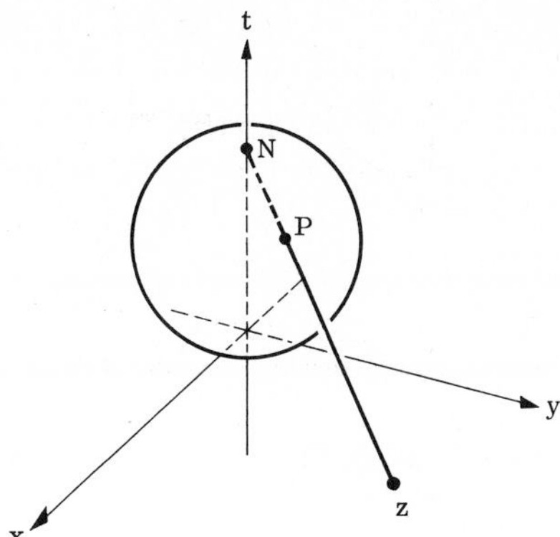

FIG. 9–15. Stereographic projection.

some $a_{-N} \neq 0$ and

$$\begin{aligned} f(z) &= \frac{a_{-N}}{z^N} + \frac{a_{-N-1}}{z^{N+1}} + \cdots \quad (|z| > R_1) \\ &= \frac{1}{z^N} g(z), \qquad\qquad (9\text{–}71) \\ g(z) &= a_{-N} + \frac{a_{-N-1}}{z} + \cdots \end{aligned}$$

Thus $g(z)$ is analytic at ∞ and $g(\infty) = a_{-N} \neq 0$. We say that $f(z)$ has a zero of order (or multiplicity) N at ∞. We can show that if $f(z)$ has a zero of order N at ∞, then $1/f(z)$ has a pole of order N at ∞, and conversely.

The significance of the complex number ∞ can be shown geometrically by the device of *stereographic projection*, i.e., a projection of the plane onto a sphere tangent to the plane at $z = 0$, as shown in Fig. 9–15. The sphere is given in xyt-space by the equation

$$x^2 + y^2 + (t - \tfrac{1}{2})^2 = \tfrac{1}{4}, \qquad (9\text{–}72)$$

so that the radius is $\frac{1}{2}$. The letter N denotes the "north pole" of the sphere, the point (0, 0, 1). If N is joined to an arbitrary point z in the xy-plane, the line segment Nz will meet the sphere at one other point P, which is the projection of z on the sphere. For example, the points of the circle $|z| = 1$ project on the "equator" of the sphere, i.e., the great circle $t = \frac{1}{2}$. As z recedes to infinite distance from the origin, P approaches N as limiting position. Thus *N corresponds to the complex number ∞*.

We refer to the z-plane plus the number ∞ as the *extended z-plane*. To emphasize that ∞ is *not* included, we refer to the *finite z-plane*.

Problems

1. For each of the following expand in a Laurent series at the isolated singularity given and state the type of singularity:

(a) $\dfrac{e^z - 1}{z}$ at $z = 0$ (b) $\dfrac{1}{z^2(z-3)}$ at $z = 0$

(c) $\dfrac{z - \cos z}{z}$ at $z = 0$ (d) $\csc z$ at $z = 0$

[*Hint for* (d): Write

$$\csc z = \frac{1}{\sin z} = \frac{1}{z - (z^3/3!) + \cdots} = \frac{a_{-1}}{z} + a_0 + \cdots$$

and determine the coefficients a_{-1}, a_0, a_1, ... so that

$$1 = (z - z^3/3! + \cdots) \cdot (a_{-1}z^{-1} + a_0 + a_1 z + \cdots).]$$

2. For each of the following find the principal part at the pole given:

(a) $\dfrac{z^2 + 3z + 1}{z^4}$ $(z = 0)$ (b) $\dfrac{z^2 - 2}{z(z+1)}$ $(z = 0)$

(c) $\dfrac{e^z \sin z}{(z-1)^2}$ $(z = 1)$ (d) $\dfrac{1}{z^2(z^3 + z + 1)}$ $(z = 0)$

3. For each of the following expand in a Laurent series at $z = \infty$ and state the type of singularity:

(a) $\dfrac{1}{1-z} = -\dfrac{1}{z}\dfrac{1}{1-(1/z)}$ (b) $\dfrac{z^2}{z+2}$ (c) $e^z + e^{1/z}$

4. Let $f(z)$ be a rational function in lowest terms:

$$f(z) = \frac{a_0 z^n + a_1 z^{n-1} + \cdots + a_n}{b_0 z^m + b_1 z^{m-1} + \cdots + b_m}.$$

The *degree* d of $f(z)$ is defined to be the larger of m and n. Assuming the fundamental theorem of algebra, show that $f(z)$ has precisely d zeros and d poles in the extended z-plane, a pole or zero of order N being counted as N poles or zeros.

5. For each of the following locate all zeros and poles in the extended plane (compare Problem 4):

(a) $\dfrac{z}{z-1}$ (b) $\dfrac{z-1}{z^2 + 3z + 2}$ (c) $\dfrac{z^3 + 3z^2 + 3z + 1}{z}$

6. Let $A(z)$ and $B(z)$ be analytic at $z = z_0$; let $A(z_0) \neq 0$ and let $B(z)$ have a zero of order N at z_0, so that

$$f(z) = \frac{A(z)}{B(z)} = \frac{a_0 + a_1(z - z_0) + \cdots}{b_N(z - z_0)^N + b_{N+1}(z - z_0)^{N+1} + \cdots}$$

has a pole of order N at z_0. Show that the principal part of $f(z)$ at z_0 is

$$\frac{a_0}{b_N}\frac{1}{(z-z_0)^N}+\frac{a_1b_N-a_0b_{N+1}}{b_N^2}\frac{1}{(z-z_0)^{N-1}}+\cdots$$

and obtain the next term explicitly. [*Hint:* Set

$$\frac{a_0+a_1(z-z_0)+\cdots}{b_N(z-z_0)^N+b_{N+1}(z-z_0)^{N+1}+\cdots}=\frac{C_{-N}}{(z-z_0)^N}+\frac{C_{-N+1}}{(z-z_0)^{N-1}}+\cdots$$

Multiply across and solve for C_{-N}, C_{-N+1}, . . .]

7. Prove *Riemann's theorem: If* $|f(z)|$ *is bounded in a deleted neighborhood of an isolated singularity* z_0, *then* z_0 *is a removable singularity of* $f(z)$. [Hint: Proceed as in Problem 6 following Section 9–11 with the aid of (9–62).]

8. Prove the *Theorem of Weierstrass and Casorati: If* z_0 *is an essential singularity of* $f(z)$, c *is an arbitrary complex number, and* $\epsilon > 0$, *then* $|f(z)-c| < \epsilon$ *for some* z *in every neighborhood of* z_0. [*Hint:* If the property fails, then $1/[f(z)-c]$ is analytic and bounded in absolute value in a deleted neighborhood of z_0. Now apply Problem 7 and conclude that $f(z)$ has a pole or removable singularity at z_0.]

Answers

1. (a) $\sum_{n=1}^{\infty}\frac{z^{n-1}}{n!}$, removable

(b) $-\frac{1}{3z^2}-\frac{1}{9z}-\sum_{n=0}^{\infty}\frac{z^n}{3^{n+3}}$, pole of order 2

(c) $-\frac{1}{z}+1+\sum_{n=0}^{\infty}\frac{(-1)^n z^{2n+1}}{(2n+2)!}$, pole of order 1

(d) $\frac{1}{z}+\frac{z}{6}+\frac{7z^3}{360}+\cdots$, pole of order 1

2. (a) $\frac{1}{z^4}+\frac{3}{z^3}+\frac{1}{z^2}$ (b) $-\frac{2}{z}$

(c) $\frac{e\sin 1}{(z-1)^2}+\frac{e(\cos 1+\sin 1)}{z-1}$ (d) $\frac{1}{z^2}-\frac{1}{z}$

3. (a) $\sum_{n=1}^{\infty}\frac{-1}{z^n}$, removable, zero of first order

(b) $z-2+\sum_{n=1}^{\infty}\frac{(-2)^{n+1}}{z^n}$, pole of order 1

(c) $2+\sum_{n=1}^{\infty}\frac{z^n}{n!}+\sum_{n=1}^{\infty}\frac{z^{-n}}{n!}$, essential

5. (a) zero: 0, pole: 1 (b) zeros: 1, ∞, poles: -1, -2
   (c) zeros: -1, -1, -1, poles: 0, ∞, ∞

6. $\dfrac{a_2 b_N^2 - a_1 b_N b_{N+1} - a_0 b_{N+2} b_N + a_0 b_{N+1}^2}{b_N^3 (z - z_0)^{N-2}}$.

9–15 Residues. Let $f(z)$ be analytic throughout a domain D except for an isolated singularity at a certain point z_0 of D. The integral

$$\oint f(z)\,dz$$

will not in general be zero on a simple closed path in D. However, the integral will have the same value on all curves C which enclose z_0 and no other singularity of f. This value, divided by $2\pi i$, is known as the *residue* of $f(z)$ at z_0 and is denoted by Res $[f(z), z_0]$. Thus

$$\text{Res}\,[f(z), z_0] = \frac{1}{2\pi i} \oint_C f(z)\,dz, \qquad (9\text{–}73)$$

where the integral is taken over any path C within which $f(z)$ is analytic except at z_0 (Fig. 9–16).

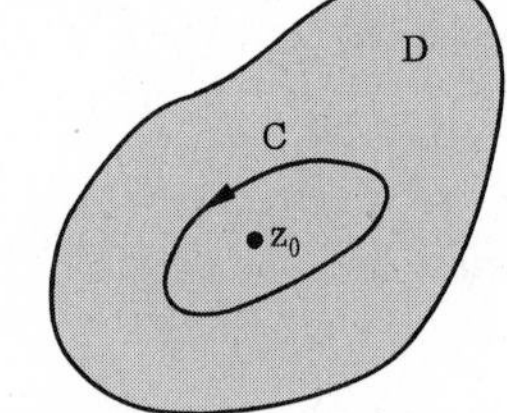

Fig. 9–16. Residue.

Theorem 27. *The residue of $f(z)$ at z_0 is given by the equation*

$$\text{Res}\,[f(z), z_0] = a_{-1}, \qquad (9\text{–}74)$$

where

$$f(z) = \cdots + \frac{a_{-N}}{(z - z_0)^N} + \cdots + \frac{a_{-1}}{z - z_0} + a_0 + a_1(z - z_0) + \cdots \qquad (9\text{–}75)$$

is the Laurent expansion of $f(z)$ at z_0.

Proof. By (9–62)

$$a_{-1} = \frac{1}{2\pi i} \oint_C f(z)\,dz,$$

where C is chosen as in the definition of residue. Hence (9–74) follows at once.

If C is a simple closed path in D, within which $f(z)$ is analytic except for isolated singularities at $z_1, \ldots, z_k$, then by Theorem 16

$$\oint_C f(z)\,dz = \oint_{C_1} f(z)\,dz + \cdots + \oint_{C_k} f(z)\,dz,$$

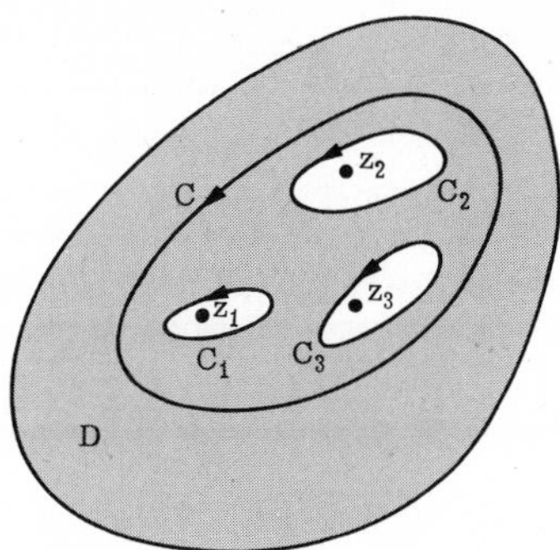

FIG. 9–17. Cauchy residue theorem.

where C_1 encloses only the singularity at z_1, C_2 encloses only $z_2, \ldots,$ as in Fig. 9–17. We thus obtain the following basic theorem:

THEOREM 28 (Cauchy's residue theorem). *If $f(z)$ is analytic in a domain D and C is a simple closed curve in D within which $f(z)$ is analytic except for isolated singularities at $z_1, \ldots, z_k$, then*

$$\oint_C f(z)\,dz = 2\pi i\,\{\text{Res}\,[f(z), z_1] + \cdots + \text{Res}\,[f(z), z_k]\}. \quad (9\text{–}76)$$

This theorem permits rapid evaluation of integrals on closed paths, whenever it is possible to compute the coefficient a_{-1} of the Laurent expansion at each singularity inside the path. Various techniques for obtaining the Laurent expansion are illustrated in the problems preceding this section. However, if we wish only the term in $(z - z_0)^{-1}$ of the expansion, various simplifications are possible. We give several rules here:

Rule I. At a simple pole z_0 (i.e., a pole of first order),

$$\text{Res}\,[f(z), z_0] = \lim_{z \to z_0} (z - z_0) f(z).$$

Rule II. At a pole z_0 of order N $(N = 2, 3, \ldots)$,

$$\text{Res}\,[f(z), z_0] = \lim_{z \to z_0} \frac{g^{(N-1)}(z)}{(N-1)!},$$

where $g(z) = (z - z_0)^N f(z)$.

Rule III. If $A(z)$ and $B(z)$ are analytic in a neighborhood of z_0, $A(z_0) \neq 0$, and $B(z)$ has a zero at z_0 of order 1, then

$$f(z) = \frac{A(z)}{B(z)}$$

has a pole of first order at z_0 *and*

$$\operatorname{Res}[f(z), z_0] = \frac{A(z_0)}{B'(z_0)}.$$

Rule IV. If $A(z)$ *and* $B(z)$ *are as in Rule III, but* $B(z)$ *has a zero of second order at* z_0, *so that* $f(z)$ *has a pole of second order at* z_0, *then*

$$\operatorname{Res}[f(z), z_0] = \frac{6A'B'' - 2AB'''}{3B''^2}, \tag{9–77}$$

where A *and the derivatives* A', B'', B''' *are evaluated at* z_0.

Proofs of rules. Let $f(z)$ have a pole of order N:

$$f(z) = \frac{1}{(z - z_0)^N}[a_{-N} + a_{-N+1}(z - z_0) + \cdots] = \frac{1}{(z - z_0)^N} g(z),$$

where

$$g(z) = (z - z_0)^N f(z), \qquad g(z_0) = a_{-N}$$

and g is analytic at z_0. The coefficient of $(z - z_0)^{-1}$ in the Laurent series for $f(z)$ is the coefficient of $(z - z_0)^{N-1}$ in the Taylor series for $g(z)$. This coefficient, which is the residue sought, is

$$\frac{g^{(N-1)}(z_0)}{(N - 1)!} = \lim_{z \to z_0} \frac{g^{(N-1)}(z)}{(N - 1)!}.$$

For $N = 1$ this gives Rule I; for $N = 2$ or higher, we obtain Rule II. Rules III and IV follow from the identity of Problem 6 following Section 9–14:

$$\begin{aligned} \frac{A(z)}{B(z)} &= \frac{a_0 + a_1(z - z_0) + \cdots}{b_N(z - z_0)^N + b_{N+1}(z - z_0)^{N+1} + \cdots} \\ &= \frac{a_0}{b_N} \frac{1}{(z - z_0)^N} + \frac{a_1 b_N - a_0 b_{N+1}}{b_N^2} \frac{1}{(z - z_0)^{N-1}} + \cdots \end{aligned}$$

For a first order pole, $N = 1$ and the residue is

$$\frac{a_0}{b_1} = \frac{A(z_0)}{B'(z_0)}.$$

For a second order pole, $N = 2$ and the residue is $[(a_1 b_2 - a_0 b_3)/b_2^2]$. Since

$$a_0 = A(z_0), \qquad a_1 = A'(z_0),$$

$$b_2 = \frac{B''(z_0)}{2!}, \qquad b_3 = \frac{B'''(z_0)}{3!},$$

this reduces to the expression (9–77).

EXAMPLE 1.

$$\oint_{|z|=2} \frac{ze^z}{z^2-1}\,dz = 2\pi i\,\{\mathrm{Res}\,[f(z), 1] + \mathrm{Res}\,[f(z), -1]\}.$$

Since $f(z)$ has first order poles at ± 1, we find by Rule I

$$\mathrm{Res}\,[f(z), 1] = \lim_{z\to 1}\,(z-1)\cdot\frac{ze^z}{z^2-1} = \lim_{z\to 1}\frac{ze^z}{z+1} = \frac{e}{2},$$

$$\mathrm{Res}\,[f(z), -1] = \lim_{z\to -1}\,(z+1)\cdot\frac{ze^z}{z^2-1} = \lim_{z\to -1}\frac{ze^z}{z-1} = \frac{-e^{-1}}{-2}.$$

Accordingly,

$$\oint_{|z|=2} \frac{ze^z}{z^2-1}\,dz = 2\pi i\left(\frac{e}{2} + \frac{e^{-1}}{2}\right) = 2\pi i\cosh 1.$$

Rule III could also have been used:

$$\mathrm{Res}\,[f(z), 1] = \left.\frac{ze^z}{2z}\right|_{z=1} = \frac{e}{2}, \qquad \mathrm{Res}\,[f(z), -1] = \left.\frac{ze^z}{2z}\right|_{z=-1} = \frac{-e^{-1}}{-2}.$$

This is simpler than Rule I, since the expression $A(z)/B'(z)$, once computed, serves for all poles of the prescribed type.

EXAMPLE 2.

$$\oint_{|z|=2} \frac{z}{z^4-1}\,dz = 2\pi i\,\{\mathrm{Res}\,[f(z), 1] + \mathrm{Res}\,[f(z), -1] + \mathrm{Res}\,[f(z), i] + \mathrm{Res}\,[f(z), -i]\}.$$

All poles are of first order. Rule III gives $A(z)/B'(z) = z/(4z^3) = 1/(4z^2)$ as the expression for the residue at any one of the four points. Moreover, $z^4 = 1$ at each pole, so that

$$\frac{1}{4z^2} = \frac{z^2}{4z^4} = \frac{z^2}{4}.$$

Hence

$$\oint_{|z|=2} \frac{z}{z^4-1}\,dz = \frac{2\pi i}{4}(1+1-1-1) = 0.$$

EXAMPLE 3.

$$\oint_{|z|=2} \frac{e^z}{z(z-1)^2}\,dz = 2\pi i\,\{\mathrm{Res}\,[f(z), 0] + \mathrm{Res}\,[f(z), 1]\}.$$

At the first order pole $z = 0$, application of Rule I gives the residue 1.

At the second order pole $z = 1$, Rule II gives

$$\operatorname{Res}[f(z), 1] = \frac{d}{dz}\left(\frac{e^z}{z}\right)\Bigg|_{z=1} = \frac{e^z(z-1)}{z^2}\Bigg|_{z=1} = 0.$$

Rule IV could also be used, with $A = e^z$, $B = z^3 - 2z^2 + z$:

$$\operatorname{Res}[f(z), 1] = \frac{6e^z(6z-4) - 2e^z \cdot 6}{3(6z-4)^2}\Bigg|_{z=1} = 0.$$

Accordingly,

$$\oint\limits_{|z|=2} \frac{e^z}{z(z-1)^2}\, dz = 2\pi i(1+0) = 2\pi i.$$

9–16 Residue at infinity. Let $f(z)$ be analytic for $|z| > R$. The *residue of* $f(z)$ *at* ∞ is defined as follows:

$$\operatorname{Res}[f(z), \infty] = \frac{1}{2\pi i}\oint\limits_C f(z)\, dz,$$

where the integral is taken in the *negative* direction on a simple closed path C, in the domain of analyticity of $f(z)$, and *outside* of which $f(z)$ has no singularity other than ∞. This is suggested in Fig. 9–18. Theorem 27 has an immediate extension to this case:

THEOREM 29. *The residue of $f(z)$ at ∞ is given by the equation*

$$\operatorname{Res}[f(z), \infty] = -a_{-1}, \tag{9–78}$$

where a_{-1} is the coefficient of z^{-1} in the Laurent expansion of $f(z)$ at ∞:

$$f(z) = \cdots + \frac{a_{-n}}{z^n} + \cdots + \frac{a_{-1}}{z} + a_0 + a_1 z + \cdots \tag{9–79}$$

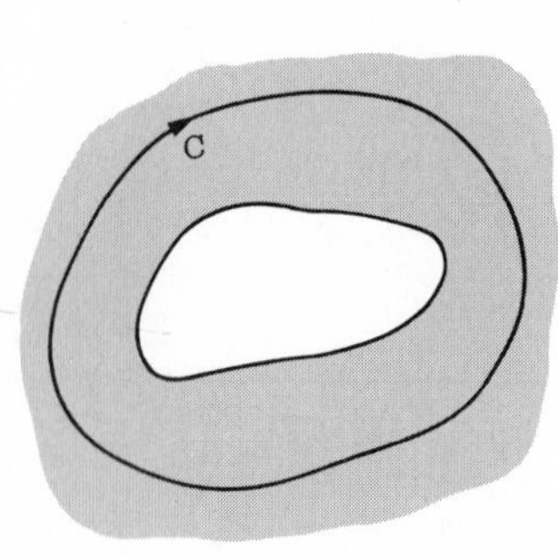

FIG. 9–18. Residue at infinity.

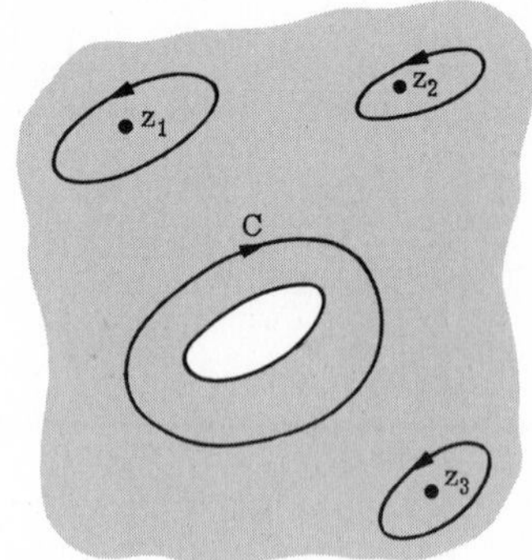

FIG. 9–19. Residue theorem for exterior.

The proof is the same as for Theorem 27. It should be stressed that presence of a nonzero residue at ∞ is not related to presence of a pole or essential singularity at ∞. That is, $f(z)$ can have a nonzero residue whether or not there is a pole or essential singularity, for the pole or essential singularity at ∞ is due to the *positive powers* of z, not to negative powers (Section 9–14 above). Thus the function $e^{1/z} = 1 + z^{-1} + (2!z^2)^{-1} + \cdots$ is analytic at ∞, but has the residue -1 there.

Cauchy's residue theorem has also an extension to include ∞:

THEOREM 30. *Let $f(z)$ be analytic in a domain D which includes a deleted neighborhood of ∞. Let C be a simple closed path in D outside of which $f(z)$ is analytic except for isolated singularities at $z_1, \ldots, z_k$. Then*

$$\oint_C f(z)\,dz = 2\pi i \,\{\text{Res}\,[f(z), z_1] + \cdots + \text{Res}\,[f(z), z_k] + \text{Res}\,[f(z), \infty]\}. \tag{9–80}$$

The proof, which is like that of Theorem 28, is left as an exercise (Problem 4, following Section 9–18). It is to be emphasized that the integral on C is taken in the *negative* direction (see Fig. 9–19) and that the *residue at ∞ must be included* on the right.

For a particular integral

$$\oint_C f(z)\,dz$$

on a simple closed path C, we have now two modes of evaluation: the integral equals $2\pi i$ times the sum of the residues inside the path (provided there are only a finite number of singularities there), and it also equals *minus* $2\pi i$ times the sum of the residues outside the path plus that at ∞ (provided there are only a finite number of singularities in the exterior domain). We can evaluate the integral both ways to check results. The principle involved here is summarized in the following theorem:

THEOREM 31. *If $f(z)$ is analytic in the extended z-plane except for a finite number of singularities, then the sum of all residues of $f(z)$ (including ∞) is zero.*

To evaluate the residues at ∞, we can formulate a set of rules like the ones above. However, the following two rules are adequate for most purposes:

Rule V. If $f(z)$ has a zero of first order at ∞, then

$$\text{Res}\,[f(z), \infty] = -\lim_{z\to\infty} zf(z).$$

If $f(z)$ has a zero of second or higher order at ∞, the residue at ∞ is zero.

Rule VI.

$$\operatorname{Res}\,[f(z),\,\infty] = -\operatorname{Res}\left[\frac{1}{z^2}f(1/z),\,0\right].$$

The proof of Rule V is left as an exercise (Problem 8, following Section 9–18). To prove Rule VI, we write

$$f(z) = \cdots + a_n z^n + \cdots + a_1 z + a_0 + \frac{a_{-1}}{z} + \frac{a_{-2}}{z^2} + \cdots \qquad (|z| > R).$$

Then for $0 < |z| < R^{-1}$,

$$f\left(\frac{1}{z}\right) = \cdots + \frac{a_n}{z^n} + \cdots + \frac{a_1}{z} + a_0 + a_{-1}z + a_{-2}z^2 + \cdots,$$

$$\frac{1}{z^2}f\left(\frac{1}{z}\right) = \cdots + \frac{a_0}{z^2} + \frac{a_{-1}}{z} + a_{-2} + \cdots$$

Hence

$$\operatorname{Res}\left[\frac{1}{z^2}f\left(\frac{1}{z}\right),\,0\right] = a_{-1},$$

and the rule follows. This result reduces the problem to evaluation of a residue at zero, to which Rules I through IV are applicable.

EXAMPLE 1. We consider the integral

$$\oint\limits_{|z|=2} \frac{z}{z^4 - 1}\,dz$$

of Example 2 in the preceding section. There is no singularity outside the path other than ∞, and at ∞ the function has a zero of order 3; hence the integral is zero.

EXAMPLE 2.

$$\oint\limits_{|z|=2} \frac{1}{(z+1)^4(z^2-9)(z-4)}\,dz.$$

Here there is a fourth order pole inside the path, at which evaluation of the residue is tedious. Outside the path there are first order poles at ± 3 and 4 and a zero of order 7 at ∞. Hence by Rule I the integral equals

$$-2\pi i\left(\frac{1}{4^4 6(-1)} + \frac{1}{(-2)^4(-6)(-7)} + \frac{1}{5^4 \cdot 7}\right).$$

9–17 Logarithmic residues; argument principle. Let $f(z)$ be analytic in a domain D. Then $f'(z)/f(z)$ is analytic in D except at the zeros of $f(z)$. If an analytic branch of $\log f(z)$ is chosen in part of D [necessarily excluding the zeros of $f(z)$], then

$$\frac{d}{dz}\log f(z) = \frac{f'(z)}{f(z)}. \tag{9–81}$$

For this reason the expression f'/f is termed the *logarithmic derivative* of $f(z)$. Its value is demonstrated by the following theorem:

THEOREM 32. *Let $f(z)$ be analytic in domain D. Let C be a simple closed path in D within which $f(z)$ is analytic except for a finite number of poles, and let $f(z) \neq 0$ on C. Then*

$$\frac{1}{2\pi i}\oint_C \frac{f'(z)}{f(z)}\,dz = N_0 - N_p,$$

where N_0 is the total number of zeros of f inside C and N_p is the total number of poles of f inside C, zeros and poles being counted according to multiplicities.

Proof. The logarithmic derivative f'/f has isolated singularities precisely at the zeros and poles of f. At a zero z_0,

$$f(z) = (z - z_0)^N g(z) \quad [g(z_0) \neq 0],$$

$$f'(z) = (z - z_0)^N g'(z) + N(z - z_0)^{N-1} g(z),$$

$$\frac{f'(z)}{f(z)} = \frac{(z - z_0)^N g'(z) + N(z - z_0)^{N-1} g(z)}{(z - z_0)^N g(z)} = \frac{g'(z)}{g(z)} + \frac{N}{z - z_0}.$$

Hence the logarithmic derivative has a pole of first order, with residue N equal to the multiplicity of the zero. A similar analysis applies to each pole of f, with N replaced by $-N$. The theorem then follows from the Cauchy residue theorem (Theorem 28), provided we show that there are only a finite number of singularities. The poles of f are finite in number by assumption, and it is easily shown that there can be only a finite number of zeros. (See Section 6–6 of the book by Kaplan listed at the end of this chapter.) Thus the theorem is proved.

Remarks. Since

$$\frac{f'(z)}{f(z)}\,dz = d\log f(z) = d\log w = d\,(\log|w| + i\arg w),$$

we have

$$\frac{1}{2\pi i}\int_C \frac{f'(z)}{f(z)}\,dz = \frac{1}{2\pi i}\int_C d\log|w| + \frac{1}{2\pi}\int_C d\arg w.$$

Since $w \neq 0$ on C, $\log|w|$ is continuous on C and the first integral on the right is zero. The second measures the total charge in $\arg w$, divided by

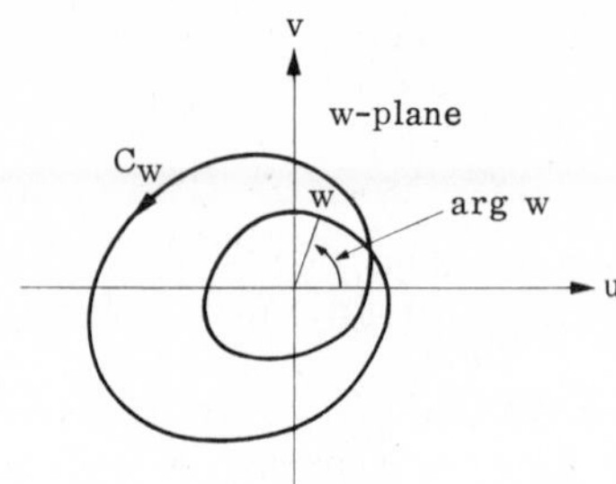

FIG. 9–20. Argument principle.

2π, as w traces the path C_w, the image of C, in the w-plane. Hence it also measures the number of times C_w winds about the origin of the w-plane; in Fig. 9–20 the number is $+2$. The statement

$$\frac{1}{2\pi}\,[\text{increase in arg } f(z) \text{ on path}] = N_0 - N_p \tag{9–82}$$

is known as the *argument principle.* This is of great value in finding zeros and poles of analytic functions.

9–18 Partial fraction expansion of rational functions. The theory of analytic functions provides a simple proof for the familiar rules for partial fraction expansions.

Let

$$f(z) = \frac{a_0 z^n + \cdots + a_n}{b_0 z^m + \cdots + a_m} \quad (a_0 \neq 0,\ b_0 \neq 0) \tag{9–83}$$

be given. We assume that $n < m$, so that f is a proper fraction. We also assume that the numerator and denominator have no common zeros. Let $z_1, z_2, \ldots, z_N$ be the *distinct* zeros of the *denominator* (no repetitions); these are the poles of $f(z)$. At z_1, $f(z)$ has a Laurent expansion:

$$f(z) = p_1(z) + g_1(z),$$
$$p_1(z) = \frac{A_{-k_1}}{(z - z_1)^{k_1}} + \frac{A_{-k_1+1}}{(z - z_1)^{k_1-1}} + \cdots + \frac{A_{-1}}{z - z_1}. \tag{9–84}$$

Here $p_1(z)$ is the principal part of $f(z)$ at z_1 and $g_1(z)$ is analytic at z_1; k_1 is the order of the pole at z_1. Similar expressions hold at the other poles.

The partial fraction expansion of $f(z)$ is now simply the identity

$$f(z) = p_1(z) + p_2(z) + \cdots + p_N(z). \tag{9–85}$$

To justify this, we let

$$F(z) = f(z) - [p_1(z) + p_2(z) + \cdots + p_N(z)].$$

Now $f(z) - p_1(z)$ has a removable singularity at z_1, while $p_2(z), \ldots, p_N(z)$ are analytic at z_1. Hence $F(z)$ has a removable singularity at z_1. In general, $F(z)$ has only removable singularities for finite z. At ∞, $f(z)$, $p_1(z), \ldots, p_N(z)$ all have zeros; hence $F(z)$ has a zero at ∞. But $F(z)$ is a rational function with no poles. Hence $F(z)$ must be a polynomial. Thus $F(z)$ has a pole at ∞ unless F is constant; since we know $F = 0$ at ∞, F must be a constant, namely zero. This proves (9–85).

If $f(z)$ has only simple poles, the principal part at each pole z_j is simply $A_j/(z - z_j)$, where A_j is the residue of f at z_j, and hence in this case

$$f(z) = \frac{A_1}{z - z_1} + \cdots + \frac{A_m}{z - z_m} \quad (A_j = \operatorname{Res}[f, z_j]) \tag{9–86}$$

is the partial fraction expansion. If we write

$$f(z) = \frac{A(z)}{B(z)},$$

then Rule III can be applied:

$$A_j = \frac{A(z_j)}{B'(z_j)}, \qquad f(z) = \sum_{j=1}^{m} \frac{A(z_j)}{B'(z_j)} \frac{1}{z - z_j}. \tag{9–87}$$

At a multiple pole z_j of order k, we can write

$$f(z) = \frac{1}{(z - z_j)^k} \phi(z)$$

The principal part $p_j(z)$ is then

$$p_j(z) = \frac{\phi(z_j)}{(z - z_j)^k} + \frac{\phi'(z_j)}{1!(z - z_j)^{k-1}} + \cdots + \frac{\phi^{(k-1)}(z_j)}{(k-1)!(z - z_j)}. \tag{9–88}$$

Hence $p_j(z)$ can be found without knowledge of the other poles.

EXAMPLE 1.

$$f(z) = \frac{z^2 + 1}{z^3 + 4z^2 + 3z} = \frac{z^2 + 1}{z(z + 1)(z + 3)}.$$

There are simple poles at 0, −1, −3. By (9–87) above,

$$f(z) = \frac{A(0)}{B'(0)} \frac{1}{z} + \frac{A(-1)}{B'(-1)} \frac{1}{z + 1} + \frac{A(-3)}{B'(-3)} \frac{1}{z + 3};$$

with $A = z^2 + 1$, $B = z^3 + 4z^2 + 3z$, $B' = 3z^2 + 8z + 3$, we find

$$f(z) = \frac{1}{3}\frac{1}{z} + \frac{2}{-2}\frac{1}{z+1} + \frac{10}{6}\frac{1}{z+3}.$$

EXAMPLE 2.

$$f(z) = \frac{z}{(z-1)^2(z^3+z+1)}.$$

At the pole $z = 1$, we write

$$f = \frac{1}{(z-1)^2}\,\phi(z) \quad \left(\phi = \frac{z}{z^3+z+1}\right).$$

Since $\phi(1) = \frac{1}{3}$, $\phi'(1) = -\frac{1}{9}$, the principal part at 1 is

$$\frac{1}{3}\frac{1}{(z-1)^2} - \frac{1}{9}\frac{1}{z-1}.$$

The cubic $z^3 + z + 1$ has one real root z_1 and two complex roots z_2, z_3. These are all simple. Hence we can write

$$f(z) = \frac{A(z_1)}{B'(z_1)}\frac{1}{z-z_1} + \frac{A(z_2)}{B'(z_2)}\frac{1}{z-z_2} + \frac{A(z_3)}{B'(z_3)}\frac{1}{z-z_3} + \frac{1}{3}\frac{1}{(z-1)^2} - \frac{1}{9}\frac{1}{z-1},$$

where $A(z) = z$, $B(z) = (z-1)^2(z^3+z+1)$. At the poles z_1, z_2, z_3, $B'(z)$ reduces to $z^2 + z + 7$, by the relation $z^3 + z + 1 = 0$.

Problems

1. Evaluate the following integrals on the paths given:

(a) $\oint \frac{z\,dz}{(z-1)(z-3)}$ $(|z| = 2)$ (b) $\oint \frac{e^{3z}\,dz}{z^2+4}$ $(|z| = 3)$

(c) $\oint \frac{\sin z}{z^4}\,dz$ $(|z| = 1)$ (d) $\oint \frac{z\,dz}{z^3+z+1}$ $(|z| = 4)$

(e) $\oint \frac{dz}{(z+1)^4(z+3)}$ $(|z| = 2)$ (f) $\oint \frac{dz}{(z+1)^5(z+3)}$ $(|z| = 4)$

(g) $\oint \frac{2z+2}{z^2+2z+2}\,dz$ $(|z| = 2)$ (h) $\oint \frac{3z^2-6z+1}{z^3-3z^2+z-3}\,dz$ $(|z| = 2)$

2. Expand each of the following in partial fractions:

(a) $\dfrac{1}{z^2-4}$ (b) $\dfrac{z+1}{(z-1)(z-2)(z-3)}$ (c) $\dfrac{z^2}{z^5+1}$ (d) $\dfrac{1}{z^n-1}$

(e) $\dfrac{z}{(z-1)^2(z+1)^2}$ (f) $\dfrac{\phi(z)}{z^n(z+1)}$ $(\phi = c_0 + c_1 z + \cdots + c_n z^n)$

(g) $\dfrac{1}{(z-1)^3(z^4+z+1)}$ (h) $\dfrac{1}{(z^2+1)(z^2+2z+2)}$

3. Prove the Fundamental Theorem of Algebra: *Every polynomial of degree at least* 1 *has a zero.* [*Hint:* Show that Res $[f'(z)/f(z), \infty]$ is not 0, but is in fact minus the degree n of the polynomial $f(z)$. Then use Theorem 32 to show that f has n zeros.]

4. Prove Theorem 30.

5. Formulate and prove Theorem 32 for integration around the boundary B of a region R in D, bounded by simple closed curves $C_1, \ldots, C_k$.

6. Prove that under the hypotheses of Theorem 32, if $g(z)$ is analytic in D and within C, then

$$\frac{1}{2\pi i}\oint_C \frac{g(z)f'(z)}{f(z)}\,dz = \sum_{k=1}^{n} g(z_k') - \sum_{l=1}^{m} g(z_l''),$$

where $z_1', \ldots, z_n'$ are the zeros of f, and $z_1'', \ldots, z_m''$ are the poles of f inside C, repeated according to multiplicity.

7. Extend Rule IV of Section 9–15 to the case in which $A(z)$ has a first order zero at z_0 and $B(z)$ has a second order zero.

8. Prove Rule V of Section 9–16.

Answers

1. (a) $-\pi i$ (b) $\pi i \sin 6$ (c) $-\pi i/3$ (d) 0 (e) $-\pi i/8$ (f) 0 (g) $4\pi i$ (h) $4\pi i$

2. (a) $\dfrac{1}{4}\dfrac{1}{z-2} - \dfrac{1}{4}\dfrac{1}{z+2}$ (b) $\dfrac{1}{z-1} - \dfrac{3}{z-2} + \dfrac{2}{z-3}$

(c) $-\dfrac{1}{5}\left[\dfrac{z_2}{z-z_1} + \dfrac{z_5}{z-z_2} + \dfrac{z_3}{z-z_3} + \dfrac{z_1}{z-z_4} + \dfrac{z_4}{z-z_5}\right]$,

$z_k = \exp\,[(2k-1)\pi i/5],\ k = 1, \ldots, 5$

(d) $\dfrac{1}{n}\left[\dfrac{z_1}{z-z_1} + \cdots + \dfrac{z_n}{z-z_n}\right],\ z_k = \exp\left(\dfrac{2k\pi i}{n}\right),\ k = 1, \ldots, n$

(e) $\dfrac{1}{4}\dfrac{1}{(z-1)^2} - \dfrac{1}{4(z+1)^2}$

(f) $\dfrac{c_0}{z^n} + \dfrac{c_1 - c_0}{z^{n-1}} + \cdots + \dfrac{c_{n-1} - c_{n-2} + \cdots + (-1)^{n-1}c_0}{z}$

$$+ (-1)^n \phi(-1)\frac{1}{z+1}$$

(g) $\frac{1}{3}\frac{1}{(z-1)^3} - \frac{5}{9}\frac{1}{(z-1)^2} + \frac{7}{27}\frac{1}{z-1} + \sum_{j=1}^{4}\frac{g(z_j)}{z-z_j}$,

where $z_1, \ldots, z_4$ are the roots of $z^4 + z + 1 = 0$ and

$$g = (-7z^3 + 5z^2 + 3z - 13)^{-1}$$

(h) $\frac{1}{10}\left[\frac{-2-i}{z-i} + \frac{-2+i}{z+i} + \frac{2-i}{z+1-i} + \frac{2+i}{z+1+i}\right]$

7. $2A'/B''$

9–19 Application of residues to evaluation of real integrals. A variety of real definite integrals between special limits can be evaluated with the aid of residues.

For example, an integral

$$\int_0^{2\pi} R(\sin\theta, \cos\theta)\, d\theta,$$

where R is a rational function of $\sin\theta$ and $\cos\theta$, is converted to a complex line integral by the substitution:

$$z = e^{i\theta}, \quad dz = ie^{i\theta}\, d\theta = iz\, d\theta,$$

$$\cos\theta = \frac{e^{i\theta} + e^{-i\theta}}{2} = \frac{1}{2}\left(z + \frac{1}{z}\right),$$

$$\sin z = \frac{e^{i\theta} - e^{-i\theta}}{2i} = \frac{1}{2i}\left(z - \frac{1}{z}\right);$$

the path of integration is the circle: $|z| = 1$.

EXAMPLE 1. $\int_0^{2\pi} \frac{1}{\cos\theta + 2}\, d\theta$. The substitution reduces this to

$$\oint_{|z|=1} \frac{-2i}{z^2 + 4z + 1}\, dz = 4\pi \operatorname{Res}[z^2 + 4z + 1, -2 + \sqrt{3}],$$

since $-2 + \sqrt{3}$ is the only root of the denominator inside the circle. Accordingly,

$$\int_0^{2\pi} \frac{1}{\cos\theta + 2}\, d\theta = \frac{2\pi}{\sqrt{3}}.$$

The substitution can be summarized in the one rule:

$$\int_0^{2\pi} R(\sin\theta, \cos\theta)\, d\theta = \oint_{|z|=1} R\left(\frac{z^2 - 1}{2iz}, \frac{z^2 + 1}{2z}\right)\frac{dz}{iz}. \qquad (9\text{–}89)$$

The complex integral can be evaluated by residues, provided R has no poles on the circle $|z| = 1$.

A second example is provided by integrals of the type

$$\int_{-\infty}^{\infty} f(x)\, dx.$$

We illustrate the procedure with an example and formulate a general principle below.

EXAMPLE 2. $\displaystyle\int_{-\infty}^{\infty} \frac{dx}{x^4+1}\cdot$ This integral can be regarded as a line integral of $f(z) = 1/(z^4+1)$ along the real axis. The path is not closed (unless one adjoins ∞), but we show that it acts like a closed path "enclosing" the upper half-plane, so that the integral along the path equals the sum of the residues in the upper half-plane.

To establish this, we consider the integral of $f(z)$ on the semicircular path C_R shown in Fig. 9–21. When R is sufficiently large, the path encloses the two poles: $z_1 = \exp(\frac{1}{4}i\pi)$, $z_2 = \exp(\frac{3}{4}\, i\pi)$ of $f(z)$. Hence

$$\oint_{CR} f(z)\, dz = 2\pi i\{\operatorname{Res}[f(z), z_1] + \operatorname{Res}[f(z), z_2]\}.$$

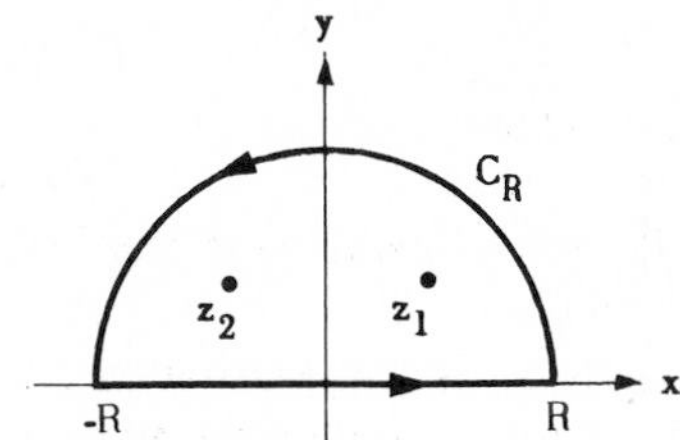

FIG. 9–21. Evaluation of $\int_{-\infty}^{\infty} f(x)\, dx$ by residues.

As R increases, the integral on C_R cannot change, since it always equals the sum of the residues times 2π. Hence

$$\oint_{CR} f(z)\, dz = \lim_{R\to\infty} \oint_{CR} f(z)\, dz = \lim_{R\to\infty} \int_{-R}^{R} \frac{dx}{x^4+1} + \lim_{R\to\infty} \int_{DR} \frac{1}{z^4+1}\, dz,$$

where D_R is the semicircle: $z = Re^{i\theta}$, $0 \leqq \theta \leqq \pi$. The limit of the first term is the integral desired (since the limits at $+\infty$ and $-\infty$ exist separately, as required; cf. Section 4–4). The limit of the second term is 0, since $|z|^4 = |z^4| = |z^4 + 1 - 1| \leqq |z^4+1| + 1$, so that on D_R

$$\left|\frac{1}{z^4+1}\right| \leqq \frac{1}{|z|^4 - 1} = \frac{1}{R^4-1}$$

and

$$\left|\int_{DR} \frac{1}{z^4+1}\, dz\right| \leqq \frac{\pi R}{R^4-1}\cdot$$

Accordingly,

$$\int_{CR} f(z)\, dz = \int_{-\infty}^{\infty} \frac{dx}{x^4+1} = 2\pi i\{\operatorname{Res}[f(z), z_1] + \operatorname{Res}[f(z), z_2]\}.$$

By Rule III above, the sum of the residues is

$$\frac{1}{4z_1^3} + \frac{1}{4z_2^3} = -\frac{1}{4}(z_1 + z_2) = -\frac{\sqrt{2}}{4}\, i$$

and hence

$$\int_{-\infty}^{\infty} \frac{dx}{x^4+1} = \frac{\pi\sqrt{2}}{2}.$$

We now formulate the general principle:

THEOREM 33. *Let $f(z)$ be analytic in a domain D which includes the real axis and all of the half-plane $y > 0$ except for a finite number of points. If*

$$\lim_{R\to\infty} \int_0^{\pi} f(Re^{i\theta})Re^{i\theta}\, d\theta = 0 \tag{9–90}$$

and

$$\int_{-\infty}^{\infty} f(x)\, dx \tag{9–91}$$

exists, then

$$\int_{-\infty}^{\infty} f(x)\, dx = 2\pi i\{\textit{sum of residues of } f(z) \textit{ in the upper half-plane}\}. \tag{9–92}$$

The proof is a repetition of the reasoning used in the above example. It is of interest to note that, even though the integral (9–91) fails to exist as improper integral, condition (9–90) implies that the principal value

$$(P)\int_{-\infty}^{\infty} f(x)\, dx = \lim_{R\to\infty} \int_{-R}^{R} f(x)\, dx \tag{9–93}$$

exists. (See Section 6–24.)

In order to apply Theorem 33, it is necessary to have simple criteria to guarantee that (9–90) holds. We list two such criteria here:

I. *If $f(z)$ is rational and has a zero of order greater than* 1 *at* ∞, *then* (9–90) *holds.*

For, when $|z|$ is sufficiently large,

$$f(z) = \frac{a_{-N}}{z^N} + \frac{a_{-N+1}}{z^{N-1}} + \cdots, \quad N > 1,$$

so that $zf(z)$ has a zero at infinity. Now, by (9–11),

$$\left|\int_0^{\pi} f(Re^{i\theta})Re^{i\theta}\, d\theta\right| \leq \int_0^{\pi} |zf(z)|\, d\theta,$$

so that the integral must converge to 0 as $R \to \infty$.

II. *If $g(z)$ is rational and has a zero of order* 1 *or greater at* ∞, *then* (9–90) *holds for $f(z) = e^{miz}g(z)$, $m > 0$.*

For a proof, and further criteria, we refer to page 115 of the treatise of Whittaker and Watson listed at the end of the chapter. The rule II makes possible the evaluation of the *Fourier integral* (Section 7–18),

$$\int_{-\infty}^{\infty} g(x)e^{mix}\, dx = \int_{-\infty}^{\infty} g(x)\cos mx\, dx + i\int_{-\infty}^{\infty} g(x)\sin mx\, dx,$$

provided both real integrals exist.

EXAMPLE 3. The integrals

$$\int_{-\infty}^{\infty} \frac{x \cos x}{x^2 + 1}\, dx \quad \text{and} \quad \int_{-\infty}^{\infty} \frac{x \sin x}{x^2 + 1}\, dx$$

both exist by the Corollary to Theorem 51 of Section 6–22. Hence

$$\int_{-\infty}^{\infty} \frac{xe^{ix}}{x^2 + 1}\, dx = 2\pi i \operatorname{Res}\left[\frac{ze^{iz}}{z^2 + 1}, i\right] = \frac{\pi i}{e}.$$

Taking real and imaginary parts, we find

$$\int_{-\infty}^{\infty} \frac{x \cos x}{x^2 + 1}\, dx = 0, \quad \int_{-\infty}^{\infty} \frac{x \sin x}{x^2 + 1}\, dx = \frac{\pi}{e}.$$

Since the first integral is an integral of an *odd* function, the value of 0 could be predicted.

Problems

1. Evaluate the following integrals:

(a) $\displaystyle\int_0^{2\pi} \frac{1}{5 + 3 \sin \theta}\, d\theta$ (c) $\displaystyle\int_0^{2\pi} \frac{1}{(\cos \theta + 2)^2}\, d\theta$

(b) $\displaystyle\int_0^{2\pi} \frac{1}{5 - 4 \cos \theta}\, d\theta$ (d) $\displaystyle\int_0^{2\pi} \frac{1}{(3 + \cos^2 \theta)^2}\, d\theta$

2. Evaluate the following integrals:

(a) $\displaystyle\int_{-\infty}^{\infty} \frac{1}{x^2 + x + 1}\, dx$ (c) $\displaystyle\int_{-\infty}^{\infty} \frac{1}{x^6 + 1}\, dx$

(b) $\displaystyle\int_{-\infty}^{\infty} \frac{1}{(x^2 + 1)(x^2 + 4)}\, dx$ (d) $\displaystyle\int_0^{\infty} \frac{1}{(x^2 + 1)^2}\, dx$

3. Evaluate the following integrals:

(a) $\displaystyle\int_{-\infty}^{\infty} \frac{\cos x}{x^2 + 4}\, dx$ (c) $\displaystyle\int_0^{\infty} \frac{x^3 \sin x}{x^4 + 1}\, dx$

(b) $\displaystyle\int_{-\infty}^{\infty} \frac{\sin 2x}{x^2 + x + 1}\, dx$ (d) $\displaystyle\int_0^{\infty} \frac{x^2 \cos 3x}{(x^2 + 1)^2}\, dx$

4. Prove that $\displaystyle\int_0^{\infty} \frac{\sin x}{x}\, dx = \tfrac{1}{2}\pi.$

[Hint: let C be a path formed of the semicircular paths D_r: $|z| = r$ and D_R: $|z| = R$, where $0 \leqq \theta \leqq \pi$ and $0 < r < R$, plus the intervals $-R \leqq x \leqq -r$, $r \leqq x \leqq R$ on the real axis. Show that

$$\lim_{r \to 0} \int_{-r \atop D_r}^{r} \frac{e^{iz} - 1}{z}\, dz = 0.$$

Use this result and II above to conclude that

$$\lim_{r \to 0} \int_{-r \atop D_r}^{r} \frac{e^{iz}}{z}\, dz = -\pi i, \quad \lim_{R \to \infty} \int_{R \atop D_R}^{-R} \frac{e^{iz}}{z}\, dz = 0.$$

Accordingly,

$$\lim_{R\to\infty}\left\{\lim_{r\to 0}\oint_C \frac{e^{iz}}{z}\,dz\right\} = \lim_{R\to\infty}\left\{\lim_{r\to 0}\left[\int_{-R}^{-r}\frac{e^{ix}}{x}\,dx + \int_{r}^{R}\frac{e^{ix}}{x}\,dx\right]\right\} - \pi i.$$

The left-hand side is zero by the Cauchy Integral Theorem. Show that the imaginary part of the right-hand side has the value $2\int_0^\infty \frac{\sin x}{x}\,dx - \pi$.]

Answers

1. (a) $\frac{\pi}{2}$, (b) $\frac{2}{3}\pi$, (c) $\frac{4\pi}{3\sqrt{3}}$, (d) $\frac{7\pi\sqrt{3}}{72}$.

2. (a) $\frac{2\pi\sqrt{3}}{3}$, (b) $\frac{\pi}{6}$, (c) $\frac{2\pi}{3}$, (d) $\frac{\pi}{4}$.

3. (a) $\frac{1}{2}\pi e^{-2}$, (b) $-2\frac{\sqrt{3}}{3}\pi e^{-\sqrt{3}}\sin 1$, (c) $\frac{1}{2}\pi e^{-\frac{1}{2}\sqrt{2}}\cos(\frac{1}{2}\sqrt{2})$, (d) $-\frac{1}{2}\pi e^{-3}$.

Suggested References

AHLFORS, LARS V., *Complex Analysis.* New York: McGraw-Hill, 1953.

BETZ, ALBERT, *Konforme Abbildung.* Berlin: Springer, 1948.

BIEBERBACH, LUDWIG, *Lehrbuch der Funktionentheorie* (2 vols.), 4th ed. Leipzig: B. G. Teubner, 1934.

CHURCHILL, RUEL V., *Complex Variables and Applications*, 2nd ed. New York: McGraw-Hill, 1960.

COURANT, R., *Dirichlet's Principle, Conformal Mapping and Minimal Surfaces.* New York: Interscience, 1950.

FRANK, P., and v. MISES, R., *Die Differentialgleichungen und Integralgleichungen der Mechanik und Physik.* Vol. 1, 2nd ed., Braunschweig: Vieweg, 1930. Vol. 2, 2nd ed., Braunschweig: Vieweg, 1935.

GOURSAT, ÉDOUARD, *A Course in Mathematical Analysis*, Vol. II, Part 1 (transl. by E. R. Hedrick and O. Dunkel). New York: Ginn and Co., 1916.

HURWITZ, A. and COURANT, R., *Funktionentheorie*, 3rd ed. Berlin: Springer, 1929.

KAPLAN, WILFRED, *Introduction to Analytic Functions.* Reading, Mass.: Addison-Wesley, 1966.

KNOPP, KONRAD, *Theory of Functions* (2 vols.), transl. by F. Bagemihl. New York: Dover, 1945.

OSGOOD, W. F., *Lehrbuch der Funktionentheorie* (2 vols.), 3rd ed. Leipzig: B. G. Teubner, 1920.

PICARD, EMILE, *Traité d'Analyse*, 3rd ed., Vol. II. Paris: Gauthier-Villars, 1922.

TITCHMARSH, E. C., *The Theory of Functions*, 2nd ed. Oxford: Oxford University Press, 1939.

WHITTAKER, E. T., and WATSON, G. N., *A Course of Modern Analysis*, 4th ed. Cambridge: Cambridge University Press, 1940.

CHAPTER 10

Partial Differential Equations

10–1 Introduction. A partial differential equation is an equation expressing a relationship between an unknown function of several variables and its derivatives with respect to these variables. For example,

$$\frac{\partial u}{\partial t} - \frac{\partial^2 u}{\partial x^2} = 0, \qquad (10\text{–}1)$$

$$\frac{\partial^2 u}{\partial x^2} + \frac{\partial^2 u}{\partial y^2} = 0 \qquad (10\text{–}2)$$

are partial differential equations.

By a *solution* of a partial differential equation is meant a particular function which satisfies the equation identically in a domain of the independent variables. For example,

$$u = e^{-t} \sin x, \quad u = x^2 - y^2$$

are solutions of (10–1) and (10–2) respectively.

Although partial differential equations have not been studied as such in the earlier chapters, they have occurred in a number of important connections. For example, the solutions of (10–2) are precisely the *harmonic* functions of x and y; such functions are studied in Chapter 2. See also Section 9–11. The Cauchy-Riemann equations: $u_x = v_y$, $u_y = -v_x$ form a system of partial differential equations; they were studied in Chapter 9. Other systems encountered earlier in this book are the following:

$$F_x = P(x, y), \quad F_y = Q(x, y);$$
$$F_x = X(x, y, z), \quad F_y = Y(x, y, z), \quad F_z = Z(x, y, z);$$
$$Z_y - Y_z = L(x, y, z), \quad X_z - Z_x = M(x, y, z), \quad Y_x - X_y = N(x, y, z).$$

The first and second sets arose in consideration of line integrals in the plane and in space (Sections 5–6 and 5–13); the third set occurred in the study of solenoidal vector fields (Section 5–13). The results of Section 2–22 also concern partial differential equations, formed from Jacobians.

In this chapter we shall not attempt to consider general methods of determining the solutions of partial differential equations, but shall rather confine ourselves mainly to one class of *linear* partial differential equations: namely, equations of the form

$$\rho \frac{\partial^2 u}{\partial t^2} + H \frac{\partial u}{\partial t} - K^2 \nabla^2 u = F(t, x, \ldots), \qquad (10\text{–}3)$$

where u is a function of t and one, two, or three space variables $x, \ldots$, ρ, H, and K^2 depend on the space variables, and F depends on t and the space variables. This equation is a natural generalization to continuous media of the equation

$$m\frac{d^2x}{dt^2} + h\frac{dx}{dt} + k^2x = F(t) \tag{10-4}$$

for forced vibrations of a spring. It will be seen that the parallel between (10–3) and (10–4) is far-reaching.

In order to make clear the relationship between (10–3) and (10–4) we shall first study the case of the motion of *two* particles attached by springs. The results obtained will then be generalized to the case of N particles. It will be seen that the equations governing the motion form a system of form

$$m_\sigma\frac{d^2u}{dt^2} + h_\sigma\frac{du_\sigma}{dt} + [\cdot\cdot\cdot] = F_\sigma(t, u_1, \ldots, u_n), \quad (\sigma = 1, \ldots, n); \tag{10-5}$$

the quantities $u_1, \ldots, u_n$ are coordinates measuring the displacements of the various particles from their *equilibrium* positions. The expression in brackets: [. . .] depends on the u_σ and, in particular, on their *differences:* $u_2 - u_1, u_3 - u_2, \ldots$. We shall show that such general systems are capable of "harmonic motion," damped vibrations, exponential decay forced motion, just as is the single mass governed by (10–4). If we let N become infinite, then n becomes infinite in (10–5) and we obtain as "limiting case" precisely the partial differential equation (10–3). The *differences* in the expression [. . .] turn into the partial *derivatives* out of which $\nabla^2 u$ is built. Finally it will be seen that the partial differential equation obtained as limit continues to exhibit all the properties observed for the systems of $1, 2, \ldots, N$ particles: harmonic motion, exponential decay, etc.

The one basic difference between the various cases is that, whereas for one particle there is only *one* frequency of oscillation, for two particles moving on a line there are *two* frequencies, for N particles moving on a line there are N frequencies, for infinitely many particles there are *infinitely* many frequencies.

10–2 Review of equation for forced vibrations of a spring. We recall briefly some basic facts (cf. Sections 8–7, 8–13) concerning the equation (10–4) above. Throughout we assume $m \geqq 0$, $h \geqq 0$, $k > 0$.

(a) *Simple harmonic motion.* Here $h = 0$ and $F(t) \equiv 0$, $m > 0$. The equation becomes

$$m\frac{d^2x}{dt^2} + k^2x = 0. \tag{10-6}$$

The solutions are sinusoidal oscillations:

$$x = A\sin(\lambda t + \epsilon), \quad \lambda = \frac{k}{\sqrt{m}}. \tag{10-7}$$

(b) *Damped vibrations.* Here $m > 0$, $F(t) \equiv 0$, and $h > 0$ but h is small: $h^2 < 4mk^2$. The equation becomes

$$m\frac{d^2x}{dt^2} + h\frac{dx}{dt} + k^2x = 0. \tag{10–8}$$

The solutions are oscillations with decreasing amplitude:

$$x = Ae^{-at}\sin(\beta t + \epsilon), \tag{10–9}$$

where $a = h/2m$ and $\beta = (4mk^2 - h^2)^{\frac{1}{2}}/2m$.

(c) *Exponential decay.* Here $m = 0$, $h > 0$, $F(t) \equiv 0$. The equation reads

$$h\frac{dx}{dt} + k^2x = 0. \tag{10–10}$$

The solutions are decaying exponential functions:

$$x = ce^{-at}, \quad a = \frac{k^2}{h}. \tag{10–11}$$

A similar result is obtained if we consider an equation (10–8) in which h is large compared to m: $h^2 > 4mk^2$; in fact, (10–10) can be considered as the limiting case: $m \to 0$ of (10–8).

(d) *Equilibrium.* We assume $F(t) = F_0$, a constant. The equation becomes

$$m\frac{d^2x}{dt^2} + h\frac{dx}{dt} + k^2x = F_0. \tag{10–12}$$

The equilibrium value of x is that for which x remains constant; hence $dx/dt = 0$, $d^2x/dt^2 = 0$. Accordingly, at equilibrium

$$x = x^* = \frac{F_0}{k^2}. \tag{10–13}$$

(e) *Approach to equilibrium.* To study the approach to equilibrium, we let

$$u = x - x^*, \tag{10–14}$$

so that u measures the difference between x and the equilibrium value. We find that

$$m\frac{d^2u}{dt^2} + h\frac{du}{dt} + k^2u = 0, \tag{10–15}$$

so that u has the form (10–7), (10–9), or (10–11). For example, when $m = 0$ and $h > 0$ [case (c)],

$$u = ce^{-at}, \quad a = \frac{k^2}{h}, \tag{10–16}$$

$$x = u + x^* = ce^{-at} + x^*; \tag{10–17}$$

x approaches its equilibrium value exponentially.

(f) *Forced motion.* We allow F to be a general function of t. If, for example, $m = 0$, $h > 0$, then

$$h\frac{dx}{dt} + k^2x = F(t). \tag{10–18}$$

Each solution consists of a transient plus a particular solution:

$$x = ce^{-at} + x^*(t), \tag{10–19}$$

$$x^*(t) = \frac{1}{h}e^{-at}\int e^{at}F(t)\,dt. \tag{10–20}$$

The solutions attempt to follow the "input" $F(t)/k^2$, are hindered by the friction.

Remark. The equations for which $m = 0$ can be realized by other physical models: for example, the cooling of a hot mass, an electric circuit containing resistance and capacitance.

10–3 Case of two particles. We consider the model illustrated in Fig. 10–1. Two particles of masses m_1, m_2 are attached to each other and to "walls" by springs. The particles move on an x axis and have coordinates x_1, x_2; the walls are at x_0, x_3. For simplicity we assume that all springs have the same natural length l and the same spring constant k^2. We assume the particles to be subject to resistances $-h_1(dx_1/dt)$, $-h_2(dx_2/dt)$ and to outside forces $F_1(t)$, $F_2(t)$. The differential equations have the form

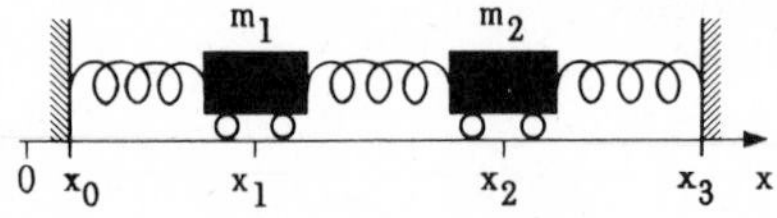

FIG. 10–1. Linear system of two masses.

$$m_1\frac{d^2x_1}{dt^2} = -k^2(x_1 - x_0 - l) + k^2(x_2 - x_1 - l) - h_1\frac{dx_1}{dt} + F_1(t),$$

$$m_2\frac{d^2x_2}{dt^2} = -k^2(x_2 - x_1 - l) + k^2(x_3 - x_2 - l) - h_2\frac{dx_2}{dt} + F_2(t).$$

They simplify to the following:

$$\begin{aligned} m_1\frac{d^2x_1}{dt^2} + h_1\frac{dx_1}{dt} - k^2(x_2 - 2x_1 + x_0) &= F_1(t), \\ m_2\frac{d^2x_2}{dt^2} + h_2\frac{dx_2}{dt} - k^2(x_3 - 2x_2 + x_1) &= F_2(t). \end{aligned} \tag{10–21}$$

We have not explicitly stated that the walls are fixed at x_0, x_3 and the differential equations remain correct even if the walls are moved in a manner controlled from outside the system. However, if x_0 and x_3 are constants x_0^*, x_3^* and $F_1(t)$, $F_2(t)$ are 0, the system has precisely one equilibrium state, namely, the solution of the equations

$$x_2 - 2x_1 + x_0^* = 0, \quad x_3^* - 2x_2 + x_1 = 0. \tag{10–22}$$

The solution is at once found to be

$$x_1 = x_1^* = x_0^* + \tfrac{1}{3}(x_3^* - x_0^*), \quad x_2 = x_2^* = x_0^* + \tfrac{2}{3}(x_3^* - x_0^*); \quad (10\text{-}23)$$

at equilibrium, the particles are equally spaced between the walls.

We now refer each particle to its equilibrium position by introducing new variables:

$$u_0 = x_0 - x_0^*, \quad u_1 = x_1 - x_1^*, \quad u_2 = x_2 - x_2^*, \quad u_3 = x_3 - x_3^*, \quad (10\text{-}24)$$

as suggested in Fig. 10-2. The differential equations (10-21) then have the appearance

$$\begin{aligned} m_1 \frac{d^2u_1}{dt^2} + h_1 \frac{du_1}{dt} - k^2(u_2 - 2u_1 + u_0) &= F_1(t), \\ m_2 \frac{d^2u_2}{dt^2} + h_2 \frac{du_2}{dt} - k^2(u_3 - 2u_2 + u_1) &= F_2(t). \end{aligned} \quad (10\text{-}25)$$

When the walls are fixed: $x_0 = x_0^*$, $x_3 = x_3^*$, u_0 and u_3 are 0; however, (10-25) allows for a general motion of the walls and, in particular, for constant nonzero values of u_0 and u_2, which would signify a shift of the equilibrium position of the walls.

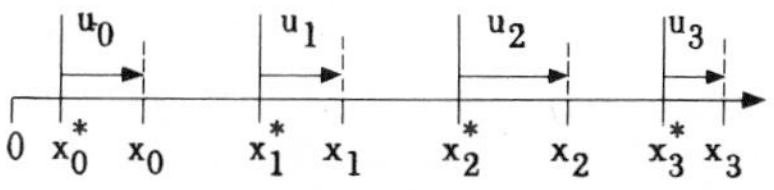

FIG. 10-2.

We now consider several special cases of (10-25), paralleling the cases (a), (c), (d), (e), and (f) of the preceding section. The discussion of the analogue of (b) is left to Prob. 10 below.

(a) *Harmonic motion.* We assume $m_1 > 0$, $m_2 > 0$, $h_1 = h_2 = 0$, $u_0 = u_3 = 0$ and $F_1(t) = F_2(t) = 0$, so that the differential equations (10-25) become

$$m_1 \frac{d^2u_1}{dt^2} - k^2(u_2 - 2u_1) = 0, \quad m_2 \frac{d^2u_2}{dt^2} - k^2(-2u_2 + u_1) = 0. \quad (10\text{-}26)$$

These are equivalent to four first-order equations:

$$\begin{aligned} \frac{du_1}{dt} &= w_1, \quad \frac{dw_1}{dt} = \frac{k^2}{m_1}(u_2 - 2u_1), \\ \frac{du_2}{dt} &= w_2, \quad \frac{dw_2}{dt} = \frac{k^2}{m_2}(-2u_2 + u_1), \end{aligned} \quad (10\text{-}27)$$

and can therefore be solved by the method of Section 8-12. However, it is simpler to work directly with the equations (10-26) as in Section 8-14. We verify that these equations can be written as

$$m_i \frac{d^2u_i}{dt^2} + \frac{\partial V}{\partial u_i} = 0, \quad i = 1, 2, \quad (10\text{-}28)$$

where $V = \frac{1}{2}\Sigma a_{ij}u_iu_j$ and $A = (a_{ij})$ is symmetric, with positive eigenvalues, so that V is positive definite (Prob. 4 below). To find the solutions, we seek the normal modes, as in the Example in Section 8–14. We set

$$u_i = A_i \sin(\lambda t + \epsilon_i), \quad i = 1, 2$$

in (10–26). We obtain the equations

$$(2k^2 - m_1\lambda^2)A_1 - k^2A_2 = 0, \quad -k^2A_1 + (2k^2 - m_2\lambda^2)A_2 = 0 \qquad (10\text{–}29)$$

and a characteristic equation

$$\begin{vmatrix} 2k^2 - m_1\lambda^2 & -k^2 \\ -k^2 & 2k^2 - m_2\lambda^2 \end{vmatrix} = 0. \qquad (10\text{–}30)$$

The roots are the four distinct real numbers $\pm\alpha$, $\pm\beta$, where

$$\begin{aligned} \alpha &= k\sqrt{p_1 + p_2 + \sqrt{p_1^2 - p_1p_2 + p_2^2}}, \\ \beta &= k\sqrt{p_1 + p_2 - \sqrt{p_1^2 - p_1p_2 + p_2^2}} \end{aligned} \qquad (10\text{–}31)$$

and $p_1 = 1/m_1$, $p_2 = 1/m_2$ (see Prob. 1 below). When $\lambda = \pm\alpha$, equations (10–29) are satisfied by $A_1 = k^2$, $A_2 = 2k^2 - m_1\alpha^2$; when $\lambda = \pm\beta$, they are satisfied by $A_1 = k^2$, $A_2 = 2k^2 - m_1\beta^2$. Accordingly, the general solution is a superposition of normal modes

$$\begin{aligned} u_1 &= C_1k^2 \sin(\alpha t + \epsilon_1) + C_2k^2 \sin(\beta t + \epsilon_2), \\ u_2 &= C_1(2k^2 - m_1\alpha^2) \sin(\alpha t + \epsilon_1) + C_2(2k^2 - m_1\beta^2) \sin(\beta t + \epsilon_2). \end{aligned} \qquad (10\text{–}32)$$

The frequencies α, β are called *resonant frequencies.*

(c) *Exponential decay.* We assume $m_1 = m_2 = 0$, $F_1 \equiv 0$, $F_2 \equiv 0$, $u_0 = u_3 = 0$, $h_1 > 0$, $h_2 > 0$. The equations (10–25) become

$$\begin{aligned} h_1\frac{du_1}{dt} - k^2(u_2 - 2u_1) &= 0, \\ h_2\frac{du_2}{dt} - k^2(-2u_2 + u_1) &= 0. \end{aligned} \qquad (10\text{–}33)$$

We seek solutions:

$$u_1 = A_1e^{\lambda t}, \quad u_2 = A_2e^{\lambda t} \qquad (10\text{–}34)$$

and proceed as in Section 8–12. The characteristic equation is of second degree and has *distinct* real roots $-a$, $-b$ which are both negative (Prob. 5 below):

$$\begin{aligned} a &= k^2(q_1 + q_2 + \sqrt{q_1^2 - q_1q_2 + q_2^2}), \\ b &= k^2(q_1 + q_2 - \sqrt{q_1^2 - q_1q_2 + q_2^2}), \end{aligned} \qquad (10\text{–}35)$$

where $q_1 = 1/h_1$, $q_2 = 1/h_2$. The general solution is found (Prob. 5 below) to be

$$\begin{aligned} u_1 &= C_1k^2e^{-at} + C_2k^2e^{-bt}, \\ u_2 &= C_1(2k^2 - h_1a)e^{-at} + C_2(2k^2 - h_1b)e^{-bt}. \end{aligned} \qquad (10\text{–}36)$$

We can say: *the general motion of the system is an exponential approach to the equilibrium state:* $u_1 = u_2 = 0$. Both "normal modes" are transients for this case, but we can consider the general motion as a linear combination of two such normal modes.

The physical model of Fig. 8–2 is not very appropriate here, since we are considering the limiting case $m_1 = 0$, $m_2 = 0$. A better model is suggested in Fig. 10–3. Here we consider a rod made of four pieces, each of which is a perfect conductor of heat, so that the temperature u is constant throughout each piece. The two end pieces are maintained at temperatures u_0 and u_3. It is assumed that heat can flow across the faces of adjacent pieces according to Newton's law, so that the rate of flow is proportional to the temperature difference. If h_1 and h_2 denote total specific heats for the middle pieces, we obtain differential equations

$$\begin{aligned} h_1 \frac{du_1}{dt} &= -k_1^2(u_1 - u_0) - k_2^2(u_1 - u_2), \\ h_2 \frac{du_2}{dt} &= -k_2^2(u_2 - u_1) - k_3^2(u_2 - u_3). \end{aligned} \tag{10–37}$$

If $k_1 = k_2 = k_3$ and $u_3 = u_0 = 0$, we obtain (10–33). It is physically clear that when the end temperatures are maintained at 0, the temperatures of the inner pieces will also gradually approach 0.

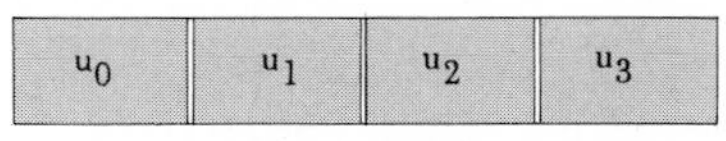

Fig. 10–3.

(d) *Equilibrium.* We assume that u_0 and u_3 have constant values u_0^* and u_3^* and that constant forces F_1 and F_2 are applied. Equations (10–25) have then an equilibrium solution, obtained by setting all derivatives with respect to t equal to 0:

$$-k^2(u_2 - 2u_1 + u_0^*) = F_1, \quad -k^2(u_3^* - 2u_2 + u_1) = F_2. \tag{10–38}$$

Accordingly, the equilibrium values of u_1, u_2 are

$$u_1^* = \frac{1}{3}\left(2u_0^* + u_3^* + \frac{2F_1}{k^2} + \frac{F_2}{k^2}\right), \quad u_2^* = \frac{1}{3}\left(u_0^* + 2u_3^* + \frac{F_1}{k^2} + \frac{2F_2}{k^2}\right). \tag{10–39}$$

(e) *Approach to equilibrium.* Let $w_1 = u_1 - u_1^*$, $w_2 = u_2 - u_2^*$, where u_1^* and u_2^* are defined by (10–39). Then substitution in (10–25), with $u_0 = u_0^*$, $u_3 = u_3^*$, $F_1 = \text{const}$, $F_2 = \text{const}$, yields the equations

$$\begin{aligned} m_1 \frac{d^2w_1}{dt^2} + h_1 \frac{dw_1}{dt} - k^2(w_2 - 2w_1) &= 0, \\ m_2 \frac{d^2w_2}{dt^2} + h_2 \frac{dw_2}{dt} - k^2(-2w_2 + w_1) &= 0. \end{aligned} \tag{10–40}$$

If, for example, $m_1 = m_2 = 0$, then these are the same as (10–33), so that

$$\begin{aligned} w_1 &= c_1k^2e^{-at} + c_2k^2e^{-bt}, \\ w_2 &= c_1(2k^2 - h_1a)e^{-at} + c_2(2k^2 - h_1b)e^{-bt}. \end{aligned} \tag{10–41}$$

Accordingly,

$$\begin{aligned} u_1 &= w_1 + u_1^* = c_1k^2e^{-at} + c_2k^2e^{-bt} + u_1^*, \\ u_2 &= w_2 + u_2^* = c_1(2k^2 - h_1a)e^{-at} + c_2(2k^2 - h_1b)e^{-bt} + u_2^*. \end{aligned} \tag{10–42}$$

Just as for one particle, the solution is an exponential approach to equilibrium.

(f) *Forced motion.* External forces can be applied both by varying the wall positions u_0, u_3 and through the forces F_1, F_2. We allow for both effects but neglect the masses by considering the equations

$$\begin{aligned} h_1\frac{du_1}{dt} - k^2[u_2 - 2u_1 + u_0(t)] &= F_1(t), \\ h_2\frac{du_2}{dt} - k^2[u_3(t) - 2u_2 + u_1] &= F_2(t). \end{aligned} \tag{10–43}$$

We now use the method of variation of parameters (Section 8–12); we replace the constants C_1, C_2 in the solutions (10–36) of the homogeneous equations by variables $v_1(t)$, $v_2(t)$:

$$\begin{aligned} u_1 &= v_1k^2e^{-at} + v_2k^2e^{-bt}, \\ u_2 &= v_1(2k^2 - h_1a)e^{-at} + v_2(2k^2 - h_1b)e^{-bt}. \end{aligned} \tag{10–44}$$

If we substitute in (10–43) and use the fact that (10–44) satisfies (10–33) when the v's are treated as constants, then we obtain equations

$$\begin{aligned} h_1(v_1'k^2e^{-at} + v_2'k^2e^{-bt}) &= F_1 + k^2u_0, \\ h_2[v_1'(2k^2 - h_1a)e^{-at} + v_2'(2k^2 - h_1b)e^{-bt}] &= F_2 + k^2u_3 \end{aligned}$$

for v_1', v_2'. The solutions are

$$\begin{aligned} &v_1 = \int e^{at}[(2k^2 - h_1b)G_1 - G_2]\,dt, \quad v_2 = -\int e^{bt}[(2k^2 - h_1a)G_1 - G_2]\,dt; \\ &G_1 = \frac{F_1(t) + k^2u_0(t)}{(a-b)h_1^2k^2}, \quad G_2 = \frac{F_2(t) + k^2u_3(t)}{h_1h_2(a-b)}. \end{aligned} \tag{10–45}$$

Substitution in (10–44) yields a solution $u_1^*(t)$, $u_2^*(t)$. The general solution is then

$$\begin{aligned} u_1 &= c_1k^2e^{-at} + c_2k^2e^{-bt} + u_1^*(t), \\ u_2 &= c_1(2k^2 - h_1a)e^{-at} + c_2(2k^2 - h_1b)e^{-bt} + u_2^*(t). \end{aligned} \tag{10–46}$$

The terms in e^{-at}, e^{-bt} can be regarded as transients. The solutions can be interpreted as following an *input*, defined by (10–39) with u_0^*, u_3^*, F_1, F_2 replaced by the given functions of t; the speed of follow-up depends on the size of h_1, h_2.

Problems

1. Show that the roots of (10–30) are given by $\pm\alpha$, $\pm\beta$ where α and β are defined by (10–31). Show that α and β are real and that $\alpha > \beta > 0$.

2. Solve (10–27) as in Section 8–12 by setting $u_1 = A_1e^{\lambda t}$, $u_2 = A_2e^{\lambda t}$, $w_1 = B_1e^{\lambda t}$, $w_2 = B_2e^{\lambda t}$. Show that (10–32) is again obtained.

3. Let $m_1 = m_2 = 1$ and $k^2 = 1$, in appropriate units, in (10–26).
(a) Write out the general solution.
(b) Obtain the particular solution for which $u_1 = u_2 = 0$ for $t = 0$ and $du_1/dt = 1$, $du_2/dt = 0$ for $t = 0$. Graph u_1 and u_2 as functions of t. Also plot the curve $u_1 = u_1(t)$, $u_2 = u_2(t)$ in the u_1u_2 plane; this is a "Lissajous figure."

4. (a) Show that Eqs. (10–26) can be written in the form (10–28) with $V = k^2(u_1^2 - u_1u_2 + u_2^2)$ and that V is positive definite.
(b) Prove that the expression $E = \frac{1}{2}m_1(du_1/dt)^2 + \frac{1}{2}m_2(du_2/dt)^2 + V(u_1, u_2)$ is constant for each solution of (10–26). [Hint: differentiate E with respect to t and use (10–26).] The first two terms give the total kinetic energy, the third term is the potential energy. E is the total energy and remains constant (Section 5–15).

5. Obtain the general solution (10–36) of (10–33) and verify that $0 < b < a$.

6. Let $h_1 = h_2 = 1$, $k^2 = 1$, in appropriate units, in (10–33).
(a) Obtain the general solution.
(b) Obtain the particular solution for which $u_1 = 1$, $u_2 = 3$ when $t = 0$. Graph u_1 and u_2 as functions of t. Also plot the curve $u_1 = u_1(t)$, $u_2 = u_2(t)$ in the u_1u_2 plane.

7. Let $m_1 = m_2 = 0$, $h_1 = h_2 = 1$, $k^2 = 1$, $F_1 = 2$, $F_2 = 3$, $u_0 = 1$, $u_3 = 4$, in appropriate units, in (10–25).
(a) Find the equilibrium state.
(b) Solve the differential equations (10–25) by *step-by-step integration* (cf. Prob. 5 following Section 8–8), starting at $u_1 = 1$, $u_2 = 2$ and using $\Delta t = 0.1$. Graph the solution obtained in the u_1u_2 plane.

8. Let $h_1 = h_2 = 1$, $k^2 = 1$, $F_1(t) \equiv 0$, $F_2(t) \equiv 0$, in appropriate units in (10–43). Let further $u_0(t) = \sin t$ and $u_3(t) \equiv 0$. Obtain the particular solution for which $u_1 = u_2 = 0$ for $t = 0$. Plot u_1 and u_2 as functions of t and also graph the curve $u_1 = u_1(t)$, $u_2 = u_2(t)$ in the u_1u_2 plane.

9. Let $m_1 = m_2 = 1$, $h_1 = h_2 = 0$, $k^2 = 1$, $u_0(t) = u_3(t) = 0$, $F_1(t) = 4 \sin t$, $F_2(t) = 4a \sin t$ in (10–25). Find a particular solution. Does resonance occur? [Hint: use (10–27) and variation of parameters.]

10. (a) Let $m_1 = m_2 = 4$, $h_1 = h_2 = 1$, $k^2 = 1$, $u_0 = u_3 = 0$, $F_1(t) \equiv F_2(t) \equiv 0$ in (10–25). Show that the solutions represent damped vibrations.
(b) Can the values of h_1, h_2 be modified so that one normal mode is undercritically damped, while the other mode is overcritically damped?

Answers

3. (a) $u_1 = c_1 \sin(\sqrt{3}t + \epsilon_1) + c_2 \sin(t + \epsilon_2)$,
$u_2 = -c_1 \sin(\sqrt{3}t + \epsilon_1) + c_2 \sin(t + \epsilon_2)$,
(b) $u_1 = \frac{1}{6}\sqrt{3} \sin \sqrt{3}t + \frac{1}{2} \sin t$, $u_2 = -\frac{1}{6}\sqrt{3} \sin \sqrt{3}t + \frac{1}{2} \sin t$.

6. (a) $u_1 = c_1e^{-t} + c_2e^{-3t}$, $u_2 = c_1e^{-t} - c_2e^{-3t}$,
(b) $u_1 = 2e^{-t} - e^{-3t}$, $u_2 = 2e^{-t} + e^{-3t}$.

7. (a) $u_1 = \frac{13}{3}$, $u_2 = \frac{17}{3}$.

8. $u_1 = 0.25e^{-t} + 0.05e^{-3t} - 0.1(3 \cos t - 4 \sin t)$,
$u_2 = 0.25e^{-t} - 0.05e^{-3t} - 0.1(2 \cos t - \sin t)$.

9. $u_1 = (1 + a)(-t \cos t), \quad u_2 = (1 + a)(-t \cos t) + (2a - 2) \sin t$. Resonance occurs except when $a = -1$.

10. (a) $u_1 = e^{-at}(c_1 \cos \beta t + c_2 \sin \beta t + c_3 \cos \gamma t + c_4 \sin \gamma t)$,
$u_2 = e^{-at}(c_1 \cos \beta t + c_2 \sin \beta t - c_3 \cos \gamma t - c_4 \sin \gamma t)$,
$a = \frac{1}{8}, \quad \beta = \sqrt{15}/8, \quad \gamma = \sqrt{47}/8$.

10–4 Case of N particles. We now consider the general case of N particles $P_1, \ldots, P_N$ of masses $m_1, \ldots, m_N$ moving on the x axis as in Fig. 10–4. The particle P_σ is attached to the particles $P_{\sigma-1}$ and $P_{\sigma+1}$ by springs; the particle P_1 is attached to a wall at P_0 and to P_2; the particle P_N is attached to P_{N-1} and to a wall at P_{N+1}. In general x_σ is the x coordinate of P_σ. For simplicity we assume that all springs have the same natural length l and spring constant k^2. We assume that P_σ is subject to a resistance $-h_\sigma \dfrac{dx_\sigma}{dt}$ and an outside force $F_\sigma(t)$. The differential equations corresponding to (10–21) are then the following:

$$m_\sigma \frac{d^2x_\sigma}{dt^2} + h_\sigma \frac{dx_\sigma}{dt} - k^2(x_{\sigma+1} - 2x_\sigma + x_{\sigma-1}) = F_\sigma(t). \tag{10–47}$$

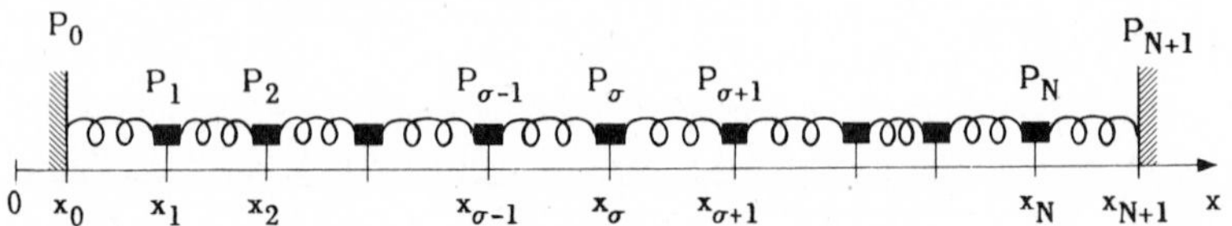

FIG. 10–4. Linear system with N particles.

If the walls are fixed: $x_0 = \overset{*}{x}_0$, $x_{N+1} = \overset{*}{x}_{N+1}$ and $F_\sigma(t) \equiv 0$ for $\sigma = 1, \ldots, N$, then there is an equilibrium state, determined by the N equations:

$$x_{\sigma+1} - 2x_\sigma + x_{\sigma-1} = 0, \quad \sigma = 1, \ldots, N. \tag{10–48}$$

Equations (10–48) can be written as follows:

$$x_\sigma = \tfrac{1}{2}(x_{\sigma+1} + x_{\sigma-1}).$$

They assert that P_σ is half-way between $P_{\sigma-1}$ and $P_{\sigma+1}$. Hence at equilibrium all particles are equally spaced between x_0 and x_{N+1}:

$$\begin{aligned} x_1 &= \overset{*}{x}_1 = \overset{*}{x}_0 + \frac{1}{N+1}(\overset{*}{x}_{N+1} - \overset{*}{x}_0), \ldots, \\ x_N &= \overset{*}{x}_N = \overset{*}{x}_0 + \frac{N}{N+1}(\overset{*}{x}_{N+1} - \overset{*}{x}_0). \end{aligned} \tag{10–49}$$

We now refer the motion of the particles to the equilibrium positions (10–49) by introducing new coordinates:

$$u_\sigma = x_\sigma - \overset{*}{x}_\sigma \quad (\sigma = 0, \ldots, N + 1). \tag{10–50}$$

The differential equations (10–47) are then replaced by the following:

$$m_\sigma \frac{d^2u_\sigma}{dt^2} + h_\sigma \frac{du_\sigma}{dt} - k^2(u_{\sigma+1} - 2u_\sigma + u_{\sigma-1}) = F_\sigma(t). \qquad (10\text{–}51)$$

A second physical model which leads to equations (10–51) is suggested in Fig. 10–5. Here the particles $P_1, \ldots, P_N$ are constrained to move along lines $x = x_1, \ldots, x = x_N$ in the xu plane. Again P_σ is attached to $P_{\sigma+1}$ and to $P_{\sigma-1}$ by springs; P_0 and P_{N+1} are points on the "walls": $x = x_0$, $x = x_{N+1}$. We assume that the lines $x = x_\sigma$ are equally spaced, at distance Δx and that all particles have the same spring constant k_0^2 and natural length l, where $l < \Delta x$. If P_σ has coordinates (x_σ, u_σ) and is subject to a resistance $-h_\sigma(du_\sigma/dt)$ and an outside force $F_\sigma(t)$, then the differential equations governing the motion are as follows:

$$m_\sigma \frac{d^2u_\sigma}{dt^2} + h_\sigma \frac{du_\sigma}{dt} = k_0^2(u_{\sigma+1} - 2u_\sigma + u_{\sigma-1}) - k_0^2 l\,(\sin\alpha_\sigma - \sin\alpha_{\sigma-1}) + F_\sigma(t), \qquad (10\text{–}52)$$

where $\sigma = 1, \ldots, N$ and α_σ is the angle from the positive x direction to $\overrightarrow{P_\sigma P_{\sigma+1}}$. If the angles α_σ remain so small that an approximation $\sin\alpha_\sigma \sim \tan\alpha_\sigma$ is justified, the equations become

$$m_\sigma \frac{d^2u_\sigma}{dt^2} + h_\sigma \frac{du_\sigma}{dt} - k^2(u_{\sigma+1} - 2u_\sigma + u_{\sigma-1}) = F_\sigma(t), \qquad (10\text{–}53)$$

where $k^2 = k_0^2[1 - (l/\Delta x)]$. The derivation of (10–52) and (10–53) is left to Prob. 1 below. Equations (10–53) are identical with (10–51).

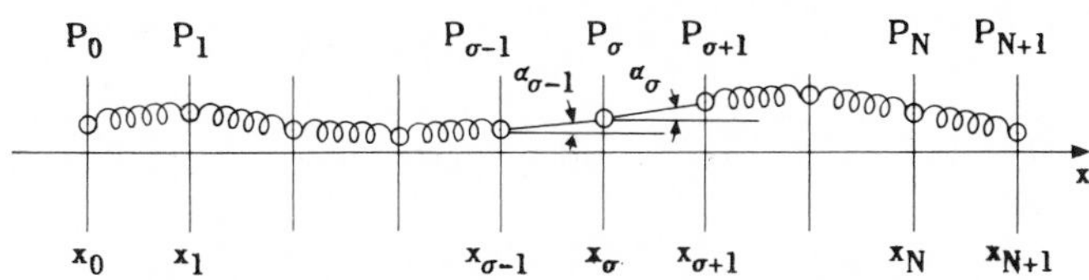

FIG. 10–5. N particle model for the vibrating string.

When $m_\sigma = 0$ for $\sigma = 1, \ldots, N$, a natural model can be devised by generalizing the heat conduction model of Fig. 10–3. Other models can be constructed, e.g., by use of electric circuits.

Equations (10–51) can be written in a form which suggests further generalizations. We write

$$V(u_1, \ldots, u_N) = k^2(u_1^2 + \cdots + u_N^2 - u_1u_2 - \cdots - u_{N-1}u_N). \qquad (10\text{–}54)$$

V is the *potential energy* associated with the system. Equations (10–51) then become

$$m_\sigma \frac{d^2u_\sigma}{dt^2} + h_\sigma \frac{du_\sigma}{dt} + \frac{\partial V}{\partial u_\sigma} = F_\sigma(t), \qquad (10\text{–}55)$$

except for $\sigma = 1$ and $\sigma = N$; the exceptional cases can be included if we modify the definitions of $F_1(t)$ and $F_N(t)$ to include $k^2u_0(t)$ and $k^2u_{N+1}(t)$ respectively.

If we allow V to be a general function of $u_1, \ldots, u_N$, instead of the special function (10–54), then an extremely broad class of physical problems is included in (10–55). Of great importance is the case in which V is a general quadratic expression:

$$V = \sum_{i=1}^{N} \sum_{j=1}^{N} a_{ij} u_i u_j. \tag{10–56}$$

This case arises in consideration of problems of "small vibrations," i.e., problems concerning the vibrations of a system of particles which remain close to equilibrium positions. The problems of Figs. 10–4 and 10–5 are of this sort. The approximation $\sin \alpha \sim \tan \alpha$ was based on the smallness of the departure from equilibrium.

We now state briefly the results corresponding to (a), ..., (f) for (10–51). Actually, many of the statements apply equally well to equations (10–55), where V is defined by (10–56) and is positive definite. This is shown for case (a) below in Section 8–14 and by similar methods in the other cases. The positive definiteness of V in (10–54) is proved in Problem 2 below.

(a) *Harmonic motion.* Here $h_\sigma = 0$, $u_0 = u_{N+1} = 0$, $F_\sigma(t) \equiv 0$. We find that there are N normal modes of vibration, with resonant frequencies $\lambda_1, \ldots, \lambda_N$, and that the general motion consists of a linear combination of these:

$$u_\sigma = \sum_{n=1}^{N} c_n A_{n,\sigma} \sin (\lambda_n t + \epsilon_n); \tag{10–57}$$

the $A_{n,\sigma}$ are certain constants and $c_1, \ldots, c_N$, $\epsilon_1, \ldots, \epsilon_N$ are arbitrary constants.

(b) *Damped vibrations.* In this case $m_\sigma > 0$, $h_\sigma > 0$, $u_0 = u_{N+1} = 0$, $F_\sigma(t) \equiv 0$. The oscillations of (a) are replaced by damped oscillations, of form $e^{-at} \sin (\lambda t + \epsilon)$, or exponential decay terms: e^{-at}, te^{-at}.

(c) *Exponential decay.* Here $m_\sigma = 0$, $F_\sigma \equiv 0$, $u_0 = u_{N+1} = 0$, $h_\sigma > 0$. We find that there are N "normal modes of decay" and that the general motion is a linear combination of the normal modes:

$$u_\sigma = \sum_{n=1}^{N} c_n A_{n,\sigma} e^{-a_n t}. \tag{10–58}$$

(d) *Equilibrium.* If $u_0 = u_n^*$, $u_{N+1} = u_{N+1}^*$, $F_\sigma = F_\sigma^*$, where the starred values are constants, then the equilibrium values: $u_\sigma = u_\sigma^*$ are equally spaced between $u_0^* + (F_1^*/k^2)$ and $u_{N+1}^* + (F_N^*/k^2)$.

(e) *Approach to equilibrium.* Under the assumptions of case (d), the general motion $u_\sigma(t)$ is a solution of the homogeneous equations corresponding to (10–51) plus the equilibrium solution u_σ^*. When the homogeneous equations are of type (c), one has therefore exponential approach to equilibrium:

$$u_\sigma = \sum_{n=1}^{N} c_n A_{n,\sigma} e^{-ant} + u_\sigma^* \quad (\sigma = 1, \ldots, N). \tag{10–59}$$

(f) *Forced motion.* Here u_0, u_{N+1}, and all F_σ are allowed to depend on t. The general motion $u_\sigma(t)$ is a solution of the homogeneous equations corresponding to (10–51) plus a particular solution $u_\sigma^*(t)$. The particular solution can always be found by variation of parameters; it can always be interpreted as a follow-up of an input. The total effect of all $F_\sigma(t)$ on the solution can be considered as a superposition of the effects of the $F_\sigma(t)$ individually. Addition of a sinusoidal term of frequency λ to one F_σ leads to addition of a sinusoidal term of frequency λ to all $u_\sigma^*(t)$, unless all h_σ are 0 and λ is a resonant frequency, in which case resonance may occur (cf. Prob. 9 following Section 10–3).

Problems

1. (a) Obtain the equations (10–52) for the model of Fig. 10–5.

(b) Show that, when the angles α_σ remain small, the equations (10–53) are justified as approximations to (10–52).

2. (a) Show that the potential energy V of (10–54) is positive definite. [Hint: Write $V = (k^2/2)[u_1^2 + (u_1 - u_2)^2 + \cdots + u_N^2]$.

(b) Show that, when $h_\sigma = 0$ and $F_\sigma(t) \equiv 0$ for $\sigma = 1, \ldots, N$, every solution of (10–55) has the property that the *total energy*

$$E = \tfrac{1}{2}m_1(du_1/dt)^2 + \cdots + \tfrac{1}{2}m_N(du_N/dt)^2 + V(u_1, \ldots, u_N)$$

is constant. [Hint: see Prob. 4 following Section 10–3.]

3. Let $N = 3$ in (10–51) and let $m_1 = m_2 = m_3 = 1$, $k^2 = 1$, $h_1 = h_2 = h_3 = 0$, $u_0 = u_4 = 0$, $F_1 = F_2 = F_3 = 0$, so that one has case (a). Seek solutions: $u_1 = A_1e^{\lambda t}$, $u_2 = A_2e^{\lambda t}$, $u_3 = A_3e^{\lambda t}$ and obtain the general solution in the form (10–57).

4. Let $N = 3$ in (10–51) and let $m_1 = m_2 = m_3 = 0$, $k^2 = 1$, $h_1 = h_2 = h_3 = 1$, $u_0 = u_4 = 0$, $F_1 = F_2 = F_3 = 0$, so that one has case (c). Seek solutions: $u_1 = A_1e^{\lambda t}$, $u_2 = A_2e^{\lambda t}$, $u_3 = A_3e^{\lambda t}$ and obtain the general solution in the form (10–58).

5. *Difference equations.* We consider functions $f(\sigma)$ of an integer variable σ: $\sigma = 0, \pm 1, \pm 2, \ldots$. Let $f(\sigma)$ be defined for $\sigma = m, \sigma = m + 1, \ldots, \sigma = n$. Then the *first difference* $\Delta_+ f(\sigma)$ is the function $f(\sigma + 1) - f(\sigma)$ $(m \leqq \sigma < n)$; the *first difference* $\Delta_- f(\sigma)$ is the function $f(\sigma) - f(\sigma - 1)$ $(m < \sigma \leqq n)$. The *second difference* is the first difference of the first difference; this could mean $\Delta_+ \Delta_- f$, $\Delta_- \Delta_+ f$, $\Delta_+ \Delta_+ f$, $\Delta_- \Delta_- f$; we shall use only

$$\Delta^2 f = \Delta_+ \Delta_- f = \Delta_- \Delta_+ f = f(\sigma + 1) - 2f(\sigma) + f(\sigma - 1);$$

cf. Section 2–18. Differences of higher order can be defined in analogous fashion. A *difference equation* is an identity to be satisfied by $f(\sigma)$ and its differences of various orders. In general, the "linear difference equation" has a theory quite analogous to that of the linear differential equation and the functions $e^{r\sigma}$ play a similar role. We consider only two cases:

(a) $\Delta^2 f(\sigma) = 0$. Show that the general solution is given by $f(\sigma) = c_1\sigma + c_2$, where c_1, c_2 are constants. Obtain the solution satisfying the *boundary conditions:* $f(0) = u_0$, $f(N + 1) = u_{N+1}$, where u_0, u_{N+1} are given constants.

(b) $\Delta^2 f(\sigma) + p^2 f(\sigma) = 0$. Show that the functions $c_1 \cos q\sigma + c_2 \sin q\sigma$ are solutions, provided $\cos q = 1 - \frac{1}{2}p^2$ and $0 < p^2 < 4$. Show that when $p^2 > 4$, distinct numbers a_1 and a_2 can be chosen so that the functions $c_1a_1^\sigma + c_2a_2^\sigma$ are solutions and that when $p^2 = 4$, the functions $c_1(-1)^\sigma + c_2\sigma(-1)^\sigma$ are solutions.

(c) Show that, if $f(0)$ and $f(1)$ are given, then the difference equation of part (b) successively determines $f(2), f(3), \ldots$. Hence for such "initial conditions" there is one and only one solution. Determine the constants c_1, c_2 for each of the cases of part (b) to match the given initial conditions. The fact that this is possible ensures that each expression gives the general solution for the corresponding value of p.

(*d*) Show that the only solutions of the difference equation of part (b) which satisfy the boundary conditions: $f(0) = 0$, $f(N+1) = 0$ are constant multiples of the N functions

$$\phi_n(\sigma) = \sin\left(\frac{n\pi}{N+1}\sigma\right) \quad (n = 1, \ldots, N),$$

and that $\phi_n(\sigma)$ is a solution only when

$$p^2 = 2\left(1 - \cos\frac{n\pi}{N+1}\right).$$

6. Show that the solution of (10–48), where $x_0 = x_0^*$, $x_{N+1} = x_{N+1}^*$ is equivalent to solution of the difference equation $\Delta^2 f(\sigma) = 0$ of Prob. 5(a), subject to boundary conditions, and compare with the solution of Prob. 5(a).

7. Let $m_1 = m_2 = \cdots = m_N = m$, $h_\sigma = 0$ and $F_\sigma(t) \equiv 0$ for $\sigma = 1, \ldots, N$ and $u_0 = u_{N+1} = 0$ in (10–51), so that one has case (a), with *equal masses*. Show that the substitution $u_\sigma = A(\sigma)\sin(\lambda t + \epsilon)$ leads to the difference equation with boundary conditions:

$$\Delta^2 A(\sigma) + p^2 A(\sigma) = 0, \quad p^2 = m\lambda^2/k^2$$
$$A(0) = 0, \quad A(N+1) = 0.$$

Use the result of Prob. 5(d) to obtain the N *normal modes*

$$u_\sigma(t) = \sin\left(\frac{n\pi}{N+1}\sigma\right)\sin(\lambda_n t + \epsilon_n),$$

$$\lambda_n = \frac{2k}{\sqrt{m}}\sin\frac{n\pi}{2(N+1)}, \quad n = 1, \ldots, N.$$

Show that $0 < \lambda_1 < \lambda_2 < \cdots < \lambda_N$.

8. For two functions $f(\sigma)$, $g(\sigma)$ defined for $\sigma = 0, 1, 2, \ldots, N+1$, we define an *inner product* (f, g) by the equation (cf. Section 7–10)

$$(f, g) = f(0)g(0) + f(1)g(1) + \cdots + f(N)g(N) + f(N+1)g(N+1);$$

the *norm* $\|f\|$ is then defined as $(f, f)^{\frac{1}{2}}$. In the following we consider *only functions which equal* 0 *for* $\sigma = 0$ *and* $\sigma = N + 1$. In particular, we use the functions $\phi_n(\sigma)$ of Prob. 5(d):

$$\phi_n(\sigma) = \sin(n\alpha\sigma), \quad \alpha = \pi/(N+1).$$

(a) Graph the functions $\phi_n(\sigma)$ for the case $N = 5$.

(b) Show that $(\phi_m, \phi_n) = 0$ for $m \neq n$ and that $\|\phi_n\|^2 = \frac{1}{2}(N+1)$. [Hint: write

$$\phi_n(\sigma) = \frac{r^\sigma - s^\sigma}{2i}, \quad r = e^{\alpha n i}, \quad s = e^{-\alpha n i}$$

and evaluate inner product and norm with the aid of the formula for sum of a geometric progression, Eq. (0–29).]

(c) Show that, if we associate with each function $f(\sigma)$ the vector $\mathbf{v} = [v_1, v_2, \ldots, v_N]$, where $v_1 = f(1)$, $v_2 = f(2), \ldots, v_N = f(N)$, then the operations $f + g$, cf, (f, g) correspond to the vector operations $\mathbf{u} + \mathbf{v}$, $c\mathbf{u}$, $\mathbf{u} \cdot \mathbf{v}$ of Section 1–8. Accordingly, the space of functions considered forms *an N-dimensional Euclidean vector space.* The vectors corresponding to the functions $\phi_n(\sigma)/\|\phi_n(\sigma)\|$ form *a system of base vectors.*

(d) Show that, if $f(\sigma)$ is defined for $\sigma = 0, \ldots, N + 1$ and $f(0) = f(N + 1) = 0$, then $f(\sigma)$ can be represented in one and only one way as a linear combination of the functions $\phi_n(\sigma)$, namely as follows:

$$f(\sigma) = \sum_{n=1}^{N} b_n\phi_n(\sigma), \quad b_n = \frac{2}{N + 1}\sum_{\sigma=0}^{N+1} f(\sigma)\phi_n(\sigma).$$

Compare with the Fourier sine series, Section 7–5.

9. Write the general solution of Prob. 7 in the form

$$u_\sigma(t) = \sum_{n=1}^{N} \phi_n(\sigma)(\alpha_n \sin \lambda_n t + \beta_n \cos \lambda_n t),$$

where α_n and β_n are arbitrary constants. Use the result of Prob. 8(d) to show that the constants α_n and β_n can be chosen in one and only one way so that $u_\sigma(t)$ satisfies given initial conditions:

$$u_\sigma(0) = f(\sigma), \quad \frac{du_\sigma}{dt}(0) = g(\sigma).$$

This shows that one has indeed obtained *all* solutions.

10. Let $h_1 = h_2 = \cdots = h_N = h$, $m_\sigma = 0$ and $F_\sigma(t) \equiv 0$ for $\sigma = 1, \ldots, N$, $u_0 = u_{N+1} = 0$ in (10–51), so that one has case (c), with *equal friction coefficients.* Show that the substitution $u_\sigma = A(\sigma)e^{\lambda t}$ leads to the difference equation with boundary conditions:

$$\Delta^2 A(\sigma) + p^2 A(\sigma) = 0, \quad p^2 = -h\lambda/k^2,$$
$$A(0) = 0, \quad A(N + 1) = 0.$$

Use the result of Prob. 5(d) to obtain the "modes of decay":

$$u_\sigma(t) = \sin\left(\frac{n\pi}{N + 1}\sigma\right)e^{-a_n t},$$
$$a_n = \frac{2k^2}{h}\left(1 - \cos\frac{n\pi}{N + 1}\right).$$

Show that $0 < a_1 < a_2 < \cdots < a_N$.

11. Prove that constants c_n can be chosen in one and only one way so that

$$u_\sigma(t) = \sum_{n=1}^{N} c_n\phi_n(\sigma)e^{-a_n t}$$

is a solution of the exponential decay problem, Prob. 10, and matches given initial conditions: $u_\sigma(0) = f(\sigma)$ (cf. Prob. 9).

12. In (10–51) let $h_1 = h_2 = \cdots = h$, $m_\sigma = 0$, for $\sigma = 1, \ldots, N$ as in Prob. 10, but let $F_\sigma(t)$ be permitted to depend on t. We assume $u_0 = 0$, $u_{N+1} = 0$, as any variation in the "walls" can be absorbed in $F_0(t)$ and $F_{N+1}(t)$. Use the

method of variation of parameters to obtain a particular solution. [Hint: the "complementary function" is given in Prob. 11. If we replace c_n by $v_n(t)$ for $n = 1, \ldots, N$ and substitute in (10–51), we obtain equations

$$h\sum_{n=1}^{N}\frac{dv_n}{dt}\phi_n(\sigma)e^{-ant} = F_\sigma(t).$$

Now use the result of Prob. 8(d) to conclude that

$$h\frac{dv_n}{dt}e^{-ant} = \frac{2}{N+1}\sum_{\sigma=1}^{N}F_\sigma(t)\phi_n(\sigma).]$$

ANSWERS

3. $u_1 = c_1 \sin(\alpha t + \epsilon_1) + c_2 \sin(\beta t + \epsilon_2) + c_3 \sin(\gamma t + \epsilon_3)$,
$u_2 = \sqrt{2}c_1 \sin(\alpha t + \epsilon_1) - \sqrt{2}c_3 \sin(\gamma t + \epsilon_3)$,
$u_3 = c_1 \sin(\alpha t + \epsilon_1) - c_2 \sin(\beta t + \epsilon_2) + c_3 \sin(\gamma t + \epsilon_3)$,
where $\alpha = (2-\sqrt{2})^{\frac{1}{2}}$, $\beta = \sqrt{2}$, $\gamma = (2+\sqrt{2})^{\frac{1}{2}}$.

4. $u_1 = c_1e^{-at} + c_2e^{-bt} + c_3e^{-ct}$, $u_2 = \sqrt{2}c_1e^{-at} - \sqrt{2}c_3e^{-ct}$,
$u_3 = c_1e^{-at} - c_2e^{-bt} + c_3e^{-ct}$, where $a = 2-\sqrt{2}$, $b = 2$, $c = 2+\sqrt{2}$.

5. (a) $u_0 + (u_{N+1} - u_0)\sigma/(N+1)$.

10–5 Continuous medium. Fundamental partial differential equation. We now consider the limiting case: $N \to \infty$. Rather than attempt to carry out a precise passage to the limit, we allow ourselves to be guided by physical intuition. The natural limiting case of the system of N particles moving on a line (Fig. 10–4) is that of a *rod* which is permitted to vibrate longitudinally, as suggested in Fig. 10–6. The individual particle is replaced by a cross section of the rod, which can be thought of as a thin layer of molecules which move together parallel to the axis of the rod. When no external forces are applied, this layer has an equilibrium position x. Just as for the particles, we can measure the displacement u of the layer from its equilibrium position x; u then becomes a function of x and t.

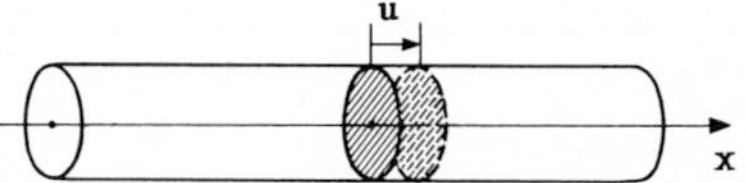

Fig. 10–6. Longitudinal vibrations of a rod.

If we pass to the limit in the model of Fig. 10–5, we obtain the *vibrating string:* for example, a violin string. To first approximation each "molecule" of such a string executes vibrations perpendicular to the line represented by the equilibrium position of the string. The displacement of the molecule at position x from its equilibrium position is measured by u, which is a function of x and t. The vibrations are assumed to take place in an xu plane; one could consider the more general case in which the vibrations are not confined to a plane.

In order to obtain a differential equation for $u(x, t)$, we return to the basic equations (10–51) and write them as follows:

$$\frac{m_\sigma}{\Delta x}\frac{d^2u_\sigma}{dt^2} + \frac{h_\sigma}{\Delta x}\frac{du_\sigma}{dt} - k^2\,\Delta x\,\frac{u_{\sigma+1} - 2u_\sigma + u_{\sigma-1}}{(\Delta x)^2} = \frac{F_\sigma(t)}{\Delta x}. \tag{10–60}$$

We assume x_0 and x_{N+1} to be fixed and let $L = x_{N+1} - x_0$, $\Delta x = L/(N+1)$. We then let N increase indefinitely. The ratio $m_\sigma/\Delta x$ represents an "average density" at position x; it is reasonable to postulate that this approaches as limit a function $\rho(x)$ representing density (mass per unit length) at x. The simplest law of friction would make h_σ proportional to m_σ, so that $h_\sigma/\Delta x$ would approach a function $H(x)$ of dimensions force per unit of length per unit of velocity. For the model of Fig. 10–4 the product $k^2\,\Delta x$ represents the tension in one of the springs when it is stretched a distance Δx. But precisely the same tension must hold in each half of the spring, which is stretched only a distance $\frac{1}{2}\,\Delta x$. Hence, if we always use springs of the same stiffness, $k^2\,\Delta x$ will approach as limit a constant force K^2. For the model of Fig. 10–5 one has indeed $k^2\,\Delta x = k_0^2(\Delta x - l)$, where k_0^2 is the actual spring constant for each spring; therefore $k^2\,\Delta x$ represents precisely the tension in each spring when all displacements u_σ are 0; the limiting value K^2 is precisely the tension in the string.

One can write

$$\frac{u_{\sigma+1} - 2u_\sigma + u_{\sigma-1}}{(\Delta x)^2} = \frac{u(x_\sigma + \Delta x, t) - 2u(x_\sigma, t) + u(x_\sigma - \Delta x, t)}{(\Delta x)^2},$$

where x_σ is the equilibrium position of P_σ. In the limit x_σ becomes a continuous variable and, as in Section 2–23, the ratio of the second difference to $(\Delta x)^2$ approaches as "limit" the derivative

$$\frac{\partial^2 u}{\partial x^2}(x, t).$$

The right-hand members we assume to approach as limit a function $F(x, t)$ representing applied force per unit length at x. We are thus led to the partial differential equation

$$\rho(x)\frac{\partial^2 u}{\partial t^2} + H(x)\frac{\partial u}{\partial t} - K^2\frac{\partial^2 u}{\partial x^2} = F(x, t). \tag{10–61}$$

This is the fundamental partial differential equation to be studied. Certain generalizations will be introduced in later sections, notably the replacement of $\partial^2 u/\partial x^2$ by the Laplacian $\nabla^2 u$:

$$\rho\frac{\partial^2 u}{\partial t^2} + H\frac{\partial u}{\partial t} - K^2\,\nabla^2 u = F(x, y, z, t). \tag{10–62}$$

This corresponds to a generalization to motion in two- or three-dimensional space. One can easily construct an N particle model for this. From the general equation (10–55) one obtains a broader class of equations in which the term $-K^2\,\nabla^2 u$ is replaced by a more complicated expression, possibly nonlinear, in u and its derivatives. Two- and three-dimensional problems can lead to *simultaneous partial differential equations.*

While the generalizations do introduce complications, the principal problems and methods reveal themselves in the equation (10–61) and, in fact, in the N particle approximation to this of the preceding section; as pointed out in the introductory section, even the single particle displays the properties which are crucial.

The limiting process by which we arrived at (10–61) was based on physical intuition and the basic test of the validity of the result is its accuracy in explaining the behavior of continuous media. This is a problem of physics, by no means simple, with which we shall not concern ourselves. However, one can ask the purely mathematical question: do the solutions of difference equations converge to the solutions of the corresponding differential equations, when the basic interval (e.g., Δx) approaches 0? This question has been made precise and answered in a generally affirmative fashion in recent research. We refer to pages 160–196 of the book by Tamarkin and Feller listed at the end of the chapter for a discussion of the problem and further references to the literature.

10–6 Classification of partial differential equations. Basic problems. Equations (10–61) and (10–62) are linear in u and its derivatives and are hence *linear partial differential equations*. They involve derivatives of u up to the second order and are hence partial differential equations of *second order*. The most general linear partial differential equation of second order in two independent variables has the form

$$A\frac{\partial^2 u}{\partial x^2} + 2B\frac{\partial^2 u}{\partial x\,\partial y} + C\frac{\partial^2 u}{\partial y^2} + D\frac{\partial u}{\partial x} + E\frac{\partial u}{\partial y} + Fu + G = 0, \qquad (10\text{–}63)$$

where $A, \ldots, G$ are functions of x and y. Equations (10–63) are classified into three types:

elliptic: $B^2 - AC < 0$, $A\xi^2 + 2B\xi\eta + C\eta^2 = 1$ is an ellipse;

parabolic: $B^2 - AC = 0$, $A\xi^2 + 2B\xi\eta + C\eta^2 + D\xi + E\eta = 0$ is a parabola;

hyperbolic: $B^2 - AC > 0$, $A\xi^2 + 2B\xi\eta + C\eta^2 = 1$ is a hyperbola.

An equation can be of one type in part of the xy plane and of another type in a second part. An analogous classification is made for equations in three or more independent variables. The three types are illustrated respectively by the equations

$$\frac{\partial^2 u}{\partial x^2} + \frac{\partial^2 u}{\partial y^2} = 0, \quad \frac{\partial u}{\partial x} - \frac{\partial^2 u}{\partial y^2} = 0, \quad \frac{\partial^2 u}{\partial x^2} - \frac{\partial^2 u}{\partial y^2} = 0.$$

The first of these is the equation $\nabla^2 u = 0$ and occurs naturally in connection with the equilibrium problem for *two* dimensions. The second corresponds to *exponential decay,* the third to *harmonic motion.*

The differential equation (10–61) was proposed as a natural one for the longitudinal oscillations of a rod, as in Fig. 10–6, or for the transverse vibrations of a string. There are other one-dimensional problems to which the equation is applicable: planar sound waves and electromagnetic waves, diffusion of heat, and other diffusion processes ($\rho = 0$). The equation can also be applied to *infinite* intervals on the x axis; while the vibration of a string infinite in length may appear to be an artificial concept, such an ideal case is of use in applications. Equation (10–62) has analogous applications in two and three dimensions, including the basic hyperbolic, parabolic, and elliptic equations:

wave equation: $$\rho \frac{\partial^2 u}{\partial t^2} - K^2 \nabla^2 u = 0,$$

heat equation: $$H \frac{\partial u}{\partial t} - K^2 \nabla^2 u = 0,$$

Laplace equation: $$\nabla^2 u = 0;$$

here ρ, H, and K^2 are usually considered to be constants. As pointed out in Section 5–15, the Laplace equation is satisfied by the velocity potential of an irrotational, incompressible fluid motion. The complete equations of hydrodynamics are simultaneous equations which are *nonlinear* (see Lamb's *Hydrodynamics*, Cambridge University Press, 1932).

The basic problems associated with (10–61) are simply the analogues for the continuous medium of the problems studied in the previous sections. For example, problem (a) concerns the case for which $\rho(x) > 0$, $H(x) = 0$, $F(x, t) = 0$, and the "walls" are fixed: $u(0, t) = 0$, $u(L, t) = 0$; we expect to show that there is one and only one solution to the initial value problem: $u = f(x)$, $\partial u/\partial t = g(x)$ for $t = 0$. Such a solution $u(x, t)$ would be defined and continuous for $0 \leqq x \leqq L$ and for $t \geqq 0$, and would be required to have partial derivatives through the second order for $0 < x < L$ and $t > 0$ and to satisfy the differential equation in the *domain* just described. One can also consider the possibility of discontinuities on the boundary; this requires care, but significant results can be obtained. The word "solution" will refer to functions $u(x, t)$ continuous for $0 \leqq x \leqq L$, $t \geqq 0$, unless otherwise indicated. Problems (b) and (c) (damped vibrations and exponential decay) are formulated in a similar manner.

The equilibrium problem (d) now becomes an *ordinary* differential equation:

$$-K^2 \frac{d^2u}{dx^2} = F(x),$$

with boundary conditions: $u = u_0$ for $x = 0$, $u = u_1$ for $x = L$. In two dimensions the analogous equation is the *Poisson equation*

$$-K^2 \nabla^2 u = F(x, y),$$

where u has given values on the boundary of a two-dimensional region; when $F \equiv 0$, this is the Dirichlet problem. In all cases, we wish to show that there is a unique equilibrium state $u^*(x)$ [in two dimensions, $u^*(x, y)$]. As for the N particle problem, the approach to equilibrium, problem (e), is described by a function $u^*(x) + u(x, t)$ where $u(x, t)$ is a solution of a *homogeneous* problem (a), (b), or (c).

The problem (f) of forced motion includes the other five problems as special cases. Boundary conditions: $u(0, t) = u_0(t)$, $u(L, t) = u_1(t)$ and initial values of u are given; we wish to show that there is one and only one corresponding solution.

The *methods* to be used are a natural extension of those used for the N particle problem. The homogeneous problems are handled by a substitution: $u(x, t) = A(x)e^{\lambda t}$, which gives the normal modes; the "general solution" is again obtained as a linear combination of normal modes. The nonhomogeneous problem of forced motion is solved by variation of parameters; the general solution is the sum of a particular solution $u^*(x, t)$ and the general solution of the homogeneous problem.

We have considered only the cases of fixed walls or walls moving in a prescribed manner. There are other natural boundary conditions; for example, one could require that $\partial u/\partial x$ be 0 for $x = 0$. For the N particle case this would correspond to the requirement that the wall P_0 move in such a fashion that $u_0 = u_1$; accordingly, the distance between P_0 and P_1 is fixed and *no energy can be transmitted*. For the heat conduction problem, this corresponds to an *insulated* boundary at $x = 0$.

10–7 The wave equation in one dimension. Harmonic motion. In the basic equation (10–61) we assume $H(x) \equiv 0$ and $F(x, t) \equiv 0$ and that ρ is constant, independent of x. The differential equation becomes

$$\rho \frac{\partial^2 u}{\partial t^2} - K^2 \frac{\partial^2 u}{\partial x^2} = 0, \quad 0 < x < L, \quad t > 0. \tag{10–64}$$

The equation is to be applied to a rod or string occupying the portion of the x axis between $x = 0$ and $x = L$. We assume the ends are fixed:

$$u(0, t) = 0, \quad u(L, t) = 0.$$

By a change of scale: $x' = \pi x/L$, we can reduce the problem to the case for which $L = \pi$. The equation becomes

$$\rho L^2 \frac{\partial^2 u}{\partial t^2} - \pi^2 K^2 \frac{\partial^2 u}{\partial x'^2} = 0.$$

For simplicity we drop the prime in the following. We introduce the abbreviation:

$$a = \frac{\pi K}{L\sqrt{\rho}}. \tag{10–65}$$

The equation and boundary conditions now read

$$\frac{\partial^2 u}{\partial t^2} - a^2 \frac{\partial^2 u}{\partial x^2} = 0, \quad 0 < x < \pi, \quad t > 0, \tag{10–66}$$

$$u(0, t) = 0, \quad u(\pi, t) = 0. \tag{10–67}$$

To determine the normal modes, we can now set

$$u(x) = A(x)e^{\lambda t} \tag{10–68}$$

in (10–66), (10–67). However, we find as in the N particle case that λ would have to be pure imaginary (Prob. 5 below) and we simplify the process by replacing (10–68) by the substitution

$$u(x) = A(x)\sin(\lambda t + \epsilon). \tag{10–68'}$$

Equations (10–66), (10–67) then become the two equations

$$-A(x)\lambda^2 - a^2A''(x) = 0, \tag{10–69}$$

$$A(0) = A(\pi) = 0. \tag{10–70}$$

The simultaneous linear equations of the N particle problem are therefore replaced by a differential equation with boundary conditions. This is anticipated in Probs. 5–12 following Section 10–4, in which it is shown that the simultaneous equations of the N particle problem can be treated as a *difference* equation with boundary conditions. The solutions obtained there are very closely related to those to be obtained for (10–69), (10–70).

The general solution of (10–69) is

$$A(x) = c_1 \sin\left(\frac{\lambda x}{a}\right) + c_2 \cos\left(\frac{\lambda x}{a}\right). \tag{10–71}$$

Equations (10–70) are satisfied only if

$$c_2 = 0, \quad \sin\left(\frac{\lambda\pi}{a}\right) = 0.$$

One obtains the *characteristic values* (resonant frequencies or *eigenvalues*)

$$\lambda_n = an \quad (n = 1, 2, \ldots), \tag{10–72}$$

and associated *characteristic functions* (or *eigenfunctions*)

$$A_n(x) = \sin nx \quad (n = 1, 2, \ldots). \tag{10–73}$$

We restrict to positive λ, since a change in sign can be absorbed in the phase constant ϵ. The normal modes are

$$\sin nx \sin(ant + \epsilon_n) \tag{10–74}$$

and constant multiples thereof; there are *infinitely many normal modes.* The set of frequencies λ_n occurring is called the *spectrum.*

We now attempt to construct the general solution $u(x, t)$ as a linear combination of normal modes:

$$u(x, t) = \sum_{n=1}^{\infty} c_n \sin nx \sin(ant + \epsilon_n). \tag{10–75}$$

However, we face a new difficulty: *the infinite series* (10–75) *may fail to converge.* Even if it does converge, it may fail to satisfy the differential equation (10–66), for this requires existence of second derivatives. Now the series (10–75) can be regarded as a Fourier sine series in x, with coefficients dependent on t. From the theory of Fourier series (Chapter 7) we easily obtain conditions on the constants c_n such that the series converge for all x and can be differentiated twice with respect to x and t.

The choice of the constants in (10–75) depends on the initial conditions, for we can write (10–75) in the form:

$$u(x, t) = \sum_{n=1}^{\infty} \sin nx[\alpha_n \sin (ant) + \beta_n \cos (ant)], \tag{10–75′}$$
$$\alpha_n = c_n \cos \epsilon_n, \quad \beta_n = c_n \sin \epsilon_n.$$

Then, if we assume the series concerned are uniformly convergent,

$$u(x, 0) = \sum_{n=1}^{\infty} \beta_n \sin nx, \quad \frac{\partial u}{\partial t}(x, 0) = \sum_{n=1}^{\infty} na\alpha_n \sin nx. \tag{10–76}$$

Thus β_n and $na\alpha_n$ are the Fourier sine coefficients of the initial displacement and velocity respectively.

THEOREM. *If the constants c_n are such that $c_n n^4$ is bounded:*

$$|c_n| < \frac{M}{n^4} \quad (n = 1, 2, \ldots), \tag{10–77}$$

then the series (10–75) *converges uniformly for all x and t and defines a solution of the wave equation* (10–66) *for all x and t. Let $f(x)$ and $g(x)$be defined for $0 \leqq x \leqq \pi$; let $f(x)$ have continuous derivatives through the fourth order and let $f(0) = f(\pi) = f''(0) = f''(\pi) = 0$; let $g(x)$ have continuous derivatives through the third order and let $g(0) = g(\pi) = g''(0) = g''(\pi) = 0$. Then there exists a solution $u(x, t)$ of the wave equation* (10–66) *with boundary conditions* (10–67) *such that*

$$u(x, 0) = f(x), \quad \frac{\partial u}{\partial t}(x, 0) = g(x); \tag{10–78}$$

namely, the series (10–75′), *where*

$$\beta_n = \frac{2}{\pi}\int_0^{\pi} f(x) \sin nx \, dx, \quad \alpha_n = \frac{2}{na\pi}\int_0^{\pi} g(x) \sin nx \, dx. \tag{10–79}$$

The solution is unique: that is, if $u(x, t)$ satisfies (10–66), (10–67), *and* (10–78), *and the partial derivatives u_{xx}, u_{tt} are continuous for $0 \leqq x \leqq \pi$, $t \geqq 0$, then $u(x, t)$ is necessarily represented by the series* (10–75′), *with coefficients given by* (10–79).

Proof. If (10–77) holds, then the Weierstrass M-test (Section 6–13) shows that the series (10–75) converges uniformly for all x and t. Similarly, the series

$$-\sum n^2 c_n \sin nx \sin (ant + \epsilon_n), \quad -\sum n^2 a^2 c_n \sin nx \sin (ant + \epsilon_n)$$

obtained by differentiating (10–75) twice with respect to x and t converge uniformly for all x and t; for by (10–77) $|n^2 c_n| < Mn^{-2}$. Accordingly, these series represent u_{xx} and u_{tt} respectively. By substitution in (10–66), we verify that the wave equation is satisfied.

If $f(x)$ and $g(x)$ satisfy the conditions stated, then (Prob. 4 below) $n^4\alpha_n$ and $n^4\beta_n$ are bounded so that

$$n^4c_n = \sqrt{(n^4\alpha_n)^2 + (n^4\beta_n)^2}$$

is bounded. Accordingly (10–77) holds, so that (10–75) or (10–75′) represents a solution $u(x, t)$ which is continuous for all x and t. When $t = 0$. the series for u and u_t reduce to the Fourier sine series of $f(x)$ and $g(x)$; these series converge to $f(x)$ and $g(x)$ (Prob. 1 following Section 7–13).

The proof of uniqueness is left to Prob. 6 below.

10–8 Properties of solutions of the wave equation. We consider the solutions in the form

$$u(x, t) = \sum_{n=1}^{\infty} \sin nx[\alpha_n \sin (nat) + \beta_n \cos (nat)]. \tag{10–80}$$

For each fixed x, the series is a Fourier series in t, with period $2\pi/a$ (Section 7–5). Accordingly *each point of the rod or string considered moves in a periodic fashion*, with period $2\pi/a$.

The normal mode for which $n = 1$ is called the *fundamental mode*. Here

$$u(x, t) = \sin x[\alpha_1 \sin (at) + \beta_1 \cos (at)].$$

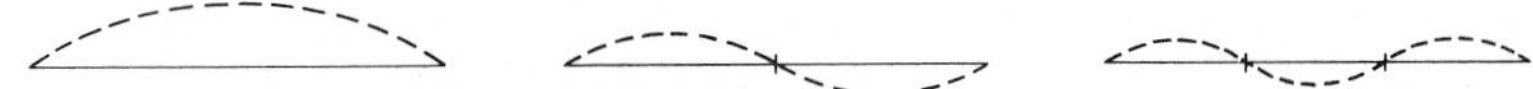

FIG. 10–7. Normal modes for vibrating string.

This is easily visualized in the case of a vibrating string, for which the displacement has the shape of a sine curve at all times (Fig. 10–7). The string is therefore vibrating in this shape with frequency $a/2\pi$ (cycles per unit time). The modes corresponding to $n = 2, 3, \ldots$ are called the *first overtone, second overtone*, etc.; musically they give the octave, octave plus a fifth, etc. These are suggested in Fig. 10–7. They are easily demonstrated on a stringed instrument, especially on low notes such as low C on a violoncello. We remark that the shapes of the normal modes are precisely the characteristic functions $A_n(x) = \sin nx$ and that these functions form a *complete orthogonal system* for the interval $0 \leqq x \leqq \pi$.

The relation between solution and initial conditions can be shown in a striking way by the following observation. Let us first assume that $g(x) \equiv 0$, so that the rod (or vibrating string) is initially at rest but has an initial displacement $f(x)$. By (10–79), $\alpha_n = 0$ for all n. We can now write

$$u(x, t) = \sum_{n=1}^{\infty} \beta_n \sin nx \cos nat = \tfrac{1}{2} \sum_{n=1}^{\infty} \beta_n[\sin n(x + at) + \sin n(x - at)].$$

Since

$$f(x) = \sum_{n=1}^{\infty} \beta_n \sin nx, \tag{10–81}$$

we can write:

$$u(x, t) = \tfrac{1}{2}[f(x + at) + f(x - at)]. \tag{10–82}$$

This representation is at first valid only for

$$0 \leqq x + at \leqq \pi, \quad 0 \leqq x - at \leqq \pi.$$

However, if we extend the definition of $f(x)$ to all x by (10–81), then (10–82) has meaning for all x and t and, under the assumptions of the theorem above, represents a solution of the wave equation for all x and t.

The term $f(x + at)$ represents the initial displacement translated at units to the left; the second term represents this displacement translated at units to the right; this is suggested in Fig. 10–8. In Fig. 10–9, $f(x)$ is chosen as a displacement confined almost entirely to an interval $\frac{1}{2}\pi - \delta < x < \frac{1}{2}\pi + \delta$, where δ is small; the solution can then be plotted as a function of x and t. The disturbance is seen to split into two disturbances which travel in opposite directions until they reach the walls, where they are reflected, with a change in sign, and move back together. This can be demonstrated experimentally in various ways: by displacing and releasing a violin string; by sound echoes.

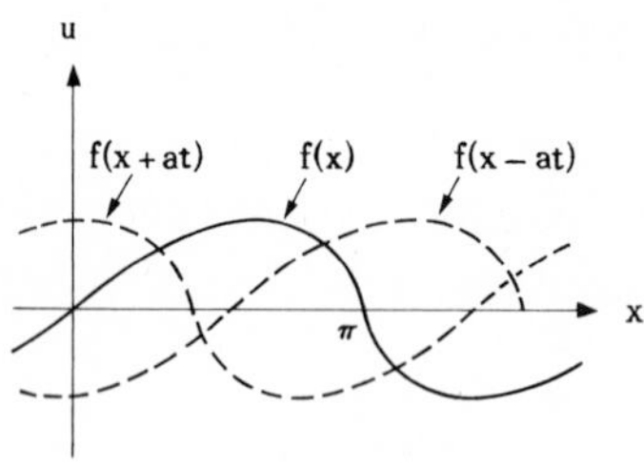

Fig. 10–8. $f(x)$ versus $f(x+at)$, $f(x-at)$.

If the initial displacement $f(x)$ has a jump discontinuity, e.g., at x_0, but is piecewise very smooth, then (10–82) continues to define a solution of the wave equation except for $x \pm at = x_0 \pm k\pi$. These lines are the paths of "propagation of discontinuities"; they are called *characteristics*.

The constant a appears as the velocity with which the disturbance or discontinuity is propagated to the left and right; it is termed the *wave velocity*. If we choose as u a single mode of vibration:

$$u(x, t) = \tfrac{1}{2}\beta_n[\sin n(x + at) + \sin n(x - at)],$$

then the initial disturbance $f(x)$ is simply $\beta_n \sin nx$, a wave repeating itself at intervals of $2\pi/n$, the *wavelength*. The oscillations at each x have a frequency of $na/2\pi$ (cycles per unit time). Hence

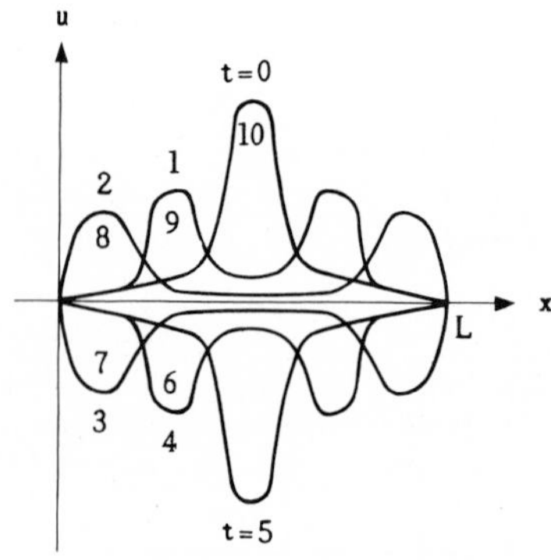

Fig. 10–9. Solution $u(x, t)$ of the wave equation. The curves show the wave forms for $t = 0, 1, \ldots, 10$. The units are chosen so that the wave velocity a is 0.2 L per unit of time.

$$(wavelength) \cdot (frequency) = \frac{2\pi}{n} \cdot \frac{na}{2\pi} = a = wave\ velocity.$$

This is one of the fundamental rules of physics.

We now consider the case when $f(x) = 0$ but the initial velocity $g(x)$ is different from zero. Now, as above,

$$g(x) = \sum_{n=1}^{\infty} na\alpha_n \sin nx. \tag{10–83}$$

Hence we can integrate (Section 7–13):

$$\int g(x)\,dx = \sum_{n=1}^{\infty} (-a\alpha_n \cos nx) + \text{const.} \tag{10–84}$$

The solution $u(x, t)$ is therefore

$$\begin{aligned} u(x, t) &= \sum_{n=1}^{\infty} \alpha_n \sin nx \sin nat = \tfrac{1}{2} \sum_{n=1}^{\infty} \alpha_n \{\cos n(x - at) - \cos n(x + at)\} \\ &= \frac{1}{2a} \int_{x-at}^{x+at} g(s)\,ds, \end{aligned} \tag{10–85}$$

where s is a dummy variable of integration. The representation (10–85) is valid for all x and t, if we use (10–83) to extend the definition of $g(x)$ to all x. Equation (10–85) can be interpreted as the *difference* of two disturbances moving to the left and right with velocity a.

If we allow for both initial displacement and initial velocity, the same reasoning as above leads to the general formula

$$u(x, t) = \tfrac{1}{2}[f(x + at) + f(x - at)] + \frac{1}{2a} \int_{x-at}^{x+at} g(s)\,ds. \tag{10–86}$$

Problems

1. Let $f(x)$ be an odd periodic function of period 2π and let $f(x) = 0$ for $0 < x < \pi/3$ and for $2\pi/3 < x < \pi$, $f(x) = 1$ for $\pi/3 < x < 2\pi/3$. Show that (10–82) defines a solution of the wave equation for a certain portion of the xt plane; analyze the solution graphically as in Figs. 10–8, 10–9. Is the Fourier series solution valid with this $f(x)$ as the initial displacement?

2. Let $g(x)$ be an odd periodic function of period 2π and let $g(x) = 0$ for $0 < x < \pi/3$ and for $2\pi/3 < x < \pi$, $g(x) = 1$ for $\pi/3 < x < 2\pi/3$. Show that (10–85) defines a solution of the wave equation for a certain portion of the xt plane; analyze the solution graphically. Is the Fourier series solution valid with this $g(x)$ as the initial velocity?

3. Show that the change of variables: $r = x + at$, $s = x - at$, converts the wave equation (10–66) into the equation

$$\frac{\partial^2 u}{\partial r\,\partial s} = 0. \tag{a}$$

Interpret the change of variables geometrically. Show that the "general solution" of (a) has the form

$$u = F(r) + G(s).$$

Discuss the relation between this representation and (10–86).

4. (a) Prove that, if $f(x)$ satisfies the conditions stated in the theorem of Section 10–7, then $\beta_n n^4$ is bounded. [Hint: use integration by parts as in Section 7–8.]

(b) Prove that, if $g(x)$ satisfies the conditions stated in the theorem of Section 10–7, then $\alpha_n n^4$ is bounded.

5. Show that the substitution (10–68) in (10–66), (10–67) leads to equations which are satisfied only when $\lambda_n = ani$ and $A = A_n(x) = c(e^{inx} - e^{-inx})$. From these expressions obtain the normal modes (10–74).

6. Prove that, under the conditions stated in the theorem of Section 10–7, the solution $u(x, t)$ satisfying initial conditions (10–78) must have the form (10–75′) and is hence uniquely determined. [Hint: under the assumptions made, $u(x, t)$ has a representation as a Fourier sine series in x:

$$u = \sum_{n=1}^{\infty} \phi_n(t) \sin nx, \quad \phi_n(t) = \frac{2}{\pi} \int_0^{\pi} u(x, t) \sin nx \, dx.$$

Differentiate the second equation twice with respect to t, using Leibnitz's Rule (Section 4–12) and integration by parts to show that $\phi_n''(t) + a^2n^2\phi_n(t) = 0$. Hence $\phi_n(t) = \alpha_n \sin(nat) + \beta_n \cos(nat)$.]

7. To raise the pitch of a violin string one can (a) increase the tension, (b) decrease the density, (c) shorten the string. Show how these conclusions follow from (10–72) and (10–65).

8. Find the general solutions of the following partial differential equations for $0 < x < \pi$, $t > 0$, with boundary conditions: $u(0, t) = u(\pi, t) = 0$:

(a) $\dfrac{\partial^2 u}{\partial t^2} - \dfrac{\partial^2 u}{\partial x^2} - u = 0$; (b) $\dfrac{\partial^2 u}{\partial t^2} - \dfrac{\partial^2 u}{\partial x^2} - 2\dfrac{\partial u}{\partial x} = 0$.

9. Find the general solution of the wave equation

$$\frac{\partial^2 u}{\partial t^2} - a^2 \frac{\partial^2 u}{\partial x^2} = 0, \quad 0 < x < 2\pi, \quad t > 0,$$

such that $u(0, t) = u(2\pi, t)$, $u_x(0, t) = u_x(2\pi, t)$.

10. Find the general solution of the wave equation

$$\frac{\partial^2 u}{\partial t^2} - a^2 \frac{\partial^2 u}{\partial x^2} = 0, \quad 0 < x < \pi, \quad t > 0,$$

such that $u(0, t) = 0$, $u_x(\pi, t) = 0$.

Answers

8. (a) $\displaystyle\sum_{n=1}^{\infty} c_n \sin nx \sin(\sqrt{n^2 - 1}\,t + \epsilon_n)$,

(b) $\displaystyle\sum_{n=1}^{\infty} c_n e^{-x} \sin nx \sin(\sqrt{n^2 + 1}\,t + \epsilon_n)$.

9. $\displaystyle\sum_{n=1}^{\infty} (a_n \cos nx + b_n \sin nx) \sin(nat + \epsilon_n)$.

10. $\displaystyle\sum_{n=1}^{\infty} c_n \sin(n + \tfrac{1}{2})x \sin[(n + \tfrac{1}{2})at + \epsilon_n]$.

10–9 The one-dimensional heat equation. Exponential decay. We return to the basic equation of Section 10–5:

$$\rho \frac{\partial^2 u}{\partial t^2} + H \frac{\partial u}{\partial t} - K^2 \frac{\partial^2 u}{\partial x^2} = F(x, t).$$

We neglect masses; i.e., we set $\rho = 0$. We assume H is a constant, that there is no outside force F, and that the "walls" are fixed. The differential equation and boundary conditions become

$$H \frac{\partial u}{\partial t} - K^2 \frac{\partial^2 u}{\partial x^2} = 0, \quad 0 < x < L, \quad t > 0; \tag{10–87}$$

$$u(0, t) = 0, \quad u(L, t) = 0. \tag{10–88}$$

Equation (10–87) is the *one-dimensional heat equation*. The three-dimensional heat equation

$$\frac{\partial T}{\partial t} - c^2 \nabla^2 T = 0 \tag{10–89}$$

was derived in Section 5–15 [Eq. (5–131)]; Eq. (10–87) can be regarded as the special case of heat conduction in an infinite slab bounded by planes $x = 0$, $x = L$ in space, the boundary conditions being such that the temperature depends only on x. One can also interpret the equation as describing conduction of heat in a thin rod, insulated except at the ends.

The transition from a system of masses attached by springs to the problem of heat conduction may at first seem artificial. However, the solutions of (10–87) do indeed display the properties of a limiting case of the system of masses as the friction increases and the total mass approaches 0. Because masses are neglected, disturbances can be propagated *instantaneously*; the wave velocity is *infinite*.

As in the preceding sections, we introduce a new variable $x' = \pi x/L$ and make the substitution:

$$c = \left(\frac{\pi^2 K^2}{L^2 H}\right)^{\frac{1}{2}}. \tag{10–90}$$

After dropping the primes, the equation and boundary conditions become

$$\frac{\partial u}{\partial t} - c^2 \frac{\partial^2 u}{\partial x^2} = 0, \quad t > 0, \quad 0 < x < \pi, \tag{10–91}$$

$$u(0, t) = 0, \quad u(\pi, t) = 0. \tag{10–92}$$

The substitution

$$u = A(x)e^{\lambda t} \tag{10–93}$$

leads to the characteristic value problem:

$$A''(x) - \frac{\lambda}{c^2} A = 0, \tag{10–94}$$

$$A(0) = A(\pi) = 0. \tag{10–95}$$

If λ is positive or zero, the only solution of (10–94), (10–95) is the trivial solution: $A(x) \equiv 0$ (Prob. 9 below). If λ is negative, the solutions are the *characteristic functions*

$$A_n(x) = \sin nx \quad (n = 1, 2, \ldots), \tag{10–96}$$

with associated *characteristic values*

$$\lambda_n = -n^2c^2. \tag{10–97}$$

Accordingly, the "normal modes" are the functions

$$\sin nx\, e^{-c^2n^2t} \tag{10–98}$$

and constant multiples thereof. We expect the general solution to be given by the linear combinations

$$u(x, t) = \sum_{n=1}^{\infty} b_n \sin nx\, e^{-c^2n^2t}, \tag{10–99}$$

where the b_n are arbitrary constants.

Again we have to be careful about convergence. The problem is simpler than that for the wave equation, for if the coefficients b_n are bounded: $|b_n| < M$ for $n = 1, 2, \ldots$, then the series (10–99) converges uniformly in x and t in each half-plane: $t \geqq t_1$, $-\infty < x < \infty$, provided $t_1 > 0$. This is a consequence of the Weierstrass M-test (Section 6–13), for

$$|b_n \sin nx\, e^{-c^2n^2t}| < Me^{-c^2n^2t_1} = M_n;$$

the series M_n is a series of constants whose convergence follows from the ratio test. In the same way we verify that the series (10–99) remains uniformly convergent in the range stated, when differentiated as often as desired with respect to x and t (Prob. 5 below). In particular,

$$\frac{\partial u}{\partial t} = \sum_{n=1}^{\infty} b_n(-c^2n^2) \sin nx\, e^{-c^2n^2t} = c^2\frac{\partial^2 u}{\partial x^2},$$

so that (10–99) is a solution of the heat equation in the range: $t > 0$, $-\infty < x < \infty$. The boundary conditions (10–92) are satisfied, since each term in the series is 0 when $x = 0$ or $x = \pi$. Every sufficiently smooth solution of the heat equation and boundary conditions in the range described must have the form (10–99) (Prob. 7 below).

For $t = 0$, the series (10–99), if convergent, reduces to

$$u(x, 0) = \sum_{n=1}^{\infty} b_n \sin nx. \tag{10–100}$$

Thus, just as for the wave equation, the initial values of u: $u(x, 0) = f(x)$, are represented by a Fourier sine series. If, for example, $f(0) = f(\pi) = 0$ and $f(x)$ has a continuous second derivative for $0 \leqq x \leqq \pi$, then its sine series will converge uniformly to $f(x)$ and the series (10–99) will converge uniformly for $0 \leqq x \leqq \pi$ and $t \geqq 0$. Accordingly, under the assumptions stated, the series (10–99) defines a function $u(x, t)$ which is continuous for

$t \geqq 0$, $0 \leqq x \leqq \pi$ and which satisfies the heat equation (10–91), the boundary conditions (10–92), and the initial condition: $u(x, 0) = f(x)$. The solution is furthermore unique (Prob. 7 below). These results show that, with minor modifications, the theorem of Section 10–7 holds for the heat equation.

If $f(x)$ is merely piecewise continuous, the constants

$$b_n = \frac{2}{\pi}\int_0^\pi f(x)\sin nx\,dx \tag{10–101}$$

are bounded, since the sequence b_n converges to 0 [Eq. (7–14), Section 7–4]. Accordingly, the series (10–99) defines a solution of the heat equation for $t > 0$. The series converges to $f(x)$ *in the mean* (Section 7–11) for $t = 0$; in fact, it can be shown that

$$\lim_{t\to 0+} u(x, t) = f(x), \quad 0 < x < \pi$$

for each x at which $f(x)$ is continuous.

10–10 Properties of solutions of the heat equation. For each fixed x, each term of the series (10–99) describes an exponential approach to 0. A similar statement applies to the sum of the series; that is,

$$\lim_{t\to\infty} u(x, t) = 0$$

(Prob. 6 below). The rate of decay varies with n; the high frequency terms in (10–99) have smaller exponential coefficients than the lower frequency terms and are accordingly damped out more rapidly. This corresponds to the observed fact that abrupt variations in temperature in an object disappear rapidly, while differences in temperature at large distances are evened out slowly.

If we write

$$\sin nx = nx - \frac{(nx)^3}{3!} + \frac{(nx)^5}{5!} - \cdots$$

in (10–99) and collect terms in powers of x, we obtain a series:

$$\sum_{k=1}^{\infty} \phi_k(t)x^{2k-1}, \quad \phi_k(t) = \sum_{n=1}^{\infty} b_n \frac{(-1)^{k-1}(n)^{2k-1}}{(2k-1)!} e^{-c^2n^2t}. \tag{10–102}$$

This operation can be justified by theorems on series or, more simply, by complex variables as in Prob. 4 below. Accordingly, for each fixed $t > 0$ the solution $u(x, t)$ can be represented by a power series in x; the power series has infinite convergence radius, so that *the series converges for all* x. The solutions $u(x, t)$ are *analytic in* x.

From this last result we deduce another property of the solutions: the infinite speed of propagation of disturbances. For example, let the initial function $f(x)$ be 0 for $0 \leqq x \leqq \pi$ and greater than 0 for $x_1 < x < x_2$, where $0 < x_1 < x_2 < \pi$. In the case of the wave equation, the solution $u(x, t)$ would remain identically 0 near $x = 0$ until the "wave" starting at $x = x_1$

could move across. For the heat equation, the solution $u(x, t)$ takes on nonzero values in every x interval for every $t > 0$. To establish this, we consider a fixed positive t. Then $u(x, t)$ can be considered as an analytic function of a complex variable x. By Theorem 26 of Section 9–13, if $u(x, t) \equiv 0$ when x ranges over a whole interval of real values, then $u \equiv 0$ for this value of t. Hence necessarily $b_n = 0$ for $n = 1, 2, \ldots$, so that $f(x) \equiv 0$, contrary to assumption. Accordingly, for each positive t, $u(x, t)$ takes on nonzero values in every interval of x. The disturbance is propagated *instantaneously*.

10–11 Equilibrium and approach to equilibrium. By analogy with case (d) of Sections 10–2 and 10–3, we allow for applied forces $F(x)$ and "wall displacements" u_0 and u_1 which do not vary with time. The differential equation and boundary conditions are as follows:

$$\rho(x)\frac{\partial^2 u}{\partial t^2} + H(x)\frac{\partial u}{\partial t} - K^2\frac{\partial^2 u}{\partial x^2} = F(x), \tag{10–103}$$

$$u(0) = u_0, \quad u(L) = u_1. \tag{10–104}$$

We seek a solution $u^*(x)$ independent of time; $u^*(x)$ then describes the equilibrium state of the system. Accordingly, we replace derivatives with respect to t by 0 in (10–103) and are led to the problem

$$-K^2\frac{d^2u}{dx^2} = F(x), \quad 0 < x < L, \tag{10–105}$$

$$u(0) = u_0, \quad u(L) = u_1. \tag{10–106}$$

Equation (10–105) is an *ordinary* differential equation whose general solution is obtained by integrating twice:

$$u = \int\left\{\int \frac{-F(x)}{K^2}\,dx\right\}dx.$$

If a particular choice $G(x)$ of the indefinite integral is made, so that $G''(x) = -F(x)/K^2$, then the general solution is

$$u = Ax + B + G(x). \tag{10–107}$$

The boundary conditions give the equations

$$B + G(0) = u_0, \quad AL + B + G(L) = u_1.$$

These are easily solved for A and B. The equilibrium state is then

$$u^*(x) = [u_1 - u_0 + G(0) - G(L)]\frac{x}{L} + u_0 - G(0) + G(x). \tag{10–108}$$

We have assumed $F(x)$ to be continuous for $0 \leqq x \leqq L$, so that the integrals above have meaning.

To describe the *approach to equilibrium*, we let $u(x, t)$ be an arbitrary solution of (10–103), (10–104) for $t > 0$, $0 < x < L$. Then

$$y(x, t) = u(x, t) - u^*(x)$$

satisfies the *homogeneous problem:*

$$\rho(x) \frac{\partial^2 y}{\partial t^2} + H(x) \frac{\partial y}{\partial t} - K^2 \frac{\partial^2 y}{\partial x^2} = 0, \tag{10–109}$$

$$y(0) = 0, \quad y(L) = 0. \tag{10–110}$$

This we verify by substitution in these equations and use of the fact that $u(x, t)$ satisfies (10–103), (10–104), while $u^*(x)$ is the solution of the equilibrium problem (10–105), (10–106). Accordingly,

$$u(x, t) = y(x, t) + u^*(x);$$

the general solution is formed of "complementary function" and particular solution in the usual manner.

If, in particular, $H(x) = 0$ and $\rho(x)$ is a positive constant ρ, then the complementary function $y(x, t)$ is a solution of the wave equation; the solutions $u(x, t)$ consist of oscillations about the equilibrium position. If H is also a positive constant, the oscillations are damped (cf. Prob. 8 below). If H is so large that ρ can be neglected, the function $y(x, t)$ is a solution of the heat equation; equilibrium is approached exponentially.

Problems

1. Determine the solution for $t \geqq 0$, $0 \leqq x \leqq \pi$ of the heat equation (10–91), with $c = 1$, such that $u(0, t) = 0$, $u(\pi, t) = 0$, $u(x, 0) = \sin x + 5 \sin 3x$. Plot the solution as a function of x and t and compare the rates of decay of the terms in $\sin x$ and $\sin 3x$.

2. Determine the solution for $t > 0$, $0 < x < \pi$ of the heat equation (10–91) which is continuous for $t \geqq 0$, $0 \leqq x \leqq \pi$ and has a continuous derivative $\partial u/\partial x$ in this region, and which furthermore satisfies the conditions: $\partial u/\partial x = 0$ for $x = 0$ and $x = \pi$, $u(x, 0) = f(x)$, where $f(x)$ has continuous first and second derivatives for $0 \leqq x \leqq \pi$. This can be interpreted as a problem in heat conduction in a slab whose faces are insulated.

3. Determine the solution, for $t > 0$, $0 < x < \pi$, of the equation

$$\frac{\partial u}{\partial t} - \frac{\partial^2 u}{\partial x^2} + 4u = 5 \sin x + 4x,$$

such that $u(0, t) = 0$, $u(\pi, t) = \pi$, $u(x, 0) = x + 2 \sin x$.

4. Prove that, if the constants b_n are bounded, then the series (10–99) can be written for each $t > 0$ as a power series in x, converging for all x. [Hint: let $t > 0$ be fixed and let

$$v(x, y) = \sum_{n=1}^{\infty} b_n \sin nx \cosh ny \, e^{-n^2c^2t}.$$

Show that the series for v converges uniformly for $-\infty < x < \infty$, $-y_1 \leqq y \leqq y_1$ by applying the M-test with

$$M_n = Me^{-\frac{1}{2}n^2c^2t}$$

for n sufficiently large. Show that the series remains uniformly convergent after differentiation any number of times with respect to x and y. Each term of the series is *harmonic* in x and y; hence conclude that $v(x, y)$ is harmonic for all x and y. By Section 9–11, $v(x, y)$ can be expanded as a power series in x and y; put $y = 0$ to obtain the desired series for $u(x, t)$.]

5. Prove that, if the constants b_n are bounded, then the series (10–99) remains uniformly convergent for $t \geqq t_1 > 0$, $-\infty < x < \infty$, after differentiation any number of times with respect to x and t.

6. Prove that, if the constants b_n are bounded: $|b_n| < M$, then the function $u(x, t)$ defined by (10–99) converges *uniformly* to 0 as $t \to \infty$; that is, given $\epsilon > 0$, a t_0 can be found such that $|u(x, t)| < \epsilon$ for $t > t_0$ and $-\infty < x < \infty$. [Hint: show that $|u(x, t)|$ is less than the sum of the geometric series $M \sum_{n=1}^{\infty} (e^{-c^2t})^n$.]

7. Let $u(x, t)$ have continuous derivatives through the second order in x and t for $t \geqq 0$ and $0 \leqq x \leqq \pi$ and let $u(0, t) = u(\pi, t) = 0$. Prove that, if $u(x, t)$ satisfies the heat equation (10–91) for $t > 0$, $0 < x < \pi$, then $u(x, t)$ has the form (10–99). [Hint: see Prob. 6 following Section 10–8.]

8. Discuss the nature of the solutions for $0 < x < \pi$, $t > 0$ of the equation

$$\rho \frac{\partial u^2}{\partial t^2} + H \frac{\partial u}{\partial t} - K^2 \frac{\partial^2 u}{\partial x^2} = 0,$$

with boundary conditions: $u(0, t) = u(\pi, t) = 0$, if ρ, H, and K are positive constants.

9. Prove that, if $\lambda \geqq 0$, the equations (10–94), (10–95) have no solution other than the trivial one: $A(x) \equiv 0$.

Answers

1. $e^{-t} \sin x + 5e^{-9t} \sin 3x$.

2. $u = \sum_{n=0}^{\infty} a_n e^{-n^2c^2t} \cos nx, \quad a_n = \frac{2}{\pi} \int_0^{\pi} f(x) \cos nx \, dx$. 3. $x + \sin x + e^{-5t} \sin x$.

8. Solutions have form $e^{-at} \sum_{n=1}^{\infty} \sin nx(\alpha_n e^{\gamma_n t} + \beta_n e^{-\gamma_n t})$, where $a = \frac{1}{2}H/\rho$, $b = K^2/\rho$, $\gamma_n = (a^2 - bn^2)^{\frac{1}{2}}$. If $\gamma_n = 0$ for $n = m$, $e^{-\gamma_m t}$ is replaced by t.

10–12 Forced motion. We now consider the general problem of type (f):

$$\rho(x) \frac{\partial^2 u}{\partial t^2} + H(x) \frac{\partial u}{\partial t} - K^2 \frac{\partial^2 u}{\partial x^2} = F(x, t), \quad 0 < x < L, \quad t > 0, \qquad (10\text{–}111)$$

$$u(0, t) = a(t), \quad u(L, t) = b(t). \qquad (10\text{–}112)$$

This is the problem of response of the one-dimensional system to outside forces varying both in position and time. The fact that the boundary conditions (10–112) are variable shows that the motion is being forced at the ends $x = 0$, $x = L$ also.

Just as in the case of the equilibrium problem of Section 10–11, we reason that, if $u(x, t)$ is a particular solution of (10–111), (10–112) for $0 < x < L$, $t > 0$, then the general solution $u(x, t)$ in this domain is

$$u(x, t) = y(x, t) + u^*(x, t), \qquad (10\text{–}113)$$

where $y(x, t)$ is a solution of the homogeneous problem (10–109), (10–110).

Accordingly, the problem is that of determining a particular solution. We can further concentrate attention on the case of fixed "walls": i.e., $a(t) = b(t) = 0$. For let

$$g(x, t) = a(t)\left(1 - \frac{x}{L}\right) + \frac{x}{L} b(t);$$

$g(x, t)$ is simply the linear function of x which interpolates between the values $a(t)$ at $x = 0$, $b(t)$ at $x = L$. Now [if we assume $a(t)$ and $b(t)$ to have the requisite derivatives]

$$\rho(x) \frac{\partial^2 g}{\partial t^2} + H(x) \frac{\partial g}{\partial t} - K^2 \frac{\partial^2 g}{\partial x^2} = G(x, t),$$

$$G(x, t) = \left(1 - \frac{x}{L}\right)[\rho a''(t) + Ha'(t)] + \frac{x}{L} [\rho b''(t) + Hb'(t)].$$

Accordingly, if $u^*(x, t)$ satisfies (10–111) and (10–112), then $w(x, t) = u^*(x, t) - g(x, t)$ satisfies

$$\rho(x) \frac{\partial^2 w}{\partial t^2} + H(x) \frac{\partial w}{\partial t} - K^2 \frac{\partial^2 w}{\partial x^2} = F(x, t) - G(x, t) = F_1(x, t), \quad (10\text{–}114)$$

$$w(0, t) = a(t) - a(t) = 0, \quad w(L, t) = 0. \quad (10\text{–}115)$$

Conversely, if $w(x, t)$ satisfies (10–114) and (10–115), then we verify that $w(x, t) + g(x, t) = u^*(x, t)$ satisfies (10–111) and (10–112). Therefore we have reduced our problem to the case of fixed walls.

The determination of a function $w(x, t)$ is now accomplished by *variation of parameters*. The applicability of this method is not dependent on the fact that the coefficients ρ, H are constant. However, we here confine attention to the case of constant coefficients; the procedure in the general case differs only slightly.

Let us suppose first that $\rho \equiv 0$ and $H = \text{const} > 0$. Then the "complementary function" is the general solution of the heat equation. We change scale and introduce the abbreviations of Section 10–9, so that the solution is

$$\sum_{n=1}^{\infty} b_n \sin nx\, e^{-c^2n^2t}.$$

We now replace the constants b_n by functions $v_n(t)$ and seek a solution

$$w(x, t) = \sum_{n=1}^{\infty} v_n(t) \sin nx\, e^{-c^2n^2t}. \quad (10\text{–}116)$$

We proceed formally and then determine conditions under which the solution obtained is valid. Substitution of (10–116) in (10–114) (with $\rho = 0$) leads to the equation

$$H \sum_{n=1}^{\infty} \{v_n'(t) e^{-n^2c^2t} \sin nx\} = F_1(x, t);$$

the other terms cancel out. This equation is simply a Fourier sine series in x for $F_1(x, t)$. Hence

$$H \frac{dv_n}{dt} e^{-n^2c^2t} = \frac{2}{\pi} \int_0^\pi F_1(x, t) \sin nx \, dx,$$
$$v_n(t) = \frac{2}{\pi H} \int e^{n^2c^2} \left\{ \int_0^\pi F_1(x, t) \sin nx \, dx \right\} dt. \tag{10–117}$$

Since we seek only one particular solution, we can choose the indefinite integral here so that $v_n(0) = 0$; that is, we choose

$$v_n(t) = \frac{2}{\pi H} \int_0^t e^{n^2c^2s} \left\{ \int_0^\pi F_1(x, s) \sin nx \, dx \right\} ds,$$

where s is a dummy variable of integration. Accordingly,

$$v_n(t) = \frac{2}{\pi H} \int_0^t \int_0^\pi F_1(x, s) e^{n^2c^2s} \sin nx \, dx \, ds, \tag{10–118}$$

and the particular solution sought is

$$w(x, t) = \frac{2}{\pi H} \sum_{n=1}^{\infty} \left\{ \sin nx \, e^{-n^2c^2t} \int_0^t \int_0^\pi F_1(r, s) e^{n^2c^2s} \sin nx \, dx \, ds \right\}. \tag{10–119}$$

We now study the validity of the result. If $F_1(x, t)$ is defined for $t \geqq 0$ and $0 \leqq x \leqq \pi$ and is continuous in both variables in this region, then $v_n(t)$ is well defined by (10–118). Furthermore, let $\partial^2 F_1/\partial x^2$ be continuous in x and t, so that $\partial^2 F_1/\partial x^2$ has a maximum $M(t_1)$ in each rectangle: $0 \leqq x \leqq \pi$, $0 \leqq t \leqq t_1$; let also $F_1(0, t) = F_1(\pi, t) = 0$. Then, as in Section 7–8, we conclude that for $0 \leqq t \leqq t_1$

$$\left| \frac{2}{\pi} \int_0^\pi F_1(x, t) \sin nx \, dx \right| \leqq \frac{2M(t_1)}{n^2}.$$

Hence by (10–118), in this rectangle,

$$|v_n(t)| \leqq \frac{2M(t_1)}{Hn^2} \int_0^t e^{n^2c^2s} \, ds = \frac{2M(t_1)}{Hn^4c^2} (e^{n^2c^2t} - 1)$$

and each term of the series (10–119) is bounded by $2M(t_1)/c^2n^4H$. The series therefore converges uniformly and continues to do so after differentiation once with respect to t or twice with respect to x. We therefore find

$$H \frac{\partial w}{\partial t} - K^2 \frac{\partial^2 w}{\partial x^2} = \frac{2}{\pi} \sum_{n=1}^{\infty} \sin nx \int_0^\pi F_1(r, t) \sin nr \, dr.$$

The right-hand side is the Fourier sine series of $F_1(x, t)$ and, under the assumptions made, converges to $F_1(x, t)$. Hence (10–114) is satisfied, as is (10–115).

Therefore (10–119) is indeed the particular solution sought. The function

$$u^*(x, t) = w(x, t) + g(x, t)$$

is then a particular solution of (10–111) and (10–112), and

$$u(x, t) = y(x, t) + u^*(x, t) = y(x, t) + w(x, t) + g(x, t)$$

is the general solution, where

$$y(x, t) = \sum_{n=1}^{\infty} b_n \sin nx \, e^{-n^2c^2t},$$

the b_n being arbitrary constants. To determine a solution satisfying the initial condition $u(x, 0) = f(x)$, we have to determine the constants b_n so that

$$\begin{aligned} f(x) &= y(x, 0) + w(x, 0) + g(x, 0) \\ &= \sum_{n=1}^{\infty} b_n \sin nx + a(0)\left(1 - \frac{x}{L}\right) + \frac{x}{L} b(0); \end{aligned}$$

that is, the series $\Sigma b_n \sin nx$ must be the Fourier sine series of

$$f(x) - a(0)\left(1 - \frac{x}{L}\right) - \frac{x}{L} b(0).$$

We have throughout assumed $\rho = 0$ and $H =$ const. If ρ is a positive constant and $H = 0$, then the complementary function is the general solution of the wave equation:

$$\sum_{n=1}^{\infty} \sin nx(\alpha_n \sin nat + \beta_n \cos nat).$$

The trial function (10–116) is now replaced by the function:

$$w(x, t) = \sum_{n=1}^{\infty} \sin nx[p_n(t) \sin nat + q_n(t) \cos nat]. \qquad (10\text{–}116')$$

Substitution in (10–114) does not yield enough conditions to determine $p_n(t)$ and $q_n(t)$, because the equation is now of *second* order in t. We choose an extra set of conditions (cf. Section 8–11):

$$p_n'(t) \sin nat + q_n'(t) \cos nat = 0, \quad (n = 1, 2, \ldots). \qquad (10\text{–}120)$$

These conditions can also be obtained by replacing (10–114) by a system of equations:

$$\frac{\partial w}{\partial t} = z, \quad \rho \frac{\partial z}{\partial t} - K^2 \frac{\partial^2 w}{\partial x^2} = F_1(x, t) \qquad (10\text{–}121)$$

(Prob. 5 below). Proceeding as above, we find

$$\begin{aligned} p_n(t) &= \frac{2}{na\pi\rho} \int_0^t \int_0^{\pi} F_1(x, s) \cos nas \sin nx \, dx \, ds, \\ q_n(t) &= \frac{-2}{na\pi\rho} \int_0^t \int_0^{\pi} F_1(x, s) \sin nas \sin nx \, dx \, ds. \end{aligned} \qquad (10\text{–}122)$$

A similar procedure applies when ρ and H are both positive constants.

Problems

1. Find the solution of the partial differential equation

$$\frac{\partial u}{\partial t} - \frac{\partial^2 u}{\partial x^2} = x^2 \cos t - 2 \sin t, \quad 0 < x < \pi, \quad t > 0,$$

satisfying boundary conditions: $u(0, t) = 0$, $u(\pi, t) = \pi^2 \sin t$, and initial conditions: $u(x, 0) = \pi x - x^2$.

2. Let $u(x, t)$ be a solution of the partial differential equation

$$\frac{\partial^2 u}{\partial t^2} - \frac{\partial^2 u}{\partial x^2} = \sin x \sin \omega t, \quad 0 < x < \pi, \quad t > 0,$$

and boundary conditions: $u(0, t) = 0$, $u(\pi, t) = 0$, $u(x, 0) = 0$, $\partial u/\partial t(x, 0) = 0$. Show that resonance occurs only when $\omega = \pm 1$ and determine the form of the solution in the two cases: $\omega = \pm 1$, $\omega \neq \pm 1$. [The other resonant frequencies 2, 3, ... are not excited because the force $F(x, t)$ is orthogonal to the corresponding "basis vectors" $\sin 2x$, $\sin 3x$, ... ; cf. Prob. 9 following Section 10–3.]

3. Let the outside force $F(x, t)$ be given as a Fourier sine series:

$$F(x, t) = \sum_{n=1}^{\infty} F_n(t) \sin nx, \quad t \geqq 0, \quad 0 \leqq x \leqq \pi.$$

Obtain a particular solution of the partial differential equation

$$H \frac{\partial u}{\partial t} - K^2 \frac{\partial^2 u}{\partial x^2} = F(x, t)$$

with boundary conditions: $u(0, t) = 0$, $u(\pi, t) = 0$, by setting

$$u(x, t) = \sum_{n=1}^{\infty} \phi_n(t) \sin nx, \qquad \phi_n(t) = 0,$$

substituting in the differential equation, and comparing coefficients of $\sin nx$ (Corollary to Theorem 1, Section 7–2). Show that the result obtained agrees with (10–119).

4. Show that the substitution of (10–116′) in (10–114), with $\rho > 0$, $H = 0$, and application of (10–120) leads to the equations (10–122) for $p_n(t)$, $q_n(t)$.

5. Show that the solution of the homogeneous problem ($F = 0$) corresponding to (10–121), with boundary conditions: $w(0, t) = 0$, $w(\pi, t) = 0$, is given by

$$w = \sum_{n=1}^{\infty} \sin nx[\alpha_n \sin nat + \beta_n \cos nat],$$

$$z = \sum_{n=1}^{\infty} \sin nx[na\alpha_n \cos nat - na\beta_n \sin nat].$$

Show that replacement of α_n by $p_n(t)$, β_n by $q_n(t)$, and substitution in (10–121) leads to the equations:

$$p_n' \sin nat + q_n' \cos nat = 0,$$

$$nap_n' \cos nat - naq_n' \sin nat = \frac{2}{\pi \rho} \int_0^{\pi} F_1 \sin nx \, dx,$$

and hence obtain (10–122).

6. Let $u_1(x, t)$, $u_2(x, t)$, $u_3(x, t)$ respectively be solutions of the problems (for $0 < x < \pi, t > 0$):

$$u_t - u_{xx} = F(x, t), \quad u(0, t) = 0, \quad u(\pi, t) = 0;$$
$$u_t - u_{xx} = 0, \quad u(0, t) = a(t), \quad u(\pi, t) = 0;$$
$$u_t - u_{xx} = 0, \quad u(0, t) = 0, \quad u(\pi, t) = b(t).$$

Show that $u_1(x, t) + u_2(x, t) + u_3(x, t)$ is a solution of the problem

$$u_t - u_{xx} = F(x, t), \quad u(0, t) = a(t), \quad u(\pi, t) = b(t).$$

This shows that the effects of the different ways of forcing the system combine by *superposition.*

Answers

1. $x^2 \sin t + \dfrac{4}{\pi} \displaystyle\sum_{n=1}^{\infty} \dfrac{1 - (-1)^n}{n^3} \sin nx \, e^{-n^2 t}$.

2. For $\omega = \pm 1$, $u = \pm\frac{1}{2} \sin x(\sin t - t \cos t)$.
For $\omega \neq \pm 1$, $u = \sin x(\sin \omega t - \omega \sin t)/(1 - \omega^2)$.

10–13 Equations with variable coefficients. Sturm-Liouville boundary value problems. In order to determine the normal modes in the problem

$$\begin{aligned} \rho(x) \frac{\partial^2 u}{\partial t^2} - K^2 \frac{\partial^2 u}{\partial x^2} &= 0, \\ u(0, t) = 0, \quad u(L, t) &= 0, \end{aligned} \tag{10–123}$$

we make the substitution:

$$u = A(x) \sin(\lambda t + \epsilon)$$

and are led to the equations:

$$\begin{aligned} K^2 A''(x) + \rho(x)\lambda^2 A(x) &= 0, \\ A(0) = A(L) &= 0. \end{aligned} \tag{10–124}$$

When $\rho(x)$ is constant, we know that the only solutions of (10–124) are the functions $A_n(x) = c \sin(n\pi x/L)$; the associated frequencies λ_n are of form an. What is the nature of the solutions when ρ is variable?

A similar question arises if we consider the problem for the heat equation with variable coefficient $H(x)$:

$$\begin{aligned} H(x) \frac{\partial u}{\partial t} - K^2 \frac{\partial^2 u}{\partial x^2} &= 0, \\ u(0, t) = 0, \quad u(L, t) &= 0. \end{aligned} \tag{10–125}$$

The substitution

$$u = A(x)e^{\lambda t}$$

leads to the equations

$$\begin{aligned} K^2 A''(x) - \lambda H(x) A(x) &= 0, \\ A(0) = 0, \quad A(L) &= 0. \end{aligned} \tag{10–126}$$

Except for a change in notation, these equations are the same as (10–124)

The problems (10–124) and (10–126) are special cases of the class of *Sturm-Liouville boundary value problems.* The general case is as follows:

$$\begin{aligned} \frac{d}{dx}\left[r(x)\frac{dy}{dx}\right] + [\lambda p(x) + q(x)]y &= 0, \\ \alpha y(a) + \beta y'(a) = 0, \quad |\alpha| + |\beta| &> 0, \\ \gamma y(b) + \delta y'(b) = 0, \quad |\gamma| + |\delta| &> 0. \end{aligned} \tag{10–127}$$

Here the function $y(x)$ is to be a solution of the differential equation for $a \leqq x \leqq b$ and is to satisfy the given boundary conditions at a and b. We further assume that $r(x)$, $p(x)$, $q(x)$ have continuous derivatives over the interval and that $r(x) > 0$, $p(x) > 0$. A value of λ for which (10–127) has a solution other than $y(x) \equiv 0$ is called a *characteristic value.* One could allow for complex characteristic values, but it can be shown that under the assumptions made this does not arise; hence we restrict to real characteristic values. For each characteristic value λ there is an associated solution $y(x)$, called a *characteristic function;* the functions $cy(x)$, where c is a constant, are also characteristic functions. It could conceivably happen that there are functions besides $cy(x)$ which have the same λ; it can be shown that under the assumptions made this cannot arise:

THEOREM. *The characteristic values of the Sturm-Liouville problem (10–127) can be numbered to form an increasing sequence:* $\lambda_1 < \lambda_2 < \cdots < \lambda_n < \cdots$. *The corresponding characteristic functions can be numbered similarly to form a sequence:* $y_n(x)$; *each* $y_n(x)$ *is determined only up to a constant multiplier. The functions* $y_n(x)$ *are orthogonal with respect to the weight function* $p(x)$:

$$\int_a^b y_m(x)y_n(x)p(x)\,dx = \begin{cases} 0, & m \neq n, \\ B_n > 0, & m = n. \end{cases}$$

The Fourier series of a function $F(x)$ *with respect to the orthogonal system* $\{\sqrt{p(x)}\,y_n(x)\}$ *converges uniformly to* $F(x)$ *for every function* $F(x)$ *having a continuous derivative for* $a \leqq x \leqq b$ *and such that* $F(a) = 0$, $F(b) = 0$.

For a proof of this theorem we refer to the books of Titchmarsh and Kamke listed at the end of this chapter.

Because of the theorem, we can be assured that, except for minor changes in form, the statements about the wave equation and heat equation in Sections 10–7 and 10–9 continue to hold when the coefficients $\rho(x)$, $H(x)$ are variable [$\rho(x) > 0$, $H(x) > 0$]. For example, the characteristic functions $A_n(x)$ of (10–124) provide normal modes:

$$A_n(x)\sin(\lambda_n t + \epsilon)$$

and constant multiples thereof. The "general solution" of (10–123) is again a series

$$\sum_{n=1}^{\infty} c_n A_n(x)\sin(\lambda_n t + \epsilon_n) = \sum_{n=1}^{\infty} A_n(x)[\alpha_n \sin\lambda_n t + \beta_n \cos\lambda_n t].$$

To satisfy initial conditions:

$$u(x, 0) = f(x), \quad \frac{\partial u}{\partial t}(x, 0) = g(x),$$

one has only to choose the constants α_n, β_n so that

$$f(x) = \sum_{n=1}^{\infty} \beta_n A_n(x), \quad g(x) = \sum_{n=1}^{\infty} \lambda_n \alpha_n A_n(x);$$

this requires expansion of the functions $\sqrt{p(x)}\, f(x)$, $\sqrt{p(x)}\, g(x)$ in Fourier series:

$$\sqrt{p(x)}\, f(x) = \sum_{n=1}^{\infty} \beta_n \sqrt{p(x)}\, A_n(x), \quad \sqrt{p(x)}\, g(x) = \sum_{n=1}^{\infty} \lambda_n \alpha_n \sqrt{p(x)}\, A_n(x).$$

By the above theorem, these expansions have the same properties as the sine series used above, so that there is no change in the results. The theory of equilibrium states, approach to equilibrium, and forced motion can also be repeated.

The only difficulty is in effective determination of characteristic values and functions. For various special equations, infinite series are effective (Section 8–15). For others one is forced to use numerical methods. These are discussed in Sections 10–16 and 10–17.

The above theorem on the Sturm-Liouville problem can be extended under appropriate assumptions to the "singular case" in which the function $r(x)$ is 0 at a or b or both, while remaining positive for $a < x < b$. This case includes in particular the important problem:

$$\frac{d}{dx}\left[(1 - x^2)\frac{dy}{dx}\right] + \lambda y = 0, \quad -1 \leqq x \leqq 1, \tag{10–128}$$

whose solutions are the Legendre polynomials $P_n(x)$, with $\lambda_n = n(n + 1)$ (Sections 7–14 and 8–15). No boundary condition is imposed at $x = \pm 1$, but it is required that the solutions remain continuous at these points. The theory can also be extended to include the problem:

$$(xy')' + \left(\lambda x - \frac{m^2}{x}\right) y = 0, \quad 0 \leqq x \leqq 1, \qquad y(1) = 0, \quad (m \geqq 0). \tag{10–129}$$

The solution is required to be continuous at $x = 0$. The solutions of (10–129) are the functions $J_m(s_{mn}x)$, where s_{mn} ranges over the positive roots of the Bessel function $J_m(x)$; the corresponding $\lambda_{mn} = s_{mn}^2$. The extensions to these cases are covered in the book of Titchmarsh listed at the end of the chapter.

10–14 Equations in two and three dimensions. Separation of variables. The generalization of our basic problem to two and three dimensions brings no basic change in the results, though the determination of normal modes is in general more complicated.

As an example we consider the wave equation for a rectangle:

$$\frac{\partial^2 u}{\partial t^2} - a^2\left(\frac{\partial^2 u}{\partial x^2} + \frac{\partial^2 u}{\partial y^2}\right) = 0, \quad 0 < x < \pi, \quad 0 < y < \pi, \tag{10–130}$$
$$u(x, y, t) = 0 \text{ for } x = 0, \quad x = \pi, \quad y = 0, \quad y = \pi.$$

The substitution

$$u(x, y, t) = A(x, y) \sin (\lambda t + \epsilon)$$

leads to the characteristic value problem

$$a^2\left(\frac{\partial^2 A}{\partial x^2} + \frac{\partial^2 A}{\partial y^2}\right) + \lambda^2 A = 0, \tag{10–131}$$
$$A(x, y) = 0 \text{ for } x = 0, \quad x = \pi, \quad y = 0, \quad y = \pi.$$

In order to determine the characteristic functions $A(x, y)$, we seek particular characteristic functions having the form of a product of a function of x by a function of y:

$$A(x, y) = X(x)Y(y).$$

Accordingly,

$$a^2[X''(x)Y + XY''(y)] + \lambda^2 X(x)Y(y) = 0,$$
$$\frac{X''}{X} + \frac{Y''}{Y} + \frac{\lambda^2}{a^2} = 0. \tag{10–132}$$

Now if we vary x, the second and third terms of the last equation cannot vary; hence the first term is a constant. Similarly, the second term is a constant:

$$X'' = -\mu X, \quad Y'' = \left(\mu - \frac{\lambda^2}{a^2}\right) Y,$$

$$X'' + \mu X = 0, \quad Y'' + \left(\frac{\lambda^2}{a^2} - \mu\right) Y = 0.$$

Because of the boundary conditions in (10–131), we are led to two new boundary value problems:

$$X'' + \mu X = 0, \quad X(0) = X(\pi) = 0;$$
$$Y'' + \left(\frac{\lambda^2}{a^2} - \mu\right) Y = 0, \quad Y(0) = Y(\pi) = 0. \tag{10–133}$$

The first is satisfied by $X_n(x) = \sin nx$, for $\mu = n^2$; for this value of μ. the second is satisfied by the functions $Y_m(y) = \sin my$, provided $(\lambda^2/a^2) - \mu = m^2$. Accordingly,

$$A_{mn}(x, y) = \sin nx \sin my$$

is a solution of (10–131), for $\lambda^2 = a^2(m^2 + n^2)$, where m and n are integers. For a given characteristic value λ there may be several characteristic functions; in fact, if $A_{mn}(x, y)$ is one function, then $A_{nm}(x, y)$ provides a second one, unless $m = n$. This causes no trouble, because we put all linear combinations into the "general solution":

$$u(x, y, t) = \sum_{m,n} c_{mn} \sin nx \sin my \sin (\lambda_{mn}t + \epsilon_n), \qquad \lambda_{mn} = a\sqrt{m^2 + n^2}. \tag{10–134}$$

The series is a "double series," to be summed over all combinations of positive integral values of m and n; the series is a Fourier series in the two variables x, y and, as pointed out in Section 7–16, can be rearranged to form a single series or summed as an "iterated" series:

$$\sum_{m=1}^{\infty} \left\{ \sum_{n=1}^{\infty} c_{mn} \sin nx \sin my \sin (\lambda_{mn}t + \epsilon_n) \right\} \cdot$$

Since only sine functions appear, the series is really a *Fourier sine series in two variables*. As in Section 7–16, the functions $\sin nx \sin my$ form a *complete orthogonal system* for the region: $0 \leqq x \leqq \pi$, $0 \leqq y \leqq \pi$. Accordingly, the results achieved in one dimension can all be generalized to two dimensions. Because of the more complicated characteristic values, the solutions are more difficult to analyze; in particular, the solutions are in general no longer periodic in t.

The crucial step above was the replacement of $A(x, y)$ by the product $X(x)Y(y)$. This led to a "separation of variables" in (10–133) and determination of particular characteristic functions $A(x, y)$ which together form a complete orthogonal system. The fact that such a procedure can be successful was already indicated in our method for determining normal modes for the problems in one dimension; the substitution $u = A(x)e^{\lambda t}$ could have been replaced by a substitution $u = A(x)T(t)$ and a separation of variables would then have led to the same results.

The method of separation of variables thus appears as a general method for attacking homogeneous linear partial differential equations. The method may in some cases provide only certain particular solutions; in a wide variety of cases these particular solutions have been shown to provide a complete set of orthogonal functions. These cases include the Laplace equation in cylindrical and spherical coordinates (Sections 2–17, 3–8); see Probs. 4 and 7 below.

The *equilibrium problem:* $\nabla^2 u = -F(x, y)$, with values of u prescribed on the boundary of a region R of the xy plane, can be attacked in several ways. It can be shown that, under appropriate assumptions on $F(x, y)$ the function (logarithmic potential)

$$u_0(x, y) = -\frac{1}{4\pi} \iint_R F(r, s) \log [(x - r)^2 + (y - s)^2] \, dr \, ds \tag{10–135}$$

satisfies the Poisson equation: $\nabla^2 u_0 = -F$ inside R and is continuous in R plus boundary. The function $v = u - u_0$ will then satisfy the condition $\nabla^2 v = -F + F = 0$ inside R and have certain new boundary values on the boundary of R. Determination of v is then a *Dirichlet* problem, which can be attacked by conformal mapping (Section 9–6). While conformal mapping is not available as a tool for the corresponding problems in three dimensions, methods based on *potential theory* can be used; see especially the book of Kellogg listed at the end of the chapter.

One can in general reduce the equilibrium problem $\nabla^2 u = -F$ to the case of zero boundary values by the following procedure (cf. Section 10–12): Let $u(x, y)$ be required to have values $h(x, y)$ on the boundary C of R. If $h(x, y)$ is sufficiently smooth, one can then find a function $h_1(x, y)$ which has continuous first and second derivatives inside R, is continuous in R plus C, and equals $h(x, y)$ on C. The function: $v = u - h_1$ is then zero on C, and $\nabla^2 v = -F - \nabla^2 h_1 = -F_1(x, y)$. The determination of v can be carried out with the aid of a *Green's function*, as indicated in Section 10–18 below.

10–15 Unbounded regions. Continuous spectrum. In many physical problems it is natural to regard the continuous medium as being *infinite* in extent. For example, in one dimension one can consider the wave equation

$$\frac{\partial^2 u}{\partial t^2} - a^2 \frac{\partial^2 u}{\partial x^2} = 0 \tag{10–136}$$

for the infinite interval: $x > 0$. If one seeks normal modes, one is led to the characteristic value problem:

$$a^2 A''(x) + \lambda^2 A(x) = 0, \quad x > 0; \quad A(0) = 0. \tag{10–137}$$

This problem has solutions for every value of λ, namely the functions $\sin \alpha x$, for $a^2\alpha^2 = \lambda^2$. Thus the resonant frequencies λ form a "continuous" set of numbers and one has a "continuous spectrum." [There also exist *unbounded* "normal modes": $u = \sinh \alpha x \, e^{a\alpha t}$. These are of less physical interest.]

One can construct linear combinations of the normal modes in order to obtain a "general solution" of the homogeneous problem. Since there is a continuous sequence of λ's, an *integration* is called for rather than a summation. For $\alpha \geqq 0$ we must integrate expressions of form

$$\sin x[p(\alpha)\cos (a\alpha t) + q(\alpha) \sin (a\alpha t)],$$

where $p(\alpha)$ and $q(\alpha)$ are "arbitrary" functions of α. We obtain the integral

$$\int_0^\infty \sin \alpha x[p(\alpha) \cos (a\alpha t) + q(\alpha) \sin (a\alpha t)] \, d\alpha.$$

For each fixed t this can be considered as a *Fourier integral* (the Fourier sine integral, Section 7–18). In particular, for $t = 0$ we obtain a Fourier integral representation of the initial displacement $u(x, 0)$:

$$\int_0^\infty p(\alpha) \sin \alpha x \, d\alpha.$$

Since the theory of the Fourier integral has been highly developed, most of the results for the finite interval can be extended to the infinite case.

Similar statements can be made for problems in two or three dimensions for unbounded regions. For more information we refer to the books of the following authors, as listed at the end of the chapter: Sneddon, Titchmarsh, Wiener, Courant and Hilbert, Frank and von Mises, and Tamarkin and Feller. The Laplace transform (Section 6–24) can also be used to represent solutions over infinite intervals. This is discussed in the texts mentioned and in the second book by Churchill.

Problems

1. (a) Let a vibrating string be stretched between $x = 0$ and $x = 1$; let the tension K^2 be $(x+1)^2$ and the density ρ be 1 in appropriate units. Show that the normal modes are given by the functions

$$A_n(x) = \sqrt{x+1}\sin\left[n\pi\frac{\log(x+1)}{\log 2}\right]\sin(\lambda_n t + \epsilon_n), \quad \lambda_n = \left(\frac{n^2\pi^2}{\log^2 2} + \frac{1}{4}\right)^{\frac{1}{2}}.$$

[Hint: make the substitution $x + 1 = e^u$ in the boundary value problem for $A_n(x)$.]

(b) Show directly that every function $f(x)$ having continuous first and second derivatives for $0 \leqq x \leqq 1$ and such that $f(0) = f(1) = 0$ can be expanded in a uniformly convergent series in the characteristic functions $A_n(x)$ of part (a). [Hint: let $x + 1 = e^u$ as in part (a). Then expand $F(u) = f(e^u - 1)e^{-\frac{1}{2}u}$ in a Fourier sine series for the interval $0 \leqq u \leqq \log 2$.]

2. Show that the general second order linear equation

$$p_0(x)y'' + p_1(x)y' + [\lambda p_2(x) + p_3(x)]y = 0,$$

where $p_0(x) \neq 0$, takes on the form of a Sturm-Liouville equation (10–127) if the equation is multiplied by $r(x)/p_0(x)$, where $r(x)$ is chosen so that $r'/r = p_1/p_0$ (cf. Section 8–6). In general, an equation of form: $(ry')' + h(x)y = 0$ is called *self-adjoint.*

3. Obtain the general solution, for $t > 0$, $0 < x < \pi$, $0 < y < \pi$, of the heat equation with boundary conditions:

$$\frac{\partial u}{\partial t} - c^2\left(\frac{\partial^2 u}{\partial x^2} + \frac{\partial^2 u}{\partial y^2}\right) = 0,$$

$$u(x, y, t) = 0 \text{ for } x = 0, \quad x = \pi, \quad y = 0, \quad y = \pi.$$

4. Show that separation of variables: $u(r, \theta) = R(r)\Theta(\theta)$ in the problem in polar coordinates for the domain $r < 1$:

$$\nabla^2 u + \lambda u \equiv \frac{1}{r^2}\left[r\frac{\partial}{\partial r}\left(r\frac{\partial u}{\partial r}\right) + \frac{\partial^2 u}{\partial \theta^2}\right] + \lambda u = 0, \quad u(1, \theta) = 0,$$

leads to the problems:

$$(rR')' + \left(\lambda r - \frac{\mu}{r}\right)R = 0, \quad R(1) = 0,$$

$$\Theta'' + \mu\Theta = 0.$$

If we require that $u(r, \theta)$ be continuous in the circle: $r \leqq 1$, then $\Theta(\theta)$ must be periodic in θ, with period 2π. Show that this implies that $\mu = m^2 \cdot (m = 0, 1, 2, 3, \ldots)$ and that $R(r) = J_m(\sqrt{\lambda_{mn}}\, r)$, for $\lambda = \lambda_{mn}$; cf. (10–129) above. Hence one obtains the characteristic functions

$$J_m(\sqrt{\lambda_{mn}}\, r)\cos m\theta, \quad J_m(\sqrt{\lambda_{mn}}\, r)\sin m\theta,$$

and linear combinations thereof. It can be shown that these characteristic functions form a complete orthogonal system for the circle: $r \leqq 1$.

5. Using the results of Prob. 4, determine the normal modes for the vibrations of a circular membrane; that is, find normal modes for the equation

$$\frac{\partial^2 u}{\partial t^2} - a^2 \nabla^2 u = 0, x^2 + y^2 < 1,$$
$$u(x, y, t) = 0 \text{ for } x^2 + y^2 = 1.$$

6. Using the results of Prob. 4, determine the general solution of the heat conduction problem:

$$\frac{\partial u}{\partial t} - c^2 \nabla^2 u = 0, \quad x^2 + y^2 < 1,$$
$$u(x, y, t) = 0 \text{ for } x^2 + y^2 = 1.$$

7. Show that the substitution: $u = R(\rho)\Phi(\phi)\Theta(\theta)$ in the problem in spherical coordinates for the domain $\rho < 1$:

$$\nabla^2 u + \lambda u \equiv \frac{1}{\rho^2 \sin^2 \phi}\left[\sin^2 \phi \frac{\partial}{\partial \rho}\left(\rho^2 \frac{\partial u}{\partial \rho}\right) + \sin \phi \frac{\partial}{\partial \phi}\left(\sin \phi \frac{\partial u}{\partial \phi}\right) + \frac{\partial^2 u}{\partial \theta^2}\right] + \lambda u = 0,$$
$$u(\rho, \phi, \theta) = 0 \text{ for } \rho = 1,$$

leads to the separate Sturm-Liouville problems:

$$(\rho^2 R')' + (\lambda \rho^2 - \alpha)R = 0, \quad R = 0 \text{ for } \rho = 1;$$
$$(\sin \phi \Phi')' + (\alpha \sin \phi - \beta \csc \phi)\Phi = 0, \quad \Theta'' + \beta\Theta = 0.$$

Here α, β and λ are characteristic values to be determined. The condition that u be continuous throughout the sphere requires Θ to have period 2π, so that $\beta = k^2$ $(k = 0, 1, 2, \ldots)$ and $\Theta_k(\theta)$ is a linear combination of $\cos k\theta$ and $\sin k\theta$. When $\beta = k^2$, it can be shown that continuous solutions of the second problem for $0 \leqq \phi \leqq \pi$ are obtainable only when $\alpha = n(n+1)$, $k = 0, 1, \ldots, n$, and Φ is a constant times $P_{n,k}(\cos \phi)$, where

$$P_{n,k}(x) = (1 - x^2)^{\frac{1}{2}n} \frac{d^k}{dx^k} P_n(x)$$

and $P_n(x)$ is the nth Legendre polynomial. When $\alpha = n(n+1)$ $(n = 0, 1, 2, \ldots)$, the first problem has a solution continuous for $\rho = 0$ only when λ is one of the roots $\lambda_{n+\frac{1}{2},1}, \lambda_{n+\frac{1}{2},2}, \ldots$ of the function $J_{n+\frac{1}{2}}(\sqrt{x})$, where $J_{n+\frac{1}{2}}(x)$ is the Bessel function of order $n + \frac{1}{2}$; for each such λ, the solution is a constant times $\rho^{-\frac{1}{2}}J_{n+\frac{1}{2}}(\sqrt{\lambda}\rho)$. Thus one obtains the characteristic functions $\rho^{-\frac{1}{2}}J_{n+\frac{1}{2}}(\sqrt{\lambda}\rho)P_{n,k}(\cos \phi) \cos k\theta$, $\rho^{-\frac{1}{2}}J_{n+\frac{1}{2}}(\sqrt{\lambda}\rho)P_{n,k}(\cos \phi) \sin k\theta$; where $n = 0, 1, 2, \ldots$, $k = 0, 1, \ldots, n$, and λ is chosen as above for each n. It can be shown that these functions form a complete orthogonal system for the spherical region $\rho \leqq 1$.

8. Show that the wave equation (10–136) for the infinite interval: $-\infty < x < \infty$ has a continuous spectrum and find the characteristic functions for the bounded normal modes.

Answers

3. $\displaystyle\sum_{m=1}^{\infty}\left\{\sum_{n=1}^{\infty} c_{mn} \sin nx \sin my e^{-c^2(m^2+n^2)t}\right\}.$

5. $J_m(\sqrt{\lambda_{mn}}r)\sin(a\sqrt{\lambda_{mn}}t+\epsilon)(c_1\cos m\theta+c_2\sin m\theta)$, where c_1 and c_2 are constants.

6. $\displaystyle\sum_{n=1}^{\infty}\left\{\sum_{m=0}^{\infty}J_m(\sqrt{\lambda_{mn}}r)e^{-c^2\lambda_{mn}t}(\alpha_{mn}\cos m\theta+\beta_{mn}\sin m\theta)\right\}.$

8. $c_1\cos\alpha x+c_2\sin\alpha x,\quad 0\leqq\alpha<\infty,\quad \lambda=a\alpha.$

10–16 Numerical methods. For problems with variable coefficients or problems in two or three dimensions concerning regions of inappropriate shape, the methods described above will in general fail to produce solutions in a form suitable for numerical applications. Similar remarks apply to classes of differential equations more general than those considered here, in particular, nonlinear equations. While the theoretical aspects of the subject are highly developed and one can often establish existence of solutions, this is not always sufficient for the needs of physics.

Accordingly, a variety of numerical methods have been devised for explicit determination of solutions satisfying given boundary conditions and initial conditions. We consider briefly some of these methods.

The first method consists simply in *a reversal of the limit process of Section* 10–5. We replace the derivative $\partial^2u/\partial x^2$ by the difference expression

$$\frac{u_{\sigma+1}-2u_\sigma+u_{\sigma-1}}{(\Delta x)^2},$$

where $u_\sigma=u(x_\sigma)$. From the differential equation

$$\rho(x)\frac{\partial^2u}{\partial t^2}+H(x)\frac{\partial u}{\partial t}-K^2\frac{\partial^2u}{\partial x^2}=F(x,t), \tag{10–138}$$

we are thus led to the system of equations

$$m_\sigma\frac{d^2u_\sigma}{dt^2}+h_\sigma\frac{du_\sigma}{dt}-k^2(u_{\sigma+1}-2u_\sigma+u_{\sigma-1})=F_\sigma(t), \tag{10–139}$$

where $\sigma=1,\ldots,N$ and

$$m_\sigma=\rho(x_\sigma)\,\Delta x,\quad h_\sigma=H(x_\sigma)\,\Delta x,\quad k^2=\frac{K^2}{\Delta x},\quad F_\sigma(t)=F(x_\sigma,t)\,\Delta x. \tag{10–140}$$

Equations (10–139) can be handled completely by the methods of Section 8–12. The tools required are basically algebraic, in particular, the solution of simultaneous equations. In order that (10–139) be an accurate approximation to (10–138), it is necessary that N be large; this makes the algebraic problems far from trivial, at least as far as time requirements are concerned.

Initial value problems. If one seeks a particular solution of (10–138) satisfying given initial conditions and boundary conditions (values of u_0 and u_{N+1}), one can set up the approximating equations (10–139) and apply the method of step-by-step integration described in Section 8–8. This can be improved on by other similar procedures, described in the books referred to at the end of that section. All these measures are even more appropriate if nonlinearity is present, e.g., if $\partial^2u/\partial x^2$ is replaced by its square; in such a case, the algebraic procedure is in general useless.

Characteristic value problems. Determination of normal modes for a wave equation or heat equation obtained from (10–138) leads in general to a Sturm-Liouville problem (10–127). This can be attacked by considering the approximating problem (10–139), for which determination of normal modes is an algebraic problem; one can also use *difference* equations, as in Probs. 5 to 10 following Section 10–4. A *variational* method is also helpful; this is described in Section 10–17 below.

One can treat the problem as an initial value problem, in the following way: to solve the equations:

$$A''(x) + \lambda\rho(x)A(x) = 0, \quad A(0) = A(1) = 0,$$

one constructs particular solutions of the initial value problem: $A(0) = 0$, $A'(0) = 1$ for different values of λ. As λ is gradually increased, the solutions vary in a simple manner and one can by trial and error determine solutions for which the condition $A(1) = 0$ is satisfied. These are precisely the characteristic functions sought.

Equilibrium problems. The equilibrium problem for (10–138) is solved in all generality in Section 10–11 above; the only difficult step is an integration, which might have to be carried out numerically as in Section 4–3. Another way of writing the solution, with the aid of a Green's function, is explained in Section 10–18.

Problems in two dimensions. If $\partial^2u/\partial x^2$ is replaced by a two-dimensional Laplacian ∇^2u in (10–138), so that one has a problem for $u(x, y, t)$ in a region R of the xy plane, an approximating system analogous to (10–139) can be devised. If R is a rectangle: $a \leqq x \leqq b$, $c \leqq y \leqq d$, one can divide R into squares (if the sides of R are commensurable) of side h. We then consider the values of u only at the corners of the squares. At each such corner (x, y), the Laplacian is computed approximately (Section 2–18) as the expression

$$\frac{u(x+h, y) + u(x, y+h) + u(x-h, y) + u(x, y-h) - 4u(x, y)}{h^2}.$$

A system of equations analogous to (10–138) is obtained. If R is not a rectangle, one can approximate R by a figure pieced together of rectangles, and proceed similarly.

The statements concerning the *initial value problem* above can now be repeated without change. The *characteristic value problem* can also be replaced by an approximating algebraic problem in the same way; the variational approach of Section 10–17 below is also useful.

The *equilibrium problem* can be attacked numerically by considering the approximating system of equations in variables u_σ as above. One has then N simultaneous linear equations; if N is large, these may be far from simple to handle. One can also regard the equilibrium problem as a special case of a *heat* equation: $u_t - K^2\nabla^2 u = F(x, y)$, with u given on the boundary of R, for all solutions of the heat equation tend exponentially to the equilibrium solution. One can assign arbitrary *initial* values and obtain a particular solution; for t large, this will approximate the equilibrium solution sought. The variational methods of Section 10–17 are also of use for the equilibrium problem.

Most of the remarks made can be generalized to problems in *three dimensions*.

The accuracy of the approximate methods described has been investigated, and in general the processes described can be carried out to yield the accuracy desired for the solutions. Some details on this are given in the books of Tamarkin and Feller, Hildebrand, and John listed at the end of the chapter. It is worth remarking that the crucial step of replacing (10–138) by (10–139) can be regarded as the *replacement of one physical model by another*. According to modern physics, both models are an oversimplification of what is observed in nature. If either one serves to describe the phenomena concerned with sufficient accuracy, then it can be regarded as a useful one.

10–17 Variational methods. The equilibrium solution for systems (10–139) and the analogous ones in two and three dimensions can be considered as problems of minimizing a function $\phi(u_1, \ldots, u_N)$. For, as remarked in Section 10–4, equations (10–139) can be written in the form

$$m_\sigma \frac{d^2u_\sigma}{dt^2} + h_\sigma \frac{du_\sigma}{dt} + \frac{\partial V}{\partial u_\sigma} = F_\sigma\ (t)\ (\sigma = 1, \ldots, N). \qquad (10\text{–}141)$$

The equilibrium problem is then the problem

$$\frac{\partial V}{\partial u_\sigma} = F_\sigma, \qquad (10\text{–}142)$$

where the F_σ are constants. The end values u_0 and u_{N+1} are also given as constants; we can consider them to be 0 by modifying the definition of F_1 and F_N. If we now let

$$\phi(u_1, \ldots, u_N) = V(u_1, \ldots, u_N) - (F_1u_1 + \cdots + F_Nu_N), \qquad (10\text{–}143)$$

then (10–142) is simply the condition that

$$\frac{\partial \phi}{\partial u_\sigma} = 0 \quad (\sigma = 1, \ldots, N). \qquad (10\text{–}144)$$

For (10–139) we have

$$\phi = k^2[u_1^2 + \cdots + u_N^2 - u_1u_2 - u_2u_3 - \cdots - u_{N-1}u_N] - F_1u_1 - \cdots - F_Nu_N, \qquad (10\text{–}145)$$

and we can verify that (10–144) has precisely one solution $u_1^*, \ldots, u_N^*$; that this critical point is a minimum point is also easily verified (Prob. 6 below). A similar statement applies to the analogous problems in two or three dimensions. Indeed, the physical picture which leads to equations of form (10–141) is almost always that of a system of particles capable of an equilibrium state, at which the potential energy V has its smallest value; equations (10–141) then describe the forced oscillations about this equilibrium state. When the applied forces are *constant*, the potential energy V is replaced by a modified one ϕ; the equations (10–141) then become

$$m_\sigma \frac{d^2u_\sigma}{dt^2} + h_\sigma \frac{du_\sigma}{dt} + \frac{\partial\phi}{\partial u_\sigma} = 0; \tag{10–146}$$

the new equilibrium state is now the minimum of ϕ.

By appropriate passage to the limit, it can be shown that the equilibrium problem for (10–138) is equivalent to minimizing the expression

$$\Phi_u = \int_0^L [\tfrac{1}{2}K^2\{u'(x)\}^2 - F(x)u]\,dx. \tag{10–147}$$

The expression Φ_u is a *functional*, that is, an expression whose value depends on the *function* $u(x)$ chosen. It can then be shown that the function $u^*(x)$ which satisfies the equilibrium conditions:

$$-K^2u''(x) = F(x), \quad u(0) = 0, \quad u(L) = 0, \tag{10–148}$$

assigns to Φ_u its smallest value attainable among all smooth functions $u(x)$ satisfying the boundary conditions (Prob. 7 below).

The general problem of minimizing functionals such as Φ_u is the subject of the *calculus of variations*. Accordingly, methods based on minimizing (or maximizing) appropriate functionals or functions are called *variational methods*.

One can attack the problem of minimizing the functional Φ_u of (10–147) by the following procedure, due to Rayleigh and Ritz. One chooses a particular function $u(x)$ depending linearly on several arbitrary constants:

$$u = c_1u_1(x) + \cdots + c_nu_n(x). \tag{10–149}$$

The functions $u_1(x), \ldots, u_n(x)$ are chosen to satisfy the boundary conditions, i.e., to be 0 at $x = 0$ and $x = L$; they are otherwise chosen as desired, though the effectiveness of the method depends greatly on the skill with which they are chosen. On substituting the expression (10–149) in (10–147) one obtains a function P whose value depends only on the constants $c_1, \ldots, c_n$. Because of the form of Φ_u, $P(c_1, \ldots, c_n)$ is also a quadratic expression and its minimum is the unique solution of the equations

$$\frac{\partial P}{\partial c_1} = 0, \ldots, \quad \frac{\partial P}{\partial c_n} = 0. \tag{10–150}$$

These are simultaneous linear equations. On solving for $c_1, c_2, \ldots, c_n$ one has found a function (10–149) which gives Φ_u a smaller value than that given by certain competing functions. If the class of competing functions is large enough, one can expect the function $u(x)$ found to be close to the true minimum of Φ_u.

For the equilibrium problem in one dimension, the procedure described is not needed, for one can solve (10–144) explicitly as in Section 10–11 above. However, for problems in two and three dimensions explicit solution is usually difficult (cf. Section 10–14) and the Rayleigh-Ritz procedure can be of considerable value. It can be shown that the solution of the equilibrium problem: $-K^2 \nabla^2 u = F$ in two and three dimensions is equivalent to minimizing the functionals

$$\iint_R [\tfrac{1}{2}K^2\{u_x^2 + u_y^2\} - F(x, y)u]\, dx\, dy,$$
$$\iiint_R [\tfrac{1}{2}K^2\{u_x^2 + u_y^2 + u_z^2\} - F(x, y, z)u]\, dx\, dy\, dz, \tag{10–151}$$

respectively.

Variational methods can also be applied to the determination of characteristic values and functions. For example, the characteristic value problem for the normal modes of (10–139), with $h_\sigma = 0$ and $F_\sigma = 0$, is the problem

$$\lambda^2 m_\sigma A_\sigma + k^2(A_{\sigma+1} - 2A_\sigma + A_{\sigma-1}) = 0 \quad (\sigma = 1, \ldots, N), \tag{10–152}$$

where $A_0 = A_{N+1} = 0$. These are the equations for minimizing the function

$$V(A_1, \ldots, A_N) = k^2(A_1^2 + \cdots + A_N^2 - A_1A_2 - \cdots - A_{N-1}A_N) \tag{10–153}$$

subject to the *side condition*

$$g(A_1, \ldots, A_N) \equiv \tfrac{1}{2}(m_1A_1^2 + \cdots + m_NA_N^2 - 1) = 0. \tag{10–154}$$

Indeed, the method of Lagrange multipliers for this problem (Section 2–20) gives the equations

$$\frac{\partial V}{\partial A_\sigma} - \lambda^2 \frac{\partial g}{\partial A_\sigma} = 0,$$

which are the same as (10–152). The side condition (10–154) fixes the constant of proportionality of the A's (up to a $\pm$ sign). The characteristic values $\lambda_1, \ldots, \lambda_N$ correspond to critical points of V on the "ellipsoid" defined by (10–154). In general two of the λ's correspond to the absolute minimum and maximum of V, when (10–154) holds.

Again a limit process leads to a variational formulation of the characteristic value problem for the continuous medium. For the equation

$$\rho(x, y)\frac{\partial^2 u}{\partial t^2} - K^2 \nabla^2 A = 0,$$

the characteristic value problem concerns the "critical points" of the functional

$$Q_A = \iint_R \tfrac{1}{2}K^2(A_x^2 + A_y^2)\,dx\,dy,$$

subject to the side condition

$$\iint_R \rho[A(x, y)]^2\,dx\,dy = 1.$$

The Rayleigh-Ritz method is applicable here in the same way as above.

For further information on this topic one is referred to the article and books by Courant, Gould, and Kantorovich and Krylov listed at the end of the chapter.

10–18 Partial differential equations and integral equations. Given the equilibrium problem

$$-k^2(u_{\sigma+1} - 2u_\sigma + u_{\sigma-1}) = F_\sigma \quad (\sigma = 1, \ldots, N), \quad u_0 = u_{N+1} = 0, \tag{10–155}$$

one could obtain the solution by the following method. One could first solve the problem for which $F_1 = 1$ and $F_2 = F_3 = F_4 = \cdots = 0$; let the solution be $u_\sigma = g_{\sigma,1}$. One could then solve for $F_2 = 1$ and all other $F_\sigma = 0$, obtaining $g_{\sigma,2}$; in general $u_\sigma = g_{\sigma,\mu}$ is the solution when $F_\mu = 1$ and $F_\sigma = 0$ for $\sigma \neq \mu$. The linear combination

$$u_\sigma = F_1 g_{\sigma,1} + F_2 g_{\sigma,2} + \cdots + F_N g_{\sigma,N} \tag{10–156}$$

is then the desired solution of (10–155); for substitution of u_σ in the left side of the first equation (10–155) gives F_1, since $g_{\sigma,1}$ gives 1 and all other g's give 0; a similar reasoning holds for the other equations. Thus the effect of all the F_σ can be built up by *superposition* of unit forces.

By a limit process, a similar result is obtained for the problem:

$$u''(x) = -F(x), \quad u(0) = u(L) = 0. \tag{10–157}$$

One finds

$$u(x) = \int_0^L g(x, s)F(s)\,ds, \tag{10–158}$$

where the $g(x, s)$ are the solutions for a force F "concentrated at a point s." The function $g(x, s)$ is called the *Green's function* for (10–157); it is 0 when $x = 0$ and $x = L$, has the value $s(L - s)/L$ when $x = s$, and is linear in x between these values. Accordingly, $g(x, s)$ has a "corner" at $x = s$, due to the concentrated force at this point, whereas $\partial^2 g/\partial x^2 = 0$ otherwise.

Similar results hold for quite general nonhomogeneous linear equations. In particular, a Green's function $g(x, y; r, s)$ can be found for the problem (Poisson equation):

$$\begin{aligned} \nabla^2 u &= -F(x, y) \quad \text{inside } R, \\ u(x, y) &= 0 \quad \text{on boundary of } R, \end{aligned} \tag{10–159}$$

for a general region R in the plane. The solutions of (10–159) are then given by a formula:

$$u(x, y) = \iint_R g(x, y; r, s)F(r, s)\, dr\, ds. \tag{10–160}$$

For each (r, s) the function g satisfies the equation $\nabla^2 g = 0$ except at $x = s$, $y = r$, where it has a discontinuity corresponding to a "point load." Also $g(x, y; r, s) = 0$ when (x, y) is on the boundary of R; therefore u is given as a "linear combination" of functions all 0 on the boundary of R, is therefore itself 0 on the boundary.

In order to solve the characteristic value problem

$$\begin{aligned} \nabla^2 u + \lambda u &= 0 \quad \text{in } R, \\ u &= 0 \quad \text{on the boundary of } R, \end{aligned} \tag{10–161}$$

one can rewrite the problem in form (10–159): $\nabla^2 u = -\lambda u$; therefore

$$u(x, y) = \lambda \iint_R g(x, y; r, s)u(r, s)\, dr\, ds. \tag{10–162}$$

This gives an implicit equation for u, the crucial operation being integration. The equation is called an *integral equation.*

A variety of other problems in partial differential equations can be restated as integral equations. A number of methods are available for obtaining solutions of integral equations and they must be considered among the most powerful ways of attacking partial differential equations. Of great importance are the following features: the theory of integral equations is much more *unified* than that of differential equations, problems in one, two, or three dimensions being treated in the same way; the treatment of *boundary values* is simpler: e.g., in (10–162) the boundary condition on u is automatically taken care of, since $g = 0$ on the boundary; the methods of solution are much better adaptable to *nonlinear* problems.

For a full discussion of integral equations and their applications, we refer to the books of Tamarkin and Feller, Frank and von Mises, Courant, Kellogg, and Mikhlin listed below.

Problems

1. Let the equilibrium problem: $\nabla^2 u(x, y) = 0$ be given for the square: $0 \leqq x \leqq 3$, $0 \leqq y \leqq 3$, with boundary values: $u = x^2$ for $y = 0$, $u = x^2 - 9$ for $y = 3$, $u = -y^2$ for $x = 0$, $u = 9 - y^2$ for $x = 3$. Obtain the solution by considering the heat equation $u_t - \nabla^2 u = 0$. Use only integer values of x, y so that only four points: (1, 1), (2, 1), (1, 2), (2, 2) inside the rectangle are concerned. Let u_1, u_2, u_3, u_4 be the four values of u at these points respectively. Using the given boundary values, show that the approximating equations are

$$\begin{aligned} u_1'(t) - (u_2 + u_3 - 4u_1) &= 0, \quad u_2'(t) = (12 + u_4 + u_1 - 4u_2), \\ u_3'(t) - (u_4 - 12 + u_1 - 4u_3) &= 0, \quad u_4'(t) - (u_3 + u_2 - 4u_4) = 0. \end{aligned}$$

Replace by difference equations in t: $\Delta u_1 = (u_2 + u_3 - 4u_1)\,\Delta t, \ldots$ and solve by step-by-step integration. Use $\Delta t = 0.1$ and initial values: $u_1 = u_2 = u_3 = u_4 = 1$. Show that for $t = 1$ the values are close to the equilibrium values: $u_1 = 0$, $u_2 = 3$, $u_3 = -3$, $u_4 = 0$.

2. The method of *relaxation* or *Liebmann's method,* as applied to Prob. 1, consists in choosing initial values of u_1, u_2, u_3, u_4, then correcting each one in turn by replacing it by the average of the *four neighboring values.* Thus at (1, 1) the value u_1 would be replaced by the average of u_2, u_3, -1, and 1. The value u_2 at (2, 1) would then be replaced by the average of 8, u_4, u_1 (new value) and 4. Apply this process repeatedly, starting with $u_1 = u_2 = u_3 = u_4 = 1$ and show that the corrected values gradually approach the equilibrium sought. This technique is discussed in the two books of Southwell listed below.

3. Let the wave equation problem: $u_{tt} - \nabla^2 u = 0$, $u(x, y, t) = 0$ on the boundary, be given for the square of Prob. 1. Determine the resonant frequencies by using difference expressions for $\nabla^2 u$ as in Prob. 1, so that one has the equations:

$$u_1''(t) - (u_2 + u_3 - 4u_1) = 0, \ldots .$$

The exact frequencies are found as in Section 10–14 to be $\frac{1}{3}\pi(m^2 + n^2)^{\frac{1}{2}}$ ($m = 1, 2, \ldots, n = 1, 2, \ldots$). Show that the four lowest frequencies are fairly well approximated.

4. Let (10–138) be the wave equation: $u_{tt} - a^2 u_{xx} = 0$ for the interval $0 < x < \pi$ as in Section 10–7. The corresponding approximating system (10–139) was considered in Prob. 7 following Section 10–4. In the notation used here, the characteristic values and functions found were

$$\lambda_n = \frac{2a(N+1)}{\pi} \sin \frac{n\pi}{2(N+1)}, \quad A_n(x_\sigma) = \sin(nx_\sigma),$$

where $\sigma = 0, 1, \ldots, N+1, n = 1, \ldots, N$; compare with the exact solutions of the wave equation. Show that for each fixed n, $\lambda_n \to an$ as $N \to \infty$.

5. Study the behavior of the solutions of the *initial value* problem:

$$u''(x) + \lambda u(x) = 0, \quad u(0) = 0, \quad u'(0) = 1,$$

as λ increases from 0 to ∞; note in particular the appearance of values of λ for which the condition: $u(1) = 0$ is satisfied. It can be shown that the same qualitative picture holds for the general Sturm-Liouville problem of Section 10–13.

6. Show that the function ϕ defined by (10–145) has precisely one critical point at which ϕ takes on its absolute minimum. [Hint: show that by proper choice of the constants $\alpha_1, \ldots, \alpha_N$, the substitution

$$w_1 = u_1 - u_2 + \alpha_1, \quad w_2 = u_2 - u_3 + \alpha_2, \ldots,$$
$$w_{N-1} = u_{N-1} - u_N + \alpha_{N-1}, \quad w_N = u_N + \alpha_N$$

transforms ϕ into an expression

$$\tfrac{1}{2}k^2[w_1^2 + \cdots + w_N^2 + (w_1 + \cdots + w_N - \alpha_1 - \cdots - \alpha_N)^2] + \text{const.}$$

This shows that $\phi \to \infty$ as $w_1^2 + \cdots + w_N^2 \to \infty$, so that ϕ has at least one critical point which gives ϕ its absolute minimum. The equations in $u_1, \ldots, u_N$ for the critical point are simultaneous linear equations. If not all the F_σ are zero, these equations are nonhomogeneous and have at most one solution. If all the F_σ are zero, then $u_1 = u_2 = \cdots = u_N = 0$ is one critical point; if $u_1^*, \ldots, u_N^*$ were a second one, then $\partial\phi/\partial u_\sigma$ would be 0 for all σ when $u_1 = u_1^* t, \ldots, u_N = u_N^* t$ and $-\infty < t < \infty$. This contradicts the fact that $\phi \to \infty$ as $w_1^2 + \cdots + w_N^2 \to \infty$. The uniqueness of the critical point can also be established by using difference equations, as in Prob. 5 following Section 10–4.]

7. Prove that the functional Φ_u defined by (10–147) attains its minimum value, among smooth functions $u(x)$ satisfying the boundary conditions $u(0) = u(L) = 0$,

when u is the solution of the equation $-K^2u''(x) = F(x)$. [Hint: take $L = \pi$ for convenience. Then express the integral in terms of Fourier sine coefficients of $u(x)$, $u'(x)$, and $F(x)$, using Theorem 14 of Section 7–13. This gives a separate minimum problem for each n, which is solved precisely when $-K^2u'' = F(x)$.]

8. (a) Determine the function $u(x)$ which minimizes

$$\int_0^1 \{[u'(x)]^2 + 6xu\}\,dx,$$

if $u(0) = u(1) = 0$.

(b) Use the Rayleigh-Ritz procedure to solve the problem of part (a), using as trial functions the functions

$$u = c_1(x - x^2) + c_2 \sin 2\pi x.$$

9. Verify that the Green's function for (10–157), as described in the text, is the following function when $L = 1$:

$$g(x, s) = x(1 - s), \quad 0 \leqq x \leqq s \leqq 1; \quad g(x, s) \equiv s(1 - x), \quad 0 \leqq s \leqq x \leqq 1.$$

Verify that the function

$$u(x) = \int_0^1 g(x, s)s\,ds$$

solves the problem: $u''(x) = -x$, $u(0) = u(1) = 0$.

Suggested References

Bateman, H., *Partial Differential Equations of Mathematical Physics.* New York: Dover, 1944.

Churchill, R. V., *Fourier Series and Boundary Value Problems*, 2nd ed. New York: McGraw-Hill, 1963.

Churchill, R. V., *Modern Operational Mathematics in Engineering.* New York: McGraw-Hill, 1944.

Courant, R., *Advanced Methods in Applied Mathematics.* Lithoprinted notes of lectures at New York University, 1941.

Courant, R., "Variational methods for the solution of problems of equilibrium and vibrations," *Bulletin of the American Mathematical Society*, Vol. 49, pp. 1–23. New York: American Mathematical Society, 1943.

Courant, R., and Hilbert, D., *Methods of Mathematical Physics.* Vol. 1, New York: Interscience, 1953. Vol. 2, New York: Interscience, 1961.

Frank, P., and v. Mises, R., *Die Differentialgleichungen und Integralgleichungen der Mechanik und Physik.* Vol. 1, 2nd ed., Braunschweig: Vieweg, 1930. Vol. 2, 2nd ed., Braunschweig: Vieweg, 1935.

Goldstein, Herbert, *Classical Mechanics.* Cambridge: Addison-Wesley Press, 1950.

Gould, S. H., *Variational Methods for Eigenvalue Problems*, 2nd ed. Toronto: University of Toronto Press, 1966.

Hildebrand, F. B., *Finite-Difference Equations and Simulations.* Englewood Cliffs, N. J.: Prentice-Hall, 1968.

John, F., *Lectures on Advanced Numerical Analysis.* New York: Gordon and Breach, 1967.

Kamke, E., *Differentialgleichungen reeller Funktionen.* Leipzig: Akademische Verlagsgesellschaft, 1933.

Kantorovich, L. V., and Krylov, V. I., *Approximate Methods of Higher Analysis,* 2nd ed., transl. by C. D. Benster. Groningen: P. Noordhoff Ltd., 1958.

Kármán, T. V., and Biot, M. A., *Mathematical Methods in Engineering.* New York: McGraw-Hill, 1940.

Kellogg, O. D., *Foundations of Potential Theory.* New York: Springer, Berlin, 1929.

Mikhlin, S. G., *Integral Equations,* transl. by A. H. Armstrong. New York: Pergamon Press, 1957.

Lord Rayleigh, *The Theory of Sound* (2 vols.), 2nd ed. New York: Dover Publications, 1945.

Sneddon, I. N., *Fourier Transforms.* New York: McGraw-Hill, 1951.

Sommerfeld, A., *Partial Differential Equations in Physics* (transl. by E. G. Straus). New York: Academic Press, 1949.

Southwell, R. V., *Relaxation Methods in Engineering Science.* Oxford: Oxford University Press, 1946.

Southwell, R. V., *Relaxation Methods in Theoretical Physics.* Oxford: Oxford University Press, 1946.

Tamarkin, J. D., and Feller, W., *Partial Differential Equations.* Mimeographed notes of lectures at Brown University, 1941.

Titchmarsh, E. C., *Eigenfunction Expansions Associated with Second-order Differential Equations.* Oxford: Oxford University Press, 1946.

Titchmarsh, E. C., *Theory of Fourier Integrals.* Oxford: Oxford University Press, 1937.

Wiener, N., *The Fourier Integral.* Cambridge: Cambridge University Press, 1933.

Index

Index